# Ion Channel Pharmacology

# Ion Channel Pharmacology

edited by
**Bernat Soria and Valentín Ceña**
Institute of Bioengineering and Institute of Neurosciences,
Universidad Miguel Hernández, Alicante, Spain

OXFORD NEW YORK TOKYO
OXFORD UNIVERSITY PRESS
1998

*Oxford University Press, Great Clarendon Street, Oxford OX2 6DP*

*Oxford New York*
*Athens Auckland Bangkok Bogota Bombay*
*Buenos Aires Calcutta Cape Town Dar es Salaam*
*Delhi Florence Hong Kong Istanbul Karachi*
*Kuala Lumpur Madras Madrid Melbourne*
*Mexico City Nairobi Paris Singapore*
*Taipei Tokyo Toronto Warsaw*
*and associated companies in*
*Berlin Ibadan*

*Oxford is a trade mark of Oxford University Press*

*Published in the United States*
*by Oxford University Press, Inc., New York*



*A catalogue record for this book is available from the British Library*

*Library of Congress Cataloging in Publication Data*
*(Data available)*

*ISBN 0 19 852360 2*

*Typeset by Hewer Text Composition Services, Edinburgh*
*Printed and bound in Great Britain by*
*Biddles Ltd, Guildford and King's Lynn*

# Preface

Ion channel research has experienced a great impulse due to the development of patch-clamp technology since the seminal papers by the Nobel prize awardees Erwin Neher and Bert Sakmann. This technique has made available for electrophysiological studies a large number of cellular types that were not suitable for it before the development of patch-clamp. The increasing number of papers studying different types of channels in a great diversity of cell types is a significant index of the success of the technique.

The advance in molecular biology techniques has also contributed to a marked increase in the identification of different channel subtypes with distinct subunit composition. Furthermore, the increased number of channel types has been matched by the discovery of toxins with a great ability to block selectively certain channel types, and by the design of drugs that could interact specifically with ion channels and that might be very useful in the treatment of certain diseases.

The present book describes how to deal with the complex world of the pharmacology of ion channels in an attempt to give the reader a precise overview of the electrophysiological properties of both voltage-dependent and receptor-operated ion channels, and the effect of the most common and selective drugs that interact with them.

*San Juan*<br>September 1997

B.S.<br>V.C.

# Contents


Contributors ix

PART ONE DEVELOPMENT

1 Neurotrophic factors and development of ionic channels in primary sensory neurones of the ear 3
F. Garcia-Diaz, C. Jiménez, E. Vázquez, J. Represa, and F. Giraldez

PART TWO SODIUM CHANNELS

2 Chemical modulation of sodium channels 23
T. Narahashi

3 Antiarrhythmic action of drugs interacting with sodium channels 74
J. Tamargo, E. Delpón, O. Pérez, and C. Valenzuela

PART THREE CALCIUM CHANNELS

4 Biophysics of voltage-dependent $Ca^{2+}$ channels: gating kinetics and their modulation 97
E. Carbone, V. Magnelli, A. Pollo, V. Carabelli, A. Albillos, and H. Zucker

5 Pharmacology of voltage-dependent calcium channels 129
B.P. Bean, I.M. Mintz, L.M. Boland, D.W.Y. Sah, and J.A. Morrill

6 Ion channels formed by amyloid β protein (AβP[1–40]). Pharmacology and therapeutic implications for Alzheimer's disease 149
H.B. Pollard, N. Arispe, and E. Rojas

PART FOUR POTASSIUM CHANNELS

7 The biophysical basis of $K^+$ channel pharmacology 167
B. Soria

8 Peptide toxins interacting with potassium channels 186
M.L. Garcia, H.-G. Knaus, R.J. Leonard, O.B. McManus, P. Munujos, W.A. Schmalofer, R.S. Slaughter, and G.J. Kaczorowski

9 Properties of the ATP-regulated $K^+$ channel in pancreatic β-cells 208
M.J. Dunne

10 Potential role of the pharmacology of $K^+$ channels in therapeutics 229
B. Soria

PART FIVE CHLORIDE CHANNELS

11 Chloride channel pharmacology 247
M.A. Valverde, S.P. Hardy, and A.S. Monaghan

PART SIX MECHANO-GATED CHANNELS

12 Drug effects on mechano-gated channels 271
O.P. Hamill and D.W. McBride, Jr

PART SEVEN INTRACELLULAR CHANNELS

13 Intracellular channels in muscle 281
E. Jaimovich

14 Basic properties of calcium release channels in neural cells 299
P. Kostyuk and A. Verkhratsky

15 Mitochondrial membrane channels and their pharmacology 321
M.-L. Campo, H. Tedeschi, C. Muro, and K.W. Kinnally

16 Modulation of a large-conductance $K^+$ channel from chromaffin granule membranes by adrenergic agonists and G proteins 343
N. Arispe, P. De Mazancourt, H.B. Pollard, and E. Rojas

PART EIGHT CYCLIC NUCLEOTIDE-GATED CHANNELS

17 Cyclic nucleotide-gated ion channels: physiology, pharmacology, and molecular biology 383
G. Matthews

PART NINE RECEPTOR-OPERATED CHANNELS

18 Excitatory amino acid-activated channels 399
J. Lerma, A.V. Paternain, N. Salvador, F. Somohano, M. Morales, and M. Casado

19 Comparison of native and recombinant NMDA receptor channels 422
P. Stern and D. Colquhoun

20 A unified theory of antidepressant action: evidence for adaptation of the NMDA receptor following chronic antidepressant treatments 438
R.T. Layer, P. Popik, G. Nowak, I.A. Paul, R. Trullas, and P. Skolnick

21 Excitatory synaptic transmission in hippocampus: modulation by purinoceptors via $Ca^{2+}$ channels 457
O. Krishtal and A. Klishin

Index 469

# Contributors

**A. Albillos** Departamento de Farmacología, Universidad Autonoma de Madrid, E-28029, Madrid, Spain.

**Nelson Arispe** Laboratory of Cell Biology and Genetics, NIDDK, National Institutes of Health, Bethesda, MD 20892, USA.
Present address: Department of Anatomy and Cell Biology, School of Medicine, Uniformed Services University, Bethesda, MD 20814, USA.

**Bruce P. Bean** Department of Neurobiology, Harvard Medical School, 200 Longwood Avenue, Boston MA 02115, USA.

**Linda M. Boland** Department of Neurobiology, Harvard Medical School, 200 Longwood Avenue, Boston MA 02115, USA.

**Maria Luisa Campo** Departamento de Bioquímica y Biología Molecular y Genética, Universidad de Extremadura, 10071, Cáceres, Spain.

**V. Carabelli** Dipartimento di Neuroscienze, Corso Raffaello 30, I-10125 Turin, Italy.

**E. Carbone** Dipartimento di Neuroscienze, Corso Raffaello 30, I-10125 Turin, Italy.

**M. Casado** Department of Neural Plasticity, Instituto Cajal, Consejo Superior de Investigaciones Científicas, Av. Doctor Arce 37, 28002 Madrid, Spain.

**David Colquhoun** Department of Pharmacology, University College London, Gower Street, London WC1E 6BT, UK.

**Eva Delpón** Department of Pharmacology, School of Medicine, Universidad Complutense, 28040 Madrid, Spain.

**Philipe De Mazancourt** Laboratory of Cell Biology and Genetics, Molecular Pathophysiology Branch, NIDDK, National Institutes of Health, Bethesda, MD 20892, USA.
Present address: Laboratory of Biochemistry, Hôpital de Poissy, rue de Champ Gaillard, F7803 Poissy, France.

**Mark J. Dunne** Department of Biomedical Science, University of Sheffield, Western Bank, Sheffield, S. Yorkshire, S10 2TN, UK.

**Maria L. Garcia** Department of Membrane Biochemistry and Biophysics, Merck Research Laboratories, P.O. Box 2000, Rahway, NJ 07065, USA.

**F. García-Diaz** Department of Physiology, Boston University School of Medicine, Boston, MA 02118, USA.

**F. Giraldez** Instituto de Biología y Genética Molecular (IBGM), Facultad de Medicina, Universidad de Valladolid-csic, 47005-Valladolid, Spain.

**Owen P. Hamill** Department of Physiology and Biophysics, The University of Texas Medical Branch at Galveston, Galveston, Texas 77555–0641, USA.

**Simon P. Hardy** Department of Pharmacy, University of Brighton, Brighton BNZ 4GJ, UK.

**Enrique Jaimovich** Departamento de Fisiologia y Biofisica, Facultad de Medicina, Universidad de Chile and Centro de Estudios Científicos de Santiago, Casilla 16244 Santiago, Chile.

**C. Jiménez** Instituto de Biología y Genética Molecular (IBGM), Facultad de Medicina, Universidad de Valladolid-csic, 47005-Valladolid, Spain.

**Gregory J. Kaczorowski** Department of Membrane Biochemistry and Biophysics, Merck Research Laboratories, P.O. Box 2000, Rahway, NJ 07065, USA.

**Kathleen W. Kinnally** Wadsworth Center for Laboratories and Research, New York State Department of Health, P.O. Box 509, Albany, New York 12201, USA.

**A. Klishin** A.A. Bogomoletz Institute of Physiology, 252024 Kiev, Ukraine.

**Hans-Gunther Knaus** Department of Membrane Biochemistry and Biophysics, Merck Research Laboratories, P.O. Box 2000, Rahway, NJ 07065, USA.
Present address: Institute for Biochemical Pharmacology, University of Innsbruck, A-6020 Innsbruck, Austria.

**Platon Kostyuk** Bogomoletz Institute of Physiology, Bogomoletz St. 4, Kiev-24, GSP 252601, Ukraine.

**O. Krishtal** A.A. Bogomoletz Institute of Physiology, 252024 Kiev, Ukraine.

**Richard T. Layer** Laboratory of Neuroscience, NIDDK, National Institutes of Health, Bethesda, MD 20892–0008, USA.

**Reid J. Leonard** Department of Membrane Biochemistry and Biophysics, Merck Research Laboratories, P.O. Box 2000, Rahway, NJ 07065, USA.

**J. Lerma** Department of Neural Plasticity, Instituto Cajal, Consejo Superior de Investigaciones Científicas, Av. Doctor Arce 37, 28002 Madrid, Spain.

**V. Magnelli** Dipartimento di Anatomia Fisiologia Umana, Corso Raffaello 30, I-10125 Turin, Italy.

**Gary Matthews** Department of Neurobiology and Behavior, State University of New York, Stony Brook, NY 11794 5230, USA.

**Don W. McBride, Jr** Department of Physiology and Biophysics, The University of Texas Medical Branch at Galveston, Galveston, Texas 77555–0641, USA.

**Owen B. McManus** Department of Membrane Biochemistry and Biophysics, Merck Research Laboratories, P.O. Box 2000, Rahway, NJ 07065, USA.

**Isabelle M. Mintz** Department of Neurobiology, Harvard Medical School, 200 Longwood Avenue, Boston MA 02115, USA.

**Alan S. Monaghan** Department of Child Health, NineWells Hospital, University of Dundee, Dundee DD1 9SY, UK.

**M. Morales** Department of Neural Plasticity, Instituto Cajal, Consejo Superior de Investigaciones Científicas, Av. Doctor Arce 37, 28002 Madrid, Spain.

**James A. Morrill** Department of Neurobiology, Harvard Medical School, 200 Longwood Avenue, Boston MA 02115, USA.

**Petraki Munujos** Department of Membrane Biochemistry and Biophysics, Merck Research Laboratories, P.O. Box 2000, Rahway, NJ 07065, USA.

**Concepción Muro** Departamento de Bioquímica y Biología Molecular y Genética, Universidad de Extremadura, 10071, Cáceres, Spain.

**Toshio Narahashi** Department of Molecular Pharmacology and Biological Chemistry, Northwestern University Medical School, 303 E. Chicago Avenue, Chicago, Illinois 60611, USA.

**Gabriel Nowak** Laboratory of Neuroscience, NIDDK, National Institutes of Health, Bethesda, MD 20892–0008, USA.

Present address: Department of Psychiatry and Human Behavior, University of Mississippi Medical Center, 2500 N. State Street, Jackson, MS 39216–4505, USA.

**A.V. Paternain** Department of Neural Plasticity, Instituto Cajal, Consejo Superior de Investigaciones Científicas, Av. Doctor Arce 37, 28002 Madrid, Spain.

**Ian A. Paul** Laboratory of Neuroscience, NIDDK, National Institutes of Health, Bethesda, MD 20892–0008, USA.
Present address: Department of Psychiatry and Human Behavior, University of Mississippi Medical Center, 2500 N. State Street, Jackson, MS 39216–4505, USA.

**Onésima Pérez** Department of Pharmacology, School of Medicine, Universidad Complutense, 28040 Madrid, Spain.

**Harvey B. Pollard** Laboratory of Cell Biology and Genetics, NIDDK, National Institutes of Health, Bethesda, MD 20892, USA.

**A. Pollo** Dipartimento di Neuroscienze, Corso Raffaello 30, I-10125 Turin, Italy.

**Piotr Popik** Laboratory of Neuroscience, NIDDK, National Institutes of Health, Bethesda, MD 20892–0008, USA.
(On leave from the Institute of Pharmacology, Polish Academy of Sciences, 12 Smetna Street, 31–343 Krakow, Poland.)

**J. Represa** Instituto de Biología y Genética Molecular (IBGM), Facultad de Medicina, Universidad de Valladolid-cisc, 47005-Valladolid-csic, Spain.

**Eduardo Rojas** Laboratory of Cell Biology and Genetics, NIDDK, National Institutes of Health, Bethesda, MD 20892, USA.

**Dinah W.Y. Sah** Department of Neurobiology, Harvard Medical School, 200 Longwood Avenue, Boston MA 02115, USA.

**N. Salvador** Department of Neural Plasticity, Instituto Cajal, Consejo Superior de Investigaciones Científicas, Av. Doctor Arce 37, 28002 Madrid, Spain.

**William A. Schmalhofer** Department of Membrane Biochemistry and Biophysics, Merck Research Laboratories, P.O. Box 2000, Rahway, NJ 07065, USA.

**Phil Skolnick** Laboratory of Neuroscience, NIDDK, National Institutes of Health, Bethesda, MD 20892–0008, USA.

**Robert S. Slaughter** Department of Membrane Biochemistry and Biophysics, Merck Research Laboratories, P.O. Box 2000, Rahway, NJ 07065, USA.

**F. Somohano** Department of Neural Plasticity, Instituto Cajal, Consejo Superior de Investigaciones Científicas, Av. Doctor Arce 37, 28002 Madrid, Spain.

**Bernat Soria** Department of Physiology and Institute of Neurosciences, P.O. Box 374, Alicante 03080, Spain.
Present address: Instituto de Bioingeniería, Universidad Miguel Hernández, Facultad de Medicina, Campus de San Juan, E-03550 San Juan (Alicante), Spain.

**Peter Stern** Department of Pharmacology, University College London, Gower Street, London WC1E 6BT, UK.
Present address: Department of Neurophysiology, Max-Planck-Institut for Brain Research, Deutschordenstrasse 46, 60528 Frankfurt, Germany.

**Juan Tamargo** Department of Pharmacology, School of Medicine, Universidad Complutense, 28040 Madrid, Spain.

**Henry Tedeschi** Department of Biological Sciences, State University of New York, Albany, New York 12222, USA.

**Ramon Trullas** Laboratory of Neuroscience, NIDDK, National Institutes of Health, Bethesda, MD 20892–0008, USA.
Present address: Neurobiology Unit, Department of Bioanalytical Medicine, CSIS, Jordi Girona 18–26, 08034 Barcelona, Spain.

**Carmen Valenzuela** Department of Pharmacology, School of Medicine, Universidad Complutense, 28040 Madrid, Spain.

**Miguel A. Valverde** Physiology Group, King's College London, The Strand, London WC2R 2LS, UK.

**E. Vázquez** Instituto de Biología y Genética Molecular (IBGM), Facultad de Medicina, Universidad de Valladolid-csic, 47005-Valladolid, Spain.

**Alexej Verkhratsky** Bogomoletz Institute of Physiology, Bogomoletz St. 4, Kiev-24, GSP 252601, Ukraine.

**H. Zucker** Max-Planck-Institut für Psychiatrie, Abt. Neurophysiologie, D-8048 Martinsried, Germany.

# 1 Development

# 1 Neurotrophic factors and development of ionic channels in primary sensory neurones of the ear

F. García-Diaz, C. Jiménez, E. Vázquez, J. Represa, and F. Giraldez

Differentiation of a particular neuronal phenotype involves the expression of a mosaic of ionic channels that provides a given neurone with characteristic functional properties. Ionic channels constitute for this reason excellent molecular and functional markers of neuronal differentiation during embryonic development. Neurones also exhibit specific connections that for sensory neurones are established with peripheral targets, the transducing elements, and with central sensory nuclei. Differentiation of neuronal pools and the organization of their connectivity occur during embryonic development on specified territories and with a fixed temporal sequence. Growing evidence support the view that cell–cell interactions via specific molecules acting during development are part of the mechanisms that allow not only differentiation and generation of connectivity, but also the possibility to secure this organization during the adult life. A prototype for these molecules, is the nerve growth factor (NGF) family of neurotrophins and their receptors, the *trk* proto-oncogenes. This family of factors show the ability to induce morphological and biochemical differentiation of sensitive neuronal populations and to support their survival. Developmental expression of neurotrophins and *trk*-receptors is restricted in time and space during development, indicating specific roles on particular neuronal populations.

We discuss here some of these issues by reviewing experiments on the development of the auditory system. We shall describe first, the temporal pattern of development of ionic currents in the auditory ganglion; secondly, the biological effects and expression pattern of NGF-related peptides and *trk*-receptors; and, finally, the possibility that trophic factors regulate ionic channel expression. Some aspects of recent literature on the effects of NGF on channel expression in other neuronal types will be summarily reviewed.

## Development of ionic currents in cochlear neurones

The expression of membrane ionic channels is subject to conspicuous changes along the various phases of neuronal differentiation. Evidence for such developmental modifications was originally obtained from examination of the changes occurring in the time course and ionic dependence of the action potential (Spitzer and Lamborghini 1976; Bader *et al.* 1983;

Mori-Okamoto *et al.*). With the introduction of the tight-seal recording technique (Hamill *et al.* 1981), it has been possible to apply voltage-clamp recording to small cells, thus affording a direct way to monitor changes in individual ionic mechanisms during the process of differentiation. It has been found that the classes of channels in an immature versus a developed cell varies from one cell type to another (Gottman *et al.* 1988; O'Dowd *et al.* 1988; Kubo 1989).

The cochlear ganglion contains the primary afferent neurones of the vertebrate auditory system and constitute the first synaptic relay after transduction of sound. Cochlear (acoustic) neurones are bipolar, with a central process that makes synapses in the brainstem, and a peripheral process that terminates on the sensory epithelium of the inner ear. Cochlear and vestibular ganglia develop in the early embryo as one single unit, the cochleovestibular ganglion (CVG), the majority of neurones arising by migration from the otic vesicle. As development proceeds, the CVG splits into distinct cochlear and vestibular ganglia. Neurones develop peripheral processes and innervate the sensory epithelium. Innervation begins around day 6 of development in the chick (14–15 gestational days in the mouse), by which time most ganglionic cells have performed their terminal mitosis. It proceeds in parallel with hair cell maturation to form the first synaptic contacts by days 8–9 and the mature synapsis by days 15–17 (gestational day 18 and post-natal in the mouse, respectively) (Anniko 1983).

## Sodium and potassium currents

Analysis of the development of membrane excitability and $N^+$ and $K^+$-currents has been carried out in neurones from the chick embryo cochlear ganglion by Valverde *et al.* (1992), using the 'patch-clamp' technique. The earliest stage explored was day 6, which corresponds to the end of the mitotic period and the beginning of the invasion of the sensory targets. The latest cells examined were large, 15–20 $\mu$m diameter neurones of day 17, exhibiting the adult mosaic of membrane currents described by Yamaguchi and Ohmori (1990). Representative current-clamp records at three different developmental stages are shown in Fig. 1.1A. Day 8 neurones showed strong rectification with sometimes a small spike activity. By day 10, spikes were clearly developed, but small. At day 14, regenerative activity with large overshooting spikes was clearly developed. Neurones recorded during these late stages showed a characteristic inflexion or 'hump' in the falling phase of the action potential and also a distinct after-hyperpolarization which could not be detected in earlier stages of development.

The general pattern of membrane currents observed throughout development is illustrated in Fig. 1.1B. Membrane current families generated by similar voltage-clamp protocols are displayed. Cells sampled on day 6 showed very small linear or weakly outwardly rectifying currents. From day 7 onwards, however, slowly inactivating outward currents were present (day $7\frac{1}{2}$ in Fig. 1.1B). These currents increased in size throughout development and by day 8 were accompanied by fast transient inward currents that also increased in magnitude over the following days. The typical current pattern of such intermediate developmental stage is illustrated by the day 10 neurone shown in Fig. 1.1B. Inward current transients, poorly resolved at the illustrated time scale, correspond to sodium currents. By day 12–14, the basic pattern of currents did not differ from that of the adult (Yamaguchi and Ohmori 1990).

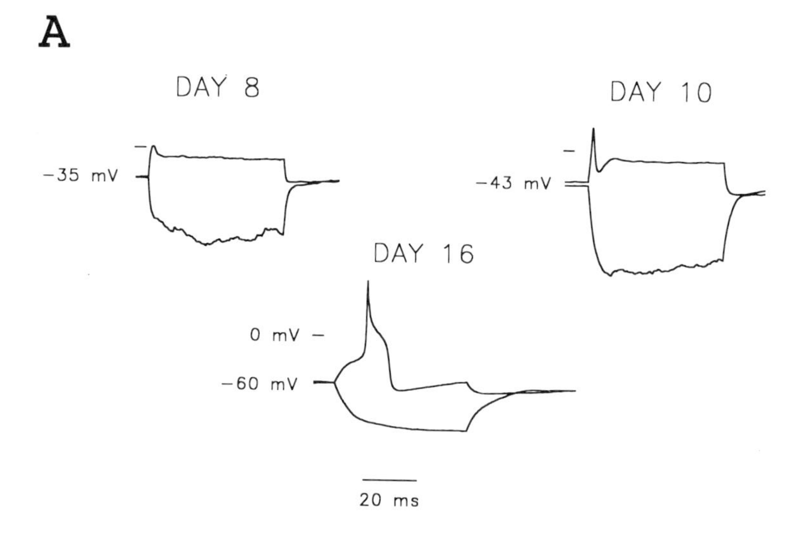


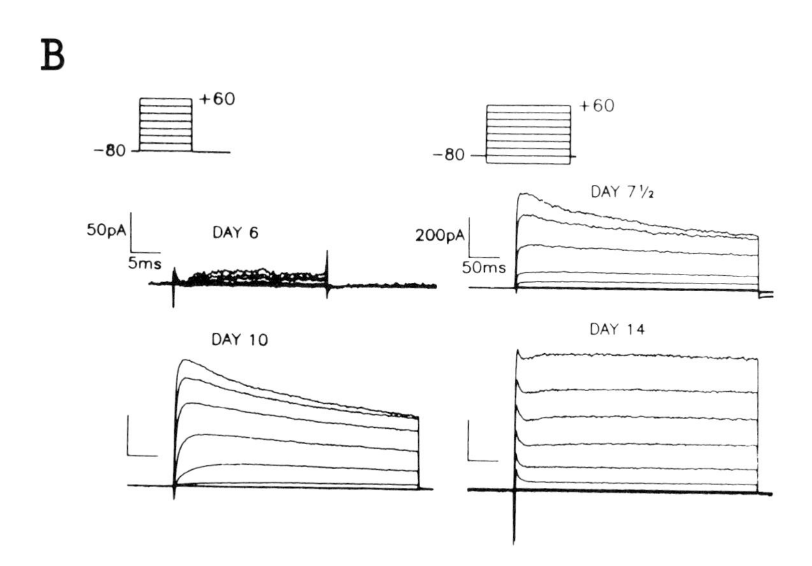


**Fig. 1.1** Development of excitability of cochlear neurones. A. Current-clamp records from isolated cochlear neurones at the stage of development indicated by the day. Positive and negative square current pulses of 20 pA from a zero-current baseline were applied and membrane potential recorded. The numbers to the left of each record indicate the zero-current potential, and hence an estimate of the resting membrane potential. B. General pattern of membrane currents recorded in isolated cochlear ganglion neurones throughout development. Cells were clamped at a holding potential of −80 mV and voltage steps applied as illustrated. Calibration of records labelled day 10 and 14 are like day $7\frac{1}{2}$. Note the presence of inward transients, poorly resolved at this time scale. Cells were bathed in a normal Ringer and patch-pipettes contained the $Ca^{2+}$-free, $K^{+}$-rich intracellular solution. (Modified from Valverde *et al.* 1992.)

Outward currents contained at least three components: (1) a non-inactivating outward current, similar to the delayed-rectifier, predominating in mature neurones; (2) a slowly-inactivating current (tau about 200 ms), most evident in early and intermediate stages (days 7 to 10); and (3) a rapidly inactivating outward current (tau about 20 ms) similar to the

A-current ($I_A$) described in other neurones, which was distinctly expressed in mature neurones.

Isolation of the fast transient outward current ($I_{to,f}$) from a differentiated neurone is illustrated in Fig. 1.2A. The activation and inactivation curves are also shown in the lower part (left). Half-inactivation ($V_{1/2}$) of $I_{to,f}$ occurs at about $-75$ mV, 90% of current being suppressed within approximately a 20 mV interval, with an activation threshold about $-60$ mV (Sheppard *et al.* 1992). In contrast to the sigmoidal activation of the non-inactivating current ($I_{ss}$), the activation curve of $I_{to,f}$ frequently showed a 'hump' between $-20$ and $+10$ mV (Fig. 1.2B). This feature is more readily apparent in the conductance versus potential plot illustrated in Fig. 1.2D, and is related to the $Ca^{2+}$-dependence of $I_{to,f}$. The effect of $Ca^{2+}$ is, at least in part, related to the influx of $Ca^{2+}$ through voltage-activated membrane channels, since both $Ca^{2+}$ removal and $Co^{2+}$ ions in the presence of extracellular

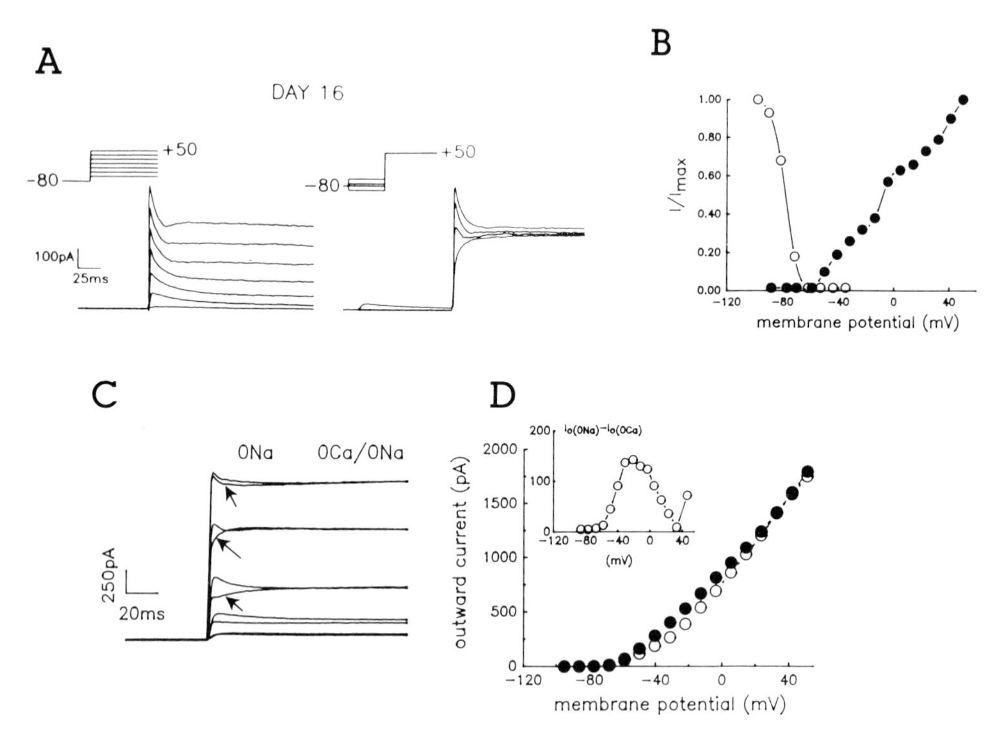


**Fig. 1.2** $Ca^{2+}$-dependent A-current type in cochlear neurones. A. Left, current family generated by depolarizing voltage steps. Right, currents generated by a pre-pulse protocol. Voltage steps are illustrated in the inset above the current traces. The calibration is the same for both. Day 16 of development. B. Activation and inactivation curve of $I_{to,f}$. Fast transient outward current measured at +50 mV voltage pulses was plotted against the pre-pulse potential (open circles). The transient outward current plotted against different command membrane potentials is also illustrated (filled circles). $I_{to,f}$ was calculated as the peak outward current minus the steady-state current measured at the end of 400 ms voltage pulses, and expressed as a proportion of the maximum current measured. Data from the same cell illustrated in a. C. Effects of $Ca^{2+}$ removal on $I_{to,f}$. Currents generated after depolarization from a holding potential of $-80$ mV. Voltage steps ilustrated are, from top to bottom, +45, +25, $-5$, $-40$ and $-60$ mV. The current family indicated by the arrowheads was obtained in the nominal absence of extracellular $Ca^{2+}$, i.e. O Na solution with no Ca salts added. (day 12 neurone). D. Current voltage plot of the experiment illustrated in C. Values of total outward current were plotted against the command potential in O Na (filled circles) and in Ca-free medium (open circles). The inset shows the subtraction between the outward current recorded in O Na and that recorded in Ca-free solution, plotted against the command potential. (Modified from Sheppard *et al.* 1992.)

$Ca^{2+}$ inhibited $I_{to,f}$. The activation of $I_{to,f}$ and its inhibition by $Co^{2+}$ showed a voltage dependence that overlaps the activation profile of $Ca^{2+}$-currents (see below). Slow and fast transient outward currents display different kinetic and pharmacological profiles that suggest different channel substrates (Sheppard *et al.* 1992).

Different types of $K^+$-currents have been observed to coexist within a single cell in several adult neurones. The slowly inactivating outward current of cochlear neurones is similar to $K^+$-currents recorded in other excitable cells, including embryonic dorsal root ganglion neurones (Valmier *et al.* 1989). The fast inactivating outward current resembles the A-current ($I_A$) characterized in molluscan neurones (Connor and Stevens 1971; Neher 1971) and thereafter found in a variety of neurones (Rogawsky 1985). $Ca^{2+}$-dependent A-currents of cochlear neurones appear as a late acquisition, coinciding with the mature neuronal phenotype, a situation also reported in other developing systems (Salkoff and Wyman 1981; Blair and Dionne 1985; Nerbonne *et al.* 1986; Aguayo 1989; Dourado and Dryer 1992; Dovrado *et al.* 1994). $K^+$-channel diversity is believed to originate by several molecular mechanisms that would allow the regulation of the expression of different channels at different developmental stages. However, the question of what is the significance of this developmental regulation and the role played by these channels in the functional development of connections between sensory receptors and primary afferents remains obscure.

From a physiological point of view, inactivating $K^+$-currents are interesting because they can modulate cell excitability by transiently hyperpolarizing the membrane. This can be of importance for fine modulation of sound transmission by acting at least at two different steps. First, at the synapse between hair cells and primary afferents, A-currents could modulate synaptic transmission in the post-synaptic membrane. If large excitatory post-synaptic potentials (EPSPs) are able to reach the threshold for A-currents, their activation may abort the depolarization induced by further EPSPs. Secondly, $I_{to}$ could control the frequency of repetitive firing by modifying the rate of decay of the after-hyperpolarization following a spike (Rogawsky 1985). This could potentially limit the frequency response of cochlear neurones as a function of hair cell receptor potential. The precise relationship between receptor potential and discharge frequency of cochlear neurones is not known, but it is well established that spike rates show strong non-linearities and shallow intensity–response functions that avoid saturation – evidence suggesting that these effects are not determined solely by the receptor potential (Weis, 1984).

## Calcium currents

Recent experiments have been carried out to study $Ca^{2+}$-channels in developing cochlear neurones. $Ca^{2+}$-and $Ba^{2+}$-currents were recorded and compared at different developmental stages. A preliminary survey of the expression of $Ca^{2+}$-currents in chick cochlear neurones during embryonic development is shown in Fig. 1.3. Currents were measured in acutely dissociated neurones using 10 mM $Ba^{2+}$ in the external solution. Open bars indicate the density (pA/pF) of the current at the end of the 200-ms voltage pulse, which is mostly due to the non-inactivating component. Hatched bars show the difference between the peak current and the current at the end of the pulse and thus indicate densities of the transient component. Although there are still few data available, the pattern indicates an increase in both

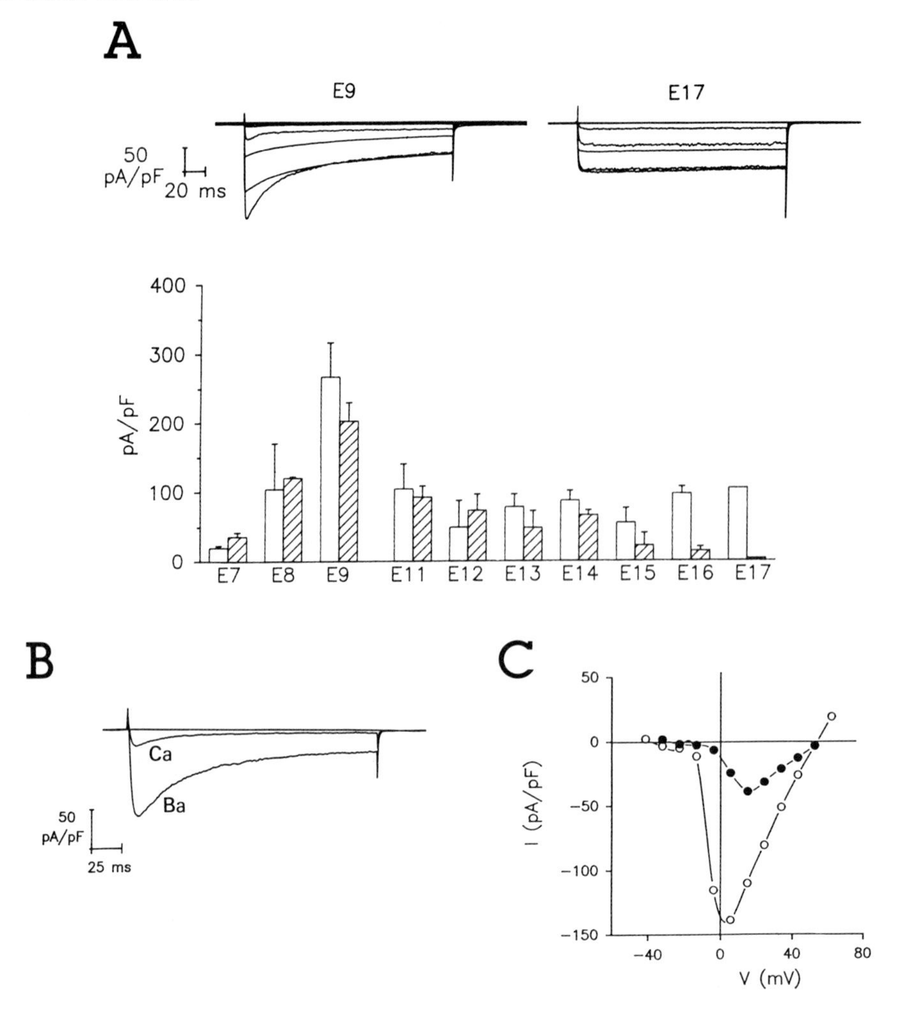


**Fig. 1.3** $Ca^{2+}$-currents in developing cochlear neurones. Temporal pattern of expression of $Ba^{2+}$ currents (10 mM $Ba^{2+}$) during embryonic development of chick cochlear ganglion neurones. Open bars indicate the current density at the end of a 200 ms pulse. Hatched bars are the peak current *minus* the steady-state value. Current density represented are maximum values obtained at voltages between −20 and +10 mV. Inset at top of figure shows characteristic currents obtained in two neurones. Holding voltage was −90 mV and depolarizations from −30 to +20 mV in 10 mV steps.

components of $Ca^{2+}$-current early in development, reaching a peak at E9. There is a subsequent decline in current density, with a more drastic reduction of the inactivating component (see insets at top of figure). These results are interesting, since they suggest that the maximum expression of $Ca^{2+}$ channels occurs during the period of early synaptogenesis (Whitehead and Morest 1985 *a* and *b*), supporting the notion that $Ca^{2+}$ channels play an important role in this process.

With 10 mM $Ba^{2+}$ as charge carrier, the maximum amplitude of inward currents was typically observed around 0 mV. $Ba^{2+}$ ions are more permeable than $Ca^{2+}$ through these channels and the IV relationship in $Ca^{2+}$ is displaced some 10 mV to more positive voltages with respect to the IV curve obtained with $Ba^{2+}$ (Fig. 1.3 B and C). Similar observations were made by Yamaguchi and Ohmori (1990) in mature chick cochlear neurones. A component of the inward current carried by $Ba^{2+}$ was sensitive to the dihydropyridine, nisoldipine, and the fraction blocked by nisoldipine exhibited a slow inactivation, probably mediated by L-type $Ca^{2+}$ channels (Bean 1992). The inorganic $Ca^{2+}$ channel blocker $Cd^{2+}$ suppressed almost completely the inward current carried by $Ba^{2+}$.

As an overview, a good correspondence can be traced between the development of membrane currents in cochlear neurones and the innervation pattern of the chick cochlea described by Whitehead and Morest (1985*a*). On days 6–7, the peripheral endings of the cochlear neurones invade the otocyst. A fraction of the cells in the ganglion concomitantly begin to show voltage-dependent currents, primarily outward currents. Early synaptic contacts are established around days 8–9, in association with continued hair cell differentiation. This period is associated with the appearance of $Na^+$-currents, slowly inactivating $K^+$-currents and distinct transient $Ca^{2+}$-currents. This picture continues through days 10 and 12, which correspond to intermediate stages of synaptogenesis. By days 14–15, the mature synaptic structure of the cochlear receptor develops. Hair cells and synaptic endings are evident and this coincides with the adult mosaic of $Na^+$-, $K^+$- and $Ca^{2+}$-currents found in mature neurones. It is interesting to note that while cochlear potentials can be evoked in chick embryos by days 10–11, central responses cannot be recorded until day 11. Following day 15, central responses become more synchronized and shorter in duration. Embryonic days 14 to 15 are also a transitional period during which behavioural responses evoked by auditory stimulation can be first elicited (Rubel 1984).

## Neurotrophic factors in ear development

Indirect evidence suggest that cochlear neurones require growth factors released from their peripheral or central targets in order to survive and differentiate (Van de Water and Ruben 1984; Ard and Morest 1985; Després and Romand 1994). Differentiating ear sensory epithelia have been shown to establish attractant fields that influence the outgrowth of CVG neurites and the setting up of inner ear innervation (Hemond and Morest 1992). Growth factors are a diverse group of polypeptides that act as extracellular messengers in the trophic and plastic growth responses of most eukaryotic cells. Nerve growth factor (NGF) was the first neurotrophic substance to be identified and characterized (Levi-Montalcini 1987) and is the prototype of a small gene family encoding functionally related proteins collectively known as neurotrophins. As in other sensory systems, the NGF family of neurotrophins emerges as a candidate to mediate cell–cell interactions during development of the auditory system (Barde 1989).

### The NGF family of neurotrophins

The NGF family of factors includes NGF, brain-derived neurotrophic factor (BDNF) and neurotrophins 3, 4 and 5 (NT-3, NT-4 and NT-5), which are structurally homologous polypeptides sharing about 60% of aminoacid identity and membrane receptors, as well as biological effects (Hallböök *et al.* 1993; Chao 1992). These neurotrophins display distinct biological effects on the developing cochlear and vestibular ganglia of mouse and avian embryos. First, they are able to stimulate cell division during early proliferative periods of CVG development (Varela *et al.* 1991; Represa *et al.* 1991*a*). This effect is coupled to the hydrolysis of a membrane glycosylphosphatidylinositol (GPI) and the generation of a biologically active inositol-phosphoglycan (IPG), that acts as a powerful mitogen for the CVG (Represa *et al.* 1991*a*). Secondly, neurotrophins are able to induce neurite outgrowth in both ganglion explants and isolated neurones (Fig. 1.4A). The effect is stage-specific, being maximal during innervation and synaptogenesis (Avila *et al.* 1993; Vazquez *et al.* 1994). Finally, neurotrophins support survival of isolated neurones in culture (Avila *et al.* 1993; Vazquez *et al.* 1994).

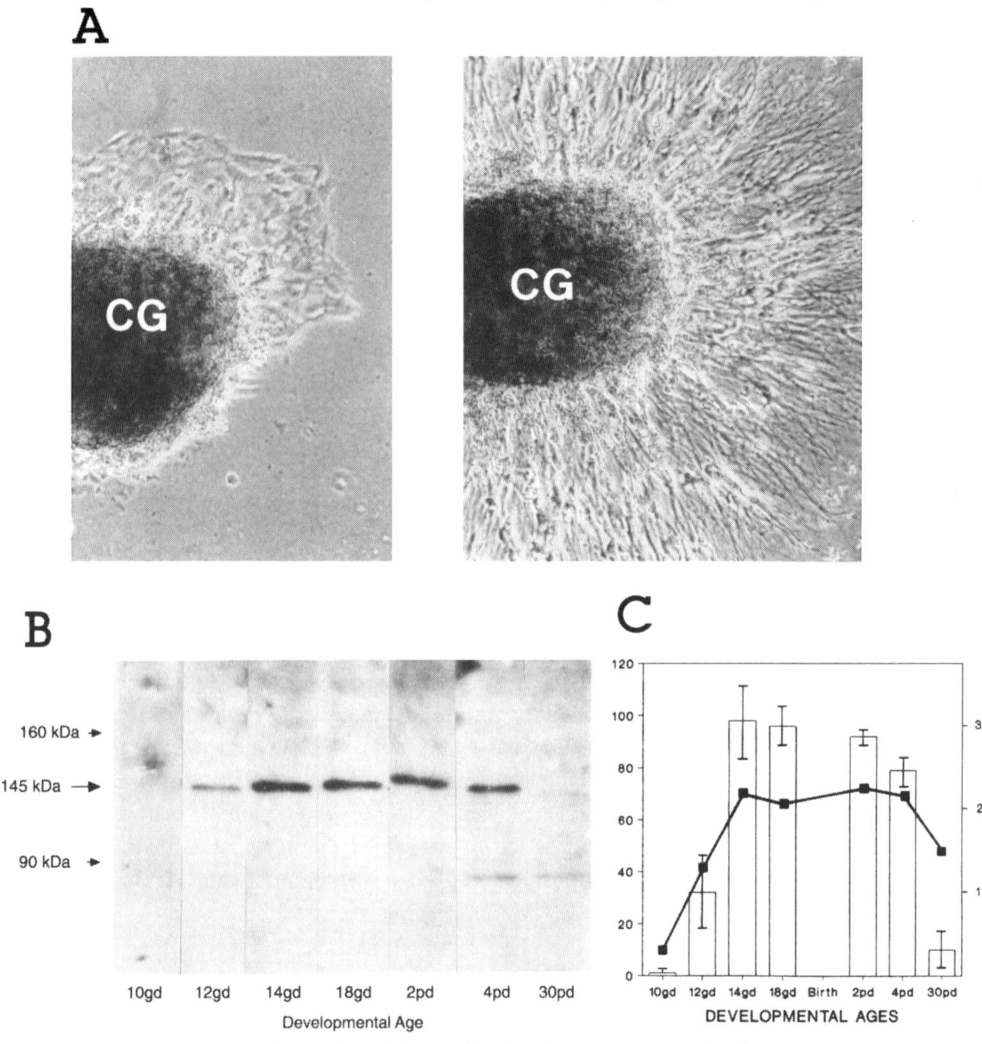


**Fig. 1.4** TrkB protein in the auditory and vestibular ganglion (CVG). A. Photomicrographs of 14 gestational days cochlear explants, after 48 h in culture with: (left) no additions of neurotrophins; and (right) with 2 ng/ml BDNF. Scale bar = 200 $\mu$m. B. Western blot analysis of trkB in the CVG throughout the developmental stages indicated by the numbers below. C. Densitometric values of trkB expression in the CVG at different stages are plotted along with the percentage of BDNF-induced cell survival on CVG neurones. Average values correspond to three experiments and the data are expressed $\pm$ S.E.M. CG, cochlear ganglion. (Modified from Vazquez *et al.* 1994.)

There are differences in the potency of the effects of the different neurotrophins throughout development. The potency of the three main neurotrophins (NGF, BDNF and NT-3) in inducing cell proliferation is comparable, although BDNF and NT-3 are one order of magnitude more potent than NGF. However, the difference is much more evident – about two orders of magnitude – when neuritogenesis and cell survival are compared (Avila *et al.* 1993; Vazquez *et al.* 1994).

Two structurally unrelated receptors have been characterized for the NGF family of neurotrophins. One of them, the $p^{75}$, serves as a low-affinity receptor for all known neurotrophins and mediates the stimulation of cell division induced by neurotrophins during the proliferative growth period of the cochlear ganglion (Represa *et al.* 1991). In agreement with that, the low-affinity $p^{75}$ receptor is differentially expressed in the auditory and vestibular system during this period (Represa *et al.* 1991*b*).

The second class of receptor is the *trk* proto-oncogene family that serves as a high-affinity receptor ($p^{145}$) and mediates the classical responses to neurotrophins (i.e. neuronal survival and neurite outgrowth). The *trk*-A, *trk*-B, and *trk*-C are high-affinity receptors for NGF, BDNF/NT4, and NT3, respectively (Meakin and Shooter 1992). The high-affinity neurotrophin receptors *trk*-B and *trk*-C, but not *trk*-A, mRNAs are expressed in rat post-natal auditory and vestibular ganglia (Ylikoski *et al.* 1993). Figure 1.4B shows the expression of trk-B protein in the cochlear–vestibular ganglion of the mouse embryo. The dominant band of about 145 kDa corresponds to the reported product of *trk*-B proto-oncogene. There is a striking parallelism between *trk*-B expression throughout different developmental stages and BDNF-induced cell survival on CVG neurones (Fig. 1.4B and C). Moreover, both events appear to be correlated with the development of inner ear afferent innervation and with BDNF mRNA expression in the developing sensory epithelium of the vestibular system (Pirvola *et al.* 1992; Hallböök *et al.* 1993). There is a good timing for the potential delivery of BDNF *in vivo*, receptor expression, and *in vitro* biological responses. This pattern applies also for NT-3 and the trk-C protein in the cochlear system. Recent analysis of mouse carrying null mutations of neurotrophic factors (BDNF or NT-3) or their receptors (trk-B −/− mutants) further confirm the strict requirement of neurotrophins for innervation of the auditory and vestibular systems (Snider 1994). The 'knock-out' of BDNF (Ernfors *et al.* 1994; Jones *et al.* 1994) and trk-B (Klein *et al.* 1993) results in serious defects in the vestibular innervation and function, and NT-3 and trk-C mutant mice show reductions in cochlear and vestibular ganglion (Fariñas *et al.* 1994; Klein *et al.* 1994). Ear innervation in *trk*-B null mutants has been recently analysed in detail by Schimmang *et al.* (1995), describing a phenotype which is basically consistent with that of BDNF (−/−). However, careful analysis throughout development has revealed that vestibular neurones are able actually to establish their peripheral connections in the absence of trk-B expression, but fail to maintain this target innervation, suggesting that neurotrophins are indispensable not for target recognition but for securing synaptic contacts.

## Other trophic factors

Experiments with organ-conditioned media have shown the presence of a factor (or factors) released by the target field of cochlear neurones, which enhances survival and neuritogenesis. Thus, the otocyst (inner ear) conditioned medium of E4–E6 chick embryos promotes neuritogenesis of chick CVG neurones (Bianchi and Cohan, 1991). Organ of Corti-conditioned medium from 5-day postpartum rat pups also has a neurotrophic effect on the survival of auditory neurones in cell culture (Lefebvre *et al.* 1992). The neurite-promoting properties of the otocyst-conditioned medium (OCM) on chick E5 neurones could not be replicated by the neurotrophins NGF, BDNF, NT-3 or NT-4, suggesting the presence of a different factor(s) in OCM (Bianchi and Cohan, 1993). The lack of response to BDNF or NT-3 seems in contradiction with the findings of Avila *et al.* (1993) mentioned above. However, the latter investigators found only a small effect of these neurotrophins at E5, the stage used by Bianchi and Cohan, but a large effect at later stages, E7 and E9. These observations underline the importance of the temporal pattern of the effects of neurotrophins. However, the possibility remains that other factors may be important in the development of cochlear neurones. A potential pitfall of experiments with OCM is the possibility of intracellular $K^+$ release from dying neurones in the cultured otocysts. This may increase the $K^+$ concentration of the OCM to a level where it will by itself promote the survival of neurones. Survival enhancement of developing neurones by high extracellular $K^+$ (20–50 mM) is a well-known phenomenon (Franklin and Johnson 1992), which in some instances is due to an autocrine effect of BDNF (Ghosh *et al.* 1994).

The fibroblast growth factors (FGFs) are a family of monomeric proteins with a 50% sequence homology and encoded by different genes (Goldfarb 1990). They primarily modulate proliferation of cells derived from the mesoderm and neuroectoderm, but also promote differentiation of certain neurones. Acidic FGF (aFGF or FGF-1) and basic FGF (bFGF or FGF-2) display similar neurotrophic effects as NGF, and their receptors also have tyrosine kinase activity (Korsching 1993). Recent experiments have shown that bFGF induces neuritogenesis in the cochlear ganglion (Lefebvre *et al.* 1990). Int-2 (FGF-3), another member of the FGF family, is required for the induction of the otic vesicle (Represa *et al.* 1991*c*) and it is expressed in regions where the sensory epithelia appears later in development (Wilkinson *et al.* 1989), suggesting that it may play a neurotrophic role during innervation of the cochlea.

Another family of growth factors, the insulin-like growth factors (IGFs), have been shown to share with NGF the capacity to enhance neurite outgrowth (Recio-Pinto *et al.* 1986; Aizenman and de Vellis 1987). Insulin acts as a co-factor in the mitogen activity of bombesin during early stages of inner ear development (Represa *et al.* 1988) and IGF-I factor and receptor are expressed throughout the development of the inner ear (León *et al.* 1995 and unpublished observations). It has been recently shown that IGF-1 causes a stimulation of N- and L-type $Ca^{2+}$ currents in neuroblastoma × glial cells (Kleppisch *et al.* 1992) and of glutamate-activated, $Na^+$- and late $K^+$-currents in cerebellar granule cells (Calissano *et al.* 1993).

## Effects of trophic factors on ionic channels

The mechanism by which ionic currents change during development, and the physiological roles that different channels may play in the course of development, are currently poorly understood. Several studies have ascribed a role for ionic fluxes to some aspects of development. In particular, calcium influx via voltage-dependent channels has been implicated in the regulation of neurite initiation and growth cone motility in several preparations (Freeman *et al.* 1985; Connor 1986; Audersik *et al.* 1990), as well as neurite elongation (Anglister *et al.* 1982; Mattson and Kater 1987) and development of a normal neural phenotype (Holliday and Spitzer 1990). Because the expression of voltage-dependent ionic channels is a crucial aspect of neuronal function, they can be used as molecular markers of neuronal differentiation. Also, since neurotrophins and other growth factors play an important role in the development of the nervous system, it is of the utmost interest to understand the function of these factors in the expression of ionic channels.

The cell line PC12, derived from a rat phaeochromocytoma, has served as a model for studies on the mechanism of action of NGF, as well as for the exploration of neuronal differentiation in general. When treated with nanomolar concentrations of NGF, these neoplastic chromaffin-like cells stop dividing and acquire the phenotype of mature sympathetic neurones. This phenotype is characterized by the extensive outgrowth of electrically excitable neurites, the ability to form functional synapses, and the acquisition of a number of biochemical markers (Green and Tischler 1976, 1982). These effects of NGF are also mimicked by bFGF. Both, NGF and bFGF, elicit an increase in the density of voltage-dependent $Na^+$ channels (Dichter *et al.* 1977; Mandel *et al.* 1988). Expression of $Na^+$ channels requires the activation of the cAMP-dependent protein kinase (Kalman *et al.* 1990; D'Arcangelo *et al.* 1993), whereas neurite outgrowth involves in part the activation of protein kinase C (Kalman *et al.* 1990). Other growth factors affect differentially neurite outgrowth and $Na^+$ channel expression (Pollock *et al.* 1990). NGF also induces the expression of a neuronal delayed rectifier $K^+$ channel (Sharma *et al.* 1993), acetylcholine-activated (Baev *et al.* 1992) and $Ca^{2+}$-currents (Usowicz *et al.* 1990) in PC12 cells. In spite of these observations in the PC12 cell line, and the stimulation by IGF-1 of the expression of certain channels mentioned before, there are hardly any studies on the involvement of the neurotrophins or other growth factors in the induction of ionic channels in developing neurones.

NGF also induces the expression of $Na^+$-currents in neurones from dorsal root ganglia (DRG), a classical model for neurotrophic factor studies. DRG neurones require NGF for survival and morphological differentiation. The effect of NGF on DRG neurones is to accelerate the acquisition and diversity of $Na^+$ channels, an effect which is evident only 2 hours after incubation but fully established after 6–7 hours (Omri and Meiri 1990).

As for the cochlear ganglion, little is known about possible effects of neurotrophic factors on ionic channel expression and function. The problem is presently under investigation once the basic developmental pattern of expression of membrane ionic currents has been clarified. Parallel to studies on the effects of incubation with different neurotrophins, we have investigated the effects of otic-conditioned media (OCM) on $Ca^{2+}$-currents. In the embryonic chick, the otocyst releases factors that can be recovered from the medium and shown to promote survival and neurite outgrowth of cochlear neurones (Lefebre *et al.* 1992; Bianchi and Cohan 1993). The nature of this factor(s) is under debate but resembles very much what is expected for BDNF and NT-3. Recent experiments suggest that $Ba^{2+}$-currents recorded from neurones cultivated in the presence of OCM for 2 days resemble those expressed later in development, suggesting that OCM may rescue the normal developing phenotype as it is able to support survival and neurite outgrowth.

Preliminary experiments have been carried out to explore whether or not NT-3 promotes the expression of ionic currents in developing cochlear ganglion neurones. Chick embryos of developmental stage E7 were chosen, since at this stage the increase in $Ca^{2+}$ channel expression begins and it is also when NT-3 exerts the maximum effect in neuronal survival and neuritogenesis (see above). Isolated neurones were plated as described and maintained in culture medium containing 2 ng/ml of human recombinant NT-3 and control cells were maintained in media supplemented with serum. Few of the control neurones survived after 2 days in culture and they showed a lack of neurite extension. In contrast, neurones maintained with NT-3 survived in larger number and exhibited long neuritic processes. Inward currents recorded with 10 mM $Ba^{2+}$ in the external solution showed an increase in current density after 2 days in culture with NT-3 (Fig. 1.4). In the few surviving control cells maintained without NT-3, $Ba^{2+}$-current density was not different or slightly lower than the initial values measured in acutely dissociated neurones at E7. The current density of neurones incubated for 2 days with NT-3, however, was lower than in acutely dissociated neurones from E9 embryos (Fig. 1.5, lower record), which suggests that some additional factor(s) may be acting *in ovo* to induce the expression of $Ca^{2+}$ channels. It must be stressed that the values shown in Fig. 1.4 are current density and thus they represent an actual increase in the expression of channels. Cell membrane area and capacitance increased appreciably in neurones cultured with NT-3. In acutely dissociated E7 neurones, the mean cell capacitance was $6.0 \pm 0.4$ pF and increased to $32.5 \pm 0.5$ pF after 2 days in culture with NT-3. In contrast, acutely dissociated E9 neurones had an average capacitance of $17.7 \pm 2.5$ pF.

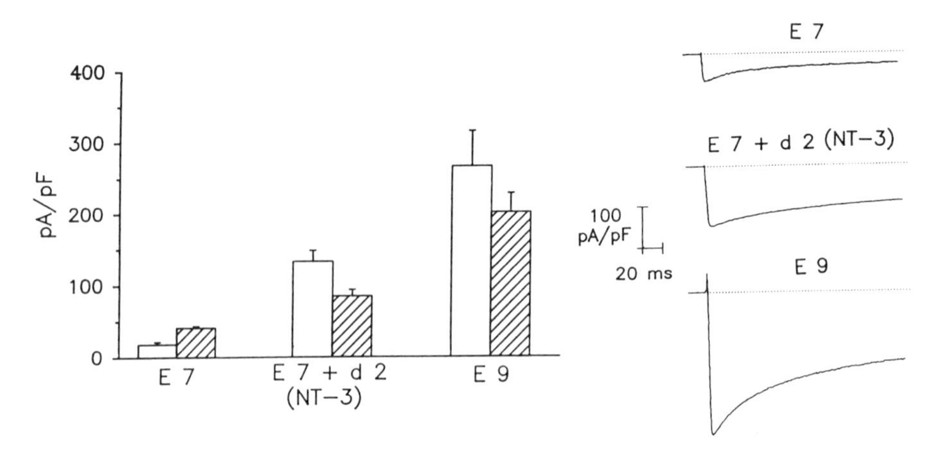


**Fig. 1.5** Effects of NT-3 on $Ca^{2+}$-currents. $Ba^{2+}$ current densities of E7 neurones acutely dissociated or maintained for 2 days in culture with 2 ng/ml NT-3, and in acutely dissociated E9 neurones. Open bars are current amplitudes at the end of the 200 ms pulse; hatched bars are the differences between the peak amplitude and the amplitude at the end of the pulse. Insets at right show examples of the currents measured following a depolarization from −90 to 0 mV. All experiments were done with 10 mM $Ba^{2+}$ as charge carrier.

E7 neurones cultured for 2 days with NT-3 also exhibited robust $Na^{+}$- and $K^{+}$-currents, typical of more advanced stages of development (Fig. 1.1). $Na^{+}$-currents are not normally seen until day 8 of embryonic development, and then they are of small amplitude. However, $Na^{+}$-currents were observed in all E7 cells cultured with NT-3 and their magnitude was comparable with those of more mature neurones (E12–E14). $K^{+}$-currents showed fast and slowly inactivating components comparable with the slow transient and the non-inactivating currents described in E8–E12 neurones by Valverde *et al.* (1992). Although preliminary, these observations seem to indicate that the neurotrophin NT-3 can support the expression of ionic channels in cochlear neurones and accelerate their mature phenotype. This would mean that NT-3 is able functionally to rescue cochlear neurones from the deprivation of normal target interactions occurring during normal development. Whether this is a specific effect of NT-3 on the induction of ionic channels, or it is a consequence of allowing cells to survive and to carry on their programme of differentiation, is not known.

In summary, we are beginning to learn about how functional properties of sensory neurones are gained throughout development, and in particular how membrane excitability and channel expression develops. On the other hand, neurotrophins and growth-promoting molecules – as well as their receptors – display biological effects and expression patterns that make them excellent candidates to regulate neuronal differentiation. At present, much of this information is being elaborated and multidisciplinary approaches are used to address the following questions: (i) do neurotrophic factors modulate ionic channel expression and function?; (ii) to what extent do ionic channels participate in development?; and (iii) are ionic channels part of the signalling mechanisms used by neurotrophic factors? These and other questions are very attractive and the subject of active research that will surely produce new answers in the near future.

## Acknowledgements

We wish to thank D.N. Sheppard, M.A. Valverde, and Thomas Schimmang for comments and co-authors of some of the published material. Financial support from DGICYT PB92-0261, FIS/1405, and Junta de Castilla y León is acknowledged. Esther Vazquez is recipient of a DGICYT studentship and C. Jiménez is an Eusko Jaularitza Student.

## References

Aguayo, L. (1989). Post-natal development of $K^+$ currents studied in isolated rat pineal cells. *Journal of Physiology*, **414**, 283–300.

Aizenman, Y. and de Vellis, J. (1987). Brain neurons develop in a serum free environment: effects of transferrin, insulin, insulin-like growth factor-I and thyroid hormone on neuronal survival, growth and differentiation. *Brain Research*, **406**, 37–42.

Anglister, L., Farber, I., Shahar, A., and Grivald, A. (1982). Localization of voltage-sensitive calcium channels along developing neurites: their possible role in regulating neurite elongation. *Developmental Biology*, **94**, 351–365.

Anniko, M. (1983). Embryonic development of vestibular sense organs and their innervation. In *Development of auditory and vestibular systems*, (ed. R. Romand), Vol. 2, pp. 375–423. Elsevier, Amsterdam.

Ard, M.D., Morest, D.K., and Hauger, S.H. (1985). Trophic interactions between the cochleovestibular ganglion of the chick embryo and its synaptic targets in culture. *Neuroscience*, **16**, 151–170.

Audersik, G., Audersik, T., Ferguson, C., Lomme, M., Shugarts, D., Rosack, J., *et al.* (1990). L-type calcium channels may regulate neurite initiation in cultured chick embryo brain neurons and N1E-115 neuroblastoma cells. *Developmental Brain Research*, **55**, 109–120.

Avila, M.A., Varela-Nieto, I., Romero, G., Mato, J.M., Giraldez, F., Van De Water, T.R., and Represa, J. (1993). Brain-derived neurotrophic factor and neurotrophin-3 support the survival and neuritogenesis response of developing cochleovestibular ganglion neurons. *Developmental Biology*, **159**, 266–275.

Bader, C.R., Bertrand, D., Dupin, E., and Kato, A.C. (1983). Development of electrical membrane properties in cultured avian neural crest. *Nature*, **305**, 808–810.

Baev, K.V., Beresovskii, V.K., Kalunov, V.N., Luschitskaya, N.I., Rusin, K.I., Vilner, B.Y., and Zavadskaya, T.V. (1992). Potential- and acetylcholine-activated ionic currents of pheochromocytoma PC12 cells during incubation with nerve growth factor. *Neuroscience*, **46**, 925–930.

Barde, Y.-A. (1989). Trophic factors and neuronal survival. *Neuron*, **2**, 1525–1534.

Bean, B.P. (1992) Whole-cell recording of calcium channel currents. *Methods in Enzymology*, **207**, 181–193.

Bianchi, L.M. and Cohan, C.S. (1991). Developmental regulation of a neurite-promoting factor influencing statoacoustic neurons. *Developmental Brain Research*, **64**, 167–174.

Bianchi, L.M. and Cohan, C.S. (1993). Effects of the neurotrophins and CNTF on developing statoacoustic neurons: comparison with an otocyst-derived factor. *Developmental Biology*, **159**, 353–365.

Blair, L.A.C. and Dionne, V.E. (1985). Developmental acquisition of $Ca^{2+}$-sensitivity by $K^{+}$ channels in spinal neurones. *Nature*, **315**, 329–331.

Calissano, P., Ciotti, M.T., Battistini, L., Zona, C. Angelini, A., Merlo, D., and Mercanti, D. (1993). Recombinant human insulin-like growth factor I exerts a trophic action and confers glutamate sensitivity on glutamate resistant cerebellar granule cells. *Proceedings of the National Academy of Sciences USA*, **90**, 8752–8756.

Chao, M.V. (1992). Neurotrophin receptors: a window into neuronal differentiation. *Neuron*, **9**, 583–593.

Connor, J. (1986). Digital imaging of free calcium changes and spatial gradients in growing processes in single, mammalian central nervous system cells. *Proceedings of the National Academy of Sciences USA*, **83**, 6179–6183.

Connor, J.A. and Stevens, C.F. (1971). Voltage clamp studies of a transient outward membrane current in gastropod neural somata. *Journal of Physiology*, **213**, 21–30.

D'Arcangelo, G., Paradiso, K., Shepherd, D., Brehm, P., Halegoua, S., and Mandel, G. (1993). Neuronal growth factor regulation of two different sodium channel types through distinct signal transduction pathways. *Journal of Cell Biology*, **122**, 915–921.

Després, R. and Romand, R. (1994). Neurotrophins and the development of cochlear innervation. *Life Sciences*, **54**, 1291–1297.

Dichter, M.A., Tischler, A.S., and Greene, L.A. (1977). Nerve growth factor-induced increase in electrical excitability and acetylcholine sensitivity of a rat pheochromocytoma cell line. *Nature*, **268**, 501–504.

Dourado, M.M. and Dryer, S.E. (1992). Changes in the electrical properties of chick ciliary ganglion neurones during embryonic development. *Journal of Physiology*, **449**, 411–428.

Dourado, M.M., Brumwell, C., Wisgirda, M.E., Jacob, M.H., and Dryer, S.E. (1994). Target tissues and innervation regulate the characteristics of $K^+$ currents in chick ciliary ganglion neurons developing in situ. *Journal of Neuroscience*, **14**, 3156–3165.

Ernfors, P., Lee, K.-F., and Jaenisch, R. (1994). Mice lacking brain-derived neurotrophic factor develop with sensory deficits. *Nature*, **368**, 147–150.

Fariñas, I., Jones, K.R., Backus, C., Wang, X.Y., and Reichardt, L.F. (1994). Severe sensory and sympathetic deficits in mice lacking neurotrophin-3. *Nature*, **369**, 658–661.

Franklin, J.L. and Johnson, E.M., Jr (1992). Supression of programmed neuronal death by sustained elevation of cytoplasmic calcium. *Trends in Neuroscience*, **15**, 501–508.

Freeman, J., Manis, P., Snipes, G., Mayes, B., Samson, P., Wikswo, J. Jr and Freeman, D. (1985). Steady growth cone currents revealed by a novel circularly vibrating probe: a possible mechanism underlying neurite growth. *Neuroscience Research*, **13**, 257–283.

Ghosh, A., Carnahan, J., and Greenberg, M.E. (1994). Requirement for BDNF in activity-dependent survival of cortical neurons. *Science*, **263**, 1618–1623.

Goldfarb, M. (1990). The fibroblast growth factor family. *Cell Growth and Differentiation*, **1**, 439–445.

Gottman, K., Dietzel, I.D., Lux, H.D., Huck, S., and Rohrer, H. (1988). Development of inward currents in chick sensory and autonomic neuronal precursor cells in culture. *Journal of Neurosciences*, **8**, 3722–3732.

Green, L.A. and Tischler, A.S. (1976). Establishment of a noradrenergic clonal line of rat adrenal pheochromocytoma cells which respond to nerve growth factor. *Proceedings of the National Academy of Sciences USA*, **73**, 2424–2428.

Green, L.A. and Tischler, A.S. (1982). PC12 pheochromocytoma cultures in neurobiological research. *Advances in Cellular Neurobiology*, **3**, 373–414.

Hallböök, F., Ibañez, C.F., Ebendal, T., and Persson, H. (1993). Cellular localization of Brain-Derived Neurotrophic Factor and Neurotrophin-3 messenger RNA expression in the early chicken embryo. *European Journal of Neuroscience*, **1**, 1–14.

Hamill, O.P., Marty, A., Neher, E., Sackman, B., and Sigworth, F.J. (1981). Improved patch-clamp techniques for high-resolution current recording from cells and cell-free membrane patches. *Pflügers Archives (European Journal of Physiology)*, **91**, 85–100.

Hemond, S.G. and Morest, D.K. (1992). Tropic effects of otic epithelium on cochleo-vestibular ganglion fiber growth in vitro. *Anatomical Record*, **232**, 273–284.

Holliday, J. and Spitzer, N. (1990). Spontaneous calcium influx and its roles in differentiation of spinal neurons in culture. *Developmental Biology*, **141**, 13–23.

Jones, K.R., Fariñas, I., Backus, C., and Reichardt, L.F. (1994). Targeted disruption of the

BDNF gene perturbs brain and sensory neuron development but not motor neuron development. *Cell*, **76**, 989–999.

Kalman, D., Wong, B., Horvai, A.E., Cline, M.J., and O'Lague, P.H. (1990). Nerve growth factor acts through cAMP-dependent protein kinase to increase the number of sodium channels in PC12 cells. *Neuron*, **4**, 355–366.

Klein, R., Smeyne, R.J., Wurst, W., Long, L.K., Auerbach, B.A., Joyner, A.L., and Barbacid, M. (1993). Targeted disruption of the trkB neurotrophin receptor gene results in nervous system lesions and neonatal death. *Cell*, **75**, 113–122.

Klein, R., Silos-Santiago, I., Smeyne, R.J., Lira, S.A., Brambilla, R., Bryant, S., *et al.* M. (1994). Disruption of the neurotrophin-3 receptor gene trkC eliminates Ia muscle afferents and results in abnormal movements. *Nature*, **368**, 249–251.

Kleppisch, T., Klinz, F.-J., and Hescheler, J. (1992). Insulin-like growth factor I modulates voltage-dependent $Ca^{2+}$ in neuronal cells. *Brain Research*, **591**, 283–288.

Korsching, S. (1993). The neurotrophic factor concept: a reexamination. *Journal of Neuroscience*, **13**, 2739–2748.

Kubo, Y. (1989). Development of ion channels and neurofilaments during neuronal differentiation of mouse embryonal carcinoma cell lines. *Journal of Physiology*, **409**, 497–523.

Lefebvre, P.P., Leprince, P., Weber, T., Rigo, J.-M., Delree, P., and Moonen, G. (1990). Neurotrophic effect of developing otic vesicle on cochleo-vestibular neurons: evidence for nerve growth factor involvement. *Brain Research*, **507**, 254–260.

Lefebvre, P.P., Weber, T., Rigo, J.-M., Staecker, H. Moonen, G., and Van de Water, T.R. (1992). Peripheral and central target-derived trophic factors effects on auditory neurons. *Hearing Research*, **58**, 185–192.

León, Y., Vazquez, E., Sanz, C., Vega, J. A., Mato, J. M., Giraldez, F., *et al.* (1995).

Levi-Montalcini, R. (1987). The nerve growth factor: thirty-five years later. *Science*, **237**, 1154–1164.

Mandel, G., Cooperman, S.S., Maue, R.A., Goodman, R.H., and Brehm, P. (1988). Selective induction of brain type II $Na^+$ channels by nerve growth factor. *Proceedings of the National Academy of Sciences USA*, **85**, 924–928.

Mattson, M.P. and Kater, S.B. (1987). Calcium regulation of neurite elongation and growth cone motility. *Journal of Neuroscience*, **7**, 4024–4033.

Meakin, S.O. and Shooter, E.M. (1992). The nerve growth factor family of receptors. *Trends in Neuroscience*, **15**, 323–331.

Mori-Okamoto, J., Ashida, M., Maru, E., and Tatsuno, J. (1983). The development of action potentials in cultures of explanted cortical neurons from chick embryos. *Developmental Biology*, **97**, 408–416.

Neher, E. (1971). Two fast transient current components during voltage clamp on snail neurons. *Journal of General Physiology*, **58**, 36–53.

Nerbonne, J.M., Gurney, A.M., and Rayburn, H.B. (1986). Development of the fast, transient outward $K^+$ current in embryonic sympathetic neurones. *Brain Research*, **378**, 197–202.

Omri, G. and Meiri, H. (1990). Characterization of sodium currents in mammalian sensory neurones cultured in serum-free defined medium with and without nerve growth factor. *Journal of Membrane Biology*, **115**, 13–29.

O'Dowd, D.K., Ribera, A.B., and Spitzer, N.C. (1988). Development of voltage-dependent calcium, sodium and potassium currents in *Xenopus* spinal neurons. *Journal of Neuroscience*, **8**, 792–805.

Pirvola, U., Palgi, J., Ylikoski, J., Lehtonen, E., Arumae, U., and Saarma, M. (1992). Brain-derived neurotrophic factor and neurotrophin 3 mRNAs are expressed in the peripheral target fields of developing inner ear ganglia. *Proceedings of the National Academy of Sciences USA*, **89**, 9915–9919.

Pollock, J.D., Krempin, M., and Rudy, B. (1990). Differential effects of NGF, FGF, EGF, cAMP, and dexamethasone on neurite outgrowth and sodium channel expression in PC12 cells. *Journal of Neuroscience*, **10**, 2626–2637.

Recio-Pinto, E., Rechler, M.M., and Ishii, D.N. (1986). Effects of insulin, insulin-like growth factor-II, and nerve growth factor on neurite formation and survival in cultured sympathetic and sensory neurons. *Journal of Neuroscience*, **6**, 1211–1219.

Represa, J.J., Miner, C., Barbosa, E., and Giráldez, F. (1988). Bombesin and other growth-factors activate cell proliferation and differentiation in chick embryo otic vesicles in culture. *Development*, **103**, 87–96.

Represa, J., Avila, M.A., Miner, C., Giraldez, F., Romero, G., Clemente, R., *et al.* (1991*a*). Glycosyl-phosphatidylinositol/ inositol phosphoglycan: a signalling system for the low affinity nerve growth factor receptor. *Proceedings of the National Academy of Sciences USA*, **88**, 8016–8019.

Represa, J., Van de Water, T.R., and Bernd, P. (1991*b*). Temporal pattern of nerve growth factor expression in developing cochlear and vestibular ganglia in quail and mouse. *Anatomy and Embryology*, **184**, 421–432.

Represa, J., León, Y., Miner, C., and Geraldez, F. *et al.* (1991*c*).

Rogawsky, M.A. (1985). The A-current: how ubiquitous a feature of excitable cells is it?. *Trends in Neurosciences*, **8**, 214–219.

Rubel, E.W. (1984). Ontogeny of auditory system function. *Annual Review of Physiology*, **46**, 213–229.

Salkoff, L. and Wyman, R. (1981). Outward current in developing *Drosophila* flight muscle. *Science*, **212**, 461–463.

Schimmang, T.S., Minichiello, L., Vazquez, E., San José, I., Giraldez, F., and Represa, J. (1995). Developing inner ear sensory neurones lacking TrkB receptors establish, but fail to maintain target innervation. *Development*, **121**, 3381–3391.

Sharma, N., D'Arcangelo, G., Kleinlaus, A., Halegoua, S., and Trimmer, J.S. (1993). Nerve growth factor regulates the abundance and distribution of $K^+$ channels in PC12 cells. *Journal of Cell Biology*, **123**, 1835–1843.

Sheppard, D.N., Valverde, M.A., Represa, J., and Giráldez, F. (1992). Transient outward currents in cochlear ganglion neurons of the chick embryo. *Neuroscience*, **51**, 631–639.

Snider, W.D. (1994). Functions of the neurotrophins during nervous system development: what the knockouts are teaching us. *Cell*, **77**, 627–638.

Spitzer, N.C. and Lamborghini, J.E. (1976). The development of the action potential mechanism of amphibian neurons isolated in culture. *Proceedings of the National Academy of Sciences USA*, **73**, 1641–1645.

Usowicz, M.M., Porzig, H., Becker, C., and Reuter, H. (1990). Differential expression by nerve growth factor of two types of $Ca^{2+}$ channels in rat phaeochromocytoma cell lines. *Journal of Physiology*, **426**, 95–116.

Valmier, J., Simonneau, M., and Boisseau, S. *et al.* (1989).

Valverde, M.A., Sheppard, D.N., Represa, J., and Giráldez, F. (1992). Development of $Na^+$ and $K^+$ currents in the cochlear ganglion of the chick embryo. *Neuroscience*, **51**, 621–630.

Van de Water, T.R. and Ruben, R.J. (1984). Neuronotrophic interaction during in vitro development of the inner ear. *Annals of Otology, Rhinology and Laryngology*, **93**, 558–564.

Varela-Nieto, I., Represa, J., Avila, M. A., Miner, C., Mato, J. M., and Giraldez, F. (1991).

Vazquez, E., Van de Water, T.R., Del Valle, M., Vega, J.A., Staecker, H., Giraldez, F., and Represa, J. (1994). Pattern of trk-B protein-like imunoreactivity and in vitro effects of brain-derived neurotrophic factor (BDNF) on developing cochlear and vestibular neurons. *Anatomy and Embryology*, **189**, 157–167.

Weis, T.F. (1984). Relation of receptor potentials of cochlear hair-cells to spike discharges of cochlear neurons. *Annual Review of Physiology*, **46**, 247–259.

Whitehead, M.C. and Morest, D.K. (1985*a*). The development of innervation patterns in the avian cochlea. *Neuroscience*, **14**, 255–276.

Whitehead, M.C. and Morest, D.K. (1985*b*). The growth of cochlear fibers and the formation of their synaptic endings in the avian inner ear: a study with the electron microscope. *Neuroscience*, **14**, 277–300.

Wilkinson, D.G., Bhatt, S., and McCahon, A.P. (1989). Expression pattern of the FGF-related proto-oncogene *int*-2 suggests multiple roles in fetal development. *Development*, **105**, 131–136.

Yamaguchi, K. and Ohmori, H. (1990). Voltage-gated and chemically gated ionic channels in the cultured cochlear ganglion neurone of the chick. *Journal of Physiology*, **420**, 185–206.

Ylikoski, J., Pirvola, U. Moshnyakov, M., Palgi, J. Arumae, U., and Saarma, M. (1993). Expression patterns of neurotrophin and their receptor mRNAs in the rat inner ear. *Hearing Research*, **65**, 69–78.

# 2 Sodium Channels

# 2 Chemical modulation of sodium channels

T. Narahashi

## Introduction

Although voltage-clamp techniques were originally developed by Cole (1949) and extensively utilized by Hodgkin, Huxley and Katz (Hodgkin and Huxley 1952 *a,b,c,d*; Hodgkin *et al.* 1952) to establish the ionic basis of nerve excitation, it was not until 1959 that the techniques were applied to the study of ion channel pharmacology. Shanes *et al.* (1959) and Taylor (1959) found that cocaine and procaine blocked both sodium and potassium currents in squid giant axons. Although these studies have established the basis for the mechanism of local anaesthesia, the effects were neither specific for one type of ion channel, nor very potent.

Shortly after that time tetrodotoxin (TTX), a puffer fish poison, was shown to block the action potential of frog skeletal muscle without changing the resting membrane potential, resting membrane resistance, or delayed rectification which is indicative of potassium channel activity (Narahashi *et al.* 1960). The results led to a hypothesis that TTX would block the sodium channel without affecting the potassium channel. The validity of this hypothesis was clearly demonstrated by voltage-clamp experiments with lobster giant axons (Narahashi *et al.* 1964). The sodium current was completely blocked by TTX while the potassium current remained unchanged. The potent and specific action of TTX in blocking the sodium channel aroused interest among neurophysiologists, and many of them started using TTX as a powerful chemical tool in the laboratory. TTX has since been used very extensively for the study of various channels and transmitter release from nerve terminals, to mention a few. For example, the potassium channel current can be recorded without being distorted by the sodium channel current in the presence of TTX. In biochemical studies, TTX was used to isolate and identify the sodium channel. The sodium channel density in various excitable cells was estimated by measuring the binding of TTX to the membrane. The mechanisms underlying transmitter release from nerve terminals were studied in the presence of TTX, which does not impair the transmitter release nor the channel activity of the post-synaptic membrane. Several review articles have been published about the action and use of TTX (Narahashi 1974, 1988a; Ritchie and Rogart 1977*c*; Catterall, 1980, 1992; Agnew, 1984; Pappone and Cahalan, 1986).

Saxitoxin (STX) is produced by dinoflagellates, and has been shown to exert almost the same sodium channel-blocking action as TTX (Narahashi *et al.* 1967*b*). STX has also been used extensively as a chemical tool in the laboratory.

The sodium channel from rat brain consists of three polypeptides: α (260 kDa), β1 (36 kDa), and β2 (33 kDa) (Hartshorne and Catterall 1981, 1984; Hartshorne *et al.* 1982; Messner and Catterall 1985). The β1 subunit is associated non-covalently with the α subunit, whereas the β2 subunit is covalently attached to the α subunit (Hartshorne *et al.* 1982; Messner and Catterall 1985). These three subunits are present in a 1 :1 :1 stoichiometry (Hartshorne and Catterall 1984).

The primary structures of sodium channels have now been identified since the pioneering works by the group of the late Professor Shosaku Numa of Kyoto University (Noda *et al.* 1984, 1986*a,b*). cDNAs encoding three distinct but highly homologous sodium channels (types I, II and III) were isolated (Noda *et al.* 1986*a*.; Kayano *et al.* 1988). The type II form is found in the embryonic and neonatal brain, whereas the alternatively spliced type IIA is present in the adult brain (Sarao *et al.* 1991; Yarowsky *et al.* 1991). The μl sodium channel α subunit is expressed primarily in adult skeletal muscle (Trimmer *et al.* 1989), whereas the hl sodium channel is found in heart and non-innervated or denervated skeletal muscle (Rogart *et al.* 1989; Kallen *et al.* 1990). The α subunit of three types of sodium channels is composed of four domains, and each domain has six transmembrane segments (S1–S6). However, the intracellular connecting loops are not well conserved. Some specific sites have been identified (Catterall, 1992). For example, cAMP-dependent phosphorylation sites are located in the inner loop between S6 of domain I and S1 of domain II. Protein kinase C phosphorylation sites and the inactivation gate are located in the inner loop between S6 of domain III and S1 of domain IV. Scorpion toxin binds to the outer loop between S5 and S6 of domain 1.

## Overview of chemicals that act on sodium channels

### Classification

The chemicals that act on the sodium channels have been classified in various ways. Catterall (1992) divided those chemicals into five groups based primarily on their binding sites. Site 1 is for TTX, STX, and μ-conotoxins; site 2 is for veratridine, batrachotoxin (BTX), aconitine, and grayanotoxin (GTX); site 3 is for α-scorpion toxins and sea anemone toxins; site 4 is for β-scorpion toxins; and site 5 is for brevetoxins and ciguatoxins. Strichartz *et al.* (1987) classified them into three large groups based primarily on their chemical properties: one group includes peptide neurotoxins such as scorpion toxins, sea anemone toxins, and μ-conotoxins; a second group includes the lipid-soluble toxins such as BTX, GTX, veratridine, aconitine, pyrethroids, and brevetoxins; and a third group includes guanidinium toxins such as TTX and STX.

Since chemicals with diverse structures may confer similar effects on the sodium channel function, and also since the chemicals that bind to a site do not necessarily exert the same effect, we here propose to classify the sodium channel agents based on their physiological effects on the channel with due consideration of their binding sites. Thus, those chemicals are classified into two large group, brokers and modulators, and the latter is further divided into several classes based on the effects on the channel activation and inactivation kinetics (Table 2.1).

**Table 2.1** Chemicals that act on sodium channels.

| Effect on channel | Chemicals | Binding Site* | References |
|---|---|---|---|
| Blockers | Tetrodotoxin | 1 | Narahashi *et al.* (1964) |
| | Saxitoxin | 1 | Narahashi *et al.* (1967b) |
| | $\mu$-Conotoxin | 1 | Kobayashi *et al.* (1986) |
| | Local anaesthetics | | Taylor (1959) |
| Modulators | | | |
| Block of inactivation | | | |
| | Pronase | | Armstrong *et al.* (1973) |
| | *N*-Bromoacetamide | | Oxford *et al.* (1978) |
| | Sea anemone toxins | 3 | Narahashi *et al.* (1969b) |
| | $\alpha$-Scorpion toxins (class 1) | 3 | Koppenhöfer and Schmidt (1968*a*,*b*) |
| | Goniopora toxin | | Muramatsu *et al.* (1985) |
| Alteration of voltage dependence of activation | | | |
| | $\beta$-Scorpion toxins (class 2) | 4 | Cahalan (1975) |
| Alteration of voltage dependence of activation and inactivation | | | |
| | $\beta$-Scorpion toxins (class 3) | 4 | Vijverberg *et al.* (1984) |
| Alteration of activation kinetics and block of inactivation | | | |
| | Batrachotoxin | 2 | Khodorov *et al.* (1975) |
| | Grayanotoxins | 2 | Seyama and Narahashi (1981) |
| | Veratridine | 2 | Ulbricht (1969) |
| | Aconitine | 2 | Schmidt and Schmitt (1974) |
| | Striatoxin | | Strichartz *et al.* (1980) |
| | Brevetoxins | 5 | Atchison *et al.* (1986) |
| | Cignatoxin | 5 | Bidard *et al.* (1984) |
| Block of inactivation and activation | | | |
| | Chloramine-T | | Ulbricht and Stole-Herzog (1984) |
| Alteration of voltage dependence and kinetics of activation and inactivation | | | |
| | Pyrethroids | 6 | Lund and Narahashi (1981*a*,*b*) |

* Catterall (1992)

## Sodium channel blockers

The chemicals that block the sodium channel comprise three groups, i.e. TTX and STX, $\mu$-conotoxin, and local anasthetics. The former three toxins have one feature in common in blocking the sodium channel selectively without affecting other channels. Local anaesthetics belong to another group, and block both sodium and potassium channels in a non-specific manner and with low potencies.

### Tetrodotoxin and saxitoxin

As described briefly earlier in this chapter, both TTX and STX block the sodium channel in a highly selective and potent manner. An example of a voltage-clamp experiment with the squid giant axon is illustrated in Fig. 2.1. Before application of TTX, a family of membrane ionic currents associated with step depolarizations to various levels is composed of initial transient sodium channel currents and late steady-state potassium channel currents. Application of

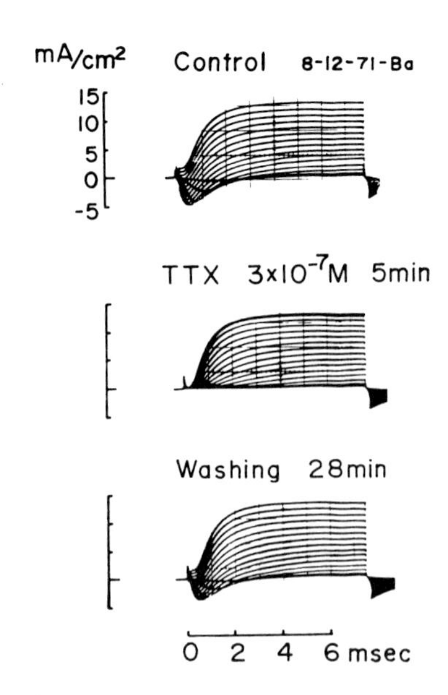


**Fig. 2.1** Block of sodium currents of a squid giant axon by tetrodotoxin (TTX). In the control, the membrane current associated with a step depolarization to various levels is composed of a transient inward or outward sodium current and a steady-state potassium current. TTX blocks the sodium currents without any effect on the potassium currents, and the effect is reversible after washing with TTX-free solution (From Narahashi 1975.)

300 nM TTX completely blocked the sodium currents with no change in the potassium currents. This effect is reversible after washing with TTX-free solutions. Single-channel patch-clamp experiments with neuroblastoma (N1E–115) cells have revealed all-or-none block of individual sodium channels by TTX (Fig.2.2). Analyses of dose–response curves indicated a one-to-one stoichiometry for TTX binding to its site with an apparent dissociation constant of 2 nM (Quandt *et al.* 1985). TTX was effective in blocking the sodium channel only when it was applied to the external surface of the nerve membrane (Narahashi *et al.* 1967*a*. These and other data led to a model that calls for one TTX/STX molecule plugging a sodium channel at its external mouth (Hille 1975). This model has since been revised by a few groups (see Narahashi, 1988*a*). However, more important questions would be to identify the binding site in the molecular structure of the sodium channel, as will be described later in connection with TTX-resistant sodium channels.

One of the remarkable examples of using TTX and STX as a chemical tool is the measurement of sodium channel density in various excitable membranes. The first experiment was performed by using a bioassay technique with lobster walking leg nerves (Moore *et al.* 1967). From the measurement of the number of TTX molecules that bound to the nerve preparations, and of the extracellular space and the total surface area of the nerve fibres contained in the preparations, the density of sodium channels was estimated to be a maximum of 13 per $\mu m^2$ of the membrane. This study was followed by several other groups with more refined techniques using [$^3$H]TTX or [$^3$H]STX, and more accurate densities were obtained. Some of these examples are given in Table 2.2. These densities are very low considering the $3 \times 5$ Å dimension of the sodium channel (Hille 1971). A density of 100 per $\mu m^2$ means that two sodium channels are separated by a distance of 1000 Å. This is roughly

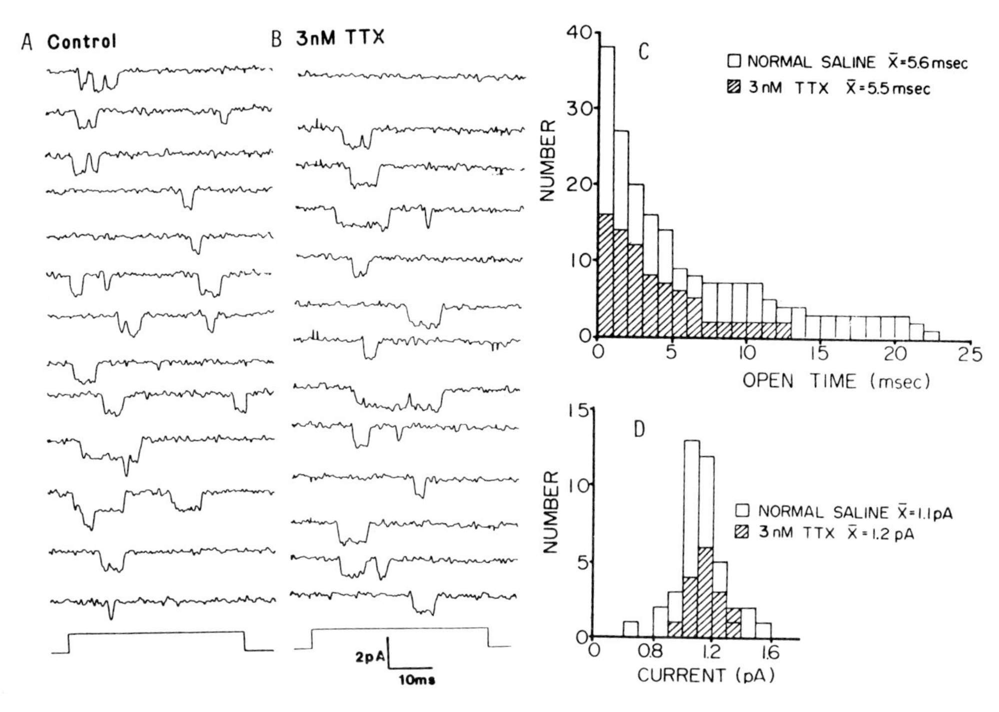


**Fig. 2.2** Tetrodotoxin (TTX) block of single sodium channel currents. Control: single channel currents recorded from an outside-out membrane patch isolated from a neuroblastoma cell (N1E-115 line) in response to depolarization from a holding potential of −90 mV to −30 mV as shown at the bottom. 3 nM TTX: after application of 3 nM TTX to the external membrane surface. Upper right: open time distributions before and after exposure to TTX. Lower right: amplitude histograms before and after TTX. Temperature 10 °C (Current records from Quandt *et al.* 1985; histograms from Narahashi 1986.)

**Table 2.2** Sodium channel densities estimated from TTX and STX binding.

| Preparation | Method | Density per $\mu m^2$ | Reference |
|---|---|---|---|
| Lobster walking leg nerve | TTX binding (bioassay) | <13 | Moore *et al.* (1967) |
| Lobster walking leg nerve | [$^3$H] STX binding | 90 | Ritchie *et al.* (1976) |
| Garfish olfactory nerve | [$^3$H] STX binding | 35 | Ritchie *et al.* (1976) |
| Squid giant axon | [$^3$H] STX binding | 290 | Keynes and Ritchie (1984) |
| Rabbit vagus nerve | [$^3$H] STX binding | 110 | Ritchie and Rogart (1977*b*) |
| Mouse neuroblastoma cell | [$^3$H] STX binding | 78 | Catterall and Morrow (1978) |
| Frog sartorius muscle | [$^3$H] STX binding | 380 | Almers and Levinson (1975) |
| Rat diaphragm muscle | [$^3$H] STX binding | 209 | Ritchie and Rogart (1977*a*) |
| Rat soleus muscle | [$^3$H] STX binding | 371 | Bay and Strichartz (1980) |
| Rabbit sciatic node | [$^3$H] STX binding | 12 000 | Ritchie and Rogart (1977*b*) |

equivalent to having golf balls (4 cm in diameter) separated by a distance of 10 m (about 30 ft), or having one golf ball in each of the 30 × 30 ft rooms, a large 900 $ft^2$ laboratory space.

TTX and STX bind to the same sodium channel site, as the binding of STX to the sodium channel was inhibited by TTX (Henderson *et al.* 1973; Ritchie and Rogart 1977*a*; Catterall and Morrow 1978). Furthermore, both TTX and STX act on the sodium channel in the cationic form (Camougis *et al.* 1967; Hille, 1968; Ogura and Mori 1968; Narahashi *et al.* 1969a). Thus, it is expected that the binding of TTX/STX is influenced by various cations present in the external solution. This was actually the case. Cations that compete with TTX/STX for binding site include $H^+$ (Benzer and Raftery 1972; Colquhoun *et al.* 1972; Henderson and Wang 1972; Ulbricht and Wagner 1975*a, b*, Bay and Strichartz 1980; Weigele and Barchi 1978); $Na^+$, $Li^+$, $K^+$, $Rb^+$, $Cs^+$, $Tl^+$, and tetramethylammonium$^+$ (Henderson *et al.* 1973, 1974; Weigele and Barchi 1978; Rhoden and Goldin 1979); and $Ca^{2+}$, $Mg^{2+}$, $Sr^{2+}$, $Ba^{2+}$, $Be^{2+}$, Be,$^{2+}$, $La^{3+}$, $Sm^{3+}$, and $Er^{3+}$ (Henderson *et al.* 1973, 1974; Weigele and Barchi 1978). However, choline was not potent in competing with STX for the site (Rhoden and Goldin 1979).

An alternative interpretation of the cation antagonism of TTX/STX binding is the change in negative surface potential in the vicinity of the sodium channel caused by cations (Grissmer 1984). The neutralization of part of the surface negative charges would decrease the TTX/STX binding. In support of this notion, the opposite effect was found by $NO^-{}_3$ and $SCN^-$, which increased the TTX/STX binding (Grissmer 1984).

Previous experiments suggested the involvement of carboxyl groups of the sodium channel in TTX binding. This was based on the observation that treatment of the crab nerve with glycine methylester HCI and 1-ethyl-3-(3-dimethylaminopropyl)carbodiimide HCl reduced TTX binding drastically (Shrager and Profera 1973). In support of this hypothesis, a similar inhibition of TTX binding was observed with trialkyloxonium salts (Baker and Robinson 1976).

The sodium channels of certain organs or cells are known to be relatively insensitive to TTX and STX. Classical examples include the sodium channels of cardiac muscle and denervated skeletal muscle (Narahashi 1974, 1988*a*). The apparent dissociation constant for TTX block of sodium channels was estimated to be 1 $\mu$M for rabbit Purkinje fibres (Cohen *et al.* 1981), 9 $\mu$M for rat myocardial cells (Matsuki and Hermsmeyer 1983), and 14 $\mu$M for pig papillary muscle (Baer *et al.* 1976). Thus, the cardiac muscles are approximately three orders of magnitude less sensitive to TTX than nerves and skeletal muscles.

TTX-R sodium channels have also been found in neuronal tissues, including bullfrog sensory neurones (Campbell 1988; Guo and Strichartz 1990), garter snake sensory neurones (Jones 1986), group C sensory neurones (Bossu and Feltz 1984), rat nodose neurones (Ikeda *et al.* 1986; Ikeda and Schofield 1987), and human and mouse dorsal root ganglion (DRG) neurones (Matsuda *et al.* 1978; Yoshida *et al.* 1978; Kostyuk *et al.* 1981; McLean *et al.* 1988; Schwartz *et al.* 1990).

TTX-S and TTX-R sodium channels of rat DRG neurones have recently been analysed in detail (Ogata and Tatebayashi 1992, 1993; Roy and Narahashi 1992; Elliott and Elliott 1993). They were different in several aspects. First, the sensitivity to TTX and STX was different between TTX-R and TTX-S sodium channels by five orders of magnitude (Fig. 2.3). Second, TTX-R sodium channels activated and inactivated more slowly than TTX-S sodium channels (Fig. 2.4). Third, TTX-R sodium channels activated and inactivated at more positive membrane potential than TTX-S sodium channels (Fig. 2.5).

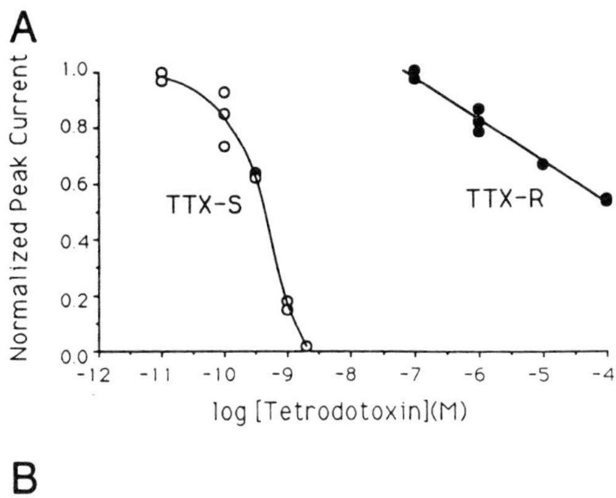


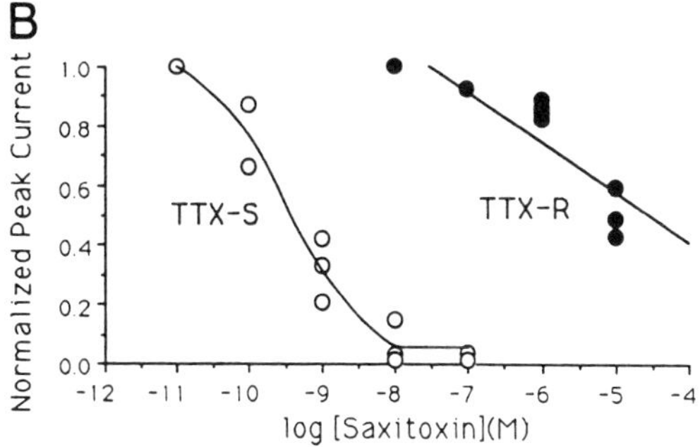


**Fig. 2.3** Dose–response relationships for TTX and STX block in rat dorsal root ganglion neurons. Cells expressing solely TTX-S ($n$=3) or TTX-R ($n$=3) currents were exposed to increasing concentrations of TTX or STX and pulsed once per minute to +10 mV to determine peak current amplitude. Steady-state peak current amplitudes reached at each concentration were normalized to control toxin-free amplitudes and plotted against toxin concentration. A. TTX dose–response curve, with $K_d$ values of 0.3 nM (TTX-S) and 100 $\mu$M (TTX-R). B. STX dose–response curve, and $K_d$ values of 0.3 nM (TTX-S) and 100 $\mu$M (TTX-R). (From Roy and Narahashi 1992.)

There also were large differences in the sensitivity to drug actions between TTX-R and TTX-S sodium channels. Lidocaine blocked TTX-S sodium channels four times more potently than were TTX-R sodium channels (Roy and Narahashi 1992). By contrast, TTX-R sodium channels were blocked by $Pb^{2+}$ and $Cd^{2+}$ more potently than were TTX-S sodium channels (Roy and Narahashi 1992). More drastic differences between TTX-S and TTX-R sodium channels were found in the action of the pyrethroid insecticides in modulating the channel kinetics (Ginsburg and Narahashi 1993; Tatebayashi and Narahashi 1994). This will be described later in more detail.

Recently, the TTX-R sodium channel of rat DRG neurones has been cloned, revealing an open reading frame of 5871 nucleotides encoding a 1957 amino acid protein, SNS, that showed 65% identity with the rat cardiac TTX-insensitive sodium channel (Akopian *et al.* 1996). This study has opened the door to the structural basis for the differential drug sensitivity of TTX-R and TTX-S sodium channels.

Striatal and hippocampal neurones of 21- to 40-day-old rats generated a slow sodium current which was insensitive to 12 $\mu$M TTX (Hoehn *et al.* 1993). This TTX-R sodium channel current activated and inactivated very slowly. Neurones isolated from the superficial layers of the medial entorhinal cortex of the rat generated sodium currents which were

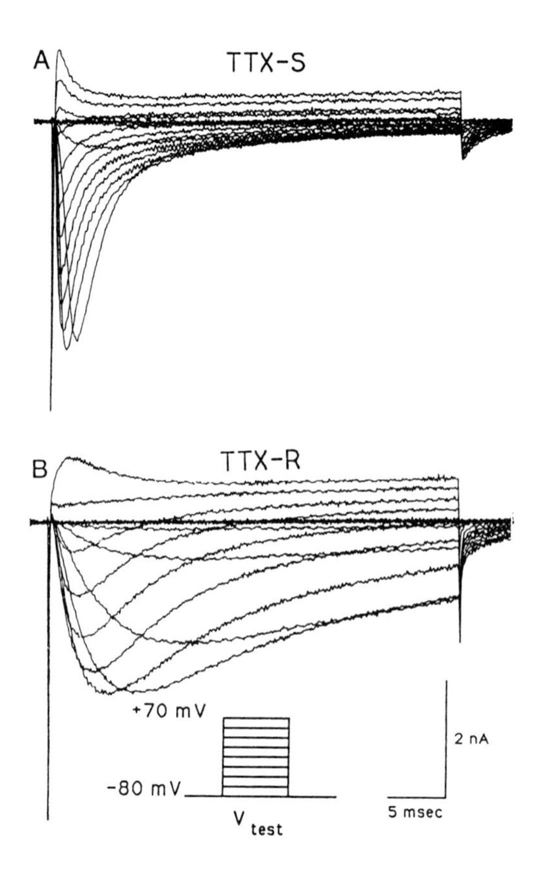


**Fig. 2.4** Families of TTX-S (a) and TTX-R (b) sodium channel currents in rat DRG neurones. Pulse protocol is shown at the bottom. (From Roy and Narahashi 1992.)

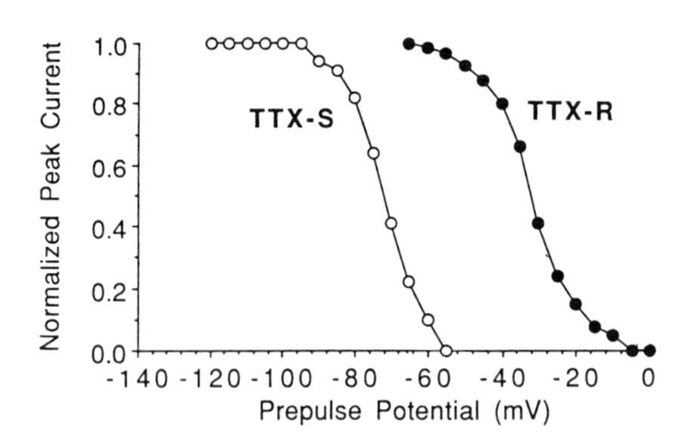


**Fig. 2.5** Steady-state inactivation curves for TTX-S and TTX-R sodium channels in rat DRG neurones. The membrane potentials at which 50% inactivation occurs are −72.5 mV and −35.4 mV for TTX-S and TTX-R channels, respectively. (From Roy and Narahashi 1992.)

relatively insensitive to TTX ($IC_{50}$ - 146 nM) (White *et al.* 1993). This TTX-insensitive sodium channel had a higher sensitivity to the blocking action of $Cd^{2+}$, $Zn^{2+}$, and $La^{3+}$ than the TTX-S sodium channel.

Sodium channels of the brain, skeletal muscle, and heart exhibit different pharmacological properties. The brain and skeletal muscle sodium channels are more sensitive to TTX and STX than the cardiac sodium channels, whereas only the skeletal muscle sodium channels are sensitive to the blocking action of $\mu$-conotoxin (Trimmer and Agnew 1989). The skeletal muscle and cardiac sodium channels have the same blocking affinity for the divalent cations, $Mg^{2+}$, $Ca^{2+}$, $Sr^{2+}$, $Ba^{2+}$, $Mn^{2+}$, $Co^{2+}$, and $Ni^{2+}$, whereas the cardiac sodium channels are more sensitive to $Cd^{2+}$ and $Zn^{2+}$ than the skeletal muscle sodium channels (Ravindran *et al.* 1991; Doyle *et al.* 1993). Competitive interactions between TTX/STX binding and cations have been shown for the alkali cations $Li^{+}$, $Na^{+}$, $K^{+}$, $Rb^{+}$, and $Cs^{+}$ (Barchi and Weigele 1979) and for $Zn^{2+}$ and $Ca^{2+}$ (Worley *et al.* 1986; Schild and Moczydlowski 1991). The concentration of $Zn^{2+}$ needed for the interaction with STX is too low to produce significant screening effect (Schild and Moczydlowski 1991).

The amino acids that are responsible for the differential actions of TTX/STX and $Zn^{2+}/Cd^{2+}$ have been studied. Iodoacetamide, a cysteine-specific alkylating agent, abolished the blocking action of $Zn^{2+}$ and modified the kinetics of STX binding in cardiac sodium channels. It was found that a cysteine group is responsible for the low sensitivity to TTX/STX and the high sensitivity to $Zn^{2+}$ (Schild and Moczydlowsk 1991), and that a tyrosine (adult skeletal muscle-specific) or phenylalanine (brain-specific) group confers on the channel high TTX and low metal sensitivity (Backx *et al.* 1992; Satin *et al.* 1994*a*). In the brain, TTX/STX and divalent cations block the sodium channel at overlapping sites (Satin *et al.* 1994*b*). Although TTX and STX have been assumed to bind to the same sodium channel site, a recent study of cardiac sodium channels using membrane-impermeant, cysteine-specific, methanethiosulphonate analogues does not support the notion that STX and TTX are interchangeable (Kirsch *et al.* 1994). A model has also been proposed to predict that TTX interacts directly with the SS1–SS2 segments of repeats I and II and that STX additionally interacts with the segments of repeats III and IV (Lipkind and Fozzard 1994).

The effect of a chemical on an ion channel may vary with the membrane potential. The voltage dependence is an important parameter in understanding the mechanism of action as it provides us with a clue to the site and mode of interaction of the chemical with the channel. For instance, it is possible to calculate the site of drug binding within the membrane potential field from the data of voltage dependence (Woodhull 1973). TTX/STX block of sodium channels was previously found to be independent of the membrane potential (Hille 1968; Almers and Levinson 1975; Jaimovich *et al.* 1976; Catterall *et al.* 1979; Krueger *et al.* 1979; Cohen *et al.* 1981; French *et al.*, 1984; Moczydlowski *et al.* 1984). However, some authors reported voltage-dependent block of sodium channels by TTX (Baer *et al.* 1976; Reuter *et al.*, 1978).

However, recent studies have cast some doubt about the notion that TTX/STX block is voltage-independent. Rando and Strichartz (1986) have shown that, although the STX block of normal sodium channels is not voltage-dependent, the STX block of the sodium channels modified by BTX is voltage-dependent. When repetitive 10 ms conditioning pulses to various voltage levels were applied, STX block was greatest at $-50$ mV and decreased with more and less negative pulse levels, resulting in a bell-shaped voltage-dependent curve. The results are interpreted as indicating that STX binding to open channels is voltage dependent, whereas STX affinity for closed and inactivated channels is voltage independent. Recent studies with rat brain sodium channels have shown an increase in affinity for STX/TTX with hyperpolarization (Satin *et al.* 1994*b*). The data are interpreted as indicating that the toxins bind to a site located 15% into the membrane.

Our study with crayfish giant axons has also shown that STX block is voltage dependent (Salgado *et al.* 1986). STX block was use dependent, and was enhanced with an increasing negativity of the holding membrane potential from −100 mV to −160 mV. On the contrary, the resting STX block was dependent on both the membrane potential and external calcium concentration in a complex manner (Fig. 2.6). With 10 mM $Ca^{2+}$ outside, no voltage dependence of STX block was observed. However, with 50 mM $Ca^{2+}$ the apparent dissociation constant for STX block was markedly increased and the block exhibited a dual voltage dependence increasing with either depolarization or hyperpolarization from −120 mV. The results are interpreted as indicating that the interaction between STX and calcium ions is modulated by the membrane potential.

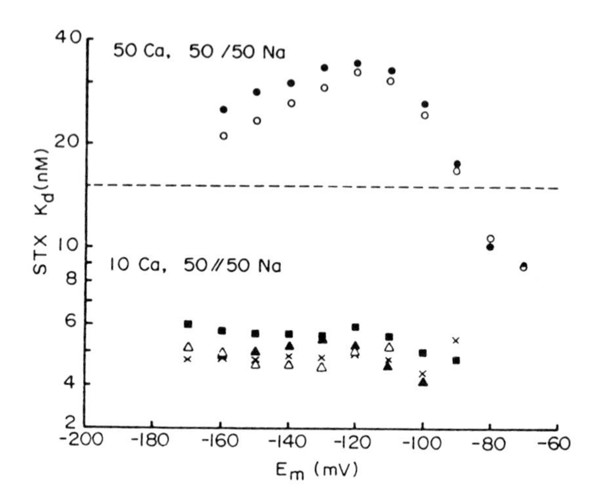


**Fig. 2.6** External $Ca^{2+}$ concentration affects the voltage dependence of resting block by STX in crayfish giant axons. Each symbol in each solution represents a different experiment. The dashed line represents the predicted $K_d$ from the change in surface potential when $Ca^{2+}$ concentration is increased from 10 mM (10 Ca) to 50 mM (50 Ca). The reduction of the negative surface potential of 13.6 mV was estimated from the degree of shift of the steady-state sodium inactivation curve when going from 10 to 50 mM $Ca^{2+}$. 50/50 Na: 50 mM external and 50 mM internal $Na^+$ concentrations. (From Salgado *et al.* 1986.)

## Conotoxins

Marine snails that belong to the genus *Conus* contain various toxins. Some of them block acetylcholine receptors, some block skeletal muscle sodium channels, and others open sodium channels. A summary of these toxins and their actions is given by Wu and Narahashi (1988), Hopkins *et al.* (1995), Martinez *et al.* (1995), Cartier *et al.* (1996), and Terlau *et al.* (1996).

Conotoxins are classified into several groups on the basis of their actions on ion channels and neuroreceptors. α-Conotoxins block acetylcholine receptors and include conotoxins GI, GIA, and GII which are isolated from *Conus geographus* (Gray *et al.* 1981; McManus *et al.* 1981;

McManus and Musick 1985, and αA-conotoxin PIVA from *c. purpurascens* (Hopkins *et al.* 1995). Conotoxins that open sodium channels include MTX from *C. magus* (Kobayashi *et al.* 1983), striatoxin from *C. striatus* (Strichartz *et al.* 1980; Kobayashi *et al.* 1981, 1982) and δ-conotoxin PVIA from *c. purpurascens* (Terlau *et al.* 1996). 314-Conotoxin (Conotoxin GVIA) from *C. geographus* (Olivera *et al.* 1984) blocks calcium channels (Kerr and Yoshikami 1984; Cruz and Olivera 1986), especially N-type calcium channels (McCleskey *et al.* 1987).

μ-Conotoxins are isolated from *Conus geographus*, and include GIIIA (gographutoxin I), GIIIB (geographutoxin II), and GIIIC. This group of toxins preferentially blocks the sodium channels of skeletal muscle and electroplax. The differential blocking action of conotoxin GIIIB on the action potentials is illustrated in Fig. 2.7. Whereas the action potential recorded from guinea pig extensor digitorum longus muscle is completely blocked by 0.5 μM conotoxin GIIIB, the action potentials from the crayfish giant axon and the papillary muscle of guinea pig heart are not affected at all at concentrations of 5 μM and 1 μM, respectively (Kobayashi *et al.* 1986). Voltage-clamp experiments with bullfrog sartorius muscle fibres clearly showed block of sodium channel currents (Kobayashi *et al.* 1986).

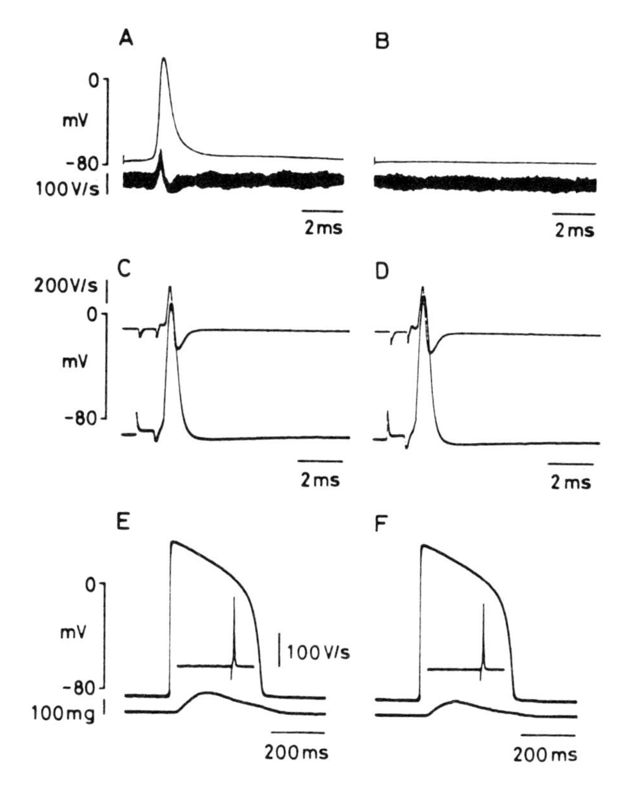


**Fig. 2.7** Effects of geographutoxin II (GTX II) on various tissues. A and B. Effect on the action potential of guinea pig extensor digitorum longus muscle. Action potential (upper trace) and its derivative (lower trace) before (A) and 6 min after (B) application of $5 \times 10^{-7}$ M GTX II. $Ca^{2+}$-free medium was used to depress muscle contraction. C and D. Effect on crayfish giant axon. Action potential (lower trace) and its derivative (upper trace) before (C) and 1 h after (D) application of $5 \times 10^{-6}$ M GTX II. E and F. Effect on guinea pig papillary muscle of the heart. Action potential (upper trace) and contraction (lower trace) before (E) and 30 min after (F) application of $1 \times 10^{-6}$ M GTX II. The derivative of the corresponding action potential shown in the inset was recorded at twice faster sweep speed. (From Kobayashi *et al.* 1986.)

$\mu$-Conotoxin GIIIA appears to bind to a site in the outer vestibule of the rat skeletal muscle sodium channel ($\mu$1) which partially overlaps with the TTX binding site (Dudley *et al.* 1995). This conclusion was based on the observations that mutation of the glutamate at position 758 to glutamine (E758Q) decreased the binding affinity of both GIIIA and TTX.

## Local anaesthetics and related blockers

Local anaesthetics are among the first chemicals that were studied by voltage-clamp techniques. Taylor (1959) and Shanes *et al.* (1959) showed that procaine and cocaine blocked both sodium and potassium currents in squid giant axons. Most local anaesthetics are tertiary amines and are ionized to become cationic forms depending on the $pK_a$ of the compound, which is in the range of 7 to 9, and the pH of the medium. Thus, a question arises as to whether the cationic form or the uncharged molecular form is responsible for conduction block. Earlier studies indicated that the uncharged molecular form was active. This conclusion was based on the observation that an increase in the pH of the medium increased the local anaesthetic potency. However, more careful studies by Ritchie and his associates have led to the conclusion that the cationic form is active (Ritchie and Greengard 1966). There still were many observations which could not be accounted for on the basis of the cationic from being active. Although they considered local anaesthetic penetration to the nerve sheath in the molecular form, the nerve membrane as an additional barrier was not taken into consideration.

We have performed extensive experiments using internally perfused squid giant axons. Two tertiary lidocaine derivatives, 6211 with a $pK_a$ of 6.3, and 6603 with a $pK_a$ of 9.8, were applied externally or internally at various pH levels, and the degree of block of the maximum rate of rise of the action potential was measured. The data were compatible only with the scheme in which the cationic form acts from inside the nerve membrane (Narahashi *et al.* 1970). Two quaternary lidocaine derivatives, QX314 and QX572, and hemicholinium-3 which is a divalent cation blocked the action potential only when applied inside the axon in a manner independent of pH (Frazier *et al.* 1970). In another study, 6211 was applied internally at various concentrations and at various pH levels so that the internally present charged cationic form remained constant. Under this condition, the degree of block was also kept constant (Frazier *et al.* 1971). Thus, it was concluded that tertiary amine local anaesthetics penetrate the nerve sheath and nerve membrane in the uncharged molecular form and are ionized to form the local anaesthetic cations which in turn block the sodium channels from inside the membrane. It is important to recognize that the charged cationic form of tertiary amine local anaesthetic and its permanently charged quaternary derivative are different in their structure, and that experiments using both tertiary and quaternary forms are crucial for drawing the conclusion.

The mechanism of local anaesthetic interaction with the sodium channels has been studied extensively by many investigators. Several important features have emerged as a result of local anaesthetic studies, including state dependence, voltage dependence, current dependence, and use dependence of block. State dependence refers to the question as to what state of the channel has a high affinity for local anaesthetics, and includes closed state, open state and inactivated state. Voltage dependence is the question of whether local anaesthetic block changes with the membrane potential. Current dependence is the question of whether local anaesthetic block depends on the intensity and direction of current flowing through the open channel in question. Use dependence refers to the question as to whether local anaesthetic block is accelerated or augmented by repetitive depolarizing pulse stimulations.

In order to understand these features of local anaesthetic block, the following over-simplified scheme for various states of the channel would be of help:

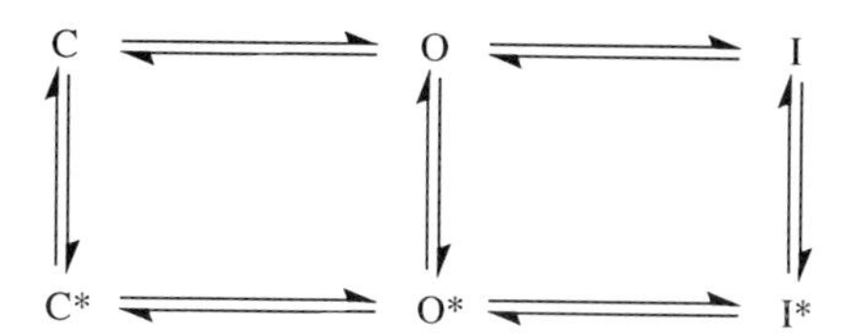


C, O and I refer to the closed, open and inactivated states of the channel, respectively, and C*, O* and I* refer to the respective drug-bound states.

A drug could have a high affinity for the closed state, open state and/or inactivated state. If it has a high affinity for the open state only, repetitive stimuli that open the channel repeatedly will facilitate block causing use-dependent block. Use-dependent block has been demonstrated for a number of blocking agents acting on the sodium channel including QX-314 (Strichartz 1973; Yeh 1982), GEA-968 (Courtney 1975), lidocaine (Hille *et al.* 1975), and chlorpromazine (Ogata *et al.* 1990) (Fig. 2.8). Open channel block is often associated with an accelerated falling phase of sodium current as the drug enters the channel as soon as it opens and curtails the current (Yeh and Narahashi 1977) (Fig. 2.9). Flickering is often observed during opening of single channels as the blocking molecule binds to and unbinds from the channel site, depending on the rates of binding and unbinding (Neher and Steinbach 1978; Yeh *et al.* 1986). High affinity of a drug for the inactivated state of the sodium channel has been shown for chlorpromazine block (Ogata *et al.* 1990). Voltage-dependent block has been shown for a number of blocking agents including pancuronium (Yeh and Narahashi 1977), chlorpromazine (Ogata *et al.* 1990), QX-314 (Strichartz 1973; Yeh 1982), 9-aminoacridine (Yeh 1979), and strychnine (Shapiro 1977). Current-dependent block has been demonstrated for the potassium channel block by tetraethylammonium (TEA) derivatives (Armstong 1969, 1971), the sodium channel block by paragracine (Seyama *et al.* 1980), and the acetylcholine channel block by methyl- and ethylguanidine (Vogel *et al.* 1984).

Development and application of molecular biology and genetics technologies to neuroreceptors and ion channels during the past 10 years have made it possible to identify the molecular site of action of local anaesthetics on the sodium channel. Site-directed mutagenesis was used to study the role of segment S6 of domain IV of the sodium channel $\alpha$ subunit from rat brain (Ragsdale *et al.* 1994). The channels were expressed in *Xenopus* oocytes. Mutation F1764A at the middle of this segment decreased the affinity of etidocaine for the open and inactivated channels and abolished both the use dependence and voltage dependence of block. Mutation N1769A increased the affinity of the resting channel. Mutation I1760A created an access pathway for local anaesthetic molecules to reach the binding site from the external side.

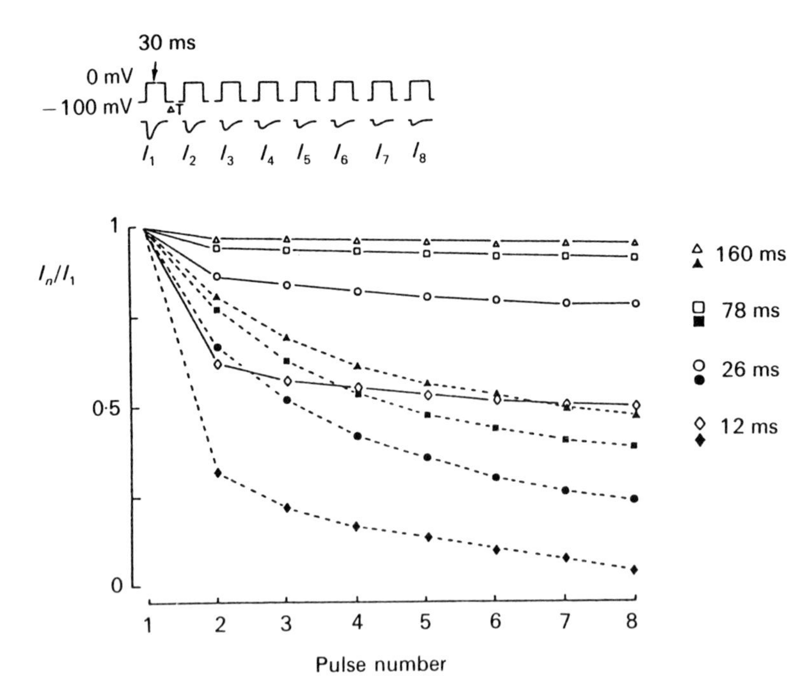


**Fig. 2.8** The time course of change in the amplitude of sodium current elicited by a train of depolarizing pulses. Eight consecutive pulses to 0 mV (30-ms duration) were delivered with various pulse intervals ranging from 12 ms to 3 s from a holding potential of −100 mV. The peak current amplitude for each pulse in the train ($I_n$) was normalized to that of the first pulse ($I_1$) and is plotted as function of the pulse number. Open symbols, control; closed symbols, in the presence of 1 $\mu$M chlorpromazine. Numerals attached to the symbols or traces indicate interpulse intervals ($\Delta$T). (From Ogata *et al.*, 1990.)

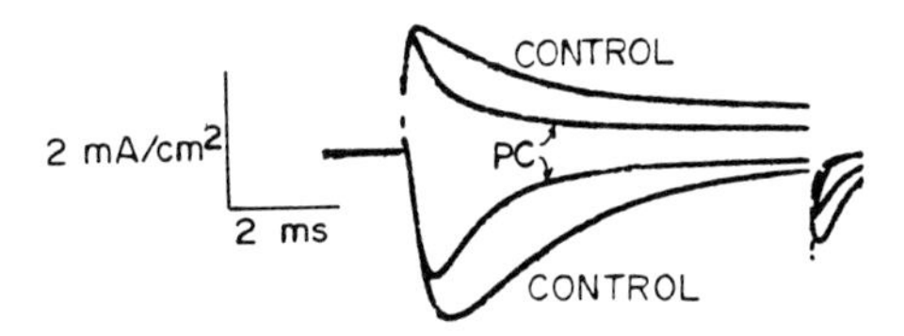


**Fig. 2.9** Effect of internal application of $1 \times 10^{-3}$ M pancuronium (PC) on sodium current in a squid giant axon internally perfused with 20 mM tetraethylammonium. The upper traces are superimposed sodium currents associated with 160 mV depolarizing pulses before and during application of PC, and the lower traces are those with 80 mV pulses. Note that the rising phase of sodium current is not affected, while its falling phase is accelerated. Temperature 6.3 °C. (From Yeh and Narahashi 1977.)

Lidocaine is known to block cardiac sodium channels more preferentially than nerve and skeletal muscle sodium channels. This difference is traditionally assumed to derive from the fact that cardiac sodium channels are in the inactivated state longer than nerve and skeletal muscle sodium channels, because local anaesthetics exhibit a higher affinity for the inactivated state than the resting state. However, experiments with the skeletal $\mu$1-$\beta_1$, sodium channel and the cardiac hH1-$\beta_1$, sodium channel expressed in frog oocytes clearly indicated that lidocaine showed a higher affinity for the hH1-$\beta_1$, sodium channel than the $\mu$1-$\beta_1$, sodium channel at their resting state (Nuss *et al.* 1995). By contrast, these two types of sodium channels at their inactivated state had the same affinity for lidocaine. The skeletal muscle sodium channel (hSkM1) and the cardiac sodium channel (hH1) expressed in human embryonic kidney cell line (HEK 293) also showed differential sensitivities to lidocaine (Wang *et al.* 1996). Tonic block, use-dependent block, and inactivated-state block by lidocaine were all greater in hH1 than in kShM1. Thus, it was concluded that the difference in lidocaine sensitivity between cardiac and skeletal muscle sodium channels is due to their intrinsic sensitivity.

It is well known that the α subunit of brain and skeletal muscle sodium channels expressed in frog oocytes generates sodium currents much slower than the currents from native sodium channels and that coexpression of the α and β1 subunits produces sodium currents with normal kinetics (Isom *et al.* 1994). However, the α subunit of cardiac sodium channels expressed alone generates currents showing the normal time course (Satin *et al.* 1992; Makielski *et al.* 1996). Unexpectedly, the sensitivity to the resting and phasic block by lidocaine was decreased by coexpression of the α and β1 subunits of cardiac sodium channels (Makielski *et al.* 1996).

## Sodium channel modulators

### Chemicals that block sodium channel inactivation

#### Pronase

Pronase is a classical example of a blocker of sodium channel inactivation (Armstrong *et al.* 1973; Rojas and Rudy 1976). It eliminates the inactivation completely when perfused internally through the squid giant axon. Therefore, pronase has been used extensively as a useful tool to remove the sodium channel inactivation, enabling accurate analysis of sodium channel activation kinetics.

#### N-bromoacetamide

*N*-Bromoacetamide and *N*-bromosuccinimide block the sodium channel inactivation completely in squid giant axons when perfused internally (Oxford *et al.* 1978). These chemicals have also been useful tools in sodium channel investigations.

#### Sea anemone toxins

Venoms from various species of sea anemones contain mixtures of polypeptide toxins. These polypeptides are classified into four groups based on their molecular weight. Two of them with lower molecular weights ($< 3000$ and 4000–6000 Da) are potent neurotoxins which modulate sodium channel gating kinetics, whereas the other two groups have no effect on ion channels (Wu and Narahashi 1988). The former two groups bind to site 3 of the sodium channel (Catterall 1992).

The selective block of the sodium channel inactivation was first demonstrated for a toxin isolated from the sea anemone *Condylactis gigantea* (Fig. 2.10) (Narahashi *et al.* 1969*b*). Since then, several other sea anemone toxins have been shown to exert the same effect as *Condylactis* toxin on sodium channels, including *Anemonia sulcata* toxin II (Romey *et al.* 1976) and anthopleurin A isolated from *Anthopleura xanthogrammica* (Low *et al.* 1979). Their amino acid sequence is similar to each other, with only seven amino acid residues out of 47 being different (Norton 1981). Interestingly, sea anemone toxins are not effective in changing the sodium channel gating kinetics in squid giant axons, although they are potent on crustacean axons, vertebrate neurones and cardiac muscles (Narahashi *et al.* 1969*b*; Romey *et al.* 1976). Binding experiments indicated that sea anemone toxins and scorpion toxins share a common site in the sodium channel (Catterall and Beress 1978; Couraud *et al.* 1978).

Because sea anemone toxins are polypeptides and bind irreversibly to the sodium channels, they have been chemically modified to produce radiolabelled, iodinated, or tritiated compounds, or labelled with fluorescent probes, phosphorescent probes, or photo-affinity probes as useful tools for the study of sodium channels (Habermann and Beress 1979; Hucho *et al.* 1979; Vincent *et al.* 1980; Stengelin *et al.* 1981; Rack *et al.* 1983).

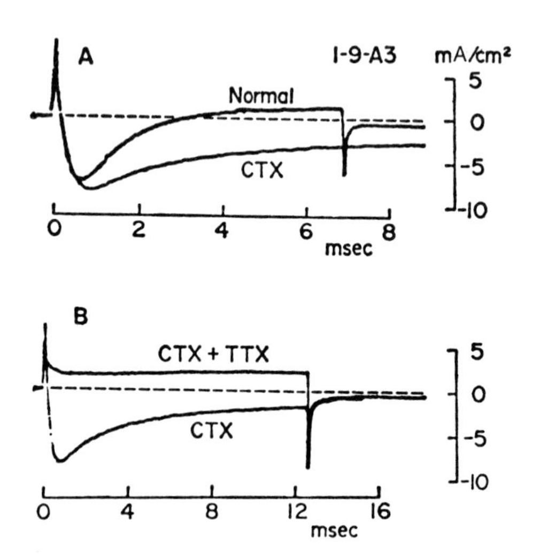


**Fig. 2.10** Membrane currents associated with step depolarization from the holding potential of −110 mV to 0 mV in a voltage-clamped crayfish giant axon. A. Before (normal) and during a longer pulse applied after washing in *Condylactis* toxin (CTX) (0.2 mg/ml) for 10 min. B. During application of CTX for 10 min and after application of CTX plus $3 \times 10^{-7}$ M tetrodotoxin (TTX) for 2 min. (From Narahashi *et al.* 1969*b*.)

## Scorpion toxins

Initial studies to demonstrate the sodium channel modulating action of scorpion toxins were performed by Koppenhöfer and Schmidt (1968*a,b*) and Narahashi *et al.* (1972). Since that time, a number of toxins (almost 100) have been isolated from various species of scorpions. Their chemical structures and physiological effects on sodium channels were reviewed by Meves *et al.* (1986), Strichartz *et al.* (1987), and Catterall (1992). Scorpion toxins may be classed into three groups based on their mechanism of modulation of sodium channel gating kinetics. Class 1 includes scorpion toxins (α -scorpion toxins) that inhibit the sodium channel inactivation. Class 2 ($\beta$-scorpion toxins) includes those that alter the sodium channel activation by shifting the activation–voltage relationship in the hyperpolarizing direction. Class 3 includes those that alter both the activation and inactivation gating kinetics. Classes 2 and 3 will be described later in this chapter.

$\alpha$ -Scorpion toxins (Class 1) that inhibit the sodium channel inactivation include: toxin V, toxin var. 1–3, and toxin II$\alpha$ from *Centruroides sculpturatus* (Hu *et al.* 1983; Wang and Strichartz, 1983; Meves *et al.* 1982); toxins $M_7$ and 2001 from *Buthus eupeus* (Mozhayeva *et al.* 1980); toxins V and XII from *Buthus tamulus* (Siemen and Vogel 1983); and toxins I and II from *Androctonus australis* (Romey *et al.* 1975; Bazan and Bernard 1984; Pichon and Pelhate 1984). In the node of Ranvier, these scorpion toxins also shift the steady-state sodium inactivation curve in the hyperpolarizing direction and elevate the curve at positive potentials so that inactivation is incomplete with deplorization (Mozhayeva *et al.* 1980; Meves *et al.* 1982; Wang and Strichartz 1983). $\alpha$ -Scorpion toxins bind to site 3 of the sodium channel, the same site as that of sea anemone toxins (Catterall 1992).

#### Goniopora toxin

The stony corals *Goniopora* contain *Goniopora* toxin. This polypeptide toxin blocks the sodium channel inactivation in much the same way as do sea anemone toxins in crayfish giant axons (Muramatsu *et al.* 1985) and in bullfrog atrial muscle (Noda *et al.* 1984).

### Chemicals that alter voltage dependence of sodium channel activation

#### Scorpion toxins

$\beta$-Scorpion toxins (Class 2) as briefly described above cause a shift of the sodium channel activation–voltage relationship in the hyperpolarizing direction without slowing the inactivation kinetics. Whole venom from *Centruroides sculpturatus* was used for the first time to demonstrate this effect (Cahalan 1975). Class 2 scorpion toxins include toxins I, III, IV, and VI isolated from *C. sculpturatus* (Meves *et al.* 1982; Hu *et al.* 1983; Simard *et al.* 1986); toxins I$\alpha$-III$\alpha$ and -III$\beta$ from the same species (Wang and Strichartz 1983); toxin II from *Centruroides suffusus suffusus* (Couraud *et al.* 1982); toxin $\gamma$ from *Tityus serrulatus* in the node of Ranvier (Zaborovskaya and Khodorov 1985). These scorpion toxins bind to site 4 of sodium channels (Catterall 1992).

### Chemicals that alter voltage dependence of both activation and inactivation of sodium channels

#### Scorpion toxins

Toxin $\gamma$ from *Tityus serrulatus*, which is described in the preceding section, alters both activation and inactivation kinetics in neuroblastoma cells (Vijverberg *et al.* 1984), but not in the node of Ranvier (Zaborovskaya and Khodorov 1985). Both activation and inactivation curves are shifted in the hyperpolarizing direction by this toxin. Toxin TsIV-5 from *Tityus serrulatus* causes similar effects as toxin $\gamma$. Toxin VII from *Centruroides sculpturatus* also shifts the activation and inactivation curves in the direction of more negative potentials (Simard *et al.* 1986). These toxins bind to site 4 of sodium channels (Catterall 1992).

## Chemicals that alter sodium channel activation kinetics and inhibit inactivation

### Batrachotoxin

Batrachotoxin (BTX) is one of the toxic components contained in the skin secretion of the Colombian poison arrow frog *Phyllobates aurotaenia* (Albuquerque *et al.* 1971; Albuquerque and Daly 1976; Khodorov 1985). BTX has a potent and irreversible depolarizing action on nerve, skeletal muscle, and cardiac muscle membranes through openings of sodium channels (Narahashi *et al.* 1971; Albuquerque *et al.* 1973). The sodium channel inactivation is removed by BTX (Khodorov *et al.* 1975; Khodorov 1978; Khodorov and Revenko 1979) which acts only when the channel is open (Fig. 2.11) (Tanguy and Yeh 1991), and the sodium channel activation curve is greatly shifted in the hyperpolarizing direction (Fig. 2.11) (Tanguy and Yeh 1991). These two effects account for membrane depolarization. BTX modifies the sodium channel in its open configuration.

Single-channel analyses have revealed that the mean open time of sodium channels is greatly prolonged by BTX (Fig. 2.12), and that, in the presence of BTX, the channels open at large negative potentials where no channels are open in normal, BTX-untreated conditions (Quandt and Narahashi 1982). The normal and BTX-modified sodium channel activities are clearly discernible in a membrane patch exposed to BTX (Fig. 2.13) (Quandt and Narahashi 1982). BTX binds to site 2 of sodium channels (Catterall 1992).

### Grayanotoxins

Grayanotoxins (GTXs) are the toxic principles isolated from the leaves of plants belonging to the family Ericaceae (*Leucothoe*, *Rhododendron*, *Andromeda*, *Kalmia*). In squid giant axons, grayanotoxin I (GTX I) and α-dihydrograyanotoxin II (α$H_2$-GTX II) cause a large membrane depolarization due to a specific increase in resting sodium permeability (Seyama and Narahashi 1973; Narahashi and Seyama 1974; Hironaka and Narahashi 1977). Voltage-clamp measurements have shown that the resting sodium permeability is increased by GTX by a factor of 10–100, depending on the concentration.

Upon step depolarization of the squid axon membrane exposed to GTX, there appears a peak sodium current which is decreased in amplitude and a secondary slow steady-state sodium current (Seyama and Narahashi 1981). The latter current represents the current flowing through the open GTX-modified sodium channels. The modified sodium channels open at large negative membrane potentials where the normal unmodified sodium channels do not open. This explains a large membrane deplorization caused by GTX. The ionic selectivity of the modified sodium channels is decreased (Frelin *et al.* 1981) and the $P_x/P_{Na}$ ratio, where $P_x$ and $P_{Na}$ refer to the permeabilities to X cations and Na ions, respectively, is greatly increased: the $P_x/P_{Na}$ ratios for hydroxylamine, formamidine, and ammonium are 1.58, 0.71, and 1.10, respectively (Seyama and Narahashi 1981). GTX binds to site 2 of sodium channels (Catterall 1992).

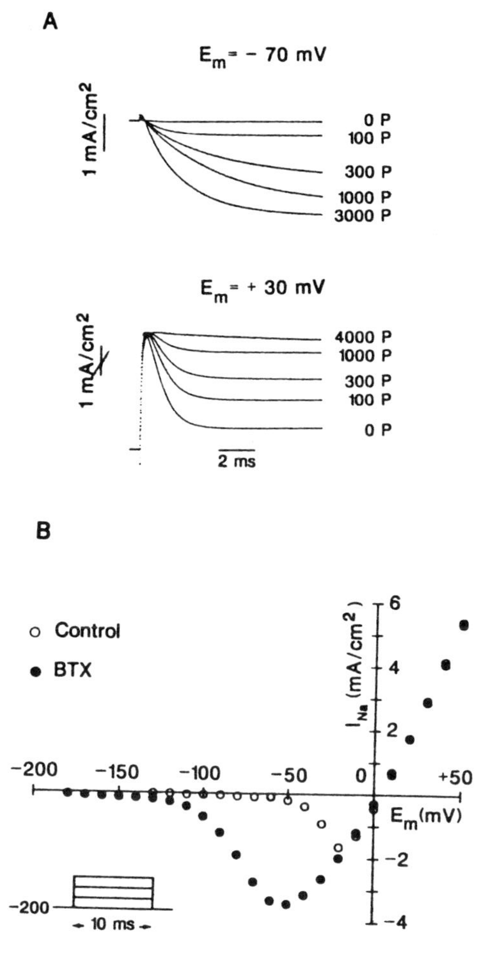


**Fig. 2.11** Batrachotoxin modification of the sodium channels of a squid giant axon. A. The membrane was repetitively depolarized by a train of 4-ms conditioning pulses during the internal perfusion with 3 $\mu$M BTX, and sodium currents were elicited by 10-ms depolarizing pulses to $-70$ mV (top) and +30 mV (bottom), before (traces labelled O P), and after a given number of conditioning pulse ($n$), as indicated for each trace. B. Current–voltage relationships for the peak sodium current in the control and the steady-state Na current after complete BTX modification by 2 $\mu$M using a train of 3200 conditioning pulses. (From Tanguy and Yeh 1991.)

### Veratridine

Veratridine is one of the veratrum alkaloids contained in plants that belong to the tribe Veratreae and the family Liliaceae. The alkaloids are found in various genera such as *Veratrum*, *Schoenocaulon*, and *Zygadenus*. Veratridine causes a membrane depolarization as a result of opening of sodium channels (Ohta *et al.* 1973). Single-channel analyses have shown that veratridine prolongs the mean open time (Garber and Miller 1987) and decreases the conductance (Yoshii and Narahashi 1984; Barnes and Hille 1988). Veratridine-modified sodium channels exhibit the ionic selectivity less than that of normal unmodified channels (Naumov *et al.* 1979; Frelin *et al.* 1981). Veratridine binds to the open sodium channels rapidly,

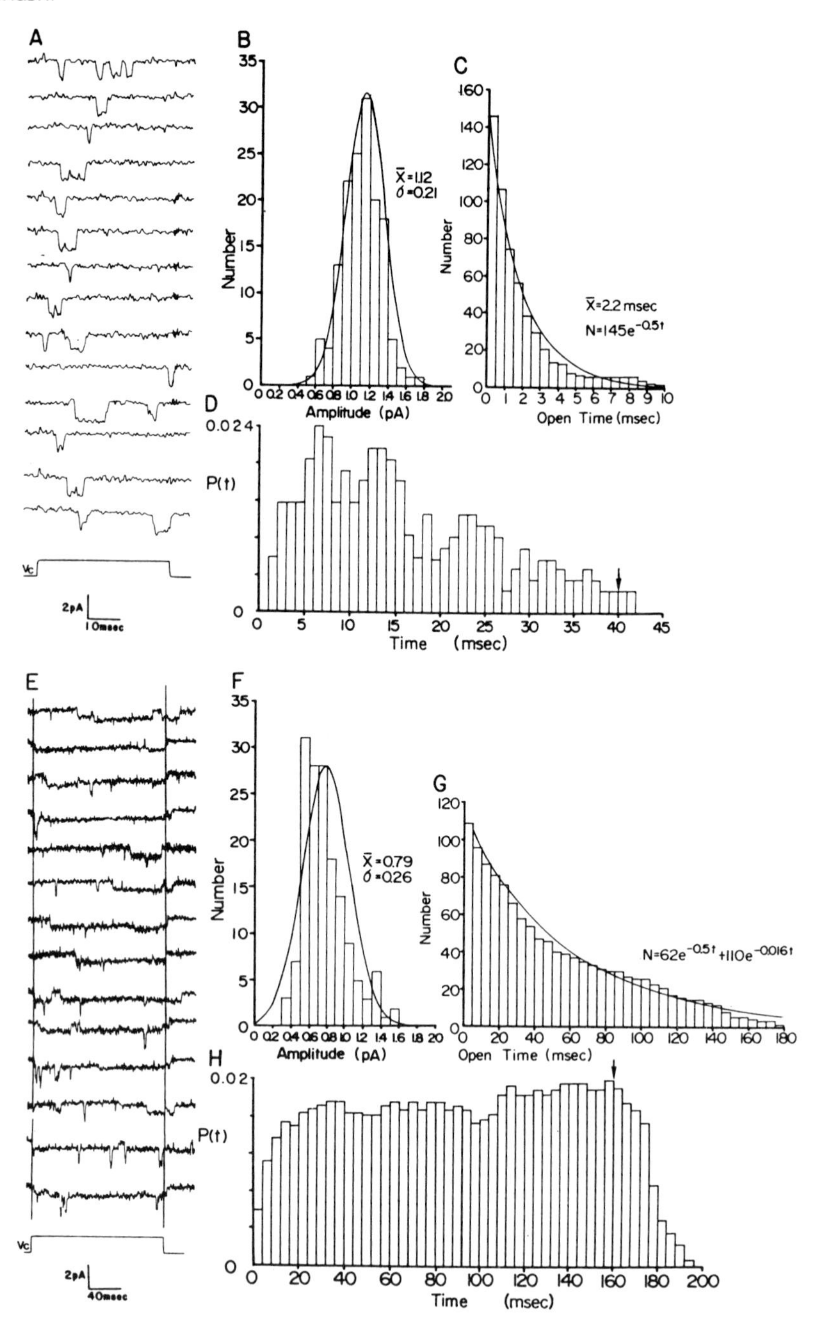


**Fig. 2.12** Single $Na^+$ channel activity in a normal neuroblastoma N1E-115 cell, as recorded by the patch–clamp method. A. Sample records of $Na^+$ channel currents (downward deflections) during step depolarization from a holding potential of −90 mV to −50 mV ($V_c$). B. Amplitude histogram of $Na^+$ channel currents. C. Poisson distribution of channel open time. D. Probability of channel openings during the step depolarization. Arrow indicates the point of step repolarization. The probability curve resembles the time course of a 'macroscopic' $Na^+$ current. E, F, G and H. Same as A, B, C and D, respectively, but in the presence of 10 $\mu$M batrachotoxin. (From Quandt and Narahashi 1982.)

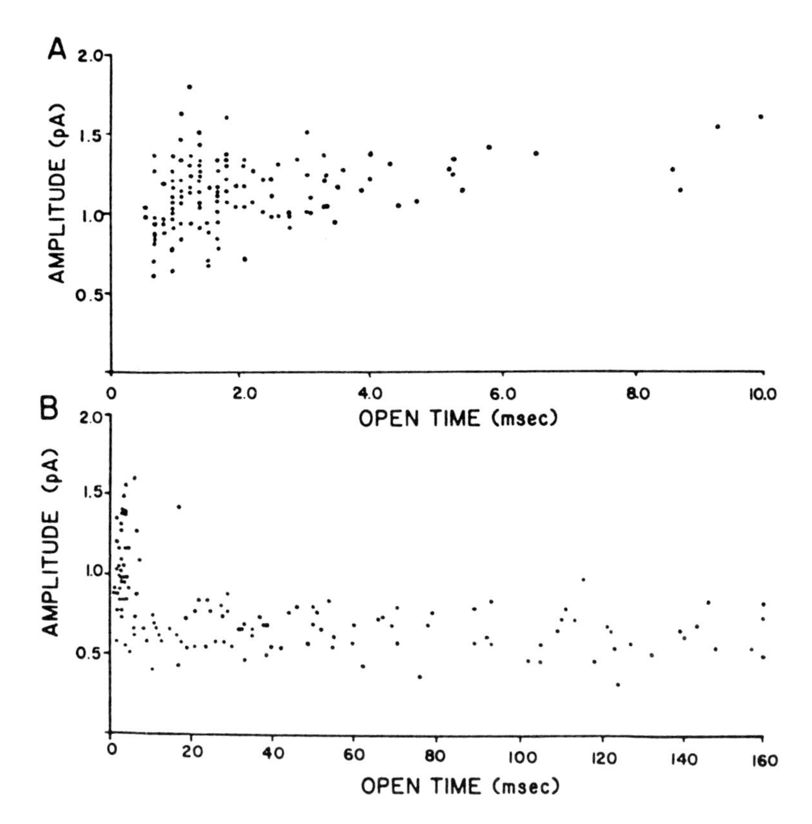


**Fig. 2.13** Two populations of open states for $Na^+$ channels observed in neuroblastoma cell membrane patches exposed to 10 $\mu$M BTX. Scattergrams of the amplitude versus the duration (open time) of each conducting state are shown. A. All channel openings in control are represented. No clear correlation between these parameters exists. Results from excised membranes exposed to BTX are shown in B. Note the longer time scale for the abscissa compared with A. The larger amplitude open states have short open times and the events with long open times all have the smallest current amplitudes. (From Quandt and Narahashi 1982.)

yet it also binds to the inactivated sodium channels slowly (Ulbricht 1969; Leicht *et al.* 1971; Sutro 1986; Rando 1989). The binding site of veratridine has been identified as site 2 of sodium channels (Catterall 1992).

### Aconitine

Aconitine is a toxic component contained in the plant *Aconitum napellus*. It causes an inhibition of the sodium channel inactivation and a shift of the sodium channel activation voltage in the hyperpolarizing direction (Schmidt and Schmitt 1974; Campbell 1982; Grishchenko *et al.* 1983). Aconitine decreases the ionic selectivity of sodium channels (Mozhayeva *et al.* 1977; Campbell 1982; Grishchenko *et al.* 1983) and binds to site 2 of sodium channels (Catterall 1992).

### Striatoxin

*Conus striatus* contains striatoxin (Kobayashi *et al.* 1982). This toxin, unlike other toxins from *Conus* such as α-conotoxin, ω-conotoxin and μ-conotoxins, inhibits the sodium channel inactivation and shifts the sodium channel activation voltage in the hyperpolarizing direction (Strichartz *et al.* 1980; Kobayashi *et al.* 1981).

### Brevetoxins

Brevetoxins are produced by the dinoflagellate *Ptychodiscus brevis* which causes red tides that kill fish (Strichartz *et al.* 1987; Wu and Narahashi 1988). Whereas the nomenclature of brevetoxins is confusing, eight components have been identified called PbTX-1 to PbTX-8 (Shimuzu 1982; Poli *et al.* 1986; Wu and Narahashi 1988). Although the potency and efficacy of the brevetoxins are different, their effects on sodium channels are characterized by the inhibition of inactivation, and the shift of activation voltage in the hyperpolarizing direction which are responsible for membrane depolarization observed in the presence of brevetoxins (Fig. 2.14) (Huang *et al.* 1984; Wu *et al.* 1985; Atchison *et al.* 1986; Sheridan and Adler 1989). Binding experiments have shown that brevetoxins bind to site 5 of sodium channels (Catterall 1992).

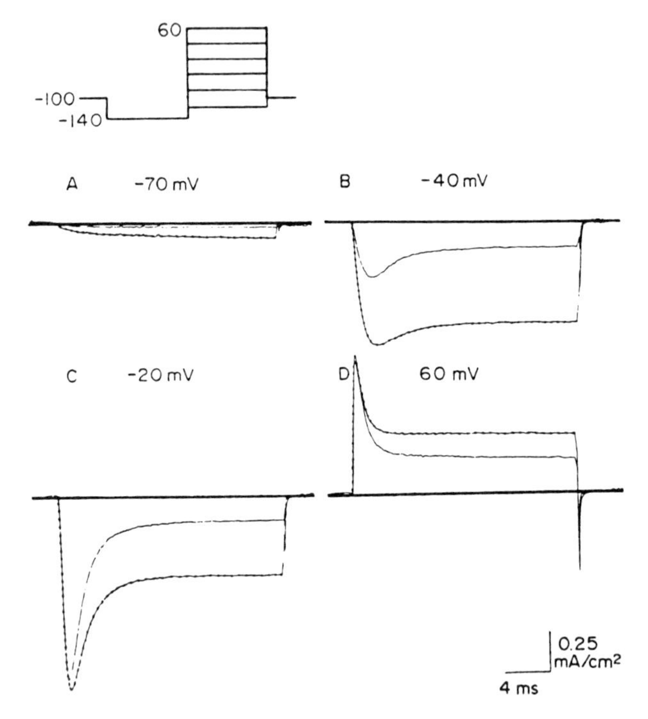


**Fig. 2.14** Effects of internally applied 30 μM brevetoxin-B on $Na^{+}$-currents recorded from a squid giant axon. Sodium currents during 20 mV step depolarizations to −70, −40, −20, and +60 mV are shown. A 20-ms prehyperpolarization to −140 mV was applied from the holding potential of −100 mV before the test step depolarizations (inset). Each of panels (A–D) shows superimposed records of control and brevetoxin-B-treated axons. The toxin caused both peak and steady-state sodium current to increase. Na concentrations in external and internal perfusates were 100 mM and 50 mM, respectively. (From Atchison *et al.* 1986.)

#### Ciguatoxin

Ciguatoxin is responsible for food poisoning from consumption of tropical fish which ordinarily are non-toxic but at times become toxic (Wu and Narahashi 1988). However, ciguatoxin is actually produced by the toxic dinoflagellate *Gambrierdiscus toxicus*. The effects of ciguatoxin on the sodium channels resemble those of brevetoxins, causing an inhibition of inactivation and a shift of activation voltage in the hyperpolarizing direction leading to a membrane depolarization (Bidard *et al.* 1984; Benoit *et al.* 1986). Ciguatoxin binds to the same site as brevetoxins, i.e. site 5 of sodium channels (Catterall 1992).

### Chemicals that inhibit sodium channel inactivation and block activation

#### Chloramine-T

Chloramine-T is an oxidant and inhibits the sodium channel inactivation irreversibly, causing a prolongation of sodium currents (Ulbricht and Stole-Herzog 1984; Wang 1984*a*,*b*; Schmidtmayer 1985; Wang *et al.* 1985; Huang *et al.* 1987). In addition, chloramine-T suppresses the amplitude of peak sodium current in a voltage-dependent manner, and this effect is reversible after washing with chloramine-T-free solutions (Huang *et al.* 1987). Interestingly, when a chloramine-T solution is kept at room temperature for 24 h, it loses the ability to inhibit sodium inactivation while maintaining the ability to block. Chloramine-T appears to contain two components, one of which undergoes chemical changes in solutions.

### Chemicals that alter voltage dependence and kinetics of activation and inactivation of sodium channels

#### Pyrethroids

Pyrethroids are synthetic derivatives of the natural toxins pyrethrin I and II which are contained in the flowers of *Chrysanthemum cinerariaefolium*. The pyrethrum insecticide was used extensively until World War II, but since the development of potent and long-lasting synthetic insecticides such as DDT and parathion, the merit of the pyrethrum insecticide was almost forgotten. Beginning in the mid-1960s, serious concerns emerged over the possible long-term health effects on humans and environmental contaminations as a result of extensive use of insecticides. A large number of pyrethrin derivatives have since been synthesized and tested for insecticidal activity and mammalian toxicity, and some two dozens of pyrethroids have proven useful insecticides due to their high potencies as insecticides, low mammalian toxicity, and biodegradability in the environment.

It has been well established that the major target site of pyrethroids is the sodium channel. The pyrethroids may be divided into two groups based on their chemical structure, those without a cyano group at the $\alpha$ position are called type I, and those with a cyano group are called type II (Fig. 2.15). Although the symptoms of poisoning caused by the pyrethroids in mammals differ somewhat between type I and type II compounds, the symptoms are basically represented by hyperactivity, and caused primarily by modulation of the sodium channels.

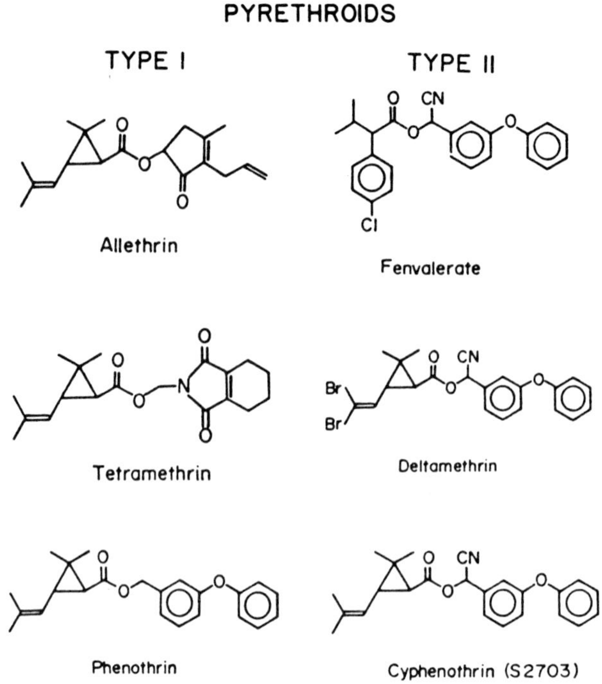

PYRETHROIDS

TYPE I

TYPE II

Allethrin

Fenvalerate

Tetramethrin

Deltamethrin

Phenothrin

Cyphenothrin (S2703)

**Fig. 2.15** Structures of type I and type II pyrethroids.

Both types of pyrethroids cause a multiple action on the gating kinetics of both the activation and inactivation of the sodium channel: (i) the activation voltage is shifted in the hyperpolarizing direction; (ii) the activation and the deactivation occur much more slowly than the normal unmodified sodium channel; (iii) the inactivation–voltage relationship is shifted in the hyperpolarizing direction; and (iv) the inactivation is inhibited (see reviews by Narahashi 1976, 1984, 1985, 1987, 1988*b*, 1989, 1992, 1994; Ruigt 1984; Soderlund and Bloomquist 1989; Vijverberg and van den Bercken 1990; Narahashi *et al.* 1995). Some of the important features of pyrethroid action are described below. DDT modifies the sodium channel in the same manner as type I pyrethroids (Lund and Narahashi 1981*c*).

Type I pyrethroids cause repetitive after-discharges to be produced by a single stimulus in nerve fibres. Depolarizing after-potential is increased and reaches the threshold for generation of action potentials (Fig. 2.16) (Narahashi 1962; Lund and Narahashi 1981*a*). This change in depolarizing after-potential is caused by an increase and prolongation of sodium currents during step depolarizing pulse and after termination of pulse (Fig. 2.17) (Lund and Narahashi 1981*a*,*b*). Thus, the sodium channel inactivation is inhibited and the sodium channel deactivation upon repolarization is greatly slowed. The steady-state sodium channel inactivation curve is shifted in the hyperpolarizing direction by pyrethroids (Fig. 2.18) (Wang *et al.* 1972; Ginsburg and Narahashi 1993; Tatebayashi and Narahashi 1994). Single-chanenel recording experiments have demonstrated that the mean open time of sodium channels is greatly prolonged and that channel opening occurs with a delay, indicating slowing of the activation kinetics (Fig. 2.19) (Yamamoto *et al.* 1983, 1984; Chinn and

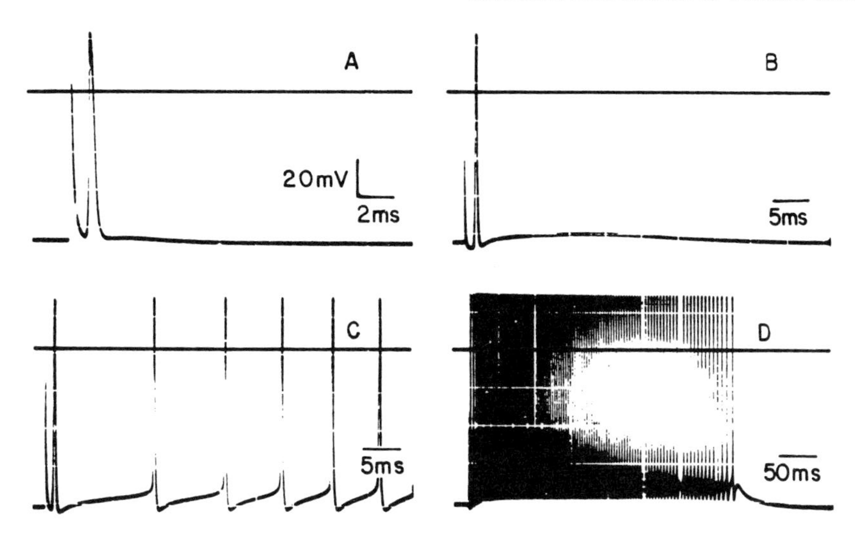


**Fig. 2.16** Repetitive discharges induced by a single stimulus in a crayfish giant axon exposed to 10 $\mu$M (+)-trans tetramethrin. Intracellular recording at 22 °C. A. Control; B. 5 min after application of tetramethrin. C and D. 10 min after tetramethrin. (From Lund and Narahashi 1981*a*.)

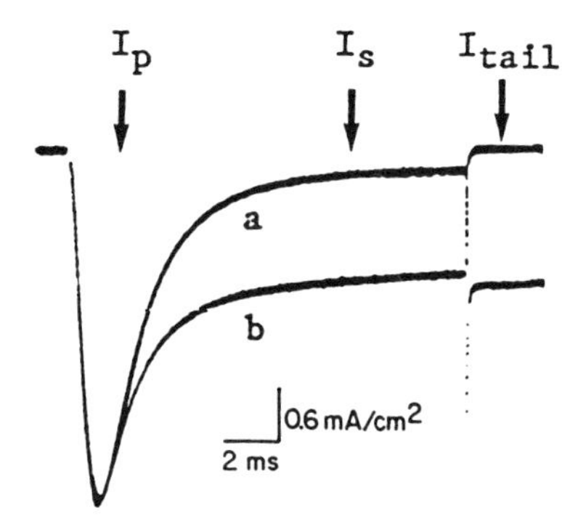


**Fig. 2.17** Membrane sodium currents in a squid giant axon before (a) and during (b) internal perfusion with 1 $\mu$M (+)-trans allethrin. Sodium current associated with a step depolarization from −100 mV to −20 mV was recorded. In the control record (a), the peak sodium current ($I_p$) is followed by a small slow sodium current ($I_s$) during a depolarizing pulse, and the tail current ($I_{tail}$) associated with step repolarization decays quickly. In the presence of allethrin (b), $I_p$ remains unchanged while $I_s$ is greatly increased in amplitude. $I_{tail}$ is also increased in amplitude and decays very slowly. (From Narahashi and Lund 1980.)

Narahashi 1986; Holloway *et al.* 1989). The slowing of sodium channel activation is also shown by applying repetitive depolarizing pulses. All sodium channels are eventually modified by pyrethroids and the sodium current develops slowly, attaining a steady state (Salgado and Narahashi 1993). Gating currents are suppressed by pyrethroids, reflecting slow kinetics of the sodium channel activation (Fig. 2.20) (Salgado and Narahashi 1993). The sodium channel activation voltage is shifted in the hyperpolarizing direction (Tatebayashi and Narahashi 1994), and this effect, when combined with the inhibition of sodium inactivation, causes membrane depolarization.

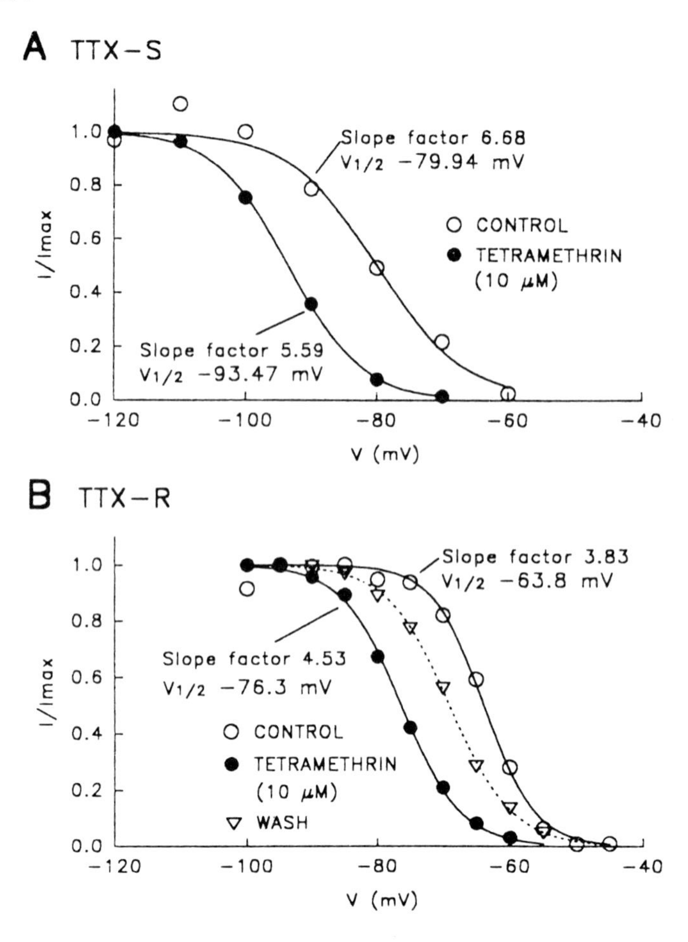


**Fig. 2.18** Effects of 10 μM tetramethrin (TM) on the steady-state inactivation curve for TTX-S and TTX-R sodium channels in rat dorsal root ganglion neurones. The membrane potential was held at various levels for 20 s and then sodium current was evoked by a step depolarization to 0 mV. A. TTX-S sodium current; the peak amplitude of sodium current during a step depolarization is plotted as a function of the conditioning voltage. ○, control; ●, 10 μM TM. B. TTX-R sodium current; ○, control; ●, 10 μM TM; ▽, washout. (From Tatebayashi and Narahashi 1994.)

Type II pyrethroids modify the sodium channels in a similar manner with some quantitative differences, and membrane depolarization rather than repetitive discharges is a predominant feature. Membrane depolarization in sensory neurones in the presence of type II pyrethroids increases discharge frequency causing paraesthesia (LeQuesne and Maxwell 1981; Tucker and Flannigan 1983; Knox *et al.* 1984; McKillop *et al.* 1987).

Vitamin E oil has been used as an effective prophylactic and therapeutic agent for pyrethroid-induced paraesthesia (Tucker *et al.* 1984; Flannigan and Tucker 1985). The mechanism underlying vitamin E action in ameliorating pyrethroid paraesthesia was totally unknown. We have recently found that vitamin E blocks pyrethroid-modified sodium channels without any effect on normal sodium channels in rat cerebellar Purkinje neurones and rat DRG neurones (Song and Narahashi 1995). Figure 2.21 shows that α-tocopherol (vitamin E) blocks tetramethrin-induced slow sodium tail currents without

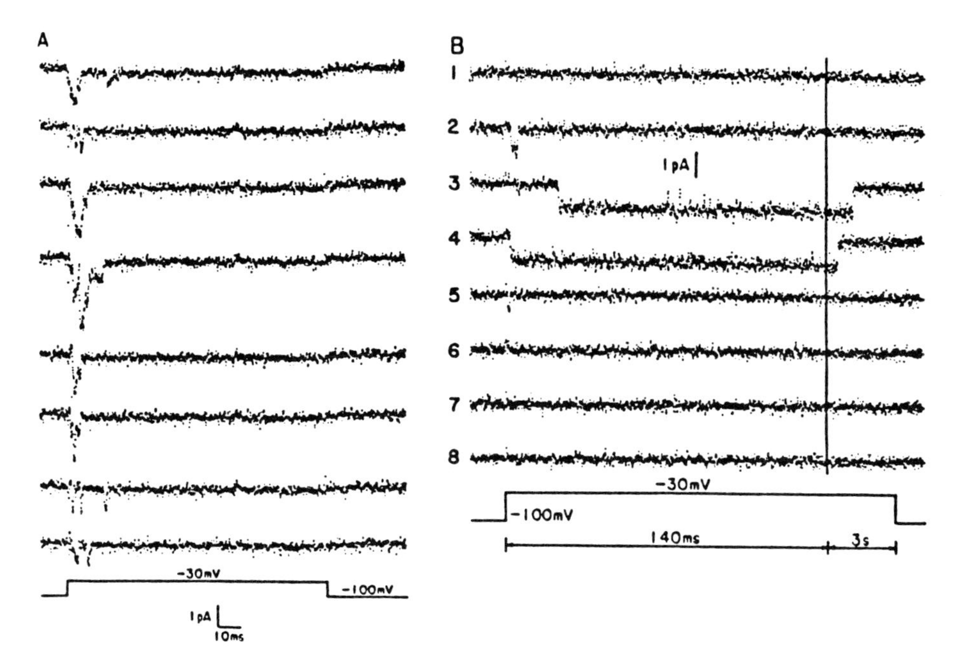


**Fig. 2.19** Effects of deltamethrin on single sodium channel currents of a neuroblastoma cell (N1E-115). A. Currents from a cell before drug treatment in response to 140-ms depolarizing steps from a holding potential of −100 to −30 mV with a 3-s interpulse interval. Records were taken at a rate of 100 μs per point. B. Currents after exposure to 10 μM deltamethrin. The membrane patch was depolarized for 3140 ms from a holding potential of −100 to −30 mV. The interpulse interval was 3 s. The time scale changes during the voltage step as indicated in the Figure. During the first 140 ms, data records were taken at a rate of 100 μs per point, and after the vertical line records were taken at a rate of 10 ms per point. (From Chinn and Narahashi 1986.)

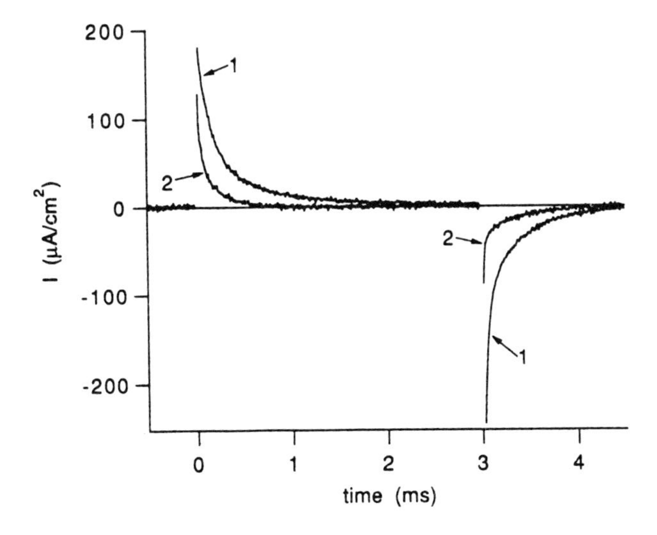


**Fig. 2.20** On-and-off asymmetry currents (gating currents) for a 3-ms depolarizing pulse to +30 mV from a hyperpolarized level of −160 mV before (trace 1) and after (trace 2) exposure to 10 μM fenvalerate, in crayfish giant axon. (From Salgado and Narahashi 1993.)

affecting peak sodium currents derived from the normal sodium channels. Repetitive after-discharges caused by tetramethrin were effectively blocked by α-tocopherol, while the first action potential induced by a single stimulus was unaffected. The mechanism of selective block of tetramethrin-modified sodium channels by vitamin E remains to be elucidated.

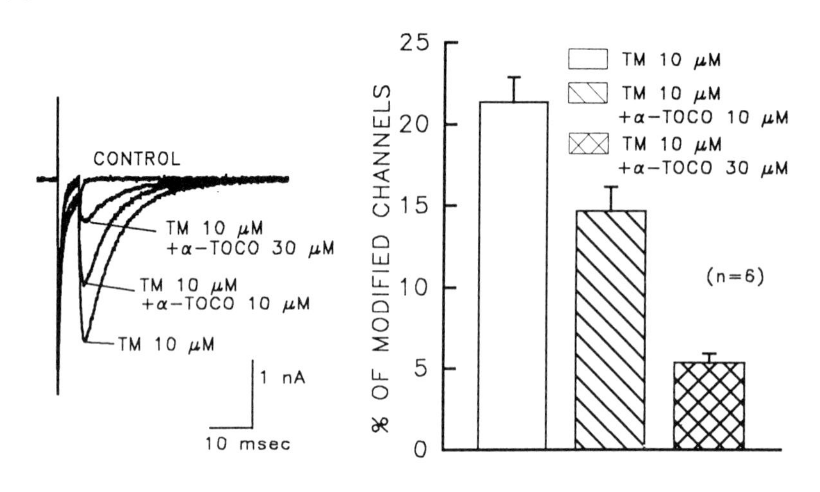


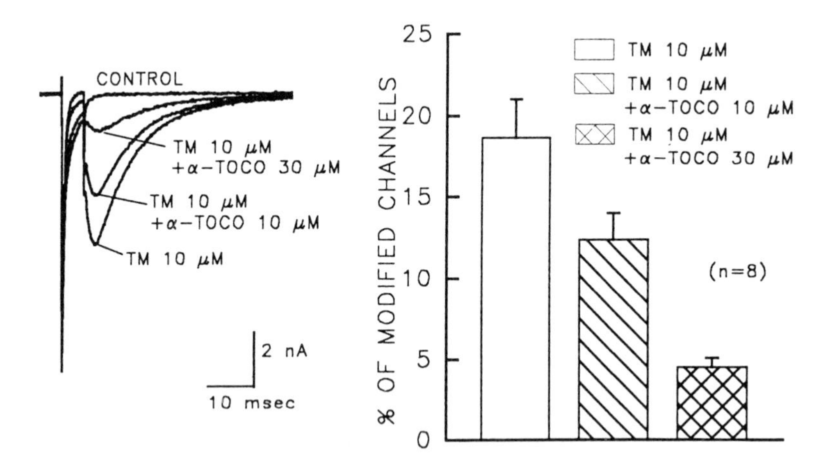


**Fig. 2.21** Suppression of 10 μM tetramethrin-induced tail currents by 10 and 30 μM (±)-α-tocopherol in TTX-S sodium channels of rat cerebellar Purkinje cells (A) and DRG cells (B). Currents were evoked by depolarizing the membrane to 0 mV for 5 ms from a holding potential of −110 mV. Cells were first treated with 10 μM tetramethrin, and then 10 or 30 μM (±)-α-tocopherol was added to the perfusion solution containing 10 μM tetramethrin. Records were taken 5 min after the addition of each chemical. The percentage of channel modification was calculated by (eqn 2.1). Mean ±S.E.M. with $n = 6$(A) and $n = 8$ (B). (From Song and Narahashi 1995.)

The question of closed versus open sodium channel modification by pyrethroids and DDT has largely been settled. Hille (1968) originally proposed open channel modification by DDT, and the concept was more recently elaborated to include allethrin (Leibowitz *et al.* 1986). We initially proposed both open and closed channel modification by tetramethrin (Lund and Narahashi 1981 *a*, *b*), but more extensive analyses using different approaches have clearly indicated that both type I and type II pyrethroids modify the sodium channel largely in its closed state before opening (de Weille *et al.* 1988; de Weille and Leinders 1989; Holloway *et al.* 1989; Brown and Narahashi 1992). However, since the sodium channels have a higher affinity for pyrethroids in the open state, more channels are modified by pyrethroids upon opening, thus explaining use-dependent modification (Lund and Narahashi 1983; Salgado and Narahashi 1993).

The rate of development of slow sodium tail currents, which are indicative of the modified channels, during depolarizing pulses is largely independent of the pyrethroid concentration, but there is a small component that shows concentration dependence (de Weille *et al.* 1988). Thus, the sodium channels are modified by pyrethroids while in the closed state, yet opening increases the affinity for pyrethroids. The increase in pyrethroid affinity by channel opening is also shown by use-dependent modification which eventually converts all channels to the modified state (Salgado and Narahashi 1993). The closed channel modification by pyrethroids has also been demonstrated by single-channel recording experiments in which the probability of channel modification by pyrethroids is independent of the duration of depolarizing pulse (Holloway *et al.* 1989). The ionic selectivity of sodium channels remains unaltered by pyrethroids, suggesting that the pyrethroid molecules do not bind to an intrachannel site (Yamamoto *et al.* 1986).

We have recently demonstrated that TTX-sensitive (TTX-S) and TTX-resistant (TTX-R) sodium channels in rat DRG neurone exhibit differential sensitivity to pyrethroids, the former being less sensitive than the latter (Ginsburg and Narahash 1993; Tatebayashi and Narahashi 1994; Song and Narahashi 1996*b*). Furthermore, TTX-S sodium channels – which are the major population of the brain sodium channels – are also less sensitive to pyrethroids than the sodium channels of invertebrate animals, including insects. This difference in pyrethroid sensitivity of sodium channels is one of the major factors that are responsible for the selective toxicity of pyrethroids between mammals and invertebrates, as will be described later in this chapter (see also Narahashi *et al.* 1995).

The effects of tetramethrin on TTX-S and TTX-R sodium channel currents of rat DRG neurone are illustrated in Fig. 2.22. In TTX-R sodium channels tetramethrin prolongs the sodium current during step depolarization and generates a large tail current that decays slowly. These changes in sodium current are very similar to those observed in invertebrate axons. In TTX-S sodium channels, however, the sodium inactivation is only slightly inhibited by tetramethrin, yet a tail current with a slow rise and fall develops. The hooked tailed current is similar in shape to that observed in vertebrate myelinated nerve in the presence of pyrethroids (Vijverberg *et al.* 1982). The hooked tail current is interpreted as follows. The sodium inactivation gate, which is almost closed during depolarizing pulse (even in the presence of tetramethrin), starts opening upon repolarization, while the activation gate remains open because its kinetics are slowed by tetramethrin. Thus, a tail current starts appearing forming a rising phase. Meanwhile, the activation gate slowly closes causing a falling phase of tail current.

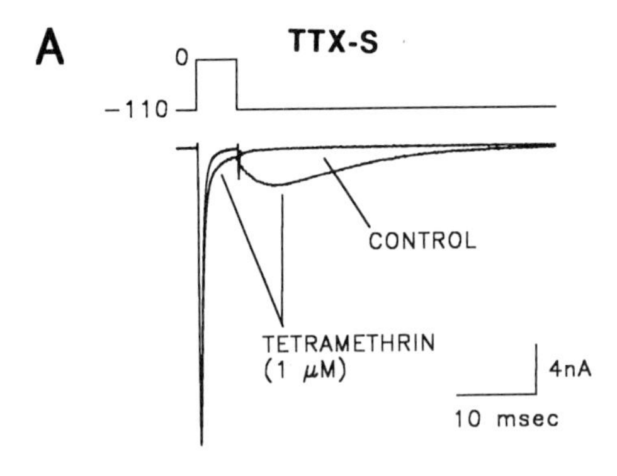


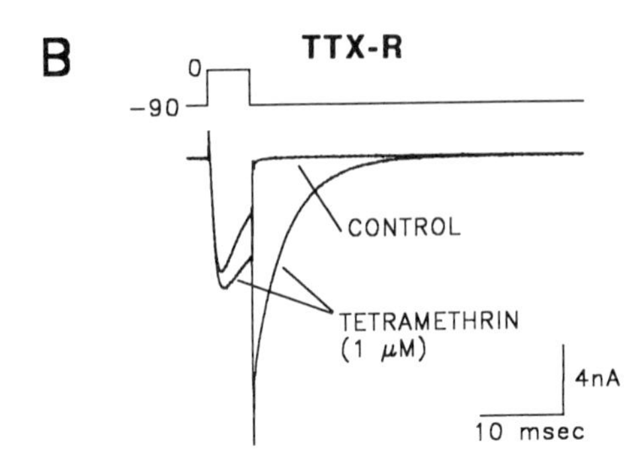


**Fig. 2.22** Effects of tetramethrin on TTX-S sodium current (A) and TTX-R sodium current (B) in rat dorsal root ganglion neurones. (From Tatebayashi and Narahashi 1994.)

The percentage of sodium channels modified by pyrethroids ($M$) can be calculated from measurements of the conductance during peak current and the conductance of slow (modified) tail current upon repolarization according to the following equation (Tatebayashi and Narahash 1994):

$$M = [\{I_{tail}/(E_h - E_{Na})\}/\{I_{Na}/(E_t - E_{Na})\}] \times 100 \qquad (2.1)$$

where $I_{tail}$ refers to the tail current amplitude obtained by extrapolation of the slowly decaying phase of the tail current to the moment of membrane repolarization, $E_h$ is the potential to which the membrane is repolarized, $E_{Na}$ is the equilibrium potential for sodium, $I_{Na}$ is the peak sodium current during depolarizing step, and $E_t$ is the potential of step depolarizing pulse. The concentration–response curves for the percentages of channel modification in TTX-S and TTX-R sodium channels are shown in Fig. 2.23, respectively. The percentages of channel modification thus obtained are given in Table 2.3. It is clearly shown that the sensitivity to tetramethrin is 30- to 100-fold different between TTX-S and TTX-R sodium channels. The level of sensitivity of TTX-R channels is in the same order of magnitude as that of invertebrate (TTX-S) sodium channels.

Since pyrethroid-modified single sodium channels display very prolonged openings, the percentage of sodium channel population can be calculated from the analysis of open time

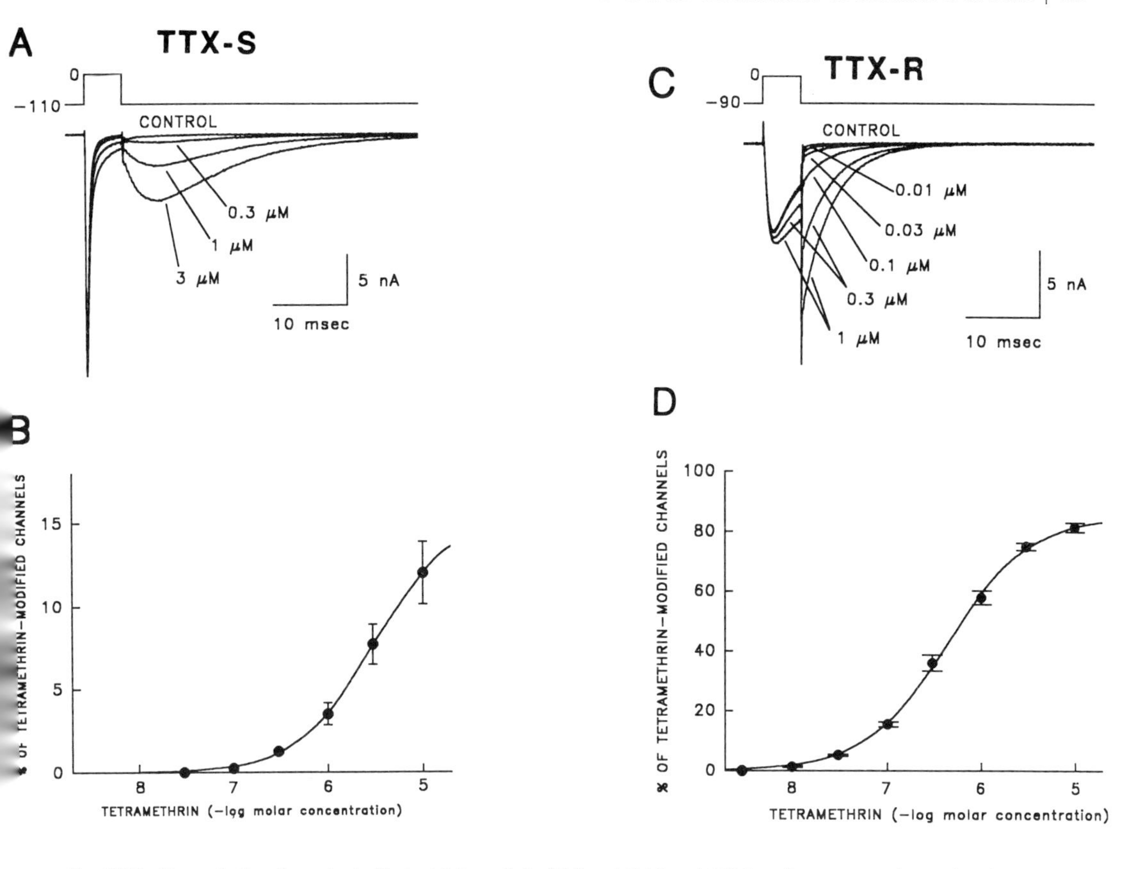


**Fig. 2.23** Concentration-dependent effect of tetramethrin (TM) on TTX-S and TTX-R sodium currents in rat dorsal root ganglion neurones. A. TTX-S currents were evoked by a 5-ms step depolarization to 0 mV from a holding potential of −110 mV under control conditions and in the presence of 0.1, 0.3, 1 and 3 μM TM. B. The dose–response relationship for induction of slow tail current. C. TTX-R currents were evoked by a 5-ms step depolarization to 0 mV from a holding potential of −90 mV under control conditions and in the presence of 0.01, 0.03, 0.1, 0.3 and 1 μM TM. D. The dose–response relationship for the induction of slow tail current. Each point indicates the mean ± S.E.M. (*n* = 4). Data were fitted by the Hill equation. (From Tatebayashi and Narahashi 1994.)

distribution. This approach has proven successful in our recent study using DRG neurones (Song and Narahashi 1996*a*). The open time distributions of TTX-S and TTX-R sodium channels in the absence and presence of 10 μM tetramethrin are illustrated in Fig. 2.24. The high concentration of tetramethrin was used in the single-channel recording experiments in order to observe sufficient numbers of modified channel openings. In the absence of tetramethrin, the open time distribution is fitted by a single exponential curve with a time constant of 1.27 ms and 1.915 ms in TTX-S and TTX-R sodium channels, respectively (Fig. 2.24 A and C). After exposure to tetramethrin, the open time distribution can be fitted by a dual exponential curve with time constants of 1.362 and 5.726 ms in TTX-S channels and 2.070 and 9.749 ms in TTX-R channels (Fig. 2.24 B and D). The shorter time constants in both TTX-S and TTX-R channels agree with the corresponding control time constant, indicating that they represent the activity of unmodified sodium channels. The longer time constants derive from the tetramethrin-modified channels. The percentages of modified

**Table 2.3** The fraction of sodium channels modified by tetramethrin at room temperature.

| Tetramethrin concentration | Cerebellar Purkinje neurone* | Dorsal root ganglion neurone† | |
|---|---|---|---|
| | TTX-sensitive | TTX-sensitive | TTX-resistant |
| (μM) | (%) | (%) | (%) |
| 0.01 | | | 1.31±0.28 |
| 0.03 | | 0 | 5.15±0.30 |
| 0.1 | 0.62±0.15 | 0.24±0.10 | 15.35±0.79 |
| 0.3 | 2.19±0.38 | 1.25±0.13 | 35.48±2.70 |
| 1 | 5.75±0.87 | 3.53±0.66 | 57.82±2.29 |
| 3 | 13.58±1.35 | 7.70±1.20 | 74.85±1.23 |
| 10 | 22.77±2.26 | 12.03±1.89 | 81.20±1.57 |
| 30 | 24.73±2.11 | | |

* From Song and Narahashi (1996*b*)
† From Tatebayashi and Narahashi (1994)

channels are 38% and 85% in TTX-S and TTX-R sodium channels, respectively. For TTX-S sodium channels, the percentage of modification is somewhat higher than 12% obtained previously in whole-cell experiments (Tatebayashi and Narahashi 1994), while for TTX-R sodium channels, the percentage of modification agrees with 81% obtained previously in whole-cell experiments (Tatebayashi and Narahashi 1994). The discrepancy for TTX-S sodium channels appears to be due to several factors that may introduce small errors (Song and Narahashi 1996*a*).

Our most recent studies have shown that Purkinje neurones of rat cerebellum are endowed only with TTX-S sodium channels and that the TTX-S channels are not highly sensitive to tetramethrin being at the same level as TTX-S sodium channels of rat DRG neurones (Table 2.3). Therefore, it is reasonable to assume that the sodium channels of rat brain neurones are less sensitive to pyrethroids than are invertebrate sodium channels.

Pyrethroids are regarded as near–ideal insecticides for three major reasons: (i) they are potent as insecticides; (ii) their mammalian toxicity is low; and (iii) they are biodegradable. The mechanism underlying the highly selective toxicity of pyrethroids in mammals and insects has been traditionally assumed, albeit with no solid experimental evidence, to be due to the difference in metabolic degradation. Our most recent studies have clearly demonstrated that the sodium channel modulation by pyrethroids is the major factor that is responsible for the selective toxicity. Differences between mammals and invertebrate animals in the various factors involved in selective toxicity of pyrethroids are listed in Table 2.4, First pyrethroids are more potent on the sodium channel at low temperature than at high temperature with a $Q_{10}$ value of 0.2 (Song and Narahashi 1996*b*). Because of the temperature difference between mammals (37 °C) and insects (ambient temperature ~25 °C), the difference in the potency of pyrethroids to modulate the sodium channel function is five-fold. Second, as described earlier, mammalian TTX-S sodium channels are less sensitive to pyrethroids than invertebrate TTX-S sodium channels, and the difference has been estimated to be at least 10-fold. Third, the mammalian sodium channels recover from the pyrethroid modulation after washout more quickly than the invertebrate sodium channels with a five-fold difference in

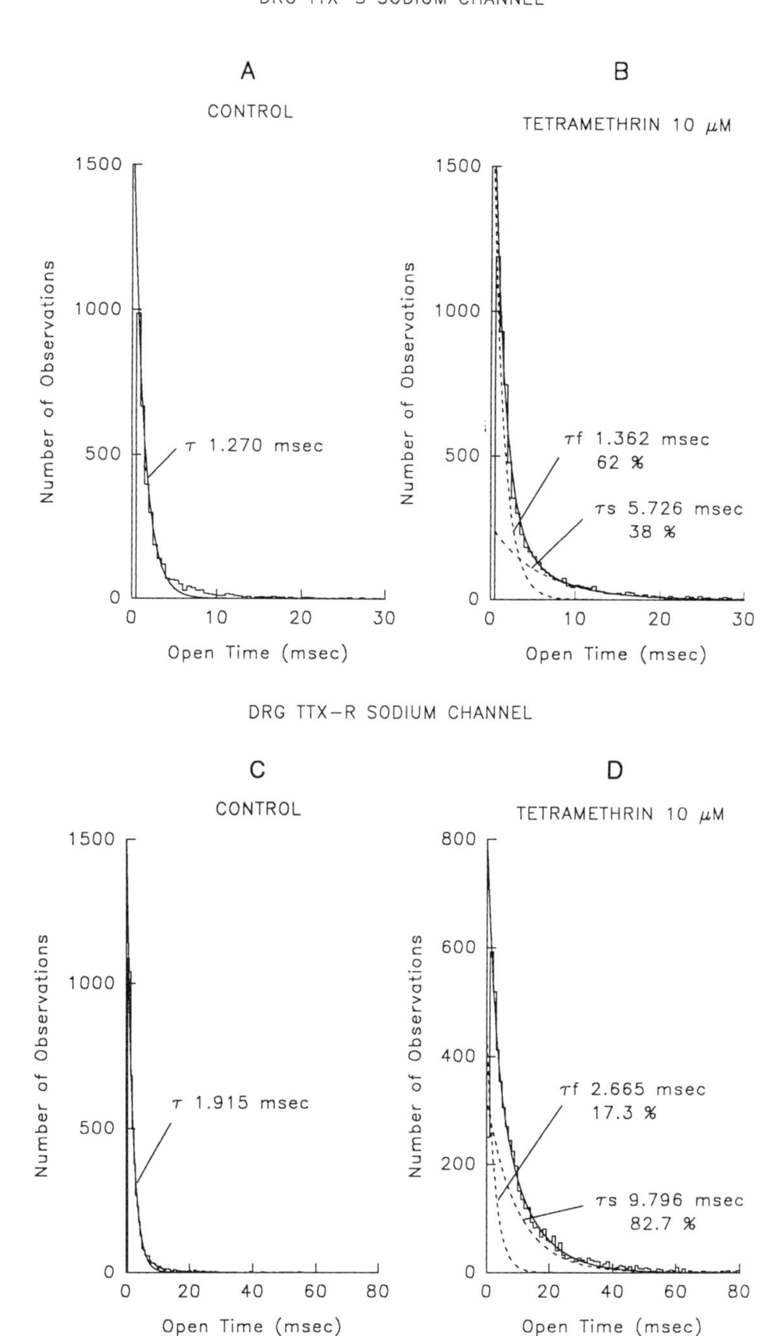


**Fig. 2.24** Open time histograms of TTX-S and TTX-R single sodium channel currents in the absence (A and C) and presence (B and D) of tetramethrin (10 $\mu$M). In controls, the histogram can be fitted by a single exponential function. In the presence of tetramethrin, the histogram can be fitted by the sum of two exponential functions. See text for further explanation. (From Song and Narahashi 1996*a*.)

**Table 2.4** Factors contributing selective toxicity of pyrethroids.

| Selectivity factor | Mammals | Insects | Differences* |
|---|---|---|---|
| Potency on nerve | | | |
| Due to temperature dependence | Low (37 °C) | High (25 °C) | 5 |
| Due to intrinsic sensitivity | Low | High | 10 |
| Recovery | Fast | Slow | 5 |
| Detoxication rate | | | |
| Due to enzymatic action | High | Low | 3 |
| Due to body size | High | Low | 3 |

* Overall difference = 2250
Source: Song and Narahashi, 1996*b*

rate. Fourth, enzymatic degradation of pyrethroids is faster in mammals than in insects due to temperature difference, and the difference in degradation rate is estimated to be three-fold. Fifth, due to the differences in body size, mammals have a greater chance than insects to detoxify pyrethroids before they can reach the target site; this difference is assumed to be three-fold. Thus, the overall difference between mammals and insects is calculated to be 2250-fold, which is in the same order of magnitude as the differences in $LD_{50}$, which range from 500-fold to 4500-fold (Wiswesser 1976; Elliott 1977; Hirai 1987; Miyamoto 1993).

The accurate estimate by (eqn 2.1) of the percentage of sodium channels modified by pyrethroids has a significant impact on the interpretation of the effective concentration of therapeutic drugs and toxic chemicals on excitable cells. We previously estimated the percentage of pyrethroid-modified sodium channels that is required to produce hyperactivity in squid giant axons (Lund and Narahashi 1982). However, we had to make several assumptions at that time because of lack of sufficient data. Nevertheless, it was estimated that less than 1% of the sodium channel population needs to be modified by tetramethrin for the depolarizing after-potential to be elevated to reach the threshold membrane potential for invitation of after-discharges. This issue has been re-assessed using rat cerebellar Purkinje neurones by directly comparing the threshold concentration of tetramethrin for initiation of repetitive after-discharges and the percentage of sodium channel modification at the threshold concentration (Song and Narahashi 1996*b*). The threshold concentration of tetramethrin to induce repetitive after-discharges was estimated to be 100 nM (Fig. 2.25). At this concentration, only 0.62% of sodium channels was modified by tetramethrin (Table 2.3). Thus, the effect of tetramethrin is greatly amplified from the channel level to the whole-cell level, and the hyperexcitation of cells is reflected in hyperactivity of animals.

The toxicity amplification described above has great impact on neuropharmacology in general. For example, a drug that inhibits an ion channel whose activity generates a slow membrane depolarization (e.g. T-type calcium channel) which in turn causes a burst of impulses may stop the burst when the channel activity is slightly suppressed to bring the slow depolarization just below the threshold for impulse discharges. This situation may apply to phenytoin, which has been shown to block T-type calcium channels (Twombly *et al.* 1988). The *in vitro* effective concentration of the drug may be far less than the *in vitro* $ED_{50}$ value, as

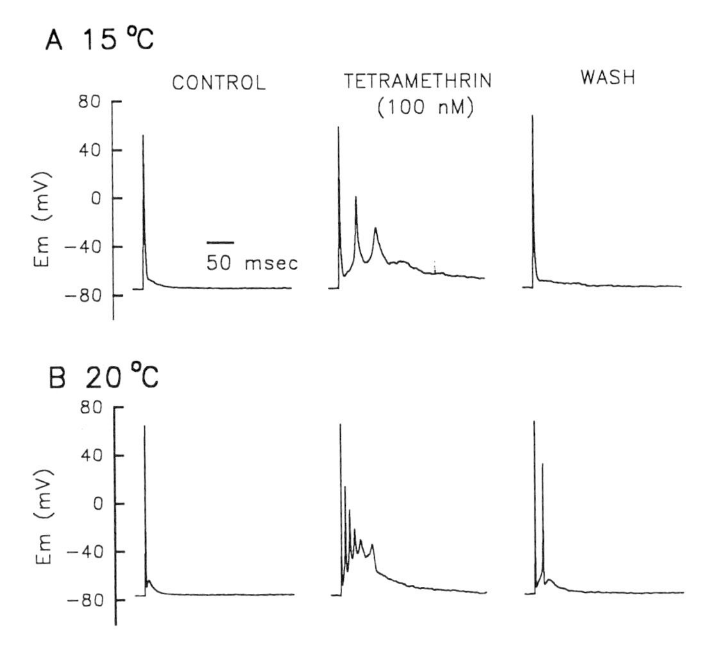


**Fig. 2.25** Effects of 100 nM tetramethrin on the action potential recorded from a rat cerebellar Purkinje neurone. This concentration is the minimum effective concentration to induce repetitive after-discharges. (From Song and Narahashi 1996*b*.)

in the case of tetramethrin. Thus, the traditional method in which *in vitro* $ED_{50}$ of a test drug is compared with the drug concentration in the patient's serum is not necessarily justified when the effect is exerted via a threshold phenomenon.

## References

Agnew, W.S. (1984). Voltage-regulated sodium channel molecules. *Annu. Rev. Physiol.*, **46**, 517–530.

Akopian, A.N., Sivilotti, L. and Wood, J.N. (1996). A tetrodotoxin-resistant voltage-gated sodium channel expressed by sensory neurons. *Nature*, **379**, 257–262.

Albuquerque, E.X. and Daly, J.W. (1976). Batrachotoxin, a selective probe for channels modulating sodium conductances in electrogenic membranes. In *The specificity and action of animal, bacterial and plant toxins. Receptors and recognition* (ed. P. Cuatrecasas), ser. B, Vol. 1, pp. 297–338, Chapman & Hall, London.

Albuquerque, E.X., Daly, J. and Witkop, B. (1971). Batrachotoxin: chemistry and pharmacology. *Science*, **172**, 995–1002.

Albuquerque, E.X., Seyama, I. and Narahashi, T. (1973). Characterization of batrachotoxin-induced deplorization of the squid giant axons. *J. Pharmacol. Exp. Ther.*, **184**, 308–314.

Almers, W. and Levinson, S.R. (1975). Tetrodotoxin binding to normal and depolarized frog muscle and the conductance of a single sodium channel. *J. Physiol. (Lond)*., **247**, 483–509.

Armstrong, C.M. (1969). Inactivation of the potassium conductance and related phenomena caused by quaternary ammonium ion injection in squid axons. *J. Gen. Physiol.*, **54**, 553–575.

Armstrong, C.M. (1971). Interaction of tetraethylammonium ion derivatives with the potassium channels of giant axons. *J. Gen. Physiol.*, **58**, 413–437.

Armstrong, C.M., Bezanilla, F. and Rojas, E. (1973). Destruction of sodium conductance inactivation in squid axons perfused with pronase. *J. Gen. Phsyiol.*, **62**, 375–391.

Atchison, W.D., Luke, V.S., Narahashi, T. and Vogel S.M. (1986). Nerve membrane sodium channels as the target site of brevetoxins at neuromuscular junctions. *Br. J. Pharmacol.*, **89**, 731–738.

Backx, P.H., Yue, D.T., Lawrence, J.H., Marban, E. and Tomaselli, G.F. (1992). Molecular localization of an ion-binding site within the pore of mammalian sodium channels. *Science*, **257**, 248–251.

Baer, M., Best, P.M. and Reuter, H. (1976). Voltage-dependent action of tetrodotoxin in mammalian cardiac muscle. *Nature*, **263**, 344–345.

Baker, P.F. and Rubinson, K.A. (1976). TTX-resistant action potentials in crab nerve after treatment with Meerwein's reagent. *J. Physiol. (Lond.)*, **266**, 3–4P.

Barchi, R.L. and Weigele, J.B. (1979). Characteristics of saxitoxin binding to the sodium channel of sarcolemma isolated from rat skeletal muscle. *J. Physiol. (Lond.)*, **295**, 383–396.

Barnes, S. and Hille, B. (1988). Veratridine modifies open sodium channels. *J. Gen. Physiol.*, **91**, 421–433.

Bay, C.M.H. and Strichartz, G.R. (1980). Saxitoxin binding to sodium channels of rat skeletal muscles. *J. Physiol. (Lond.)*, **300**, 89–103.

Bazan, M. and Bernard, P. (1984). Effect of α-neurotoxins from scorpion venom on sodium current in neuroblastoma cells. *J. Biophys. Med. Nucl.*, **8**, 3–8.

Benoit, E., Legrand, A.M. and Dubois, J.M. (1986). Effects of ciguatoxin on current and voltage clamped frog myelinated nerve fibre. *Toxicon*, **24**, 357–364.

Benzer, T.I. and Raftery, M.A. (1972). Partial characterization of a tetrodotoxin-binding component from nerve membrane. *Proc. Natl Acad. Sci. USA*, **69**, 3634–3637.

Bidard, J.-N., Vijverberg, H.P.M., Frelin, C., Chungue, E., Legrand, A.-M., Bagnis, R. and Lazdunski, M. (1984). Ciguatoxin is a novel type of $Na^+$ channel toxin. *J. Biol. Chem.*, **259**, 8353–8357.

Bossu, J.L. and Feltz, A. (1984). Patch-clamp study of the tetrodotoxin-resistant sodium current in group C sensory neurones. *Neurosci. Lett.*, **51**, 241–246.

Brown, L.D. and Narahashi, T. (1992). Modulation of nerve membrane sodium channel activation by deltamethrin. *Brain Res.*, **584**, 71–76.

Cahalan, M.D. (1975). Modification of sodium channel gating in frog myelinated nerve fibres by *Centruroides sculpturatus* scorpion venom. *J. Physiol. (Lond.)*, **244**, 511–534.

Camougis, G., Takman, B.H. and Tasse, J.R.P. (1967). Potency difference between the zwitterion form and the cation forms of tetrodotoxin. *Science*, **156**, 1625–1627.

Campbell, D.T. (1982). Modified kinetics and selectivity of sodium channels in frog skeletal muscle fibers treated with aconitine. *J. Gen. Physiol.*, **80**, 713–731.

Campbell, D.T. (1988). Differential expression of Na channel subtypes in two populations of sensory neurons. *Biophys. J.*, **53**, 15a.

Cartier, G.E., Yoshikami, D., Gray, W.R., Luo, S., Olivera, B.M. and McIntosh, J.M. (1996). A new α-conotoxin which targets α3β2 nicotinic acetylcholine receptors. *J. Biol. chem.* **271**, 7522–7528.

Catterall, W.A. (1980). Neurotoxins that act on voltage-sensitive sodium channels in excitable membranes. *Annu. Rev. Pharmacol. Toxicol.*, **20**, 15–43.

Catterall, W.A. (1992). Cellular and molecular biology of voltage-gated sodium channels. *Physiol. Rev.*, **72**(4) Suppl., S15–S48.

Catterall, W.A. and Beress, L. (1978). Sea anemone toxin and scorpion toxin share a common receptor site associated with the action potential sodium ionophore. *J. Biol. Chem.*, **253**, 7393–7396.

Catterall, W.A. and Morrow, C.S. (1978). Binding of saxitoxin to electrically excitable neuroblastoma cells. *Proc. Natl Acad. Sci. USA*, **75**, 218–222.

Catterall, W.A., Morrow, C.S. and Hartshorne, R.P. (1979). Neurotoxin binding to receptor sites associated with voltage-sensitive sodium channels in intact, lysed, and detergent-solubilized brain membranes. *J. Biol. Chem.*, **254**, 11379–11387.

Chinn, K. and Narahashi, T. (1986). Stabilization of sodium channel states by deltamethrin in mouse neuroblastoma cells. *J. Physiol. (Lond.)*, **380**, 191–207.

Cohen, C.J., Bean, B.P., Colatsky, T.J. and Tsien, R.W. (1981). Tetrodotoxin block of sodium channels in rabbit Purkinje fibers. *J. Gen. Physiol.*, **78**, 383–411.

Cole, K.S. (1949). Dynamic electrical characteristics of the squid axon membrane. *Arch. Sci. Physiol.*, **3**, 253–258.

Colquhoun, D., Henderson, R. and Ritchie, J.M. (1972). The binding of labelled tetrodotoxin to non-myelinated nerve fibres. *J. Physiol. (Lond.)*, **227**, 95–126.

Couraud, F., Rochat, H. and Lissitzky, S. (1978). Binding of scorpion and sea anemone neurotoxins to a common site related to the action potential $Na^+$ ionophore in neuroblastoma cells. *Biochem. Biophys. Res. Commun.*, **83**, 1525–1530.

Couraud, F., Jover, E., Dubois, J.M. and Rochat, H. (1982). Two types of scorpion toxin receptor sites, one related to activation, the other to inactivation of the action potential sodium channel. *Toxicon*, **20**, 9–16.

Courtney, K.R. (1975). Mechanism of frequency-dependent inhibition of sodium currents in frog myelinated nerve by the lidocaine derivative GEA 968. *J. Pharmacol. Exp. Ther.*, **195**, 225–236.

Cruz, L.J. and Olivera, B.M. (1986). Calcium channel antagonists. *J. Biol. Chem.*, **261**, 6230–6233.

de Weille, J.R. and Leinders, T. (1989). The action of pyrethroids on sodium channels in myelinated nerve fibres and spinal ganglion cells of the frog. *Brain Res.*, **482**, 324–332.

de Weille, J.R., Vijverberg, H.P.M. and Narahashi, T. (1988). Interactions of pyrethroids and octylguanidine with sodium channels of squid giant axons. *Brain Res.*, **445**, 1–11.

Doyle, D.D., Guo, Y., Lustig, S.L., Satin, J., Rogart, R.B. and Fozzard, H.A. (1993). Divalent cation competition with [$^3$H]saxitoxin binding to tetrodotoxin-resistant and -sensitive sodium channels. A two-site structural model of ion/toxin interaction. *J. Gen. Physiol.*, **101**, 153–182.

Dudley, S.C., Jr, Todt, H., Lipkind, G. and Fozzard, H.A. (1995). A $\mu$-conotoxin-insensitive $Na^+$ channel mutant: possible localization of a binding site at the outer vestibule. *Biophys. J.*, **69**, 1657–1665.

Elliott, A.A. and Elliott, J.R. (1993). Characterization of TTX-sensitive and TTX-resistant sodium currents in small cells from adult rat dorsal root ganglia. *J. Physiol. (Lond.)*, **463**, 39–56.

Elliott, M. (1977). Synthetic pyrethroids. In *Synthetic Pyrethroids*, ACS Symposium Series 42, (ed. M. Elliott), pp. 1–28. American Chemical Society, Washington, DC.

Flannigan, S.A. and Tucker, S.B. (1985). Variation in cutaneous sensation between synthetic pyrethroid insecticides. *Contact Dermatitis*, **13**, 140–147.

Frazier, D.T., Narahashi, T. and Yamada, M. (1970). The site of action and active form of local anesthetics. II. Experiments with quaternary compounds. *J. Pharmacol. Exp. Ther.*, **171**, 45–51.

Frazier, D.T., Murayama, K. and Narahashi, T. (1971). Comparison of the blocking potency of local anesthetics applied at different pH values. *Experientia*, **28**, 419–420.

Frelin, C., Vigne, P. and Lazdunski, M. (1981). The specificity of the sodium channel for monovalent cations. *Eur. J. Biochem.*, **119**, 437–442.

French, R.J., Worley, J.F. and Krueger B.K. (1984). Voltage-dependent block by saxitoxin of sodium channels incorporated into planar lipid bilayers. *Biophys. J.*, **45**, 301–310.

Garber, S. and Miller, C. (1987). Single $Na^+$ channels activated by veratridine and batrachotoxin. *J. Gen. Physiol.*, **89**, 459–480.

Ginsburg, K.S. and Narahashi T. (1993). Differential sensitivity of tetrodotoxin-sensitive and tetrodotoxin-resistant sodium channels to the insecticide allethrin in rat dorsal root ganglion neurons. *Brain Res.*, **627**, 239–248.

Gray, W.R., Luque, A., Olivera, B.M., Barrett, J. and Cruz, L.J. (1981). Peptide toxins from *Conus geopgraphus* venom. *J. Biol. Chem.*, **256**, 4734–4740.

Grishchenko, I.I., Naumov, A.P. and Zubov, A.N. (1983). Gating and selectivity of aconitine modified sodium channels in neuroblastoma cells. *Neuroscience*, **9**, 549–554.

Grissmer, S. (1984). Effect of various cations and anions on the action of tetrodotoxin and saxitoxin on frog myelinated nerve fibers. *Pflügers Arch.*, **402**, 353–359.

Guo, X. and Strichartz. G. (1990). Differential gating of TTX-sensitive and -resistant Na currents in bullfrog sensory neurons. *Biophys. J.*, **57**, 107a.

Habermann, E. and Beress, L. (1979). Iodine labelling of sea anemone toxin II and binding to normal and denervated diaphragm. *Naunyn-Schmiedeberg's Arch. Pharmacol.*, **309**, 165–170.

Hartshorne, R.P. and Catterall, W.A. (1981). Purification of the saxitoxin receptor of the sodium channel from rat brain. *Proc. Natl. Acad. Sci. USA*, **78**, 4620–4624.

Hartshorne, R.P. and Catterall, W.A. (1984). The sodium channel from rat brain. Purification and subunit composition. *J. Biol. Chem.*, **259**, 1667–1675.

Hartshorne, R.P., Messner, D.J. Coppersmith J.C. and Catterall, W.A. (1982). The saxitoxin receptor of the sodium channel from rat brain. Evidence for two nonidentical beta subunits. *J. Biol. Chem.*, **257**, 13888–13891.

Henderson, E. and Wang, J.H. (1972). Solubilization of a specific tetrodotoxin-binding component from garfish olfactory nerve membrane. *Biochemistry*, **11**, 4565–4569.

Henderson, R., Ritchie, J.M. and Strichartz G. (1973). The binding of labelled saxitoxin to the sodium channels in nerve membranes. *J. Physiol. (Lond.)*, **235**, 783–804.

Henderson, R., Ritchie, J.M. and Strichartz, G.R. (1974). Evidence that tetrodotoxin and saxitoxin act at a metal cation binding site in the sodium channels of nerve membrane. *Proc. Natl Acad. Sci. USA*, **71**, 3936–3940.

Hille, B. (1968). Pharmacological modification of the sodium channels of frog nerve. *J. Gen. Physiol.*, **51**, 199–219.

Hille, B. (1971). The permeability of the sodium channel to organic cations in myelinated nerve. *J. Gen. Physiol.*, **58**, 599–619.

Hille, B. (1975). The receptor for tetrodotoxin and saxitoxin. A structural hypothesis. *Biophys. J.*, **15**, 615–619.

Hille, B., Courtney, K. and Dum, R. (1975). Rate and site of action of local anesthetics in myelinated nerve fibers. In *Molecular mechanisms of anesthesia*, Vol. 1, (ed. B.R. Fink), pp. 13–20. Raven Press, New York.

Hirai, H. (1987). ETOC, a new pyrethroid. *Sumitomo Pyrethroid World* No. 9 (Autumn), 2–4.

Hironaka, T. and Narahashi, T. (1977). Cation permeability ratios of sodium channels in normal and grayanotoxin-treated squid axon membranes. *J. Membr. Biol.*, **31**, 359–381.

Hodgkin, A.L. and Huxley, A.F. (1952*a*). Currents carried by sodium and potassium ions through the membrane of the giant axon of *Loligo*. *J. Physiol. (Lond.)*, **116**, 449–472.

Hodgkin, A.L. and Huxley, A.F. (1952*b*). The components of membrane conductance in the giant axon of *Loligo*. *J. Physiol. (Lond.)*, **116**, 473–496.

Hodgkin, A.L. and Huxley, A.F. (1952*c*). The dual effect of membrane potential on sodium conductance in the giant axon of *Loligo*. *J. Physiol. (Lond.)*, **116**, 497–506.

Hodgkin, A.L. and Huxley, A.F. (1952*d*). A quantitiative description of membrane current and its application to conduction and excitation in nerve. *J. Physiol. (Lond.)*, **117**, 500–544.

Hodgkin, A.L. and Huxley, A.F. and Katz, B. (1952). Measurement of current–voltage relations in the membrane of the giant axon of *Loligo*. *J. Physiol. (Lond.)*, **116**, 424–448.

Hoehn, K., Watson T.W.J. and MacVicar, B.A., (1993). A novel tetrodotoxin-insensitive, slow sodium current in striatal and hippocampal neurons. *Neuron*, **10**, 543–552.

Holloway, S.F., Narahashi, S.F., Salgado, V.L. and Wu C.H., (1989). Kinetic properties of single sodium channels modified by fenvalerate in mouse neuroblastoma cells. *Pflügers Arch.*, **414**, 613–621.

Hopkins, C., Grilley, M., Miller, C., Shon, K.-J., Cruz, L.J., Gray, W.R., Dykert, J., Rivier, J., Yoshikami, D. and Olivera, B.M. (1995). A new family of *Conus* peptides targeted to the nicotinic acetylcholine receptor. *J. Biol. Chem.*, **270**, 22361–22367.

Hu, S.L., Meves, H., Rubly, N. and Watt, D.D. (1983). A quantitative study of the action of *Centruroides sculpturatus* toxins III and IV on the Na currents of the node of Ranvier. *Pflügers Arch.*, **397**, 90–99.

Huang, J.M.C., Wu, C.H. and Baden, D.G. (1984). Depolarizing action of a red-tide dinoflagellate brevetoxin on axonal membranes. *J. Pharmacol. Exp. Ther.*, **229**, 615–621.

Huang, J.M.C., Tanguy, J. and Yeh, J.Z., (1987). Removal of sodium inactivation and block of sodium channels by chloramine-T in crayfish and squid giant axons. *Biophys. J.*, **52**, 155–163.

Hucho, F., Beress, L. and Stengelin, S. (1979). Covalently reacting derivatives of neurotoxins and ion channel ligands. In *Function and molecular aspects of biomembrane transport*, ed. E. Quagliariello, F. Palmieri, S. Papa and M. Klinkenberg), pp. 43–51. Elsevier, Amsterdam, The Netherlands.

Ikeda, S.R. and Schofield, G.G. (1987). Tetrodotixin-resistant sodium current of rat nodose neurones: monovalent cation selectivity and divalent cation block. *J. Physiol. (Lond.)*, **389**, 255–270.

Ikeda, S.R., and Schofield, G.G. and Weight F.F. (1986). $Na^+$ and $Ca^{2+}$ currents of acutely isolated adult rat nodose ganglion cells. *J. Neurophysiol.*, **55**, 527–539.

Isom, L.L., De Jongh K.S. and Catterall, W.A. (1994). Auxiliary subunits of voltage-gated ion channels. *Neuron*, **12**, 1183–1194.

Jaimovich, E., Venosa, R.A., Shrager, P. and Horowicz, P. (1976). Density and distribution of tetrodotoxin receptors in normal and detubulated frog sartorius muscle. *J. Gen. Physiol.*, **67**, 399–416.

Jones, S.W. (1986). Two sodium currents in dissociated bullfrog sympathetic neurons. *Soc. Neurosci. Abstr.*, **12**, 1512.

Kallen, R.G., Sheng. Z.H., Yang, J., Chen, L.Q., Rogart, R.B. and Barchi, R.L. (1990). Primary structure and expression of a sodium channel characteristic of denervated and immature rat skeletal muscle. *Neuron*, **4**, 233–242.

Kayano, T., Noda, M., Flockerzi, V., Takahashi, H. and Numa, S. (1988). Primary structure of rat brain sodium channel III deduced from the cDNAa sequence. *FEBS Lett.*, **228**, 187–194.

Kerr, L.M. and Yoshikami, D. (1984). A venom peptide with a novel presynaptic blocking action. *Nature*, **308**, 282–284.

Keynes, R.D. and Ritchie, J.M. (1984). On the binding of labelled saxitoxin to the squid giant axon. *Proc. R. Soc. Lond. (Biol.)*, **222**, 147–153.

Khodorov, B.I. (1978). Chemicals as tools to study nerve fiber sodium channels; effects of batrachotoxin and some local anesthetics. In *Membrane transport processes*, (ed. D.C. Tosteson, A.O. Yu and R. Latorre), Vol. 2, pp. 153–174. Raven, New York.

Khodorov, B.I. (1985). Batrachotoxin as a tool to study voltage-sensitive sodium channels of excitable membranes. *Prog. Biophys. Mol. Biol.*, **45**, 57–148.

Khodorov, B.I. and Revenko, S.V. (1979). Further analysis of the mechanism of action of batrachotoxin on the membrane of myelinated nerve. *Neuroscience*, **4**, 1315–1330.

Khodorov, B.I., Peganov, E.M., Revenko, S.V. and Shishkova, L.D. (1975). Sodium currents in voltage clamped nerve fiber of frog under the combined action of batrachotoxin and procaine. *Brain Res.*, **84**, 541–546.

Kirsch, G.E. Alam, M. and Hartmann, H.A. (1994). Differential effects of sulfhydryl reagents on saxitoxin and tetrodotoxin block of voltage-dependent Na channels. *Biophys. J.*, **67**, 2305–2315.

Knox, J.M. II, Tucker, S.B. and Flannigan, A. (1984). Paresthesia from cutaneous exposure to a synthetic pyrethroid insecticide. *Arch. Dermatol.*, **120**, 744–746.

Kobayashi, J., Nakamura, H. and Ohizumi, Y. (1981). Biphasic mechanical responses of the guinea-pig isolated ileum to the venom of the marine snail *Conus striatus*. *Br. J. Pharmacol.*, **73**, 583–585.

Kobayashi, J., Nakamura, H., Hirata, Y. and Ohizumi, Y. (1982). Isolation of a cardiotonic glycoprotein, striatoxin, from the venom of the marine snail *Conus striatus*. *Biochem. Biophys. Res. Commun.*, **105**, 1389–1395.

Kobayashi, J., Nakamura, H. and Ohizumi, Y. (1983). Excitatory and inhibitory effects of a myotoxin from *Conus magus* venom on the mouse diaphragm, the guinea-pig atria, taenia caeci, ileum and vas deferens. *Eur. J. Pharmacol.*, **86**, 283–286.

Kobayashi, M., Wu, C.H., Yoshii, M., Narahashi, T., Nakamura, H., Kobayashi, J. and Ohizumi, Y. (1986). Preferential block of skeletal muscle sodium channels by geographutoxin II, a new peptide toxin from *Conus geographus*. *Pflügers Archiv.*, **407**, 241–243.

Koppenhöffer, E. and Schmidt, H. (1968*a*). Die Wirkung von Skorpiongift auf die Ionenstrome des Ranvierscher Schnurrings. I. Die Permeabilitäten $P_{Na}$ and $P_K$. *Pflügers Arch.*, **303**, 133–149.

Koppenhöffer, E. and Schmidt, H. (1968*b*). Die Wirkung von Skorpiongift auf die Ionenstrome des Ranvierschen Schnurrings. II. Unvollstandige Natrium-Inaktivierung. *Pflügers Arch.*, **303**, 150–161.

Kostyuk, P.G., Veselovsky, N.S. and Tsyndrenko, A.Y. (1981). Ionic currents in the somatic membrane of rat dorsal root ganglion neurons-I. Sodium currents. *Neuroscience*, **6**, 2423–2430.

Krueger, B.K., Ratzlaff, R.W., Strichartz, G.R. and Blaustein, M.P. (1979). Saxitoxin binding to synaptosomes, membranes, and solubilized binding sites from rat brain. *J. Membr. Biol.*, **50**, 287–310.

Le Quesne, P.M. and Maxwell, I.C. (1981). Transient facial sensory symptoms following exposure to synthetic pyrethroids: a clinical and electrophysiological assessment. *Neurotoxicology*, **2**, 1–11.

Leibowitz, M.D., Sutro, J.B. and Hille, B. (1986). Voltage-dependent gating of veratridine-modified Na channels. *J. Gen. Physiol.*, **87**, 25–46.

Leicht, R., Meves, H. and Wellhoner, H.H. (1971). The effect of veratridine on *Helix promatia* neurones. *Pflügers Arch.*, **323**, 50–62.

Lipkind, G.M. and Fozzard, H.A. (1994). A structural model of the tetrodotoxin and saxitoxin binding site of the $Na^+$ channel. *Biophys. J.*, **66**, 1–13.

Low, P.A., Wu, C.H. and Narahashi, T. (1979). The effect of anthopleurin-A on crayfish giant axon. *J. Pharmacol. Exp. Ther.*, **210**, 417–421.

Lund, A.E. and Narahashi, T. (1981*a*). Modification of sodium channel kinetics by the insecticide tetramethrin in crayfish giant axons. *Neurotoxicology*, **2**, 213–229.

Lund, A.E. and Narahashi, T. (1981*b*). Kinetics of sodium channel modification by the insecticide tetramethrin in squid axon membranes. *J. Pharmacol. Exp. Ther.*, **219**, 464–473.

Lund, A.E. and Narahashi, T. (1981*c*). Interaction of DDT with sodium channels in squid giant axon membranes. *Neuroscience*, **6**, 2253–2258.

Lund, A.E. and Narahashi, T. (1982). Dose-dependent interaction of the pyrethroid isomers with sodium channels of squid axon membranes. *NeuroToxicology*, **3**, 11–24.

Lund, A.E. and Narahashi, T. (1983). Kinetics of sodium channel modification as the basis for the variation in the nerve membrane effects of pyrethroids and DDT analogs. *Pesticide Biochem. Physiol.*, **20**, 203–216.

Makielski, J.C., Limberis, J.T., Chang, S.Y., Fan, Z. and Kyle, J.W. (1996). Coexpression of $\beta 1$ with cardiac sodium channel $\alpha$ subunits in oocytes decreases lidocaine block. *Mol. Pharmacol.*, **49**, 30–39.

Martinez, J.S., Olivera, B.M., Gray, W.R., Craig, A.G., Groebe, D.R., Abramson, S.N. and and McIntosh, J.M. (1995). α-Conotoxin EI, a new nicotinic acetylcholine receptor antagonist with novel selectivity. *Biochemistry*, **34**, 14519–14526.

Matsuda, Y., Yoshida, S. and Yonezawa, T. (1978) Tetrodotoxin sensitivity and Ca component of action potentials of mouse dorsal root ganglion cells cultured *in vitro*. *Brain Res.*, **154**, 69–82.

Matsuki, N. and Hermsmeyer, K. (1983). Tetrodotoxin-sensitive $Na^+$ channels in isolated single cultured rat mycardial cells. *Am. J. Physiol.*, **245**, C381–C387.

McCleskey, E.W., Fox, A.P., Feldman, D., Cruz, L.J., Olivera, B.M., Tsien, R.W. and Yoshikami, D. (1987). ω-Conotoxin: direct and persistent blockade of specific types of calcium channels in neurones but not muscle. *Proc. Natl Acad. Sci. USA*, **84**, 4327–4331.

McKillop, C.M., Brock, J.A.C., Oliver, G.J.A. and Rhodes, C. (1987). A quantitiative assessment of pyrethroid-induced paraesthesia in the guinea-pig flank model. *Toxicol. Lett.*, **36**, 1–7.

McLean, M.J., Bennett, P.B. and Thomas, R.M. (1988). Subtypes of dorsal root ganglion neurons based on different inward currents as measured by whole-cell voltage clamp. *Mol. Cell. Biochem.*, **80**, 95–107.

McManus, O.B. and Musick, J.R. (1985). Postsynaptic block of frog neuromuscular transmission by conotoxin GI. *J. Neurosci.*, **5**, 110–116.

McManus, O.B., Musick, J.R. and Gonzalez, C. (1981). Peptides isolated from the venom of *Conus geographus* block neuromuscular transmission. *Neurosci. Lett.*, **25**, 57–62.

Messner, D.J. and Catterall, W.A. (1985). The sodium channel from rat brain. Separation and characterization of subunits. *J. Biol. Chem.*, **260**, 10597–10604.

Meves, H., Rubly, N. and Watt, D.D. (1982). Effect of toxins isolated from the venom of the scorpion *Centruroides sculpturatus* on the Na currents of the node of Ranvier. *Pflügers Arch.*, **393**, 56–62.

Meves, H., Simard, J.M. and Watt, D.D. (1986). Interactions of scorpion toxins with the sodium channel. *Ann. N.Y. Acad. Sci.*, **479**, 113–132.

Miyamoto, J. (1993). A risk assessment of household insecticides. *Sumitomo Pyrethroid World*, No. 20 (Spring), 14–19.

Moczydlowski, E., Hall, S., Garber, S.S., Strichartz, G.S. and Miller, C. (1984). Voltage-dependent blockade of muscle $Na^+$ channels by guanidinium toxins. Effect of toxin charge. *J. Gen. Physiol.*, **84**, 687–704.

Moore, J.W., Narahashi, T. and Shaw, T.I. (1967). An upper limit to the number of sodium channels in nerve membrane? *J. Physiol. (Lond.)*, **188**, 99–105.

Mozhayeva, G.N., Naumov, A.P., Negulyaev, Y.A. and Nosyreva, E.D. (1977). The permeability of actonitine-modified $Na^+$ channels to univalent cations in myelinated nerve. *Biochim. Biophys. Acta*, **466**, 461–473.

Mozhayeva, G.N., Naumov, A.P., Nosyreva, E.D. and Grishin, E.V. (1980). Potential-dependent interaction of toxin from venom of the scorpion *Buthu is eupeus* with sodium channels in myelinated fibre. *Biochim. Biophys. Acta*, **597**, 587–602.

Muramatsu, I., Fujiwara, M., Miura, A. and Narahashi, T. (1985). Effects of *Goniopora* toxin on crayfish giant axons. *J. Pharmacol. Exp. Ther.*, **234**, 307–315.

Narahashi, T. (1962). Effect of the insecticide allethrin on membrane potentials of cockroach giant axons. *J. Cell. Comp. Physiol.*, **59**, 61–65.

Narahashi, T. (1974). Chemicals as tools in the study of excitable membranes. *Physiol. Rev.*, **54**, 813–889.

Narahashi, T. (1975). Mode of action of dinoflagellate toxins on nerve membranes. In *Proceedings of the First International Conference on Toxic Dinoflagellate Blooms*. (ed. V.R. LoCicero), pp. 395–402. Massachusetts Science and Technology Foundation. Wakefield, Massachusetts.

Narahashi, T. (1976). Effects of insecticides on nervous conduction and synaptic transmission. In *Insectcide biochemistry and physiology*, (ed. C.F. Wilkinson), pp. 327–352. Plenum Publ. Corp., New York.

Narahashi, T. (1984). Nerve membrane sodium channels as the target of pyrethroids. In *Cellular and molecular neurotoxicology*, (ed. T. Narahashi), pp. 85–108. Raven Press, New York.

Narahashi, T. (1985). Nerve membrane ionic channels as the primary target of pyrethroids. *NeuroToxicology*, **6**(2), 3–22.

Narahashi, T. (1986). Nerve membrane ionic channels as the target of toxicants. Toxic interfaces of neurones, smoke and genes. *Arch. Toxicol. Suppl.*, **9**, 3–13.

Narahashi, T. (1987). Neuronal target sites of insecticides. In *Sites of action for neurotoxic pesticides*. (ed. R.M. Hollingworth and M.B. Green), pp. 226–250. American Chemical Society Symposium Series. No. 356, ACS, Washington, DC.

Narahashi, T. (1988*a*). Mechanism of tetrodotoxin and saxitoxin action. In *Handbook of natural toxins, Vol. 3, Marine toxins and venoms*, (ed. A.T. Tu), pp. 185–210. Marcell Dekker, New York.

Narahashi, T. (1988*b*). Molecular and cellular approaches to neurotoxicology: past, present and future. In *Neurotox '88. Molecular basis of drug and pesticide action*, (ed. G.G. Lunt), pp. 269–288. Elsevier, Amsterdam.

Narahashi, T. (1989). The role of ion channels in insecticide action. In *Insecticide action: from molecule to organism*, (ed. T. Narahashi and J.E. Chambers), pp. 55–84. Plenum Press, New York.

Narahashi, T. (1992). Nerve membrane $Na^+$ channels as targets of insecticides. *Trends in Pharmacological Sciences*, **13**, 236–241.

Narahashi, T. (1994). Role of ion channels in neurotoxicity. In *Principles of neurotoxicology*, (ed. L.W. Chang), pp. 609–655. Marcel Dekker, New York.

Narahashi, T. and Lund, A.E., (1980). Giant axons as models for the study of the mechanism of action of insecticides. In *Insect neurobiology and pesticide action (Neurotox 79)*, pp. 495–505. Soc. Chem. Industry, London.

Narahashi, T. and Seyama, I. (1974). Mechanism of nerve membrane depolarization caused by grayanotoxin I. *J. Physiol. (Lond.)*, **242**, 471–487.

Narahashi, T., Deguchi, T., Urakawa, N. and Ohkubo, Y. (1960). Stabilization and rectification of muscle fiber membrane by tetrodotoxin. *Am. J. Physiol.*, **198**, 934–938.

Narahashi, T., Moore, J.W. and Scott, W.R. (1964). Tetrodotoxin blockage of sodium conductance increase in lobster giant axons. *J. Gen. Physiol.*, **47**, 965–974.

Narahashi, T., Anderson, N.C. and Moore, J.W. (1967*a*). Comparison of tetrodotoxin and procaine in internally perfused squid giant axons. *J. Gen. Physiol.*, **50**, 1413–1428.

Narahashi, T., Hass, H.G. and Therrien, E.F., (1967*b*). Saxitoxin and tetrodotoxin: comparison of nerve blocking mechanism. *Science*, **157**, 1441–1442.

Narahashi, T., Moore, J.W. and Frazier, D.T. (1969*a*). Dependence of tetrodotoxin blockage of nerve membrane conductance on external pH. *J. Pharmacol. Exp. Ther.*, **169**, 224–228.

Narahashi, T., Moore, J.W. and Shapiro, B.I. (1969*b*). *Condylactis* toxin: interaction with nerve membrane ionic conductances. *Science*, **163**, 680–681.

Narahashi, T., Frazier, D.T. and Yamada, M. (1970). The site of action and active form of local anesthetics. I. Theory and pH experiments with tertiary compounds. *J. Pharmacol. Exp. Ther.*, **171**, 32–44.

Narahashi, T., Albuquerque, E.X., and Deguchi, T. (1971). Effects of batrachotoxin on membrane potential and conductance of squid giant axons. *J. Gen. Physiol.*, **58**, 54–70.

Narahashi, T., Shapiro, B.I., Deguchi, T., Scuka, M. and Wang, C.M. (1972). Effects of scorpion venmom on squid axon membranes. *Am. J. Physiol.*, **222**, 850–857.

Narahashi, T., Frey, J.M., Ginsburg, K.S., Nagata, K., Roy, M.L. and Tatebayashi, H. (1995). Sodium channels and GABA-activated channels as the target sites of insecticides. ACS Symposium Series, *Molecular Action and Pharmacology of Insecticides on Ion Channels*, (ed. J.M. Clark), pp. 26–43, Am. Chem. Soc., Washington, DC.

Naumov, A.P., Negulayer, Y.A. and Nosyreva, E.D., (1979). Changes of selectivity of sodium channels in membrane of nerve treated with veratrine. *Zytologia*, **21**, 692–696.

Neher, E. and Steinbach, J.H. (1978). Local anaesthetics transiently block currents through single acetylcholine-receptor channels. *J. Physiol. (Lond.)*, **277**, 153–176.

Noda, M., Ikeda, T., Kayano, T., Suzuki, H., Takeshima, H., Kurasaki, M., Takahashi, H., Kuno, M. and Numa, S. (1986*a*). Existence of distinct sodium channel messenger RNAs in rat brain. *Nature*, **320**, 188–192.

Noda, M., Ikeda, T., Suzuki, T., Takeshima, H., Takahashi, T., Kuno, M. and Numa, S. (1986*b*). Expression of functional sodium channels from cloned cDNA. *Nature*, **322**, 826–828.

Noda, M., Shimizu, S., Tanabe, T., Takai, T., Kayano, T., Ikeda, T., *et al.* (1984). Primary structure of *Electrophorus electricus* sodium channel deduced from cDNA sequence. *Nature*, **312**, 121–127.

Noda, M., Muramatsu, I. and Fujiwara, M. (1984). Effects of *Goniopora* toxin on the membrane currents of bullfrog atrial muscle. *Naunyn-Schmiedeberg's Arch. Pharmacol.*, **327**, 75–80.

Norton, T.R. (1981). Cardiotonic polypeptides from *Anthopleura xanthogrammica* (Brandt) and *A. elegantissima* (Brandt). *Fed. Proc.*, **40**, 21–25.

Nuss, H.B., Tomaselli, G.F. and Marbán, E. (1995). Cardiac sodium channels (hH1) are intrinsically more sensitive to block by lidocaine than are skeletal muscle ($\mu$1) channels. *J. Gen. Physiol.*, **106**, 1193–1209.

Ogata, N. and Tatebayashi, H. (1992). Slow inactivation of tetrodotoxin-insensitive $Na^+$ channels in neurons of rat dorsal root ganglia. *J. Membr. Biol.*, **129**, 71–80.

Ogata, N. and Tatebayashi, H. (1993). Kinetic analysis of two types of $Na^+$ channels in rat dorsal root ganglia. *J. Physiol. (Lond.)*, **466**, 9–37.

Ogata, N., Yoshii, M. and Narahashi, T. (1990). Differential block of sodium and calcium channels by chlorpromazine in mouse neuroblastoma cells. *J. Physiol. (Lond.)*, **420**, 165–183.

Ogura, Y. and Mori, Y. (1968). Mechanism of local anesthetic action of crystalline tetrodotoxin and its derivatives. *Eur. J. Pharmacol.*, **3**, 58–67.

Ohta, M., Narahashi, T. and Keeler, R.F., (1973). Effects of veratrum alkaloids on membrane potential and conductance of squid and crayfish giant axons. *J. Pharmacol. Exp. Ther.*, **184**, 143–154.

Olivera, B.M., McIntosh, J.M., Cruz, L.J., Luque, F.A. and Gray, W.R. (1984). Purification and sequence of a presynaptic peptide toxin from *Conus geographus* venom. *Biochemistry*, **23**, 5087–5090.

Oxford, G.S., Wu, C.H. and Narahashi, T. (1978). Removal of sodium channel inactivation in squid giant axons by *N*-bromoacetamide. *J. Gen. Physiol.*, **71**, 227–247.

Pappone, P.A. and Cahalan, M.D. (1986). Ion permeation in cell membranes. In *Physiology of membrane disorders*, (ed. T.E. Andreoli, J.F. Hoffman, D.D. Fanestil, and S.G. Schultz), pp. 249–272. Plenum Press, New York.

Pichon, Y. and Pelhate, M. (1984). Effect of toxin 1 from *Androctonus australis* Hector on sodium currents in giant axons of *Loligo forbesi*. *J. Physiol. (Paris)*, **79**, 318–326.

Poli, M.A., Mende, T.J. and Baden, D.G. (1986). Brevetoxins, unique activators of voltage-sensitive sodium channels, bind to specific sites in rat brain synaptosomes. *Mol. Pharmacol.*, **30**, 129–135.

Quandt, F.N. and T. Narahashi, (1982). Modification of single $Na^+$ channels by batrachotoxin. *Proc. Natl. Acad. Sci. USA* **79**, 6732–6736.

Quandt, F.N., Yeh, J.Z. and Narahashi, T. (1985). All or none block of single $Na^+$ channels by tetrodotoxin. *Neurosci. Lett.*, **54**, 77–83.

Rack, M., Meves, H., Beress, L. and Grunhagen, H.H. (1983). Preparation and properties of fluorescence labeled neuro- and cardiotoxin II from the sea anemone (*Anemonia sulcata*). *Toxicon*, **21**, 231–237.

Ragsdale, D.S., McPhee, J.C. Scheuer, T. and Catterall, W.A. (1994). Molecular determinants of state-dependent block of $Na^+$ channels by local anesthetics. *Science*, **265**, 1724–1728.

Rando, T. (1989). Rapid and slow gating of vertridine-modified sodium channels in frog myelinated nerve. *J. Gen. Physiol.*, **93**, 43–65.

Rando, T.A. and Strichartz, G.R. (1986). Saxitoxin blocks batrachotoxin-modified sodium channels in the node of Ranvier in a voltage-dependent manner. *Biophys. J.*, **49**, 785–794.

Ravindran, A., Schild, L. and Moczydlowski, E. (1991). Divalent cation selectivity for external block of voltage-dependent $Na^+$ channels prolonged by batrachotoxin. $Zn^{2+}$ induces discrete substates in cardiac $Na^+$ channels. *J. Gen. Physiol.*, **97**, 89–115.

Reuter, H., Baer, M. and Best, P.M. (1978). Voltage dependence of TTX action in mammalian cardiac muscle. In *Biophysical aspects of cardiac muscle*, (ed. M. Morad), pp. 129–142. Academic Press, New York.

Rhoden, V.A. and Goldin, S.M. (1979). The binding of saxitoxin to axolemma of mammalian brain. Cooperative competition between saxitoxin and sodium ion. *J. Biol. Chem.*, **254**, 11199–11201.

Ritchie, J.M. and Greengard, P. (1966). On the mode of action of local anesthetics. *Annu. Rev. Pharmacol.*, **6**, 405–430.

Ritchie, J.M. and Rogart, R.B. (1977*a*). The binding of labelled saxitoxin to the sodium channels in normal and denervated mammalian muscle, and in amphibian muscle. *J. Physiol. (Lond.)*, **269**, 341–354.

Ritchie, J.M. and Rogart, R.B. (1977*b*). Density of sodium channels in mammalian myelinated nerve fibers and nature of the axonal membrane under the myelin sheath. *Proc. Natl Acad. Sci. USA*, **74**, 211–215.

Ritchie, J.M. and Rogart, R.B. (1977*c*). The binding of STX and TTX to excitable tissue. *Rev. Physiol. Biochem. Pharmacol.*, **79**, 1–50.

Ritchie, J.M., Rogart, R.B. and Strichartz, G.R. (1976). A new method for labelling saxitoxin and its binding to non-myelinated fibres of the rabbit vagus, lobster walking leg, and garfish olfractory nerves. *J. Physiol. (Lond.)*, **261**, 477–494.

Rogart, R.B., Cribbs, L.L., Muglia, L.K. Kephart, D.D. and Kaiser, M.W. (1989). Molecular cloning of a putative tetrodotoxin-resistant rat heart sodium channel isoform. *Proc. Natl Acad. Sci. USA*, **86**, 8170–8174.

Rojas, E. and Rudy, B. (1976). Destruction of the sodium conductance inactivation by specific protease in perfused nerve fibers from *Loligo*. *J. Physiol. (Lond.)*, **262**, 501–531.

Romey, G., Chicheportiche, R., Lazdunski, M., Rochat, H., Miranda, F. and Lissitzky, S. (1975). Scorpion neurotoxin – a presynaptic toxin which affects both $Na^+$ and $K^+$ channels in axons. *Biochem. Biophys. Res. Commun.*, **64**, 115–121.

Romey, G., Abita, J.-P., Schweitz, H., Wunderer, G. and Lazdunski, M. (1976). Sea anemone toxin: a tool to study molecular mechanisms of nerve conduction and excitation–secretion coupling. *Proc. Natl Acad. Sci. USA*, **73**, 4055–4059.

Roy, M.-L. and Narahashi, T. (1992). Differential properties of tetrodotoxin-sensitive and tetrodotoxin-resistant sodium channels in rat dorsal root ganglion neurons. *J. Neurosci.*, **12**, 2104–2111.

Ruigt, G.S.F. (1984). Pyrethroids. In *Comprehensive insect physiology, biochemistry and pharmacology*, Vol. 12, (ed. G.A., Kerkut, and L.I. Gilbert) pp. 183–263. Pergamon Press, Oxford.

Salgado, V.L. and Narahashi, T. (1993). Immobilization of sodium channel gating charge in crayfish giant axons by the insecticide fenvalerate. *Mol. Pharmacol.*, **43**, 626–634.

Salgado, V.L., Yeh, J.Z. and Narahashi, T. (1986). Use- and voltage-dependent block of sodium channel by saxitoxin. *Ann. N. Y. Acad. Sci.*, **479**, 84–95.

Sarao, R., Gupta, S.K., Auld, V.J. and Dunn, R.J. (1991). Developmentally regulated alternative RNA splicing of rat brain sodium channel mRNAs. *Nucleic Acids Res.*, **19**, 5673–5679.

Satin, J., Kyle, J.W., Chen, M., Rogart R.B. and Fozzard, H.A. (1992). The cloned cardiac sodium channel α subunit expressed in *Xenopus* oocytes show gating and blocking properties of native channels. *J. Membr. Biol.*, **130**, 11–22.

Satin, J., Kyle, J.W., Fan, Z., Rogart, R., Fozzard, H.A. and Makielski, J.C. (1994*a*). Post-repolarization block of cloned sodium channels by saxitoxin: the contribution of pore-region amino acids. *Biophys. J.*, **66**, 1353–1363.

Satin, J., Limberis, J.T., Kyle, J.W., Rogart, R.B. and Fozzard, H.A. (1994*b*). The Saxitoxin/tetrodotoxin binding site on cloned rat brain IIa Na channels is in the transmembrane electric field. *Biophys. J.*, **67**, 1007–1014.

Schild, L. and Moczydlowski, E. (1991). Competitive binding interaction between $Zn^{2+}$ and saxitoxin in cardiac $Na^+$ channels. Evidence for a sulfhydryl group in the $Zn^{2+}$/saxitoxin binding site. *Biophys. J.*, **59**, 523–537.

Schmidt, H. and Schmitt, O. (1974). Effect of aconitine on the sodium permeability at the node of Ranvier. *Pflügers Arch.*, **349**, 133–148.

Schmidtmayer, J. (1985). Behavior of chemically modified sodium channels in frog nerves supports a three-state model of inactivation. *Pflügers Arch.*, **404**, 21–28.

Schwartz, A., Palti, Y. and Meiri, H. (1990). Structural and developmental differences between three types of Na channels in dorsal root ganglion cells of newborn rats. *J. Membr. Biol.*, **116**, 117–128.

Seyama, I. and Narahashi, T. (1973). Increase in sodium permeability of squid axon membranes by α-dihydrograyanotoxin II. *J. Pharmacol. Exp. Ther.*, **184**, 299–307.

Seyama, I. and Narahashi, T. (1981). Modulation of sodium channels of squid nerve membranes by grayanotoxin I. *J. Pharmacol. Exp. Ther.*, **219**, 614–624.

Seyama, I., Wu, C.H. and Narahashi, T. (1980). Current-dependent block of nerve membrane sodium channels by paragracine. *Biophys. J.*, **29**, 531–537.

Shanes, A.M., Freygang, W.H., Grundtest, H. and Amatniek, E. (1959). Anesthetic and calcium action in the voltage clamped squid giant axon. *J. Gen. Physiol.*, **42**, 793–802.

Shapiro, B.I. (1977). Effects of strychnine on the sodium conductance of the frog node of Ranvier. *J. Gen. Physiol.*, **69**, 915–926.

Sheridan, R.E. and Adler, M. (1989). The actions of a red tide toxin from *Ptychodiscus brevis* on single sodium channels in mammalian neuroblastoma cells. *FEBS Lett.*, **247**, 448–452.

Shimizu, Y. (1982). Recent progress in marine toxin research. *Pure Appl. Chem.*, **54**, 1973–1980.

Shrager, P. and Profera, C. (1973). Inhibition of the receptor for tetrodotoxin in nerve membranes by reagents modifying carboxyl groups. *Biochim. Biophys. Acta*, **318**, 141–146.

Siemen, D. and Vogel, W. (1983). Tetrodotoxin interferes with the reaction of scorpion toxin (*Buthus tamulus*) at the sodium channel of the excitable membrane. *Pflügers Arch.*, **397**, 306–311.

Simard, J.M., Meves, H. and Watt, D.D. (1986). Effects of toxins VI and VII from the scorpion *Centruroides sculpturatus* on the Na currents of the frog node of Ranvier. *Pflügers Arch.*, **406**, 620–628.

Soderlund, D.M. and Bloomquist, J.R. (1989). Neurotoxic actions of pyrethroid insecticides. *Annu. Rev. Entomol.*, **34**, 77–96.

Song, J.-H. and Narahashi, T. (1995). Selective block of tetramethrin-modified sodium channels by (±)-α-tocopherol (Vitamin E). *J. Pharmacol. Exp. Ther.*, **275**, 1402–1411.

Song, J.-H. and Narahashi, T. (1996*a*). Differential effects of the pyrethroid tetramethrin on tetrodotoxin-sensitive and tetrodotoxin-resistant single sodium channels. *Brain Res.*, **712**, 258–264.

Song, J.-H. and Narahashi, T. (1996*b*). Modulation of sodium channels of rat cerebellar Purkinje neurons by the pyrethroid tetramethrin. *J. Pharmacol. Exp. Ther.*, **277**, 445–453.

Stengelin, S., Rathmayer, W., Wunderer, G., Beress, L. and Hucho, F. (1981). Radioactive labeling of toxin II from *Anemonia sulcata*. *Anal. Biochem.*, **113**, 277–285.

Strichartz, G. (1973). The inhibition of sodium currents in myelinated nerve by quaternary derivatives of lidocaine. *J. Gen. Physiol.*, **62**, 37–57.

Strichartz, G.R., Wang, G.K., Schmidt, J., Hahin, R. and Shapiro, B.I. (1980). Modification of ionic currents in frog nerve by crude venom and isolated peptides of the mollusc *Conus striatus*. *Fed. Proc.*, **39**, 2065.

Strichartz, G., Rando, T. and Wang, G.K. (1987). An integrated view of the molecular toxinology of sodium channel gating in excitable cells. *Annu. Rev. Neurosci.*, **10**, 237–267.

Sutro, J.B. (1986). Kinetics of veratridine action on Na channels of skeletal muscle. *J. Gen. Physiol.*, **87**, 1–24.

Tanguy, J. and Yeh, J.Z. (1991). BTX modification of Na channels in squid axons. *J. Gen. Physiol.*, **97**, 499–519.

Terlau, H., Shon, K.-J., Grilley, M., Stocker, M., Stühmer, W. and Olivera, B.M. (1996). Strategy for rapid immobilization of prey by a fish-hunting marine snail. *Nature*, **381**, 148–151.

Tatebayashi, H. and Narahashi, T. (1994). Differential mechanism of action of the pyrethroid tetramethrin on tetrodotoxin-sensitive and tetrodotoxin-resistant sodium channels. *J. Pharmacol. Exp. Ther.*, **270**, 595–603.

Taylor, R.E. (1959). Effect of procaine on electrical properties of squid axon membranes. *Am. J. Physiol.*, **196**, 1071–1078.

Trimmer, J.S. and Agnew, W.S. (1989). Molecular diversity of voltage-sensitive Na channels. *Annu. Rev. Physiol.*, **51**, 401–418.

Trimmer, J.S., Cooperman, S.S., Tomiko, S.A., Zhou, J.Y., Cream, S.M., Boyle, M.B., Kallen, R.G., Sheng, Z.H., Barchi, R.L., Sigworth, F.J., Goodman, R.H., Agnew, W.S. and Mandel, G. (1989). Primary structure and functional expression of a mammalian skeletal muscle sodium channel. *Neuron*, **3**, 33–49.

Tucker, S.B. and Flannigan, S.A. (1983). Cutaneous effects from occupational exposure to fenvalerate. *Arch. Toxicol.*, **54**, 195–202.

Tucker, S.B., Flannigan, S.A. and Ross, C.E. (1984). Inhibition of cutaneous paresthesia resulting from synthetic pyrethroid exposure. *Internat. J. Dermatol.*, **23**, 686–689.

Twombly, D.A., Yoshii, M. and Narahashi, T. (1988). Mechanisms of calcium channel block by phenytoin. *J. Pharmacol. Exp. Ther.*, **246**, 189–195.

Ulbricht, W. (1969). The effects of veratridine on excitable membranes in nerve and muscle. *Ergeb. Physiol. Biol. Chem. Exp. Pharmakol.*, **61**, 17–71.

Ulbricht, W. and Stole-Herzog, M. (1984). Distinctly different rates of benzocaine action on sodium channels of Ranvier nodes kept open by chloramine-T and veratridine. *Pflügers Arch.*, **402**, 439–445.

Ulbricht, W. and Wagner, H.-H. (1975*a*). The influence of pH on equilibrium effects of tetrodotoxin on myelinated nerve fibres of *Rana esculenta*. *J. Physiol. (Lond.)*, **252**, 159–184.

Ulbricht, W. and Wagner, H.-H. (1975*b*). The influence of pH on the rate of tetrodotoxin action on myelinated nerve fibres. *J. Physiol. (Lond.)*, **252**, 185–202.

Vijverberg, H.P.M. and van den Bercken, J. (1990). Neurotoxicological effects and the mode of action of pyrethroid insecticides. *Crit. Rev. Toxicol.*, **21**(2), 105–126.

Vijverberg, H.P.M., van der Zalm, J.M. and van den Bercken, J. (1982). Similar mode of action of pyrethroids and DDT on sodium channel gating in myelinated nerves. *Nature*, **295**, 601–603.

Vijverberg, H.P.M., Pauron, D. and Lazdunski, M. (1984). The effect of *Tityus serrulatus* scorpion toxin $\gamma$ on Na channels in neuroblastoma cells. *Pflügers Arch.*, **401**, 297–303.

Vincent, J.P., Balerna, M., Barhanin, J., Fosset, M. and Lazdunski, M. (1980). Binding of sea anemone toxin to receptor sites associated with gating system of sodium channel in synaptic nerve endings *in vitro*. *Proc. Natl Acad. Sci. USA*, **77**, 1646–1650.

Vogel, S.M., Watanabe, S., Yeh, J.Z., Farley, J.M. and Narahashi, T. (1984). Current-dependent block of end-plate channels by guanidine derivatives. *J. Gen. Physiol.*, **83**, 901–918.

Wang, C.M., Narahashi, T. and Scuka, M. (1972). Mechanism of negative temperature coefficient of nerve blocking action of allethrin. *J. Pharmacol. Exp. Ther.*, **182**, 442–453.

Wang, D.W., Nie, L., George, A.L. Jr. and Bennett, P.B. (1996). Distinct local anesthetic affinities in $Na^+$ channel subtypes. *Biophys. J.*, **70**, 1700–1708.

Wang, G.K., Brodwick, M.S. and Eaton, D.C. (1985). Removal of sodium channel inactivation in squid axon by the oxidant chloramine-T. *J. Gen. Physiol.*, **86**, 289–302.

Wang, G.K. (1984*a*). Irreversible modification of sodium channel inactivation in toad myelinated nerve fibers by the oxidant chloramine-T. *J. Physiol. (Lond.)*, **346**, 127–141.

Wang, G.K. (1984*b*). Modification of sodium channel inactivation in single myelinated nerve fibers by methionine-reactive chemicals. *Biophys. J.*, **46**, 121–124.

Wang, G.K. and Strichartz, G.R. (1983). Purification and physiological characterization of neurotoxins from venoms of the scorpions *Centruroides sculpturatus* and *Leiurus quinquestriatus*. *Mol. Pharmacol.*, **23**, 519–533.

Weigele, J.B. and Barchi, R.L. (1978). Saxitoxin binding to the mammalian sodium channel. Competition by monovalent and divalent cations. *FEBS Lett.*, **95**, 49–53.

White, J.A., Alonso, A. and Kay, A.R. (1993). A heart-like $Na^+$ current in the medical enthorhinal cortex. *Neuron*, **11**, 1037–1047.

Wiswesser, W.J. (1976). *Pesticide Index*, 5th edn, Entomological Society of America, College Park, MD.

Woodhull, A.M. (1973). Ionic blockage of sodium channels in nerve. *J. Gen. Physiol.*, **61**, 687–708.

Worley, J.R., III, French, R.J. and Krueger, B.K. (1986). Trimethyloxonium modification of single batrachotoxin-activated sodium channels in planar bilayers. Changes in unit conductance and in block by saxitoxin and calcium. *J. Gen. Physiol.*, **87**, 327–349.

Wu, C.H. and Narahashi, T. (1988). Mechanism of action of novel marine neurotoxins on ion channels. *Annu. Rev. Pharmacol. Toxicol.*, **28**, 141–161.

Wu, C.H., Huang, J.M.C., Vogel, S.M., Luke, V.S., Atchison, W.D. and Narahashi, T. (1985). Actions of *Ptychodiscus brevis* toxins on nerve and muscle membranes. *Toxicon*, **23**, 481–487.

Yamamoto, D., Quandt, F.N. and Narahashi, T. (1983). Modification of single sodium channels by the insectidice tetramethrin. *Brain Res.*, **274**, 344–349.

Yamamoto, D., Yeh, J.Z. and Narahashi, T. (1984). Voltage-dependent calcium block of normal and tetramethrin-modified single sodium channels. *Biophys. J.*, **45**, 337–344.

Yamamoto, D., Yeh, J.Z. and Narahashi, T. (1986). Ion permeation and selectivity of squid axon sodium channels modified by tetramethrin. *Brain Res.*, **372**, 193–197.

Yarowsky, P.J., Krueger, B.K., Olson, C.E., Clevinger, E.C. and Koos, R.D. (1991). Brain and heart sodium channel subtype mRNA expression in rat cerebral cortex. *Proc. Natl Acad.Sci. USA*, **88**, 9453–9457.

Yeh, J.Z. (1979). Dynamics of 9-aminoacridine block of sodium channels in squid axon. *J. Gen. Physiol.*, **73**, 1–21.

Yeh, J.Z. (1982). A pharmacological approach to the structure of the Na channel in squid axon. In *Proteins in the nervous system: structure and function*, (ed. B. Haber, J. Prez-Polo, and J. Coulter), pp. 17–49, Liss, New York.

Yeh, J.Z. and Narahashi, T. (1977). Kinetic analysis of pancuronium interaction with sodium channels in squid axon membranes. *J. Gen. Physiol.*, **69**, 293–323.

Yeh, J.Z., McCarthy, W.A. Jr, Quandt, F.N. and Yamamoto, D. (1986). Single-channel analysis of the action of Na channel blockers 9-aminoacridine and QX-314 in neuroblastoma cells. In *Molecular and cellular mechanisms of anesthetics*, (ed. S.H. Roth and K.W. Miller), pp. 227–241. Plenum, New York.

Yoshida, S., Matsuda, Y. and Samejima, A. (1978). Tetrodotoxin-resistant sodium and calcium components of action potentials in dorsal root ganglion cells of the adult mouse. *J. Neurophysiol.*, **41**, 1096–1106.

Yoshii, M. and Narahashi, T. (1984). Patch clamp analysis of veratridine-induced sodium channels. *Biophys. J.*, **45**, 184a.

Zaborovskaya, L.D. and Khodorov, B.I. (1985). Effect of *Tityus* $\gamma$ toxin on the activation process in sodium channels of frog myelinated nerve. *Gen. Physiol. Biophys.*, **4**, 101–104.

# 3 Antiarrhythmic action of drugs interacting with sodium channels

J. Tamargo, E. Delpón, O. Pérez, and C. Valenzuela

## Voltage-gated sodium channels

### Introduction

Cardiac cells display a complex action potential which is generated by a sequence of ionic movements across the cell membrane. In most cardiac cells (atrial and ventricular muscle, His–Purkinje system), the rapid depolarizing phase of the cardiac action potential which supports excitability and conduction velocity result from the rapid entry of $Na^+$ ions through voltage-sensitive $Na^+$ channels[1–4] $Na^+$ channels are characterized by voltage-dependent activation, rapid inactivation and high selectivity for $Na^+$ ions over other monovalent cations. In addition, the finding that tetrodotoxin (TTX) and local anaesthetics of the lidocaine type, at concentrations which had no effect on the maximum upstroke velocity of the action potential, markedly shortened the action potential duration suggested that $Na^+$ channels also play an important role in the plateau phase of the cardiac action potential.[5]

Antiarrhythmic drugs are compounds that effectively suppress or prevent tachyarrhythmias at concentrations at which they should not exert adverse effects on a normal heart beat. Antiarrhythmic therapy has focused on electrical cardiac activity and thus, ion channels involved in the regulation of the cardiac action potential represent the therapeutic target for antiarrhythmic drugs. Disturbances in $Na^+$ channel function play a central role in the slow intracardiac conduction and block leading to re-entrant arrhythmias. $Na^+$ channel blockers, e.g. class I antiarrhythmic drugs (Table 3.1), comprise a large number of drugs with disparate chemical structures and distinct electrophysiological and pharmacological properties that share the ability to inhibit the influx of $Na^+$ ions through voltage-gated channels. [6–9] $Na^+$ channels are also the site of action of a wide variety of toxins used as pharmacological tools for a better characterization of channel structure and function [CTX and saxitoxin (STX) for their selective blocking effects; veratridine, batrachotoxin (BTX), scorpion, and sea anemone toxins for their agonistic effects].

Even when many voltage-dependent properties of $Na^+$ channels in heart and nerve are similar, functional, molecular, and pharmacological differences have been reported. [1–4, 10, 11] In neuronal and skeletal muscle, TTX binds to $Na^+$ channels at nanomolar

**Table 3.1** Classification of $Na^+$ channel blockers, i.e. class I antiarrhythmic agents.

| Subclasses | IA | IB | IC |
|---|---|---|---|
| $I_{Na}$ at normal cardiac rates | Depressed | Little effect | Markedly depressed |
| QRS | Modest prolongation | Little effect | Marked prolongation |
| Onset–offset kinetics of UDB | Intermediate | Fast | Slow |
| Drugs | Quinidine<br>Procainamide<br>Disopyramide<br>Imipramine<br>Ajmaline | Lidocaine<br>Mexiletine<br>Tocainide<br>Etmozine<br>Phenytoin<br>Pirmenol | Propafenone<br>Flecainide<br>Encainide<br>Lorcainide<br>Diprafenone<br>Recainam<br>Indecainide |

UDB: use-dependent $I_{Na}$ block

concentrations, whereas the predominant mammalian cardiac voltage-gated $Na^+$ channels are resistant to TTX (TTX-R, $K_D \approx 0.3$–$1\ \mu M$) and $\mu$-conotoxin blocks skeletal muscle $Na^+$ channels but has no effect on neuronal or cardiac $Na^+$ channels. [2–4] Conversely, local anaesthetics block cardiac TTX-R channels with 50–1000 times greater potency than their blockage of neuronal $Na^+$ channels. [2–4, 12, 13] Two characteristics distinguish cardiac TTX-R from TTX-S $Na^+$ channels. First, the blockade induced by TTX is potentiated by membrane depolarization and repetitive high-frequency stimulation. [2–4] The use-dependent block of $I_{Na}$ occurred because TTX dissociates considerably more slowly from inactivated channels than from open channels.[14] Secondly, TTX-R channels show an increased sensitivity to blockade by group IIB divalent cations ($Cd^{2+}$, $Zn^{2+}$). Divalent cations also competitively inhibit TTX and STX binding, which suggests that they share a common binding site on $Na^+$ channels.

## $Na^+$ channel $\alpha$-subunits

The predominant isoform in adult heart is a TTX-R $Na^+$ channel that is responsible for most of the $I_{Na}{}^+$ observed in mammalian cardiac myocytes.[1–4] Additionally, there is evidence that both the rat brain I isoform and another partially characterized $Na^+$ channel cDNA sequence are also present in the heart[15]. The cDNA of voltage-gated human atrial and ventricular $Na^+$ channels (hH1), encodes a 2016-amino acid protein that bears 90% identity to the TTX-R $Na^+$ channel characteristics of rat heart and of immature and denervated skeletal muscle.[13] Northern blot analysis demonstrated a ~9-kilobase transcript expressed in cardiac tissue that when expressed in Xenopus oocytes exhibits rapid kinetics, similar to native cardiac $Na^+$ channels. The $Na^+$ channel contains four homologous internal domains (I–IV) surrounding a central ion-selective pore acting as the conduction pathway.[2,3,16] Each domain has six transmembrane hydrophobic segments (S1–S6) with a high $\alpha$-helical potential and a S5–S6 (or H5 region) linker that may partly span the membrane as a hairpin and is involved in controlling ion selectivity. [2,3,17] High-affinity binding toxins (TTX) and site-directed mutagenesis provide strong evidence that the highly conserved H5 region is the $Na^+$ channel pore. The S4 region which contains positively charged amino acids [X-X-(arginine or lysine)]$_n$ at every third position has been proposed to serve as a voltage sensor. In fact, reducing the net positive charge in the IS4 region causes a decrease in apparent gating charge, as manifested by a decrease in the steepness of the potential dependence of activation. The segments preceding and following each domain are on the cytoplasmic side of the membrane, while the segment following S5 is in the extracellular space and presents several N-glycosylation sites. In addition to the main $\alpha$-subunit, two $\beta$ subunits ($\beta_1$ and $\beta_2$) have been described in nerve cells. The $\beta_1$ subunit has multiple effects on $Na^+$ channel function: increased peak current, accelerated activation, and inactivation and altered voltage dependence of inactivation.[18] However, the coexpression of $\alpha$ and $\beta_1$ subunits of cardiac $Na^+$ channels evoked $Na^+$ currents with similar characteristics as those exhibited by the expression of the $\alpha$ subunit[1].

In rat heart homogenates, approximately 20% of the binding sites for STX exhibit high affinity, suggesting the presence of a $Na^+$ channel isoform other than the predominant TTX-R channel.[1–4, 15] A TTX and $Mn^{2+}$-insensitive slow $Na^+$ channel has also been described in the myocardial membranes of genetically cardiomyopathic hamsters. In addition, voltage-clamp recordings of normal heart cells from a Variety of species have demonstrated the existence of a class of low-amplitude, $Na^+$- and voltage-dependent, slow inward currents with much slower activation and inactivation kinetics and a different TTX sensitivity than those responsible for the rapid depolarization phase of the cardiac action potential.[1–5, 15] These channels may contribute to the plateau phase of the cardiac action potential and may also participate in certain diseases.[15] This suggests the existence of voltage-, $Na^+$-conducting channels that exhibit functional properties very distinct from the predominant cardiac TTX-R channels. All of the mammalian cDNA sequences predict proteins which exhibit stricking similarity to one another ($>80\%$ overall amino acid identity) and appear to comprise a single multi-gene family.[15] There is little information about subtypes of cardiac $Na^+$ channels, even when changes in kinetics and TTX sensitivity during development have been reported in rat cardiomyocytes.[1, 19] It would be of interest to know how the prevalence of certain isoforms of cardiac $Na^+$ channels may be altered under physiopathological conditions.

Very recently, the cDNAs that encode an $Na^+$ channel $\alpha$-subunit have been cloned from adult human heart and fetal skeletal muscle.[20] The 7.2 Kb sequence, named hNav2.1 (2.1 refers to the first member of a second gene sub-family of $Na^+$ channels) encodes a 1682-amino acid protein that bears 46% overall identity with $Na^+$channels cloned from rat heart. The length of hNav2.1 and the size of the corresponding mRNA transcript are significantly smaller than those of other $Na^+$ channels (1819–2018 residues, 8.5–9.0 Kb) because their carboxy terminus is 38–67 amino acids shorter in hNav2.1 than in other $Na^+$ channels. But the most striking difference between hNav2.1 and other $Na^+$ channels exists in the S4 segments, which are believed to function as voltage sensors. The absolute number of positively charged residues (arginine or lysine) is less in repeat domains I, III and IV than in other $Na^+$ channels. In addition, in IVS4, histidines (residues 1355 and 1367) replace two of the highly conserved arginines. Tissue distribution of hNav2.1 mRNA transcripts indicated that this $Na^+$ channel is prominently expressed in adult human heart and uterus, moderately in skeletal muscle, and is absent or expressed at very low level in cerebral cortex, kidney, vascular smooth muscle, liver, and spleen. High-affinity binding toxins (TTX, STX) and site-directed mutagenesis experiments provide strong evidence that the highly conserved H5 region is the $Na^+$ channel pore. [2,3,17] TTX and STX have a guanidine residue and because guanidine is a permeant ion, it enters and binds to a receptor site located near the extracellular mouth of the pore of the $Na^+$ channel and lodges there. Studies of chimeric (TTX-sensitive and TTX-R voltage-gated) $Na^+$ channels suggest that TTX sensitivity is determined by a 22-amino acid stretch that forms part of the external loop connecting the S5–S6 region of domain I.[2,3] Neutralization of negatively charged amino acids in the short segment (glutamate-387 by glutamine, aspartate-384 by asparagine, tyrosine-401 by cysteine) renders the channel insensitive to TTX and SXT, reduces single-channel conductance, shifts the voltage dependence of activation toward more positive potentials, and alters cation sensitivity.[3,21,22] Complete loss of TTX sensitivity was observed in rat heart after mutation of tyrosine-374 for cysteine, while the reverse mutation markedly increased sensitivity to TTX[15]. In TTX-sensitive (brain, skeletal muscle) and TTX-insensitive (cardiac) channels, glutamate-387 is conserved, while the adjacent asparagine residue that is present in all TTX-sensitive $Na^+$ channel isoforms is replaced by arginine in cardiac TTX-R channels.[13] This charge difference could repel the positively charged guanidium group of TTX, conferring a lower sensitivity to the cardiac $Na^+$ channel. Mutation of one to three positively charged amino acid residues in segment S4 of domains I and II from arginine to lysine, which retains a positive charge, or to non-charged glutamate, markedly inhibited TTX binding, decreased steepness in the voltage dependence of activation, and shifted the current–voltage relationship towards more positive values than expected if these positively charged amino acid residues serve as gating charges.[17,22] Mutation of a hydrophobic residue in IIS4 from leucine to phenylalanine also causes a 20-mV shift in the voltage dependence of gating to more positive membrane potentials. In addition, mutations of lysine-1422 and alanine-1714 to glutamate in the H5 region of domains III and IV, respectively, reduced selectivity for $Na^+$ and the double mutant changed the selectivity of the channel from $Na^+$ selective to $Ca^{2+}$-selective, [21] which indicated that both residues form part of the selectivity filter of the channel.

## Kinetics of cardiac $Na^+$ channels

Drug-induced blockade of cardiac $Na^+$ channels is responsible for the suppression of cardiac arrhythmias. The effectiveness of $Na^+$ channel blockers in depressing $I_{Na}$ is voltage- and time-dependent, and is additionally determined by their physicochemical properties, so that marked differences in the degree of inhibition would be observed among them. Single-channel studies have stressed that antiarrhythmic drugs exert their blocking action in an all-or-none fashion.[19] Blockade prevents the $Na^+$ channel from attaining the conducting configuration, while unblocked channels show apparently normal elementary properties, including unitary current size and open-state kinetics.

Single-channel recordings from isolated cardiac cells have been used to characterize activation and inactivation behaviour of $Na^+$ channels using a simple Markov chain model of channel stales. The basic three-state model chosen is schematically presented below:

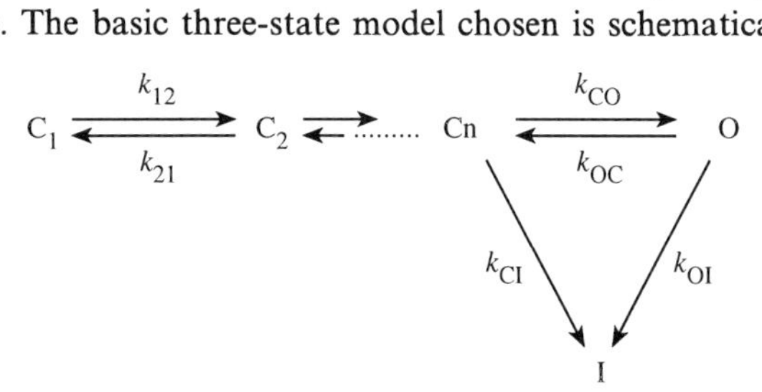

States C are closed, rested states (but available to open), O is an open state and I is a closed, absorbing, inactive state (not available to open); k indicated the rate constants.[1,2,4,23,24] All these states of the $Na^+$ channel are candidates for drug binding. The normal gating transition of the $Na^+$ channel includes several closed states before the opening transition of the channel ($C_1 \ldots C_n \rightarrow O$), one open state and one absorbing inactivated state. At normal membrane potentials, $Na^+$ channels are mostly in the closed state but are available to be open and provoke a $Na^+$ current on depolarization.

Upon depolarization, some voltage-gated $Na^+$ channels inactivate directly without even opening (C → l), and others transiently enter into the open state (C → O) after a variable delay with a sigmoid time course which indicates that individual channels must move into multiple (4–6) non-conducting states before activation. This indicates that only a fraction of the cardiac $Na^+$ channels may contribute to the rapid depolarization (phase 0) of the action potential.[1] When open, the channels inactivate (O → l) or close (O → C), but if the channel returns to the closed state, it can reopen. The probability of at least one opening is $k_{CO}/(k_{Cl} + k_{CO})$ and the probability of inactivation without opening (C → l) is $k_{cl}/k_{cl} + k_{CO})$. The increase in $Na^+$ permeability reached a maximum after a few milliseconds, but then the current spontaneously decreases, even during a maintained depolarization as the channels inactivate. Because the inactivated state is absorbing, once the channel inactivates, it remains inactivated until the membrane is repolarized. The inactivated state predominates at depolarized levels, i.e. during the plateau phase of the action potential and in depolarized cardiac tissues. Thus, all situations which prolong the cardiac action potential and depolarize the membrane potential (ischaemia, hyperkalaemia) will shift the equilibrium of channels to the inactivated state. Calculation of the rates for simple models indicates that the main voltage dependence of the transition rates is related to the transitions between the closed states and from the closed state to the open state. On the other hand, the transition rate from open to inactivated state is only weakly voltage-dependent.[1,4,10,11,23]

Channels must recover from inactivation (l → R or reactivation) during diastole before they can be available for reopening. The recovery from inactivation is a fast ($\tau_{re} < 35$ ms) and voltage-dependent process[7,8,25] that takes place at the beginning of diastole, so that in sinus rhythm (between 60 and 70 beats per min) $Na^+$ channel availability is complete at the time the next action potential arrives.

Although cardiac $Na^+$ channels exhibit many of the properties similar to those reported for nerve cells,[4,10,11] cardiac $Na^+$current kinetics is slower and inactivation occurs with more than one time constant.[4,10,26] Figure 3.1 shows whole-cell currents recorded from a guinea pig ventricular myocyte (upper panel) and the current-voltage relationship (lower panel). Membrane current rises rapidly and reaches a peak within a few milliseconds and then gradually declines as the channel inactivates. The rates of activation and inactivation increase with membrane depolarization. With larger depolarizations, more channels are activated, with an apparent increase in both rate of activation and inactivation. The normal $l_{Na}$ begins to activate at −70 mV, reaches a maximum peak at −20 mV and between −60 and −10 mV shows a steep voltage dependence which is responsible for the all-or-none feature of the propagated action potential.

At the single-channel level, depolarization into the activating voltage range produces one of following patterns: (i) a single opening after a variable delay; (ii) a brief burst of openings

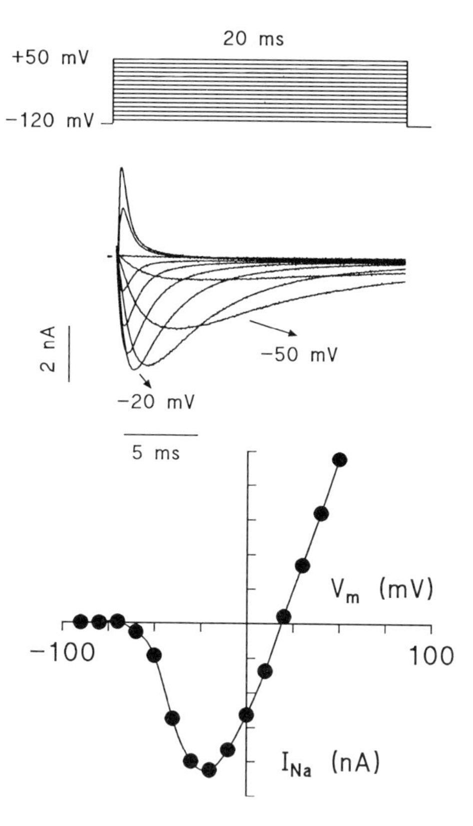


**Fig. 3.1** Whole-cell $Na^+$-currents recorded in guinea-pig ventricular myocytes. Upper panel: family of $Na^+$-currents evoked by depolarizing test pulses from a holding potential of −120 mV. Lower panel: peak $Na^+$-current is plotted against test pulse potential (peak current–voltage relationship).

at the beginning of the depolarizing step; or (iii) no opening at all. In all three cases the channel can be considered to have inactivated, because there are no further channel openings if one continues to hold the membrane at the depolarized potential. The mean values obtained from currents recorded at different voltages can be plotted as I–V relationships (Fig. 3.2). Under many conditions, at least over most of the voltage ranges in which $Na^+$ channels are physiologically active, the relationship is ohmic, and thus, the slope of this relation represents the single-channel conductance ($\gamma$) for those conditions. The $\gamma$ values reported for cardiac $Na^+$ channels ranged between 10 and 25 pS.[1–4] Subconductance states have also been described under normal conditions and more frequently in chemically modified channels, e.g. when the fast inactivation was removed by DPI 201–106,[19,27] although their physiological role is still uncertain. There has been controversy as to whether the opening times of cardiac $Na^+$ channels are voltage-dependent, but most of the experimental data has provided evidence that they are.[1–4,23,28] Whereas neuronal $Na^+$ channels usually open only once before 'irreversibly' entering the inactivated state,[29] cardiac $Na^+$ channels have a greater tendency to reopen at membrane potentials between −60 and −40 mV.[1,4,10,11,23,28]

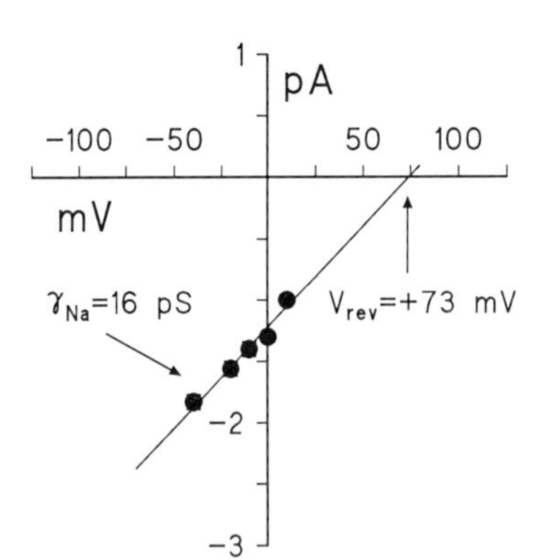


**Fig. 3.2** Current voltage (I–V) relationship. Data are means obtained from six experiments with two times the S.E.M. indicated in each direction. The dashed line has a slope of 16 pS and an x-intercept (reversal potential) of +73 mV. (From Valenzuela and Bennett, 1994[28].)

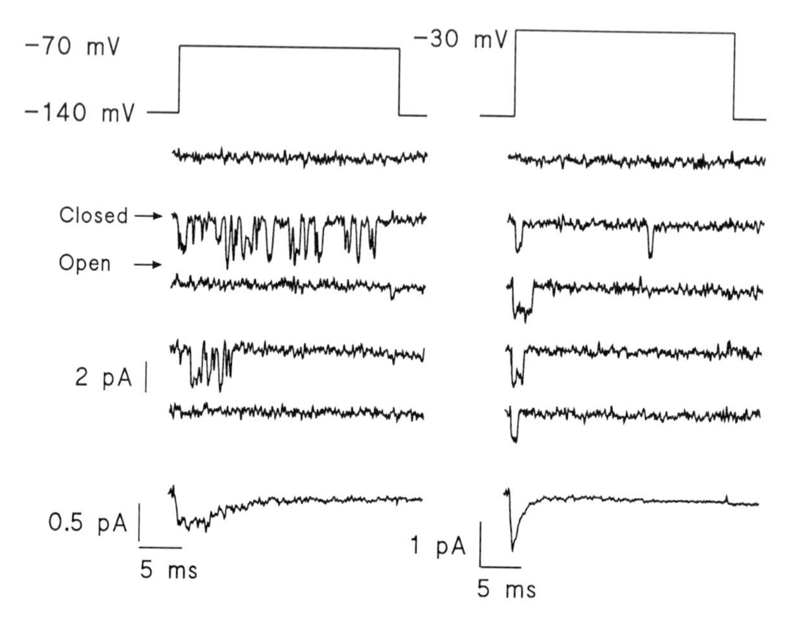


**Fig. 3.3** Single $Na^+$ channel currents and ensemble averages (bottom) in an inside-out patch after applying depolarizing pulses to −70 mV (left panel) and to −30 mV (right panel) during 30 ms. The patch was held at −140 mV for 1 s between pulses either to −70 or −30 mV.

At membrane potentials close to the threshold potential (−70 mV), the probability of inactivation without opening is high, so that a significant number of resting channels (about 50%) directly inactivate without opening, because $k_{CI}$ is larger than $k_{CO}$ (Fig. 3.3). During depolarizing steps to −60 and −70 mV, channel openings occur throughout the step, and the first opening often occurs well after the start of depolarization. At these negative potentials, channels also show a substantial probability of reopening, so that a single channel can open several times before inactivating and thus, the time course of the ensemble average current decay becomes slower than at more positive depolarizations. Thus, the time course of the macroscopic current decay is the result of two processes, inactivation of the channel and channel reopening. Reopening is a consequence of oscillations between the open state and the last closed state ($O \leftrightarrows C_n$) in preference to progression to the inactivated state, because near the threshold, $k_{OC}$ and $k_{CO}$ are similar and faster than $k_{OI}$, so that every time a channel opens, it very likely returns to C. Thus, $Na^+$ channels will produce short bursts of openings before they finally inactivate. As the membrane is depolarized towards more positive potentials, the situation is reversed, so that $k_{CO}$ increases more rapidly. However, since $k_{OI}$ predominates over $k_{OC}$, nearly all open channels are far more likely to exit to the inactivated state with little chance of reopening.[1,11]

At strongly depolarized potentials (more positive than −30 mV), the probability of inactivation without reopening is less than that observed at the threshold potential. During a depolarizing step to 0 mV, most of the $Na^+$ channels open and then rapidly inactivate and, in contrast to negative potentials, the probability of reopening is very low, since $k_{OC}$ is much slower than $k_{OI}$. Thus, reopening is an important source of $Na^+$ current for the cardiac cell at voltages near threshold, accounting for as much as 39% of the current, whereas at more depolarized potentials this is reduced dramatically, contributing only 5–10% of the current[1,4,10,23] (Fig. 3.3).

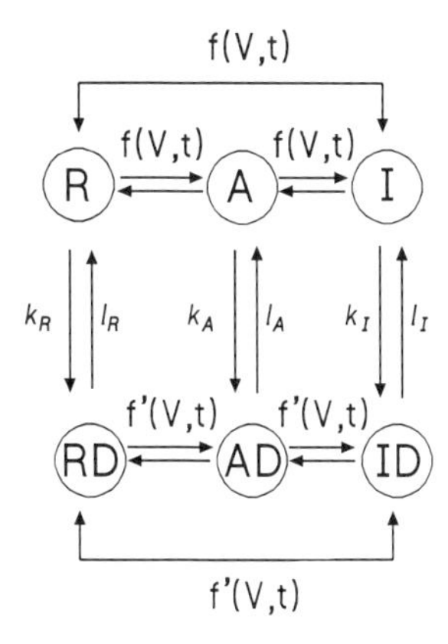


**Fig. 3.4** Representation of modulated receptor hypothesis for antiarrhythmic drug action. Sodium channels are depicted as existing in three states: rested (R), activated-open (O) and inactivated (I). RD, AD and ID represent drug-associated states. Binding and unbinding of antiarrhythmic drugs with each state are governed by association (*k*) and dissociation (*l*) rate constants. Transitions between states are governed by a function which depends on time and voltage [f(*V*, *t*)] and is modified in the presence of the antiarrhythmic drug [f'(*V*, *t*)].

The inactivation process of cardiac $Na^+$ channels consists of a fast, and a more slow phase, although the majority of the relaxation is related to the fast one.[1–4,10,11] Internal protelolysis with α-chymotrypsin removes fast inactivation but leaves slow inactivation intact, which suggests that these two processes are distinct and involve different domains in $Na^+$ channel structure.[28] The slow inactivation of $Na^+$-current may play an important role in cardiac cells because: (i) it may contribute to maintain the plateau phase of the cardiac action potential and in fact, TTX and lidocaine shorten the action potential duration; (ii) certain agents that suppress cardiac excitability bind to the slow inactivated channels; and (iii) it may mediate suppressed excitability in regions that are depolarized for prolonged periods (i.e. ischaemia[28]). In addition, the slow component of $Na^+$ current may also be responsible for the shortening of the action potential duration (APD) observed at high frequencies.[26]

### Inactivation

When the membrane is depolarized, $Na^+$ channels are activated and the influx of $Na^+$ ions increases but then slowly decreases (inactivation), despite the fact that the membrane voltage is maintained at the positive value that induced channel openings. The inactivation process can be removed or slowed after intracellular application of toxins or endopeptidases and is associated with little gating charge movement. This leads to the proposal of the 'ball and chain' model,[30] which assumes that the channel contains on its cytoplasmic surface a positively charged inactivation particle (ball) which is tethered to the cytoplasmic mouth of the channel by a protease-cleavage chain. During depolarization the activation gate opens, the ball moves into its receptor located in the ion pore, and inactivation occurs.

The cytoplasmic loop connecting homologous domains III and IV ($L_{III/IV}$) is involved in the fast inactivation of $Na^+$ channels.[1–4] Application of polyclonal antibodies against this region inhibits fast inactivation, whereas molecular cuts and small insertions at this level slowed inactivation.[1–4,17,31] The effect of the antibodies was voltage-dependent, being more rapid and marked at negative membrane potentials, when the channel is not inactivated, than at more positive membrane potentials, when the channel is inactivated. Phosphorylation of serine-1506 in $L_{III/IV}$ by protein kinase C slows inactivation and reduces $I_{Na}$, suggesting that activation of protein kinase C may be involved in the modulation of $Na^+$ channels by neurotransmitters and hormones. $L_{III/IV}$ contains 12 positively charged amino acids and three negatively charged residues. Neutralization of six of the 12 charged residues to either asparagine or glutamate resulted in a faster inactivation than the native channel.[3] Deletion of the first 10-amino acid segment at the amino-terminal end of $L_{III/IV}$ suppressed the inactivation. Mutation of the three-residue hydrophobic cluster isoleucine-1488, phenylalanine-1489, methionine-1450 to glutamine (IFMQ3) completely removed $Na^+$ channel inactivation.[3] Phenylalanine-1489 is the critical residue, since its mutation (F1489Q) nearly completely prevents inactivation and decreases peak conductance, while mutants 1448Q and M1450Q exhibit a slower $Na^+$ channel inactivation versus the wild-type. Therefore, these residues are an essential part of the fast inactivation particle which occludes the intracellular ion pore of the $Na^+$ channel.

## Modulated receptor hypothesis

The voltage- and frequency-dependent blockade of $Na^+$ channels produced by local anaesthetics and antiarrhythmic drugs in nerve and cardiac muscle, can be explained in the framework of the modulated receptor hypothesis.[25] This proposes that $Na^+$ channel blockers bind to a specific receptor located on or near the $Na^+$ channel but their binding affinity and the onset and offset rate constants are modulated by the conformational state of the channel (rested-R, open-O, inactivated-I)(Fig. 3.4). The concept of the specific receptor site on the channel is supported by the finding that $Na^+$ channel blockers inhibit $I_{Na}$ and non-competitively inhibit $^3$H-BTX binding, both effects being saturable and stereospecific.[1,5,32] At therapeutic concentrations ($\mu$M range), $Na^+$ channel blockers show a much higher affinity for either the open and/or inactivated channels than for the rested channels ($K_{DR}$ >0.1 mM).[7,8,25] Drug-associated $Na^+$ channels fail to conduct in response to membrane depolarization, exhibit time- and voltage-dependent transitions between the RD, OD and ID states, and their voltage dependence is shifted to more negative potentials, indicating a greater inhibition of $Na^+$ channels at depolarized levels. In addition, $Na^+$ channel blockers prolong the $\tau_{re}$, the degree of this prolongation being specific for each drug (Table 3.2).[7,9,4,25,33]

**Table 3.2** Relative affinity and recovery time constants ($\tau_{re}$) of different sodium channel blockers.

| Drug | Active ($\mu$M) | Inactive ($\mu$M) | $\tau_{re}$ | Subgroup |
|---|---|---|---|---|
| Quinidine | 100 | >100.000 | 4.7 s | |
| Procainamide | 28 | 1.0 | 2.3 s | IA |
| Disopyramide | 300 | >100.000 | 2.2 s | |
| Lidocaine | 50 | 1.2 | 230 ms | |
| Mexiletine | 200 | 0.7 | 470 ms | IB |
| Amiodarone | 1000 | 0.2 | 1.4 s | |
| Propafenone | 1 | 0.6 | 8.5 s | |
| Flecanide | 15 | >100.000 | 15.5 s | IC |
| Encainide | 10 | >100.000 | 20.3 s | |

Source: Modified from Tamargo *et al.*[7] and Snyders *et al.*[24]

Recovery from block is a slow diastolic process usually related to the dissociation from closed channel states (RD → R and ID → I → R)(Fig. 3.4). Recovery from inactivation is favoured at negative membrane potentials and in fact, the recovery from block for lidocaine, quinidine, mexiletine, tocainide, and procainamide is accelerated by hyperpolarization.[7,8,24,25,33] This can be explained by assuming that hyperpolarization shifts the distribution between inactivated- and rested-bound channels (ID → RD) and therefore, the RD → R pathway is favoured. These features explain why antiarrhythmic drugs produced a greater inhibition of $I_{Na}$ at faster heart rates and in depolarized tissue.

In addition to this slow recovery from block measured at negative membrane potentials, there has also been described a fast dissociation following $I_{Na}$ block with some open-channel blockers (quinidine, disopyramide, penticainide).[34,35] This has been explained assuming that these drugs bind to the open state and are trapped when the channel returns from the open to the rested state (i.e. activation trapping). Since drug dissociation requires channel opening and the possibility of opening or activating the channel is less the more negative the membrane potential, a slowing of recovery from block is expected.

This activation or use-dependent unblocking is related to the ID → RD → OD → O pathway[24] and can be an important mechanism of dissociation of antiarrhythmic drugs that have very little rest recovery (Fig. 3.4). Following strong hyperpolarizations, channels switch from ID to RD and subsequent depolarization can cause the transition to drug-associated open channels (OD → O) when the block exceeds the open state equilibrium ([OD] > > [O]). These characteristics explain why antiarrhythmic drugs produced a greater inhibition of $I_{Na}$ at faster heart rates and in depolarized tissue.

## Tonic and use-dependent $I_{Na}$ block

In the presence of $Na^+$ channel blockers, two types of $I_{Na}$ block can be observed.[7,8,24,25,33] *Tonic block*, is the inhibition of $I_{Na}$ observed during the first depolarization after a long rest period at sufficiently negative potentials. Most $Na^+$ channel blockers produced a small tonic block ($\acute{K}_{RD}$ in the mM range), suggesting that at negative resting potentials most channels remained unblocked.

Under physiological conditions, $Na^+$ channels are open-activated and inactivated during each cardiac action potential and the reactivation occurs very fast ($\tau_{re} < 35$ ms) at the beginning of diastole, so that $Na^+$ channel availability is complete when the next action potential arrives. In the presence of antiarrhythmic drugs during repetitive stimulation (tachycardia) the $I_{Na}$ decreases beat by beat until a new steady state is obtained.[7–9,24,25,33,36] The accumulation of $I_{Na}$ inhibition produced by $Na^+$ channel blockers during repetitive stimulation is called *use(frequency)-dependent block* (UDB). Repetitive depolarization induced the transition from the rested state to channel states (open and inactivated) for which $Na^+$ channel blockers exhibit higher affinity and the diastolic time for reactivation is shortened, so when the next regular beat arrives some channels are still drug-bound. This would lead to a progressive increase of non-conducting (ID) channels, which slowly return to the rested state. Steady-state UDB will be achieved when the block accumulated during depolarization is matched by that dissipated during the diaslolic interval. UDB explains why $Na^+$ channel blockers would inhibit $I_{Na}$ to a greater extent during tachyarrhythmia than at normal heart rates. The UDB is voltage-dependent and is enhanced at a less negative holding potential, so that drug molecules may dissociate from $Na^+$ channel receptors after the drug-bound inactivated state (ID) is converted to the drug-bound rested state (RD).

Most $Na^+$ channel blockers are weak bases ($pK_a = 7–10$), and therefore, the proportion of charged-hydrophilic and uncharged-lipophilic forms will vary with pH. Drugs are thought to bind at a site located at the channel.[2] The charged forms of tertiary amines and permanently charged quaternary derivatives do not block $Na^+$ channels when applied from outside and preferentially access the $Na^+$ channel receptor through a hydrophilic pathway from the inside only when the channel is open, and thus, they cause a pronounced UDB[2]. The uncharged blockers gain access to its receptor via the hydrophobic pathways and interact with the channel by lateral diffusion through the lipid layer, so that drug access is not strongly affected by the channel state but exerts a tonic block. Membrane depolarization, acidosis and hyperkalaemia would favour the interaction of the charged form with its receptor and markedly slow the $\tau_{re}$.[8,25] In contrast, at high external pH and hyperpolarized potentials, most drug molecules are uncharged and the $\tau_{re}$ is expected to speed up. This can explain the effectiveness of alkalosis-inducing salts in the treatment of cardiotoxicity induced by $Na^+$ channel blockers.[7,8,24,25,33]

## Subclassification of $Na^+$ channel blockers

According to the rates of onset and offset of UDB, $Na^+$ channel blockers have been subclassified as fast (IB), slow (IC) and intermediate (IA) kinetics (Table 3.1).[6,9]

### Fast kinetic (class IB) blockers

These bind very rapidly to the inactivated state of the $Na^+$ channel and unbind quickly during repolarization ($\tau_{re} < 0.5$ s). The fast offset kinetics mean that at normal heart rates in normally polarized cardiac tissues, the diastole is long enough to enable most channels to have recovered from inactivation (ID $\rightarrow$ R) by the time the next action potential arrives.[7,8,25,33,36] These antiarrhythmic drugs would exert minimal, if any, effect on intracardiac conduction velocity (as reflected in the lengthening of the HV interval on His bundle recordings) and do not alter the QRS duration of the surface electrocardiogram at the slow rates of sinus rhythm, whereas they selectively depress conduction at faster rates (tachycardia) or when ventricular conduction is previously depressed (ischaemia, marked hyperkalaemia). Because of their fast kinetics, steady-state UDB is reached within five beats after a sudden increase in heart rate, so that they may be very effective in the prevention of a sudden increase in heart rate (early premature ventricular extrasystoles and tachycardia onset), but less effective than class IA and IC drugs against slower tachycardias or late premature beats[7,25,36].

### Intermediate (IA) and slow kinetic antiarrhythmic (IC) drugs

These exhibit higher affinity for the open state and because of their slow onset and offset kinetics (IA: $1\ \text{s} < \tau_{re} < 5$ s; IC $\tau_{re} > 8$ s) the UDB reached a steady-state level after $> 10$ beats.[6–9,25,33,36] Moreover, in the presence of slow kinetics $Na^+$ channel blockers, a considerable number of channels are still drug-associated (ID) when the next action potential occurs, so that they slow conduction and prolong the QRS duration not only during the tachyarrhythmia but also at normal heart rates, even when these effects are more marked at increasing heart rates. Thus, $Na^+$ channel blockers with long $\tau_{re}$ values (IA and particularly IC drugs) will not discriminate between normal heart rates and tachycardia and, therefore, cannot selectively suppress the tachycardia without also slowing conduction of normal heart beats.[7,36] Decreased excitability and slowed conduction may suppress re-entry by further depressing conduction, thus converting regions of unidirectional into bidirectional conduction block. Slowed conduction may, however, facilitate or aggravate re-entrant arrhythmias, so that these drugs may produce proarrhythmic effects and the risk would increase with the prolongation of the $\tau_{re}$ value.[36] Because slowed intracardiac conduction is much more likely in patients with pre-existing conduction disturbances, slow (and intermediate) kinetic $Na^+$ channel blockers may produce incessant monomorphic ventricular tachycardia associated with fast heart rates and excessive slowing of conduction.

The onset kinetics of the UDB also explain the effects of $Na^+$ channel blockers on cardiac effective refractory periods (ERP).[6,9] The ERP is determined by interpolating premature extrastimuli at various coupling intervals from the basic stimuli,[9] so that the increase in refractoriness indicates the ability of the drug to produce an additional inhibition of $Na^+$ channels in response to a sudden increase in heart rate. Fast kinetics drugs prolong the ERP relative to the APD because they produce a rapid additional depression of $Na^+$ channels in response to a sudden increase in heart rate. In contrast, because of their slow onset of block, slow and intermediate kinetics blockers do not allow a rapid additional blocking effect and produce moderate or minor increases in the ERP relative to the APD, respectively.

## Open and inactivated $Na^+$ channel blockers

There are important clinical implications depending on the affinity of $Na^+$ channel blockers for a particular state of the $Na^+$ channel. In general, class IA and IC antiarrhythmic drugs exhibit a high affinity for the open state, while class IB drugs exhibit a higher affinity for the inactivated state of the channel, even when many antiarrhythmic drugs exhibit affinity for the open and the inactivated state of the channel.[7,8,24,25]

### Open channel blockers

These produce a marked increase of block during action potential upstroke, but little further block is observed after the closure of the channel.[24,34] These drugs can produce an inhibition of $I_{Na}$ following a single brief pulse (3–5 ms), which activates $Na^+$ channels without inducing much inactivation and a UDB following a train of short-duration pulses (2–10 ms). The degree of $I_{Na}$ block does not increase when lengthening the duration of the depolarizing pulse which augmented the duration of the inactivated state or at more negative holding potentials when the channels are shifted to the inactivated state, while the block may be partly reversed by increasing $[Na^+]_o$. An open channel blocker would be expected to produce an abbreviation of the channel open times if block occurs on a time scale similar to channel gating, while if the blocking kinetics are much faster than the open time of the channel, it may cause an apparent decrease in single-channel amplitude because of the limited bandwidth of the recording. In single-channel studies, blockade occurring at the beginning of the depolarizing pulse following a long rest period is termed *open-state block* and includes block of the open state and of those preceding the opening. Open-channel blockers of unmodified $Na^+$ channels like ethacizin, propafenone, diprafenone, prajmaline, quinidine, amiodarone, and disopyramide caused use-dependent block of neonatal rat $Na^+$ channels and shorten the mean open time of the channel without decreasing single-channel conductance.[34,37] The shortening of the mean open time can be explained by a faster closure rate to inactivated or closed states for drug-bound channels.

The short mean open time ($<1$ ms) and the difficulty in separating the capacitative transient and ionic current due to the limited bandwidth of the recording have limited the analysis of the cardiac $Na^+$ current in general and particularly of open-channel blockers, even at room temperature.[1] One approach to isolate the closed–open transition ($C \rightarrow O$) is to modify $Na^+$ channel gating by removing (pronase, trypsin, DPI201–106),[19,28] or markedly slowing inactivation (scorpion and coelenterate peptide $\alpha$ toxins, BTX).[14] In rat ventricular myocytes, BTX slows the time course of activation, removes fast inactivation, shifts the voltage dependence of activation by almost 20 mV in the negative direction, and reduces the selectivity of the channel for $Na^+$.[14,36] After removal or slowing of fast inactivation with BTX or DPI 201-106, two types of activity sweeps can be observed at the single channel level, one with low ($P_o < 0.1$) and another with a high open probability ($P_o > 0.1$) which is seen as repetitive openings over the entire depolarizing pulse. In cell-attached patches from rat neonatal myocytes, the open state of DPI-modified $Na^+$ channels is highly sensitive to antiarrhythmic drugs. Propafenone, diprafenone, prajmalium, disopyramide, and to a lesser extent lidocaine, produce a *flicker block*, that is, long-lasting openings observed under control conditions were chopped into a large number of short and grouped openings.[19,34,37] This drug-induced shortening of the open state was paralleled by a large increase in the number of rapid transitions between a drug-associated, blocked state and a drug-free, conducting state. However, the unitary conductance of this drug-free state (between 12–20 pS) was very similar to that obtained in the absence of antiarrhythmic drugs. Thus, the flicker block can be modelled in the simplest scheme by:

$$\text{closed} \underset{\beta}{\overset{\alpha}{\rightleftarrows}} \text{open} \underset{d}{\overset{a[D]}{\rightleftarrows}} \text{blocked}$$

so that DPI-modified $Na^+$ channels may attain, after leaving their open configuration, two non-conducting states, a closed state or, after interaction with the antiarrhythmic drug, a blocked state. The association (*a*) and dissociation (*d*) rate constants that govern the transition from one state to another in the presence of some antiarrhythmic drugs are shown in Table 3.3. The values for propafenone and prajmalium are in the same order of magnitude as those obtained with several local anaesthetics in acetylcholine-activated channels. The flicker block kinetics are voltage-dependent and the affinity of the channel-associated receptor increases ($K_D$ value decreases) on membrane depolarization.

Open channel blockers are more effective at faster rates (tachycardia), in cardiac tissues with short action potentials firing at high rates (e.g. atrial flutter) and in hyperpolarized tissues, but may be partly reversed by increasing the $[Na^+]_o$ and in depolarized tissues, where

**Table 3.3** Values for the apparent association (*a*) and dissociation (*d*) rate constants for drugs that block the open state of the channel.

| Drug | a ($mol^{-1} s^{-1}$) | d ($s^{-1}$) |
|---|---|---|
| Propafenone | $6.1 \times 10^7$ | 833 |
| Prajmalium | $1.5 \times 10^7$ | 409 |
| Disopyramide | $3.71 \times 10^7$ | 286 |
| Lidocaine | $10^4$ | |
| Penticainide | $3.74 \times 10^6$ | 286 |

Source: references 19,34,35

the inactivated state predominates. Since the duration of the upstroke is almost the same in all cardiac tissues, the inhibition of $I_{Na}$ produced by these drugs is independent of APD and it can be predicted to be effective in both supraventricular as well as ventricular tachyarrhythmias.[7,25,36] In addition, they also exhibit use-dependent unblocking. Because $Na^+$ channels are open for a few ms during each action potential, steady-state UDB would be achieved after a certain number of beats (> 10 beats).

## Inactivated channel blockers

These antiarrhythmic drugs (i.e. lidocaine, mexiletine, tocainide, aprindine) produce a continuous increase in block after the open state is no longer available until a steady-state is reached.[24,34] Both in multicellular preparations and in whole-cell voltage-clamp studies, the degree of block increased with the duration of the depolarizing pulse and at more depolarized membrane potentials, so that they shifted the steady-state $Na^+$ channel voltage–availability relationship toward more negative potentials. In voltage-clamped Purkinje fibres, the concentration lidocaine reducing the $I_{Na}$ by 50% varied from 10 $\mu$M at a membrane potential of $-50$ mV, when inactivation was nearly complete, to $>0.3$ mM at a membrane potential of $-120$ mV.[12]

The block produced by lidocaine, an inactivated $Na^+$ channel blocker, disappears after removal of fast inactivation with $\alpha$-chymotrypsin or toxins.[37] In single-channel studies, the exposure of the intracellular side of the excised patch to lidocaine does not modify the latency to the first opening or the mean opening time of $\alpha$-chymotrypsin-modified $Na^+$ channels. Mutations of three amino acids in the III–IV interdomain responsible for the fast inactivation process (IFM by QQQ) leads to a complete lost of fast inactivation and causes loss of lidocaine block at therapeutic concentrations[41]. These results suggest that the molecular site of action of this drug is the structural region of the channel that is responsible for inactivation.

The degree of $I_{Na}$ block will increase under circumstances where the inactivated state predominates.[7,8,24,25,36] First, in tissues with longer (ventricular muscle and Purkinje fibres) than in those with shorter APD (atrial muscle cells), so that under circumstances which lengthen the APD (i.e. synergistic combination with $K^+$ channel blockers) the amount of block increases. This may explain, at least in part, the negligible effect of these agents on supraventricular tachyarrhythmias. The affinity for the inactivated state of the channel would be enhanced by lengthening the APD, so that the effects of some class IB antiarrhythmic drugs may be potentiated when given in combination with such drugs that lengthen the APD. However, fast kinetics drugs can also shorten the APD by decreasing window $Na^+$-current or by increasing $K^+$ conductance,[7] which results in shortening the time that the $Na^+$ channel spends in the inactivated state and decreases the opportunity for the drug to inhibit the $I_{Na}$. Second, at faster rates (tachycardia), because cardiac tissues spend more time in the depolarized state and there is less time to recover from block during diastole. Third, in partially depolarized tissues (ischaemia, hyperkalaemia). Myocardial ischaemia increases extracellular $K^+$, decreases intracellular pH and depolarizes the resting membrane potential, leading to partial inactivation of the $I_{Na}$. Because membrane depolarization shifted $Na^+$ channels from the rested to the inactivated state during diastole and prolonged the $\tau_{re}$, the inactivated-state blockers will be able to bind to a greater fraction of channels and thus, more $I_{Na}$ block will occur (i.e. voltage-dependent block).[7,8,25] This is the basis for a selectivity of $Na^+$ channel blockers, where the acidosis increases the charged form, to suppress cardiac arrhythmias in ischaemic cardiac tissues compared with normally polarized myocardium. In addition, less drug-associated channels move from the ID to the RD state (Fig. 3.4), so that less activation unblocking (RD → OD → O) will occur. Increased inactivation blocking and decreased activation unblocking would result in selective depression of $I_{Na}$ in depolarized tissue compared with normally polarized tissue by antiarrhythmic drugs. However, the possible selectivity of inactivated channel blockers may be counteracted because in ischaemic-depolarized cells, the $\tau_{re}$ is lengthened.[7,8,25]

## Competition between Na⁺ channel blockers

The rationale for combination drug therapy in the treatment of cardiac arrhythmias is to enhance the antiarrhythmic effectiveness when patients fail to or only partly respond to either agent alone, and to reduce the incidence of dose-related toxic side effects. If $Na^+$ channel blockers bind to a specific receptor site in the $Na^+$ channel, when given in combination they can compete at this level, thus altering each other's efficacy.

Combinations of low concentrations of antiarrhythmic drugs with fast-IB (lidocaine, mexiletine) and intermediate-IA (quinidine, disopyramide) or slow-IC (propafenone, flecainide) kinetics would be expected to produce a higher $I_{Na}$ block along the diastole, i.e. synergistic combination (see Table 3.4 for a review).[7,25,36,38] Figure 3.5 shows that this is the case when combining mexiletine (fast) and flecainide (slow). In the presence of mexiletine, recovery from block was completed after 500 ms, so that it would not be very effective in suppressing extrasystoles late in diastole. However, in the presence of flecainide, the percentage of block early during diastole was not much greater than late in diastole because flecainide slows the recovery from block. When both drugs were combined, many more $Na^+$ channels were blocked than could be achieved by flecainide alone, while at the end of diastole many more channels remained blocked than could be blocked by mexiletine alone. Thus, the combination of IA and IC drugs provides a means to achieve a high level of block during diastole which cannot be obtained by each drug alone.

The combination of intermediate and slow (IA + IC) or two slow kinetics drugs would not be expected to produce an important benefit and, by further depression of conduction velocity, would increase the risk of proarrhythmic effects.[7,34,36] When an intermediate/slow recovery kinetics drug achieves a high degree of block, then the addition of a fast kinetics drug can occupy the channel during depolarization. During diastole, the rapid offset of the fast drug may reduce the amount of drug-bound channels and produce, at least at certain driving rates, an increase in the available $I_{Na}$ at the time of the next excitation (antagonistic combination (Fig. 3.5).[7,36] However, if the $\tau_{re}$ of the antiarrhythmic drug is too long (class IC), little displacement would be induced by a fast drug. The competitive interaction of fast $Na^+$ channel blockers has been used to potentiate the antiarrhythmic effects and reverse drug-induced cardiotoxicity produced by slow kinetics drugs.

Most antiarrhythmics of the local anaesthetic type are extensively metabolized to electrophysiologically active metabolites. These are closely related structurally to the parent

**Table 3.4** Competition between $Na^{+}$ channel blockers.

| Fast drug (displacing) | Slow drug (displaced) | Reference |
|---|---|---|
| Lidocaine | Bupivacaine | Clarkson, C.W. and Hondeghem, L.M. (1985). *Circ. Res.*, **56**, 496–506. |
| | Propafenone | Kohlhardt, M. and Seifert, C. (1985). *Naunyn Schmiedeberg's Arch. Pharmacol.*, **315**, 55–62. |
| | Aprindine | Kodama, I. *et al.* (1987). *J. Pharmacol. Exp. Ther.*, **241**, 1065–1071. |
| | Propoxyphene | Whitcomb, D.C. *et al.* (1989). *J. Clin. Invest.*, **84**, 1629–1636. |
| Lidocaine | Ethacizin | Nesterenko *et al.* (1991). |
| Benzocaine | Lidocaine | Sánchez Chapula, J. (1985). *J. Mol. Cell. Cardiol.*, **17**, 495–503. |
| Glycylxylidide | Lidocaine | Bennett, P.B. *et al.* (1988). *Circulation*, **78**, 692–700. |
| Mexiletine | Quinidine | Valenzuela, C. and Sánchez Chapula, J. (1989). *J. Cardiovasc.* |
| | Ropitoin | *Pharmacol.*, **14**, 783–789. |
| | Propafenone | Valenzuela, C. *et al.* (1989). *J. Cardiovasc. Pharmacol.*, **14**, 351–357. |
| | Flecainide | Delpón, E. *et al.* (1991). *Br. J. Pharmacol.*, **103**, 1411–1416. |

Source: Grant and Wendt (1992)[34] and Tamargo *et al.* (1992)[7]

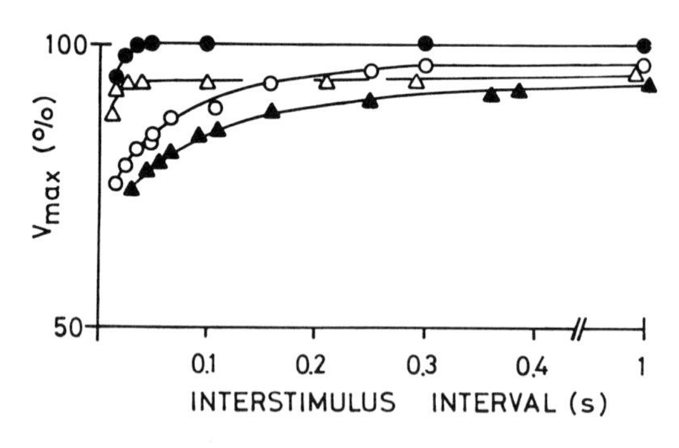


**Fig. 3.5** First 1000 ms of the time course of recovery of the maximum upstroke velocity ($V_{max}$) of the action potentials recorded in guinea pig papillary muscles driven at a basal rate of 1 Hz. Data are plotted as percentage of the maximal value of $V_{max}$ under control conditions. The recovery time constant from use-dependent $V_{max}$ block ($\tau_{re}$) was determined by applying test stimuli every 10 basic stimuli. ●: controls; ○: mexiletine, $10^{-5}$M; △: flecainide $10^{-6}$M; ▲: flecainide plus mexiletine. (See Table 4 of Delpón *et al.* 1991[40].)

compound, so that both may bind and compete at the $Na^{+}$ channel level, altering the efficacy of the parent drug. The steady-state effect of the combination, parent compound and metabolite, will depend on the relative concentrations of the individual drugs, their affinities for the receptor and the onset–offset kinetics of the binding.[3,24,36] Metabolites with slower $\tau_{re}$ may exert synergistic interactions, while those with a faster $\tau_{re}$ exert antagonistic interactions.[7,36] If the metabolites are equally or even more potent than the parent compound to inhibit $I_{Na}$, they may be responsible for the antiarrhythmic/cardiodepressant effects previously attributed to the parent drug. We have recently described that desipramine and 5-hydroxypropafenone, the main metabolites of imipramine (IA) and propafenone (IC) respectively, produced a more marked UDB and exhibited a slower onset and offset kinetics than the parent compounds, i.e. both metabolites were class IC antiarrhythmic drugs.[39,40] Sometimes, the metabolites exhibit different electrophysiological properties, so are responsible, at least in part, for the antiarrhythmic effectiveness of the parent compound. *N*-Acetylprocainamide (NAPA), the main metabolite of procainamide, inhibited $K^+$-currents and prolonged the APD, an effect which is not prominent with procainamide.[38] This prolongation of the APD may potentiate the effects of procainamide which exhibits affinity for the inactivated state of the channel.[25] This explains why patients whose arrhythmias responded to intravenous procainamide (before any NAPA accumulates) may fail to respond to oral therapy with NAPA, and vice versa.[38] The activity of cytochrome P-450 2D6 is responsible for the metabolism of some antiarrhythmic drugs (i.e. propafenone, flecainide, mexiletine, encainide, ajmaline). This activity is genetically determined, with 93% and 7% of the population having the extensive and the poor metabolizer phenotype, respectively. Therefore, the metabolic phenotype of the patient is a determinant of the types of antiarrhythmic action that can be acting to suppress an arrhythmia.

Because opening precedes inactivation of $Na^+$ channels, among $Na^+$ channel blockers with similar $\tau_{re}$, open channel blockers would be expected to possibly prevent the binding of inactivated blockers.[25] The accumulation of the lidocaine metabolite glycylxylidide, that blocks primarily activated $Na^+$ channels, may be the mechanism that reduces clinical effectiveness and excessive $I_{Na}$ block (cardiodepression, proarrhythmia) produced by the parent compound lidocaine, an inactivated $Na^+$ blocker (Table 3.4). This could be a possible strategy for reversing the excessive $I_{Na}$ block produced by drugs with high affinity for the inactivated state of the channel.

Some $Na^+$ channel blockers are racemic mixtures of at least two enantiomers which may differ in their potency and onset–offset kinetics to inhibit $I_{Na}$. For tocainide, disopyramide and mexiletine, the (R)-enantiomer is a more potent inhibitor of $I_{Na}$ than the S-form.[7,38] Moreover, each enantiomer may exhibit other properties unrelated to the blockade of $I_{Na}$ that may be responsible at least partly for the antiarrhythmic effectiveness and/or some of the side effects of the racemic mixture.

## Future directions

The examination of the blocking mode of drugs at the single channel level have improved our understanding of the cellular mechanisms of arrhythmias, the actions of $Na^+$ channel blockers (i.e. class I antiarrhythmic drugs), and the interaction between class I antiarrhythmic drugs and $Na^+$ channels at the molecular level. New information regarding the structure and function of $Na^+$ channels has recently been obtained by the application of molecular biology and immunology techniques. The channel has been cloned and sequenced and the use of selective antibodies against specific domains and site-directed mutagenesis yields new information on receptor structure. Consequently, we hope that, in the near future, the sequences that can be considered as receptor sites for antiarrhythmic drugs can be identified. This information will help to improve the design of more specific and effective drugs, so that therapy can be more effectively tailored to a particular arrhythmia.

However, there is still much to learn about the $Na^+$ channel and the mechanism of action of $Na^+$ channel blockers. Initially, the effects of antiarrhythmic drugs were inferred from healthy cardiac tissues. However, there is evidence that the effects of $Na^+$ channel blockers in cardiac tissues often differed markedly in healthy compared with diseased tissues. Moreover, the effects may differ in different animal species; therefore, extrapolation of the observed effects is often difficult. Since severe cardiac arrythmias appeared in patients with previous myocardial infarction, basic research into the biochemical links between ischaemia, membrane currents, and arrhythmias is absolutely essential before more specific drugs can be designed. Finally, traditionally unappreciated aspects of $Na^+$ channel blockers, such as stereochemical differences in $Na^+$ channel blocking activity, the effects of active metabolites and their interactions with the parent compounds, and the effects of the chemical environment of the $Na^+$ channel receptor (i.e. changes in pH) on the drug–receptor interaction must receive further attention in the future.

## Acknowledgements

This research was supported by CICYT Grant (SAF96–0042).

## References

1. Fozzard, H.A. and Hank, D.A. (1992). Sodium channels. In *The heart and cardiovascular system*, (ed. H.A. Fozzard, E. Haber, R. Jennings, A.M. Katz, and H.E. Morgan), pp. 1091–1119. Raven Press, New York.
2. Hille, B. (1992). *Ionic channels of excitable membranes.* Sinauer, Sunderland. MA.
3. Catterall, W.A. (1992). Cellular and molecular biology of voltage-gated sodium channels. *Physiol. Rev.*, **72**, S15–S48.
4. Kirsch, G.E. and Brown, A.M. (1990). Cardiac sodium channels. In *Cardiac electrophysiology: from cell to bedside*, (ed. D.P. Zipes and J. Jalife), pp. 1–10. Saunders, Philadelphia.
5. Coraboeuf, E., Deroubaix, E. and Coulombe, A. (1979). Effect of tetrodotoxin on action potentials of the conducting system in the dog heart. *Am. J. Physiol.*, **236**, H561–H567.
6. Vaughan Williams, E.M. (1984). A classification of antiarrhythmic actions reassessed after a decade of new drugs. *J. Clin. Pharmacol.*, **24**, 129–147.
7. Tamargo, J., Valenzuela, C. and Delpón, E. (1992). New insights into the pharmacology of sodium channel blockers. *Eur. Heart J.*, **13**(suppl. F), 2–12.
8. Grant, A.O., Starmer, C.F. and Strauss, H.C. (1984). Antiarrhythmic drug action: blockade of the inward sodium current. *Circ. Res.*, **55**, 427–439.
9. Campbell, T.J. (1983). Kinetics of onset of rate-dependent effects of Class I antiarrhythmic drugs are important in determining their effects on refractoriness in guinea-pig ventricle, and provide a theoretical basis for their subclassification. *Cardiovasc. Res.*, **17**, 344–352.
10. Kunze, D.L., Lacerda, A.E., Wilson, D.L. and Brown, A.M. (1985). Cardiac Na currents and the inactivating, reopening and waiting properties of single cardiac Na channels. *J. Gen. Physiol.*, **86**, 697–719.

11. Patlak, J.B. and Ortiz, M. (1986). Slow currents through single sodium channels of adult rat heart. *J. Gen. Physiol.*, **86**, 613–642.

12. Bean, B.P., Cohen, C.J. and Tsien, R.W. (1983). Lidocaine block of cardiac sodium channels. *J. Gen. Physiol.*, **81**, 613–642.

13. Gellens, M.E., George, A.L. Jr, Chen, L., Chahine, M., Horn, R., Barchi, R.L. and Kallen, R.G. (1992). Primary structure and functional expression of the human cardiac tetrodotoxin-insensitive voltage-dependent sodium channel. *Proc. Natl. Acad. Sci. USA*, **89**, 554–558.

14. Huang, L.M., Yatani, A. and Brown, A.M. (1987). The properties of batrachotoxin-modified cardiac Na channels, including state-dependent block by tetrodotoxin. *J. Gen. Physiol.*, **90**, 341–360.

15. Tamkum, M.M., Knittle, T.J., Keal, K.K., House, M.H., Roberds, S.L., Po, S., Bennett, P.B., George, A.L. Jr. and Snyders, D. (1994). Molecular physiology of voltage-gated potassium and sodium channels: ion channel diversity within the cardiovascular system. In *Ion channels in the cardiovascular system. Function and dysfunction*, (ed. P.M. Spooner, A.M. Brown, W.A. Catterall, G.J. Kaczorowski and H.C. Strauss), pp. 287–316. Futura Publishing, Armonk, New York.

16. Lombet, A. and Lazdunski, M. (1984) Characterization, solubilization, affinity labelling and purification of the cardiac $Na^+$ channel using *Tityus* toxin $\gamma$. *Eur. J. Biochem.*, **141**, 651–660.

17. Stühmer, W. (1991). Structure-function of voltage-gated ion channels. *Annu. Rev. Biophys. Chem.*, **20**, 65–78.

18. Isom, L.L., De Jongh, K.S. and Catterall, W.A. (1994) Auxiliary subunits of voltage-gated ion channels. *Neuron*, **12**, 1183–1194.

19. Kohlhardt, M., Frobe, U. and Herzig, J.W. (1981). Properties of normal and non-inactivating single cardiac $Na^+$ channels. *Proc. R. Soc. Ser. B.*, **232**, 71–93.

20. George, A.L. Jr, Knittle, T.J. and Tamkun, M.M. (1992). Molecular cloning of an atypical voltage-gated sodium channel expressed in human heart and uterus: evidence for a distinct gene family, *Proc. Natl. Acad. Sci USA*, **89**, 4893–4897.

21. Heineman, S.H., Terlau, H., Stühmer, W., Imoto, K. and Numa, S. (1992). Calcium channel characteristics conferred on the sodium channel by single mutations *Nature*, **356**, 441–443.

22. Terlau, H., Heinemann, S.H., Stühmer, W., Pusch, M., Conti, F., Imoto, K. and Numa, S. (1991). Mapping the site of block by tetrodotoxin and saxitoxin of sodium channel II. *FEBS Lett.*, **293**, 93–96.

23. Scanley, B.E., Hanck, D.A., Chay, T. and Fozzard, H.A. (1990) Kinetic analysis of single sodium channels from canine cardiac Purkinje cells. *J. Gen. Physiol.*, **95**, 411–435.

24. Snyders, D.J., Bennett, P.B., and Hondeghem, L.M. (1992). Mechanisms of drug-channel interaction. In *The heart and cardiovascular system*, (ed. H.A. Fozzard, E. Haber, R. Jennings, A.M. Katz, and H.E. Morgan). pp. 2165–2193. Raven Press, New York.

25. Hondeghem, L.M. and Katzung, B.G. (1984). Antiarrhythmic agents: the modulated receptor mechanism of action of sodium and calcium channel-blocking drugs. *Annu. Rev. Pharmacol. Toxicol.*, **24**, 387–423.

26. Carmeliet, E. (1987). Slow inactivation of sodium current in rabbit cardiac Purkinje fibres. *Pflügers Arch.*, **408**, 18–26.

27. Nilius, B. and Carmeliet, E. (1989). Different conductance states of the bursting Na channel in guinea-pig ventricular myocytes. *Pflügers Arch.*, **413**, 242–248.

28. Valenzuela, C. and Bennett, P.B. (1994). Gating of cardiac $Na^+$ channels in excised membrane patches after modification by α-chymotrypsin. *Biophys. J.*, **67**, 161–171.

29. Aldrich, R.W., Corey, D.P. and Stevens, C.F. (1983). A reinterpretation of mammalian sodium channel gating based on single channel recording. *Nature*, **306**, 436–441.

30. Armstrong, C.M. and Bezanilla, F. (1977). Inactiation of the sodium channel. II. Gating current experiments. *J. Gen. Physiol.*, **70**, 567–590.

31. Patton, D.E., West, J.W., Catterall, W.A. and Goldin, A.L. (1992). Amino acid residues required for fast $Na^+$-channel inactivation. Charge neutralizations and deletions in the III-IV linker. *Proc. Natl Acad. Sci. USA*, **89**, 10905–10909.

32. Sheldon, R.S., Cannon, N.J. and Duff, H.J. (1987). A receptor for type I antiarrhythmic drugs associated with rat cardiac sodium channels. *Circ. Res.*, **61**, 482–497.

33. Courtney, K.R. (1987). Review: quantitative structure/activity relations based on use-dependent $I_{Na}$ block and repriming kinetics in myocardium. *J. Mol. Cell. Cardiol.*, **19**, 318–330.

34. Grant, A.O. and Wendt, D.J. (1992). Block and modulation of cardiac $Na^+$ channels by antiarrhythmic drugs, neurotransmitters and hormones. *TIPS*, **13**, 352–358.

35. Carmeliet, E. (1988). Activation block trapping of penticainide, a disopyramide analogue, in the $Na^+$ channel of the rabbit cardiac Purkinje fibres. *Circ. Res.*, **63**, 50–60.

36. Hondeghem, L.M. (1990). Molecular interactions of antiarrhythmic agents with their receptor sites. In *Cardiac electrophysiology: from cell to bedside*, (ed. D.P. Zipes and J. Jalife), pp. 865–812. Saunders Co, Philadelphia.

37. Kohlhardt, M., Fichtner, H., Froebe, U. and Herzig, J.W. (1989). On the mechanism of drug-induced blockade of $Na^+$ currents: interaction of antiarrhythmic compounds with DP1-modified single cardiac $Na^+$ channels. *Circ. Res.*, **64**, 867–881.

38. Woosley, R.L. (1991). Antiarrhythmic drugs. *Annu. Rev. Pharmacol. Toxicol.*, **31**, 427–455.

39. Valenzuela, C., Delpón, E. and Tamargo, J. (1988). Tonic and phasic $V_{max}$ block induced by 5-hydroxypropafenone in guinea pig ventricular muscles. *J. Cardiovasc. Pharmacol.*, **12**, 423–431.

40. Delpón, E., Valenzuela, C., Pérez, O., and Tamargo, J. (1993). Electrophysiological effects of the combination of imipramine and desipramine in guinea pig papillary muscles. *J. Cardiovasc. Pharmacol.*, **21**, 13.

41. Bennett, P., Valenzuela, C., Chen, L. Q. and Kallen, R. (1995). On the molecular nature of the lidocaine receptor of cardiac $Na^+$ channels. Modification of block by alterations in the α-subunit III–IV interdomain. *Circ. Res.*, **77**, 584–592.

# 3 Calcium Channels

# 4 Biophysics of voltage-dependent $Ca^{2+}$ channels: gating kinetics and their modulation

E. Carbone, V. Magnelli, A. Pollo, V. Carabelli, A. Albillos, and H. Zucker

## Introduction

Voltage-dependent $Ca^{2+}$ channels represent one of the membrane pathways through which cytoplasmic $Ca^{2+}$ levels are effectively regulated. As the increase of intracellular $Ca^{2+}$ represents the triggering event of many biological functions, the identification of $Ca^{2+}$ channel subtypes and the characterization of their gating properties has become a crucial task for understanding the genesis of cell activities (Bean 1989; Tsien *et al.* 1991). Regulation of $Ca^{2+}$ entry through voltage-activated $Ca^{2+}$ channels may occur by varying the number of functioning channels, the constraints of ion permeation through open pores, or by affecting their gating characteristics. Modification of channel gating can be as effective as changing the number of functioning channels. On a short time scale, for instance, $Ca^{2+}$ ions and activated G proteins may alter the time course of $Ca^{2+}$ channel activation and inactivation, thus regulating the $Ca^{2+}$ entry into the cell (Carbone and Swandulla 1989). Alternatively, $Ca^{2+}$ channel modulation may proceed through a cascade of slow events mediated by activated membrane receptors and cytoplasmic second messengers, in which either the probability of channel opening or the activation–inactivation kinetics can be significantly modified (Hille 1994).

$Ca^{2+}$ channel gatings are primary characteristics of the channel. They help to identify the class and the role of the channel in the genesis of many cell activities. Given the number of channel subtypes and the possible couplings among second messenger pathways, $Ca^{2+}$ channel gating modifications induced by first messengers are expected to undergo a complex series of molecular events. Such complexity, however, may result in an effective mechanism controlling hormone and neurotransmitter release. In neurones and neurosecretory cells, for instance, $Ca^{2+}$ channel gating can be down-modulated by the released neurotransmitter itself, thus originating a molecular feedback loop that may finely control the secreted material through the activation of membrane autoreceptors. Whether self-regulated or stimulated by endogenous ligands, the receptor-mediated modulation of $Ca^{2+}$ channel gating is expected to be one of the major routes through which inhibition or facilitation of exocytotic responses may take place at the presynaptic terminals and in neuroendocrine glands.

In this article, we review the gating properties of classical and newly described $Ca^{2+}$ channel subtypes, focusing on the most relevant discoveries on $Ca^{2+}$ channel modulation and their basic implications. In doing this, we have chosen to discuss the arguments in an informal way, giving illustrative examples of the complex phenomena rather than an exhaustive list of contributions to the field. This might be less rigorous but is certainly more intuitive.

## $Ca^{2+}$ channel gatings

### Activation–deactivation kinetics

Voltage-dependent $Ca^{2+}$ channels are multisubunit proteins composed of a main channel-forming $\alpha_1$ subunit interacting with $\beta$ and $\alpha_2\delta$ auxiliary subunits. The $\alpha_1$ subunit confers the main characteristics to the channel, i.e. the $Ca^{2+}$ selectivity, the voltage dependency of open–closed kinetics, and the sensitivity to most pharmacological agents. The voltage sensitivity of $Ca^{2+}$ channel openings derives from localized charged groups of the $\alpha_1$ subunits that move or reorient upon changes of the intramembrane electric field, making the channel permeable to ion passage. Voltage-dependent probability of $Ca^{2+}$ channel opening implies that the rate constants regulating the transitions between the closed (non-conductive, C) and the open (conductive, O) forms of the channel are, at different degrees, sensitive to voltage. Monotonic increases of open channel probability ($p_o$) with increasing voltages can be produced by either increasing the forward rate ($\alpha$) from closed to open ($C \rightarrow O$) or decreasing the backward rate ($\beta$) from open to closed ($C \leftarrow O$). For a $Ca^{2+}$ channel with several sequential closed states and one open state, the rate constants for opening and closing can be easily estimated assuming that $Ca^{2+}$ channels obey an Hodgkin–Huxley (H–H) formalism in which all the transition rates are taken as multiples of the two rates $\alpha$ and $\beta$. Under these conditions, macroscopic current measurements at constant membrane potentials in whole-cell clamp configuration (Hamill *et al.* 1981) furnish a complete set of $\alpha$ and $\beta$ values versus voltage that allows a full description of $Ca^{2+}$ channel activation ($\tau_{act}$), deactivation ($\tau_{deact}$) and $p_o(V)$. The determination of rate constants is more complex if the $Ca^{2+}$ channel is assumed not to obey an H-H scheme. Transitions rates are all independent variables and their evaluation becomes far more complicated. For most $Ca^{2+}$ channels, however, activation and deactivation are sufficiently well described by a minimal sequential model with two closed and one open state ($C_1 \leftrightarrow C_2 \leftrightarrow O$) in which the transitions rates $C_1 \leftrightarrow C_2$ and $C_2 \leftrightarrow O$ can be univocally determined by measuring the mean open and mean closed times and the mean bursts duration of single-channel current recordings at different potentials (Fenwick *et al.* 1982). For several $Ca^{2+}$ channels activating at relatively high membrane voltages, transitions $C_2 \rightarrow O$ and $C_2 \leftarrow O$ are usually very frequent (500–1500 $s^{-1}$), weakly voltage-dependent and responsible for the fast flickering of the channel during prolonged depolarizations above 0 mV. In contrast, the rate of transition $C_1 \rightarrow C_2$ is low (50–250 $s^{-1}$) at potentials near the threshold of activation (−20 mV in 5 mM $Ba^{2+}$) while the backward rate ($C_1 \leftarrow C_2$) is high. Transitions $C_1 \leftrightarrow C_2$, together with the fast reaction $C_2 \leftrightarrow O$, are mainly responsible for the sigmoidal time course of $Ca^{2+}$ channel activation.

In whole-cell recordings, $Ca^{2+}$ channel activation causes a fast sigmoidal rise of the inward $Ca^{2+}$ current that reaches peak values at shorter times with increasing membrane depolarizations. The size of the current is proportional to the number of total available channels at rest (N), the single-channel conductance ($\gamma$), the probability of channel opening ($p_o$) and the driving force ($V_m - E_{Ca}$), where $V_m$ is the membrane potential and $E_{Ca}$ the Nernst equilibrium potential for $Ca^{2+}$. Channel deactivation on return to resting potentials gives rise to instantaneous inward current jumps (due to the increased negative driving force) followed by fast exponential decays ('tails'). Time constants of the tails change from fractions of milliseconds to some milliseconds at $-60$ mV in 10 mM $Ba^{2+}$ and, in some cases, help to identify the $Ca^{2+}$ channel subtype (Swandulla and Armstrong 1988).

A main feature of $Ca^{2+}$ channels is their voltage range of activation which is most simply determined by current amplitude measurements at different voltages (I–V curve). In 5 mM $Ca^{2+}$, the I–V relationships of $Ca^{2+}$ channels possess a characteristic region of negative slope at negative potentials, a peak value and a more or less linear region with positive slope at positive potentials. The region of negative slope reflects the voltage range of channel activation that is usually identified by the voltage of half-maximal current ($V_{½}$) and by the slope at $V_{½}$ (voltage sensitivity). In most neurones possessing low-voltage-activated (LVA) and high-voltage-activated (HVA) $Ca^{2+}$ channels, the I–V relationship is formed by two pharmacologically separable curves with distinct peak amplitude, $V_{½}$ value, slope at $V_{½}$ and sensitivity to holding potentials.

## Inactivation kinetics

Time-dependent inactivation is a gating feature common to most voltage-dependent ion channels. It may develop on a broad time scale (from a few milliseconds to several seconds) and can be used to identify ion channel subtypes. For $Ca^{2+}$ and other ion channels, inactivation is defined as the molecular events leading an open or closed channel into a temporarily inactive form (non-conductive) different from the deactivated (closed) channel. The time constant of $Ca^{2+}$ channel inactivation ($\tau_h$) may depend either on membrane voltage (voltage-dependent), or on the levels of intracellular $Ca^{2+}$ ($Ca^{2+}$-dependent) or on other intracellular factors and may be coupled to a variable degree to channel activation (Eckert and Chad 1984; Gutnick *et al.* 1989). At the single-channel level, the inactivation of open channels is undistinguishable from channel closing except for the appearance of prolonged interruptions of channel activity during sustained depolarizations. In current recordings from patches containing only one active channel, inactivation is usually responsible for the last channel closure after repeated channel activity. Averaged recordings of this form result in decaying ensemble currents with inactivation time constants comparable with the $\tau_h$ derived from whole-cell recordings.

As for the activation gatings, voltage-dependent inactivation implies that the rate of channel inactivation ($1/\tau_h$) is a steep function of membrane voltage. When increasing the voltage by 10 to 20 mV, $\tau_h$ decreases an e-fold or even more. Quick voltage-dependent inactivation (tens of ms) is characteristic of some LVA channels which turn on transiently during low depolarizations in excitable cells (Bean 1989). Slower inactivation time courses ($\tau_h > 100$ ms) are characteristic of the HVA channels or related $\alpha_1$ subunits cloned from neuronal tissues (Tsien *et al.* 1991). Voltage-dependent inactivation may occur also very slowly (tens of seconds) when changing the holding potential to more positive values. Slow $Ca^{2+}$ channel inactivation near resting potentials is common to several LVA and HVA channels and leads to a steady decrease of available channels at rest. This property is routinely exploited to selectively reduce the contribution of specific channel subtypes, thus dissecting macroscopic $Ca^{2+}$ currents into multiple components.

A peculiarity of some $Ca^{2+}$ channels is the sensitivity of their inactivation gatings to the cytoplasmic free $Ca^{2+}$ levels (Eckert and Chad 1984). $Ca^{2+}$-dependent inactivation is an effective mechanism to limit intracellular $Ca^{2+}$ elevations during prolonged cell activity. The process is vital for the survival of many cells and is the best example of self-controlled ion fluxes through open channels. $Ca^{2+}$-dependent inactivation is characterized by three main features: (i) the rate of $Ca^{2+}$ current inactivation depends on the current size. It increases with increasing the amplitude of $Ca^{2+}$ currents; (ii) in double-pulse experiments in which $Ca^{2+}$ currents are measured after a preconditioning pulse to different potentials, the amplitude of the test current decreases (inactivates) proportionally to the amount of the current flowing during the pre-pulse (Eckert and Chad 1984; Plant 1988). The corresponding inactivation curve versus voltage appears U-shaped rather than monotonic, with maximal inactivation at potentials of maximal $Ca^{2+}$ currents (+10, +20 mV in 10 mM $Ca^{2+}$); (iii) $Ba^{2+}$ and $Na^{+}$ currents through $Ca^{2+}$ channels inactivate less than do $Ca^{2+}$ currents and the corresponding U-shaped inactivation curve appears much less pronounced; and (iv) inactivation is strongly relieved in the presence of internal $Ca^{2+}$ chelators. $Ca^{2+}$ current inactivation in non-dialysed cells (perforated-patch configuration) is usually more prominent than in cells dialysed (whole-cell configuration) with an effective intracellular $Ca^{2+}$ buffer (10 mM EGTA). $Ca^{2+}$-dependent inactivation is specific for some HVA $Ca^{2+}$ channels (see p. 105) and may occur in combination with the voltage-dependent inactivation (Gutnick *et al.* 1989).

## Gating modes

Single-channel recordings have brought new insights into the gating properties of voltage-dependent $Ca^{2+}$ channels. Classical kinetic models predict that the estimated parameters of $Ca^{2+}$ channel gatings (open times, close times, $p_o$ and $\gamma$) are statistically distributed around mean values which univocally identify the gating feature of the channel. In a series of single-channel recordings, a classical model predicts channel activities having a homogeneous gating pattern characterized by one set of kinetic parameter values (*gating mode*). Single-channel recordings in cardiac and neuronal tissues, however, show that $Ca^{2+}$ channel activity may derive from a discrete number of gating modes, each characterized by its own set of distinct kinetic parameters (Hess *et al.* 1984; Delcour *et al.* 1993). $Ca^{2+}$ channel gating modes can be distinguished according to their probability of openings (high-, medium- or low-$p_o$), their inactivation kinetics (fast or slow), their unitary current amplitude (large or small) and voltage-dependence (strong or weak). Transitions from one mode to the other are infrequent, with mean sojourns of about 10 s in each mode. This implies that the channel undergoes thousands of brief open–closed transitions in one mode before switching to a different mode. Mode switching occurs preferentially during prolonged repolarization to resting potentials and is rarely observed within short sweeps of channel activity lasting less than 1 s. Thus, transitions between modes are better envisaged from a series of consecutive recordings in which a sequence of sweeps with similar channel activity is followed by one or more sweeps of markedly different gating characteristics.

Gating mode switching has been observed for both the cardiac L-type channel (Hess *et al.* 1984) and the neuronal N-type channel (Delcour *et al.* 1993). Although the kinetic scheme ($C_1 \leftrightarrow C_2 \leftrightarrow O$) within a mode and the number of modes is similar for the two channels, the set of rate constants for each mode are significantly different. The L-type channel has been shown to possess three gating modes (Hess *et al.* 1984): (i) mode 0, with no activity ($p_o \sim 0$); (ii) mode 1, characterized by brief ($\sim$ 1 ms) openings occurring in rapid bursts and low $p_o$; and (iii) mode 2, with relatively long lasting channel openings, brief closing and high $p_o$ ($>0.7$). Modes 1 and 0 occur most frequently in normal cells. Mode 2 is only occasionally observed in control conditions but is more often visible in the presence of dihydropyridine (DHP) agonists. This accounts for the large increase of L-type $Ca^{2+}$ currents induced by these drugs in most tissues.

Three gating modes with significantly different mean $p_o$ values have been reported also for the N-type channel of bullfrog sympathetic neurones (Delcour *et al.* 1993). However, the picture is likely to be complicated by the presence of other gating modes with different unitary current amplitudes and inactivation kinetics, as shown for the N-type channel of rat sympathetic neurones (Rittenhouse and Hess 1994). The detailed work of Delcour *et al.* shows that the three gating modes with high, medium and low $p_o$ have sharply different voltage-dependence of activation, but comparable unitary current amplitude. The high $p_o$ mode has the steepest voltage-dependence and the lowest threshold of activation ($\geqslant -20$ mV in 100 mM $Ba^{2+}$). The medium $p_o$ has threshold of activation similar to the high-$p_o$ but less-steep voltage-dependence of activation. The low $p_o$ mode has high threshold of activation ($> -10$ mV) and weak voltage-dependence. The most surprising aspect of these data is that the high and medium $p_o$ modes, which are responsible for most of the macroscopic N-type current, activate at relatively low membrane potentials ($\geqslant -20$ mV) and reach maximal $p_o$ slightly above 0 mV in 100 mM $Ba^{2+}$. In 5 mM $Ba^{2+}$ this would correspond to a threshold of activation of $-50$ mV and a peak current at around $-20$ mV, if allowance is made for a $-30$ mV shift due to the unscreening of membrane negative charges upon reducing extracellular [$Ba^{2+}$] from 100 to 5 mM. These parameter values are typical of low-threshold rather than N-type channels. Only the low $p_o$ mode seems to possess all the prerequisites of the high-threshold channels. It activates positive to 0 mV and reaches maximal $p_o$ above $+30$ mV in 100 mM $Ba^{2+}$. It is likely, therefore, that rather than representing three gating modes of the same channel, the high, medium, and low $p_o$ modes belong to two distinct $Ca^{2+}$ channels: one activating at low membrane potentials (high and medium $p_o$) and one activating at higher voltages (low-$p_o$). The low $p_o$ would correspond to the 'true' N-type channel that is down-modulated in a voltage-dependent manner by external neurotransmitters in most neurones while the high and medium $p_o$ modes belong very likely to the recently described 19 pS low-threshold channel of bullfrog sympathetic neurones (Elmslie *et al.* 1994, see p. 103).

## Channel gatings help to distinguish $Ca^{2+}$ channels subtypes

### Low-threshold channels

By definition, the LVA channel family includes $Ca^{2+}$ channels that activate at relatively low membrane potentials ($-50$ to $-30$ mV in 5 mM $Ba^{2+}$) (Bean 1989). Since a decade, these

**Table 4.1** Biophysical and pharmacological properties of LVA $Ca^{2+}$ channels.*

| **Property** | **T-type** vertebrate sensory neurones† | **R-type** cerebellum granule cells‡ | **rbE-II** $\alpha_1$ subunit from rat brain § | **doe-1** $\alpha_1$ subunit from marine ray** | **'Novel'** bullfrog sympathetic neurone†† |
|---|---|---|---|---|---|
| Range of activation (mV) | $\geqslant -40$ | $\geqslant -40$ | $\geqslant -40$ | $\geqslant -30$ | $\geqslant -40$ |
| Inactivation at $-50$ mV | Strong | Strong | Strong | Strong | Strong |
| Inactivation kinetics | Fast | Fast | Medium | Medium | Medium |
| Ca/Ba permeability | $P_{Ca} \leqslant P_{Ba}$ | $P_{Ca} \leqslant P_{Ba}$ | Not studied | $P_{Ca} \leqslant P_{Ba}$ | Not studied |
| Single-channel conductance (pS) | 8 | Not studied | Not studied | 14.4 | 19 |
| Sensitivity to DHP and $\omega$-peptides | Insensitive | Insensitive | Insensitive | Insensitive | Insensitive |
| $Ni^{2+}$ block ($IC_{50}$) ($\mu$M) | 15–30 | 66 | 28 | 33 | (‡‡) |
| $Cd^{2+}$ block ($IC_{50}$) ($\mu$M) | 3–5 | 1 | ~3 | 1 | (‡‡) |

* With the exception of the single-channel conductance values, all data are derived from experiments in 2–5 mM [$Ba^{2+}$] or [$Ca^{2+}$]
† Carbone and Lux 1987
‡ Sather *et al.* 1993
§ Soong *et al.* 1993
** Ellinor *et al.* 1993
†† Elmslie *et al.* 1994
‡‡ Data available only at high $Ba^{2+}$ concentration (90 mM)

channels are thought to represent a homogeneous class of $Ca^{2+}$ channels with fast activation and inactivation kinetics. The T-type (transient) channel of peripheral and central neurones is the best example of $Ca^{2+}$ channel that inactivates fully during short depolarizations in a voltage-dependent manner ($\tau_h$ 20 ms at 0 mV in 5 mM $Ba^{2+}$). The fast and voltage-dependent inactivation of T-type channels is well preserved also at the single-channel level, where unitary T-type current events can be measured in both cell-attached and excised membrane patches with either $Ca^{2+}$ or $Na^+$ as charge carrier (Carbone and Lux 1987). With the exception of sympathetic neurones (Marchetti *et al.* 1986; Wanke *et al.* 1987), chromaffin cells (Artalejo *et al.* 1994) and human small-cell lung carcinoma cells (Codignola *et al.* 1993), T-type channels are expressed in most neurones, muscle and secretory cells (Tsien *et al.* 1991). A peculiarity of T-type channels is their relatively equal permeability to $Ca^{2+}$ and $Ba^{2+}$ and, compared with HVA channels, their weaker block by $Cd^{2+}$ ($IC_{50}$ 3–5 $\mu$M in 5 mM $Ba^{2+}$), higher sensitivity to $Ni^{2+}$ ($IC_{50}$ 15–30 $\mu$M) and $Ca^{2+}$-independent inactivation. T-type channels are also partially blocked by high doses of amiloride, and largely insensitive to dihydropyridines and most animal $\omega$-toxin peptides (see chapter 5).

The T-type channel, however, does not seem any longer to be the only member of the LVA channel family (Table 4.1). Although well conserved, the different kinetics and pharmacology of LVA channels in various cells is suggestive of the existence of multiple LVA channels with strong homology. For instance, the R-type channel expressed in cerebellar granules has impressive similarities to the T-type channel (Zhang *et al.* 1993) in that (i) it activates between $-40$ and $-30$ mV in 5 mM $Ba^{2+}$; (ii) it is almost equally permeable to $Ca^{2+}$ and $Ba^{2+}$; (iii) it is insensitive to most $Ca^{2+}$ antagonists selective for the HVA channels; (iv) it is reprimed by very negative holding potentials (−80 mV); (v) it inactivates quickly and in a voltage-dependent manner ($\tau_{inact}$ 29 ms in 5 mM $Ba^{2+}$ at 0 mV); and (vi) it is blocked by micromolar amounts of $Ni^{2+}$ ($IC_{50}$ 66 $\mu$M). The only significant difference from T-type channels seems to be the high sensitivity of the R-type channel for $Cd^{2+}$ ($IC_{50}$ 1 $\mu$M).

Despite the strong similarities with T-type channels, the cerebellar granule R-type channel is classified in the HVA family (Zhang *et al.* 1993) and it is proposed to be the counterpart of two cloned neuronal $Ca^{2+}$ channels: the $\alpha_1$ subunit doe-1 cloned from the marine ray forebrain (Ellinor *et al.* 1993) and the $\alpha_1$ subunit rbE-II derived from the rat brain (Soong *et al.* 1993). Indeed, although structurally related to the HVA class, the rbE-II channel displays kinetics and pharmacological characteristics similar to the T-type and is therefore proposed as a member of the LVA channel family (Soong *et al.* 1993). $Ba^{2+}$ currents through the rbE-II channel activate at significantly negative membrane potentials (−50 mV) and peak around −10 mV in 4 mM $Ba^{2+}$. Inactivation is somehow slower than T- and R-type channels ($\tau_h$ 99 ms at −10 mV in 4 mM $Ba^{2+}$) but increases significantly with increasing membrane depolarization. Like the T- and R-type channels, the pharmacological profile of the rbE-II $\alpha_1$ subunit is also distinct from that of HVA channels. rbE-II is either insensitive or weakly affected by most $Ca^{2+}$ channel antagonists selective for the HVA channels and is effectively blocked by $Ni^{2+}$ ($IC_{50}$ 28 $\mu$M). However, unlike the T-type but similar to the R-type and the doe-1 channel, the rbE-II channel is potently blocked by $Cd^{2+}$ ($IC_{50}$ 3 $\mu$M; see Table 4.1).

Finally, a low-threshold channel with kinetics and pharmacological characteristics similar to rbE-II and doe-1 has been recently described in bullfrog sympathetic neurones (Elmslie *et al.* 1994). The channel is best revealed at high $Ba^{2+}$ concentrations. In 90 mM $Ba^{2+}$ it activates above −30 mV, inactivates less rapidly and deactivates more quickly than T-type channels. Like the T-type, however, it steadily inactivates at low holding potentials (−50 mV), it is insensitive to DHPs, $\omega$-CTx-GVIA and has opposite sensitivities to $Ni^{2+}$ and $Cd^{2+}$ compared with the HVA channels. This channel has therefore significant similarities to rbE-II and doe-1 $\alpha_1$ subunits, including the low-threshold range of activation. Indeed, if allowance is made for a −30 mV voltage shift, the activation threshold of this 'novel' $Ca^{2+}$ channel would occur around −60, −50 mV in 5 mM $Ba^{2+}$. As the channel possesses a single-channel conductance comparable with that of HVA channels (19 pS) and is not clearly detectable at low $Ba^{2+}$ concentrations (2–3 mM), its unexpected presence in high $Ba^{2+}$ solutions may cause serious misinterpretations of single-channel recordings at low voltages (−20 to 0 mV). Thus, single-channel analysis in 100 mM $Ba^{2+}$ should always consider the possibility that a 19 pS low-threshold $Ca^2$ channel may be fully active below 0 mV and that the channel can be easily confused with an HVA channel subtype activating at low voltages (Elmslie *et al.* 1994).

## High-threshold channels

High-voltage-activated $Ca^{2+}$ channels form a broad class of voltage-dependent channels with a threshold of activation more positive than LVA channels (> −20 mV in 5 mM $Ba^{2+}$) (Tsien *et al.* 1991). HVA channels have apparently similar voltage range of activation and inactivation (particularly in $Ba^{2+}$ solutions). For this reason, their separation in channel subtypes is not unequivocal and is usually achieved by means of specific ligands (see Chapter 5). Dihydropyridines (DHPs) and animal toxins (cono- and aga-peptides) are commonly used to dissect macroscopic $Ca^{2+}$ currents into multiple components of different size, depending on the cell type and function.

Following these criteria, HVA channels are commonly identified as: (i) DHP-sensitive (L-type), that are blocked by micromolar concentrations of DHP antagonists (nifedipine, nitrendipine, nimodipine) or potentiated by low doses of DHP agonists (Bay K 8644, (+)-(S)-202–791, (+)-PN200–110). These channels are ubiquitous but predominate in muscle (Bean 1989), secretory cells (Plant 1988; Lopez *et al.* 1994) and some neurones (Carbone and Swandulla 1989); (ii) $\omega$-Conotoxin (CTx)-GVIA-sensitive channels (N-types), that are quickly and potently blocked by micromolar concentrations of $\omega$-CTx-GVIA applied in low $Ca^{2+}$ solutions (Sher and Clementi 1991). They predominate in neuronal tissues (Hirning *et al.* 1988; Carbone *et al.* 1990; Regan *et al.* 1991; Kasai and Neher, 1992) but are expressed at lower densities also in neurosecretory cells (Codignola *et al.*; Pollo *et al.* 1993; Albillos *et al.* 1994*a*, Artalejo *et al.* 1994); (iii) $\omega$-Aga-IVA-sensitive channels (P-type), that are slowly blocked by submicromolar concentrations of the toxin (Mintz *et al.* 1992). Block of P-type channels by $\omega$-Aga-IVA is voltage-dependent. It is weak and reversible at very positive potentials, but largely irreversible below 0 mV. P-type channels are insensitive to DHP and $\omega$-CTx-GVIA and represent the predominant $Ca^{2+}$ channel subtype expressed by cerebellar Purkinje neurones (Llinas *et al.* 1992). P-type channels are reported also to be expressed in neurosecretory cells (Codignola *et al.*. 1993; Albillos *et al.* 1994*b*; Artalejo *et al.* 1994) and in other neurones (see Chapter 5); and (iv) the Q-type channel represents the last new entry in the list of HVA channels (Randall *et al.* 1993; Sather *et al.* 1993). To date, the channel is not yet pharmacologically identified by any selective ligand. It is insensitive to DHPs and $\omega$-CTx-GVIA and is weakly affected by $\omega$-Aga-IVA. Micromolar concentrations of $\omega$-CTx-MVIIC are sufficient to block irreversibly the channel, but this toxin is also shown to block irreversibly the P-type channel and reversibly the N-type (Hillyard *et al.* 1992). Thus, the pharmacological separation of Q- from P-type channels appears subtle and suggests close relation between the two channels. Nevertheless, the two channels differ noticeably with respect to their voltage range of activation, sensitivity to holding potential and time course of inactivation. The P-type channel activates at relatively low membrane potentials ($\geqslant -30$ mV in 5 mM $Ba^{2+}$), inactivates very little during prolonged depolarizations, and persists to low membrane potentials ($V_h > -50$ mV). The Q-type channel in cerebellar granule cells activates at more positive potentials ($\geqslant -20$ mV in 5 mM $Ba^{2+}$), inactivates more rapidly and is steadily inactivated at $V_h > -50$ mV. In contrast, however, recordings of single $Ca^{2+}$ channels with pharmacological profile similar to the Q-type show weak time-dependent inactivation during pulses of 800 ms duration in cerebellar granules (Forti *et al.* 1994). The Q-type channel is highly expressed in the central nervous system and is proposed to support synaptic transmission in CA3 and CA1 hippocampal neurones (Wheeler *et al.* 1994). HVA channels similar to the Q-type are also shown to control catecholamine secretion in bovine chromaffin cells (Lopez *et al.* 1994) and to be expressed in rat insulin–secreting cells (Magnelli *et al.* 1995).

Despite some unavoidable controversies, the universally accepted separation of voltage-dependent $Ca^{2+}$ channels into LVA and HVA families still seems adequate. Most of the presently identified $Ca^{2+}$ channels appear to fulfil the basic requirements of each subset. Increasing molecular homologies between the two families, however, suggest some caution in rigidly assigning some channels to a specific class and it would not be surprising if separation of new $Ca^{2+}$ channels into LVA and HVA will soon appear more semantic than substantial.

## Activation and inactivation kinetics of high-threshold channels

Close inspections of the kinetics and voltage-range of activation of $Ca^{2+}$ currents often reveal small but significant differences among HVA $Ca^{2+}$ channel subtypes that, in combination with specific ligands, help to better characterize $Ca^{2+}$ channel diversity in excitable cells. We will review here some kinetics peculiarities that, in our experience, can be helpful for the routine work on $Ca^{2+}$ channel identification. L-type channels, for instance, possess two distinctive features. They inactivate markedly with increasing intracellular $Ca^{2+}$ level (Plant 1988; Kasai and Neher 1992; Pollo *et al.* 1993; Marchetti *et al.* 1994) and activate at slightly more negative potentials than other HVA channels (Carbone *et al.* 1990; Regan *et al.* 1991; Kasai and Neher 1992; Pollo *et al.* 1993).

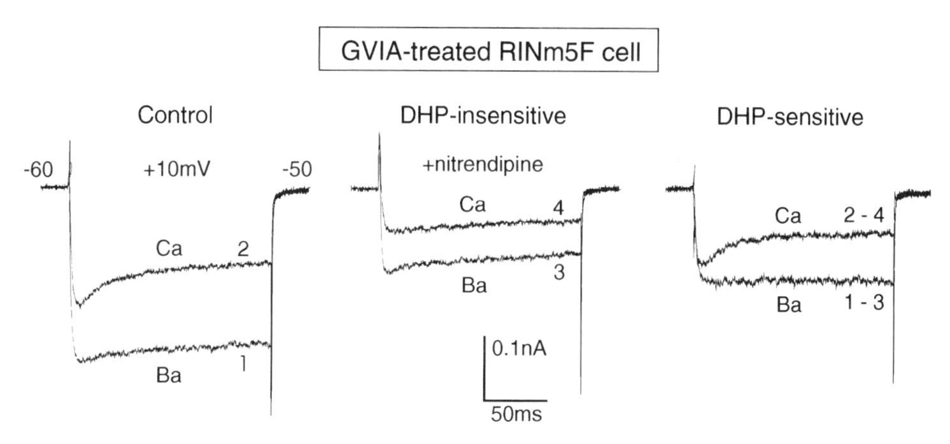


**Fig. 4.1** Effects of $Ca^{2+}/Ba^{2+}$ substitution on DHP-sensitive and DHP-insensitive currents in a RINm5F cell. The current traces were recorded at +10 mV in 10 mM $Ba^{2+}$ or $Ca^{2+}$ from a cell pre-treated with ω-CTx-GVIA (3.2 μM for 20 min), before (trace 1 and 2, left) and after (trace 3 and 4, middle) addition of 10 μM nitrendipine. The right-hand panel shows the corresponding DHP-sensitive currents obtained by subtracting trace 3 from 1 and trace 2 from 4. Holding potential −60 mV. Repolarizations to −50 mV. (From Pollo *et al.* 1993.)

### $Ca^{2+}$-dependent inactivation of the L-type channel

The sharp $Ca^{2+}$ sensitivity of L-type channel inactivation is illustrated in Fig. 4.1. The HVA $Ba^{2+}$ current recorded at +10 mV from a rat insulin-secreting RINm5F cell inactivates slowly and little during pulses of 150 ms duration (trace 1, left panel). Replacement of $Ba^{2+}$ with $Ca^{2+}$ causes a marked peak amplitude reduction and a fast $Ca^{2+}$ current inactivation ($\tau_h$ 30 ms) (trace 2, left panel). The inactivating phase of the current is removed after addition of 10 μM nitrendipine (trace 4, middle panel) suggesting that most of the $Ca^{2+}$-dependent inactivation is associated to DHP-sensitive channels (trace 1–3 and 2–4 right panel), whose contribution to the total current is determined by subtracting the resistant component (middle panel) from the control current (left panel). From Fig. 4.1 it is also evident that DHP-resistant (non-L-type) channels exhibit little or no inactivation during pulses of 150 ms, independently of whether $Ca^{2+}$ or $Ba^{2+}$ is the current-carrying ion. $Ca^{2+}$-dependent inactivation is better highlighted during double-pulse experiments on DHP-sensitive currents in insulin-secreting cells (Plant *et al.* 1988; Marchetti *et al.* 1994) and peripheral neurones (Kasai and Neher 1992). In both cases, the size of $Ca^{2+}$ currents after pre-pulse is markedly current-dependent rather than voltage-dependent and the corresponding I–V relationships show clear U-shaped inactivation characteristics (see p. 100).

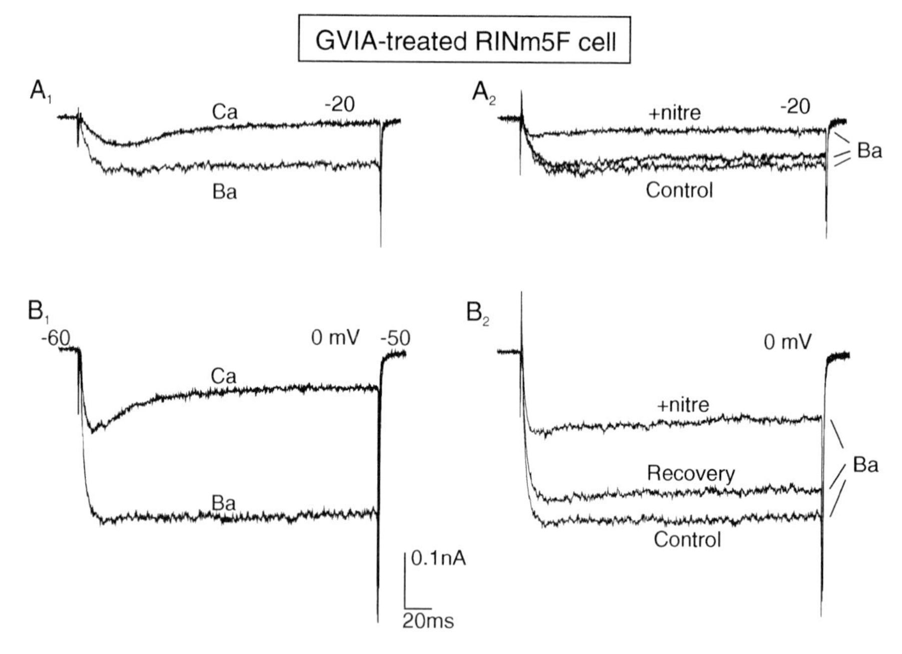


**Fig. 4.2** $Ca^{2+}$-dependent inactivation of L-type channels activated at low membrane potentials in a RINm5F cell. $Ca^{2+}$ and $Ba^{2+}$ currents at $-20$ mV ($A_1$) and 0 mV ($B_1$). Panel $A_2$ and $B_2$ show the blocking potency of nitrendipine (10 $\mu$M) on $Ba^{2+}$ currents at $-20$ and 0 mV, respectively. Traces recorded under same conditions as Fig. 4.1.

#### Voltage-dependence of L-type versus non-L-type channels

The second distinctive property of L-type channels is their more negative voltage range of activation compared with some high-threshold non-L-type channels, such as the N- and Q-type. In 5 mM $Ba^{2+}$, L-type channels activate around $-30$ mV, i.e. at about 10 to 15 mV more negative voltages than other non-L-type channels. Activation of L-type channels in some excitable cells is so low that, with external $Ca^{2+}$ solutions, the time course of their currents can be easily confused with the fast inactivating T-type current reported in most neurones, muscles, and secretory cells (Carbone and Swandulla 1989). This is evident in Fig. 4.2, in which $Ca^{2+}$ currents recorded from a rat insulinoma RINm5F cell at $-20$ and 0 mV are illustrated (panels $A_1$ and $B_1$). Similar to the T-type channel, the $Ca^{2+}$ current at $-20$ mV inactivates quickly and almost completely during the voltage pulse. However, unlike the T-type channel, the inactivation is largely removed when external $Ca^{2+}$ is replaced by $Ba^{2+}$. Above $-20$ mV, $Ba^{2+}$ current inactivation is nearly absent and addition of nitrendipine (10 $\mu$M) abolishes a large fraction of the current. In line with the idea that L-type channels predominate at low voltages, nitrendipine action appears more potent at $-20$ mV than at 0 mV (79% versus 57% block; Fig. 4.2, panels $A_2$ and $B_2$). This implies that, like other cells, RINm5F cells possess subclasses of HVA channels: one activating at more negative potentials (L-type) and one activating at somewhat more positive values (non-L-type). This point is proved when studying the combined action of $Ca^{2+}$ channel agonists and antagonists at different membrane potentials (Fig. 4.3) or, even better, when comparing the blocking action of DHPs and $\omega$-CTx-MVIIC that blocks selectively non-L-type channels (Figs. 4.4 and 4.5). Block by nifedipine and potentiation of $Ba^{2+}$ currents by Bay K 8644 are significantly more effective at $-30$ mV than at $-10$ mV (Fig. 4.3). At $-30$ mV, nifedipine blocks about 80% of the current while Bay K 8644 causes a four-fold increase of their size and a 10-fold prolongation of its monoexponential deactivation kinetics on return to $-50$ mV (tail 1), suggesting a predominance of active L-type channels at $-30$ mV. Above this range, nifedipine and Bay K 8644 actions are both weaker. Nifedipine block drops to about 30% and Bay K 8644 induces only a two-fold increase of the current size. Tail currents following step depolarizations to $-10$ mV are clearly bi-exponential (tail 2), due to the mixed presence of slowly deactivating Bay K 8644-modified channels and fast deactivating non-L-type channels activated at $-10$ mV.

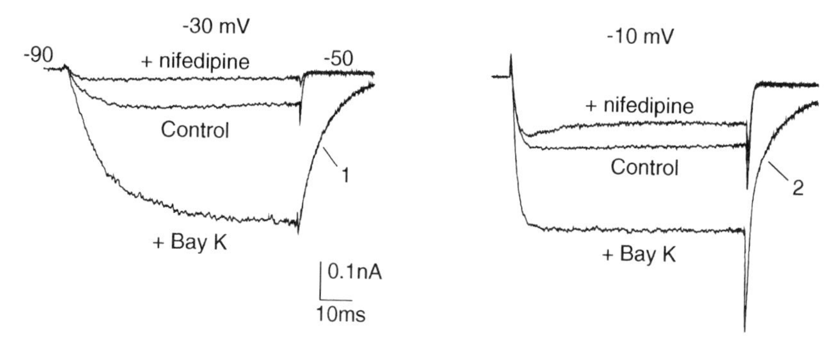


**Fig. 4.3** Nifedipine inhibition and Bay K 8644 potentiation of HVA currents is more effective at low membrane potentials. The current traces were recorded in 10 mM $Ba^{2+}$ at $-30$ mV (left) and $-10$ mV (right) from the same RINm5F cell. The sequence of the recordings was: control, nifedipine (10 $\mu$M) and Bay K 8644 (1 $\mu$M). Holding potential $-90$ mV. Tail currents on repolarization to $-50$ mV were sampled at 10 $\mu$s per point. The two slow tails (1 and 2) of Bay K 8644-modified currents were fitted by double exponentials with markedly different time constants ($\tau_{fast}$, $\tau_{slow}$). Tail 1 was dominated by $\tau_{slow}$ (5.05 ms), while tail 2 was best fitted by a fast (0.45 ms) and a slow component (5.65 ms) of comparable size. (From Pollo *et al.* 1993.)

The separation of L- from non-L-type currents is better achieved by testing the action of DHP antagonists and $\omega$-CTx-MVIIC, that is shown to block N-, P-, and Q-type channels in neurons (Hillyard *et al.* 1992). Indeed, contrary to nifedipine, $\omega$-CTx-MVIIC blocks more effectively the HVA $Ba^{2+}$ currents of RINm5F cells above $-10$ mV (Magnelli *et al.* 1995). The block by $\omega$-CTx-MVIIC is large and partly irreversible at $+30$ mV, while it is weak and

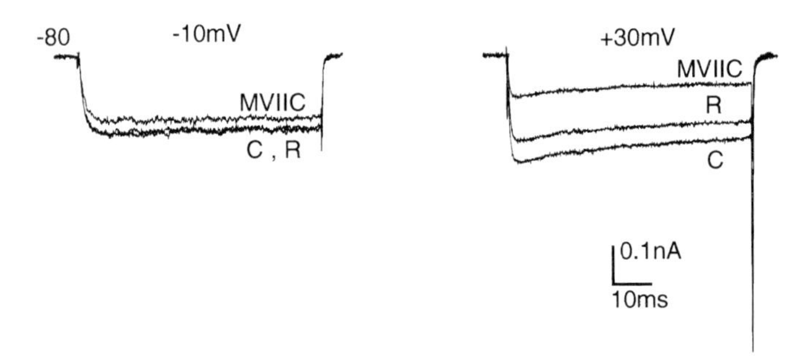


**Fig. 4.4** The blocking action of $\omega$-CTx-MVIIC at different voltages. The records were obtained before (C), during (MVIIC) and after (R) acute application of 3 $\mu$M $\omega$-CTx-MVIIC to a RINm5F cell at $-10$ mV (left) and +30 mV (right). Contrary to the DHPs (Figs. 4.2 and 4.3), the block by $\omega$-CTx-MVIIC (3 $\mu$M) is larger at +30 mV (74%) than at $-10$ mV (15%). Holding potential $-80$ mV. (From Magnelli *et al.* 1995.)

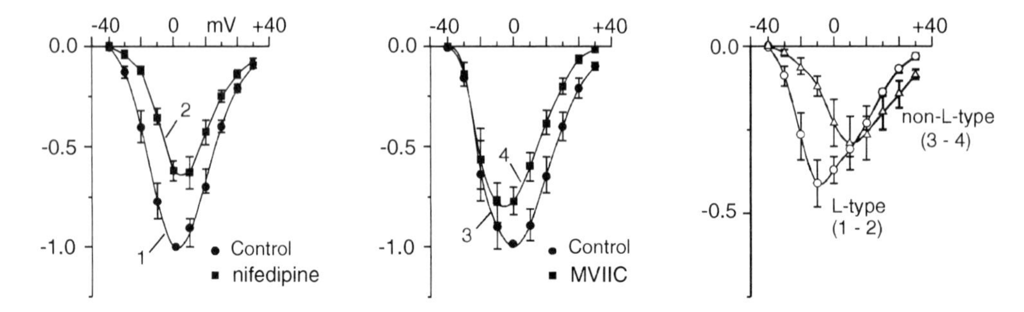


**Fig. 4.5** Distinct activation of L- and non-L-type channels in RINm5F cells. Nifedipine (5 $\mu$M) and $\omega$-CTx-MVIIC (3 $\mu$M) induce opposite voltage shifts to the normalized I–V curves at control (•). The right-hand panel show the I–V curves derived by subtracting curve 2 from 1 and curve 4 from 3. They represent, respectively, the I–V characteristics of L-type (○) and non-L-type channels (△). (From Magnelli *et al.* 1995.)

reversible at $-10$ mV (Fig. 4.4). The different voltage ranges of nifedipine and $\omega$-CTx-MVIIC action are illustrated in Fig. 4.5. Nifedipine block induces a slight shift to the right of the I/V characteristics (curve 2) while $\omega$-CTx-MVIIC shifts the I/V curve in the opposite direction (curve 4). The corresponding blocked currents obtained by subtracting curve 2 from 1 and curve 4 from 3 gives well-separated I/V curves for L- (curve 1–2) and non-L-type channels (curve 3–4). Recordings of nearly pure L-type channels by using a combination of $\omega$-conopeptides to irreversibly block N-, P-, and Q-type channels confirm nicely that L-type channels activate at potentials positive to $-30$ mV in 5 mM $Ba^{2+}$ and that, under these conditions, nifedipine action is potent and homogeneous at all the potentials tested (Fig. 4.6) (Magnelli *et al.* 1995).

L-type and N-type channels have been shown to activate at well-separated voltage ranges in a number of neurones (Carbone *et al.* 1990; Regan *et al.* 1991; Kasai and Neher, 1992) and neurosecretory cells (Pollo *et al.* 1993; Codignola *et al.* 1993; Albillos *et al.* 1994*a*). This property now includes other non-L-type channels that are also distinct from the N-type channels. Insulin-secreting cells, chromaffin and human small-cell lung carcinoma cells have recently been shown to possess L-, N-, P-, and Q-type channels (Codignola *et al.* 1993;

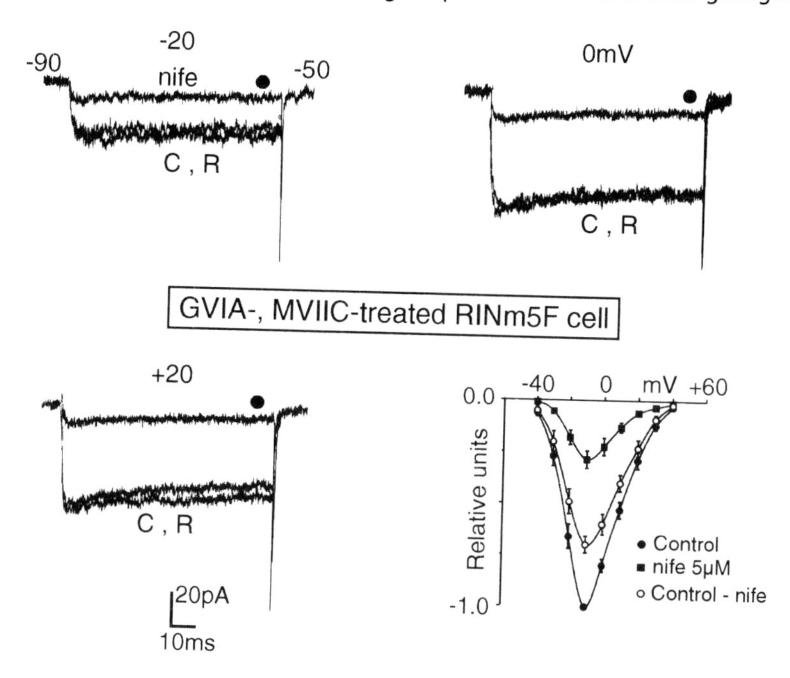


**Fig. 4.6** Predominance of DHP-sensitive channels in RINm5F cells chronically-treated with ω-CTx-GVIA and ω-CTx-MVIIC. Current traces were obtained in 5 mM $Ba^{2+}$ at control (C), with nifedipine (5 μM, ●) and after washing (R) at −20, 0 and +20 mV. The block by nifedipine was 85%, 83%, and 87%, respectively. The I–V relationships at the bottom-right are derived from the HVA currents at control (●), with nifedipine (■) and from their difference (○). Holding potential −80 mV. (From Magnelli *et al.* 1995.)

Albillos *et al.* 1994*b*; Artalejo *et al.* 1994; Marchetti *et al.* 1994; Gandia *et al.* 1995). In all these cells L-type channels activate at somewhat more negative potentials than non-L-type channels.

#### Functional role of $Ca^{2+}$ channels based on their threshold of activation and inactivation kinetics

The threshold of activation and the time course of inactivation are important parameters for the voltage-dependent $Ca^{2+}$ channels involved in the control of $Ca^{2+}$-dependent cell activities. $Ca^{2+}$ channels activating at low voltages (LVA) are usually assumed to be involved in the initiation of electrical oscillations in excitable cells (Bean 1989), while channels activating at more positive potentials (HVA) are shown to be conditional for the maintenance of sustained cell functions (Tsien *et al.* 1991). The different thresholds of activation of HVA channels introduce a further degree of flexibility in the role that voltage-dependent $Ca^{2+}$ channels play in the regulation of $Ca^{2+}$ dependent cell functions. For instance, the relatively low threshold of activation of L-type channels in most secretory cells and neurones, confers to these channels a critical role in exocytosis. If, as in pancreatic β-cells, the oscillatory electrical bursts controlling insulin secretion are critically dependent on the threshold of activation of a slowly inactivating $Ca^{2+}$ conductance (−40 mV in 3 mM $Ca^{2+}$) (Satin and Smolen 1994), the L-type channel will result conditional to the initiation of cell burstings and to the subsequent insulin release during plasma glucose increases. On the other hand, since inactivation of the L-type channel in pancreatic $\beta$-cells is fast ($\tau_h$ 50 ms) and $Ca^{2+}$-dependent (Plant 1988), the L-type channel alone cannot account for the slowly inactivating $Ca^{2+}$ conductance ($\tau_h$ 2–3 s) that underlies the prolonged plateau potentials and is responsible for bursts durations. The slowly inactivating $Ca^{2+}$ conductance is thus likely to be provided by another set of slowly inactivating non-L-type channels whose activation, however, may critically depend on the L-type channel activating at lower voltages (Magnelli *et al.* 1995). In conclusion, the expression of various $Ca^{2+}$ channels with different threshold of activation and time course of inactivation offers an unquestionable powerful means to modulate the $Ca^{2+}$-dependent physiological functions of excitable cells.

## Modulation of high-threshold $Ca^{2+}$ channels

On a short time scale, the inhibition of high-threshold channels induced by neurotransmitters and hormones represents the best example of $Ca^{2+}$ channel modulation in neurones and neurosecretory cells. The phenomenon was first reported in chick sensory neurones (Dunlap and Fischbach 1978) and subsequently extended to most peripheral and central neurones by an impressive number of reports (Swandulla *et al.* 1991). The main 'ingredients' of neurotransmitter action are well established and attributed to a receptor-mediated reaction modulated by either pertussis toxin-sensitive (Dolphin and Scott 1987; Wanke *et al.* 1987) or cholera toxin-sensitive G-proteins (Zhu and Ikeda 1994*b*). The activated βγ-subunit of the G protein down-modulates reversibly the HVA $Ca^{2+}$ channels by reducing and delaying their probability of opening (Ikeula 1996). An interesting aspect of this phenomenon is that membrane voltage controls with rapid kinetics a significant fraction of $Ca^{2+}$ channel inhibition. Strong and short depolarizations remove the inhibition, while negative potentials help to recover the depression (Marchetti *et al.* 1986; Grassi and Lux 1989; Elmslie *et al.* 1990). A number of recent papers have focused on the basic mechanisms underlying the voltage-dependence, voltage-independence and selectivity of G protein action (Hille 1994). Some points are universally accepted, while others are still questioned. Here, we will review the converging issues and discuss the contrasting ones, highlighting the role that the fast $Ca^{2+}$ channel modulation by G proteins may play in the control of neurotransmitter and hormone release in synaptic terminals and secretory cells.

### Selective action of neurotransmitters on $Ca^{2+}$ channel subtypes

Neurotransmitter action on $Ca^{2+}$ channels is reversible and mediated by membrane receptors. In human neuroblastoma IMR32 cells (Pollo *et al.* 1992) and bullfrog sympathetic neurones (Boland and Bean 1993) the block of HVA currents by receptor agonists is concentration-dependent and develops within few seconds at saturating doses of the agonist. Wash-out develops independently of neurotransmitter concentration, being complete between 6 and 10 s. The degree of neurotransmitter inhibition is related to the density of membrane receptors expressed by the cell. For instance, noradrenergic, muscarinic and $\delta$- and $\mu$-opioid receptors coexist at different concentrations in IMR32 cells. In these cells, the action of noradrenaline (NA), oxotremorine (OXO) and [D-Pen$^2$, D-Pen$^5$]-Enkephalin (DPDPE; $\delta$-

opioid selective agonist) is usually much more pronounced than that of the μ-opioid selective agonist [D-Ala$^2$, N-MePhe$^4$, Gly-ol$^5$] enkephalin (DAMGO) (Fig. 4.7). The inhibition appears more pronounced at the beginning of the step depolarization and is partially relieved toward the end of the pulse. Neurotransmitter action is prevented by receptor antagonists (Fig. 4.7C,D; Fig. 4.8A) and, in most cells, the effect is limited to non-L-type channels. In differentiated IMR32 cells possessing 20% L-type and 80% N-type channels (Carbone *et al.* 1990), the action of OXO, NA and DPDPE is nearly absent in cells pre-treated with ω-CTx-GVIA and thus deprived of N-type channels (Fig. 4.8B). The same is true for most peripheral neurones (Elmslie *et al.* 1990; Kasai 1992; Boland and Bean 1993). L-type channels appear little inhibited by neurotransmitters. Neurotransmitter action is always additive to that of L-type blockers, suggesting a distinct action of neurotransmitters on non-L-type channels. In Fig. 4.8C, nifedipine abolishes the same amount of current independently of whether oxotremorine is present or absent. In both cases, the DHP reduces the size without affecting the time course of the current. In contract, OXO produces a marked change to the activation-inactivation kinetics either with or without the DHP.

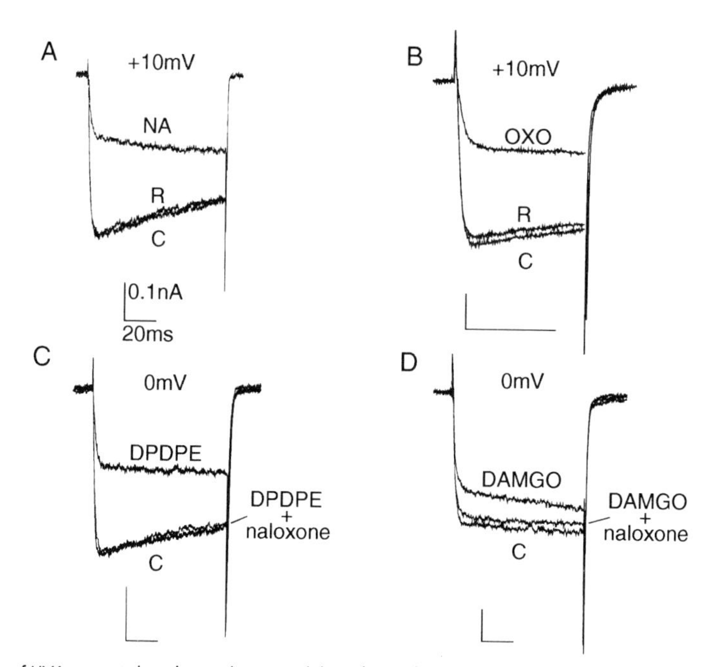


**Fig. 4.7** Inhibition of HVA currents by adrenergic, muscarinic and opioid agonists in human neuroblastoma IMR32 cells. A, B. Reversible effects of noradrenaline (NA, 10 μM) and oxotremorine (OXO, 10 μM) on HVA currents at +10 mV in 10 mM $Ba^{2+}$. The traces were recorded before (C) and after (R) short exposures (1 min) of the cell to the neurotransmitters. Notice the appearance of a slow phase of activation and the absence of inactivation during application of the neurotransmitter. Holding potential −90 mV. C, D. Inhibitory effects of the δ-selective opioid agonist DPDPE (1 μM) and μ-selective agonist DAMGO (1 μM) on two IMR32 cells. The different action of the two agonists was prevented by the unspecific opioid antagonist, naloxone. Test depolarizations to 0 mV in 10 mM $Ba^{2+}$. Holding potential −90 mV.

At variance with HVA channels, neurotransmitters have marginal effects on LVA channels in most neurones. An example of a weak action of neurotransmitter on LVA channels is shown in Fig. 4.9. LVA and HVA currents recorded using a ramp command at control (C)

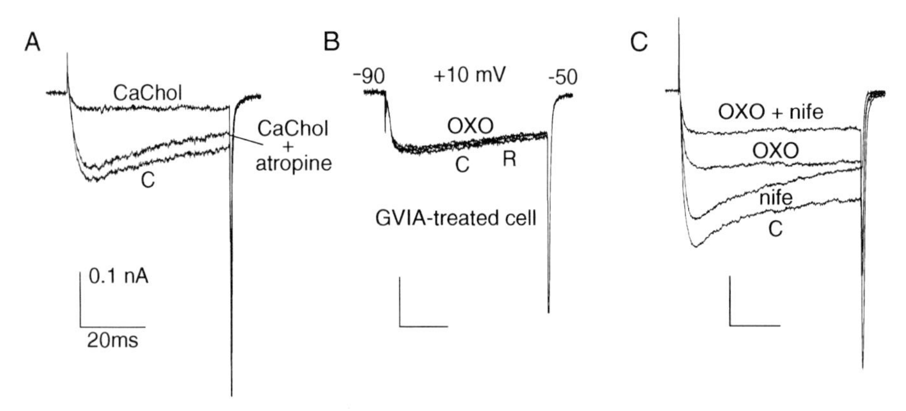


**Fig. 4.8** Muscarinic inhibition of HVA currents in IMR32 cells is selective for ω-CTx-GVIA-sensitive (N-type) channels. A. The inhibitory action of carbachol (CaChol) (1 μM) on HVA currents is antagonized by addition of 1 μM atropine to the solution containing the agonist. Test depolarization to +10 mV from −90 mV in 10 mM $Ba^{2+}$. B. 10 μM oxotremorine (OXO) has no effect on HVA currents persisting after cell pre-treatment with ω-CTx-GVIA (3.2 μM for 15 min in 2 mM $Ca^{2+}$). Same conditions as in panel A. C. Oxotremorine action is additive to the blocking effects of nifedipine. The sequence of recordings was: control (C) nifedipine (nife, 10 μM), oxotremorine (OXO, 10 μM) and oxotremorine + nifedipine (OXO + nife). Same conditions as in panels (A) and (B).

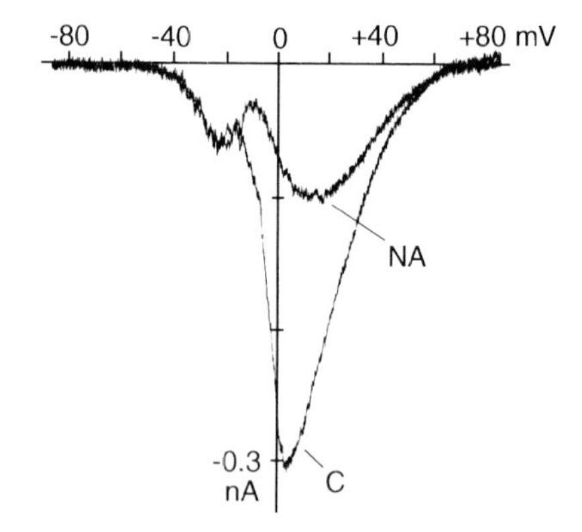


**Fig. 4.9** Selective action of noradrenaline (NA) on HVA currents. The current traces are recorded in response to ramp commands with a slope of 0.55 mV $ms^{-1}$ from −90 mV holding potential in 10 mM $Ba^{2+}$. The resulting I–V relationships shows an early peak at around −25 mV (LVA) that is unaffected by 10 μM NA. On the contrary, the second peak at about +5 mV (HVA) is largely depressed by the neurotransmitter. The recordings were corrected for $Cd^{2+}$-insensitive currents persisting after addition of 200 μM $Cd^{2+}$.

and during noradrenaline application (NA, 10 μM) are differently affected by the neurotransmitter. NA causes a marked depression of the peak current activating at high voltages (+5 mV) preserving the early small current originating at low voltages (−25 mV). Small effects on LVA channels have been reported in several other neurones using most neurotransmitters (Cox and Dunlap 1992), but there are also clear exceptions to this rule (Marchetti *et al.* 1986; Kasai 1992).

## Voltage-dependent versus voltage-independent modulation

Neurotransmitter inhibition of HVA currents is time- and voltage-dependent. The depression is marked at low voltages and partially relieved at higher depolarizations. Figure 4.9 shows that the NA inhibition in IMR32 cells is more potent at 0 mV (75%) than at +50 mV (28%) and the modified I–V curve is shifted to the right. The time-dependent removal of inhibition with voltage is illustrated in Fig. 4.10. At −10 mV, depression by OXO is strong (87%) and remains nearly unchanged during the 30 ms step depolarization. At +10 mV, the current reduction is maximal soon after the onset of the pulse, but is significantly relieved at the end of the pulse. Tail currents on return to −40 mV are depressed by only 67%. At +50 mV, the time course of the inhibited current is hardly separable from the control trace and the inhibition is largely relieved during the 30 ms step depolarization (facilitation). The corresponding tail current is reduced by only 27%.

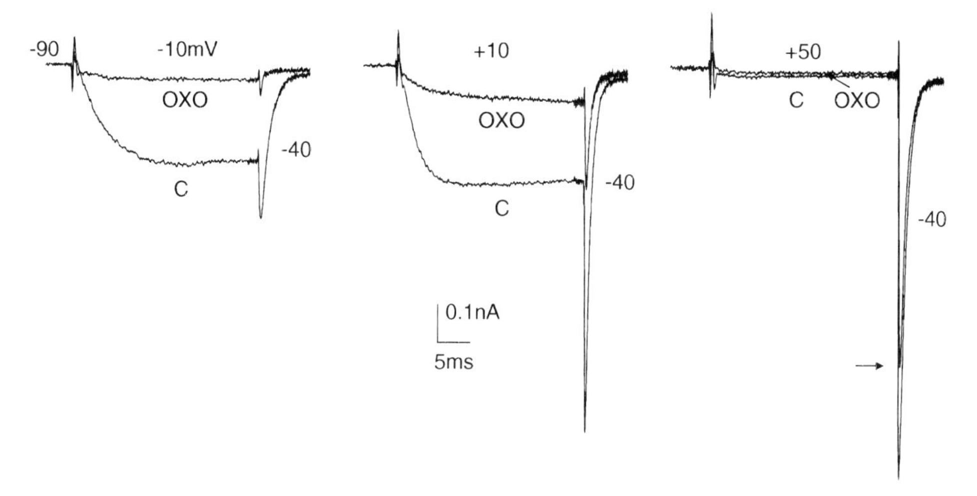


**Fig. 4.10** Time- and voltage-dependent removal of oxotremorine (OXO) inhibition of HVA $Ba^{2+}$ currents in IMR32 cells. The traces were recorded before (C) and during application of 10 $\mu$M oxotremorine in 10 mM $Ba^{2+}$ at the potentials indicated. Step repolarizations to −40 mV; holding potential −90 mV. Tail currents were recorded at a faster sampling rate (20 $\mu$s/point) with respect to the current relaxation at test potentials (50 $\mu$s/point). The arrow on the right-hand panel indicates the peak amplitude of the 'facilitated' tail current with OXO.

Voltage-induced 'facilitation' may be significant (>80% at +9 mV) but is usually partial, leaving a variable amount of residual voltage-independent depression. This raises several important questions. Are the voltage-dependent and voltage-independent inhibitions controlled by different mechanisms (G proteins or second messengers)? Are they associated with different $Ca^{2+}$ channels? How accurately can the percentage of voltage-dependent facilitation be measured? Figure 4.11A shows a typical voltage protocol used to determine the amount of voltage-dependent facilitation of $Ca^{2+}$ currents exposed to saturating doses of neurotransmitter (Grassi and Lux 1989; Elmslie *et al.* 1990). The current recorded during the test pulse (+10 mV) from the holding potential (trace 1) is compared with the current relaxation at the same potential preceded by a 50 ms step depolarization to +90 mV (trace

2). In control conditions, the two current relaxations (1 and 2) are coincident if allowance is made for trace 1 to reach its peak value. With 10 $\mu$M OXO, the two traces (3 and 4) deviate markedly (Fig. 4.11B). There is 71% inhibition when comparing the two traces without pre-pulses at the peak (vertical arrows on traces 1 and 3), but there is only 16% residual inhibition when comparing the amplitude of traces 2 and 4 at the end of the facilitating pre-pulse (horizontal arrows on traces 2 and 4). This is a relatively high degree of facilitation with pre-pulses and, in our view, is the most accurate way to evaluate the percentage of voltage-dependent modulation. Protocols using a short repolarization (5–10 ms) to $-90$ mV interposed between the pre conditioning and the test pulse usually underestimate the percentage of facilitation for two reasons. Channel re-inhibition after facilitation develops in a concentration-dependent manner with a time constant ($\tau_{reinh}$) of 30–50 ms at the holding potential (Lopez and Brown 1991; Golard and Siegelbaum 1993; Elmslie and Jones 1994) and may cause a 28% to 18% channel re-inhibition after 10 ms, respectively. In addition, channel activation at test potentials around 0 mV further delays the estimate of the current facilitation by 3–5 ms, i.e. the time required to reach the peak. Thus, in double-pulse protocols with a 10 ms repolarization to $-90$ mV and a test to $+10$ mV (trace 5 in Fig. 4.11C), the measurement of channel facilitation at the peak of the facilitated current is delayed by about 15 ms after the end of the facilitating stimulus. With a ($\tau_{reinh}$) of 30–50 ms, this causes a 40% to 25% amplitude reduction of the facilitated current with a consequent proportional underestimate of the voltage-dependent inhibition. By comparing traces 2 and 4 in Fig. 4.11C the residual (voltage-independent) inhibition is 16% while by comparing traces

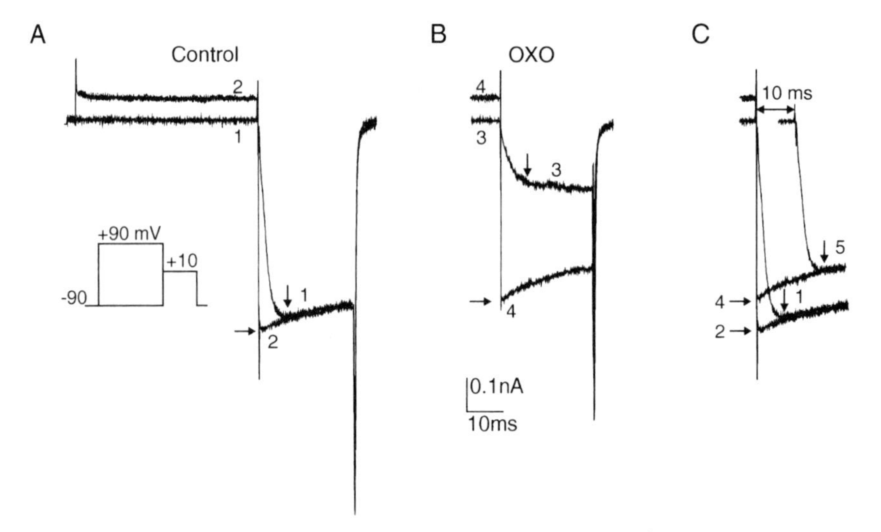


**Fig. 4.11** Prepulse-induced facilitation of muscarinic inhibition in IMR32 cells. The two overlapping traces in A and B were recorded without (trace 1 and 3) and with (trace 2 and 4) a 50 ms conditioning pre-pulse to +90 mV. Test depolarizations to +10 mV; holding potential $-90$ mV. The double-pulse protocol was delivered before (A) and during (B) application of 10 $\mu$M oxotremorine (OXO), with 3 s intervals between pulses. The vertical arrows on traces 1 and 3 indicate the time at which OXO-inhibition is estimated. The horizontal arrows on traces 2 and 4 indicate the amplitude of the facilitated current relaxation to +10 mV. Traces 1, 2 and 4 are shown also in C, together with trace 5. This latter is obtained by scaling and shifting trace 1 in order to mimic a hypothetical facilitated current recorded after 10 ms repolarization to $-90$ mV. The percentage of voltage-independent inhibition estimated from the ratio of trace 2 and 4 and trace 1 and 5 is 16% and 23%, respectively.

1 and 5, the depression increases to 23% and can be larger if channel inactivation and reinhibition become faster. In conclusion, the voltage-dependent $Ca^{2+}$ channel facilitation is more accurately estimated at the end of the facilitating pre-pulse, and the neurotransmitter-induced depression persisting after pre-pulse represents the best estimate of the voltage-independent depression not recovered by voltage.

Variable degrees of voltage-independent inhibition of $Ca^{2+}$ currents have been reported on most peripheral and central neurones. The steady-state depression insensitive to voltage is commonly associated with N-type channels (Luebke and Dunlap 1994), but there is also evidence for a similar action on L-type channels in neurosecretory cells (Pollo *et al.* 1993; Albillos *et al.* 1996), peripheral and central neurones (Bley and Tsien 1990; Amico *et al.* 1995) as well as on T-type (Marchetti *et al.* 1986; Kasai 1992) and P-type channels (Swartz 1993). In insulin-secreting RINm5F cells, for instance, NA induces both voltage-independent and voltage-dependent inhibition that can appear either in isolation or in combination. Voltage-independent inhibition in RINm5F cells is largely removed by applications of saturating doses of nifedipine, which however spare the voltage-dependent component, mainly associated with the N-type channel (Pollo *et al.* 1993). The same is true for the GABA-induced depression of L-type currents in cerebellar granule cells (Amico *et al.* 1995) and for the opioid inhibition of DHP-sensitive currents in bovine chromaffin cells (Albillos *et al.* 1996).

Steady-state inhibition by neurotransmitters has been mostly investigated on N-type currents. It is preserved in the presence of DHP-antagonists and contributes significantly in neurones expressing a small percentage of L-type channels (Cox and Dunlap 1992). So far, separation of the voltage-dependent and voltage-independent depression was mostly based on qualitative grounds. Both pathways are controlled by G proteins, as intracellular GTP-$\gamma$-S mimics the two actions and GDP-$\beta$-S prevents them. In our experience, however, internal GDP-$\beta$-S is extremely effective in preventing the slowing of channel activation, but spares some steady-state inhibition by addition of neurotransmitters. This, very likely, is the first direct evidence that the two modulations develop partially through distinct pathways. The second evidence is the lack of correlation between the two effects. Although the presence of channel inactivation does not often allow a clear separation of the slow activation phase from steady-state inhibition, the two modulations may be separately estimated by the conditioning pre-pulse method described above. Other evidences are: (i) the two phenomena have sharply different voltage-dependence (Luebke and Dunlap 1994); (ii) different neurotransmitters cause either more voltage-dependent or voltage-independent inhibitions in the same cell (Formenti *et al.* 1993); and (iii) voltage-dependent and voltage-independent modulations are regulated by different G protein subunits (Diversé-Pierluissi *et al.* 1995). Kinetic slowing induced by $\alpha_2$-adrenergic and $GABA_B$ agonists in chick sensory neurones seems controlled by $G_o$ $\alpha$ subunits with no involvement of protein kinase C (PKC) (but see p. 121 Swartz 1993; Zhu and Ikeda 1994*a*), while steady-state inhibition through $\alpha_2$-adrenergic receptors is mediated by $G_i$ $\beta\gamma$ subunits and PKC activation. In conclusion, there is evidence for the existence of two modulatory pathways which may subserve different cellular functions.

The existence of separate voltage-dependent and voltage-independent down-modulations of $Ca^{2+}$ currents increases the degree of control of $Ca^{2+}$-dependent cell functions by endogenous or exogenous membrane receptor agonists. Voltage-dependent inhibitions are expected to be attenuated upon repeated cell stimulation, and may be useful in the phasic or use-dependent control of cell function. Voltage-independent depression may be effective in the tonic inhibition of neurotransmitter and hormone secretion, even in the presence of high-frequency stimulations or prolonged cell depolarizations.

## Models of 'Voltage-dependent' inhibition and facilitation

The voltage-dependent down-modulation of HVA currents by neurotransmitters is usually ascribed to changes in channel gating mediated by activated G proteins. Internal GTP-$\gamma$-S mimics the current depression by external neurotransmitters, while GDP-$\beta$-S prevents this action. Identified G proteins (Dolphin 1991) are postulated to interact with $Ca^{2+}$ channel gating in a voltage-dependent manner. Low negative potentials stabilize the binding of the G protein to the channel, while strong positive depolarizations favour the normal gating conditions. It is still unclear how the voltage can affect this 'protein–protein' interaction. It could be that membrane voltage affects directly the coupling and uncoupling of the two macromolecules by acting on a voltage sensor on the G protein (Swandulla *et al.* 1991) or that the 'voltage-dependence' results from a permanent modification of $Ca^{2+}$ channels, in slow equilibrium with unmodified channels (Kasai 1992; Pollo *et al.* 1992; see scheme 1 and 2 in Fig. 4.12). In the first cases, the modified ($M^*$) channels are facilitated by the voltage-dependent rate constants $\gamma$ and $\delta$ that regulate the modified gating mode. $k_1$ and $k_{-1}$ are the rate constants regulating the slow equilibrium between normal and modulated channels (see legend Fig. 4.12). In the second case, the facilitation of inhibited channels ($H^*$) to the normal gating mode ($C \leftrightarrow O$) is further favoured by fast state transitions with voltage-independent rate constants ($\mu$ and $\upsilon$) (Pollo *et al.* 1992). State transitions are assumed to occur most favourably from open ($4\upsilon$) rather than from closed ($\upsilon$) modified channels (Elmslie *et al.* 1990). 'Voltage-dependence' may derive also from the reduced affinity of the G protein for open channels and would be, therefore, a direct consequence of the voltage-sensitivity of channel gatings (Boland and Bean 1993; Golard and Siegelbaum 1993). In this case, the inhibited channel is thought to shift from a 'reluctant' to a 'willing' gating mode during membrane depolarizations (scheme 3). The 'willing' mode has increased occupational probability as the G protein dissociates either from the closed or from the open state. The off-rate of G protein unbinding ($\lambda$) is independent of the activated G protein concentration, [G], but increases by a factor 64 at each sequential activation of one of the four channel gating domains, to reach a maximum of $64^4\,\lambda$ when the channel is fully open ($O^*$). The 'reluctant' mode is favoured by G protein binding through the on-rate $\kappa$ [G] as the channel deactivates on return to resting potentials.

All the above models explain some, but not all, of the features of voltage-dependent inhibition and facilitation. For instance, schemes 1 and 2 account nicely for the time course of facilitation at different neurotransmitter concentrations, for the recovery of channel activation after pre-pulse and for the concentration- and voltage-independent re-inhibition of open channels. Models 1 and 2 are particularly suitable to account for the slow equilibrium kinetics ($k_1$, $k_{-1}$) regulating the partial block of N-type channels at saturating doses of neurotransmitters. As they are, however, the two schemes have some shortcomings. The most serious one is that channel activation must always follow channel facilitation through transition $M^* \rightarrow O^*$ (scheme 1) or $H^* \rightarrow O^*$ (scheme 2). In other words, activation cannot

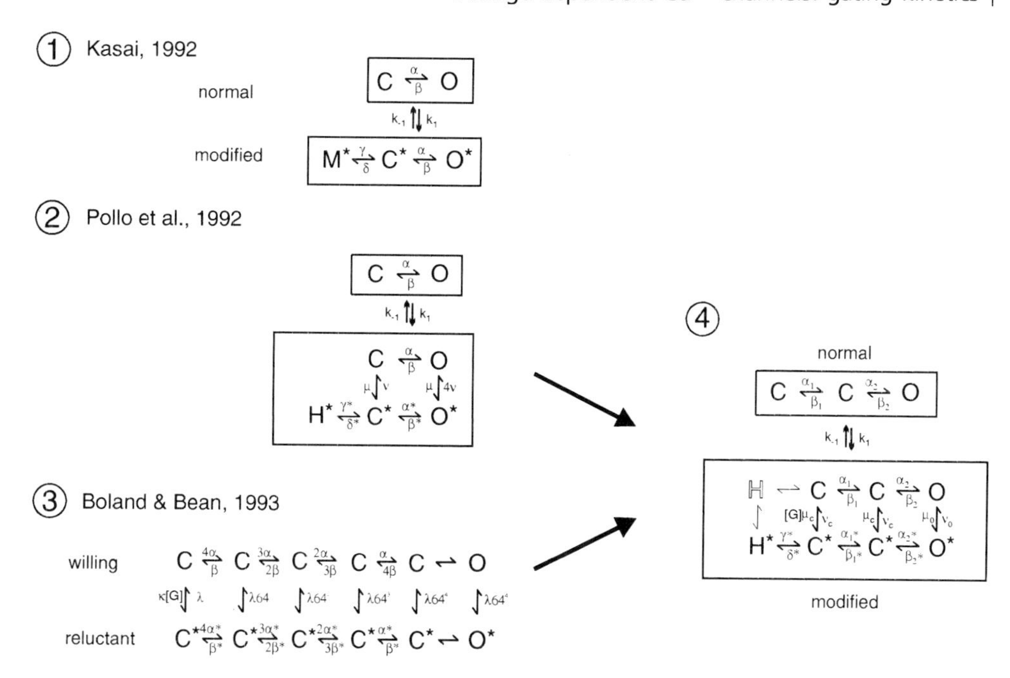


**Fig. 4.12** Kinetic models for the $Ca^{2+}$ channel gating modulation by neurotransmitters. In scheme 1, C and O represent the closed and open states of the normal channel which are in equilibrium through the voltage-dependent rate constants $\alpha$ and $\beta$ (Kasai 1992). $C^*$ and $O^*$ are the corresponding closed and open states of the modified channel. $M^*$ is the modified (non-conductive) state of the channel and $\gamma$ and $\delta$ are the rate constants producing the slow activation of the channel during depolarization. Scheme 2 is from Pollo *et al.* 1992. At variance with model 1, the inhibited state of the channel ($H^*$) is allowed to equilibrate with the normal gating mode ($C \leftrightarrow O$) through the fast transition rate constants $\mu$ and $\upsilon$ taken from Elmslie *et al.* (1990). $k_1$ and $k_{-1}$ are the rate constants regulating the slow equilibrium between normal and modified channels. They are estimated from the onset and offset of $Ca^{2+}$ channel inhibition during application of saturating doses of neurotransmitters. Scheme 3 is derived from Boland and Bean (1993). The 'willing' and the 'reluctant' gating modes are in equilibrium through slow equilibrium rate constants as indicated (see text). The activation kinetics in each mode obey a Hodgkin–Huxley formalism with four gating subunits ($m^4$). Scheme 4 is an alternative model derived from schemes 2 and 3 and should account for the various drawbacks of these two schemes. With respect to scheme 2, the new scheme should be able to allow channels to activate before facilitating at very positive potentials. State H may not be strictly necessary, but it is included to show the close relation between schemes 3 and 2. With respect to scheme 3, scheme 4 accounts for the partial depression induced by the neurotransmitters and should furnish a better fit of the slow activation and fast re-inhibition after facilitation.

be faster than facilitation, while channel activation may develop five times faster than facilitation at very positive potentials (+130 to +150 mV) ($\tau_{act}$ 1 ms and $\tau_{facil}$ 5 ms; Boland and Bean 1993). This can be partly overcome in scheme 2 by re-designing the on–off rates $\gamma$ and $\delta$, thus allowing the channels in state $H^*$ to reach state O and $O^*$ in 1 ms and then wait for the facilitating transition $O^* \rightarrow O$ to occur with 5 ms time constant (Elmslie *et al.* 1990). Scheme 3 overcomes this drawback and accounts for most of the kinetic features of the voltage-dependent inhibition and facilitation, but fails to mimic the partial block of N-type channels at saturating doses of neurotransmitters and the concentration-independent time constant of slow activation at different doses of neurotransmitters (Kasai 1992; Pollo *et al.* 1992; Elmslie and Jones 1994). While giving a detailed and reasonable account of the G-protein–$Ca^{2+}$ channel interaction, the Boland and Bean model is unable to match the slow equilibrium kinetics between receptor activation and channel inhibition developing within seconds and the fast relaxations between activated G proteins and inhibited $Ca^{2+}$ channels occurring within tens of ms.

New experimental findings have further complicated the overall picture of voltage-dependent modulation and have made difficult an updated description of the phenomenon in terms of kinetic schemes. Thus, before proposing new models, it seems more reasonable to summarize all the relevant features that should be satisfied:

1. Voltage-dependent depression by neurotransmitter is only partial (60–75% maximal inhibition) and concentration-dependent.
2. Activation of modified channels is slow at around $-10$ mV and accelerates steeply with voltage (Marchetti *et al.* 1986). The activation kinetics of modified channels can be very slow at some potential ($\tau_{act}$ 40–70 ms) and cannot be attributed to a simple voltage shift of the normal channel gating.
3. The amplitude *but not the time constant* of the slow activation is concentration-dependent (Kasai 1992; Pollo *et al.* 1992; Elmslie and Jones 1994). This implies that, although the current depression develops through a slow dose-dependent equilibrium reaction, the fast voltage-dependent removal of inhibition is paradoxically independent of [G].
4. Facilitation is rate limiting the activation of $Ca^{2+}$ channels at potentials below $+60$ mV (Pollo *et al.* 1992). Above $+70$ mV, activation develops faster ($\tau_{act} \approx 1$ ms) than facilitation ($\tau_{fac} \approx 5$ ms) (Elmslie and Jones 1990; Boland and Bean 1993). This implies that modified and unmodified channels first activate and then facilitate, supporting a model in which G proteins dissociate more easily from open than from closed channels (Boland and Bean 1993).
5. Channel re-inhibition after facilitation is fast ($\tau_{reinh}$ 30–50 ms) and [G]-independent at potentials around 0 mV where channels are preferentially open (Kasai 1992; Pollo *et al.* 1992), but is [G]-dependent at very negative potentials ($-80$ mV) where channels are mostly closed (Golard and Siegelbaum 1993; Elmslie and Jones 1994). This point introduces two complications that none of the proposed models satisfy. Equilibrium of facilitated and re-inhibited channels occurs on a faster time scale (tens of ms) compared with the slow onset of inhibition during application of external neurotransmitters. Onset and offset of inhibition occur on a slower time scale independently of whether the channel is preferentially closed (0.5 s) or open (0.3 s) (see Fig. 4 in Pollo *et al.* 1992). This suggests that the reactions leading to G-proteins activation by receptor occupation is rate limiting the true G protein–$Ca^{2+}$ channel interaction that occurs on a faster time scale when the G protein is activated. The [G]-dependent re-inhibition at $-80$ mV implies that the on-rate of G protein binding is driven by mass law and is either favoured by negative voltages or by the closed state of the channel (Elmslie and Jones 1994). Alternatively, the [G]-independent re-inhibition around 0 mV may be favoured by the less negative potential or by the open state of the channel.

The last point sets strict constraints to the model accounting for the 'voltage-dependent' modulation of HVA channels. As none of the proposed models satisfies all the above conditions, it is reasonable to believe that, with proper modifications, they could converge to a more general one in which concentration-dependence, voltage-dependence and time course of inhibition, facilitation and re-inhibition are all well fitted. Scheme 4, derived from schemes 2 and 3 could for example be a sufficiently general one for simulating most of the observed kinetic features. It would certainly require further improvements when tested, but represents an interesting challenge for future studies.

## A single-channel approach to $Ca^{2+}$ channel modulation

The voltage-dependent and voltage-independent inhibition by neurotransmitters should also find a correspondence at the single-channel level. To date, cell-attached recordings in high $Ba^{2+}$ solutions have shown that neurotransmitters cause a marked reduction of open channel probability interpreted as a shift from a high- (or medium-) to a low-$p_o$ mode mediated by the activated G protein (Delcour and Tsien 1993, see also p. 100). The absence of effects on single-channel conductance supports the view that neurotransmitters mainly affect channel gatings rather than channel permeability. The mode-switching model proposed by Tsien's group satisfies most of the observations on the voltage-independent inhibition of macroscopic currents, but is unable to explain the fast kinetics of the voltage-dependent facilitation and reinhibition ($\tau \approx 20$ ms) observed in whole-cell recordings at positive voltages (Fig. 4.7). If mode switching is conditioned by G proteins binding and unbinding, and the average sojourn of the channel in one mode is about 10 s, it seems unlikely that the fast voltage-dependent unbinding of the G protein from the channel in the low $p_o$ mode causes a shift to the high $p_o$ mode and produces a current facilitation within 20 ms at positive potentials. Indeed, we were able to prove that, in cell-attached patches containing 20 μM NA and 1 μM DPDPE, N-type channel openings are preferentially delayed with a significant increase of the latency to first openings. In addition, test depolarizations to +20 mV preceded by a 50 ms step depolarization to +90 mV allow full recovery of single-channel activity by drastically reducing the first latency of openings (Carabelli *et al.* 1996). This suggests that, despite the technical difficulties, also at the single-channel level, voltage-dependent facilitation and reinhibition occur on a fast time scale as predicted by most kinetic models. The absence of serious discrepancies between whole-cell and single-channel measurements should stimulate further comparisons of the results by the two methodologies in other neurones; this could possibly overcome some inherent limitations of cell-attached recordings that do not allow to run test and control experiments on the same channel.

## Gating versus ion permeation modulation

Most groups agree that channel inhibition and facilitation by neurotransmitters is a consequence of channel gating modulation. A recent report, however, proposes that neurotransmitter inhibition may derive from a partial reduction of channel permeability to divalent cations (Kuo and Bean 1993). Activated G proteins are thought not only to interfere with the channel subunits responsible for channel gating, but also to modify the energy profile controlling ion passage through open channels. Evidence in favour of this idea is based on the following observations: (i) G protein inhibition of N-type currents is less effective when the channel carries either inward $Na^+$ or outward $Cs^+$ instead of $Ba^{2+}$; (ii) inhibition is attenuated also when $[Ba^{2+}]$ increases from 3 to 150 mM; and (iii) in the presence of intracellular GTP-$\gamma$-S, the single-channel conductance estimated by noise analysis on tail currents to $-40$ mV is 25% smaller if measured before a strong facilitating prepulse rather than after. Kuo and Bean propose that the activated G protein at the inner site of the membrane produces a conformational change of the negative protein charges at the outer site of the channel where two high-affinity binding sites for $Ba^{2+}$ (or $Ca^{2+}$) are located (Kuo and Hess 1993). The two binding sites are close enough to produce high fluxes of divalent cations by ion–ion repulsion with millimolar $[Ca^{2+}]_o$ or block of $Na^+$ currents by micromolar $[Ca^{2+}]_o$ (Almers and McCleskey 1984). In low $[Ba^{2+}]$, the strong binding of divalent cations to the negative sites is partly destabilized by the conformational change induced by the G protein. This would reduce the effectiveness of ion–ion interactions and consequently the size of $Ba^{2+}$ current. The conformational change induced by the G protein is apparently less critical when either $Na^+$ or high $[Ba^{2+}]$ are involved. In the first case, $Na^+$ or $Cs^+$ ions are expected to be loosely bound to the negative sites of the pore and ion–ion interaction is less affected by the displaced charges. In the second case, both sites would be occupied by $Ba^{2+}$ ions and the ligand displacement induced by the G protein would result less determinant. $Ba^{2+}$ fluxes in 150 mM $Ba^{2+}$ would be less depressed by the G protein in this condition.

The results and interpretation of Kuo and Bean are very attractive, but contrast with the evidence of no marked effects of G proteins on $Ba^{2+}$ (or $Ca^{2+}$) ion permeation. Single-channel experiments in 100 mM $Ba^{2+}$ show no changes on single-channel conductance in the presence or absence of neurotransmitter, despite a significant depression of macroscopic currents (Delcour and Tsien 1993). The same is true when comparing single-channel current amplitude before and after facilitating prepulses in the presence of saturating doses of neurotransmitter (Carabelli *et al.* 1996). In addition, in IMR32 cells we find that $Ba^{2+}$ current depression induced by 20 μM NA in 100 mM $Ba^{2+}$ is only 10–15% smaller than that in 5 mM $Ba^{2+}$, while NA inhibition of $Na^+$ currents through $Ca^{2+}$ channels is far more attenuated (V. Carabelli, M. Lovallo, and E. Carbone, unpublished results). Thus, it seems that for G protein–$Ca^{2+}$ channel interaction, it is more relevant whether the channel conducts $Na^+$ or $Ba^{2+}$ rather than how many $Ba^{2+}$ ions flow at low or high $[Ba^{2+}]$. How this can occur remains to be studied. There are several alternatives that can be tested before drawing any conclusion. One is that $Na^+$ and $Ba^{2+}$ permeation through open channels may be regulated by an intramembrane binding site whose energy profile depends more critically on the type of ion flowing (Lux *et al.* 1990) than the energy profile of the two binding sites located at the outer site of the pore (Almers and McCleskey 1984; Kuo and Hess 1993). A differential G protein-induced conformational change of the intramembrane binding site for $Na^+$ and $Ba^{2+}$ would also explain the reduced inhibition of $Na^+$ currents by neurotransmitter. Alternatively, it might be that the weaker inhibition of $Na^+$ current by G proteins derives from an increased gating activity of the channel in low $Ca^{2+}$ solutions. If the G protein–$Ca^{2+}$ channel interaction depends on the state of the channel (open or closed), a possible increased flickering of the channel with $Na^+$ as the main monovalent cation will facilitate the unbinding of the G protein from the more frequently open channel, with consequent relieved inhibition. These and other possibilities need to be tested at the single-channel level with different $Na^+$ and $Ba^{2+}$ concentrations (10 to 100 mM) that allow accurate estimates of $Ca^{2+}$ channel permeability.

## G Proteins, intracellular modulators and voltage-dependent facilitation

Given the well-established role of G proteins in controlling the voltage-dependent inhibition of high-threshold $Ca^{2+}$ channels by activation of neurotransmitter receptors, a related question is whether other intracellular messengers are involved in the G protein–$Ca^{2+}$ channel coupling. Early works on the action of several second messengers have shown that neither cAMP, nor cGMP or IP3 are involved in HVA channel modulation (Hille 1994). To date, only protein kinase C (PKC) seems to be the only effective intracellular modulator able to interfere with neuronal $Ca^{2+}$ channels (but see Surmeier *et al.* 1995). Early works on the PKC action on HVA channels have shown that PKC activation may cause either up- or down-regulation of $Ca^{2+}$ currents, without any significant specificity for a channel subtype (Anwyl 1991). In spite of this, recent studies on the N-type channel of rat central and peripheral neurones have proved clearly that PKC activation effectively up-regulates the N-type current by removing the voltage-dependent inhibition induced by endogenous GTP or by neurotransmitter receptor activation (Swartz 1993, Zhu and Ikeda, 1994*a*). This, however, seems not to be the general rule, since up-regulation of N- and L-type currents in frog sympathetic neurones occurs independently of HVA channel inhibition by neurotransmitters (Yang and Tsien 1993) and removal of inhibition by PKC activation is not followed by current increases in chick sympathetic neurones (Golard *et al.* 1993). Nevertheless, there seems at least to be a converging view that PKC interferes with the voltage-dependent machinery of $Ca^{2+}$ channel down-modulation. Whether PKC interferes with the receptor–G protein coupling (Golard *et al.* 1993) or with the G protein–$Ca^{2+}$ channel interaction (Swartz 1993; Zhu and Ikeda 1994*a*) may depend on the cell type, neurotransmitter receptor and G protein subtype. The same variability is expected for the PKC regulation of the voltage-independent $Ca^{2+}$ channel inhibition (or facilitation) by neurotransmitters, that may play as well a critical role in the control of cell functions.

There remains, however, the physiological significance that these findings anticipate. Removal of $Ca^{2+}$ channel inhibition induced by released neurotransmitter is an effective mechanism by which presynaptic terminals may enhance their efficiency. Whether this occurs through an increased nerve activity (voltage-dependent facilitation) or through the activation of PKC seems irrelevant to the final result, but underlines the possibility that multiple regulatory mechanisms with different genesis and time courses may variably lead to the common goal of increased synaptic efficacy.

An issue of interest concerns the role that cell dialysis may play in the maintenance of the voltage-dependent kinetic slowing of $Ca^{2+}$ channels by neurotransmitters in whole-cell clamped neurones. Somatostatin-induced inhibition of $Ca^{2+}$ currents is reported to be voltage-dependent, with little kinetic slowing when tested on a ciliary ganglion neurone in perforated-patch conditions (Meriney *et al.* 1994). Somatostatin action is increased and the kinetic slowing is more prominent when $Ca^{2+}$ currents are recorded from whole-cell clamped neurones. The interpretation of Meriney *et al.* is that kinetic slowing and voltage dependency are unrelated phenomena and that kinetic slowing, but not the voltage-dependence, depends on the wash-out of some intracellular factor (a cGMP-dependent protein kinase) that prevents $Ca^{2+}$ channel inhibition in intact cells. In contrast to this, however, $Ca^{2+}$ current recordings in perforated patches show strong voltage-dependence and a significant slow down in the presence of external neurotransmitters (Sher *et al.* 1996; Luebke and Dunlap 1994). Also, single-channel recordings in cell-attached conditions display clear kinetic slowing of channel activation and a significant voltage-dependent recruitment of inhibited channels by conditioning prepulses (Carabelli *et al.* 1996), suggesting that cell dialysis may not be as critical as indicated. In our view, kinetic slowing depends critically on the time course of channel inactivation, that is faster and more complete in perforated-patch than in whole-cell conditions, due to the absence of strong intracellular $Ca^{2+}$ buffers. Fast inactivation may easily distort the slowing down of channel activation that would appear as a steady-state inhibition during neurotransmitter application. Under these conditions it would be wise to normalize the time course of modified currents to the time course of control currents (Kasai 1992; Elmslie and Jones 1994). What remains unresolved, in all cases, is the mutual interference of activation and inactivation gating with $Ca^{2+}$ channel modulators and the possible back action of modulators on the coupling of activation–inactivation gatings.

Finally, it should be mentioned that parallel to the fast modulation of $Ca^{2+}$ channel gatings, other forms of presynaptic facilitation may cause net increases of cytoplasmic $Ca^{2+}$ by recruiting functional $Ca^{2+}$ channels at the plasmalemma. Several neurones and neuroendocrine cells are shown to possess an intracellular pool of N-type $Ca^{2+}$ channels available to recruitment at the cell surface by membrane depolarization and PKC activation (Passafaro *et al.* 1994). The same two kinds of stimuli facilitating $Ca^{2+}$ channel gatings within seconds, are thus able to stimulate N-type $Ca^{2+}$ channels translocation to the cell surface on a slower time scale (minutes), further proving the pathways complexity through which $Ca^{2+}$ channel modulation may occur.

## Conclusions

Much has been achieved in the past few years in the biophysics and physiology of $Ca^{2+}$ channels. Many groups have contributed to the identification of new mammalian $Ca^{2+}$ channels in neurones and neurosecretory cells, and to a better understanding of $Ca^{2+}$ channel gating modulation. Other groups have provided a detailed molecular description of $Ca^{2+}$ channel subunits and a more precise spatial localization of $Ca^{2+}$ channels in different areas of the central nervous system. Much, however, remains to be done. Little is known, for instance, on the role that colocalized $Ca^{2+}$ channel subtypes exert in the control of cell exocytosis, on the $Ca^{2+}$-dependent modulatory responses of some central neuronal networks and on the relationship between molecular structure and $Ca^{2+}$ channel gatings. However, given the rapid progress in the field it is predictable that some of these goals will soon be attained.

## Acknowledgements

We wish to thank our close collaborators: Drs E. Sher, M. Lovallo, G. Aicardi, A. Garcia, L. Gandia and A. Artalejo for the many stimulating discussions that contributed to our present understanding on $Ca^{2+}$ channels.

The work was supported by the European Neuroscience Programme (grant No. 125, Italy–Spain) and by NATO (grant No. CRG 870576 to E.C.).

## References

Albillos, A., Artalejo, A.R., López, M.G., Gandía, L., García, A.G., and Carbone, E. (1994*a*). Calcium channel subtypes in cat chromaffin cells. *Journal of Physiology*, **477**, 197–213.

Albillos, A., Garcia, A.G., and Gandia, L. (1994*b*). ω-Agatoxin-IVA-sensitive calcium channels in bovine chromaffin cells. *FEBS Letters*, **336**, 259–262.

Albillos, A., Carbone, E., Gandía, L., García, A.G., and Pollo, A. (1996). Opioiol inhibition of $Ca^{2+}$ channel subtypes in bovine chromaffin cells: selectivity of action and voltage-dependence. *European Journal of Neuroscience*, **8**, 1561–1570.

Almers, W. and McCleskey, E.W. (1984). Non-selective conductance in calcium channels of frog muscle: calcium selectivity in a single-file pore. *Journal of Physiology*, **353**, 585–608.

Amico, C., Marchetti, C., Nobile, M., and Usai, C. (1995). Pharmacological types of calcium channels and their modulation by baclofen in cerebellar granules. *Journal of Neuroscience*, **15**, 2839–2848.

Anwyl, R. (1991). Modulation of vertebrate neuronal calcium channels by transmitters. *Brain Research Review*, **16**, 265–281.

Artalejo, C.M., Adams, M.E., and Fox, A.P. (1994). Three types of $Ca^{2+}$ channel trigger secretion with different efficacy in chromaffin cells. *Nature*, **367**, 72–76.

Bean, B.P. (1989). Classes of calcium channels in cardiac muscle, vascular muscle, and neurons. *Annu. Rev. Physiol.*, **51**, 367–384.

Bley, K.R. and Tsien, R.W. (1990). Inhibition of $Ca^{2+}$ and $K^{+}$ channels in sympathetic neurons by neuropeptides and other ganglionic transmitters. *Neuron*, **4**, 379–391.

Boland, L. and Bean, B.P. (1993). Modulation of N-type calcium channels in bullfrog sympathetic neurons by luteinizing hormone-releasing hormone: kinetics and voltage-dependence. *Journal of Neuroscience*, **13**, 516–533.

Carabelli, V., Lovallo, M., Magnelli, V., Zucker, H., and Carbone, E. (1996). Voltage-dependent modulation of single N-type $Ca^{2+}$ channel kinetics by receptor agonists in IMR32 cellls. *Bioplynical Journal*, **70**, 2744–2154.

Carbone, E. and Lux, H.D. (1987). Single low-voltage-activated calcium channels in chick and rat sensory neurons. *Journal of Physiology*, **386**, 571–601.

Carbone, E. and Swandulla, D. (1989). Neuronal calcium channels: kinetics, blockade and modulation. *Prog. Biophys. Mol. Biol.*, **54**, 31–58.

Carbone, E., Sher, E., and Clementi, F. (1990). Ca currents in human neuroblastoma IMR32 cells: kinetics, permeability and pharmacology. *Pflügers Archiv*, **416**, 170–179.

Codignola, A., Tarroni, P., Clementi, F., Pollo, A., Lovallo, M., Carbone, E., and Sher, E. (1993). Calcium channel subtypes controlling serotonin release from human small cell lung carcinoma cell lines. *Journal of Biological Chemistry*, **268**, 26240–26247.

Cox, D. and Dunlap, K. (1992). Pharmacological discrimination of N-type from L-type calcium current and its selective modulation by transmitters. *Journal of Neuroscience*, **12**, 906–914.

Delcour, A.H. and Tsien, R.W. (1993). Altered prevalence of gating modes in neurotransmitter inhibition of N-type calcium channels. *Science*, **259**, 980–984.

Delcour, A.H., Lipscombe, D., and Tsien, R.W. (1993). Multiple gating modes of N-type $Ca^{2+}$ channel activity distinguished by differences in gating kinetics. *Journal of Neuroscience*, **13**, 181–194.

Diversé-Pierluissi, M., Goldsmith, P.K., and Dunlap, K. (1995). Transmitter-mediated inhibition of N-type calcium channels in sensory neurons involves multiple GTP-binding proteins and subunits. *Neuron*, **14**, 191–200.

Dolphin, A.C. (1991). Regulation of calcium channel activity by GTP binding protein and second messengers. *Biochimica et Biophysica Acta*, **1091**, 68–80.

Dolphin, A.C. and Scott, R.H. (1987). Calcium channel currents and their inhibition by (-)-baclofen in rat sensory neurones. *Journal of Physiology*, **386**, 1–17.

Dunlap, K. and Fischbach, G.D. (1978). Neurotransmitters decrease the calcium component of sensory neurone action potentials. *Nature*, **276**, 837–839.

Eckert, R. and Chad, J.E. (1984). Inactivation of Ca channels. *Prog. Biophys. Mol. Biol.*, **44**, 215–267.

Ellinor, P.T., Zhang, J.-F., Randall, A.D., Zhou, M., Schwarz, T.L., Tsien, R.W., and Horne, W.A. (1993). Functional expression of a rapidly inactivating neuronal calcium channel. *Nature*, **363**, 455–458.

Elmslie, K.S. and Jones, S.W. (1994). Concentration dependence of neurotransmitter effects on calcium current kinetics in frog sympathetic neurones. *Journal of Physiology*, **481**, 35–46.

Elmslie, K.S., Zhou, W., and Jones, S.W. (1990). LHRH and GTP-$\gamma$-S modify calcium current activation in bullfrog sympathetic neurons. *Neuron*, **5**, 75–80.

Elmslie, K.S., Kammermeier, P.J., and Jones, S.W. (1994). Reevaluation of $Ca^{2+}$ channel types and their modulation in bullfrog sympathetic neurons. *Neuron*, **13**, 217–228.

Fenwick, E.M., Marty, A., and Neher, E. (1982). Sodium and calcium channels in bovine chromaffin cells. *Journal of Physiology*, **331**, 599–635.

Formenti, A., Arrigoni, E., and Mancia, M. (1993). Two distinct modulatory effects on calcium channels in adult rat sensory neurons. *Biophysical Journal*, **64**, 1029–1037.

Forti, L., Tottene, A., Moretti, A., and Pietrobon, D. (1994). Three types of voltage-dependent calcium channels in rat cerebellar neurons. *Journal of Neuroscience*, **14**, 5243–5256.

Gandia, L., Borges, R., Albillos, A., and Garcia, A. (1995). Multiple calcium channel subtypes in isolated rat chromaffin cells. *Pflügers Archiv*, **430**, 55–63.

Golard, A. and Siegelbaum, S.A. (1993). Kinetic basis for the voltage-dependent inhibition of N-type calcium current by somatostatin and norepinephrine in chick sympathetic neurons. *Journal of Neuroscience*, **13**, 3884–3894.

Golard, A., Role, L.W., and Siegelbaum, S.A. (1993). Protein kinase C blocks somatostatin-induced modulation of calcium current in chick sympathetic neurons. *Journal of Neurophysiology*, **70**, 1639–1643.

Grassi, F. and Lux, H.D. (1989). Voltage-dependent GABA-induced modulation of calcium currents in chick sensory neurons. *Neuroscience Letters*, **105**, 113–119.

Gutnick, M.J., Lux, H.D., Swandulla, D., and Zucker, H. (1989). Voltage-dependent and calcium-dependent inactivation of calcium channel current in identified snail neurones. *Journal of Physiology*, **412**, 197–220.

Hamill, O.P., Marty, A., Neher, E., Sakmann, B., and Sigworth, F.J. (1981). Improved patch-clamp techniques for high-resolution current recording from cells and cell-free membrane patches. *Pflügers Archiv*, **391**, 85–100.

Hess, P., Lansman, J.B., and Tsien, R.W. (1984). Different modes of Ca channel gating behaviour favoured by dihydropyridine $Ca^{2+}$ agonists and antagonists. *Nature*, **311**, 538–544.

Hille, B. (1994). Modulation of ion-channel function by G-protein-coupled receptors. *Trends in Neurosciences*, **17**, 531–532.

Hillyard, D.R., Monje, V.D., Mintz, I.M., Bean, B.P., Nadasdi, L., Ramachandran, J., Miljanich, G., Azimi-Zoonooz, A., McIntosh, J.M., Cruz, L.J., Imperial, J.S., and Oliveira, B.M. (1992). A new conus peptide ligand for mammalian presynaptic $Ca^{2+}$ channels. *Neuron*, **9**, 69–77.

Hirning, L.D., Fox, A.P., McCleskey, E.W., Olivera, B.M., Thayer, S.A., Miller, R.J., and Tsien, R.W. (1988). Dominant role of N-type $Ca^{2+}$ channels in evoked release of norepinephrine from sympathetic neurons. *Science*, **239**, 57–61.

Ikeda, S.R. (1996). Voltage-dependent modulation of N-type calcium channels by G-proteins $\beta\gamma$ sub units. *Nature*, **380**, 255–258.

Kasai, H. (1992). Voltage- and time-dependent inhibition of neuronal calcium channels by a GTP-binding protein in a mammalian cell line. *Journal of Physiology*, **448**, 189–209.

Kasai, H. and Neher, E. (1992). Dihydropyridine-sensitive and $\omega$-conotoxin-sensitive calcium channels in a mammalian neuroblastoma-glioma cell line. *Journal of Physiology*, **448**, 161–188.

Kuo, C.-C. and Bean, B.P. (1993). G-protein modulation of ion permeation through N-type calcium channels. *Nature*, **365**, 258–262.

Kuo, C.-C. and Hess, P. (1993). Characterization of the high-affinity $Ca^{2+}$ binding sites in the L-type $Ca^{2+}$ channel pore in rat phaeochromocytoma cells. *Journal of Physiology*, **466**, 657–682.

Llinas, R., Sugimori, M., Hillman, D.E., and Cherksey, B. (1992). Distribution and functional significance of the P-type, voltage-dependent $Ca^{2+}$ channels in the mammalian central nervous system. *Trends in Neurosciences*, **15**, 351–356.

Lopez, H.S. and Brown, A.M. (1991). Correlation between G protein activation and reblocking kinetics of $Ca^{2+}$ channel current in rat sensory neurons. *Neuron*, **7**, 1061–1068.

López, M.G., Villaroya, M., Lara, B., Martinez-Sierra, R., Albillos, A., García, A. G., and Gandia, L. (1994). Q- and L-type $Ca^{2+}$ channels dominate the control of secretion in bovine chromaffin cells. *FEBS Letters*, **349**, 331–334.

Luebke, J.I. and Dunlap, K. (1994). Sensory neuron N-type calcium currents are inhibited by both voltage-dependent and -independent mechanisms. *Pflügers Archiv*, **365**, 258–262.

Lux, H.D., Carbone E., and Zucker, H. (1990). $Na^+$ currents through low-voltage-activated $Ca^{2+}$ channels of chick sensory neurones: block by external $Ca^{2+}$ and $Mg^{2+}$. *Journal of Physiology*, **430**, 159–188.

Magnelli, V., Pollo, A., Sher, E., and Carbone, E. (1995). Block of non-L, non-N-type $Ca^{2+}$ channels in rat insulinoma RINm5F cells by $\omega$-agatoxin IVA and $\omega$-conotoxin MVIIC. *Pflügers Archiv*, **429**, 762–771.

Marchetti, C., Carbone, E., and Lux, H.D. (1986). Effects of dopamine and noradrenaline on Ca channels of cultured sensory and sympathetic neurons of chick. *Pflügers Archiv*, **406**, 104–111.

Marchetti, C., Amico, C., Podestà, D., and Robello, M. (1994) Inactivation of voltage-dependent calcium current in an insulinoma cell line. *European Biophysics Journal*, **23**, 52–58.

Meriney, S.D., Gray, D.B., and Pilar, G.R. (1994). Somatostatin-induced inhibition of neuronal $Ca^{2+}$ current modulated by cGMP-dependent protein kinase. *Nature*, **369**, 336–339.

Mintz, I.M., Adams, M.E., and Bean, B.P. (1992). P-type calcium channels in rat central and peripheral neurons. *Neuron*, **9**, 85–95.

Passafaro, M., Clementi, F., Pollo, A., Carbone, E., and Sher, E. (1994). $\omega$-Conotoxin and $Cd^{2+}$ stimulate the recruitment to the plasmamembrane of an intracellular pool of voltage-operated $Ca^{2+}$ channel. *Neuron*, **12**, 317–326.

Plant, T.D. (1988). Properties and calcium-dependent inactivation of calcium currents in cultured mouse pancreatic $\beta$-cells. *Journal of Physiology*, **404**, 731–747.

Pollo, A., Lovallo, M., Sher, E., and Carbone, E. (1992). Voltage-dependent noradrenergic modulation of $\omega$-conotoxin-sensitive $Ca^{2+}$ channels in human neuroblastoma IMR32 cells. *Pflügers Archiv*, **422**, 75–83.

Pollo, A., Lovallo, M., Biancardi, E., Sher, E., Socci, C., and Carbone, E. (1993). Sensitivity to dihydropyridines, $\omega$-conotoxin and noradrenaline reveals multiple high-voltage activated $Ca^{2+}$ channels in rat insulinoma and human pancreatic $\beta$-cells. *Pflügers Archiv*, **423**, 462–471.

Randall, A.D., Wendland, B., Schweizer, F., Miljanich, G., Adams, M.E., and Tsien, R.W. (1993). Five pharmacologically distinct high-voltage-activated $Ca^{2+}$ channels in cerebellar granule cells. *Society for Neuroscience Abstracts*, **19**, 1478.

Regan, L.J., Sah, D.W.Y., and Bean, B.P. (1991). $Ca^{2+}$ channels in rat central and peripheral neurons: high threshold current resistant to dihydropyridine blockers and $\omega$-conotoxin. *Neuron*, **6**, 269–280.

Rittenhouse, A.R. and Hess, P. (1994). Microscopic heterogeneity in unitary N-type calcium currents in rat sympathetic neurons. *Journal of Physiology*, **474**, 87–99.

Sather, W.A., Tanabe, T., Zhang, J.–F., Mori, Y., Adams, M.E., and Tsien, R.W. (1993). Distinctive biophysical and pharmacological properties of class A (B1) calcium channel $\alpha_1$ subunits. *Neuron*, **11**, 291–303.

Satin, L.S. and Smolen, P.D. (1994). Electrical bursting in β-cells of the pancreatic islets of Langerhans. *Endocrine*, **2**, 677–687.

Sher, E. and Clementi, F. (1991) $\omega$-Conotoxin-sensitive voltage-operated calcium channels in vertebrate cells. *Neuroscience*, **42**, 301–307.

Sher, E., Cesare, P., Codignola, A., Clementi, F., Tarroni, P., Pollo, A., Magnelli, V. and Carbone, E. (1996). Activation of $\delta$-opioda receptors inhibits neuronal-like calcium channels and dirtal steps of $Ca^{2+}$-dependent secretion in human small-cell lung carcinoma cells. *Journal of Neuroscience*, **16**, 3672–3684.

Soong, T.W., Stea, A., Hodson, C.D., Dubel, S.J., Vincent, S.R., and Snutch, T.P. (1993). Structure and functional expression of a member of the low voltage-activated calcium channel family. *Science*, **260**, 1133–1136.

Surmeier, D.J., Bargas, J., Hemmings, H.C. Jr., Nairn, A.C., and Greengard, P. (1995). Modulation of calcium currents by a $D_1$ dopaminergic protein kinase/phosphatase cascade in rat neostriatal neurons. *Neuron*, **14**, 385–397.

Swandulla, D. and Armstrong, C.M. (1988). Fast deactivating calcium channels in chick sensory neurons. *Journal of General Physiology*, **92**, 197–218.

Swandulla, D., Carbone, E., and Lux, H.D. (1991). Do calcium channel classifications account for neuronal calcium channel diversity? *Trends in Neurosciences*, **14**, 46–51.

Swartz, K.J. (1993). Modulation of $Ca^{2+}$ channels by protein kinase C in rat central and peripheral neurons: disruption of G protein-mediated inhibition. *Neuron*, **11**, 1–20.

Tsien, R.W., Ellinor, P.T., and Horne, W.A. (1991). Molecular diversity of voltage-dependent $Ca^{2+}$ channels. *Trends in Pharmacological Science*, **12**, 349–354.

Wanke, E., Ferroni, A., Malgaroli, A., Ambrosini, A., Pozzan, T., and Meldolesi, J. (1987). Activation of muscarinic receptor selectively inhibits a rapidly inactivated $Ca^{2+}$ current in rat sympathetic neurons. *Proc. Natl. Acad. Sci USA*, **84**, 4313–4317.

Wheeler, D.B., Randall A., and Tsien, R.W. (1994). Roles of N-type and Q-type $Ca^{2+}$ channels in supporting hippocampal synaptic transmission. *Science*, **264**, 107–111.

Yang, J. and Tsien, R.W. (1993). Enhancement of N- and L-type calcium channel currents by protein kinase C in frog sympathetic neurons. *Neuron*, **10**, 127–136.

Zhang, J.–F., Randall, A.D., Ellinor, P.T., Horne, W.A., Sather, W.A, Tanabe, T., Schwarz, T.L., and Tsien, R.W. (1993). Distinctive pharmacology and kinetics of cloned neuronal

$Ca^{2+}$ channels and their possible counterparts in mammalian CNS neurons. *Neuro-pharmacology*, **32**, 1075–1088.

Zhu, Y. and Ikeda, S.R. (1994*a*). Modulation of $Ca^{2+}$-channel currents by protein kinase C in adult rat sympathetic neurons. *Journal of Neurophysiology*, **72**, 1549–1560.

Zhu, Y. and Ikeda, S.R. (1994*b*). VIP inhibits N-type $Ca^{2+}$ channels of sympathetic neurons via a pertussis toxin-insensitive but cholera toxin-sensitive pathway. *Neuron*, **13**, 657–669.

# 5 Pharmacology of voltage-dependent calcium channels

B. P. Bean, I. M. Mintz, L. M. Boland, D. W.Y. Sah, and J. A. Morrill

## Introduction

Voltage-dependent calcium channels are present in the cell membranes of many cells and serve a variety of functions. In a few cell types, calcium channels play an important role in electrical activity. In many cells, though, the most important role of calcium channels is to transduce the electrical signal of an action potential into the intracellular chemical signal of an increase in calcium (Hille 1992). The increase in calcium resulting from calcium entry through calcium channels mediates many cellular functions that are triggered by depolarization, from contraction of smooth and cardiac muscle to synaptic transmission between neurones. In the past decade, it has become clear that there are multiple types of voltage-dependent channels in mammalian cells. All share the fundamental property of being closed at normal resting potentials and being opened by depolarization of the cell, but they differ in details of voltage dependence, kinetics, and single-channel properties. Different types of calcium channels also differ in their pharmacology, and selective drugs and toxins have become valuable for recognizing distinct classes of channels. Many neurones, in particular, possess multiple types of calcium channels, and pharmacological agents have been especially useful in distinguishing between different types of calcium channels in neurones (reviewed by Bean 1989; Hess 1990; Tsien *et al.* 1991; Miller 1992).

Recently, distinctions between classes of calcium channels have also been made based on molecular cloning (reviewed by Tsien *et al.* 1991; Miller 1992; Snutch and Reiner 1992). So far, there is only partial overlap between classes of calcium channel identified by molecular biology and those characterized in native cells. Pharmacological profiles have been crucial in establishing links between cloned channels and native channels.

Many calcium channel blocking drugs are useful as therapeutic agents (Table 5.1). The existence of multiple types of calcium channels offers opportunities for the development of new pharmaceutical agents. By exploiting differences between the pharmacology of calcium channels in different cell types, it may be possible to develop new drugs with more selective effects.

**Table 5.1** Calcium channel pharmacological agents and details of their suppliers.

| Substance | Channel types | Effect | Reference |
|---|---|---|---|
| ω-Conotoxin GVIA | N-type | Block | *Pflugers. Arch.*, **414**, 150 (1989) |
| ω-Conotoxin MVIIC | N-type, P-type, others | Block | *Neuron*, **9**, 69 (1992) |
| ω-Aga-IIIA | N-type, L-type, P-type | Block | *Proc. Natl Acad. Sci. USA*, **88**, 6628 (1991) |
| ω-Aga-IVA | P-type | Block | *Neuron*, **9**, 1 (1992) |
| ω-Aga-IVB | P-type | Block | *Mol. Pharmacol.*, **44**, 681 (1994) |
| FTX | P-type | Block | *Proc. Natl Acad. Sci. USA*, **86**, 1689 (1989) |
| Fluspirilene | | N-type, P-type | *Mol. Pharmacol.*, **45**, 84 (1994) |
| | | T-type, L-type | *Mol. Pharmacol.*, **42**, 34 (1992) |
| Pimozide | | N-type, P-type | *Mol. Pharmacol.*, **45**, 84 (1994) |
| | | T-type | *J. Neurophysiol.*, **68**, 213 (1992) |
| | | L-type | *Mol. Pharmacol.*, **37**, 752 (1990) |
| Thioridazine | | N-type, P-type | *Mol. Pharmacol.*, **45**, 84 (1994) |
| Flunarizine | | T-type | *J. Neurophysiol.*, **68**, 213 (1992) |
| FPL 64176 | L | Enhancement | *J. Pharm. Exp. Ther.*, **259**, 982 (1991) |
| Bay K 8644 | L | Enhancement | *Nature*, **311**, 538 (1984) |
| Amiloride | T | Block | *Science*, **240**, 213 (1988) |
| Phenytoin | T | Block | *Science* **235**, 680 (1987) |
| Ethosuximide | T | Block | *Ann. Neurol.*, **25**, 582 (1989) |
| Methoxuflurane | T-type | Block | *J. Neurophysiol.*, **68**, 213 (1992) |

*Suppliers details*

| | |
|---|---|
| ω-Conotoxin GVIA | Research Biochemicals International<br>Customer Service Department, One Strathmore Road, Natick, MA 01760-2418, USA<br>Phone: 508-651-8151 Fax: 508-655-1359 |
| | Peninsula Laboratories Inc.<br>611 Taylor Way, Belmont, CA 94002-4041, USA<br>Phone: 415-592-5392 Fax: 415-595-4071 |
| | BACHEM California<br>3132 Kashiwa Street, Torrance, CA 90505, USA<br>Phone: 310-539-4171 Fax: 310-530-1571 |
| | Peptides International, Inc.<br>P.O. Box 24658, Louisville, KY 40224<br>Phone: 502-266-8787 Fax: 502-267-1329 |
| ω-Conotoxin MVIIC | BACHEM, Peptides International |
| ω-Conotoxin MVIIA | BACHEM |
| ω-Aga-IVA | Pfizer Inc. Groton, CT |
| Fluspirilene | Research Biochemicals Incorporated |
| Pimozide | Research Biochemicals Incorporated |
| Thioridazine | Research Biochemicals Incorporated |
| Flunarizine | Research Biochemicals Incorporated |
| Bay K 8644 | Research Biochemicals Incorporated |
| Amiloride | Research Biochemicals Incorporated |
| Phenytoin | Research Biochemicals Incorporated |
| Ethosuximide | Research Biochemicals Incorporated |

This article reviews the pharmacology of voltage-dependent calcium channels in mammalian cells. The emphasis is on drugs and toxins that have been useful in making distinctions between different kinds of calcium channels, especially those in neurones. We will consider in turn each of the classes of calcium channels that are presently distinguished one from another.

## T-type calcium channels

T-type calcium channels have properties of voltage dependence and kinetics that clearly distinguish them from the other classes of well-characterized calcium channels (with the possible exception of the recently recognized 'R-type'). T-type calcium channels have four distinguishing characteristics that led to their recognition as a distinct class of channels (Carbone and Lux 1984, 1987; Fox *et al.* 1987*a*,*b*; see Bean 1989, and Hess 1990 for reviews). First, they can be activated by small depolarizations from the resting potential, so they are also known as low voltage-activated or low-threshold calcium channels. Second, they inactivate more rapidly (tens of milliseconds) than other types of calcium channel. Third, they are more susceptible to being inactivated by steady depolarization of the resting potential. Often, for example, T-type calcium channels are entirely inactivated by steady resting potentials positive to −60 mV or so. At the single-channel level, T-type calcium channels are distinguished by having a smaller single-channel conductance than other calcium channel types.

Compared with other types of calcium channels, T-type calcium channels have a restricted distribution. They are prominent in many cells that are capable of spontaneous pacemaking, including cells in the sinoatrial node of the heart (Hagiwara *et al.* 1988), and thalamic neurones in the brain (Coulter *et al.* 1989*a*; Huguenard and Prince 1992).

Potent, selective blockers for T-type calcium channels do not yet exist. A number of drugs have some selectivity for blocking T-type calcium channels over other types (Herrington and Lingle 1992), but none is very potent and none is highly selective. Amiloride is able to inhibit T-type calcium channels in cardiac muscle and in neurones and shows some selectivity for blocking T-type channels compared with other types (Tang *et al.* 1988). However, amiloride is not very potent, since it takes about 1 mM concentration to inhibit T-type current almost completely. It also affects many non-calcium channels, being first recognized as a blocker of epithelial sodium channels. Still, it currently may be the most useful tool for separating components of T-type current from those carried by other channels. It will be interesting to see if it is possible to make derivatives of amiloride with more potent and selective actions on T-type channels.

Petit mal anticonvulsants such as ethosuximide are capable of blocking T-type calcium channels (Coulter *et al.* 1989*b*). These partially inhibit T-type calcium current in thalamic relay neurones while having less effect on other channel types in the same neurones. Interestingly, ethosuximide has very little effect on T-type current in neuroendocrine $GH_3$ cells (Herrington and Lingle 1992), suggesting that the drug may have selectivity for different subtypes of T-type calcium channels.

## L-type calcium channels

L-type calcium channels are both the most widespread and the best understood class of calcium channels. They produce most of the overall calcium current in skeletal, cardiac, and smooth muscle cells and a significant fraction of the current in most neurones and neuroendocrine cells. L-type channels should be regarded as a class or family of channels rather than a single type, since there are obvious differences in the kinetics and voltage dependence of the L-type current in different cell types. For example, L-type current in cardiac cells inactivates much more rapidly than that in neurones. The L-type calcium current in skeletal muscle activates very slowly (time constant of 10s to 100s of milliseconds) compared with that in other cells (time constant of several millisecond). Nevertheless, even before the structure of the channels was known, it was reasonable to regard them as members of the same family based on two common properties: (i) sensitivity to dihydropyridine drugs; and (ii) a large single-channel conductance (Fox *et al.* 1987*a*,*b*). Since the early definition of L-type calcium channels, it has been appreciated that some other calcium channel types can have single-channel conductances nearly as large as that of L-type channels (Plummer *et al.* 1989), so that sensitivity to dihydropyridine drugs has become the principal distinguishing characteristic of L-type channels. The view that there are distinct types of channels within the general class of L-type channels has been supported by recent results from molecular biology studies (Snutch and Reiner 1992).

Dihydropyridine (DHP) calcium channel blockers were developed for clinical use against hypertension and angina pectoris. The most common DHP blocking drugs are nifedipine, nitrendipine, and nimodipine. The clinical effects of the drugs are most likely primarily due to inhibition of calcium channels in the smooth muscle of blood vessels, causing vasorelaxation. However, the drugs are potent inhibitors of L-type calcium channels in cardiac muscle, skeletal muscle, and neurones, as well as those in smooth muscle. Figure 5.1 shows inhibition of calcium channel current in a myocardial cell from the right atrium of a rabbit heart. At a concentration of 1 IuM, nimodipine completely inhibited the calcium channel current elicited by a depolarization of the membrane from −70 mV to −20 mV. At similar concentrations, nimodipine has almost no effect on P-type calcium channels in rat cerebellar Purkinje neurones or N-type calcium channels in rat sympathetic neurones (Mintz *et al.* 1992*a*; I.M. Mintz and B.P. Bean, unpublished results), suggesting a high degree of specificity for L-type channels over other channel types.

Dihydropyridine drugs block L-type calcium channel current more potently when the current is elicited from depolarized holding potentials (where the channels are partially inactivated) than from hyperpolarized holding potentials (Bean 1984; Sanguinetti and Kass 1984; Uehara and Hume 1985). This may indicate that the drugs bind with very high affinity ($K_d$ < 1 nM) to the inactivated state of the channel, but more weakly ($K_d$10 nM to 1 uM) to the resting state of the channel (Bean 1984).

The voltage dependence of dihydropyridine block is probably important for their clinical usefulness. At nanomolar concentrations of drug, many cells with normal, negative resting potentials would be little affected. However, calcium entry in cells that are depolarized (perhaps due to anoxia or damage) would be more potently blocked.

There are two other classes of clinically used calcium channel blockers that are effective

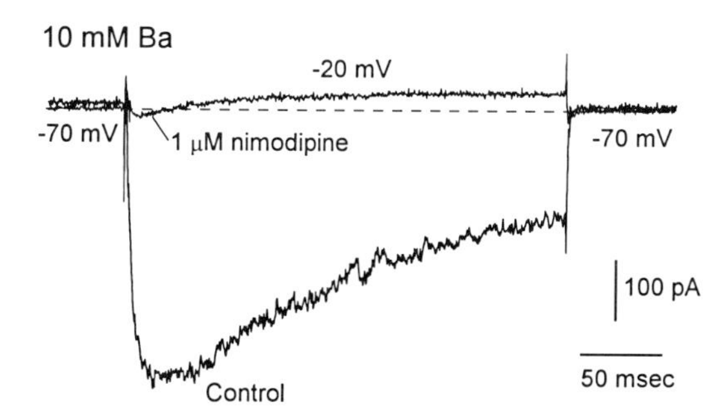


**Fig. 5.1** Nimodipine block of L-type current in a rat atrial myocyte. Internal solution: 115 mM CsCl, 5 mM MgATP, 0.3 mM GTP (Tris salt), 5 mM creatine phosphate (diTris salt), 10 mM BAPTA, 10 mM HEPES, pH 7.4 with CsOH. External solution: 10 mM $BaCl_2$, 160 mM tetraethylammonium (TEA) Cl, 10 mM HEPES, 3 μM tetrodotoxin, pH adjusted to 7.4 with TEA OH. (Adapted with permission from Bean 1991.)

blockers of L-type channels: phenylalkylamines, including verapamil; and benzothiazepines, including diltiazem. Both bind to high-affinity binding sites on L-type calcium channels, and these binding sites are distinct from high-affinity dihydropyridine binding sites. There have been relatively few studies with verapamil and diltiazem on other types of calcium channels, so the degree of specificity for L-type channels is not yet well established.

Some dihydropyridine molecules enhance, rather than inhibit, current through L-type calcium channels. The best studied of these 'calcium channel agonists' is Bay K 8644, a mixture of two optical isomers, and another dihydropyridine molecule, (+)-(S)-202–791. The dihydropyridine agonists increase the probability of channel opening by inducing or stabilizing unusually long-lasting openings of the channels (Hess *et al.* 1984; Nowycky *et al.* 1985), and they are often used as tools to obtain enhanced single-channel activity.

## N-type calcium channels

In contrast to L-type channels, N-type calcium channels seem restricted to neurones and to neuroendocrine cells such as pituitary cells and adrenal chromaffin cells. N-type calcium channels in neurones were first distinguished from L-type channels by different characteristics of voltage dependence, kinetics, and single-channel behaviour in recordings from chick sensory neurones in tissue culture (Carbone and Lux 1984, 1987; Fox *et al.* 1987*a,b*). Subsequent work in a variety of peripheral and central neurones has confirmed the distinction between L-type and N-type calcium channels in neurones, but it has also shown that the differences in kinetics and single-channel size are sometimes less clear than originally recognized (Plummer *et al.* 1989).

The clearest distinction between current components carried by L-type and N-type channels in neurones can usually be made by their different pharmacology. N-type calcium channels are not blocked by dihydropyridine blockers, and they are not enhanced by dihydropyridine agonists (Fox *et al.* 1987*a,b*, Aosaki and Kasai 1989). The most effective blocker of N-type channels known so far is a peptide toxin called ω-conotoxin GVIA (ω-CgTx GVIA), originally isolated from the venom of a sea snail, *Conus geographus* (Olivera *et al.* 1985; McCleskey *et al.* 1987). ω-CgTx GVIA is now widely available as a synthetic peptide. It has been shown to be a highly selective blocker of N-type calcium channels, with no effect on L-type calcium channels in many neurones (Aosaki and Kasai 1989; Plummer *et al.* 1989; Regan *et al.* 1991). Interestingly, however, there are a number of reports of particular neurones where the toxin does block L-type channels (Aosaki and Kasai 1989; Mynlieff and Beam 1992). A cloned L-type calcium channel from human brain was found to be weakly blocked by ω-CgTx GVIA (Williams *et al.* 1992). It will be interesting to determine the difference between the rarely encountered L-type channels that are blocked by the toxin from the more common L-type channels that are completely resistant.

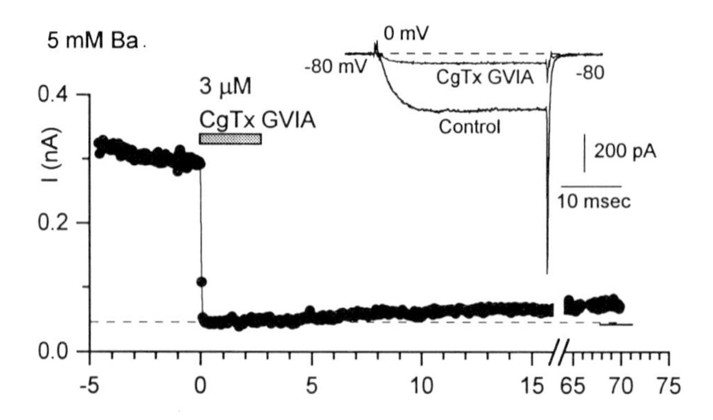


**Fig. 5.2** Block by 3 μM ω-CgTx GVIA of calcium channel current in a rat sympathetic neurone. Current was elicited by a depolarization from −80 mV to 0 mV, delivered every 3 s. Current was measured at the end of the test pulse after correcting for leak current (calculated from a step from −80 mV to −70 mV) and is plotted versus experimental time (minutes). Inset: leak-corrected currents in control and after exposure to ω-CgTx GVIA for 12 s. External solution: 5 mM $BaCl_2$, 160 mM TEACl, 10 mM HEPES, 3 μM nimodipine, pH 7.4 with TEAOH. Internal solution: 108 mM CsCl, 9 mM HEPES, 9 mM EGTA, 4.5 mM $MgCl_2$, 4 mM MgATP, 0.3 mM GTP (Tris salt), pH 7.4 with CsOH. (Adapted with permission from Boland *et al.* 1994.)

Figure 5.2 shows the effect of ω-CgTx GVIA on high-threshold calcium channel current in a sympathetic neurone isolated from the superior cervical ganglion of a rat. At 3 μM, the toxin rapidly (< 6 s) inhibited the majority of the calcium channel current in the neurone. In this neurone, a small component of L-type calcium current was first blocked by 3 μM nimodipine, which was present in all the solutions. ω-CgTx GVIA inhibited about 85% of the remaining current. Dose–response experiments show a half-maximally effective concentration of about 0.7 nM (Boland *et al.* 1994), so the toxin evidently binds with very high affinity to the N-type calcium channel. Consistent with high-affinity block, reversal of block is extremely slow, with only minimal reversal over 1 hour of washing in the experiment shown in Fig. 5.2. Interestingly, reversal of block by ω-CgTx is considerably faster in sympathetic neurones from bullfrogs, suggesting a somewhat lower-affinity binding than for rat N-type calcium channels.

ω-CgTx GVIA block of N-type channels appears to be all-or-none, so that when a channel is inhibited by toxin it is completely silent (Boland *et al.* 1994). The block is not obviously voltage-dependent, either with changes in holding potential or with applications of large depolarizing pulses. Current is blocked equally well for all voltage steps, and current carried by both divalent and monovalent cation can be inhibited. Figure 5.3 shows currents activated by depolarizations over a wide range of voltages in a frog sympathetic neurone. Moderate depolarizations, in the range of $-40$ to $+50$ mV, activate inward currents, carried by 2 mM Ba in the external solution. These are almost completely blocked by $\omega$-CgTx GVIA. Large depolarizations, positive to the reversal potential of $+60$ mV, activate outward currents, which are carried by intracellular Cs ions flowing outward through the channels. These outward currents are also inhibited by the toxin.

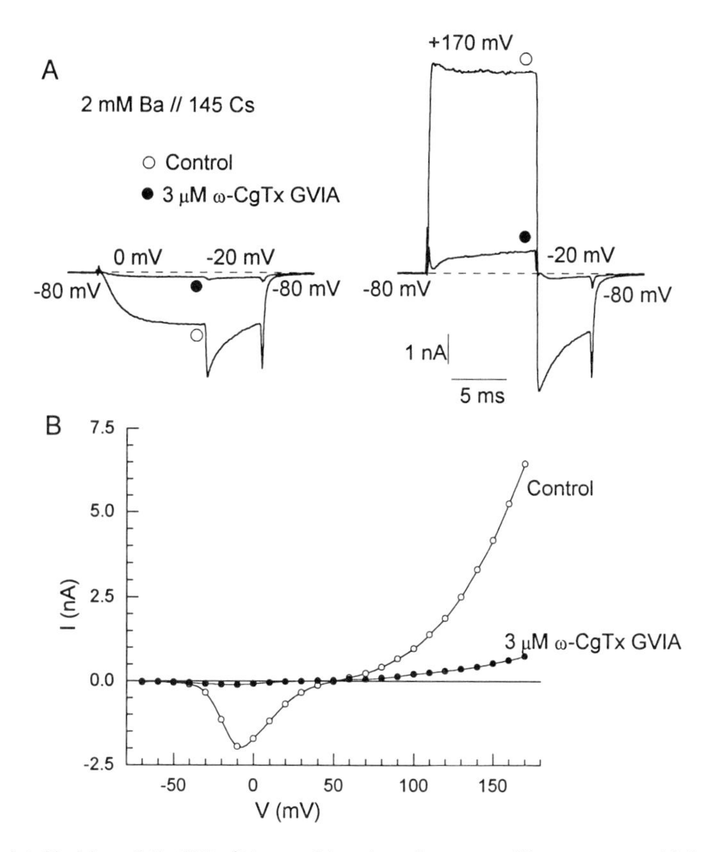


**Fig. 5.3** Complete block by $\omega$-CgTx GVIA of N-type calcium channel current at different test potentials in a frog sympathetic neurone. Currents were elicited by 10-ms steps from $-80$ mV to test potentials from $-70$ to $+170$ mV. A. Leak-corrected currents at 0 mV (left) and +170 mV (right) in control and after application of 3 $\mu$M $\omega$-CgTx GVIA for 4 min. B. Current–voltage relationship for test pulse current recorded as in A. External solution: 2 mM $BaCl_2$, 160 mM TEACl, 10 mM HEPES, 3 $\mu$M nimodipine, 3 $\mu$M TTX, pH 7.4 with TEAOH. Internal solution as in Fig. 5.2. (Adapted with permission from Boland *et al.* 1994.)

A striking feature of block of N-type channels by $\omega$-CgTx GVIA is that the rate of block depends strongly on the concentration of divalent cations in the extracellular solution (McCleskey *et al.* 1987). As the concentration of Ca or Ba is increased, the development of block by a given concentration of $\omega$-CgTx GVIA becomes slower. Figure 5.4A shows examples in bullfrog sympathetic neurones. At 3 μM, $\omega$-CgTx GVIA blocks with a time constant of 29 s in 5 mM Ba and with a time constant of 354 s in 30 mM Ba. If the rate of block is expressed as a first-order rate constant, the rate of block is depends on Ba concentration with an inverse linear relationship. This is the relationship that would be expected if $\omega$-CgTx GVIA binds at a site that can also be occupied by divalent ions with high affinity, so that the rate of successful collisions would depend on the fraction of time that the site is unoccupied by a divalent ion. A particularly interesting possibility is that this hypothetical site is the high-affinity 'set' of divalent binding sites that is believed to underlie the divalent selectivity of the calcium channel (Kuo and Hess 1993).

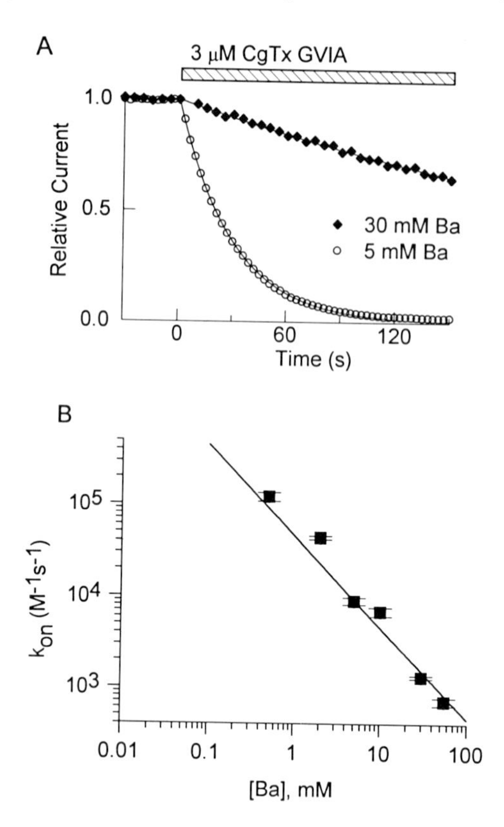


**Fig. 5.4** Effect of Ba concentration on blocking rate. A. The time course of 3 $\mu$M $\omega$-CgTx GVIA block in 30 mM Ba (▲) or 5 mM Ba (○), in two frog sympathetic neurones. Currents are normalized to the size at the time of toxin application. Solid lines are single exponentials decaying to zero, with time constants of 354 s for 30 mM Ba and 29 s for 5 mM Ba. B. Rate constant of block versus. Ba concentration in frog neurones. Data points are the inverse of the time constant of block (mean $\pm$ S.E.M., $n$ = 4–35). The solid line has a slope of $-1$, the relation expected if $\omega$-CgTx GVIA competes with Ba for a binding site to which Ba binds with high affinity. External solutions: 0.5–55 mM $BaCl_2$, 160 mM TEACl (80 mM TEACl with 55 mM $BaCl_2$), 0.1 mM EGTA, 10 mM HEPES, pH adjusted to 7.4 with TEAOH. Internal solution as in Fig. 5.2. (Adapted with permission from Boland *et al.* 1994.)

## P-type calcium channels

In cerebellar Purkinje neurones, the predominant current is a high-threshold current that has very little contribution from L-type or N-type calcium channels. As illustrated in Fig. 5.5A, both dihydropyridines and $\omega$-CgTx GVIA block only a very small fraction of the high-threshold current in rat cerebellar Purkinje neurones (Regan 1991; Regan *et al.* 1991). These observations support an earlier suggestion that calcium channels in the dendrites of cerebellar

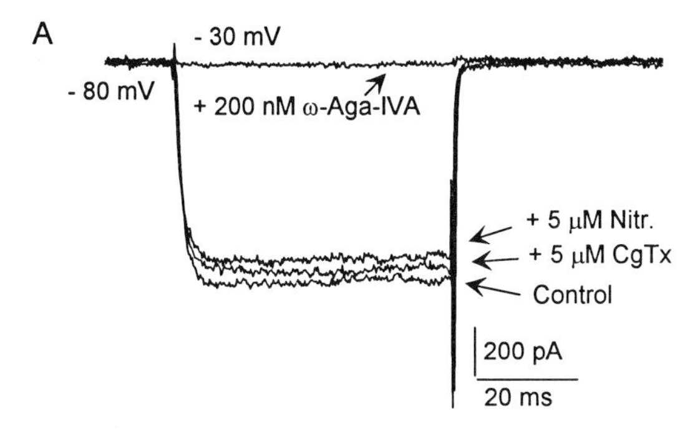


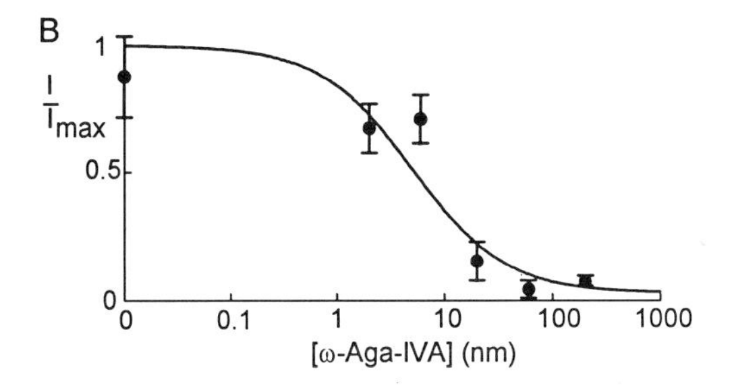


**Fig. 5.5** ω-Aga-IVA block of high-threshold current in cerebellar Purkinje neurones. A. Complete block by ω-Aga-IVA of high-threshold calcium current resistant to ω-CgTx GVIA and nitrendipine. High-threshold current in control, in the presence of ω-CgTx GVIA (5 μM), ω-CgTx GVIA plus nitrendipine (5 μM), and ω-CgTx GVIA, nitrendipine, and ω-Aga-IVA (200 nM). Internal solution: 108 MCs-methanesulphonate, 4 mM $MgCl_2$, 9mM EGTA, 9mM HEPES, 4mM MgATP, 14mM creatine phosphate (Tris salt), 1mM GTP (Tris salt), pH 7.4. External solution: 5mM $BaCl_2$, 160mM TEACl, 0.1mM EGTA, 10mM HEPES, pH 7.4. (Reprinted with permission from Mintz *et al.* 1992*a*.) B. Collected dose–response data for block of high-threshold current in Purkinje neurones by ω-Aga-IVA. (Reprinted with permission from Mintz *et al.* 1992*b*.)

Purkinje cells are pharmacologically different from L-type and N-type channels, and the name 'P-type' channels was proposed for the predominant calcium channels in Purkinje neurones (Llinas *et al.* 1989).

The venom of *Agelenopsis aperta*, a funnel web-weaving spider living in the American south-west, was found to contain a 48-amino acid peptide toxin, named ω-Aga-IVA, that is a highly effective blocker of P-type calcium channels in cerebellar Purkinje neurones (Mintz *et al.* 1992*a*,*b*). Figure 5.5A shows an example in which ω-Aga-IVA completely blocked the high-threshold current in a rat Purkinje neurone that was left after addition of nitrendipine and ω-conotoxin. ω-Aga-IVA is very potent, inhibiting P-type calcium channels with a $K_d$ of a few nM (Fig. 5.5B). The initial characterization of ω-Aga-IVA was done using peptide purified from *Agelenopsis aperta* venom. Subsequently, the peptide has been successfully synthesized based on the deduced amino acid structure of the purified toxin. The synthetic peptide, which is available in larger quantities, has almost identical properties of potency and selectivity as the purified toxin. The only difference is that the synthetic toxin blocks about twice more slowly at a given concentration and has a slightly higher estimated $K_d$ for block (Mintz and Bean 1993). Since the toxin has four disulphide bridges that are likely to be important for determining the active structure of the molecule, one possibility is that not all of the synthetic peptide forms the pattern of disulphide bridges required for the most active form.

In either the purified or the synthetic form, $\omega$-Aga-IVA is highly selective for blocking P-type channels but not other channel types (Mintz *et al.* 1992a; Mintz and Bean 1993). At 100 nM, well above saturation for P-type channel block, the toxin has no effect on T-type channels in dorsal rat ganglion neurones (Fig. 5.6B). It also has no effect on L-type current in neurones (which in Fig. 5.6C is enhanced by Bay K 8644) and also does not block L-type channels in rat cardiac myocytes. Rat sympathetic neurones have primarily current carried by N-type calcium channels, and $\omega$-Aga-IVA has no effect on this current (Fig. 5.6D).

Block of P-type channels by $\omega$-Aga-IVA reverses only very slowly with wash-out of the toxin, as shown in Fig. 5.7. However, the rate of recovery can be dramatically enhanced by giving a short series of large depolarizations. In the example shown in Fig. 5.7, a series of 25

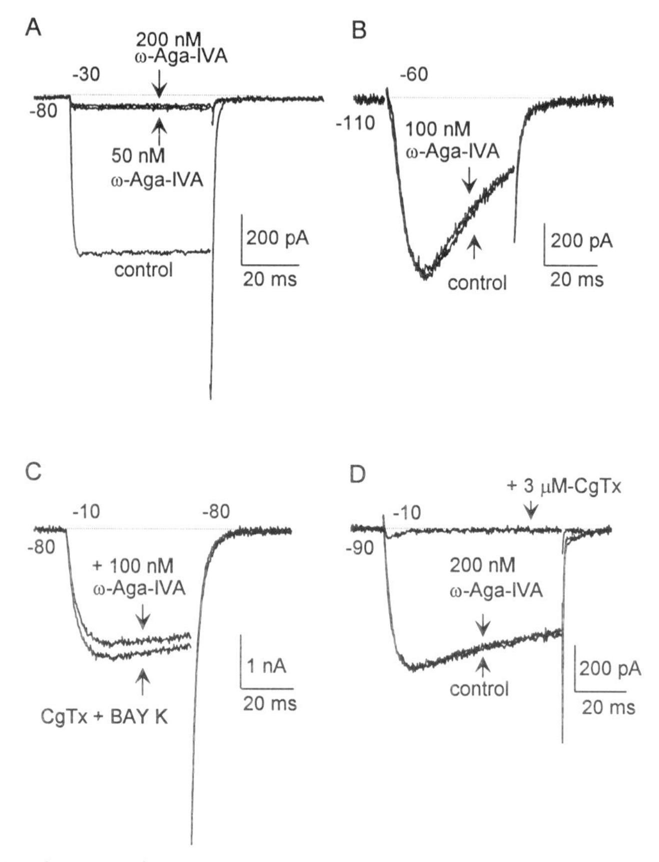


**Fig. 5.6** Discrimination of $\omega$-Aga-IVA for P-type channels. A. 50 nM $\omega$-Aga-IVA maximally inhibits the high-threshold current in Purkinje neurone with no further effect of 200 nM. Current was elicited by a step from −80 to −30 mV. B. T-type current in a rat DRG neurone is not blocked by 100 nM $\omega$-Aga-IVA. The current was elicited by a step from −110 to −60 mV; it had the rapid inactivation and slow deactivation typical of T-type channels, and it was eliminated by holding at − 70 mV. C. Neuronal Bay-K 8644-enhanced L-type current is insensitive to 100 nM $\omega$-Aga-IVA. High-threshold currents were recorded in a rat DRG neurone in the continuous presence of $\omega$-CgTx GVIA (2 $\mu$M) and Bay K 8644 (1 $\mu$M). Current was elicited by a step from −80 to −10 mV. $\omega$-Aga-IVA slightly reduced the current elicited at −10 mV and had no effect on the slowly deactivating tail current at −80 mV induced by Bay-K 8644. D. Calcium channel current in a rat sympathetic neurone was unaffected by $\omega$-Aga-IVA (200 nM) but was abolished by $\omega$-CgTx GVIA (3 $\mu$M). Solutions as in Fig. 5.5. (Redrawn with permission from Mintz *et al.*, 1992*a*.)

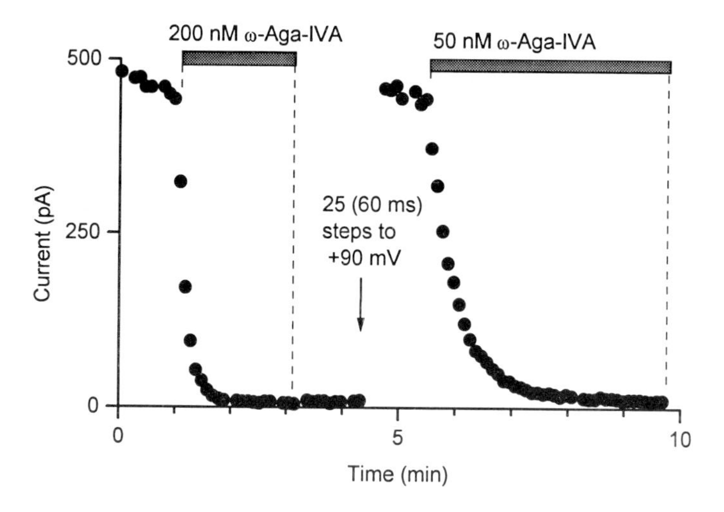


**Fig. 5.7** Depolarizing pulses alleviate ω-Aga-IVA-induced current block of P-type current in a cerebellar Purkinje neurone. Application of 200 nM ω-Aga-IVA produced complete inhibition of the current, with very little recovery in a minute following removal of the toxin. Data points represent currents elicited by 60-ms steps from −90 mV to −30 mV, applied every 6 s. After 1 min of wash-out, a burst of 25 separate 60-ms steps from −80 to +90 mV (repeated at 1-s intervals within the burst) was delivered (arrow). This series of short depolarizations produced complete relief of block. The current was then re-blocked by addition of 50 nM ω-Aga-IVA. Solutions as in Fig. 5.5. (Reprinted with permission from Bean and Mintz 1994.)

separate 60-ms depolarizations to +90 mV (delivered at 1 Hz) resulted in complete recovery of the current (which was then re-blocked by 50 nM ω-Aga-IVA). Quantitatively, it seems that the unbinding rate of toxin is accelerated by a factor of about $10^4$ at +90 mV compared with −90 or −80 mV. If the binding rate of toxin is not affected by voltage (which has not yet been tested carefully), the affinity of toxin for the channel would be $10^4$ times lower at very positive voltages than near the resting potential. The molecular basis of this effect is not yet clear. An interesting possibility is that it reflects the conformational change that the channel undergoes when it opens in response to depolarization.

Many other types of neurones, in addition to Purkinje neurones, possess calcium channels that are blocked by ω-Aga-IVA, but in most other neurones a much smaller fraction of the overall high-threshold current is blocked by ω-Aga-IVA. The fraction of current sensitive to ω-Aga-IVA usually coexists with other fractions sensitive to ω-CgTx GVIA and to dihydropyridines. Figure 5.8A shows the action of these different agents on the overall high-threshold calcium current in a pyramidal neurone from the CA1 region of the hippocampus. ω-CgTx GVIA blocked about 45% of the current, and addition of 2 μM nitrendipine (with ω-CgTx GVIA still present) blocked an additional 15%. Addition of 100 nM ω-Aga-IVA inhibited most of the remaining current (about 30% of the original current). Increasing the concentration of ω-Aga-IVA to 200 nM had no further effect (Fig. 5.8B), suggesting that 100 nM was sufficient to produce saturating block of the channels sensitive to the toxin.

There is an interesting difference in the rate of block by ω-Aga-IVA in different types of neurones. Block in hippocampal neurones is substantially slower than that by the same concentrations of toxin in Purkinje neurones. For example, block by 100 nM ω-Aga-IVA develops with a time constant of about 2 min in the hippocampal neurone shown in Fig. 5.8,

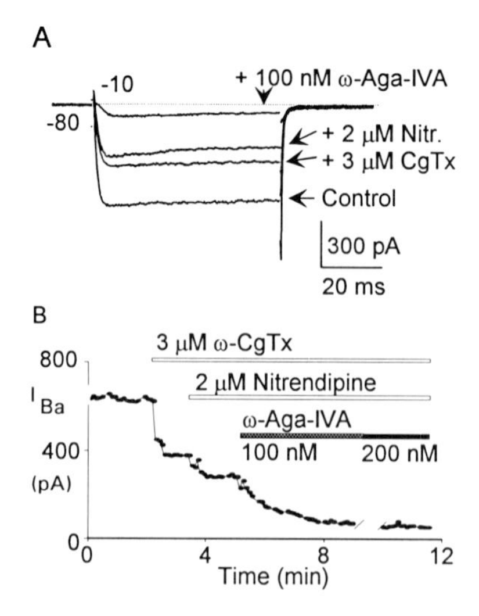


**Fig. 5.8** Block by $\omega$-Aga-IVA of a fraction of current in a hippocampal CA1 neurone. Currents were elicited by 60-ms depolarizations from −80 mV to −10 mV, delivered every 7 s. Drugs were applied with cumulative addition of 3 μM $\omega$-CgTx GVIA, 2 μM nitrendipine, 100 nM $\omega$-Aga-IVA, and 200 nM $\omega$-Aga-IVA. Solutions as in Fig. 5.5. (Redrawn with permission from Mintz *et al.* 1992*a*.)

while a two-fold lower concentration of toxin blocks with a time constant of about 1 min in the Purkinje cell shown in Fig. 5.7. This is a consistent difference, and suggests that there may be differences in the channels sensitive to $\omega$-Aga-IVA in different cell types. Block in spinal cord neurones is rapid (Fig. 5.9A), and similar to that in Purkinje neurones; in contrast, block in hippocampal neurones and dorsal root ganglion neurones is slower (Mintz *et al.* 1992*a*). This might reflect different channel types being blocked by the toxin in the different types of neurones. It is not yet clear whether the potency of block is lower in hippocampal neurones than in spinal cord neurones or Purkinje neurones. However, it cannot be very much lower, since 100 nM produces saturating block, even in hippocampal neurones (Fig. 5.8).

## Other types of calcium channels

In almost all neurones other than Purkinje neurons, there is some calcium channel current that cannot be assigned to T-type, L-type, N-type, or P-type channels. In many neurones with no T-type current, application together of saturating concentrations of $\omega$-Aga-IVA, $\omega$-CgTx GVIA, and dihydropyridine blockers fails to completely block the high-threshold current. Examples are shown for a hippocampal neurone in Fig. 5.8 and for a spinal cord neurone in Fig. 5.9. The properties of the remaining current have not been completely described in most cases. One case where there is some information about such an 'unclassified' current is in rat sympathetic neurones. Particularly when studied with external solutions containing high Ba concentrations, there is significant current that remains unblocked by $\omega$-CgTx GVIA, and dihydropyridines (Ikeda 1991; Zhu and Ikeda 1993; Boland *et al.* 1994). This current is not affected by $\omega$-Aga-IVA, suggesting that it does not correspond to P-type channels. It differs from the predominant N-type current in the same neurones by being activated by smaller depolarizations and in inactivating somewhat more rapidly (Fig. 5.10).

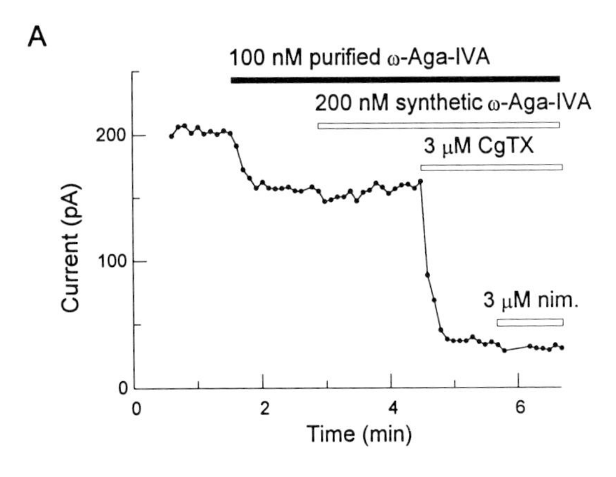


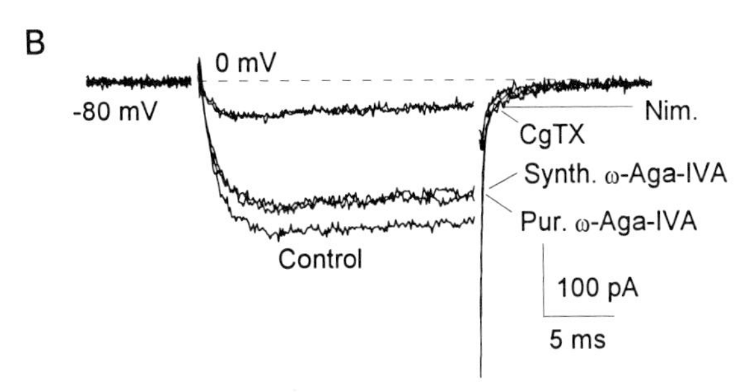


**Fig. 5.9** High-threshold calcium current in a spinal cord neurone resistant to ω-CgTx GVIA, nimodipine, and ω-Aga-IVA. A. Purified ω-Aga-IVA at 100 nM occluded further block by 200 nM synthetic ω-Aga-IVA. ω-CgTx GVIA blocked substantially more current, and nimodipine little more. B. Currents recorded in the various conditions. Solutions as in Fig. 5.5. (Redrawn with permission from Mintz and Bean 1993.)

In some central neurones, the current remaining in ω-CgTx GVIA, ω-Aga-IVA, and dihydropyridines can be partially blocked by a peptide derived from the cone snail *Conus magus*, ω-CgTx MVIIC (Hillyard *et al.* 1992). Unfortunately, the peptide is not a selective blocker of these channels, since ω-CgTx MVIIC also blocks both N-type and P-type calcium channels (Hillyard *et al.* 1992). Initially, it seemed that ω-CgTx MVIIC blocked N-type channels with high potency and P-type channels more weakly (Hillyard *et al.* 1992), but more recent experiments show that block of P-type calcium channels also occurs with high potency but very slowly, so that micromolar concentrations are necessary to produce significant block within several minutes (K.J. Swartz, I.M. Mintz, and B.P. Bean, unpublished results), although nanomolar concentrations will block if applied for very long times.

ω-CgTx MVIIC is useful in drawing connections between cloned calcium channels and their counterparts in native cells. Sather and colleagues (1993) have shown that cloned calcium channels of the Class A variety, when expressed in *Xenopus* oocytes, are relatively resistant to most other blockers, but are blocked by ω-CgTx MVIIC. These channels may correspond to part of the 'unclassified' current component in native neurones. Specifically, it has been proposed that a class of 'Q-type' calcium channels exists, that is resistant to ω-CgTx

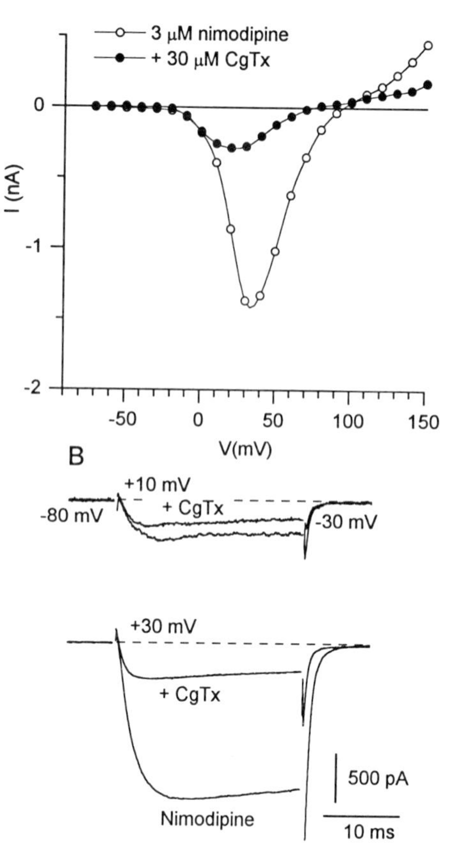


**Fig. 5.10** Current–voltage relation for ω-CgTx GVIA- and nimodipine-resistant current in a rat sympathetic neurone. 30 μM ω-CgTx GVIA was applied in a solution containing 110 mM $BaCl_2$, 3 μM nimodipine, 3 μM TTX, 10 mM HEPES (pH adjusted to 7.4 with $Ba(OH)_2$). A. Peak current versus voltage before and after application of ω-CgTx GVIA. B. Currents at +10 mV (top) and +30 mV (bottom) before and after application of ω-CgTx GVIA. The tail current without ω-CgTx GVIA is truncated to save space. Internal solution as in Fig. 5.2. (Reprinted with permission from Boland *et al.* 1994.)

GVIA and nimodipine, is weakly blocked by ω-Aga-IVA, and is sensitive to block by ω-CgTx MVIIC, and that the channels contribute to synaptic transmission in the mammalian brain (Wheeler *et al.* 1994).

Ellinor and colleagues (1993) and Zhang and colleagues (1993) have described additional components of calcium current in cerebellar granule neurones that persist in the presence of ω-CgTx MVIIC, ω-CgTx GVIA, and nimodipine. These 'R-type' channels differ from 'Q-type' channels in being resistant to ω-CgTx MVIIC. They may correspond to the E class of cloned calcium channels (Ellinor *et al.* 1993; Zhang *et al.* 1993).

## Non-selective organic blockers

Although there are peptide toxins that selectively block N- and P-type calcium channels, there are not yet small synthetic molecules that are selective blockers of both calcium channel types.

However, there are a number of agents which block all types of calcium channels, including P- and N-type channels, with roughly equal potency.

One such compound is fluspirilene, a diphenylbutylpiperidine. Figure 5.11 shows that fluspirilene at 30 μM completely inhibits the calcium channel current in a cerebellar Purkinje neurone (predominantly P-type current). Studied with a moderate rate of depolarization from a resting potential of −80 mV, fluspirilene inhibits P-type channels with an $IC_{50}$ of about 6 μM. It inhibits N-type channels in sympathetic neurones with similar potency; the overall high-threshold current in rat sympathetic neurones is blocked with an $IC_{50}$ of about 2 μM (Fig. 5.12).

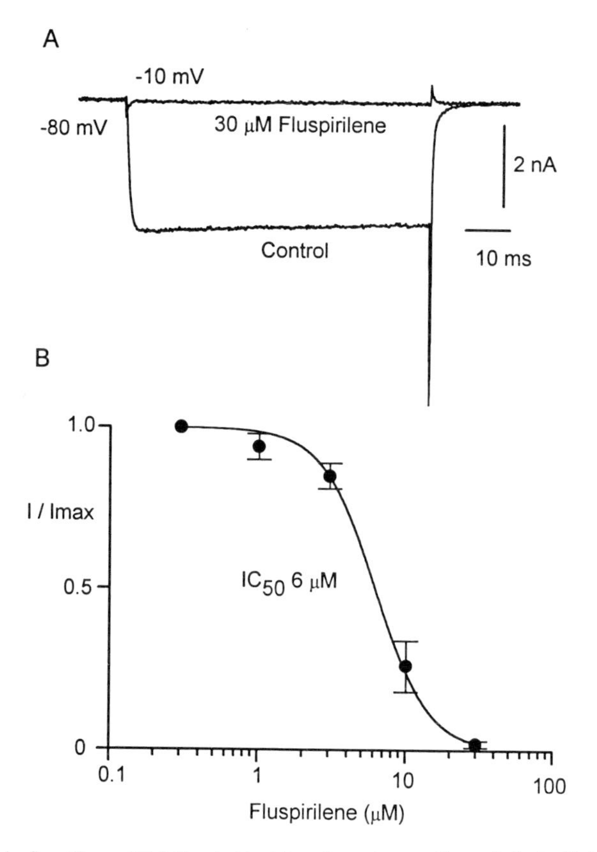


**Fig. 5.11** A. Inhibition by fluspirilene of high-threshold calcium channel current in cerebellar Purkinje neurones. Current was elicited by a 50-ms depolarization from −80 mV to −10 mV delivered every 5 s. B. Dose-dependent inhibition by fluspirilene of high-threshold calcium channel current in cerebellar Purkinje neurones. Each point represents the mean ± S.E.M. for 3–11 cells. Smooth curves represent the best fit to $1/(1+([Drug]/EC50)^n)$ with $EC_{50} = 6.1$ μM and $n = 2.4$. External solution: 160 mM TEA-Cl, 5 mM $BaCl_2$, 10 mM HEPES, pH 7.4 with TEA-OH, with 3 μM tetrodotoxin. Internal solution as in Fig. 5.2. (Redrawn with permission from Sah and Bean 1994.)

Fluspirilene is not selective for P-type and N-type channels, since it has previously been found to inhibit both L-type and T-type calcium channels (Galizzi *et al.* 1986; Enyeart *et al.* 1992).

Like many organic blockers of voltage-dependent ion channels, the potency of fluspirilene

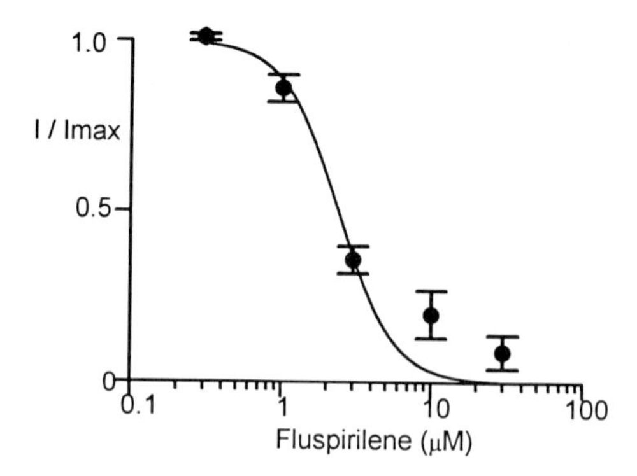


**Fig. 5.12** Dose-dependent inhibition by fluspirilene of N-type calcium channel current in rat sympathetic neurones. Solutions and protocols as in Fig. 5.11. $EC_{50} = 2.4\ \mu M$, $n = 2.3$. (Redrawn with permission from Sah and Bean 1994.)

block of P- and N-type channels depends on the pattern of activation of the channels. Depolarized voltages that activate (and inactivate) the channels tend to promote more potent block. This is illustrated in Fig. 5.13 for fluspirilene block of P-type calcium channels. Applied at 7 μM, the drug produces relatively little inhibition if the cell is stimulated infrequently from a negative holding potential, but block is greatly enhanced when the cell is depolarized to a holding potential of −40 mV. This characteristic, which is reminiscent of dihydropyridine block of L-type calcium channels, may be a useful property for a clinically

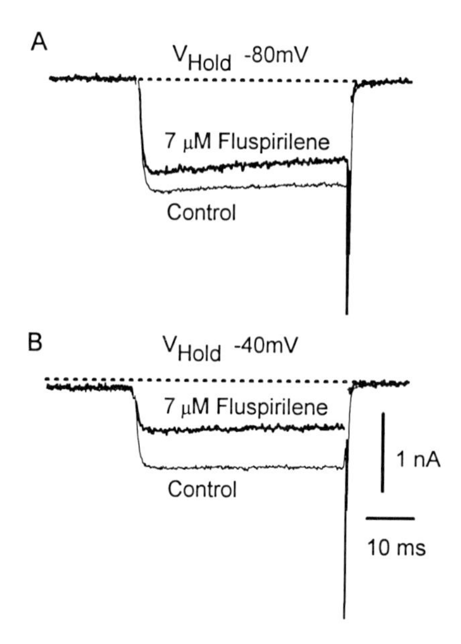


**Fig. 5.13** Voltage-dependence of fluspirilene block of high-threshold calcium channel current in a cerebellar Purkinje neurone. Current was elicited by a 50-ms depolarization to −10 mV delivered from a holding potential of −80 mV (A) or after the holding potential was changed to −40 mV for 10 s (B). Fluspirilene (7 μM) was applied at a holding potential of −80 mV while current was elicited every 20 s with short (3-ms) depolarizations to −10 mV; after 2 min the effects of fluspirilene reached steady state and the records shown were taken. Tail currents were truncated to save space. Solutions as in Fig. 5.11. (Redrawn with permission from Sah and Bean 1994.)

useful drug. There may be little block with 'normal' activation of channels, but steady depolarization of a cell due to damage would greatly potentiate the ability of the drug to block calcium entry. Pathologically high amounts of calcium entry might therefore be preferentially inhibited. Drugs that have this property and are capable of selectively blocking N-type, P-type, Q-type, or R-type calcium channels could well be useful in treating various diseases of the central nervous system. It will be interesting to see whether such compounds can be discovered.

## References


Aosaki, T. and Kasai, H. (1989). Characterization of two kinds of high-voltage-activated calcium-channel currents in chick sensory neurons. Differential sensitivity to dihydropyridines and omega-conotoxin GVIA. *Pflugers. Arch.*, **414**, 150–156.

Bean, B.P. (1984). Nitrendipine block of cardiac calcium channels: high-affinity binding to the inactivated state. *Proc. Natl Acad. Sci. USA* **81**, 6388–6392.

Bean, B.P. (1989). Classes of calcium channels in vertebrate cells. *Annu. Rev. Physiol.* **51**, 367–384.

Bean, B.P. (1991). Pharmacology of calcium channels in cardiac muscle, vascular muscle, and neurons. *Am. J. Hypertension*, **4**, 406S–411S.

Bean, B.P. and Mintz, I.M. (1994) Pharmacology of different types of calcium channels in rat neurons in *Handbook of membrane channels*, (ed. C.Peracchia), Academic Press, San Diego, pp. 199–210.

Boland, L.M., Morrill, J.A., and Bean, B.P. (1994). $\omega$-Conotoxin block of N-type calcium channels in frog and rat sympathetic neurons. *J. Neurosci.* **14**, 5011-5027.

Carbone, E. and H.D. Lux. (1984). A low-voltage-activated, fully inactivating Ca channel in vertebrate sensory neurons. *Nature* **310**, 501–502.

Carbone, E. and H.D. Lux. (1987). Single low-voltage-activated calcium channels in chick and rat sensory neurones. *J. Physiol. (Lond.)* **386**, 571–601.

Coulter, D.A., Huguenard, J.R., and Prince, D.A. (1989*a*). Calcium currents in rat thalamocortical relay neurones; kinetic properties of the transient, low-threshold current. *J. Physiol. (Lond.)*, **414**, 587–604.

Coulter, D.A., Huguenard, J.R., and Prince, D.A. (1989*b*). Characterization of ethosuximide reduction of low-threshold calcium current in thalamic relay neurons. *Ann. Neurol.* **25**, 582–593.

Ellinor, P.T., Zhang J.-F., Randall, A.D., Zhou, M., Schwarz, T.L., Tsien, R.W., and Horne, W.A. (1993). Functional expression of a rapidly inactivating neuronal calcium channel. *Nature*, **363**, 455–458.

Enyeart, J.J., Biagi, B.A. and Mlinar, B. (1992). Preferential block of T-type calcium channels by neuroleptics in neural crest-derived rat and human C cell lines. *Mol. Pharmacol*, **42** 364–372.

Fox, A.P., Nowycky, M.C., and Tsien R.W. (1987*a*). Single-channel recordings of three types of calcium channels in chick sensory neurones. *J. Physiol. (Lond.)*, **394**, 173–200.

Fox, A.P., Nowycky, M.C., and Tsien R.W. (1987*b*). Kinetic and pharmacological properties distinguishing three types of calcium currents in chick sensory neurones. *J. Physiol. (Lond.)*, **394**, 149–172.

Galizzi, J.-P., Fosset M., Romey, G., Laduron P., and Lazdunski M. (1986). Neuroleptics of the diphenylbutylpiperidine series are potent calcium channel inhibitors. *Proc. Natl Acad. Sci. USA*, **83**, 7513–7517.

Hagiwara, N., Irisawa, H. and Kameyama, M. (1988). Contribution of two types of calcium channels to the pacemaker potentials of rabbit sinoatrial node cells. *J. Physiol. (Lond.)*, **395**, 233–252.

Herrington, J. and Lingle, C.J. (1992). Kinetic and pharmacological properties of low voltage-activated $Ca^{2+}$ current in rat clonal ($GH_3$) pituitary cells. *J. Neurophysiol.*, **68**, 213–232.

Hess, P. (1990). Calcium channels in vertebrate cells. *Annu. Rev. Neurosci.*, **13**, 337–356.

Hess, P., Lansman, J.B., and Tsien, R.W. (1984). Different modes of calcium channel gating behaviour favoured by dihydropyridine agonists and antagonists. *Nature*, **311**, 538–544.

Hille, B. (1992). *Ionic channels of excitable membranes*. Sinauer, Sunderland, MA.

Hillyard, D.R., Monje, V.D., Mintz, I.M., Benan, B.P., Nadasdi, L., Ramachandran, J., Miljanich, G., Azimi-Zoonooz, A., McIntosh, J.M., Cruz, L.J., Imperial, J.S., and Olivera, B.M. (1992). A new Conus peptide ligand for mammalian presynaptic $Ca^{2+}$ channels. *Neuron*, **9** 69–77.

Huguenard, J.R. and Prince, D.A. (1992). A novel T-type current with slowed inactivation contributes to prolonged $Ca^{2+}$-dependent burst firing in GABAergic neurons of rat thalamic reticular nucleus. *J. Neurosci*, **12**, 3804–3817.

Ikeda, S.R. (1991). Double-pulse calcium channel facilitation in adult rat sympathetic neurons. *J. Physiol. (Lond.)*, **439** 181–214.

Kuo, C.-C. and Hess, P. (1993). Ion permeation through the L-type $Ca^{2+}$ channel in rat phaeochromocytoma cells: two sets of ion binding sites in the pore. *J. Physiol.*, **466**, 629–655.

Llinas, R., Sugimori, M., Lin, J.W., and Cherksey, B. (1989). Blocking and isolation of a calcium channel from neurons in mammals and cephalopods utilizing a toxin fraction (FTX) from funnel-web spider poison. *Proc. Natl. Acad. Sci. USA*, **86**, 1689–1693.

McCleskey, E.W., Fox, A.P., Feldman, D.H., Cruz, L.J., Olivera, B.M., Tsien, R.W., and Yoshikami, D. (1987). Omega-conotoxin: direct and persistent blockade of specific types of calcium channels in neurons but not muscle. *Proc. Natl Acad. Sci. USA*, **84**, 4327–4331.

Miller, R.J. (1992). Voltage sensitive $Ca^{2+}$ channels. *J. Biol. Chem.*, **267**, 1403–1406.

Mintz, I.M. and Bean, B.P. (1993). Block of calcium channels in rat neurons by synthetic $\omega$-Aga-IVA. *Neuropharmacology*, **32**, 1161–1169.

Mintz, I.M., Adams, M.E., and Bean, B.P. (1992*a*). P-type calcium channels in central and peripheral neurons. *Neuron*, **9**, 1–20.

Mintz, I.M., Venema, V.J., Swiderek, K., Lee, T., Bean, B.P., and Adams, M.E. (1992). P-type calcium channels blocked by the spider toxin ω-Aga-IVA. *Nature* **355**, 827–829.

Mynlieff, M. and Beam, K.G. (1992). Characterization of voltage-dependent calcium currents in mouse motorneurons. *J. Neurophysiol*, **68**, 85–92.

Nowycky, M.C., Fox, A.P., and Tsien, R.W. (1985). Long-opening mode of gating of neuronal calcium channels and its promotion by the dihydropyridine calcium agonist Bay K 8644. *Proc. Natl Acad. Sci. USA*), **82**, 2178–2182.

Olivera, B.M., Gray, W.R., Zeikus, R., McIntosh, J.M., Varga, J., Rivier, J., de Santos, V., and Cruz, L.J. (1985). Peptide neurotoxins from fish-hunting cone snails. *Science*, **230**, 1338–1343.

Plummer, M.R., Logothetis, D.E., and Hess, P. (1989). Elementary properties and pharmacological sensitivities of calcium channels in mammalian peripheral neurons. *Neuron*, **2**, 1453–1463.

Regan, L.J. (1991). Voltage-dependent calcium currents in Purkinje cells from rat cerebellar vermis. *J. Neurosci*, **11**, 2259–2269.

Regan, L.J., Sah, D.W., and Bean, B.P. (1991). $Ca^{2+}$ channels in rat central and peripheral neurons: high-threshold current resistant to dihydropyridine blockers and omega-conotoxin. *Neuron*, **6**, 269–280.

Sanguinetti, M.C. and Kass, R.S. (1984). Voltage-dependent block of calcium channel current in the calf cardiac Purkinje fiber by dihydropyridine calcium channel antagonists. *Circ. Res.*, **55**, 336–348.

Sather, W.A., Tanabe, T., Zhang, J.F., Mori, Y., Adams, M.E., and Tsien, R.W. (1993). Distinctive biophysical and pharmacological properties of class A(BI) calcium channels αl subunits. *Neuron*, **11**, 291–303.

Soong, T.W., Stea, A., Hodson, C.D., Dubel, S.J., Vincent, S.R., and Snutch, T.P. (1993). Structure and functional expression of a member of the low voltage-activated calcium channel family. *Science*, **260**, 1133–1136.

Snutch, T.P. and Reiner, P.B. (1992). $Ca^{2+}$ channels: diversity of form and function. *Curr. Opin. Neurobiol.*, **2**, 247–253.

Tang, C.M., Presser, F., and Morad, M. (1988). Amiloride selectively blocks the low threshold (T) calcium channel. *Science*, **240**, 213–215.

Tsien, R.W., Lipscombe, D., Madison, D.V., Bley, K.R., and Fox, A.P. (1988). Multiple types of neuronal calcium channels and their selective modulation. *Trends Neurosci*, **11**, 431–438.

Tsien, R.W., Ellinor, P.T. and Horne, W.A. (1991). Molecular diversity of voltage-dependent $Ca^{2+}$ channels. *Trends Pharmacol. Sci.*, **12**, 349–354.

Uehara, A. and Hume, J.R. (1985). Interactions of organic calcium channel antagonists with calcium channels in single frog atrial cells. *J. Gen. Physiol.*, **85**, 621–647.

Wheeler, D.B., Randall, A., and Tsien, R.W. (1994). Roles of N-type and Q-type $Ca^{2+}$ channels in supporting hippocampal synaptic transmission. *Science*, **264**, 107–111.

Williams, M.E., Feldman, D.H., McCue, A.F., Brenner, R., Velicelebi, G., Ellis, S.B., and Harpold, M.M. (1992). Structure and function of $\alpha 1$, $\alpha 2$, and $\beta$ subunits of a novel human neuronal calcium channel subtype. *Neuron*, **8**, 71–84.

Zhang J.-F., Randall, A.D., Ellinor, P.T., Horne, W.A., Sather, W.A., Tanabe, T., Schwarz, T.L., and Tsien, R.W. (1993). Distinctive pharmacology and kinetics of cloned neuronal $Ca^{2+}$ channels and their possible counterparts in mammalian CNS neurons. *Neuropharmacology*, **32**, 1075–1088.

Zhu, Y. and Ikeda, S.R. (1993). Adenosine modulates voltage-gated $Ca^{2+}$ channels in adult rat sympathetic neurons. *J. Neurophysiol.*, **70**, 610–620.

# 6 Ion channels formed by amyloid β protein (AβP[1–40]). Pharmacology and therapeutic implications for Alzheimer's disease

H. B. Pollard, N. Arispe, and E. Rojas

## Introduction

Alzheimer's disease (AD) is a chronic dementia which is becoming increasingly prevalent as the population ages.[1–4] It has been generally estimated that as many as one-third of those reaching 85 years of age will suffer from a dementia of some kind, and Alzheimer's disease will afflict nearly half of these.[5] Inasmuch as modern medicine has made it very likely that many humans now alive will live to see their one hundredth year, attention to the quality of those lives suggests that Alzheimer's disease portends to become a very significant public health problem.

The cause of Alzheimer's disease is not known, although a hint has come from a consideration of the pathological findings in the brains of those afflicted. The most common observations include the occurrence of extracellular amyloid plaques, as well as intraneuronal neurofibrillary tangles, and concomitant vascular and neuronal damage.[1, 4, 6–11] In addition, reactive microglia are often found surrounding the plaques.[12] However, the major proteinaceous component of the plaque is a 38-to-42-residue peptide, termed amyloid β protein (AβP, βA/4).[4, 10, 13] This peptide is a proteolytic product of the widely distributed amyloid precursor protein ($APP_{751}$) defined by a locus on chromosome 21. AβP has been circumstantially linked to the toxic principle causing cell damage in the disease, although the mechanism has remained obscure.[4, 7, 14–17]

The importance of AβP has been substantiated indirectly by the fact that in Down's syndrome, where chromosome 21 is trisomic, those patients who survive into their late twenties suffer massive deposition of amyloid plaques and Alzheimer's disease.[6, 7] Furthermore, several types of familial Alzheimer's disease have been traced to mutations on the amyloid precursor protein.[18] However, there are other genetic loci statistically associated with the disease. One such associated locus is on chromosome 14.[19–21, 72] Another, the APO-E4 locus, is on chromosome 19.[22] Another is on chromosome 1.[73] Nonetheless, in all of these cases of Alzheimer's disease, the common denominator is a substantial deposition of characteristic amyloid. Thus, whatever the mechanism by which these different mutations increase the likelihood of the development of AD, AβP remains a central actor on the stage of disease.[23, 85]

In the remainder of this article we will consider the structure of the AβP peptide and will review the evidence that the peptide itself might be a toxic principle responsible for cell death in the disease. We will further develop the new concept that the mechanism of amyloid toxicity might lie in the cation channel properties which this peptide exhibits in bilayer and cellular membranes. The basic hypothesis to be developed is that the channel might be responsible for holes in the membranes of target cells, and that these holes might be responsible for the death of the cells. As will be shown, the amyloid channel preferentially conducts calcium, indicating that oxidative stress due to increments in intracellular calcium concentration could be a possible ultimate mechanism of cell death. Finally, if amyloid channels were to be the basis of neuronal cell death, we will develop the concept that amyloid channel blocking drugs ought to be useful as potential agents to arrest cell loss by this process.

## Structure and molecular biology of amyloid β protein

### Properties of the amyloid precursor protein and proteolytic production of AβP

While the amyloid precursor protein (APP) constitutes as much as 2.5% of the total brain mRNA, its function is not known. Neuronal injury, either from trauma,[24] or defined chemical insults such as kainic acid,[25, 26] cause acute increases in the high molecular weight APP. This result has led to the concept that the production of APP may be a normal response to local injury. Consistently, loss of subcortical innervation causes induction of APP in the cerebral cortex.[27] The association of APP with stress is further emphasized by the recent report of APP in adult bovine chromaffin cells,[28] and by our own observations (G. Lee and H.B. Pollard, unpublished results), that immunoreactive APP is an intrinsic constituent of purified chromaffin secretory granule membranes. The implications of this line of reasoning about the origins of APP are that physical or other stresses may be viewed as predisposing to the development of Alzheimer's disease. Real-time studies on APP metabolism in the brains of control and Alzheimer's disease patients have only recently become possible with the development by Wolozin and colleagues [29] of the olfactory neuroblast system.

In the brain, APP occurs naturally as a membrane-bound protein of ca. 100 KDa. Kang *et al.*[30] have suggested, on the basis of molecular considerations, that APP could span the membrane through a single C-terminal hydrophobic α-helical domain. A soluble portion of the APP (sAPP) can be released (or 'secreted') by action of a proteolytic enzyme termed 'secretase.' The function of this large secreted APP fragment (illustrated in Fig. 6.1), is not known. The remaining residues constitute the putative transmembrane domain and different portions of the N- and C-terminal flanking residues of APP. The common cleavage site on the APP is at lysine-16 (K16) in the numbering scheme for AβP(1–40) shown in Fig. 6.1. The remaining peptide is not thought to be toxic. However, when cleavage occurs at aspartic acid-1 ('$D_1$'; see Fig. 6.1) the toxic AβP is produced.

Until only recently AβP [1–40] and its close variants were thought to be produced and accumulated only in the diseased state. However, these amyloid peptides have now been

$H_2N$-$D_1$-$A_2$-$E_3$-$F_4$-$R_5$-$H_6$-$D_7$-$S_8$-$G_9$-$Y_{10}$-$E_{11}$-$V_{12}$-$H_{13}$-$H_{14}$-$Q_{15}$-$K_{16}$-$L_{17}$-

$V_{18}$-$F_{19}$-$F_{20}$-$A_{21}$-$E_{22}$-$D_{23}$-$V_{24}$-$G_{25}$-$S_{26}$-$N_{27}$-$K_{28}$-$G_{29}$-$A_{30}$-$I_{31}$-$I_{32}$-$G_{33}$-

$L_{34}$-$M_{35}$-$V_{36}$-$G_{37}$-$G_{38}$-$V_{39}$-$V_{40}$-OH.

**Fig. 6.1** Sequence of the amyloid *β* protein (A*β*5P(1–40)] from amyloid precursor protein (APP). The letters are the single letter codes for the 20 amino acids. K16 (lysine-16) is the site of normal inactivational cleavage by the *α*-secretase. Besides C-terminal cleavage at V40 (valine-40) to yield the major A*β*P(1–40) species, other cleavages occur in this region to yield a truncated amyloid peptide with 39 residues [A*β*P(1–39)], or an extended amyloid peptide with up to 43 residues [e.g. A*β*P(1–41, –42, or –43)].

shown to occur naturally in the cerebrospinal fluid of both Alzheimer's disease patients and apparently normal controls, in the supernatant of cultured cells, and also in lower mammals.[31–35]

## Structure of A*β*P in solution

Synthetic AβP forms stable dimers in aqueous solution,[36] or dimers, trimers, and tetramers.[37] Natural A*β*P from AD brain associates in fibrils which have a cross *β*-sheet conformation,[38] and even shorter segments from the A*β*P peptide contain the *β*-sheet conformation.[36,39–42] Hilbich *et al.*[36] report that synthetic A*β*P has a central *β*-turn, while Barrow *et al.*[37] emphasize that synthetic A*β*P can express different proportions of *α*-helix and *β*-sheet, depending on physiologically relevant environmental variables such as ionic strength, pH, and hydrophobicity. Further details of the sensitivity of the structure to pH have been described,[43] and models based on X-ray diffraction have been proposed.[44] Kang *et al.* have also suggested, on the basis of molecular considerations, that A*β*P could span the membrane through a C-terminal hydrophobic *α*-helical domain, leaving free the N-terminal *β*-sheet domain. Thus, both the spectroscopic and molecular data on natural and synthetic A*β*P emphasize the structural flexibility of the A*β*P molecule, and the possibilities of membrane interactions.

Further insights to conformational possibilities come from a closer examination of the specific sequences in different parts of the peptide. For example, within the putative N-terminal *β*-sheet domain, alternate residues [1, 3, 5, 7, (9), 11, 13, and 15,] are mostly charged or neutral, and that these residues are separated by mostly hydrophobic or neutral residues [2, 4, (6), 8,10,12, and (14)]. This is typical of amphipathic *β*-sheet structures and is consistent with above-described biophysical measurements. This amphipathic *β*-sheet character was instrumental in our decision to examine the amyloid peptide for ion channel activity, since this motif is characteristic of the pore structure of a number of ion channels. The best defined example is from the porin/VDAC system, where both modelling,[45] genetics,[46] and, eventually, X-ray crystallography, verified that the channel was composed of an amphipathic *β*-barrel structure. In the shaker $K^+$ channel system, molecular biology experiments have demonstrated that the transmembrane α-helical structures only serve to hold an amphipathic 19 residue β-hairpin structure in place as the actual pore[47]. The isolated pore sequence binds to phospholipids,[48] and the isolated hairpin 26 mer sequence has been shown to form ion channels in planar lipid bilayers.[49] Derivative modelling and experiments have implicated such structures in voltage-gated sodium, calcium, and potassium channels,[50, 51] and the channels for annexin I, [52, 74], annexin V, [75, 76], and annexin VII, [77, 78, 79].

### Amyloid peptides cause death of cells in culture

The C-terminal domain of APP, containing the (1–40) domain, has been reported to be toxic to neurones in culture.[9,53] The mechanism is not well understood, but a number of investigators have suggested that the toxic event may involve potentiation of other neurotoxins.[54–56] For example, glutamate toxicity has been shown to be potentiated in neuroblasts from embryonic hippocampus.[55] Toxicity independent of excitotoxins has also been demonstrated.[57] It has been suggested that amyloid disrupts calcium homoeostasis in target cells,[4] and direct measurements of intracellular $[Ca^{2+}]$ have shown that exposure to AβP[1–40] causes an elevation over time.[55] However, the details of how the calcium enters the cells have not been elucidated. It was concern for this question which led to our discovery that amyloid itself could provide an intrinsic entry pathway for this ion.[58,59]

## Amyloid β protein forms calcium channels in membranes

### Amyloid channels conduct cations in planar lipid bilayers

Planar bilayers were formed by applying a suspension of palmityloleoyl phosphatidylethanolamine (POPE) and phosphatidylserine (PS), 1 : 1, 50 mg/ml, each in *n*-decane, to a hole of ca. 100–120 μm in diameter in a thin Teflon film separating two compartments that contained salt solutions.[60] In our initial experiments,[58] channels were incorporated from a sonicated suspension of pure phosphatidylserine liposomes and AβP[1–40] peptide (Bachem, Torrance, CA). The liposome adduct was added to the *cis* chamber in small aliquots, and incorporation occurred directly from the experimental solutions. In later experiments the amyloid peptide was diluted into the solution in the chamber, and channels observed to incorporate directly into the bilayer.[59] In either case, the channels observed were very similar.

When symmetric solutions of 40 mM KCl were placed in each chamber, ion channel activity could be observed which lasted for minutes. In the specific case described by Arispe *et al.*,[58] direct closures to the zero current level were observed over a wide range of voltages. In symmetric ionic solutions, the reversal potential was at O V, and the apparent slope conductance was 320 pS. With asymmetric solutions of KCl in the *cis* and *trans* chambers, the selectivity of the amyloid channels for $K^+$ over $Cl^-$ was found to be 12.3 : 1.

The nature of the ion selectivity of amyloid channels was further studied using bi-ionic conditions. In summary, the amyloid channel was found to conduct $Cs^+$, $Li^+$, $K^+$, and $Na^+$ as well as $Ca^{2+}$, in the following order of preference:

$$[Cs^+ > Li^+ > Ca^{2+} = K^+ > Na^+]$$

### Amyloid channels interact specifically with calcium ions

As described by Arispe *et al.*,[58] the amyloid channels also treated calcium in a manner very reminiscent of typical calcium channels. For example, Arispe *et al.*[58] showed that when A$\beta$P[1–40] channels were incorporated into a membrane separating symmetric solutions of 200 mM CsCl, a robust conductance of $Cs^+$ could be elicited from the system over the voltage range of $\pm$ 60 mV. However, upon addition of 10 mM $CaCl_2$ to the *cis* chamber, channel activity became only barely detectable, and then only at extremes of voltage. Thus, while the amyloid channel will conduct monovalent cations, the channel is blocked down to a lower conductance when simultaneously given access to calcium. Apparently, the amyloid channel has calcium binding sites within or around the pore which control the flow of ions through the pore.

### Multiple conductance states of amyloid channels

Whether added to the bilayer complexed in liposomes or free in solution, amyloid channels frequently exhibited multiple conductance levels.[58] This was particularly evident as the voltage was raised. Since complete closures were observed from almost all levels of conductance, we conclude that single amyloid channels are able to undergo conformational transitions. We noted that conductance levels ranged from hundreds of pS into the range of 4–5 nS.[58,59]

An example of this interconversion between conductance states is shown in Fig. 6.2. These data are from an experiment in which amyloid was added as the free peptide to a bilayer system composed of symmetric CsCl. The figure depicts currents at $-2$ and $\pm 4$ mV with a conductance range between 0.134 to 3.565 nS. These data also illustrate the capacity of the amyloid channel system to be open and stable for long periods, and occasionally to make minor switches in conductance in the middle of these prolonged open states. In addition, the data shown in the fifth trace from the top illustrate another aspect of the amyloid channel system, with an abrupt episode of relatively rapid gating and switches to higher conductances being shown to occur.

## Pharmacology of amyloid ion channels

### Blockade of amyloid channels by aluminium

Nearly everyone aware of the problem of Alzheimer's disease is sensitive to the possibility that aluminium might in some way contribute to the onset or severity of the disease.[61,62] One reason is that the amyloid plaques associated with the disorder have been reported to contain high concentrations of aluminium. This finding is itself presently an issue, but it was this possibility which led us to investigate the question of whether aluminium might have some action on the amyloid channels.

A bi-ionic system was constructed, composed of (*cis*) 40 mM NaCl and (*trans*) 40 mM KCl (Fig. 6.3). At O V, an upward going current can be observed. The chemical potentials of $Na^+$ and $K^+$ are equal, but since $P_k > P_{Na}$, the direction of the current shows that there is

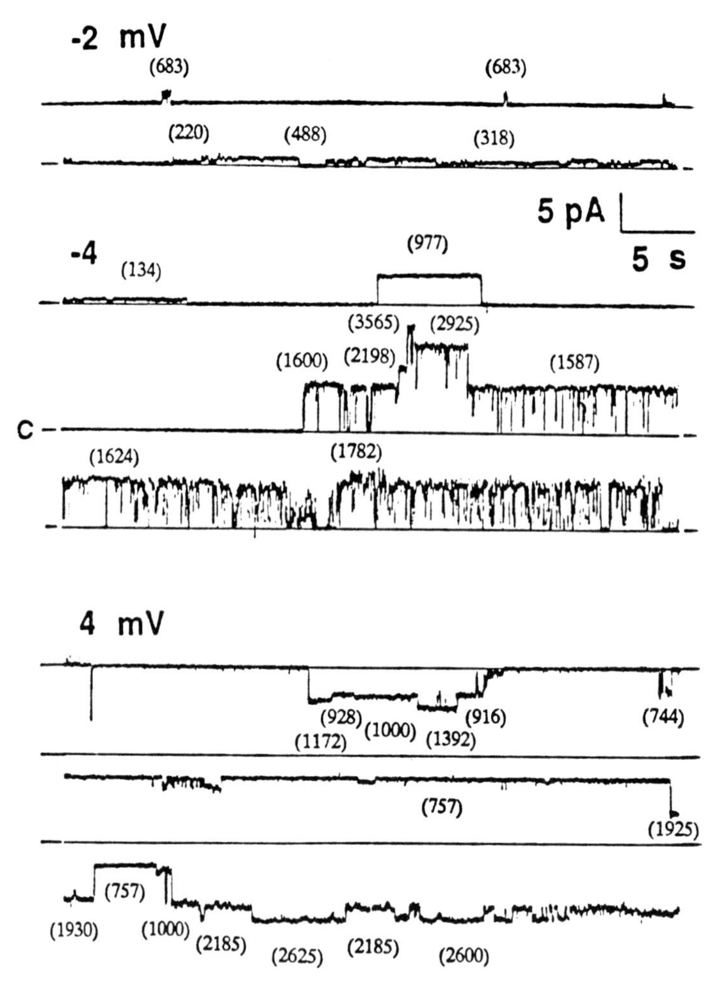


**Fig. 6.2** Multiple conductance behaviour of amyloid $\beta$ channel protein (A$\beta$P) in planar lipid bilayers. Amyloid $\beta$ protein [A$\beta$P(1–40)] was incubated at room temperature for 9 weeks in 50 mM NaHEPES buffer, pH 7.4. The peptide was diluted in chamber containing the same buffer, and allowed to insert directly into a bilayer composed of POPE : PS (1 : 1). Ionic current was recorded at various potentials (viz. −2, −4 and 4 mV). Conductances (values in parentheses) were observed within the range of ca. 200 pS to ca. 4 nS. Closures occurred between many of the current levels and zero.

net $K^+$ flux from *trans* to *cis*. When 1 mM $AlCI_3$ was added to the *cis* chamber, the current promptly went to zero. A small amount of current could be elicited at $\pm 60$ mV, the effect of $Al^{3+}$ being slightly more profound for current flowing from *cis* to *trans*.

The effect of aluminium is also detectable at much lower concentrations. As an illustration, we show in Fig.6.4 the blocking action of 20 $\mu$M $AlCl_3$ in the *cis* compartment on a bi-ionic system of (*cis*) 37.5 mM CsCI and (*trans*) 25 mM $CaCI_2$. The upper record obtained at $-40$ mV (*cis*) depicts an upward-going current, consistent with the flow of $Ca^{2+}$ from *trans* to *cis*. When $Al^{3+}$ is added, the current is blocked to the extent that, even at $-60$ mV, the movement of $Ca^{2+}$ from the *trans* to *cis* compartments was virtually zero. However, at $-100$ mV instances of higher conductance could be detected intermittently. This result

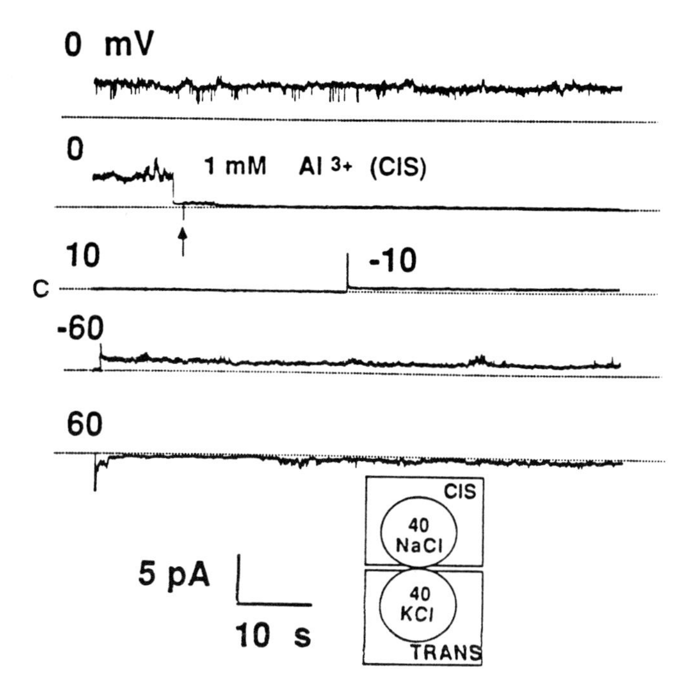


**Fig. 6.3** Influence of 1 mM $AlCl_3$ on [AβP(1–40)] channels. Amyloid channels were formed in an asymmetric system of 40 mM NaCl (*cis*) and 40 mM KCl (*trans*). 1 mM $AlCl_3$ was added to the *cis* chamber and blocking effects on current documented at different transmembrane potentials. (Data are from Arispe *et al.*[58].)

indicates that $Al^{3+}$ might be blocking the channel, but that it could be 'kicked off' by a substantial voltage gradient. In a separate experiment with the same bi-ionic system, with only 10 μM $AlCl_3$, similar profound blocking action could be detected (Fig. 6.5). More recent experiments have shown that $Zn^{2+}$can block amyloid channels both in bilayers[62] and neuronal plasma membranes[80].

## Blockade of amyloid channels by tromethamine

Tromethamine, also known as THAM or TRIS, is also able to block amyloid channels. As shown in Fig. 6.6, addition of 10 mM TRIS to a bi-ionic system of (*cis*) 37.5 mM CsCl and (*trans*) 25 mM $CaCl_2$ is able to block the movement of $Ca^{2+}$ from *trans* to *cis*, or the movement of $Cs^+$ in the opposite direction. A similar result has been obtained in an entirely different experimental conditions. Typical amyloid channel activity can be observed in a solvent-free PS bilayer constructed on the tip of a patch pipette. This activity can also be abolished by addition of tromethamine to the bathing solution.[59] These data thus indicate two important conclusions. First, the amyloid channel formation is not related in any way to the solvent (*n*-decane) in the planar lipid bilayer system. Second, the tromethamine inhibition is a general effect, and independent of the method used to observe the amyloid channels.

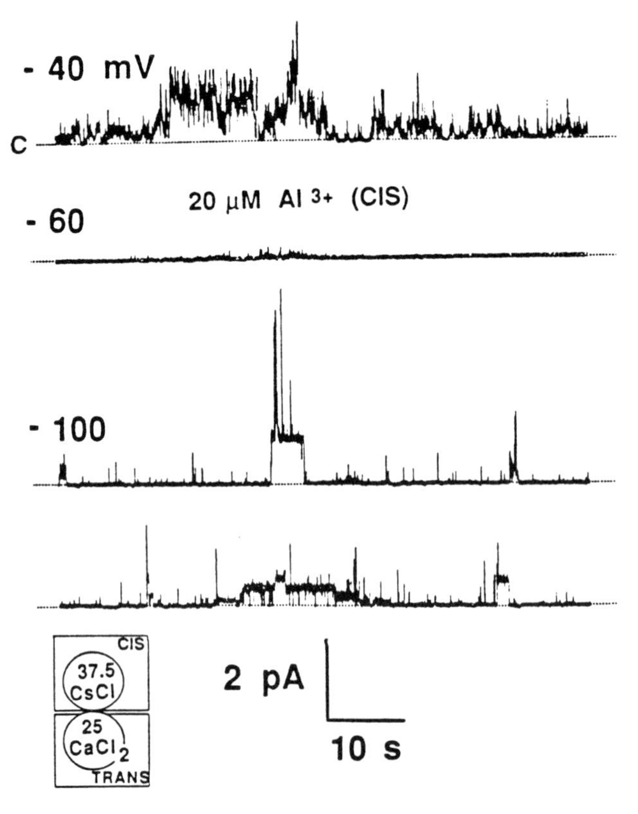


**Fig. 6.4** Influence of 20 $\mu$M $AlCl_3$ on [A$\beta$P(1–40)] channel current. Amyloid channels were formed in an asymmetric system of 37.5 mM CsCl (*cis*) and 25 mM $CaCl_2$ (*trans*). The difference in concentrations is designed to maintain an osmotic balance. 20 $\mu$M $AlCl_3$ was added to the *cis* compartment and the blocking effects on ionic current generated by different potentials. (Data are from Arispe *et al.*[58].)

## Amyloid channel activity is insensitive to nitrendipine

Drugs such as nitrendipine, which principally block L-type calcium channels,[81] have been previously considered as possible therapeutics for Alzheimer's disease. In addition, it has been recently reported that amyloid peptide toxicity *in vitro* is attenuated by drugs in this category.[63] The fact that the amyloid channel was also a calcium channel therefore led us to test whether nitrendipine could also block this channel. As shown in Fig. 6.7, A$\beta$P[1–40] was introduced into a bilayer system, and a set of voltages between $\pm$40 mV applied. Nitrendipine (10 $\mu$M) was then added first to the *cis* chamber and then to the *trans* chamber of the system, and the same programme of voltage pulses applied systematically. No obvious effect was observed, as born out in the I–V curves in Fig. 6.8. From these data we conclude that however the amyloid channel is structured, it is not some sort of L-type calcium channel.

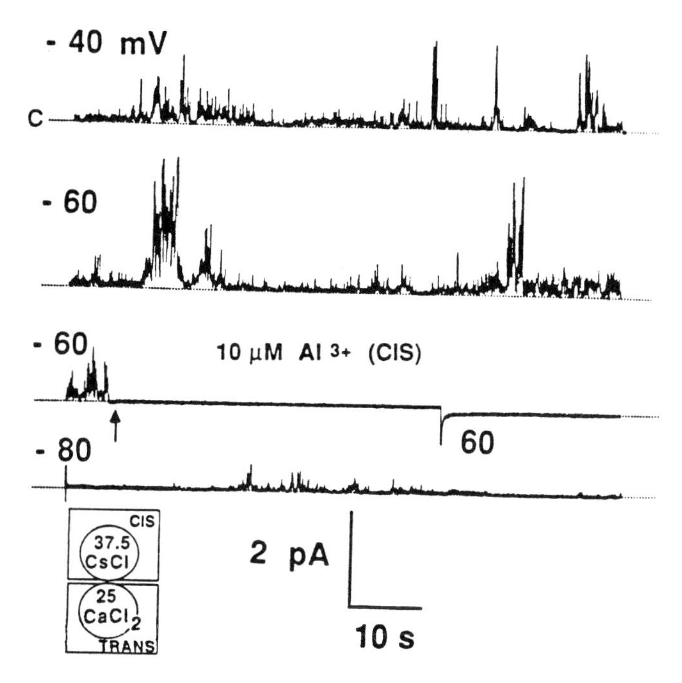


**Fig. 6.5** Influence of 10 $\mu$M $AlCl_3$ on [A$\beta$P(1–40)] channels. Amyloid channels were formed in an asymmetric system of 37.5 mM CsCl (*cis*) and 25 mM $CaCl_2$ (*trans*). 10 $\mu$M $AlCl_3$ was added to the *cis* compartment and the blocking effects on the conductance documented with driving forces up to $-80$ mV. (Data are from Arispe *et al.*[58].)

## Amyloid channel activity is insensitive to tacrine

Tacrine (9-amino-1,2,3,4-tetrahydroacridine) is also known as Cognex®, the only drug presently approved by the United States Food and Drug Administration for the treatment of Alzheimer's disease. The concept behind the use of this drug is that it blocks acetylcholine esterase at submicromolar concentrations, and thereby might be expected to elevate acetylcholine levels in the brain. Destruction of cholinergic neurones does occur in Alzheimer's disease, and thus any agent able to spare the acetylcholine has received attention as a possible therapeutic. From our own perspective, tacrine attracted our attention because of a recent report that this compound inhibited the three major types of calcium channel.[64] The $IC_{50}$ values for T-type channels was 125 $\mu$M, and 80 $\mu$M for N- and L-type channels. However, as shown in Fig. 6.9, tacrine did not appear to affect amyloid channels over the voltage range of $-20$ to 30 mV. This conclusion was born out by the I–V curves for these data shown in Fig. 6.10.

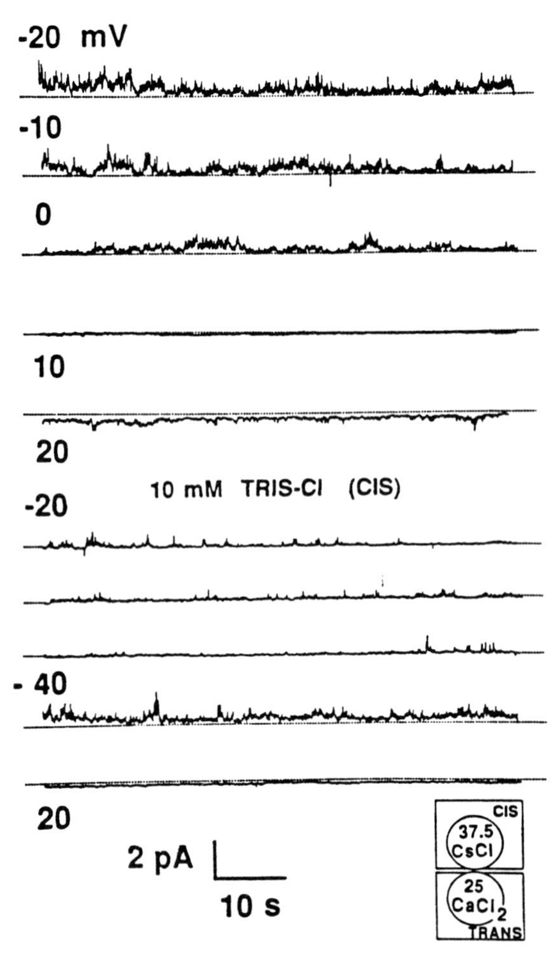


**Fig. 6.6** Influence of tromethamine (THAM, TRIS) on [AβP(1–40)] channels. Amyloid channels were formed in an asymmetric system of 37.5 mM CsCl (*cis*) and 25 mM $CaCl_2$ (*trans*). 10 mM TRIS was added to the *cis* compartment. Blocking effects on $Cs^+$ and $Ca^{2+}$ currents are shown for different transmembrane potentials. (Data are from Arispe *et al.*[58].)

## Therapeutic implications of amyloid channels for Alzheimer's disease

### Amyloid channel hypothesis for neurotoxicity

These data have led us to consider the possibility that the ion channel activity of amyloid might form the basis of the neurotoxic properties of this molecule. As summarized in Fig. 6.11, the origin of the amyloid is the amyloid precursor protein (APP), which is depicted on the far left-hand side of the figure. Proteolytic cleavages release the large soluble portion to the extra-cellular compartment. A portion of the membrane-spanning domain, shown as the the blackened segment, becomes the amyloid β protein (AβP). In its soluble form, the AβP can aggregate to form various lower-order polymers. These polymers can further aggregate to form the extensive polymers found in amyloid plaques, or can enter accessible membranes to form channels (upper portion of the figure). $Ca^{2+}$, or $K^+$, or other ions can then enter or exit the cell, with toxic consequences. For example, if one amyloid channel were placed in a cell of

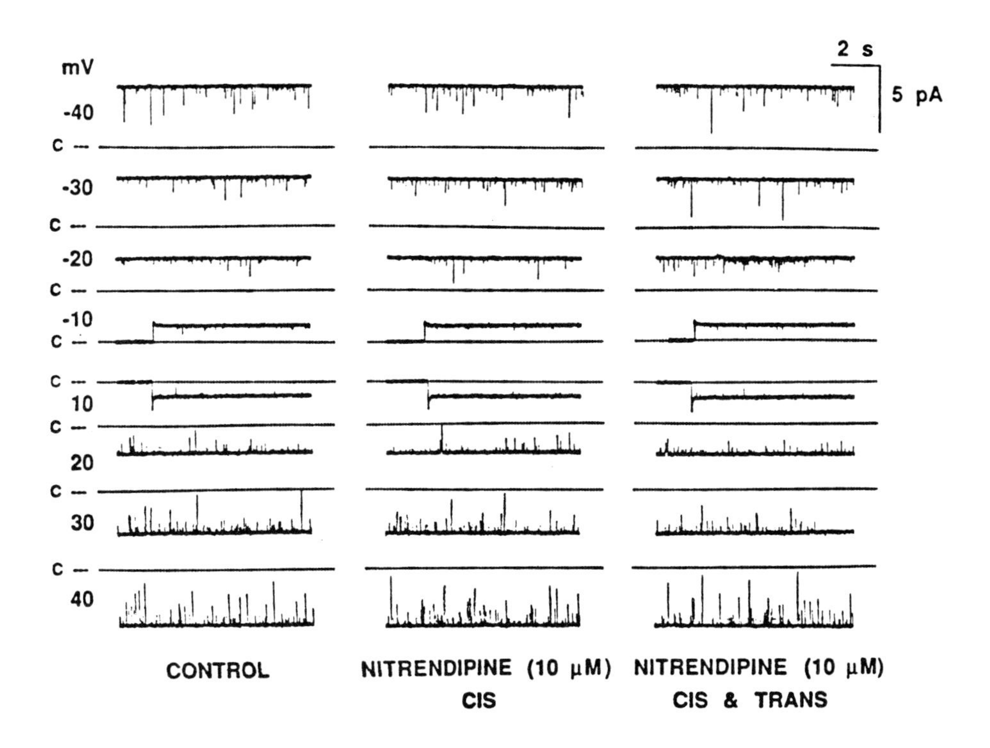


**Fig. 6.7** Amyloid [AβP(1–40)] channels in the presence of 10 μM nitrendipine. The channels are studied in the symmetric system of 50 mM CsCl. The left-hand set of current records are the control case. The middle set of current records are with 10 μM nitrendipine in the *cis* chamber. The right-hand set of current records are with 10 μM nitrendipine in both the *cis* and *trans* chambers. (Data are from Arispe *et al.*[59].)

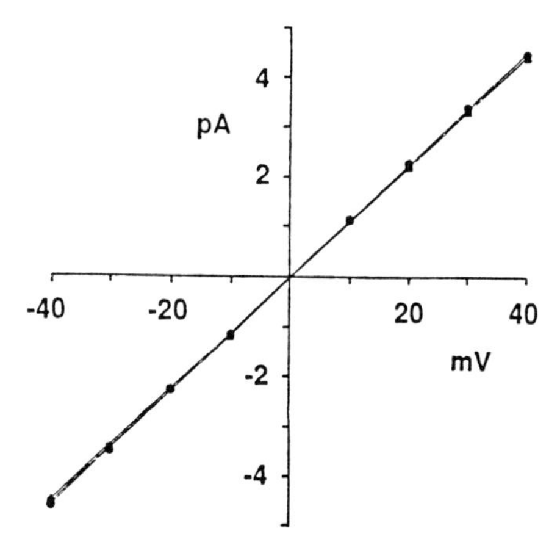


**Fig. 6.8** I–V curve for amyloid [AβP(1–40)] channels in the presence and absence of 10 μM nitrendipine. The data are from Fig. 6.7. Data are from the control (*triangle*), drug in the *cis* chamber (*square*), and in both chambers (*circle*).

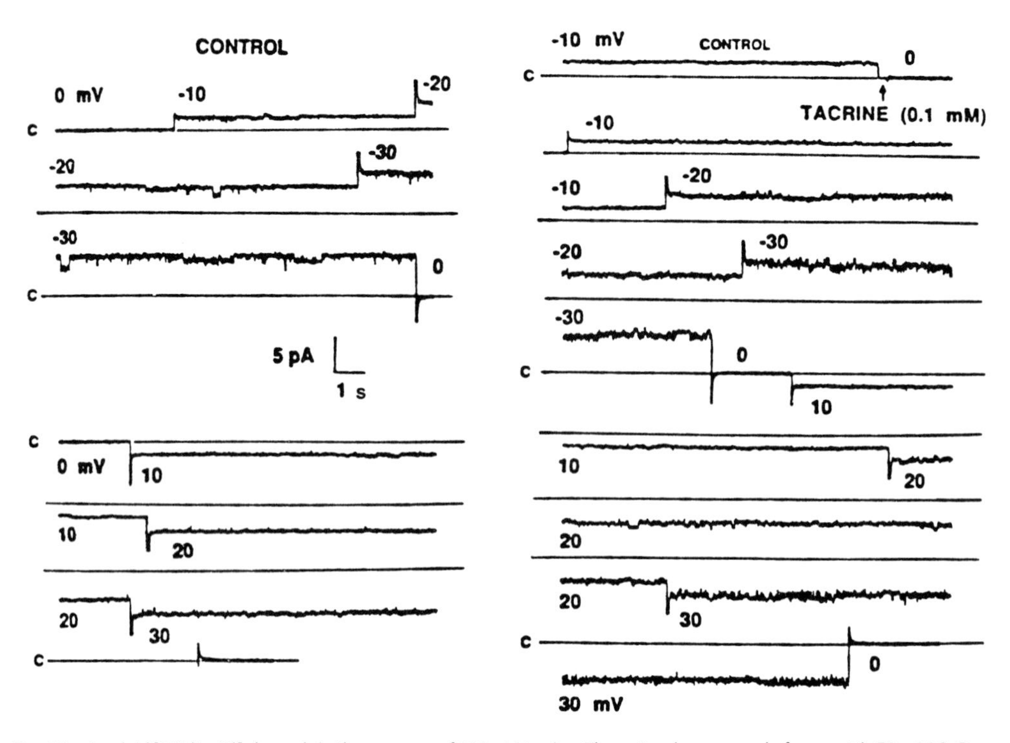


**Fig. 6.9** Amyloid [AβP(1–40)] channels in the presence of 100 $\mu$M tacrine. The system is composed of *symmetric 50 mM* CsCl. Tacrine is also known as Cognex®, a drug approved by the FDA in 1993 for the treatment of Alzheimer's disease. The effective concentration in patients is said to be in the range of 100 nM.

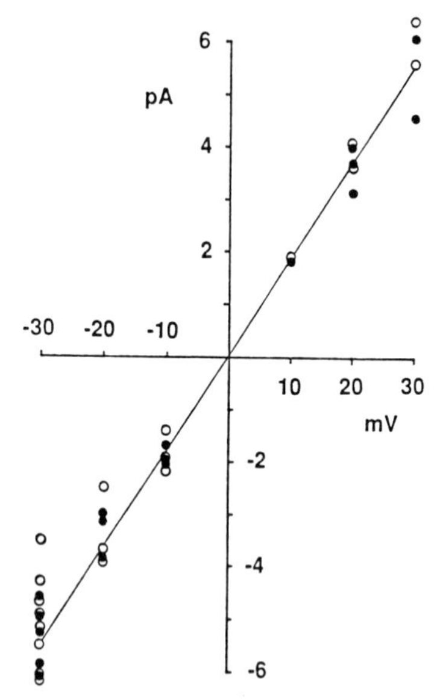


**Fig. 6.10** I–V curve for amyloid [AβP(1–40)] channels in the presence and absence of tacrine. The data are from Fig. 6.6. The open symbols are from the control case. The filled symbols are in the presence of 100 $\mu$M tacrine. No significant distinction can be made between the two cases.

25 $\mu$m in diameter, the ensuing flux of the less-permeable $Na^+$ would lead to a change of 10 $\mu$M/s. Ions such as $Ca^{2+}$ or $K^+$, with greater permeability, would obviously permeate the channel at a greater rate. $Ca^{2+}$ need only reach the millimolar range for a few minutes to be lethal. Amyloid plaques themselves have sometimes been considered to be toxic,[65] although the matter is a subject of some controversy. However, dissociation to intrinsically toxic amyloid channel oligomers could also occur. We should mention at this juncture that the soluble amyloid can also potentiate the action of other receptor systems, such as NMDA,[55] or modify the expression of other channels, such as $K^+$ channels in fibroblasts.[66]

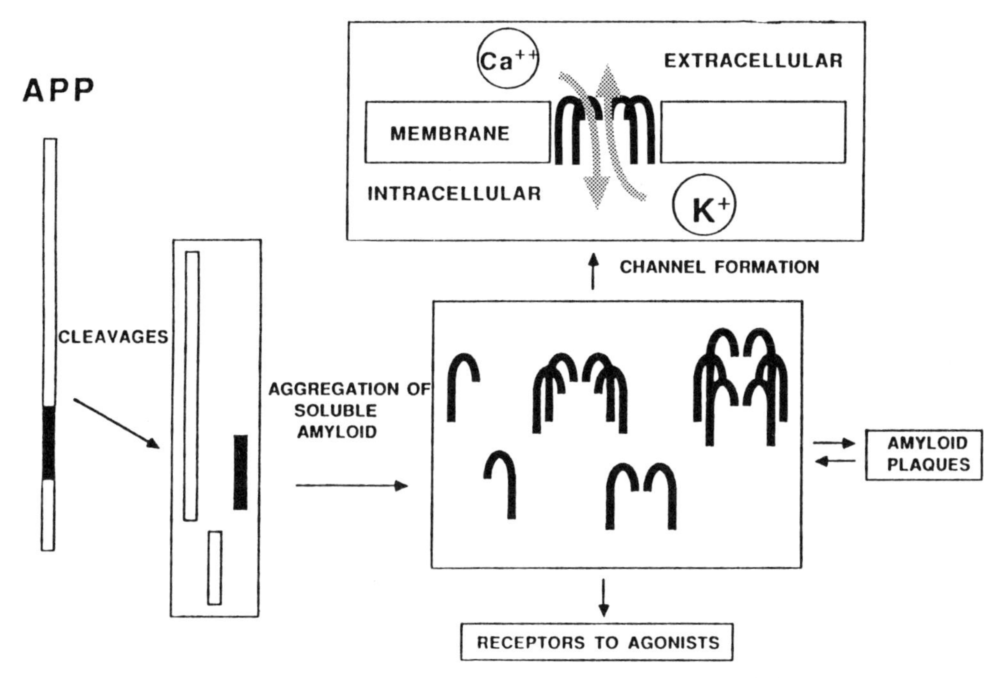


**Fig. 6.11** Amyloid channel hypothesis for neurotoxicity. (Figure reproduced from Arispe *et al.*[70,71].)

## Amyloid channel blocking drugs as potential therapeutic agents

If the ion channel hypothesis for the neurotoxicity of amyloid proves to be correct, and if this neurotoxic principle is the cause of cell death in Alzheimer's disease, then drugs which block the amyloid channel could prove useful in preventing or at least retarding the progression of the disorder. From the data at hand we at least know that there do exist some agents able to block the amyloid channel. Aluminium could be problematic in that toxicity is a well-known phenomenon.[67] However, the normal concentration of aluminium in the blood is 0.170 g/l, and soft tissues contain between 0.2 and 0.6 ppm.[68] It is thus possible that the origin of aluminium in amyloid plaques, if true, could be due in part to the intrinsic affinity of amyloid for this naturally occurring substance.[58] Zinc is much less toxic, and has been reported to be metabolically associated with ageing and Alzheimer's disease[82–84].

Tromethamine, on the other hand, has certain advantages in terms of administration to humans.[69] Known medicinally as THAM, it has been administered routinely to humans as part of the preoperative preparation for cardiac bypass surgery. The equilibrating concentration is ca. 16 mM, and its function is to prevent local acidosis when the circulation is arrested. However, there is no evidence to suggest that tromethamine has any action on Alzheimer's disease, and for the present we can only use this information to verify that non-toxic compounds do exist which interfere with amyloid channel function. No doubt others can be found, and the amyloid channel system may well turn out to be a useful drug discovery vehicle for this important problem.[70,71]

## References


1. Reisberg, B. (1983). *Alzheimer's disease*. The Free Press, New York.
2. Evans, D.A., Funkstein, H.H., Albert, M.S., *et al.* (1989). *JAMA*, **262**, 2551–2556.
3. Jorm, A.F. (1990). *The epidemiology of Alzheimer's disease and related disorders*. Chapman & Hall, London.
4. Hardy, J.A. and Higgins, G.A. (1992). *Science*, **256**, 184–185.
5. Skoog, I., Nilsson, L., Palmertz, B., Andreasson, L.-A., and Svanborg, A. (1993). *N. Engl. J. Med.*, **328**, 153–158.
6. Glenner, G.G. and Wong, C.W. (1984). *Biochem. Biophys. Res. Commun*, **122**, 1131–1135.
7. Masters, C.L., Simms, G., Weinman, N.A., Multhaup, G., McDonald, B.L., and Beyreuther, K (1985). *Proc. Natl. Acad. Sci. USA*, **82**, 4245–4249.
8. Fraustchy, S.A., Baird, A., and Cole, G.M. (1991). *Proc. Natl. Acad. Sci. USA*, **88**, 8362–8366.
9. Neve, R.L., Dawes, L.R., Yankner, B.A., Benowitz, L.L., Rodriguez, W., and Higgins, G.A. (1990) *Prog. Brain Res.*, **86**, 257–267.
10. Selkoe, D.J. (1991) *Neuron*, **6**, 487–498.
11. Selkoe, D.J. (1993). *Trends in Neuroscience*, **16**, 403–409.
12. Dickson, D.W. and Rogers, J. (1992). *Neurobiol. of Aging*, **13**, 793–798.
13. McKee, A.C., Kosik, K.S., and Kowall, N.W. (1991). *Ann Neurol*, **30**, 156–165.
14. Roth, M., Tomlinson, B.E., and Blessed, G. (1966). *Nature*, **209**, 109–110.
15. Blessed, G., Tomlinson, B.E., and Roth, M. (1968). *Br. J. Psychol.*, **114**, 797–811.
16. Terry, R.D. and Peña, C. (1965). J. *Neuropathol. Exp. Neurol.*, **24**, 200–210.
17. Joachim, C.L., Duffy, L.K., and Selkoe, D.J. (1988). *Brain Res.*, **474**, 100–111.
18. Goat, A., Chartier-Harlin, M.C., Mullan, M. *et al.* (1991). *Nature*, **349**, 704–706.
19. Schellenberg, G., Bird, T., Wijsman, E., *et al.* (1992) *Science*, **258**, 668–671.
20. St George-Hyslop, P., Rogaev, J., Montilla, M., *et al.* (1992) *Nature Genetics*, **2**, 330–334.
21. Van Broeckhoven, C., Backhovens, H., Cruts, M., De Winter, G., Bruyland, M., Cras, P., and Martin, J. (1992). *Nature Genetics* **2**, 335–339.
22. Strittmatter, W., Saunders, A., Schmeckel, D., Pericak-Vance, M., Enghild, J., Salvesen, G., and Roses, A. (1993). *Proc. Natl. Acad. Sci. USA*, **90**, 1977–1981.
23. Rosenberg, R.N. (1993). *Neurology*, **43**, 851–856.
24. Roberts, G.W., Gentleman, S.M., Lynch, A., and Graham, D.I. (1991). *Lancet*, **338**, 1422–1423.
25. Siman, R., Card, P., Nelson, R.B., and Davis, L.G. (1989). *Neuron*, **3**, 275–285.
26. Kawarabayashi, T., Shoji, M., Harigaya, Y., Yamaguchi, H., and Hirai, S. (1991). *Brain Res.*, **563**, 334–338.

27. Wallace, W. (1993). *Proc. Natl. Acad. Sci. USA*, **90**, 8712–8716.
28. Takeda, M., Tanaka, S., Daikoku, S., Oka, M., Sakai, K., and Katunuma, N. (1994). *Neurosci. Lett.*, **168**, 57–60.
29. Wolozin, B., Lesch, K., Lebobics, R., and Sunderland, T. (1993). *Biol. Psychol.*, **34**, 824–838.
30. Kang, J., Lemaire, H.-G., Unterbeck, A., *et al. Nature*, **325**, 733–736.
31. Shoji, M., Golde, T., Ghiso, J., *et al. Science*, **258**, 126–129.
32. Seubert, P., Oltersdorf, T., Lee, M.G., *et al.* (1993). *Nature*, **361**, 260–263.
33. Haass, C., Schlossmacher, M., Hung, A., *et al.* (1992). *Nature*, **359**, 322–327.
34. Busciglio, J., Gabuzda, D.H., Matsudaira, P., and Yankner, B.A. (1993). *Proc. Natl. Acad. Sci. USA*, **90**, 2092–2096.
35. Oosawa, T., Shohi, M., Harigaya, Y., Cheung, T., Shaffer, L., Younkin, S,G., Younkin, S., and Hirai, S. (1993). *Soc. Neurosci. Abst.*, **19**, 1038.
36. Hilbich, C., Kisters-Woike, B., Reed, J., Master, C., and Beyreuther, K. (1991). *J. Mol. Biol.*, **218**, 149–163.
37. Barrow, C.J., Yasuda, A., Kenny, T.M., and Zagorsky, M.G. (1992). *J. Mol. Biol.*, **225**, 1075–1093.
38. Kirshner, D.A., Abraham, C., and Selkoe, D.J. (1986). *Proc. Natl. Acad. Sci. USA*, **83**, 503–507.
39. Burdick, D., Soreghan, B., Kwon, M., Kosmoski, J., Knauer, M., Henschen, A., Yates, J., Cotman, C., and Glabe, C. (1992). *J. Biol. Chem.*, **267**, 546–554.
40. Gorevic, P.D., Castano, E.M., Sarma, K., and Frangione, B. (1987). *Biochem. Biophys. Res. Commun.*, **147**, 854–862.
41. Castano, E.M., Ghiso, J., Prelli, F., Gorevic, P.D., Migheli, A., and Frangione, B. (1986). *Biochem,. Biophys. Res. Commun.*, **141**, 782–789.
42. Kirshner, D.A., Inouye, H., Duffy, L.K., Sinclair, A., Lind, M., and Selkoe, D.J. (1987). *Proc. Natl. Acad. Sci. USA*, **84**, 6953–6957.
43. Fraser, P.E., Nguyen, J.T., Swiewicz, W.K., and Kirschner, D.A. (1992). *Biophys. J.*, **60**, 1190–1201.
44. Inouye, H., Fraser, P., and Kirschner, D.A. (1993). *Biphys. J.*, **64**, 502–519.
45. Forte, M, Guy, H.R., and Mannella, C. (1987). *J. Bioenergetics Biomembr.*, **19**, 341–350.
46. Blachly-Dyson, E., Peng, S.Z., Colombini, M. and Foret, M. (1989). *J. Bioenergetics Biomembr.*, **21**, 471–483.
47. Yellen, G., Jurman, M.E., Abramson, T., and MacKinnon, R. (1991). *Science*, **251**, 939–942.
48. Peled, H. and Shai, Y. (1993). *Biochemistry*, **32**, 7879–7885.
49. Shinozaki, K., Anzai, K., Kirino, Y., Lee, S., and Aoyagi, H. (1994). *Biochem. Biphys. Res. Commun.*, **198**, 445–450.
50. Guy, H.R. and Conti, F. (1990). *Trends in Neuroscience*, **13**, 3867–3874.
51. Guy, H.R., Durrell, S., and Raghunathan, G. (1992). *Biophys. J.*, **62**, 243–251.
52. Pollard, H.B., Guy, H.R., Arispe, N., *et al.* (1992). *Biophys. J.*, **62**, 15–18.
53. Yankner, B.A., Duffy, L.K., and Kirshner, D.A. (1990). *Science*, **250**, 279–282.
54. Koh, J.Y., Yang, L.L., and Cotman, C.W. (1990). *Brain Res.*, **533**, 315.
55. Mattson, M.P., Barger, S.W., Cheng, B., Lieberberg, I., Smith-Swintosky, V.L., and

Rydel, R.E. (1993). *Trends in Neuroscience*, **16**, 409–414.
56. Pike, C., Walencewicz, A.J., Glabe, C.G., and Cotman, C.W. (1991). *Brain Res.*, **563**, 311–314.
57. Busciglio, J., Yeh, J., and Yankner, B.A. (1993). *J. Neurochem.*, **61**, 1565–1568.
58. Arispe, N., Rojas, E., and Pollard, H.B. (1993*a*). *Proc. Natl. Acad. Sci. USA*, **90**, 567–571.
59. Arispe, N., Pollard, H.B., and Rojas, E. (1993*b*). *Proc. Natl. Acad. Sci. USA*, **90**, 10573–10577.
59. Wisniewski, H.M. and Wen, G.Y. (1992). *Ciba Foundation Symposium*, **169**, 142–154, and Discussion 154–164.
60. Arispe, N., Pollard, H.B., and Rojas, E. (1992). *J. Membr, Biol.*, **130**, 191–202.
61. Mantyh, P.W., Ghilardi, J.R., Rogers, S., DeMaster, E., Allen, C., Stimson, E.R. and Maggio, J.E. (1993). *J. Neurochem.*, **61**, 1171–1174.
62. Arispe, N., Pollard, H.B., and Rojas, E. (1996). *Proc. Natl. Arad. Sci. USA*, **93**, 1710–1715
63. Weiss, J.H., Pike, C.J., and Cotman, C.W. (1994). *J. Neurochem.*, **62**, 372–375.
64. Kelly, K.M., Gross, R.A., and Macdonald, R.L. (1991). *Neurosci. Lett.*, **132**, 247–252.
65. Jarrett, J.T. and Lansbury, P.T. (1993). *Cell*, **73**, 1055–1058.
66. Etcheberrigaray, R., Ito, E., Kim, C.S., and Alkon, D.L. (1994). *Science*, **264**, 276–279.
67. Doull, J., Klaassen, C.D., and Amdur, M.O. (ed.) (1980) *Toxicology*, 2nd edn. MacMillan Publishing Co., New York.
68. Underwood, W.J. (1971). *Trace elements in human and animal nutrition*. Academic Press, Inc., New York.
69. Nahas, G.G. (1962). *Pharmacol. Rev.*, **14**, 447–472.
70. Arispe, N., Pollard, H.B., and Rojas, E. (1994*a*). *Molecular and Cellular Biochem*, **140**, 119–125.
71. Arispe, N., Pollard, H.B., and Rojas, E. (1994*b*). *Ann. N.Y. Acad. Sci.*, **747**, 256–266.
72. Sherrington, R., Rosqev, E.I., Liang, Y. *et al.* (1995). *Nature*, **375**, 754-760.
73. Levy-Lahad, E., Wasco, W., and Poorkaj, P. (1995). *Science*, **269**, 973–977.
74. Cohen, E.B., Lee, G., Arispe, N., and Pollard, H.B. (1995). *FEBS Lett*, **377**, 444–450.
75. Rojas, E., Pollard, H.B., Haigler, H.T., Panro, C. and Burns, A.L. *J. Bide. Chem.*, **265**, 21207–21215.
76. Rojas, E., Arispe, N., Haigler, H., Burns, A.L., and Pollard, H.B. (1992). *Bone and Mineral*, **17**, 214–218.
77. Pollard, H.B., and Rojas, E. (1988). *Proc. Natl Acad. Sci. USA*, **85**, 2974–2978.
78. Pollard, H.B., Burns, A.L. and Rojas, E. (1990). *J. Membr. Biol.*, **117**, 101–112.
79. Raynal, P. and Pollard, H.B. (1994). *BBA Biomembranes*, **1197**, 63–93.
80. Kawahava, M., Arispe, N., Kuroda, Y., and Rojas, E. (1997). *Biophys. J.*, **73**(1), (in press).
81. Tsien, R.W., Hess, P., McClesky, W. and Rosenberg, R. (1987). *Annu. Rev of Biophys and Biophys. Chem.*, **16**, 265–290.
82. Bush, A.L., Pettingel, W.H., Multhaup, G. *et al.* (1994). *Science*, **265**, 1464–1767.
83. Kaiser, J. (1994). *Science*, **265**, 1365.
84. Tully, C.L., Snowdon, D.A., and Markesbery, W.R. (1995). *Neuroreport*, **6**, 105–108.
85. Hardy, J. (1997). *Trends Neurosci.*, **20**, 154–159.

# 4 Potassium Channels

# 7 The biophysical basis of $K^+$ channel pharmacology

B. Soria

## Introduction

$K^+$ channels are found in all cells, where they have an important role in maintaining transmembrane potential. Closure of these channels leads to membrane depolarization, which can be followed by cell-specific activity such as the contraction of vascular smooth muscle, or secretion of insulin from pancreatic $\beta$-cells. The central physiological feature is that the activation of $K^+$ permeability drives the transmembrane voltage towards the $K^+$ equilibrium potential. $K^+$ channel modulators change the microscopic properties of the channels, thus leading to changes in membrane potential or intracellular ionic activity as a direct consequence of other voltage-activated ion channels opening. $K^+$ channel defects may result in electrophysiological changes, and diseases such as hypertension, heart arrhythmias or diabetes are, or could be, treated with drugs which modulate the activity of $K^+$ channels. It is not surprising therefore that a number of drugs have been introduced which influence $K^+$ channels by either blocking or opening them. For example, it is customary to treat type 2 (non-insulin-dependent) diabetes mellitus with sulphonylurea derivatives, which exert their insulinotropic effect by closing the $K_{ATP}$ channels of the pancreatic $\beta$-cell. This has stimulated the pharmaceutical industry to explore the structure – function relationship of $K^+$ channels and $K^+$ channel modulators in an attempt to design more effective and specific therapeutic weapons against such disorders.

Pharmacological studies on wild-type and cloned $K^+$ channels have yielded important insights into $K^+$ channel properties and structure. Receptor sites for $K^+$ channel blockers have been located at the pore-forming structures of voltage-gated $K^+$ channels. The pore region is highly conserved in all the $K^+$ channel types; hence, 'pore-binding' drugs will not be specific for channel subtypes or locations. A detailed knowledge of the structure and function of other channel regions will provide invaluable information for rational drug design in the future.

## Physiological role of potassium channels

Upon activation, $K^+$ channels drive the membrane potential to the $K^+$ equilibrium potential, whereas blocking them will transfer control of the equilibrium potential to other permeabilities (for example, depolarizing permeabilities as non-specific cation conductances).

As a general rule, $K^+$ channel blockade leads to membrane depolarization, whereas $K^+$ channel activation hyperpolarizes the cell membrane. Each cell is endowed with a particular set of $K^+$ channels which may vary both in functional properties and location. Thus, the spatiotemporal integration of their activity results in a particular set of physiological characteristics for the cell. All ionic species distribute unequally across the cell membrane and an equilibrium potential or 'battery' can be calculated for each ion from the Nernst equation. The main determinant of equilibrium potential, at a given temperature, is the ratio between the internal and the external ion concentrations. In mammalian cells the $K^+$ equilibrium potential is close to $-90$ mV.

Before describing the individual properties of each $K^+$ channel type, it is worth describing in some detail both input conductance and the I–V curve for $K^+$ permeabilities. Input conductance determines not only the zero current potential (ZCP), but also the range over which membrane potential can be modulated by changes in the opening probability of membrane ion channels. The input conductance consists usually of one or more outward components carried by $K^+$, together with inward currents carried by $Na^+$ and/or $Ca^{2+}$ at depolarized membrane potentials (more positive than $-35$ mV) or by $K^+$ at membrane potentials more negative than the $E_K$ ($K^+$ equilibrium potential). The concept of ZCP, or the potential at which the net current crossing the membrane is zero, is a useful parameter, more so than the so-called resting membrane potential which is difficult to define and is at most times misleading. In spontaneously active tissues (e.g. the pancreatic *B*-cell at physiological nutrient concentrations) there is not a true resting potential. Furthermore, the membrane potential can step from one 'resting' value to another, depending on the cells, physiological condition.

The current-to-voltage relationship (I–V curves) (Fig.7.1) of excitable cells measured using the whole-cell configuration shows that the net input conductance (dI/dV) in the potential range between $-35$ mV and the $K^+$ equilibrium potential is usually small. An overall conductance of 1 nS can be measured (this means that a change in the ionic current of 10 pA would cause a membrane potential change of 10 mV). These quite large potential changes as a consequence of small current changes should be kept in mind when judging any result with a putative $K^+$ channel modulator.

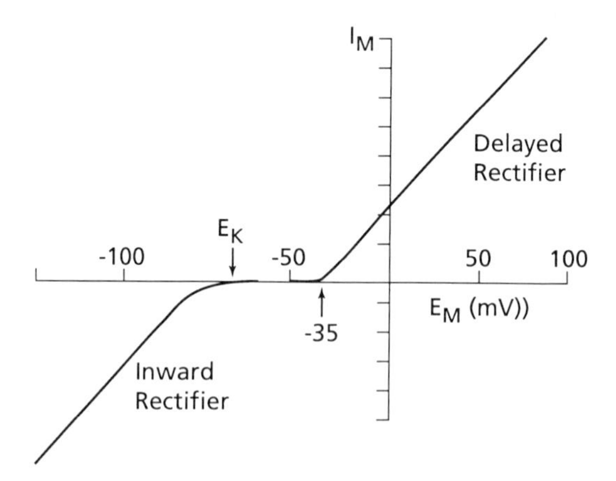


**Fig. 7.1** Current intensity versus membrane potential (I–V curves) for voltage-activated $K^+$ channels.

Outside these membrane potential limits ($-35$ mV and $E_K$), the input conductance increases several fold when the membrane potential is either depolarized (outward rectifier) or hyperpolarized (inward rectifier) (Fig. 7.1). The opening of $K^+$ channels will drive the membrane potential to the potassium equilibrium potential, thus keeping the membrane potential within discrete limits.

## Biophysical properties and nomenclature of $K^+$ channels in excitable cells

When compared with the number of $Ca^{2+}$ channel types described to date (Chapter 4), $K^+$ channels in excitable tissue represent a much larger group. More complementary cDNAs have been cloned for voltage-dependent $K^+$ channels than for any other voltage-dependent ion channel. Purely functional studies anticipated this result, since $K^+$-currents are far more diverse than voltage dependent $Na^+$, $Ca^{2+}$ or $Cl^-$ currents.[1,2]

### Functional classification of $K^+$ channels

Based on their single-channel microscopic properties, gating mechanisms, kinetics, voltage sensing, and pharmacology, $K^+$ channels can be divided into five different groups (Table 7.1).

**Table 7.1** Functional subdivision of K channels.

| |
|---|
| Voltage-dependent |
| Delayed rectifier ($K_{DR}$) |
| Transient outward currents ($K_A$) |
| Inward rectifier ($K_{IR}$) |
| Calcium-activated |
| High-conductance ($BK_{Ca}$) |
| Small-conductance ($S_{Ca}$) |
| Intermediate conductance ($I_{Ca}$) |
| Sodium-activated channels ($K_{Na}$) |
| Receptor-coupled |
| M-currents ($K_M$) |
| Muscarinic-Activated ($K_{ACh}$) |
| Serotonin-Inactivated (S5-HT) |
| Metabolically directed |
| ATP-modulated ($K_{ATP}$) |
| $O_2$-modulated ($K_{O2}$) |
| Fatty acid-activated ($K_{FA}$) |

This classification, based solely on functional criteria, may be incomplete. A class of cyclic nucleotide-gated channels could be added. Additionally, $O_2$-regulated and fatty acid-activated $K^+$ channels might be considered not as distinct classes, but to represent a more general effect on ion channels by $O_2$ and fatty acids.

## Structural classification of $K^+$ channels

Molecular cloning of the cDNAs for different $K^+$ channels, coupled with detailed electrophysiological studies, have greatly expanded our knowledge of $K^+$ channel function. Based on their amino acid sequences, these $K^+$ channels fall into three groups: (i) shaker-like channels, (ii) minimal $K^+$ (MinK) voltage-dependent $K^+$ channels: and (ii) a grouping of ATP-sensitive and vertebrate inward rectifier channels. Various cDNAs and genomic DNAs have been cloned which allow a correlation to be made between functional and structural aspects of voltage-gated channels when expressed in oocytes or cell lines which lack native $K^+$ channels.

### Shaker-like $K^+$ channels

A group of $K^+$ channel genes from different subfamilies have now been isolated and shown to share sequence similarity to that of the *Shaker* gene family. Current evidence suggests that each subunit of the *Shaker*-like $K^+$ channel contains several putative transmembrane fragments (S1 to S6). The pore segment (P), also called H5 region (between S5 and S6), is thought to be involved in the actual ion conduction, whereas voltage sensitivity and cooperativity in channel opening are associated with charged amino acids present in the S4 and S5 segments and the S4–S5 linker. Inactivation involves the amino terminus acting as a 'ball' which ocludes the pore (N-type inactivation). Another type of inactivation (C-type inactivation) is linked to the ion permeation process.[3,4]

P(H5) designates a hydrophobic region that is highly conserved among $K^+$ channels. It has been proposed that this region may enter and exit the lipid bilayer from the extracellular side. The N-terminus of some $K^+$ channels contains an amphipathic structure (illustrated by the shaded ball of Fig 7.2) which may occlude the pore of the channel. It has been proposed that the subunit assembly signal is also located in the N-terminal region (it is assumed that four subunits assemble to make a functional $K^+$ channel).[5,6] On the other hand, the C-termini contain ligand-binding domains ($Ca^{2+}$, cGMP), together with a phosphorylation site. Voltage-activated outward $K^+$ channels and calcium-activated $K^+$ currents fall into this group.

### MinK channels

MinK are very slow, voltage-dependent $K^+$ channels that differ from Shaker-like channels in their structural organization. Each subunit (130 residues) appears to contain a single transmembrane segment; however, it is not clear how many subunits are needed to form a functional $K^+$ channel. Channel activity and gating is affected by mutations in the amino acid sequence, implying that the minK protein itself exhibits channel activity. Membrane potential changes result in oligomer formation in a manner reminiscent of pore-forming antibiotics such as alamethicin.[7]

### ATP regulated and inward rectifier channels

Inward rectifier $K^+$ channels have two transmembrane segments (M1 and M2) separated by a P loop equivalent to an H5 region. A central feature of these channels is the existence of a

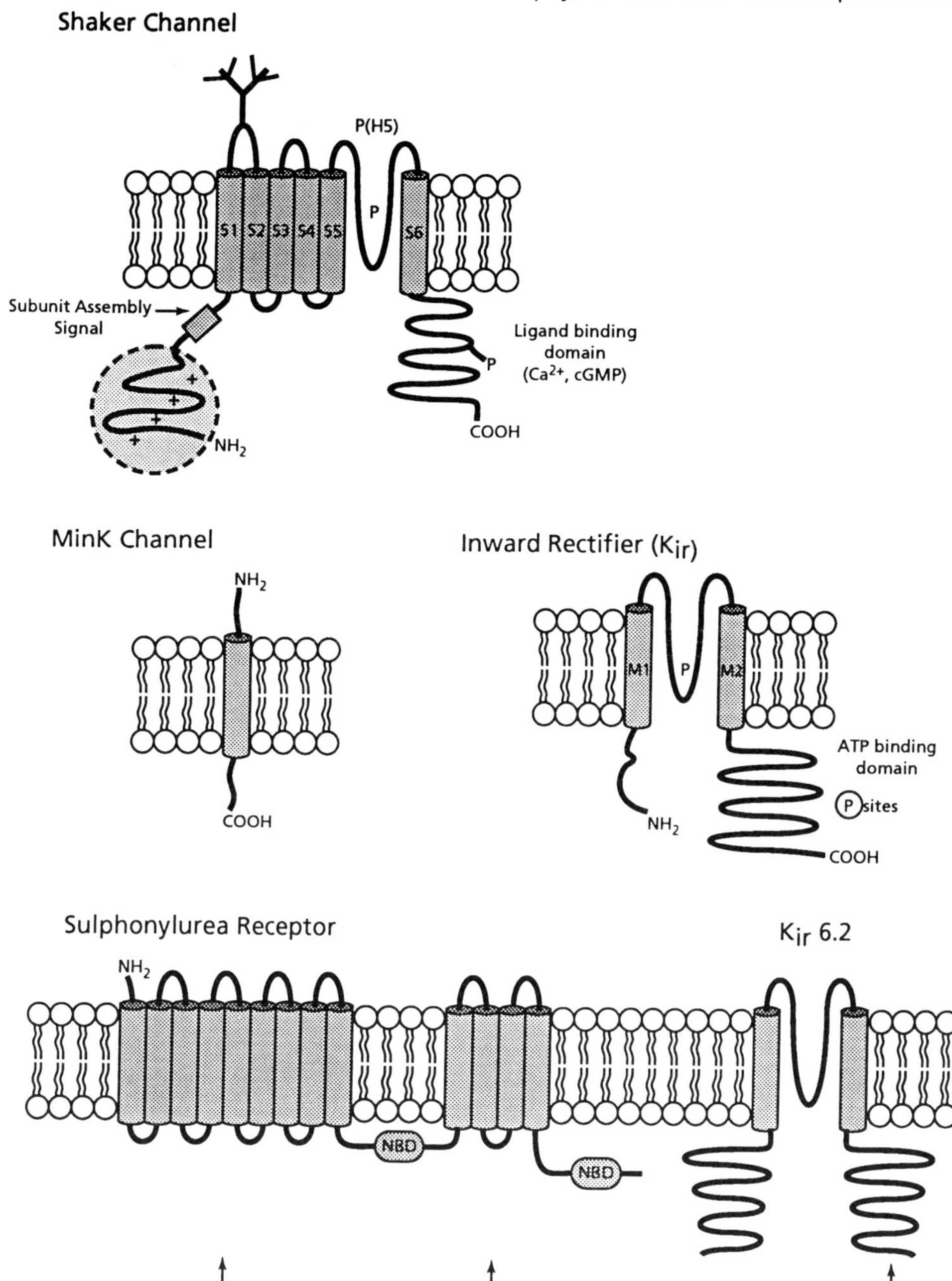


**Fig. 7.2** Structure of $K^+$ channels. NBD: Nucleotide binding domain.

large C-terminal domain, which somehow contributes to the pore structure and function. This domain contains sites for phosphorylation and regulatory molecules (ions, nucleotides, G protein subunits). As with *Shaker*-type $K^+$ channels, inward rectifier channels form tetrameric structures in a similar manner. It should be noted that some $K^+$ channel subtypes may require heteromultimeric assembly. Functional reconstitution of $K_{ATP}$ channels co-expressing Kir 6.2 (a member of the inward rectifier family), together with the sulphonylurea receptor (a member of the ATP-binding cassette superfamily), results in the full functional reconstitution of *in vivo* inward rectifier activities and properties.[8] The resultant channel has a conductance of 76 pS, can be blocked by ATP and sulphonylureas, and is activated by diazoxide. Expression of either sulphonylurea receptor or Kir 6.2 alone does not result in any measurable channel activity.

Other $K^+$ channels have not yet been characterized at the molecular level. $O_2$-regulated and $Na^+$-activated $K^+$ channels of the plasmalemma and $K^+$ channels present in intracellular membranes may come to represent additional gene families.

## Voltage-activated K⁺ channels

In voltage-dependent channels the probability of being open depends on the membrane potential. Voltage-dependent $K^+$ currents from a variety of eukaryotic cells are currently classified as: (i) delayed (outward) rectifier; [9–11] (ii) transient outward (A current).[12] and (iii) inward (anomalous) rectifier $K^+$ currents[13]. A new current type represented by the sarcoplasmic reticulum potassium channel has been recently added to the voltage-dependent channel group.

**Table 7.2** Properties of voltage-activated potassium channels.

| Property | Delayed rectifier ($K_{DR}$) | Transient outward ($K_A$) | Inward rectifier ($K_{IR}$) | Sarcoplasmic reticulum ($K_{SR}$) |
|---|---|---|---|---|
| Single-channel conductance (pS) | 50–60 | <1–20 | 5–30 | 180 |
| Blockers | TEA >>> 4-AP<br>Forskolin<br>Phencyclidine<br>9-Aminoacridine<br>Margatoxin<br>Imperator toxin<br>Zn2⁺ | 4-AP >>> TEA<br>Dendrotoxin<br>Quinidine<br>Phencyclidine<br>Tetra-aminoacridine<br>MCD (mast cell degranulating peptide) | Polyamines<br>Gaboon viper venom<br>$Sr^{2+}$<br>LY 97241 | Decamethonium<br>Hexamethonium<br>$Cs^+$ |
| Regulation | Delayed activation<br>Slow Inactivation | Fast activation and inactivation | Mg 2⁺-sensitive | Strong voltage dependence<br>Low $Na^{+/K}+$ selectivity |
| Physiological role | Action potential repolarization | Space-repetitive responses | Resting membrane potential | |

4-AP: 4-aminopyndine; TEA: tetraethyl ammonium

## Delayed rectifier current ($K_{DR}$)

This is described by Hodgkin and Huxley in their classical set of papers.[9–11] It consists of a current that turns on after a brief delay following the onset of a membrane depolarization, and persists while the depolarization is maintained for periods of $<100$ ms. This current, which contributes to the repolarization of the action potential (to keep action potentials short), is widely distributed in all excitable cells. For longer periods of depolarization ($>$ 100 ms) this current shows a slow (or steady-state) inactivation as observed in various tissues. Steady-state inactivation has been suggested as one of the targets for $K^+$ channel modulation.[14]

Delayed rectifier channels show single-channel conductances which vary between 5 and 60 pS. In these channels, tetraethyl ammonium (TEA) is a more effective blocker than 4-aminopyridine (4-AP).[15] They are also blocked by forskolin, phencyclidine, phalloidin, 9-aminoacridine, margatoxin, imperator toxin, and $Zn^{2+}$. TEA in millimolar quantities blocks all delayed rectifier channels when applied to the intracellular side. Armstrong and colleagues (for a review, see ref. 16) showed that the internal receptor for blockers lies in the pore region of the channel, and is only accessible when the channel is opened by depolarization. All these studies have been confirmed by molecular cloning of the channel (see below).

## Transient outward current ($K_A$)

Alternatively known as A-current ($K_A$, this is also activated by depolarization but shows fast inactivation when depolarization is maintained (Fig. 7.3). The current is partially inactivated at the resting membrane potential; thus, the amount of permeability is maximal when activated by a depolarizing step imposed after hyperpolarizing pulse.[12,15] The role of A-current is to space repetitive responses. Encoding membranes fire at a rate which reflects the stimulus intensity. Steady-state gating parameters for this current (activation and inactivation curves) explain why these channels serve as clampers in the interspike interval, spacing successive action potentials much more effectively than a standard combination of $Na^+$, $K^+$ (delayed rectifier) and leak channels could alone. Contrary to what happens with delayed rectifier currents, transient outward currents are more effectively blocked by 4-AP than by TEA (Fig. 7.4).[12,15]

## Inward (anomalous) rectifier K current ($K_{IR}$)

This current mirrors the characteristics of the delayed rectifier current in that it passes a larger inward current than an outward direction. Inward rectifier K channels were first discovered in $K^+$ depolarized muscle by Katz,[13] and so far have been elucidated in muscle tissue, endothelium, eggs, etc. Katz[13] used the term 'anomalous' rectification to contrast with the properties of 'normal' delayed rectification (predicted by the Goldman current equation). The function of both inward and outward rectifier currents is to keep the membrane potential between discrete limits (Fig. 7.1). In some cells, the inward rectifier current permits long depolarizing pulses.

The ability of inward rectifier currents to stabilize the membrane potential in excitable tissues such as heart, neurones, and blood vessels may be crucial to some physiological

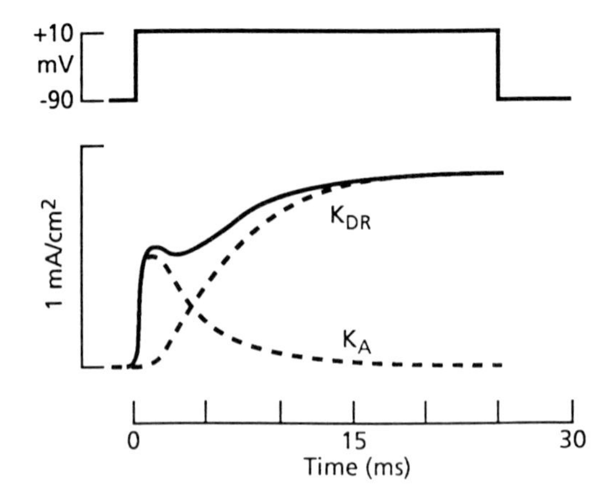


**Fig. 7.3** Two K-components in the crab giant axon.

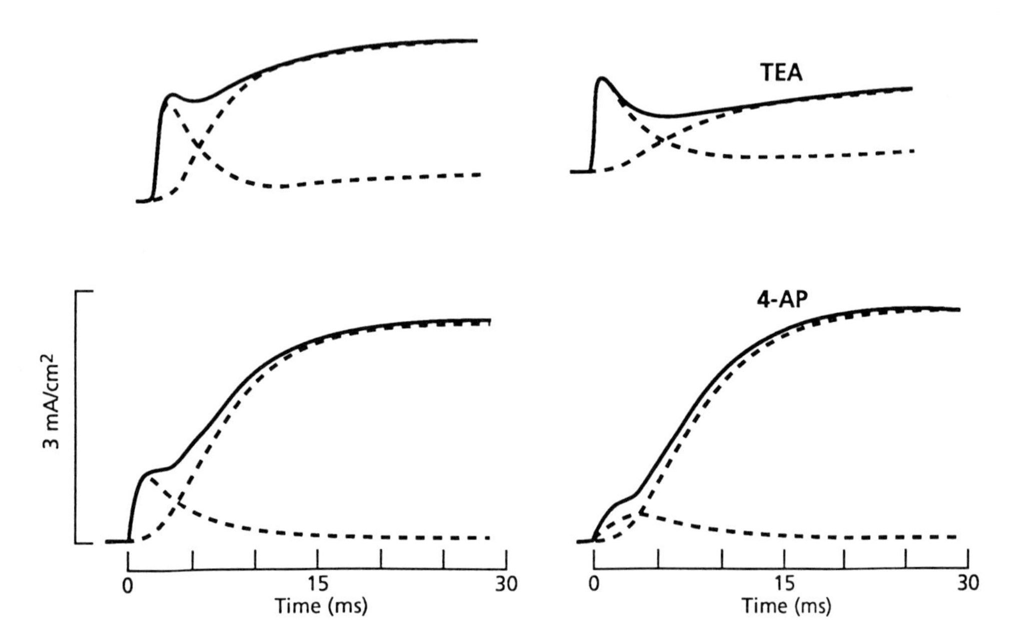


**Fig. 7.4** Differential blockade of two types of outward $K^+$-currents by tetraethy lammonium (TEA) and 4-aminopyridine (4-AP).

processes. It has been suggested that cerebral vascular $K_{IR}$ channels mediate vasodilation by their activation in response to extracellular $K^+$ (1–15 mM) released as a consequence of neuronal activity. Likewise, polyamines (putrescine, spermidine, and spermine) have been shown to be responsible for inward rectification of a class of $K^+$ channels. This is of particular significance, since rectification is lost in excised patches, a phenomenon which cannot be attributed solely to the absence of a voltage-dependent $Mg^{2+}$ block on the channel. An 'endogenous rectifier factor' was suggested therefore. Interestingly, polyamines restore rectification when applied to the cytosolic side of inside-out patches. Cerebral hypoxia, ischaemia, or hypoglycaemia are all associated with elevated $K^+$ levels. Subsequently, the altered polyamine metabolism observed with such abnormalities could affect extracellular $K^+$-induced vasodilation in these tissues.[17]

### Sarcoplasmic reticulum channel ($K_{sr}$)

This is a high-conductance (180 pS) ion channel with a $K^+/Na^+$ selectivity lower than in other voltage-dependent $K^+$ channels. It shows a strong voltage dependence and is blocked by $Cs^+$, decamethonium, and hexamethonium.

## Calcium-activated $K^+$ channels

Although this current was first described in neurones, the introduction of patch-clamp methodology unexpectedly resulted in it being found in nearly every excitable cell type. In neurones, $Ca^{2+}$-activated $K^+$ channels produce long, hyperpolarizing pauses. Action potentials in neurones are followed by hyperpolarization, which can last for several seconds. This hyperpolarization has several phases which are mediated by the activation of distinct types of $Ca^{2+}$-activated $Ca^{2+}$ channels.[18]

**Table 7.3** Properties of calcium – Activated potassium channels.

| Property | High conductance ($BK_{Ca}$) | Small conductance ($SK_{Ca}$) | Intermediate conductance ($IK_{Ca}$) | Others |
|---|---|---|---|---|
| Single-channel conductance (pS) | 100–250 | 6–14 | 18–50 | 3–7 |
| Blockers | TEA (< 1 mM)<br>Charybdotoxin<br>Iberiotoxin<br>Noxiustoxin<br>Kaliotoxin<br>Soyasaponins<br>d-Tubocurarine<br>$Ba^{2+}$, $Na^+$, $Cs^+$ | TEA (> 20 mM)<br>Apamin<br>d-Tubocurarine<br>Quinine<br>Mepacrine | TEA<br>Clotrimazole<br>Ceteidil<br>Nitrendipine<br>Trifluoroperazine<br>Haloperidol | TEA (> 20 mM)<br>Ryanodine |
| Openers | NS 004<br>NS 1619<br>DHS-1 | | | |
| Voltage sensitivity | +++ | – | + | – |
| $Ca^{2+}$-sensitivity near the resting membrane potential (−50, −70 mV) | 1–10 $\mu$M | 100–400 nM | | |
| Physiological role (current involved) | Action potential Repolarization (IC) | Afterhyperpolarization (IAHP) | | Afterhyperpolarization (IAHP) |

Single-channel studies have revealed two families of $Ca^{2+}$-activated $K^+$ channels (high-conductance ($BK_{Ca}$) and small-conductance ($SK_{Ca}$), which differ in their physiological role, conductance, and pharmacology (Table 7.3). Some authors also differentiate an intermediate conductance ($IK_{Ca}$) channel. Their biophysical properties suggest that the physiological role of repolarization or hyperpolarization of cells when their internal free $Ca^{2+}$ rises above a certain level.

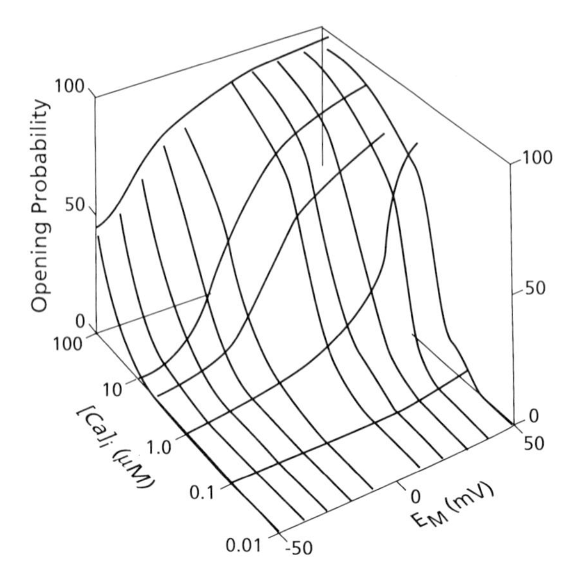


**Fig. 7.5** Voltage- and calcium-dependence in calcium activated $K^+$ channels.

## High-conductance (calcium-activated $K^+$ ($BK_{Ca}$ or Maxi-K)

These channels posses a large unitary conductance (100–250 pS), are ubiquitously distributed in different cell types and tissues, and play an important role in cellular processes as distinct as repolarization of the action potential or regulation of arterial tone. Despite its large conductance, this channel is highly specific for $K^+$ions. $BK_{Ca}$ channels are essentially impermeable to $Na^+$ and $Cs^+$ and conduct $K^+$ 10 times more effectively than $Rb^+$.

### Channel gating

$BK_{Ca}$ is activated by cytosolic $Ca^{2+}$ and by depolarizing voltages, both of which lead to an increase in opening probability (Fig. 7.5). In this sense, $BK_{Ca}$ is a voltage-dependent $K^+$ channel which is also activated by $Ca^{2+}$. Thus, under physiological conditions, voltage-dependence is controlled by the entry of $Ca^{2+}$. The channel kinetics are complex. In addition to the open and closed states, this channel shows at least six conductance substates and changes between modes of high and small open probability (at constant $Ca^{2+}$), as well as flickering modes.[19]

#### Channel modulation

*Blockers*

The main pharmacological characteristic of $BK_{Ca}$ channels is their sensitivity to the scorpion toxin, charybdotoxin (ChTX).[20] Nevertheless, not all $BK_{Ca}$ channels are sensitive to this toxin and, most importantly, ChTX also blocks $K^+$ channels that are not $BK_{Ca}$ (some $SK_{Ca}$ and some $Ca^{2+}$-independent $K^+$-channels). Regarding toxin specificity, another scorpion toxin, iberiotoxin (IbTX), has proved to be more specific for $BK_{Ca}$ channels than ChTX (see Chapter 8). Quaternary ammonium ions such as TEA and nonyltrimethylammonium are able to block $BK_{Ca}$, both from the internal and the external side. Nonetheless, the affinity of the internal site is 150-fold lower than that of the external site.

In addition to toxins (ChTX, IbTX), $BK_{Ca}$ can be blocked by ATP, angiotensin II, and thromboxane $A_2$. ATP decreases the open probability of $BK_{Ca}$ in rabbit aorta, trachea, and pig coronary artery with an apparent dissociation constant in the range of 0.2–0.6 nM. However, ATP does not modify the activity of $BK_{Ca}$ in skeletal muscle and smooth muscle membranes, indicating the existence of channel subtypes. Angiotensin II and thromboxane $A_2$, a metabolite of arachidonic acid, are potent vasoconstrictors which subsequently act as potent inhibitors of $BK_{Ca}$.

*Activators*

Endothelin-1, a vasoactive peptide, increases the open probability of $BK_{Ca}$ in smooth muscle cells of the porcine coronary artery, at concentrations as low as 1 nM. At higher concentrations, it may act as a blocker. Endothelin has some tissue and/or species specificity. In the rabbit portal vein, endothelin is a potent activator but in other tissues it is not. Niflumic acid, known for its inhibitory action on the $Ca^{2+}$-activated $Cl^-$-channels increases the opening probability of $BK_{Ca}$ channels from coronary arteries incorporated into lipid bilayers. Finally, three organic compounds present in the crude extract of a medicinal herb have proven to be potent activators of $BK_{Ca}$ channels.[19] The compounds were identified as triterpenoid glycosides (dehydrosoyasaponin, soyasaponin I, and soyasaponin III).

### Small-conductance calcium-activated $K^+$ channels ($SK_{Ca}$) and other Ca-activated potassium conductances

Ionic currents underlying the hyperpolarization (AHP) that follows the action potential in many neurones can be divided into two distinct types. One current peaks rapidly (1–5 ms) following the action potential and decays with a time constant of 100–200 ms. This current is voltage-insensitive, and is blocked by the bee-venom toxin apamin and high concentrations of D-tubocurarine, but not by low concentrations of TEA ($<1$ nM). The other type of current rises to a peak in 0.5 s and decays with a time constant of 1–2 s. This current, also voltage-insensitive, is not blocked by apamin. This current cannot be classified as $BK_{Ca}$, but as a small-conductance $Ca^{2+}$ activated $K^+$ channel ($SK_{Ca}$) either apamin-sensitive or insensitive. $SK_{Ca}$ is much more sensitive to $Ca^{2+}$ than $BK_{Ca}$ (100–400 nm $Ca^{2+}$ activates $SK_{Ca}$, whereas $BK_{Ca}$ requires 1–10 $\mu$M $Ca^{2+}$ near the resting membrane potentials (in the range of $-50$ to

−70 mV). The apamin-insensitive $Ca^{2+}$-activated $K^+$ channels are blocked by ryanodine (20 $\mu$M) and are modulated by neurotransmitters, such as noradrenaline (10 $\mu$M).

#### $Ca^{2+}$-activated $K^+$ channels are $Ca^{2+}$ binding proteins

To better understand many physiological and pharmacological properties associated with these channels, it is important to remember that $Ca^{2+}$-activated $K^+$-channels are $Ca^{2+}$-binding proteins. The divalent cation selectivity sequence found for the $BK_{Ca}$ of skeletal muscle follows the same order as that found for $Ca^{2+}$ binding proteins (such as calmodulin, troponin C, or parvalbumin). For example, the potency of divalent cations in opening $SK_{Ca}$ of human erythrocytes follows the sequence:

$$Pb^{2+} > Cd^{2+} > Ca^{2+} > Co^{2+} > Mg^{2+},\ Fe^{2+}$$

which is similar to the selectivity sequence of the $Ca^{2+}$ site in parvalbumin. The $BK_{Ca}$ selectivity sequence is closer to that found in the $Ca^{2+}$ binding sites of calmodulin or troponin

$$Ca^{2+} > Cd^{2+} > Hg^{2+} > Si^{2+} > Mn^{2+} > Zn^{2+} > Pb^{2+} Co^{2+} > Mg^{2+},\ Ni^{2+},\ Ba^{2+}$$

## Sodium-activated $K^+$-current ($K_{Na}$)

The central property of this current was found to be its activation by cytosolic $Na^+$ first described in ventricular myocytes by Kaneyama *et al.*[21] and has also been shown in crayfish neurones, avian ganglionic neurones, brainstem neurones, and quail trigeminal neurones.[22] The unitary current conductance measured from these channels is relatively high (100–200 pS). However, their physiological role remains unresolved. Kaneyama *et al.* concluded that this channel might only become active under pathological conditions (it needs over 25 mM $Na^+$ in the cytosol to become activated). Other authors postulate that this channel may already be activated under physiological conditions. With $[Na^+]_i$ close to 10 mM, the $K_{Na}$ results in low-frequency openings, which may contribute to determining the resting potential in quail sensory neurones. It has also been suggested that this current accelerates the repolarization phase of action potentials. $Na^+$-activated $K^+$ channels are blocked by 4-AP. Neurotransmitters such as noradrenaline and muscarine may also affect this current in some tissues.

## Receptor-operated $K^+$-channels

$K^+$-channels may open not only in response to changes in the membrane potential (voltage-activated) or by direct effect of a ligand ($Ca^{2+}$-activated, $Na^+$-activated, ATP- and $O_2$-regulated), but also through a mechanism somehow coupled to a receptor for an extracellular messenger (neurotransmitter, hormone). Receptor-operated $K^+$-channels include: M-current, atrial muscarinic-activated, and serotonin-inactivated $K^+$-channels (Table 7.4).

**Table 7.4** Properties of receptor-operated potassium channels.

| Property | Muscarinic inactivated ($K_M$) | Atrial muscarinic activated ($K_{ACh}$) | Serotonin inactivated ($K_{5\text{-}HT}$) |
|---|---|---|---|
| Single-channel conductance (pS) | 5–18 | 7–50 | 55 |
| | $Ba^{2+}$ | $Ba^{2+}$ | $Ba^{2+}$ |
| Blockers | LH<br>Substance P<br>Bradykinin | $Cs^+$<br>4-AP<br>TEA<br>Quinine | $Cs^+$<br>4-AP<br>TEA |
| Openers | Somatostatin<br>β-Adrenergic agonists | Adenosine | FMRF-amide |
| Regulation | Time- and voltage-dependent<br>Slow activation | Inward rectifying<br>Voltage activated | Weakly voltage dependent |

4-AP: 4-aminopyridine; LH: luteinizing hormone; TEA: tetraethylammonium; FMRF-amide: Phe-Met-Arg-Phe-$NH_2$.

## M-current ($K_M$)

A time- and voltage-dependent channel with a single unitary conductance of 5–18 pS, this was first described in bullfrog sympathetic ganglion[23] where muscarinic receptor activation led to a decrease in a voltage-dependent $K^+$ conductance. It has been identified in many tissues, including visceral (but not in vascular) smooth muscle and is inhibited by muscarinic activation (acetylcholine, carbachol) and by luteinizing hormone (LH), substance P, $Ba^{2+}$, etc. It seems that muscarinic agonists suppressed the M-current by acting at a locus downstream from regulation of cAMP levels by adenylcyclase and phosphodiesterase. $K_M$ is activated by somatostatin.

## Atrial muscarinic-activated ($K_{ACh}$)

This is a voltage-activated current with inward rectifying properties and a conductance of 7–50 pS. It is blocked by $Ba^{2+}$, $Cs^+$, 4-AP, TEA, and quinine.

## Serotonin-inactivated ($K_{5\text{-}HT}$)

This is weakly dependent on voltage and with a higher conductance (55 pS) than other receptor-operated $K^+$ channels. This channel is opened by FMRF-amide and blocked by $Ba^{2+}$, $Cs^+$, $Cs^+$, 4-AP, and TEA. Serotonin (type 1A receptor) may negatively control a strongly rectifying IRK-type inwardly rectifying $K^+$ channel involved in the control of neuronal excitability in the mammalian brain. IRK1 channel activity results by inhibited direct protein kinase A-mediated phosphorylation.[24]

## Metabolically steered K-channels

### ATP regulated $K^+$-Channels ($K_{ATP}$)

Described for the first time in cardiac myocytes,[25] this current is also present in the pancreatic $\beta$-cell where it plays a key role in the nutrient-induced release of insulin[26] as well as in other tissues such as muscle and neurones. Its presence in mitochondria has been recently reported. In response to stimulatory glucose concentrations (> 7mM), the pancreatic $\beta$-cell membrane depolarizes, initiating a characteristic pattern of electrical activity and insulin release.[27] There is considerable evidence that this depolarization results from a decrease in the resting $K^+$ permeability of the $\beta$-cell as a consequence of nutrient metabolism.[28,29] Glucose metabolism is believed to exert an inhibitory action on $K_{ATP}$ channels by increasing the cytoplasmic ATP/ADP ratio. A novel class of glucose-induced intracellular messengers, the diadenosine polyphosphates has been recently hypothesized.[30] Diadenosine polyphosphates (diadenosine tri- and tetraphosphate, $Ap_3A$, $Ap_4A$) are a group of low-molecular weight compounds that increase 30- to 70-fold in concentration after exposure to glucose at concentrations that induce insulin release. $Ap_3A$ and $A_4A$ are effective inhibitors of the ATP-regulated $K^+$-channels when applied to the intracellular side of excised membrane patches from $\beta$-cell cultures. In the ischaemic myocardium, opening of this channel (by the reduction of ATP production) will result in a hyperpolarization which precludes $Ca^{2+}$ entry (and $Ca^{2+}$ overload) during ischaemic and/or reperfusion periods (Table 7.5).

**Table 7.5** Properties of metabolically steered potassium channels.

| Property | ATP-regulated ($K_{ATP}$) | $O_2$-regulated ($K_{O_2}$) | Fatty-acid-activated ($K_{FA}$) |
|---|---|---|---|
| Single-channel conductance (pS) | 5–90 | 20 | 23 |
| Blockers | Tolbutamide<br>Glibenclamide<br>Phentolamine<br>Ciclazindol<br>Lidocaine<br>TEA, 4-AP<br>$Cs^+$, $Ba^{2+}$ | Sulphydryl-reducing agents (gluthatione, dithiothreitol) | |
| Openers | Chromakalim<br>Diazoxide<br>Pinacidil<br>Aprikalim | | Fatty acids (10–100 $\mu$M) arachidonic, myristic, linoleic, palmitoleic, etc. |
| Regulation | Inhibited by: ATP, diadenosine polyphosphates<br>Activated by: ADP | Inhibited by $O_2$<br>Activated by voltage | |

4-AP: 4-amino-pyridine; TEA: tetraelthyl ammonium.

The ion permeating subunit, comes from a new family of genes encoding inward rectifying channels (Kir 6.2).[8] Its transmembrane topology, unique among ion channels, seems to include only two membrane-spanning domains, neither of which is sensitive to sulphonylureas, as seen in native $\beta$-cell $K_{ATP}$ channels. Sulphonylurea sensitivity is associated with a separate protein, the sulphonylurea receptor, which belongs to the ATP-binding cassette group, and contains two nucleotide binding sites on the cytosolic side (see Fig. 7.2).

### $O_2$-regulated $K^+$ channels ($K_{O_2}$)

The pioneering work of De Castro (1928) and Heymans (1930) demonstrated that the carotid bodies, strategically located in the bifucartion of the carotid artery, sense reductions in arterial $PO_2$, and stimulate the brainstem respiratory centres to produce a hyperventilatory response to hypoxia. Glomus cells are electrically excitable and exhibit an $O_2$-sensitive K-current.[31,32] Low $PO_2$ selectively and reversibly inhibits the activity of a specific $K^+$ channel, which confers the cell's chemoreceptive properties. As with delayed-rectifier $K^+$-currents, $O_2$-sensitive $K^+$-currents open upon membrane depolarization and have a unitary conductance of 20 pS (Table 7.5). This $K^+$-current is also present in the neuroepithelial bodies of the lung (small innervated organs distributed throughout the airways mucosa that may be important in the regulation of respiratory activity after birth) and in lung smooth muscle. From the biophysical point of view, $O_2$-sensitive $K^+$ channels belong to the family of delayed rectifier $K^+$ channels in that they are affected by the same type of drugs. On the other hand, sulphydryl-reducing agents (such as glutathione and dithiothreitol) produce an inhibition of $K^+$ channel inhibition similar to that observed upon exposure to hypoxic solutions.[32]

### Fatty acid-activated $K^+$-channels ($K_{FA}$)

Smooth muscle cells from toad stomach possess a $K^+$ channel ($K_{FA}$) which is activated by low concentrations (10–100 $\mu$M) of arachidonic, myristic, linoleic, palmitoleic and other types of fatty acids.[33] Its unitary channel conductance is 23 pS. $K_{FA}$ is active in the absence of $Ca^{2+}$ and nucleotides, and does not show a pronounced voltage dependence. Fatty acids modulate the opening probability of this channel by acting directly on it. The mechanism of action for these fatty acids is unknown. Palmitoleic and myristic acids are not substrates for cyclo- or lipooxygenase, and eicosatetraenoic acid blocks cyclo- and lipooxygenases, as well as cytochrome $P_{450}$. At higher concentrations, this can alter the fluidity of the membrane bilayer and thus the properties of the proteins embedded there.

## Structural basis for $K^+$ channel modulation

There are two types of extracellular blockers for voltage-gated $K^+$ channels: organic compounds (usually positively charged: TEA, 4-AP, and quinine); and peptide toxins (charybdotoxin, dendrotoxin, and mast-cell degranulating peptide).

## Mechanisms of charybdotoxin block

Charybdotoxin and other small peptide scorpion toxins specifically interact with the external mouth of the pore of some Shaker-like $K^+$ channels, thereby blocking ion conduction. Binding and unbinding rate constants can be modified by mutating both the channel and the toxin.[34] Neutralization of selected positively charged residues in the toxin molecule cause a large decrease in the affinity of the toxin.

## Mechanisms of TEA block

TEA block of channels is associated with an amino acid in the P(H5) segment of Shaker-like voltage-dependent $K^+$ channels.[35] Point mutation in this region has shown that when an aromatic residue (tyrosine or phenylalanine) is present in this region the channels are blocked with high affinity (< 1 mM) to TEA, but with a decreased voltage dependence. These mutations also reinforce the hypothesized tetrameric structure of the channels. Several workers have generated heteromultimeric Shaker-like $K^+$ channels made up of different combinations of wild-type subunits and subunits with a mutation in the external TEA-binding site. Despite this heterogeneity, the TEA sensitivity of the heteromultimers fits that of single site binding isotherm more consistent with a uniform population of channels.

### Internal TEA-binding site

Chimeric $K^+$ channels have provided compelling evidence that the P(H5) region not only plays a major role in determining the external TEA-binding site, but also that of the internal site and the conducting pore. A single mutation in the P(H5) region of the channel changes not only internal TEA binding, but also the selectivity properties of the pore. Molecular observations are consistent with more classical functional observations which locate the TEA-binding site at the mouth of the pore.[16] It also will explain the possibility of TEA being removed by a $K^+$ ion stream flowing in the opposite direction (for example, outward current for external blockade) as in the 'uncorking' of a champagne bottle (Fig. 7.6).

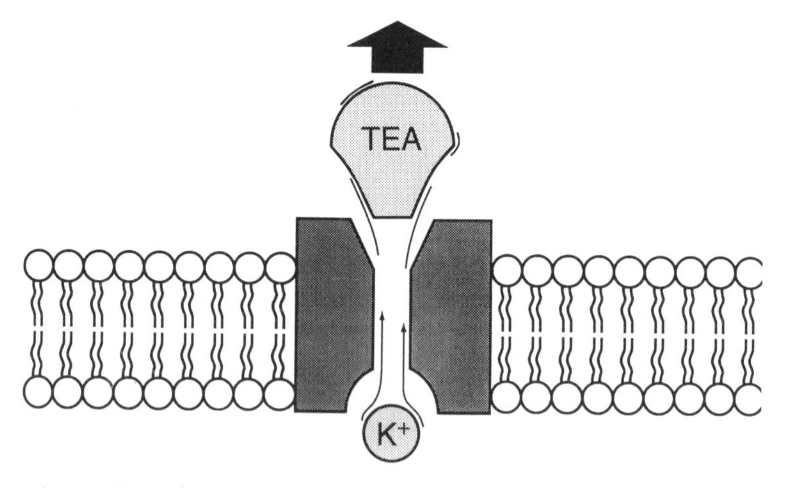


**Fig. 7.6** 'Uncorking' mechanism for the removal of TEA block.

### Dendrotoxin binding site

Dendrotoxin (DTX) is a snake venom produced by the Eastern green mamba, *Dendroaspis augusticeps*. This peptide induces epileptiform activity in hippocampal CA1 neurones. The main target for DTX action appears to be a subtype of outwardly rectifying $K^+$ channels. The unitary conductance of DTX-sensitive $K^+$ channels is in the order of 10 pS. The DTX binding site is located on the extracellular surface of $K^+$ channels, as in the TEA- and CTX-binding sites. Although an extensive mutational analysis of the DTX binding is still needed, it is already clear from the observation that CTX- and DTX-binding sites are not identical. In this sense, mutations of the P(H5) region that affect TEA and CTX binding do not similarly affect DTX binding. For example, substitution of lysine at position 19 of the P(H5) region by tyrosine produces a TEA-sensitive RCK4 channel. This mutant is not very sensitive to DTX. Other negatively charged residues located in the N-terminal side of the P(H5) region appear to be important for DTX binding.

### Mechanism for polyamine-induced rectification

The properties and mechanism for polyamine-dependent rectification have been studied using amine analogues and mutant channels. Polyamine derivatives containing bulky carboxyl groups or ring structures were ineffective as rectifier – inducers. This, together with other chemical modifications, suggests that a negatively charged binding site lying in a narrow, confined space may be responsible for the high specificity of the effect. From the structural view point, the second transmembrane domain (M2) fulfils the binding site requirements. A single negatively charged residue (Asp-172) appears to be crucial for a substantial proportion of polyamine-dependent rectifier activity. Another residue (Glu-244) located in the C-terminal domain seems to contribute to this effect in a manner independent from the other region. Thus, inward rectifier channels possess two independent binding sites for polyamines.[17]

### Nucleotide binding sequences and phosphorylation binding sites

Ion channels contain many consensus sequences for nucleotide binding, phopsphorylation, or G protein modulation. Point mutation analysis is revealing the exact regions which participate in each type of modulation. For example, protein kinase C eliminates fast inactivation in an A-type $K^+$ channel (Kv 3.4), by phosphorylating two serine residues in the N-terminal domain. The ROMK1 channel contains a putative nucleotide binding site similar to that found in cyclic nucleotide-gated channels and cystic fibrosis transmembrane (conductance) regulator (CFTR) within its C-terminal domain, while the sulphonylurea receptor by contrast contains the ATP-binding sites in a place far from the pore region.[8]

In **summary**, knowledge of the molecular structure of cloned $K^+$ channels has made it increasingly clear than the primary structure of the external pore region is highly conserved. Thus, external blockers of voltage-dependent $K^+$ channels are unlikely to be very specific for one type of $K^+$ channel and may thus be of limited therapeutic use. In such case, there is a great need for the development and design of $K^+$ channel blockers that are targeted away from the external pore region, to sequences that are less conserved and unique to the channel of interest.

## Aknowledgements

The author has been supported by Grants from the Ministry of Health (FIS 94–0014–01 and FIS 96–1994–01), Generalitat Valenciana (GV-3117–95) and European Commission (Contract ERB-SCI 920833).

## References

1. Hille, B. (1991). *Ionic channels in excitable membranes*. Sinauer Assoc. Inc., Sunderland, Massachusets..
2. Pallotta, B.S. and Wagoner, P.K. (1992). *Physiol. Rev.*, **72**(4), S49–S67.
3. MacKinnon, R. (1991). *Nature*, **350**, 232–234.
4. Hoshi, T. and Zagotta, W.N. (1993). *Curr. Opin. Neurobiol.*, **3**, 283–290.
5. Pongs, O. (1992). *Trends Pharmacol. Sciences*, **13**, 359–365.
6. Pongs, O. (1992) *Physiol. Rev.*, **72**(4), S69–S88.
7. Breitwieser, G.E. (1996). *J. Membr. Biol.*, **152**, 1–11.
8. Aguilar-Bryan, L., Nichols, C.G., Wechsler, S.W., Clement, J.P., Boyd A.E. III, Gonzalez, G., Herrera-Sosa, H., Nguy, K., Bryan, J., and Nelson, D.A. (1995). *Science*, **268**, 423–425.
9. Hodgkin, A.L. and Huxley, A.F. (1952). *J. Physiol. (Lond.)*, **116**, 449–472.
10. Hodgkin, A.L. and Huxley, A.F. (1952). *J. Physiol. (Lond.)*, **116**, 473–496.
11. Hodgkin, A.L. and Huxley, A.F. (1952) *J. Physiol. (Lond.)*, **117**, 500–544.
12. Connor, J.A. and Stevens, C.F. (1971). *J. Physiol. (Lond.)*, **213**, 21–30.
13. Katz, B. (1949). *Arch. Sci. Physiol.*, **3**, 285–299.
14. Sala, S. and Soria, B. (1991). *Eur. J. Neurosci.*, **3**, 462–472.
15. Soria, B., Arispe, N., Quinta-Ferreira, M.E., and Rojas, E. (1985). *J. Membr. Biol.*, **84**(2), 117–126.
16. Armstrong, C.M. (1992). *Physiol. Rev.*, **72**, S5–S13.
17. Johnson, T.D. (1996). *Trends in Pharmacol. Sci.*, **17**, 22–27.
18. Sah, P. (1996). *Trends in Neurosciences*, **19**, 150–154.
19. Latorre, R. (1994) In: *Handbook of membrane channels. Molecular and cellular physiology*, (ed. C. Peracchia), pp 79–102. Academic Press, London.
20. Miller, C. (1988). *Neuron*, **1**, 1003–1006.
21. Kaneyama, M., Kakei, M., Sato, R., Shabasaki, T., Matsuda, H., and Irisawa, H. (1984). *Nature*, **309**, 354–356.

22. Dryer, S.E. (1994). *Trends in Neurosciences*, **17**, 155–160.
23. Brown, D.A. and Adams, P.R. (1980). *Nature*, **283**, 673–676.
24. Wischmeyer, E. and Karschin, A. (1996). *Proc. Natl Acad. Sci. USA*, **93**(12), 5819–5823.
25. Noma, A. (1983). *Nature*, **305**, 147–148.
26. Ashcroft, F.M., Harrison, D., and Ashcroft, S.J.H. (1984). *Nature*, **312**, 446–448.
27. Atwater, I., Rojas, E., and Soria, B. (1986). *Biophysics of the pancreatic B-cell*. Plenum Press, New York.
28. Valdeolmillos, M., Nadal, A., Contreras, D., and Soria, B. (1992). *J. Physiol.*, **455**, 173–186.
29. Martin, F. and Soria, B. (1995). *J. Physiol. (Lond.)*, **486**, 361–371.
30. Ripoll, C., Martin, F., Rovira, J.M., Pintor, J., Miras-Portugal, M.T., and Soria, B. (1996). *Diabetes*, **45**, 1431–1434.
31. Lopez-Barneo, J. (1994). *Trends in Neurosciences*, **17**, 133–135.
32. Lopez-Barneo, J. (1996). *Trends in Neurosciences*, **19**, 435–440.
33. Ordway, R.W., Walsh, J.V., and Singer, J.J. (1989). *Science*, **244**, 1176–1179.
34. Goldstein, S.A. and Miller, C. (1992). *Biophys. J.*, **62**, 5–7.
35. MacKinnon, R. and Yellen, G. (1990). *Science*, **250**, 276–279.

# 8 Peptide toxins interacting with potassium channels

M. L. Garcia, H.-G. Knaus, R. J. Leonard, O. B. McManus, P. Munujos, W. A. Schmalhofer, R. S. Slaughter, and G. J. Kaczorowski.

## Introduction

Potassium channels are integral membrane proteins that allow the transmembrane movement of ions at high rates through aqueous-pores with high selectivity for $K^+$. This family of proteins controls electrical excitability, as well as the resting membrane potential, in many different cell types. Gating of these channels occurs through conformational changes that are controlled by membrane potential and/or ligand binding.

Structural information has accumulated during the past several years on voltage-dependent $K^+$ channels due to the successful cloning and functional expression of the pore-forming subunit of a large number of these proteins (Jan and Jan 1992; Pongs 1992). Interestingly, the predicted secondary structure of these proteins resembles one of the four repeated regions present in $Na^+$ and $Ca^{2+}$ channels. In fact, it has been demonstrated that functional $K^+$ channels are formed from the association of either four identical or four related subunits (Christie *et al.* 1990; Isacoff *et al.* 1990; Ruppersberg *et al.* 1990; MacKinnon 1991. This phenomenon could explain the wide diversity of $K^+$ channels found in different tissues. Recently, another degree of complexity has emerged in terms of the diversity of $K^+$ channels based on the molecular cloning of a complementary subunit that copurifies with the pore-forming subunit of a voltage-gated $K^+$ channel from rat brain (Scott *et al.* 1994). Coexpression of the two subunits reveals a dramatic effect of the newly identified protein on the inactivating properties of the pore-forming subunit (Rettig *et al.* 1994).

Within the subfamily of ligand-gated channels, $Ca^{2+}$-activated $K^+$ channels are perhaps the most extensively studied. Within this group, three subtypes of channels have been classified based on unitary single channel conductance; large-conductance (100–300 pS), intermediate-conductance (40–100 pS), and small-conductance (10–35 pS) channels (Latorre *et al.* 1989). Two types of high-conductance, $Ca^{2+}$-activated $K^+$ (maxi-K) channels, the *Drosophila slowpoke* (Adelman *et al.* 1992) and the mammalian *mSlo* (Butler *et al.* 1993), have recently been cloned and the predicted secondary structure of these proteins is expected to be similar to that of voltage-gated $K^+$ channels, although the former channels possess a large C-terminus that could form additional membrane-spanning regions. Whereas the diversity of voltage-gated $K^+$ channels appears to be due to the association in a tetrameric structure of different subunits, each one derived from a single gene product, the diversity of maxi-K channels is due to alternative RNA splicing of a single gene. The recent cloning of maxi-K channels from human brain has led to the identification of four different spliced sites within the C-termini of the protein (Tseng-Crank *et al.* 1994). All possible combinations of these exons could lead to 48 different maxi-K channel variants with distinct biophysical properties. In addition, a complementary subunit has been identified and cloned after its copurification with the pore-forming subunit of the maxi-K channel from bovine tracheal smooth muscle (Knaus *et al.* 1994a). Coexpression experiments in *Xenopus* oocytes with the two subunits leads to a very profound effect on the biophysical and pharmacological properties of the pore-forming subunit (McManus *et al.* 1995; Meera *et al.* 1995).

Despite the extensive structural information on $K^+$ channels that has been derived from molecular biology approaches, there is limited knowledge as to which particular subtype of channel is expressed in given tissues, and, more importantly, what is the physiological role of those channels in target tissues of interest. In this respect, the discovery over the past few years of peptidyl toxin molecules, that in some cases interact very specifically with a given type of $K^+$ channel, has allowed one to gain some insight into the role that $K^+$ channels play in cell function. It is important to note that these probes also represent unique tools with which to attempt the purification of channel proteins from native tissues. They can also be used in the discovery of small organic molecule ion channel modulators for possible therapeutic application in those pathophysiological conditions that involve $K^+$ channel function.

Since the number of peptidyl toxin molecules targeting $K^+$ channels has grown considerably over the past 10 years, and their targets are very diverse, for consistency they can be subdivided into three major groups: small-size toxins (18–31 amino acids long) from various organisms that block small-conductance, $Ca^{2+}$-activated $K^+$ channels (e.g. apamin, leiurustoxin I); intermediate-size toxins (37–39 amino acids long) derived from scorpion venoms that target voltage-dependent and $Ca^{2+}$-activated $K^+$ channels; large-toxin molecules (larger than 50 amino acids long) derived from snake venoms that inhibit voltage-dependent $K^+$ channels (e.g. dendrotoxins).

In this review, we will concentrate on the second group of scorpion toxins and illustrate how their discovery has helped to advance our knowledge of the structure and function of $K^+$ channels. Within this group a further subdivision can be made based on amino acid sequence homologies between the various members of this family. Thus, a class is formed by charybdotoxin, iberiotoxin, limbatustoxin, and *Leiurus quinquestriatus* toxin 2. The second class comprises noxiustoxin, margatoxin, and *Centruroides limpidus limpidus* toxin I, while a third class is defined by kaliotoxins and the three recently identified agitoxins. The primary amino acid sequences of these toxins are presented in Fig. 8.1. These classes will be discussed in that order.

## Charybdotoxin, iberiotoxin, limbatustoxin, and L. quinquestriatus toxin 2

Venom of the scorpion *L. quinquestriatus* contains several peptide toxins that alter the gating kinetics of $Na^+$ channels. These toxins have been widely used in biochemical studies to study

| | | Sequence |
|---|---|---|
| I. | ChTX | Z F T N V S **C** T T S K E **C** W S V **C** Q R L H N T S R G K **C** M N K K **C** R **C** Y S |
| | IbTX | Z F T D V D **C** S V S R E **C** W S V **C** K D L F G V D R G K **C** M G K K **C** R **C** Y Q |
| | $Lq_2$ | Z F T Q E S **C** T A S N Q **C** W S I **C** K R L H N T N R G K **C** M N K K **C** R **C** Y S |
| II. | NxTX | T I I N V K **C** T S P K Q **C** S K P **C** K E L Y G S S A G A K **C** M N G K **C** K **C** Y N N |
| | MgTX | T I I N V K **C** T S P K Q **C** L P P **C** K A Q F G Q S A G A K **C** M N G K **C** K **C** Y P H |
| | C.I.I. I | I T I N V K **C** T S P Q Q **C** L R P **C** K D R F G Q H A G G K **C** I N G K **C** K **C** Y P |
| III. | $AgTX_1$ | G V P I N V K **C** T G S P Q **C** L K P **C** K D A G M R F G K **C** I N G K **C** H **C** T P K |
| | $AgTX_2$ | G V P I N V S **C** T G S P Q **C** I K P **C** K D A G M R F G K **C** M N R K **C** H **C** T P K |
| | $AgTX_3$ | G V P I N V P **C** T G S P Q **C** I K P **C** K D A G M R F G K **C** M N R K **C** H **C** T P K |
| | KTX | G V E I N V K **C** S G S P Q **C** L K P **C** K D A G M R F G K **C** M N R K **C** H **C** T P K |
| | $KTX_2$ | V R I P V S **C** K H S G Q **C** L K P **C** K D A G M R F G K **C** M N G K **C** D **C** T P K |

**Fig. 8.1** Comparison of amino acid sequences of charybdotoxin (ChTX), iberiotoxin (IbTX), *L. quinquestriatus* toxin 2 (Lq2), noxiustoxin (NxTX), margatoxin (MgTX), *C. limpidus limpidus* toxin I (C.1.1. I), agitoxin I ($AgTX_1$), agitoxin II ($AgTX_2$), agitoxin III ($AgTX_3$), kaliotoxin (KTX), and kaliotoxin 2 ($KTX_2$). The sequences have been aligned with respect to the six cysteine residues which are in boldface type.

$Na^+$ channel function (Catterall 1980). This venom has also provided a rich source of toxins targeted against $K^+$ channels. The first such activity discovered in this venom was named charybdotoxin (ChTX) by Miller *et al.* (1985). This group found that addition of the crude scorpion venom produced a reversible block of maxi-K channels present in mammalian skeletal muscle t-tubules that were reconstituted into bilayers. This activity was manifested only when added to the external face of the channel. Addition of ChTX causes the appearance of silent periods, due to binding of a single toxin molecule to the channel to block ion conduction, interspersed between bursts of normal channel activity which represent times at which the toxin has dissociated from the channel. Toxin binding occurs through a simple bimolecular reaction, since the average blocked time is independent of toxin concentration, while the duration of the burst periods is inversely proportional to toxin concentration (Smith *et al.* 1986).

Soon after its discovery, ChTX was purified to homogeneity using ion-exchange and reversed-phase chromatographies, and its complete amino acid sequence was determined in a number of different laboratories (Gimenez-Gallego *et al.* 1988; Lucchesi *et al.* 1989; Schweitz *et al.* 1989; Strong *et al.* 1989). ChTX is a 37-amino acid peptide containing six cysteines that form three disulphide bridges, and it has its N-terminal amino acid blocked in the form of pyroglutamine. The peptide has a high content of positively charged residues, four lysines, three arginines, and a single histidine, which confers a net charge of +5. Some of these residues have been shown to be very important for the mechanism of $K^+$ channel inhibition (see below).

The primary amino acid sequence of ChTX has been confirmed after chemical synthesis of the peptide using solid-phase FMOC pentafluorophenyl ester methodologies. After its synthesis, the reduced peptide could be oxidized to yield biologically active material which was found to be indistinguishable from native ChTX (Sugg *et al.* 1990). After enzymatic digestion of either native or synthetic oxidized ChTX, identical peptide fragments were obtained, suggesting the same folding for both entities. From these studies, it was possible to assign the disulphide bonds in ChTX as: Cys-7–Cys-28, Cys-13–Cys-33, and Cys-17–Cys-35.

Large quantities of biologically active ChTX can also be produced by recombinant techniques. In this approach, the toxin is produced in *E. coli* as part of a fusion protein with the $T_7$ gene 9 protein. After purification of the fusion protein, the toxin can be folded and enzymatically cleaved using either factor Xa protease, enterokinase, or trypsin. After conversion of the N-terminal residue glutamine to pyroglutamic acid by acid treatment, the resulting peptide can be purified to homogeneity by reversed-phase chromatography (Park *et al.* 1991). By this means, not only large quantities of protein can be produced (i.e. $\sim 3$ mg $l^{-1}$ of *E. coli* culture), but also specific residues in the molecule can be modified for structure–function studies. These studies have been guided by the knowledge of the three-dimensional structure of ChTX in solution, as determined by NMR techniques (Lambert *et al.* 1990; Bontems *et al.* 1991*a*, *b*).

ChTX is a $20 \times 20 \times 25$ Å globular structure formed by a three-turn $\alpha$-helix lying upon a small three-strand antiparallel $\beta$-sheet that is maintained rigid by three disulphide bonds. The peptide's entire internal volume is filled by the six cysteine residues. All non-cysteine side chains lie on the surface and project into the solvent. All residues exposed to the solvent have been subjected to mutagenesis and analysed for maxi-K channel blocking properties (Stampe *et al.* 1994). Out of all residues, eight were found to be crucial based on the fact that their modification leads to a very significant increase in toxin dissociation from the channel. These residues comprise Ser-10, Trp-14, Arg-25, Lys-27, Met-29, Asn-30, Arg-34, and Tyr-36. Lys-27 is a particularly interesting residue. It had been shown previously that ChTX blockade of maxi-K channels is voltage-dependent, but this dependence is not due to a direct action of voltage on the toxin (MacKinnon and Miller 1988). Rather, it appears to be caused by $K^+$ entering the channel from the internal medium and binding at a site located along the ion conduction pathway to destabilize ChTX binding at its receptor. This idea is supported by data showing that toxin dissociation is enhanced by channel-permeant ions present at the inner mouth of the channel, such as $K^+$, $Rb^+$ and $NH_4{}^+$, while impermeant ions, such as $Li^+$, $Na^+$, $Cs^+$ and arginine, do not cause the effect. When position 27 in ChTX contains a positively charged residue, internal $K^+$ increases the dissociation rate of toxin in a voltage-dependent manner (Park and Miller 1992). However, if a neutral Asn or Gln is substituted at this position, the effect of either internal $K^+$ or applied voltage on the toxin dissociation rate is abolished. The specificity of the voltage dependence for position 27 is indicated by the fact that charge-neutral substitutions at other positions fail to eliminate the $K^+$ destabilization phenomenon. These results strongly suggest that when ChTX is bound to the channel, Lys-27 lies physically close to a $K^+$-specific binding site at an external site of the ion conduction pathway. The occupancy of this site by $K^+$ causes a destabilization of bound ChTX through direct electrostatic repulsion with the $\varepsilon$-amino group of Lys-27.

The non-hydrogen van der Waals surface of ChTX, as calculated from NMR-determined coordinates, indicates that the eight crucial residues are spatially separated from unimportant residues on the molecular surface of ChTX. When viewed by a potassium ion emerging from the conduction pathway of the channel, six of these residues form a flat triangular surface that is predicted to make close contact with a complementary surface on the channel. The interaction surface covers about 25% of ChTX's molecular surface. Making several assumptions, a model of the complementary receptor site in the external vestibule of the maxi-K channel has been deduced (Stampe *et al.* 1994). Complementary mutagenesis studies between toxin and channel residues should afford determination of distances between residues and a more complete picture of the maxi-K channel ChTX receptor site.

Although ChTX was first described as an inhibitor of maxi-K channels, soon after its discovery it was found that it could also inhibit other types of voltage-dependent and $Ca^{2+}$-activated $K^+$ channels. In particular, ChTX blocks with high affinity the $K_v1.3$ type of $K^+$ channel that is present in human T-lymphocytes and neuronal tissue (Garcia *et al.* 1991). Such data supporting the promiscuous behaviour of this toxin indicate that ChTX is not the most suitable tool for defining the physiological role of specific $K^+$ channels. For this purpose, more selective toxins are needed.

Initially, ChTX was also reported to be a high-affinity inhibitor of native *Shaker* $K^+$ channels (MacKinnon *et al.* 1988). Although it was later discovered that the blocking activity was due to a contaminant present in the ChTX preparation, and the inhibitory peptides identified as agitoxins I-III (see below), *Shaker* $K^+$ channels have become a useful tool to study the receptor site of ChTX. This has been possible by the discovery that a single amino acid substitution at position 425 (Gly for Phe) in the wild-type *Shaker* $K^+$ channel, confers high-affinity ChTX sensitivity ($K_d$ of 75 pM). The features of ChTX interaction with *Shaker* are similar to those found with the maxi-K channel. Thus, ChTX binds through a bimolecular reaction in which Lys-27 interacts with $K^+$ in the ion conduction pathway (Goldstein and Miller 1993). The interaction surface of the toxin with the *Shaker* channel has also been deduced and five residues were found to be critical for the toxin's interaction (Goldstein *et al.* 1994). These residues are also well separated from unimportant ones and lie together in an area of 530 $Å^2$, which represents 17% of the toxin's molecular surface. The shape of this surface suggest that the narrow $K^+$ conduction pore widens at its external face to a 25 × 35 Å plateau. Complementary mutagenesis of both ChTX and *Shaker* indicates that Phe-425 in the channel makes contact with an area close to Thr-8 and Thr-9 on ChTX, causing the toxin to bind poorly. Toxin binding can be enhanced by mutating these two threonine residues to smaller ones. When Phe-425 is made smaller, the destabilizing contact is lost, the toxin binds strongly, and binding is no longer sensitive to the size of the residue on the toxin. Given this interaction, Phe-425 in *Shaker* must be located at a 20 Å radial distance from the pore axis, and 10–15 Å above the receptor floor (Goldstein *et al.* 1994). Complementary mutagenesis has also identified a pair of residues, Lys-427 in *Shaker*, and Lys-11 in ChTX, that sense each other via through-space electrostatic forces (Stocker and Miller 1994). These studies suggest that Lys-427 in *Shaker* is located 10–15 Å above the wide plateau that forms the channel's ChTX interaction surface, at the same elevation as Phe-425. These two *Shaker* residues appear to form a low wall enveloping the floor of the receptor.

Since ChTX contains a single tyrosine residue, the molecule can be iodinated to high specific activity with $^{125}I$ and the resulting monoiodotyrosine-ChTX ([I]ChTX) separated by reversed-phase chromatography (Vazquez *et al.* 1989). Sequence analysis confirmed that a single iodine is incorporated into Tyr-36 of the peptide. [I]ChTX reversibly blocks maxi-K channels in isolated outside-out patches of cultured bovine aortic smooth muscle cells by a mechanism identical to that of native toxin, although with ca. 10-fold reduced potency. When [$^{125}$IChTX is incubated with purified bovine smooth muscle sarcolemmal membranes derived from either aortic or tracheal tissue – particularly rich sources of maxi-K channels – there is time and concentration-dependent association of toxin with these membranes (Slaughter *et al.* 1988; Vazquez *et al.* 1989). Toxin binding is a freely reversible bimolecular reaction, and displays predictable characteristics based on the functional interaction of ChTX with the channel. Thus, binding is highly sensitive to the ionic strength of the incubation media, and is blocked by a number of metal ions that are known to bind with high affinity to sites located along the ion conduction pathway of the channel. In addition, tetraethylammonium (TEA) blocks ChTX binding by an apparently competitive mechanism with a $K_i$ value identical to that measured for inhibition of maxi-K channels in functional assays. Therefore, the receptor site identified for [$^{125}$I]ChTX in these smooth muscle membrane preparations has all the characteristics expected of a maxi-K channel.

Recently, a ChTX mutant containing Cys-19 has been produced and shown to have the same biological activity as native toxin (Shimony *et al.* 1994). Interestingly, this ChTX mutant can be reacted with thiol alkylating reagents such as [$^{3}$H]*N*-ethylmaleimide or fluorescence derivatives of maleimide to yield a labelled ChTX molecule with identical biological activity as native toxin.

Using [$^{125}$I]ChTX binding to bovine aortic sarcolemma as a monitor of maxi-K channels, a variety of scorpion venoms were tested for their ability to modulate the binding reaction. Venom of the scorpion *Buthus tamulus* contains such an activity and a peptide that blocked [$^{125}$I]ChTX binding was purified to homogeneity. This peptide was named iberiotoxin (IbTX), and determination of its primary sequence reveals 70% homology with ChTX (Galvez *et al.* 1990). Like ChTX, the N-terminus of IbTX is blocked in the form of a pyroglutamic acid residue, and it also contains six cysteine residues. However, IbTX is a much less basic peptide than ChTX due to the presence of four more acidic amino acids. The identity of the peptide was also confirmed by solid-phase synthesis and by its production using recombinant techniques. The three-dimensional solution structure of IbTX has been deduced by NMR techniques (Johnson and Sugg 1992). Its backbone structure is identical to that of ChTX, as is the position of those residues critical for ChTX's interaction with the maxi-K channel. In single-channel recordings of maxi-K channels incorporated into planar lipid bilayers, IbTX reversibly inhibits channel activity by a mechanism identical to that of ChTX (Candia *et al.* 1992; Giangiacomo *et al.* 1992). However, IbTX causes the appearance of much longer silent periods than those observed in the presence of ChTX, suggesting that, once it is bound to the channel, IbTX must overcome a much higher energy barrier to dissociate from its receptor site. The molecular determinants for the tighter association of IbTX are unknown, but it is possible that more than a single residue could contribute to the stabilization of the toxin bound state.

One of the most interesting features of IbTX is its high selectivity for maxi-K channels. IbTX does not affect other ChTX-sensitive $K^+$ channels, such as $K_v1.3$ and small-conductance $Ca^{2+}$-activated $K^+$ channels (Leonard *et al.* 1992; Giangiacomo *et al.* 1993). Therefore, IbTX represents a selective and high-affinity probe with which to explore the physiological role of maxi-K channels in different tissues.

In an effort to identify those determinants in ChTX and IbTX which are responsible for each respective type of channel selectivity, hybrid molecules were synthesized in which the N-terminal 19 amino acid residues of one peptide were coupled to the final 18 C-terminal residues of the other peptide, and vice versa. In addition, a ChTX deletion mutant in which the first six residues were not present was also synthesized. Single-channel recordings of smooth muscle maxi-K channels, or whole-cell current recordings of $K_v1.3$ channels expressed in *Xenopus* oocytes indicate that the C-terminus of each peptide appears to dictate receptor specificity. Moreover, the N-terminus of ChTX appears to be critical for high-affinity interaction with the maxi-K, but not with the $K_V1.3$ channel (Giangiacomo *et al.* 1993).

Another peptide that has similar characteristics to IbTX in terms of selectivity and potency has been isolated and characterized from venom of the new world scorpion, *Centruroides limbatus*. Limbatustoxin blocks maxi-K channels with similar kinetics as IbTX, and displays a net overall charge intermediate between IbTX and ChTX, since it possess two less negatively charged residues than IbTX (Novick *et al.* 1991).

From venom of the scorpion, *L. quinquestriatus*, another peptide, Lq2 toxin, has been purified and characterized (Lucchesi *et al.* 1989). Lq2 toxin blocks maxi-K channels, although with lower potency than ChTX. In addition, this agent also blocks the *Shaker* H4 channel expressed in *Xenopus* oocytes. The wide spectrum of activities of Lq2 toxin makes this peptide of limited utility for characterization of $K^+$ channels in physiological studies.

The functional role of the maxi-K channel has been studied in pharmacological experiments employing both IbTX and ChTX, and a number of different smooth muscle tissues. The myogenic activity of guinea pig bladder and ileum smooth muscle tissues is markedly enhanced by both IbTX and ChTX. This is an indication that maxi-K channels in these tissues represent a major pathway for cell repolarization after $Ca^{2+}$ entry following action potential generation (Suarez-Kurtz *et al.* 1991). However, the myogenic activity of guinea pig uterine smooth muscle appears to be quite insensitive to the presence of either toxin. Since maxi-K channels are known to be present in this tissue, these data indicate that, in the guinea pig, other types of $K^+$ channels play a more predominant role in controlling uterine contraction. The role of maxi-K channels in smooth muscle is not only tissue-dependent, but also species-dependent, since different effects have been found with smooth muscle tissues from rat (Winquist *et al.* 1989).

When guinea pig airway smooth muscle is contracted by carbachol, relaxation of the preparation can be achieved by a number of agents such as $\beta$-agonists, phosphodiesterase inhibitors, and dibutyryl-cAMP. In the presence of either IbTX or ChTX, the relaxation dose–response curves for these agents are shifted to the right and under some experimental conditions, (e.g. in the presence of $\beta$-agonists) abolished (Jones *et al.* 1990, 1993). These findings may suggest that part, or all, of the relaxation mechanism of these agents involves activation of maxi-K channels, and it would be consistent with the observation that the open probability of this channel is enhanced by protein kinase A-dependent phosphorylation (Kume *et al.* 1989). However, abolition of relaxation to $\beta$-agonists by IbTX is prevented in the presence of a $Ca^{2+}$-entry blocker, indicating that functional antagonism might also account for some of these results (Huang *et al.* 1993). Apparently, the threshold for opening voltage-gated $Ca^{2+}$ channels is achieved due to the depolarization induced in a carbachol-contracted preparation by IbTX. $Ca^{2+}$ entry through voltage-gated $Ca^{2+}$ channels may be sufficient to overcome the $Ca^{2+}$-lowering mechanism(s) activated by $\beta$-agonists and prevent relaxation by these agents. Thus, functional studies have to be interpreted with caution in trying to characterize the role of maxi-K channels in trachea.

Stimulation of the sensory neurones in guinea pig airways leads to the release of tachykinins, such as the neurokinins and substance P, that can bind to receptors in the smooth muscle to cause contraction. Tachykinins have been implicated in the pathophysiology of asthma, since they can also produce mucus hypersecretion and plasma extravasation, which are typical features of asthmatic diseases. Release of tachykinins from the sensory neurones can be prevented by $\mu$-opiod agonists, $\alpha$-agonists, and neuropeptide Y. However, the effects of these agents is greatly diminished in the presence of either IbTX or ChTX. These data suggest that, at the sensory nerve terminals, these three dissimilar agents prevent the release of tachykinins by activation of maxi-K channels (Stretton *et al.* 1992), and, therefore, an activator of maxi-K channels could have utility in the treatment and prevention of the neurogenic inflammatory component of asthmatic diseases.

As stated above, the development of specific peptide probes for $K^+$ channels provides a tool with which to identify other pharmacologically relevant modulators of $K^+$ channels. In this respect, three compounds were isolated and characterized from extracts of the plant, *Desmodium adscendens*, based on their ability to inhibit [$1^{25}$I]ChTX binding to tracheal smooth muscle sarcolemmal membranes (McManus *et al.* 1993). It is interesting to note that alcoholic extracts of this plant are used in Ghana for the chronic treatment of asthma and other diseases associated with smooth muscle contraction. These compounds were identified as dehydrosoyasaponin I (DHS-I), soyasaponin I, and soyasaponin III. The most potent of these compounds in binding experiments is DHS-I. This agent is a partial inhibitor of [$1^{25}$I]ChTX binding, causing 62% maximum inhibition with a $K_i$ of 120 nM. DHS-I also displays non-competitive behaviour in saturation experiments, and enhances toxin dissociation kinetics, suggesting that it is an allosteric modulator of [$1^{25}$I]ChTX binding. In single-channel recordings, DHS-I causes a marked increase in channel open probability when applied to the inner membrane surface, the side at which $Ca^{2+}$ binds to activate the channel, but this agent has no effect when added at the external surface of the channel. Activation of maxi-K channel activity can be observed at concentrations as low as 10 nM, and the effect is fully reversible. The chemical differences in the soyasaponins isolated indicate a defined structure–activity relationship for channel activation with contributions from both the sugar and triterpene moieties of the molecule. DHS-I applied at the inner channel surface destabilizes ChTX bound in the external vestibule of the channel, as was previously observed in binding experiments. These data are a clear demonstration of an allosteric interaction between two different binding sites on the channel and predict, that in analogy with $Na^+$ and $Ca^{2+}$ channels, maxi-K channels will also be a multi-drug receptor. Unfortunately, the site of action of these compounds, together with their membrane impermeant properties, has precluded their study in pharmacological experiments necessary to determine the utility of a maxi-K channel agonist in intact tissue evaluations.

A series of maxi-K channel antagonists has also been identified because of their ability to modulate [$1^{25}$I]ChTX binding to smooth muscle membranes (Knaus *et al.* 1994*c*). These agents consist of a family of tremorgenic indole diterpenes which have been isolated from fungal microorganisms and which are known to produce intermittent tremors in animals. In *in vitro* studies, members of this structural family enhance neurotransmitter release from different synaptosomal preparations. Although the mechanism of action of these compounds has not been elucidated, these *in vitro* effects could be related to maxi-K channel blockade. In fact, these compounds have now been shown to interact very potently with this channel. In [$1^{25}$I]ChTX binding experiments, the indole diterpenes display a complex interaction with the ChTX receptor. Agents such as paxilline, verruculogen, and paspalicine cause a marked stimulation of toxin binding, which is due to an increase in toxin affinity resulting from decreases in toxin dissociation kinetics. Other structurally related members of this family, such as paspalitrem A and C, penitrem A, aflatrem, and paspalinine, inhibit ChTX binding with a defined rank order of potency, and inhibition is incomplete in most cases. Importantly, the inhibition produced by these compounds is reversible after washing the membranes, and paxilline reverses aflatrem inhibition of toxin binding while aflatrem reverses paxilline stimulation of toxin binding. These latter data suggest that both stimulators and inhibitors of the binding reaction share a common receptor site from where they allosterically modulate toxin bound to a distinct receptor in the mouth of the channel.

Despite the differential modulation of ChTX binding by these compounds, all are potent inhibitors of the maxi-K channel. Agents such as penitrem A and aflatrem cause complete inhibition of channel activity at concentrations as low as 10 nM. Blockade of the maxi-K channel appears specific, since members of this structural class display marginal activity at $\mu$M concentrations against other types of $K^+$ channels, and they do not display $Ca^{2+}$-entry blocker activity or modulate $Na^+$ channels. The tremorgenic activity of these compounds is known to be associated with the presence of a hydroxyl group at position 19 of the diterpene nucleus. The only structural difference between paspalicine and paspalinine is the presence of the C-19 hydroxyl group, which results in paspalinine being tremorgenic, while its deshydroxy analogue, paspalicine, is non-tremorgenic. However, paspalicine is also an effective blocker of maxi-K channels. These data indicate that tremorgenicity is not strictly associated with maxi-K channel blockade. It is possible, however, that some of the *in vitro* properties of these compounds which are the result of enhanced neurotransmitter release could be related to maxi-K channel blockade, since this action could lead to a broadening of the action potential and enhanced $Ca^{2+}$ entry at the nerve terminal (Robitalle *et al.* 1993). It is interesting to speculate that agents with such properties could have potential therapeutic application in the treatment of neuronal degenerative diseases where neurotransmitter release is impaired.

The discovery of peptidyl toxins that interact with specific ion channels has also led to the isolation and characterization of the molecular components of those channels. Although molecular biological techniques have afforded the identification of the pore-forming subunit of many channels, the discovery of auxiliary subunits of these channels has only been accomplished after purification of the channel complex to homogeneity. In the case of $Na^+$ and $Ca^{2+}$ channels, these proteins modulate the biophysical and pharmacological properties of their respective channels (Isom *et al.* 1994). There is biochemical and immunological evidence for the existence of other subunits that are associated with pore-forming $K^+$ channel subunits (Rehm and Lazdunski *1988*; Parcej and Dolly *1989*; Trimmer 1991). Recently, the first $\beta$-subunit of a voltage-gated $K^+$ channel was identified based on its copurification with the $\alpha$-dendrotoxin receptor from bovine brain (Scott *et al.* 1994). Coexpression of $\beta$- and the pore-forming subunit of members of the *Shaker* family lead to marked functional effects on the behaviour of these channels (Rettig *et al.* 1994).

As stated above, [$^{125}$I]ChTX binds to a single class of receptors in aortic and tracheal smooth muscle sarcolemma that are associated with maxi-K channels. Therefore, these preparations represent an appropriate source of starting material for the purification of this protein. To accomplish this, tracheal sarcolemmal membrane vesicles were solubilized in the presence of digitonin (Garcia-Calvo *et al.* 1991), and the solubilized material was then subjected to fractionation using a combination of conventional chromatographic techniques and sucrose density gradient centrifugation (Garcia-Calvo *et al.* 1994). These procedures yielded a final preparation that is enriched ca. 2000-fold over the initial starting material and that displays a specific activity of ca. 1 nmol [$^{125}$I]ChTX binding sites per mg protein. Importantly, all the pharmacological properties for toxin interaction that have been observed with intact membranes are preserved in the purified preparation. When fractions from the last step in purification are subjected to SDS–PAGE, silver staining of the preparation reveals the presence of a protein with an $M_r$ of 62 000 Da that comigrates with binding activity. However, SDS–PAGE after [$^{125}$I]Bolton-Hunter labelling of these same fractions indicates the presence of two subunits with $M_r$s of 62 000 Da ($\alpha$) and 31 000 Da ($\beta$) that copurify with binding activity. The $\beta$-subunit is heavily glycosylated by N-linked sugars, which may explain the difficulties in staining the protein. After enzymatic deglycosylation, the $M_r$ of the core protein is reduced to 22 000 Da. This subunit represents the protein of the maxi-K channel complex to which [$^{125}$I]ChTX is specifically and covalently attached in the presence of the bifunctional crosslinking reagent, disuccinimidyl suberate (Garcia-Calvo *et al.* 1991; Knaus *et al.* 1994*c*).

When the purified maxi-K channel preparation is reconstituted into liposomes, single maxi-K channel activity can be readily observed after fusion of this material with artificial lipid bilayers (Garcia-Calvo *et al.* 1994). These channels are voltage- and $Ca^{2+}$- dependent. Thus, the channel open probability is enhanced by membrane depolarization and by raising intracellular $Ca^{2+}$ levels. A plot of the channel open probability versus voltage at different $Ca^{2+}$ concentrations indicate that this ion shifts the mid-point of activation to more hyperpolarized potentials. This property, as well as the ion selectivity, single-channel conductance, and modulation by pharmacological agents, suggests that the two structural components of the purified ChTX receptor preparation are sufficient to reconstitute a native maxi-K channel complex.

In order to obtain amino acid sequence information from these proteins to accomplish their molecular cloning, the subunits were separated by SDS–PAGE, electroeluted, and subjected to proteolytic digestion. Individual fragments were separated by RP–HPLC and their amino acid sequence determined by Edman degradation. Seven sequences were obtained from the $\alpha$-subunit (Knaus *et al.* 1994*b*). These all display a very high degree of homology or are identical with sequences from the recently cloned maxi-K channel, *mSlo*, from mouse brain and skeletal muscle or *hslo*, from human brain. The obtained sequences are located in the S4 segment, the link between the S5 and pore region, the S6 segment, and different regions in the large C-terminal domain of *mSlo*. Under denaturing conditions, the $\alpha$-subunit from the purified maxi-K channel preparation can be specifically immunoprecipitated by a site-directed antibody raised against one of the obtained sequences. However, under nondenaturing conditions, both the $\alpha$- and $\beta$-subunits of the maxi-K channel complex are specifically immunoprecipitated, suggesting that they are tightly associated. The predicted mass for the *mSlo* protein is ca. 140 000 Da, but the purified $\alpha$-subunit of the maxi-K channel displays an $M_r$ of 62 000 Da. Using a battery of site-directed antibodies raised against different domains of *mSlo*, it has been shown that the size of the tissue-expressed *mSlo* is ca. 115 000 Da, and that during purification a specific proteolytic cleavage takes place (Knaus *et al.* 1995).

Although the resulting fragments remain associated in the complex, they can be resolved after SDS–PAGE under reducing conditions to yield the major product of 62 000 Da.

From the β-subunit of the bovine tracheal maxi-K channel, a unique 28-amino acid sequence was obtained that was used to synthesize oligonucleotide probes against both ends of the molecule. These were used to amplify a PCR product that, after sequencing, confirmed the identity of the peptide. This product was then used to screen cDNA libraries from both bovine tracheal and aortic smooth muscle which led to the isolation of a full-length cDNA coding for the β-subunit (Knaus *et al.* 1994*a*). The deduced amino acid sequence predicts a protein of 22 000 Da that contains two α-helical transmembrane domains connected by a large extracellular loop with two putative sites for N-linked glycosylation, in excellent agreement with the known biochemical properties of this protein. Both the N- and the C-terminus are predicted to be cytoplasmic, and the N-terminus contains a consensus sequence for phosphorylation by cAMP-dependent protein kinase A, close to the first transmembrane domain. Under denaturing conditions, both the [$^{125}$I]ChTX-crosslinked and [$^{125}$I]Bolton-Hunter labelled β-subunit can be specifically immunoprecipitated by site-directed antibodies raised against two sequences of the putative extracellular domain of the β-subunit, suggesting that the cloned subunit represents the protein to which [$^{125}$I]ChTX is covalently attached in the presence of a bifunctional crosslinking reagent. Under non-denaturing conditions, these antibodies immunoprecipitate both α- and β-subunits of the maxi-K channel complex, further demonstrating that the maxi-K channel exists *in vivo* as a heterodimer of α- and β-subunits. Injection of mRNA translated from the β-subunit into *Xenopus* oocytes does not lead to the appearance of ion channel activity. However, coexpression of the β-subunit with the pore-forming subunit of the maxi-K channel leads to marked alterations in the biophysical and pharmacological properties of the later subunit (McManus *et al.* 1995; Meera *et al.* 1995). Thus, the voltage dependence of channel activation is shifted by more than 50 mV in the hyperpolarized direction, and the channel becomes sensitive to the maxi-K channel agonist DHS-I. It will be interesting to determine if the presence of a β-subunit represents a physiological mechanism for maxi-K channel regulation.

## Noxiustoxin, margatoxin, and Centruroides limpidus limpidus toxin I

This second class of scorpion toxin peptides displays significant sequence homology with both ChTX, IbTX, limbatustoxin, and Lq2 toxin. However, they appear to constitute a distinct group of toxins based on their selectivity against voltage-dependent $K^+$ channels.

Noxiustoxin (NxTX) was the first toxin identified to be targeted against $K^+$ channels (Possani *et al.* 1982). It is a 39-amino acid peptide containing six cysteine residues that was isolated from venom of the scorpion, *Centruroides noxius*. Initially, it was shown that NxTX inhibited the delayed rectifier $K^+$ channel in squid axon, although high concentrations were needed to produce this blockade (Carbone *et al.* 1982). Later, it was found that NxTX could also block maxi-K channels from skeletal muscle t-tubule membranes incorporated into planar lipid bilayers (Valdivia *et al.* 1988). Again, the interaction of NxTX with maxi-K channels was not of high affinity in that it displayed a *K*d of 450 nM. It was through the use of ChTX, which displays a broader selectivity against $K^+$ channels, that the specificity of NxTX was elucidated. As stated in the previous section, ChTX blocks $K_V1.3$ channels in neuronal and human T-lymphocyte preparations. In rat brain membranes, [$^{125}$I]ChTX binds to a single class of receptor sites with pharmacological properties expected for an interaction with $K_v1.3$ channels (Vazquez *et al.* 1990). In particular, [$^{125}$I]ChTX binding in this preparation is modulated by low concentrations of both NxTX and α-dendrotoxin, but it is not sensitive to the presence of high concentrations of IbTX. This suggested that the high-affinity receptor for NxTX was neither the delayed rectifier $K^+$ channel, nor the maxi-K channel, but rather $K_v1.3$. Consistent with this idea, NxTX has no affect on either [$^{125}$I]ChTX binding to maxi-K channels in smooth muscle membranes, or maxi-K channel activity in smooth muscle cells. However, it does inhibit with high affinity, $K_d = 0.2$ nM, $K_v1.3$ type $K^+$ channels in Jurkat cells, as well as $K_v1.3$ channels expressed in *Xenopus* oocytes (Sands *et al.* 1989; Swanson *et al.* 1990). In addition, [$^{125}$I]ChTX binding to $K_v1.3$ channels in both human T-lymphocytes and Jurkat cells is inhibited by low concentrations of NxTX (Deutsch *et al.* 1991; Slaughter *et al.* 1991). Recently, it has been shown that NxTX inhibits the dendrotoxin-sensitive $K_v1.2$ channel stably transfected in mammalian cells with similar affinity as it blocks $K_v1.3$ (Grissmer *et al.* 1994). Unlike ChTX, few mechanistic studies have been performed with NxTX, but it is also predicted to function as a channel pore blocker. NxTX, and the newly discovered margatoxin (MgTX), have been useful probes for defining the physiologic role that $K_v1.3$ channels play in human T-lymphocytes (see below).

In the search for other selective toxins directed against the $K_v1.3$ type of $K^+$ channel, the [$^{125}$I]ChTX binding reaction in rat brain membranes was used to screen different scorpion venoms for blocking activity. Crude venom of the scorpion *Centruroides margaritatus* inhibits this binding reaction with an $IC_{50}$ value of 4.5 ng/ ml crude venom. Much higher concentrations of the crude venom are needed to elicit inhibition of toxin binding to maxi-K channels in smooth muscle, suggesting that this extract could contain an activity(s) specifically directed against $K_v1.3$ channels. Fractionation of the crude venom by ion-exchange and RP–HPLC led to the purification and characterization of MgTX (Garcia-Calvo *et al.* 1993). This novel peptide consists of 39 amino acids and six cysteine residues, and displays high homology (79% identity) in its primary amino acid sequence with NxTX. MgTX is a selective and high-affinity blocker of $K_v1.3$ channels, displaying a $K_d$ of 50 pM in electrophysiological experiments. It has no activity whatsoever on maxi-K channels at concentrations up to 1 $\mu$M (Garcia-Calvo *et al.* 1993), nor does it block other types of ChTX-sensitive small-conductance $Ca^{2+}$-activated $K^+$ channels that are present in human T-lymphocytes (Leonard *et al.* 1992). Several different voltage-dependent $K^+$ channels were assessed for their sensitivity to MgTX after their expression in *Xenopus* oocytes. The only other *Shaker*-like channels sensitive to MgTX are $K_v1.1$ and $K_v1.2$; $K_v1.6$ is also sensitive but it is blocked with a 100-fold lower potency, whereas $K_v1.5$ and $K_v3.1$ are completely insensitive to the toxin.

The amino acid sequence of MgTX was confirmed by producing the peptide in *E. coli* by recombinant techniques, as well as by solid-phase synthesis (Garcia-Calvo *et al.* 1993; Bednarek *et al.* 1994). Both approaches offer the possibility of producing MgTX mutants with which to define the contact points of the toxin with the pore of the channel. The three-dimensional structure of MgTX has been determined by $^{1}$H, $^{13}$C, $^{15}$N triple-resonance NMR spectroscopy (Johnson *et al.* 1994). The global structure is very similar to that of ChTX and IbTX. A helix is present from residues 11 to 20 and includes two Pro residues at positions 15 and 16. There is a two-strand antiparallel sheet from residues 25 to 38, with a turn at residues

30–33. The additional two residues in MgTX extend the $\beta$-sheet by one residue relative to ChTX and IbTX. This longer sheet could have implications for channel selectivity.

MgTX has been radiolabelled in biologically active form and its interaction with rat brain synaptic membrane vesicles studied (Koch *et al.* 1995). In this system, [$^{125}$I]MgTX binds to a single class of receptor sites that display a *K*d of 0.1 pM, as determined by either ligand or receptor saturation studies, kinetics of ligand association and dissociation, or competition binding experiments with native MgTX. MgTX, therefore, represents the highest affinity ligand for any membrane receptor or ion channel identified to date. In human peripheral T-lymphocytes and Jurkat plasma membranes, [$^{125}$I]MgTX binds to receptor sites that display lower affinity (ca. 10-fold lower) than in brain membranes (Felix *et al.* 1995). Since the receptor site in either T-lymphocytes or Jurkat cells is presumed to be a homotetramer of $K_V1.3$, the high-affinity interaction found in brain must represent toxin binding to a heteromultimeric receptor, or yet some other unidentified channel.

Given the fact that, in human T-cells, both NxTX and MgTX appear to inhibit only $K_V1.3$ channels, the role of this channel in T-lymphocyte function was investigated. It has been proposed that $K^+$ channels play a role in the physiology of T-cell activation based on results which demonstrated that either raising extracellular $K^+$ concentration, or addition of several small-molecule $K^+$ channel inhibitors, will prevent mitogen-induced elevations in cytosolic $Ca^{2+}$ concentration, as well as the synthesis and release of IL–2. This would result in blockade of T-cell proliferation. Unfortunately, the agents that were employed are not very selective at the high concentrations needed to block $K^+$ channels. Later, it was shown that ChTX would inhibit mitogen-induced proliferation of human peripheral T-lymphocytes, and that this inhibition could be reversed by addition of exogenous IL–2 (Price *et al.* 1989). However, ChTX not only blocks $K_V1.3$ channels, but also the small-conductance $Ca^{2+}$ – activated $K^+$ channels present in these cells (Leonard *et al.* 1992). Therefore, the relative contribution of block of either type of channel to account for the antiproliferative properties of ChTX was difficult to assess. With the use of NxTX and MgTX, this question could now be easily addressed.

When the membrane potential of non-activated peripheral human T-lymphocytes is monitored with the lipophilic cation [$^3$H]tetraphenylphosphonium ion, the cells display a resting potential of approximately −50mV. In the presence of increasing concentrations of either MgTX, NxTX, or ChTX, the cells depolarize to a maximum value of −30 mV, and the concentrations of the peptides required to produce this effect correlate with their ability to block the $K_V1.3$ channel (Leonard *et al.* 1992). These data indicate that the resting potential in nonactivated human T-cells is determined by the activity of $K_V1.3$, and not the small-conductance $Ca^{2+}$-activated $K^+$ channels.

Mitogen-induced T-cell activation consists of a number of processes that lead to production of IL–2, which serves as an autocrine factor to initiate T-cell proliferation. The initial step in T-cell activation is a rise in cytoplasmic $Ca^{2+}$ levels due to release of this ion from intracellular stores and the influx of extracellular $Ca^{2+}$ through depleted-stores-activated $Ca^{2+}$ channels. The sustained entry of $Ca^{2+}$ through these channels appears to be necessary for T-cell proliferation. This rise in intracellular $Ca^{2+}$ is inhibited by MgTX, NxTX, and ChTX and, as expected, these peptides also inhibit lymphokine production (Lin *et al.* 1993). These data suggest that membrane depolarization caused by inhibition of $K_V1.3$ in human T-cells is sufficient to prevent the rise in $Ca^{2+}$ that is required for T-cell activation. It is possible that an internal negative membrane potential contributes to the driving force for influx of $Ca^{2+}$ into T-lymphocytes, and that this process becomes limited by depolarization. Although the precise mechanism by which membrane depolarization leads to alterations in $Ca^{2+}$ homeostasis is not well understood, the above data suggest that selective $K_v1.3$ blockers could represent a novel approach to immunosuppresant therapy.

Recently, two novel toxins have been purified from the venom of the Mexican scorpion, *Centruroides limpidus limpidus*, using an immunoassay based on antibodies raised against NxTX (Martin *et al.* 1994). The primary structure of one of these toxins, *C. l. limpidus* toxin I, has been determined and shown to have 74% and 64% homology with MgTX and NxTX, respectively. Both toxins were able to inhibit binding of radiolabelled NxTX to rat brain synaptosomes, and to inhibit a transient voltage-dependent $K^+$ current in cultured rat cerebellar granule cells. The specificity of these toxins against other types of $K^+$ channels has not been determined.

## Kaliotoxins and agitoxins

A third class of peptidyl scorpion toxins directed against $K^+$ channels can be defined based on the primary amino acid sequence of kaliotoxins (KTX) and the agitoxins (AgTX). Despite their high degree of homology, there appears to be a large difference in the selectivity of members of this class of toxins.

KTX is a 38-amino acid peptide isolated from venom of the scorpion, *Androctonus mauretanicus*. In the first attempt to determine the sequence of this peptide, only 37 residues were identified (Crest *et al.* 1992). In a later publication, the last residue of the peptide was characterized and the sequence corrected (Romi *et al.* 1993). The identity of the peptide was confirmed after its synthesis by solid-phase methodologies. KTX specifically suppresses the whole-cell $Ca^{2+}$ activated $K^+$ current from *Helix pomatia* without affecting voltage-gated $K^+$ currents in that preparation. This $Ca^{2+}$ –activated $K^+$ current appears to correspond to a channel that displays a single unitary conductance of 30–35 pS, under asymmetrical $K^+$ conditions. However, an interaction of KTX with voltage-gated $K^+$ channels has been inferred since iodinated KTX 1–37 binds to a class of receptor sites in brain membranes that appear to have pharmacological properties related to voltage-gated $K^+$ channels. Moreover, KTX blocks $K_v1.3$ stably transfected in a mammalian cell line with very high affinity (Grissmer *et al.* 1994). Another related toxin, $KTX_2$, has been isolated from *Androctonus australis* scorpion venom based on its ability to compete with [$^{125}$ I] KTX 1–37 for binding to rat brain synaptosomes (Laraba-Djebari *et al.* 1994). $KTX_2$ consists of 37 amino acids and has 76% identity with KTX. $KTX_2$ displays lower affinity than KTX as an inhibitor of either the *H. pomatia* $Ca^{2+}$ –activated $K^+$ current or of binding of KTX to rat brain synaptosomes. Recently, the three-dimensional structure of KTX 1–37 has been determined by NMR techniques (Fernandez *et al.* 1994). It was found that the helical region of this toxin is shorter when compared with ChTX and distorted, and that the N-terminal strand and the $\alpha$-helix interact with opposite sides of the $\beta$-sheet, whereas in ChTX they interact with the same face of the $\beta$-sheet. The implications of these structural differences for channel inhibition remain to be determined. Since KTXs are not high-affinity blockers of maxi-K channels, the specificity of these toxins is different from that of the other two classes of toxins previously described.

Agitoxins represent a group of three closely related peptides that also display high homology with KTX. These three newly identified peptides isolated from *L. quinquestriatus* venom (Garcia *et al.* 1994) block *Shaker* $K^+$ channels with very high affinity, as well as the mammalian homologues of *Shaker*. They do not, however, inhibit maxi-K channels, and their specificity is clearly different from that of KTX. The AgTXs are a very interesting class of peptides since they represent the blocking activity detected against *Shaker* that was previously attributed to ChTX (MacKinnon *et al.* MacKinnon 1991; 1988, 1990). The primary amino acid sequence of these peptides was confirmed after their production in biologically active form by recombinant techniques. Given the conserved placement of the cysteine residues in the amino acid sequence of this class of toxins with that of ChTX, IbTX, and MgTX, it is expected that they will have a similar three-dimensional structure as the first group of toxins.

Five out of the eight critical residues of ChTX are not conserved in this class of toxins. For example, Met 29, when substituted by Ile in ChTX, decreases the affinity of toxin for the maxi-K channel by 1000-fold (Stampe *et al.* 1994), but Ile is present in the corresponding position in one of the AgTXs. Clearly, the specific interactions between residues on the contact surface of these toxins with their target channels will be different.

Recently, $AgTX_2$ has been used in a study to transfer toxin sensitivity from $K_v1.3$ to the toxin-insensitive potassium channel, $K_v2.1$. This was accomplished by transferring the stretch of amino acids between transmembrane domains $S_5$ and $S_6$. These studies indicate that this region represents the only part of the ion channel that makes direct contact with the bound toxin. The presence of two residues in $K_v2.1$, Lys356 and Lys386, was found to be the reason for this channel's insensitivity to $AgTX_2$ (Gross *et al.* 1994).

The number of peptidyl scorpion toxins shown to be directed against $K^+$ channels continues to grow. It is very likely that more toxins will be identified in the future since scorpion venoms are a rich source of such agents. These peptides are clearly magnificent tools for studying at all levels the structure and function of $K^+$ channels. A summary of the pharmacological properties of several $K^+$ channels is presented in Table 8.1.

## References

Adelman, J.P., Shen, K.-Z., Kavanaugh, M.P., Warren, R.A., Wu, Y.-N., Lagrutta, A. Bond, C.T., and North R.A. (1992). Calcium-activated potassium channels expressed from cloned complementary DNAs. *Neuron*, **9**, 209–216.

Bednarek, M.A., Bugianesi, R.M., Leonard, R.J., and Felix, J.P., (1994). Chemical synthesis and structure-function studies of margatoxin, a potent inhibitor of voltage-dependent potassium channel in human T lymphocytes. *Biochem. Biophys. Res. Commun.*, **198**, 619–625.

Bontems, F., Roumestand, C., Boyot, P., Gilquin, B., Dolijansky, Y., Menez, A., and Toma, F. (1991*a*). Three-dimensional structure of natural charybdotoxin in aquous solution by $^1H$ – NMR. Charybdotoxin possesses a structural motif found in other scorpion toxins. *Eur. J. Biochem.*, **196**, 19–28.

Bontems, F., Roumestand, C., Gilquin, B., Menez, A., and Toma, F. (1991*b*). Refined structure of charybdotoxin: common motifs in scorpion toxins and insect defensins. *Science*, **254**, 1521–1523.

**Table 8.1** Sensitivity of $K^+$ channels to various pharmacological agents.

| | TEA (mM) | 4-AP (mM) | α-DTX (nM) | ChTX (nM) | IbTX (nM) | NxTx (nM) | MgTX (nM) | KTX (nM) | $AgTX_1$(nM) | $AgTX_2$(nM) |
|---|---|---|---|---|---|---|---|---|---|---|
| $K_V1.1$ | 0.3–0.6[a,b] | 0.3–1[a,b] | 12–20[a,b] | 1500[c] | 0.14[K] | >25[a] | 0.14[K] | 41[a] | 136[c] | 0.044[c] |
| $K_V1.2$ | 129–560[a,b] | 0.6–0.8[a,b] | 4–17[a,b] | 14[a] | N.D. | 2[a] | 0.15[k] | >1000[a] | N.D. | N.D. |
| $K_V1.3$ | 10–50[a,b] | 0.2–1.5[a,b] | >200[a,b] | 0.19–2.6[a,b,c] | >100[d] | 1[a] | 0.03[e] | 0.65[a] | 1.7[c] | 0.004[c] |
| $K_V1.4$ | >100[b] | 12.5[b] | >200[b] | >40[b] | N.D. | N.D. | N.D. | N.D. | N.D. | N.D. |
| $K_V1.5$ | 330[a] | 0.27[a] | >1000[a] | >100[a] | >100[f] | >25[a] | >200[e] | >1000[a] | N.D. | N.D. |
| $K_V1.6$ | 7[b] | 1.5[b] | 20[b] | 22[c] | N.D. | N.D. | 5[e] | N.D. | 149[c] | 0.037[c] |
| maxi-K | 0.14–0.29[g] | N.D. | N.D. | 3[d] | 1[h] | 450[i] | >1000 | >300[j] | >200[c] | >200[c] |

Sources: The inhibition constant, $K_i$, is reported for each case.
[a] Grissmer *et al.* (1994)
[b] Pongs (1992)
[c] Garcia *et al.* (1994)
[d] Giangiacomo *et al.* (1993)
[e] Garcia-Calvo *et al.* (1993)
[f] R. Swanson, personal communication
[g] Latorre (1994)
[h] Giangiacomo *et al.* (1992)
[i] Valdivia *et al.* (1988)
[j] Laraba-Djebari *et al.* (1994)
[k] Unpublished information.
N.D.: not determined.

Butler, A., Tsunoda, S., McCobb, D.P., Wei, A., and Salkoff, L. (1993). mSlo, a complex mouse gene encoding 'maxi' calcium-activated potassium channels. *Science*, **261**, 221–224.

Candia, S., Garcia, M.L., and Latorre, R. (1992). Mode of action of iberiotoxin, a potent blocker of the large conductance $Ca2^+$-activated $K^+$ channel. *Biophys. J.*, **63**, 583–590.

Carbone, E., Wanke, E., Prestipino, G., Possani, L.D., and Melicke, A. (1982). Selective blockage of voltage-dependent $K^+$ channels by a novel scorpion toxin. *Nature*, **296**, 90–91.

Catterall, W.A. (1980). Neurotoxins that act on voltage-sensitive sodium channels in excitable membranes. *Annu. Rev. Pharmacol. Toxicol.*, **20**, 15–43.

Christie, M.J., North, R.A., Osborne, P.B., Douglas, J., and Adelman, J.P. (1990). Heteropolymeric potassium channels expressed in *Xenopus* oocytes from cloned subunits. *Neuron*, **4**, 405–411.

Crest, M., Jacquet, Gola, M., Zerrouk, H., Benslimane, A., Rochat, H., Mansuelle, P., and Martin-Eauclaire, M.-F. (1992). Kaliotoxin, a novel peptidyl inhibitor of neuronal BK-type $Ca^{2+}$-activated $K^+$ channels characterized from *Androctonus mauretanicus mauretanicus* venom. *J. Biol. Chem.*, **267**, 1640–1647.

Deutsch, C., Price, M., Lee, S., King, V.F., and Garcia, M.L. (1991). Characterization of high affinity binding sites for charybdotoxin in human T lymphocytes: evidence for association with the voltage-gated $K^+$ channel. *J. Biol. Chem.*, **266**, 3668–3674.

Felix, J.P., Bugianesi, R.M., Abramsom, A.A., and Slaughter, R.S. (1995). Binding of mono- and di-iodinated margatoxin to human peripheral T lymphocytes and to Jurkat plasma membranes. *Biophys. J.*, **68**, 267A.

Fernandez, I., Romi, R., Szendeffy, S., Martin-Eauclaire, M.F., Rochat, H., Ban Rietschoten, J., Pons, M., and Giralt, E. (1994). Kaliotoxin (1–37) shows structural differences with related potassium channel blockers. *Biochemistry*, **33**, 14256–14263.

Galvez, A., Gimenez-Gallego, G., Reuben, J.P., Roy-Contancin, L., Feigenbaum, P., Kaczorowski, G.J., and Garcia, M.L. (1990). Purification and characterization of a unique, potent, peptidyl probe for the high conductance calcium-activated potassium channel from venom of the scorpion *Buthus tamulus*. *J. Biol. Chem.*, **265**, 11083–11090.

Garcia, M.L., Galvez, A., Garcia-Calvo, M., King, V.F., Vazquez, J., and Kaczorowski, G.J. (1991). Use of toxins to study potassium channels. *J. Bioenerg. Biomembr.*, **23**, 615–646.

Garcia, M.L., Garcia-Calvo, M., Hidalgo, P., Lee, A., and MacKinnon, R. (1994). Purification and characterization of three inhibitors of voltage-dependent $K^+$ channels from *Leiurus quinquestriatus* var. *hebraeus* venom. *Biochemistry*, **33**, 6834–6839.

Garcia-Calvo, Knaus, H.-G., McManus, O.B., Giangiacomo, K.M., Kaczorowski, G.J., and Garcia, M.L. (1994). Purification and reconstitution of the high-conductance calcium-activated potassium channel from tracheal smooth muscle. *J. Biol. Chem.*, **269**, 676–682.

Garcia-Calvo, M., Vasquez, Smith, M. Kaczorowski, G.J. and Garcia M.L. (1991). Characterization of solubilized charybdotoxin receptor from bovine aortic smooth muscle. *Biochemistry*, **30**, 11157–11164.

Garcia-Calvo, M., Leonard, R.J. Novick J., Stevens, S.P., Schmalhofer, Kaczorowski, G.J. and Garcia M.L. (1993). Purification, characterization, and biosynthesis of margatoxin, a component of *Centruroides margaritatus* venom that selectively inhibits voltage-dependent potassium channels. *J. Biol. Chem.*, **268**, 18866–18874.

Giangiacomo, K.M., Garcia, M.L. and McManus, O.B. (1992). Mechanism of iberiotoxin block of the large-conductance calcium-activated potassium channel from bovine aortic smooth muscle. *Biochemistry*, **31**, 6719–6727.

Giangiacomo, K.M., Sugg, E.E. Garcia-Calvo, M. Leonard, R.J. McManus, O.B., Kaczorowski, G.J. and Garcia, M.L. (1993). Synthetic charybdotoxin-iberiotoxin chimeric peptides define toxin binding sites on calcium-activated and voltage-dependent potassium channels. *Biochemistry*, **32**, 2363–2370.

Gimenez-Gallego, G., Navia, M.A. Reuben, J.P. Katz, G.M. Kaczorowski, and G.J. and Garcia, M.L. (1988). Purification, sequence, and model structure of charybdotoxin, a potent selective inhibitor of calcium-activated potassium channels. *Proc. Natl Acad. Sci. USA*, **85**, 3329–3333.

Goldstein, S.A.N. and Miller, C. (1993). Mechanism of charybdotoxin block of a voltage-gated $K^+$ channel. *Biophys.* J., **65**, 1613–1619.

Goldstein, S.A.N., Pheasant, D.J. and Miller, C. (1994). The charybdotoxin receptor of a *Shaker* $K^+$ channel: peptide and channel residues mediating molecular recognition. *Neuron*, **12**, 1377–1388.

Grissmer, S., Nguyen, A.N. Aiyar, J. Hanson, D.C. Mather, R.J. Gutman, G.A., Karmilowicz, M.J. Auperin, D.D., and Chandy, K.G. (1994). Pharmacological characterization of five cloned voltage-gated $K^+$ channels, types $K_v$ 1.1, 1.2, 1.3, 1.5, and 3.1, stably expressed in mammalian cell lines. *Mol. Pharmacol.*, **45**, 1227–1234.

Gross, A., Abramson, T. and MacKinnon, R. (1994). Transfer of the scorpion toxin receptor to an insensitive potassium channel. *Neuron*, **13**, 961–966.

Huang, J.-C., Garcia, M.L., Reuben, J.P. and Kaczorowski, G.J. (1993). Inhibition of $\beta$-adrenoceptor agonist relaxation of airway smooth muscle by Ca2+-activated $K^+$ channel blockers. *Eur. J. Pharmacol.*, **235**, 37–43.

Isacoff, E.Y., Jan, Y.N., and Jan, L.Y. (1990). Evidence for the formation of heteromultimeric potassium channels in *Xenopus* oocytes. *Nature*, **345**, 530–534.

Isom, L.L., De Jongh, K.S. and Catterall, W.A. (1994). Auxiliary subunits of voltage-gated ion channels. *Neuron*, **12**, 1183–1194.

Jan, L.Y. and Jan, (1992). Structural elements involved in specific $K^+$ channel functions. *Annu. Rev. Physiol.*, **54**, 537–555.

Johnson, B.A. and Sugg, E.E. (1992). Determination of the three-dimensional structure of iberiotoxin in solution by $^1H$ nuclear magnetic resonance spectroscopy. *Biochemistry*, **31**, 8151–8159.

Johnson, B.A., S.P. Stevens, and Williamson, J.M. (1994). Determination of the three-dimensional structure of margatoxin by $^1H$, $^{13}C$, $^{15}N$ triple-resonance nuclear magnetic resonance spectroscopy. *Biochemistry*, **33**, 15061–15070.

Jones, T.R., Charette, M.L Garcia, and Kaczorowski, G.J. (1990). Selective inhibition of relaxation of guinea-pig trachea by charybdotoxin, a potent $Ca^{2+}$-activated $K^+$ channel inhibitor. *J. Pharmacol. Exp. Ther.* **255**, 697–705.

Jones, T.R., Charette, M.L., Garcia, M.L. and Kaczorowski, G.J. (1993). Interaction of iberiotoxin with beta adrenoceptor agonists and sodium nitroprusside on guinea pig trachea. *J. Appl. Physiol.*, **74**, 1879–1884.

Knaus, H.-G., Folander, K., Garcia-Calvo, M., Garcia, M.L., Kaczorowski, G.J., Smith, M., and Swanson, R. (1994a). Primary sequence and immunological characterization of the $\beta$-subunit of the high-conductance $Ca^{2+}$-activated $K^+$ channel from smooth muscle. *J. Biol. Chem.*, **269**, 17274–17278.

Knaus, H.-G., Garcia-Calvo, M., Kaczorowski, G.J., and Garcia, M.L. (1994*b*). Subunit composition of the high conductance calcium-activated potassium channel from smooth muscle, a representative of the *mSlo* and *slowpoke* family of potassium channels. *J. Biol. Chem.*, **269**, 3921–3924.

Knaus, H.-G., McManus, O.B., Lee, S.H., Schmalhofer, W.A., Garcia-Calvo, M., Helms, L.M.H., Sanchez, M., Giangiacomo, K., Reuben, J.P., Smith, A.B. III, Kaczorowski, G.J., and Garcia, M.L. (1994*c*). Tremorgenic indole alkaloids potently inhibit smooth muscle high-conductance $Ca^{2+}$-activated $K^+$ channels. *Biochemistry*, **33**, 5819–5828.

Knaus, H.-G., Eberhart, A., Munujos, P., Kaczorowski, G.J., Schmalhofer, W.A., Warmke, J.W., and Garcia, M.L. (1995). Characterization of the $\alpha$-subunit of the high-conductance $Ca^{2+}$-activated $K^+$ channel from tracheal smooth muscle. *Biophys. J.*, **68**, 267A.

Koch, R.O.A., Eberhart, A., Seidl, C.V., Saria, A., Kaczorowski, G.J., Slaughter, R.S., Garcia, M.L., and Knaus, H.-G. (1995). [$^{125}$I]Margatoxin, an extraordinary high-affinity ligand for voltage-gated $K^+$ channels in mammalian brain. *Biophys. J.*, **68**, 266A.

Kume, H., Tokuno, H., and Tomita, T. (1989). Regulation of $Ca^{2+}$-dependent $K^+$-channels in trachael myocytes by phosphorylation. *Nature*, **341**, 152–154.

Lambert, P., Kuroda, H., Chino, N., Watanabe, T.X., Kimura, T., and Sakakibara, S. (1990). Solution synthesis of charybdotoxin (ChTX), a $K^+$ channel blocker. *Biochem, Biophys. Res. Commun.*, **170**, 684–690.

Laraba-Djebari, F., Legros, C., Crest, M., Ceard, B., Romi, R., Mansuelle, P., Jacquet, G., van Rietschoten, J., Gola, M., Rochat, H., Bougis, P.E., and Martin-Eauclaire, M.-F. (1994). The kaliotoxin family enlarged. *J. Biol. Chem.*, **269**, 32835–32843.

Latorre, R. (1994). The molecular workings of large conductance (maxi) $Ca^{2+}$-activated $K^+$ channels. In *Membrane channels: molecular and cellular physiology*, (ed. C. Peracchia), pp. 79–101. Academic Press, New York.

Latorre, R., Oberhauser, A., Labarca, P., and Alvarez, O. (1989). Varieties of calcium-activated potassium channels. *Annu. Rev. Physiol.*, **51**, 385–399.

Leonard, R.J., Garcia, M.L., Slaughter, R.S., and Reuben, J.P. (1992). Selective blockers of voltage-gated $K^+$ channels depolarize human T lymphocytes: mechanism of the antiproliferative effect of charybdotoxin. *Proc. Natl. Acad. Sci. USA*, **89**, 10094–10098.

Lin, C.S., Boltz, R.C., Blake, J.T., Nguyen, M., Talento, A., Fischer, P.A., Springer, M.S., Sigal, N.H., Slaughter, R.S., Garcia, M.L., Kaczorowski, G.J., and Koo, G.C. (1993). Voltage-gated potassium channels regulate calcium-dependent pathways involved in human T lymphocyte activation. *J. Exp. Med.*, **177**, 637–645.

Lucchesi, K., Ravindran, A., Young, H., and Moczydlowski, E. (1989). Analysis of the blocking activity of charybdotoxin homologs and iodinated derivatives against $Ca^{2+}$-activated $K^+$ channels. *J. Membr. Biol.*, **109**, 269–281.

MacKinnon, R. (1991). Determination of the subunit stoichiometry of a voltage-activated potassium channel. *Nature*, **350**, 232–235.

MacKinnon, R., and Miller, C. (1988). Mechanism of charybdotoxin block of the high-conductance, $Ca^{2+}$-activated $K^+$ channel. *J. Gen. Physiol.*, **91**, 335–349.

MacKinnon, R., Heginbotham, L., and Abramson, T. (1990). Mapping the receptor site for charybdotoxin, a pore-blocking potassium channel inhibitor. *Neuron.*, **5**, 767–771.

Martin, B.M., Ramirez, A.N. Gurrola, G.B. Nobile, M. Prestipino G., and Possani, L. D. (1994). Novel $K^+$-channel-blocking toxins from the venom of the scorpion *Centruroides limpidus limpidus* Karsch. *Biochem. J.*, **304**, 51–56.

McManus, O. B., Helms L.M.H., Pallanck, L., Ganetzky, B., Swanson, R, and Leonard, R. J. (1995). The beta subunit of maxi-K channels modifies gating and pharmacology of expressed mslo channels. *Neuron*, **14**, 1–20.

McManus, O. B., Giangiacomo, K. L., Harris, G. H., Addy, M. E., Reuben, J. P, Kaczorowski G., and Garcia, M.L., (1993). A maxi-K channel agonist isolated from a medicinal herb. *Biophys. J.*, **64**, 3A.

Meera, P., Wallner, M., Ottolia M., Adelman, J.P., Kaczorowski, G. Garcia, M.L., and Toro L., (1995). Differential effects of the maxi-K $\beta$ subunit on the $\alpha$ subunit of human and *Drosophila* expressed in *Xenopus laevis* oocytes. *Biophys. J.*, **68**, 3A.

Miller, C., Moczydlowski, Latorre, R., and Phillips, M. (1985). Charybdotoxin, a protein inhibitor of single $Ca^{2+}$-activated $K^+$ channels from mammalian skeletal muscle. *Nature*, **313**, 316–318.

Novick, J., Leonard, R. J., King, V. F., Schmalhofer, W., Kaczorowski, G.J., and Garcia, M. L. (1991). Purification and characterization of two novel peptidyl toxins directed against $K^+$ channels from venom of new world scorpions. *Biophys. J.*, **59**, 78A.

Parcej, D. N. and Dolly J. O., (1989). Dendrotoxin acceptor from bovine synaptic plasma membranes: binding properties, purification and subunit composition of a putative constituent of certain voltage-activated $K^+$ channels. *Biochem. J.*, **257**, 899–903.

Park, C.-S. and Miller, C., (1992). Interaction of charybdotoxin with permeant ions inside the pore of a $K^+$ channel. *Neuron*, **9**, 307–313.

Park, C. S., Hausdorff, S. F., and Miller, C., (1991). Design, synthesis, and functional expression of a gene for charybdotoxin, a peptide blocker of $K^+$ channels. *Proc Natl Acad. Sci. USA*, **88**, 2046–2050.

Pongs, O. (1992). Structural basis of voltage-gated $K^+$ channel pharmacology. *Trends Pharm. Sci.*, **13**, 359–365.

Possani, L.D., Martin, B.M., and Svendsen, I.B., (1982). The primary structure of Noxiustoxin: a $K^+$ channel blocking peptide, purified from the venom of the scorpion *Centruroides noxius* Hoffmann. *Carlsberg Res. Commun*, **47**, 285–289.

Price, M., Lee, S.C., and Deutsch, C. (1989). Charybdotoxin inhibits proliferation and IL2 production in human peripheral blood lymphocytes. *Proc. Natl Acad. Sci. USA*, **86**, 10171–10175.

Rehm, H. and Lazdunski, M., (1988). Purification and subunit structure of a putative $K^+$ channel protein identified by its binding properties for dendrotoxin I. *Proc. Natl Acad. Sci. USA*, **85**, 4919–4923.

Rettig, J., Heinemann, S.H., Wunder, F., Lorra, C., Parcej, D.N., Dolly, J.O., and Pongs, O. (1994). Inactivation properties of voltage-gated $K^+$ channels altered by presence of $\beta$ subunit. *Nature*, **369**, 289–294.

Robitalle, R., Garcia, M.L., Kaczorowski, G.J., and Charlton, M.P. (1993). Functional colocalization of calcium and calcium-gated potassium channels in control of transmitter release. *Neuron*, **11**, 645–655.

Romi, R., Crest M., Gola, M., *et al.* (1993). Synthesis and characterization of kaliotoxin. *J. Biol. Chem.*, **268**, 26302–26309.

Ruppersberg, J.P., Schroeter, K.H., Sakmann, B., Stocker, M., Sewing, S., and Pongs, O. (1990). Heteromultimeric channels formed by rat brain potassium-channel proteins. *Nature*, **345**, 535–537.

Sands, S.B., Lewis, R.S. and Cahalan, M.D. (1989). Charybdotoxin blocks voltage-gated $K^+$ channels in human and murine T lymphocytes. *J. Gen. Physiol.*, **93**, 1061–1074.

Schweitz, H., Bidard, J.-N., Maes, P., and Lazdunski, M. (1989). Charybdotoxin is a new member of the $K^+$ channel toxin family that includes dendrotoxin I and mast cell degranulating peptide. *Biochemistry*, **28**, 9708–9714.

Scott, V.E.S., Rettig, J., Parcej, D.N., Keen, J.N., Findlay, J.B.C., Pongs, O., and Dolly, J.O. (1994). Primary structure of a $\beta$ subunit of $\alpha$-dendrotoxin-sensitive $K^+$ channels from bovine brain. *Proc. Natl Acad. Sci. USA*, **91**, 1637–1641.

Shimony, E., Sun, T., Kolmakova-Partensky, L., and Miller, C. (1994). Engineering a uniquely reactive thiol into a cysteine-rich peptide. *Protein Engineering*, **7**, 503–507.

Slaughter, R.S., Kaczorowski, G.J., and Garcia, M.L. (1988). Charybdotoxin binds with high affinity to a single class of sites in bovine tracheal smooth muscle sarcolemmal membrane vesicles. *J. Cell Biol.*, **107**, 143A.

Slaughter, R.S., Shevell, J.L., Felix, J.P., Lin, C.S., Sigal, N.H., and Kaczorowski, G.J. (1991). Inhibition by toxins of charybdotoxin binding to the voltage-gated potassium channel of lymphocytes: correlation with block of activation of human peripheral T-lymphocytes. *Biophys. J.*, **59**, 213A.

Smith, C., Phillips, M. and Miller, C. (1986). Charybdotoxin, a specific inhibitor of the high conductance $Ca^{2+}$-activated $K^+$ channel. *J. Biol. Chem.*, **261**, 14607–14613.

Stampe, P., Kolmakova-Partensky, L., and Miller, C. (1994). Intimations of $K^+$ channel structure from a complete functional map of the molecular surface of charybdotoxin. *Biochemistry*, **33**, 443–450.

Stocker, M. and Miller, C. (1994). Electrostatic distance geometry in a $K^+$ channel vestibule. *Proc. Natl Acad. Sci. USA*, **91**, 9509–9513.

Stretton, D., Motohiko, M. Belvesi, M.G., and Barnes, P.J. (1992). Calcium-activated potassium channels mediate prejunctional inhibition of peripheral sensory nerves. *Proc. Natl Acad. Sci. USA*, **89**, 1325–1329.

Strong, P.N., Weir, S.W., Beech, D.J., Heistand, P., and Kocher, H.P. (1989). Effects of potassium from *Leiurus quinquestriatus hebraes* venom on responses to cromakalim in rabbit blood vessels. *Br. J. Pharmacol.*, **98**, 817–826.

Suarez-Kurtz, G., Garcia, M.L., and Kaczorowski, G.J. (1991). Effects of charybdotoxin and iberiotoxin on the spontaneous motility and tonus of different guinea pig smooth muscle tissues. *J. Pharmacol. Exp. Ther.*, **259**, 439–443.

Sugg, E.E., Garcia, M.L., Reuben, J.P., Patchett, A.A., and Kaczorowski, G.J. (1990). Synthesis and structural characterization of charybdotoxin, a potent peptidyl inhibitor of the high conductance $Ca^{2+}$-activated $K^+$ channel. *J. Biol. Chem.*, **265**, 18745–18748.

Swanson, R., Marshall, J., Smith, J.S. *et al.* (1990). Cloning and expression of cDNA and genomic clones encoding three delayed rectifier potassium channels in rat brain. *Neuron*, **4**, 929–939.

Trimmer, J.S. (1991). Immunological identification and characterization of a delayed rectifier $K^+$ channel polypeptide in rat brain. *Proc. Natl Acad. Sci. USA*, **88**, 10764–10768.

Tseng-Crank, J., Foster, C.D., Krause, J.D., Mertz, R., Godinot, N., DiChiara, T.J., and Reinhart, P.H. (1994). Cloning, expression, and distribution of functionally distinct $Ca^{2+}$-activated $K^+$ channel isoforms from human brain. *Neuron*, **13**, 1315–1330.

Valdivia, H.H., Smith, J.S., Martin, B.M., Coronado, R., and Possani, L.D. (1988). Charybdotoxin and noxiustoxin, two homologous peptide inhibitors of the $K^+(Ca^{2+})$ channel. *FEBS Lett.*, **2**, 280–284.

Vazquez, J., Feigenbaum, P., Katz, G., King, V.F., Reuben, J.P., Roy-Contancin, L., Slaughter, R.S., Kaczorowski, G.J., and Garcia, M.L. (1989). Characterization of high affinity binding sites for charybdotoxin in sarcolemmal membranes from bovine aortic smooth muscle: evidence for a direct association with the high conductance calcium-activated potassium channel. *J. Biol. Chem.*, **264**, 20902–20909.

Vazquez, J., Feigenbaum, P., King, V.F., Kaczorowski, G.J., and Garcia, M.L. (1990). Characterization of high affinity binding sites for charybdotoxin in synaptic plasma membranes from rat brain: evidence for a direct association with an inactivating, voltage-dependent, potassium channel. *J. Biol. Chem.*, **256**, 15564–15571.

Winquist, R.J., Heaney, L.A., Wallace, A.A., Baskin, E.P., Stein, R.B., Garcia, M.L., and Kaczorowski, G.J. (1989). Glyburide blocks the relaxation response to BRL34915 (cromakalim), minoxidil sulfate and diazoxide in vascular smooth muscle. *J. Pharmacol. Exp. Ther.*, **248**, 149–156.

# 9 Properties of the ATP-Regulated $K^+$ channel in pancreatic $\beta$-cells

M. J. Dunne

## Introduction

The maintenance of a constant blood glucose concentration is a challenging, highly complex, integrated process and one that is principally controlled by the hormone insulin. When the concentration of glucose in the plasma is elevated, insulin is released by the process of exocytosis from the $\beta$-cells of the pancreatic islets of Langerhans. It is now widely accepted that the key intracellular regulator of secretion is an increase in the cytosolic concentration of free calcium ions ($[Ca^{2+}{}_i]$), a rise that is mainly dependent on the presence of calcium outside the cell.[1,2]

It was first demonstrated in the late 1960s that, when islets of Langerhans are challenged with a number of compounds known to elicit insulin secretion, marked changes were brought about in the electrical behaviour of the cells. In these experiments, using glass microelectrodes, Dean and Matthews (in 1968)[3] were able to show that carbohydrate secretagogues caused a depolarization of the cell membrane and the generation of spike potentials. Later, in 1975, Matthews and Sakamoto[4] further demonstrated that these voltage-gated spikes were caused by the inward movement of $Ca^{2+}$. Moreover, the link between carbohydrate metabolism and the initiation of the membrane depolarization was subsequently shown to be associated with a decrease in the permeability of the membrane to potassium ions.[5–7] However, it was not until nearly 10 years later, through the use of the improved patch-clamp technique to study single-channel currents that these and other ionic conductance pathways were identified.

In intact resting pancreatic $\beta$-cells, only one particular type of $K^+$-selective ion channel is operational – a channel that is closed when cells are challenged with carbohydrate secretagogues.[8,9] Since this channel appears to be exactly the same as the one shown to be directly inhibitable by intracellularly applied adenosine 5′-trisphosphate (ATP),[10], we now understand that closure of ATP-sensitive $K^+$ channels in intact insulin-secreting cells initiates a depolarization of the cell membrane potential. Patch-clamp experiments have shown, using a number of sources of insulin-secreting cells, from human to rodent tissues, that there are close similarities in many of the properties of this channel (i.e. single-channel conductance, sensitivity to ATP, etc.) and that under normal conditions they show remarkably little in the way of species differentiation (Table 9.1). This is fortunate, since inaccessibility to human tissue, and the requirement of high numbers of insulin-secreting cells for biochemical determinations, have meant that insulin-secreting cell lines and rodent $\beta$-cells are used widely in this field. Finally, these channels have now been cloned and the components of the molecular architecture of the complex elucidated. This has turned up the very surprising observation that $K_{ATP}$ channels result from the associations of two very discrete proteins – which include an ATP-binding cassette (ABC) protein and an inward rectifier $K^+$ channel.

**Table 9.1** Published values for the general properties of ATP-sensitive $K^+$ channels in insulin-secreting cells.

| Tissue | Inward current conductance (pS) | Approximate $K_I$ for ATP ($\mu$M)* | Hill coefficient |
|---|---|---|---|
| Human $\beta$-cells | 60 | 10 | 1.2 |
| Rat (adult) $\beta$-cells | 58 | 26 | 1 |
| Rat (neonatal) $\beta$-cells | 54 | 15 | 1.2 |
| Mouse $\beta$-cells | 56 | 18 | 1.8 |
| RINm5F cells | 50 | 50 | 1.8 |
| CRI-G1 cells | 56 | 13 | 1.2 |
| HIT-T15 cells | 52 | 60 | ? |

* $K_I$ values for 50% inhibition of $K_{ATP}$ channel have been obtained using the inside-out membrane patch configuration in the presence of symmetrical 140 mM KCl-rich solutions.

## Biophysical properties of $K_{ATP}$ channels in $\beta$-cells

Early reports indicated that pancreatic $\beta$-cells possessed two types of ATP-sensitive $K^+$ ($K_{ATP}$) channel.[11] However, as these channels always appeared under the same experimental conditions, and differed only in their unit conductance or size, it is hard to say with conviction that the 'two channels' are independent conductances rather than sub-states of one complex. In intact insulin-secreting cells at rest, the frequency of openings of $K_{ATP}$ channels is typically extremely low. However, when either detergents, such as digitonin or saponin, are used to make holes in the plasma membrane (outside of the isolated patch area from which the recording is made), thereby equilibrating the cell interior with the bath solution ('open' or permeabilized cell) or when a patch of membrane is excised from the cell, many more channel open events become apparent, with much longer open times.[12,13] Experiments of this type suggest that, in intact cells, ATP-sensitive potassium channels are tonically inhibited by some endogenous factor. Since the channels are inhibited by ATP in cell-free patches, this 'endogenous factor' is likely to be intracellular ATP.[12,13] Estimates of the number of $K_{ATP}$ channels per cell vary considerably between 5000–10 000 in number,[14,15] and although an average value of approximately 12 channels per patch has been reported, it is not unusual to record currents from greater than 40 channels in an individual patch of membrane[16] which might indicate – as with voltage-gated $Ca^{2+}$ channels,[17] – that $K_{ATP}$ channels are 'clustered' at the plasma membrane.

$K_{ATP}$ channels are highly selective for external potassium ions relative to $Na^+$ ($P_{Na}/P_K$ = 0.007)[18] which explains why in the presence of physiological solutions containing 140 mM $Na^+$ the zero-current transmembrane potential is invariably found to be close to the $K^+$ equilibrium potential. The channels will allow rubidium ions to move, but the $Rb^+$ conductance is low ($P_{RB}/P_K$ = 0.73), and $Rb^+$ will also act as a permeable blocker of the channel.[18] $K_{ATP}$ channels are also inhibited by barium ions, which will produce both a time-and voltage-dependent block of $K_{ATP}$ channels that is increased by increasing degrees of hyperpolarization of the cell membrane potential.[19]

Current–voltage relationship plots obtained from either intact cells or excised patches in the presence of quasi-physiological cation gradients, suggest that the single-channel current conductance of $K_{ATP}$ channels is within the region of approximately 60 pS for the inward current (Table 9.1). $K_{ATP}$ channels, however, express a marked degree of inward current rectification and the current amplitude increases very little at potentials more positive than +20 mV. This indicates that these channels preferentially allow $K^+$ to enter the cell while restricting its efflux.[8,12,20] Rectification probably results from the intracellular concentrations of either or both $Mg^{2+}$ and $Na^+$.[21]

There has been a recent suggestion that the endogenous gating of $K_{ATP}$ channels in $\beta$-cells also plays a key role in governing the oscillatory responses of insulin-secreting cells to glucose. This is potentially very important, since oscillations in electrical activity are the gear to drive oscillations in the cytoplasmic free $Ca^{2+}$ concentration, and ultimately pulsatile insulin release. There are several possibilities that might account for the establishment of oscillations in electrical activity, but fluctuations in the endogenous gating of $K_{ATP}$ channels observed in single cells in the presence glucose could account for the initiation of the pulsative responses.[22] Furthermore, these oscillations in $K_{ATP}$ channel activity persist during glucose metabolism and are sensitive to changes in the metabolic status of the cells.

## Intracellular regulation of $K_{ATP}$ channel function

### Changes in cell membrane potential and $[Ca^{2+}]_i$

The frequency of $K_{ATP}$ channel openings and closures is relatively unaffected by changes in the cell membrane potential or by changes in the concentration of free ionized intracellular calcium ($[Ca^{2+}]_i$) see Dunne and Petersen[14] and Rorsman and Trube[15].

### Effects of intracellular pH changes

Both Cook and Hales[10] and Misler and his collaborators,[9] originally reported that, over the pH range 6.6 to 7.3, changes in the internal pH had no significant effect upon $K_{ATP}$ channel activity. However, in the presence of cytosolic ATP or in intact cells, changes in $pH_i$ appear to influence $K_{ATP}$ channels, with increases in cellular acidification tending to lead to an increase in the frequency. These data might indicate that changes in $pH_i$ do not directly influence $K^+$ channels, but rather that its actions are mediated through the availability of $ATP^{4-}$. This was subsequently confirmed by Proks and colleagues[23] who showed that while the $K_i$ for ATP-evoked channel inhibition was reduced from 18 to 32 $\mu$M when $pH_i$ was lowered, there was no effect of changes in $pH_i$ on the $ATP^{4-}$-mediated closure of channels. By comparison, $K_{ATP}$ channels in other tissues are far more sensitive to changes in intracellular pH than in β-cells; decreases in $pH_i$ have been found to inhibit $K_{ATP}$ channels in skeletal muscle and cardiac tissue in the absence of cytosolic ATP, whereas in the presence of ATP the opposite effect occurs.[24,25]

## Effects of ATP

The principal action of ATP when added to the inside of the membrane is to evoke channel closure.[10] This effect is concentration-dependent, with $K_i$ values, corresponding to half-maximal inhibition, between 10 μM and 60 μM (Table 9.1). Estimates of the Hill coefficient for channel blockade vary from approximately 1 to 1.8, making it unclear as to whether several molecules of ATP are able to bind to the channel in order to bring about closure in a cooperative manner.[10,27,27] Studies of $K_{ATP}$ channels in isolation from the cells are complicated by the problem of 'run-down' which is associated with the time-dependent loss of channel activity.[28] Interestingly, this process is prevented by the maintained presence of ATP, and since run-down can be reversed by exposing the cytosolic face of the membrane to ATP and then removing the nucleotide – so-called 'refreshment'[13] – this demonstrates that ATP has more than one mechanism of $K_{ATP}$ channel modulation; namely inhibition and activation.

Hydrolysis of ATP that might lead to phosphorylation of the protein complex, appears not to be involved in closing $K_{ATP}$ channels since non- and partially hydrolysable analogues of ATP will still inhibit channel activity,[26,29], and ATP will close channels in the complete absence of internal $Mg^{2+}$.[26,30,31] Indeed, the effects of $Mg^{2+}$-free solutions on $K_{ATP}$ channels are particularly interesting. In the absence of internal ATP ($[ATP]_i$), removing $Mg^{2+}$ from the inner membrane causes an increase in channel activity.[30] This activation is characterized by an increase in the single-channel conductance,[21,30] an increase in the number of channel openings[21,30] and by the disappearance of channel run-down.[31] These effects contrast markedly with the response of $K_{ATP}$ channels to the removal of $Mg^{2+}$ in the presence of ATP, which instead of activating $K_{ATP}$ channels, actually increases the potency of ATP inhibition.[30,32] These findings have been taken as evidence that it is the 'free' form of ATP (non-$Mg^{2+}$-bound) – probably $ATP^{4-}$ – that determines the degree of $K^+$ channel inhibition.[30,32] ATP-induced channel closure involves a decrease both in the mean open time of the channel and the number of open events per burst of activity;[32] there is no direct effect of ATP on the single-channel conductance and the actions of ATP are independent of the holding potential applied to the membrane.

In contrast to the inhibitory effects of ATP, the recovery of channel run-down is strictly dependent upon ATP hydrolysis and is not supported by ATP analogues or $ATP^{4-}$.

## Effects of ADP

The modulatory effects of ADP on $K_{ATP}$ channels are probably as important as those of ATP. In general, relatively low concentrations of ADP ($< 500$ μM), when added to the inside of the membrane, tend to open channels, whereas higher concentrations will inhibit.[33,34]

However, when 'inhibitory concentrations' of ADP are used at the intracellular domain of the cell membrane in the presence of ATP, instead of potentiating the actions of ATP, ADP causes an increase in channel activity. This unusual finding is also very important, since it is a key piece of evidence that has been used to explain not only how glucose metabolism can result in a depolarization of the cell membrane, but also how $K_{ATP}$ channels can operate in the intact cell. Since the earliest reports of $K_{ATP}$ channels in $\beta$-cells it has been long recognized that these channels are open in intact resting cells and that they provide the $K^+$ leak current used to establish the resting membrane potential. However, for several years there was no apparent way to explain how these channels could be operational, since electrophysiological data consistently revealed that the channels have a relatively very high sensitivity to intracellular ATP ($K_i = 10$–$60\ \mu M$; Table 9.1), yet estimates of the internal concentration of ATP vary between 4 and 6 mM. As ADP will offset the inhibition by ATP, and ADP is present at reasonably high concentrations in $\beta$-cells ($\sim 1$ mM), these findings provide a logical explanation for this discrepancy.[33,34] Furthermore, $K_{ATP}$ channels are also remarkably sensitive to ADP, which will activate ATP-inhibited channels even if the concentration of ATP is some 30-fold in excess, and small changes in the ratio of ATP : ADP markedly affect the gating of $K_{ATP}$ channels.[29] Despite the actions of ADP, it has been estimated that between 95% and 99% of the channels will be closed under normal resting conditions, and this has been argued as energetically favourable, since if glucose acts by closing channels to evoke a membrane depolarization, it is convenient for the cell to have already in the resting situation, the majority of channels inhibited ('spare channel hypothesis' [35,36]) The ADP regulation of ATP-inhibited channels also provides a rational explanation for the coupling of glucose metabolism to the generation of electrical activity in $\beta$-cells, and this may very well be a 'tight' cooperation, since the intracellular ATP : ADP ratio will also oscillate during metabolism with a similar periodicity to the cycling of electrophysiological changes in the membrane potential.[37–39]

The mechanism of ADP-evoked modulation of ATP-inhibited channels has also been addressed. Several pieces of data reveal that while the direct activation of channels by ADP may involve protein phosphorylation, the recovery of ATP-inhibited channels by ADP does not: ADP$\beta$S, a non-hydrolysable analogue of ADP, closes channels directly but will activate $K_{ATP}$ channels in the presence of either ATP, or ATP analogues (ATP$\gamma$S, AMP-PNP, or AMP-PCP). The implication of these findings is that ADP also has multiple binding sites on the $K_{ATP}$ channel complex which result from both competitive interactions with nucleotide binding site(s) on the channel and from protein phosphorylation-dependent mechanisms.

### Effects of other nucleotides

The reduced- and non-reduced forms of the pyridine nucleotides also influence $K_{ATP}$ channel behaviour. High concentrations ($> 500\ \mu M$) of each of the nucleotides will evoke inhibition, whereas lower concentrations ($< 200\ \mu M$) tend to open channels. These effects occur both directly and in the presence of ATP/ADP.[16] In $Mg^{2+}$-free solutions, the stimulatory effects of the pyridine nucleotides were lost, while the inhibitory actions are retained (Dunne, unpublished results). As the intracellular concentrations of $NADH^+$, NADPH, and NADP have been estimated to lie between 30 and 100 μM,[16] it is possible that these nucleotides contribute to the background activation of channels along with intracellular ADP. The concentration of $NAD^+$, on the other hand is thought to be considerably higher at approximately 350 $\mu$M. Since, $NAD^+$ at this concentration will inhibit $K_{ATP}$ channels in the presence of quasi-physiological ATP/ADP concentrations, $NAD^+$ might also be another candidate in the intact cell to bring about the tonic closure of $K_{ATP}$ channels.

## Regulation by G proteins

Until recently, the evidence to support a role for the direct regulation of $K_{ATP}$ channels by G proteins was somewhat ambiguous. First, GTP, GDP, GDP$\beta$S, and GTP$\gamma$S were all shown to be able evoke a reversible, concentration-dependent (10 $\mu$M–2 mM) increase in channel openings,[40] which were $Mg^{2+}$ dependent.[21] However, one of the complications in interpreting these data is that several nucleotides will affect the activity of $K_{ATP}$ channels in $\beta$-cells through a 'nucleotide-dependent binding site' rather than a specific G protein-dependent mechanism. Secondly, the alumino-fluoride complex $AlF_4^-$, which is known to have effects on a number of G protein-regulated systems, dose-dependently (1 to 30 mM) activates $K_{ATP}$ channels.[41] However, these effects are not influenced by either cholera toxin or pertussis toxin pre-treatment of cells, they will also occur without exogenous nucleotides being added to the cell membrane, and they can be observed in isolated cell-free patches many, many minutes after the formation of inside-out patches; conditions that are not likely to preserve G protein function.

More palpable evidence to support a role for the G protein-dependent gating of $K_{ATP}$ channels in insulin-secreting cells is available, but this has so far only been observed in the insulinoma cell lines, RINm5F and HIT T15. Both galanin and somatostatin are inhibitors of insulin secretion and in RINm5F cells it has been shown that these agents will mediate their actions through a hyperpolarization of the cell membrane potential, and this is brought about by the selective activation of $K_{ATP}$ channels. The evidence to favour a G protein involvement is based upon the fact that the response are abolished by pre-treatment of cells with pertussis toxin,[41,42] and by the fact that the isolated $\alpha$-subunits of the GTP-binding proteins $G_o$ and $G_i$ also caused the activation of $K_{ATP}$ channels.[43,44]

## Regulation by protein kinase C

There is no direct evidence that protein kinase C will alter the activity of $\beta$-cell $K_{ATP}$ channels, but there are several studies of the effects of modulators of PKC and the actions of these compounds on ATP-sensitive $K^+$ channels. Phorbol esters, such as phorbol myristate acetate (PMA) are potent activators of protein kinase C, that can both initiate secretion and potentiate glucose-induced insulin release from $\beta$-cells.[45,46] Using RINm5F cells, there have been several attempts to resolve the mechanisms of action of PMA on $K_{ATP}$ channels. Wollheim and collaborators[47] originally showed that PMA caused the closure of $K_{ATP}$ channels and led to a depolarization of the cell membrane potential. However, exactly the opposite effect on $K_{ATP}$ channels and the cell membrane potential were subsequently reported by two other groups.[48,49] More recent data shows that PMA has a dual control of $K_{ATP}$ channels – closing the channels in the short term and increasing their activity over longer periods of time.[50] One explanation for this is that there may be more than one type of protein kinase C associated with the channel. This is supported by the fact that cell-permeable diacyglycerols, such as dideconylglycerol ($DC_{10}$), can both activate and inhibit.[47] $K_{ATP}$ channels under different experimental conditions.

## Extracellular control of $K_{ATP}$ channel function

### Effects of carbohydrate secretagogues

ATP-regulated $K^+$ channels are the only $K^+$ channels seen under resting conditions in intact cells, and it is now clear that these channels play a very important role in initiating a complex pattern of electrical responses following glucose metabolism. This culminates in waves or oscillations of depolarization/repolarization of the membrane that are used to initiate the activity of voltage-gated $Ca^{2+}$ channels, which lead to a rise in the $[Ca^{2+}]_i$ and the release of insulin by exocytosis. The link that couples metabolism to channel closure most likely involves changes in the concentrations of ATP and ADP close to the cell membrane interior, but a regulatory role of other molecules (e.g. NAD(P)H/NAD(P)), fatty acid derivatives, or the importance of protein kinase-mediated channel phosphorylation cannot be excluded. Yet more novel intracellular compounds that may be also be implicated, such as diadenosine tetraphosphates which have been recently shown to block $K_{ATP}$ channels in $\beta$-cells and bring about channel 'refreshment' following run-down.[51]

### Effects of hormones and neurotransmitters

Galanin is a 29-amino acid polypeptide that will abolish insulin secretion under a number of experimental conditions, both *in vivo* and in the isolated perfused pancreas. At least part of these inhibitory effects involve $K_{ATP}$ channels, since electrophysiological recordings have demonstrated that galanin will hyperpolarize RINm5F cells, terminate the cell electrical activity[42] and activate $K_{ATP}$ channels.[41,42] Since these actions were abolished by overnight pretreatment of cells with pertussis toxin, these data suggest a G protein-dependent mechanisms of action. Somatostatin, another potent inhibitor of insulin secretion has also been shown to mediate a similar effect on $K_{ATP}$ channels in RINm5F cells.[49]

Also using RINm5F cells it has been shown that agents which potentiate glucose-induced insulin secretion will do so by closing $K_{ATP}$ channels and facilitating a depolarization of the cell membrane potential; these include vasopressin,[52] glucagon,[43] and the purinoceptor agonist, ATP.[53]

## Pharmacology of $K_{ATP}$ channels

As $K_{ATP}$ channels have a pivotal role to play in stimulus–response coupling events in $\beta$-cells, and may be considered as an 'on-off' switch for Ca-dependent secretory events, the pharmacology of these channels can have important medicinal implications for the control of blood glucose levels. Inhibitors of $K_{ATP}$ channels will tend towards hypoglycaemia through promoting insulin secretion, while channel activators will inhibit insulin release and lead to elevated serum glucose levels.[54–56]

## $K_{ATP}$ channel activators

In general, this aspect of the pharmacology of ATP-sensitive $K^+$ channels is more widely associated with potassium channel openers (KCOs) and the applicability of these compounds as anti-hypertensive agents and as smooth muscle relaxants: cromakalim, pinacidil, and nicorandil are typical compounds in this context. However, the benzothiadiazine diazoxide was the first compound to be shown to have the capacity to activate or 'open' $K^+$ channels, and it was also the first agent to be shown to be selective for this action on the $K_{ATP}$ channel. Other types of $K^+$ channel, such as delayed rectifier- , and calcium- and voltage-gated $K^+$ channel are not activated by diazoxide. Several $K^+$ channel openers are able to activate $K_{ATP}$ channels in $\beta$-cells, and while the mechanisms of action are similar, there are very marked differences in the selectivity of these compounds for $K_{ATP}$ channels in different tissues. Thus, while cromakalim, pinacidil, nicorandil, and RP49356 are typically very potent vasodilators and activators of $K_{ATP}$ channels in smooth/cardiac muscle, they are only weakly effective in the pancreas (Table 9.2). By contrast, diazoxide is a very effective agonist of $\beta$-cell $K_{ATP}$ channels, but inhibits $K_{ATP}$ channels in cardiac tissue.[57] Interestingly, the most potent group of compounds known to activate $K_{ATP}$ channels in $\beta$-cells are the pyridothiadiazine derivatives[58] that include BPDZ-44 and BPDZ-62.[59] These compounds are structurally similar to both diazoxide and pinacidil, but are much more effective than either compound at inhibiting insulin secretion and opening $K_{ATP}$ channels.

**Table 9.2** Typical synthetic potassium channel openers and their actions on $K_{ATP}$ channels in pancreatic $\beta$-**cells.**

| Chemical group | Drug | Action on $K_{ATP}$ channels |
|---|---|---|
| Benzothiadiazine | Diazoxide | Potent agonist |
| Benzopyran | Cromakalim<br>SDZ PCO-400 | Weak agonist<br>Weak antagonist |
| Nicotinamide | Nicorandil | Weak agonist |
| Cyanoguanidine | Pinacidil | Moderate agonist |
| Thioformamide | RP 49356 | Weak agonist |
| Pyridothiadiazine | BPDZ-44/-62 | Potent agonist |

The mechanisms of action of diazoxide (and other KCOs) has been widely documented. Diazoxide does not alter cAMP accumulation, nor does it seem to have any direct effect upon intracellular $Ca^{2+}$ homoeostasis. Rather, diazoxide causes the activation of $K_{ATP}$ channels and this in turn will lead to a hyperpolarization of the $\beta$-cell membrane potential, terminating voltage-gated $Ca^{2+}$ influx. From electrophysiological studies it is clear that diazoxide will operate when added directly to either the inside (inside-out patches) or the outside (outside-out patches) face of the plasma membrane, indicating that there is no specific requirement for a soluble cytosolic component to mediate the actions of the drug. In addition, diazoxide will also open channels in cell-attached patches when added to the bath solution. Since in these experiments diazoxide is added to the area of the cell membrane that is outside of that from where the recording is being made, this suggests that the compound can penetrate the cell membrane and bind to a 'receptor site' located in the lipid domain of the plasma membrane, rather than bind only to a cytosolic site. The nature of the interaction of diazoxide with the $K_{ATP}$ channels has recently been investigated by molecular biology studies. In these experiments, it was elegantly shown that the recombinant channel complex formed from the subunits $K_{IR}6.2$ and SUR1 was diazoxide-sensitive[60] and that SUR1 can also confer the capacity for diazoxide to open other inward rectifier $K^+$ channel clones.[61] The implications of this work are that the binding site for diazoxide is located on the sulphonylurea receptor and that there is unlikely to be a separate protein complex associated with the channels.

From studies of native $K_{ATP}$ channels we know that diazoxide and KCOs mediate their actions in $\beta$-cells by increasing the channel opening frequency, and that they do not appear to increase the number of operational channels in the membrane, nor do they enhance the permeability of potassium ion flow through the channel pore. One major characteristic of these compounds is that their ability to open channels is acutely dependent upon the availability of cytosolic nucleotides, in particular ATP and ADP.[30,62] Using cell-free patches of cell membrane it has been shown that there are no effects of KCOs on channels in the absence of internal ATP, or when the cytosolic face of the cell membrane is exposed to non-hydrolysable ATP analogues, or Mg-free ATP-containing solutions.[30,63,64] Indeed, under these conditions diazoxide will decrease the activity of $K_{ATP}$ channels by reducing the amplitude of the single-channel current event.[30, 65] Although ATP and ATP hydrolysis are a prerequisite for effective channel opening, increases in the ATP concentration will also reverse the actions of the $K^+$ channel openers by inducing a marked rightward shift in the concentration–response relationship at elevated ATP levels.[30] One implication of this is that these drugs will be relatively more effective on damaged or diseased tissue where the ATP levels are low, rather than in normal tissue.

From an experimental perspective, the ADP-dependency of diazoxide-induced $K_{ATP}$ channel activation is intriguing. We and others have shown clearly that, under some experimental conditions, one can manipulate the intracellular environment to show that while diazoxide no longer activates ATP-inhibited channels to work, ADP availability will lead to $K^+$ channel opening. The experimental conditions used are patches in which run-down of channels is extensive, which therefore reflects the inability of the ion channels to support phosphorylation. Under these experimental conditions diazoxide will no longer activate $K_{ATP}$ channels in the presence of ATP, but if ADP is delivered to the same patch of cell membrane, $K^+$ channel activation results. These findings suggest that ADP and diazoxide share a similar binding site on SUR1, and that the hydrolysis of ATP to ADP (which would not occur in these experiments) is acutely required for diazoxide's actions under normal conditions. This is supported by the fact that hydrolysis of ATP is a requirement of diazoxide-evoked activation, and by the fact that hydrolysis of ATP to ADP is the critical step in this process; diazoxide has no significant effect on channels in the presence of monophosphate derivatives of the nucleotide.

Several years ago there was a great deal of interest in $K^+$ channel agonist pharmacology and this was an important time for the generation of ligands with different chemical structures to explore the sensitivity of $K_{ATP}$ channel modulation and the mechanisms of action of the drugs. One disappointing aspect of the KCO field, is that many of these types of compounds showed limited selectivity for their target tissue, and this has led to difficulties in taking the drugs to the clinical situation. However, we now know much more about the function and the regulatory differences in the 'superfamily' of ATP-sensitive $K^+$ channels, and this is backed up by molecular initiatives that have already begun to unravel the architecture of $K_{ATP}$ channels. The β-cell channel, for instance, is a herteromultimeric complex of SUR1 and $K_{IR}6.2$[60] the cardiac $K_{ATP}$ channel appears to be composed of SUR2A and $K_{IR}6.2$ (a channel with high sensitivity to pinacidil and cromakalim, but not diazoxide),[66] and the smooth muscle $K_{ATP}$ channel is a complex formed from SUR2B and $K_{IR}6.2$.[67] With these novel and innovative approaches to ion channels, more specific and more potent compounds will become available that may have a range of medicinal benefits.

The only routine clinical situation that currently relies upon the use of diazoxide to alleviate insulin hypersecretion is in the management of the neonatal disorder, persistent hyperinsulinaemic hypoglycaemia of infancy (PHHI).[68] The disorder is characterized by inappropriate insulin secretion in the face of hypoglycaemia and diagnosed by demonstrating hypoketotic hypo-fatty acidaemic hypoglycaemia in association with hyperinsulinism.[68] PHHI has a variable clinical phenotype, usually presenting within the first few hours/days of birth, and the combination of hypoketotic hypoglycaemia carries with it a substantial risk of neurological damage if not recognized and treated promptly and adequately. Short-term medical therapy for the disorder involves the provision of an increased carbohydrate intake to meet the elevated requirement and usually one or more drugs that inhibit insulin secretion. One such agent is diazoxide, which was first introduced to treat hyperinsulinism in the 1960s.[69] In the clinical environment, the sensitivity of children with PHHI to medical therapy is unfortunately highly variable, with a spectrum from extreme sensitivity through to total drug resistance. In normal pancreatic β-cells, the actions of diazoxide are well described, but in PHHI β-cells the situation is somewhat complicated by the fact that this disease arises from a loss of function of $K_{ATP}$ channels which occurs in association with gene mutations that encode the SUR1 subunit of the channel complex.[70] The absence of functional $K_{ATP}$ channels in β-cells from PHHI patients will explain why many children fail to respond to therapy and require partial pancreatectomy to alleviate the symptoms of hypoglycaemia. However, the presence of other (novel?) diazoxide-sensitive $K^+$ channels in β-cells from these patients also explains why some children will respond to medical therapy, although this in many cases is inappropriate to maintain normoglycaemia without the necessity for surgery.

## Inhibition of $K_{ATP}$ channels by sulphonylureas

Sulphonylurea compounds, such as glibenclamide, tolbutamide, etc., are a class of hypoglycaemia-inducing drugs that have been used for many years to successfully treat Type II or non-insulin-dependent diabetes (NIDDM).[71] These compounds are currently the most widely used of all agents in the therapy of Type II diabetes. It has been known for a long time that these agents will decrease the $K^+$ permeability of the β-cell membrane – which, as with glucose, will then lead to a deplorization of the cell, the initiation of voltage-gated $Ca^{2+}$ influx, and the release of insulin. Patch-clamp studies have now shown conclusively that the decrease in $K^+$ permeability results from the closure of $K_{ATP}$ channels, and that the actions of sulphonylureas are highly selective for these channels in $\beta$-cells.[72]

Sulphonylureas have undergone several generations in the development of their functional potency and a number of compounds are now available that have different degrees of effectiveness in their inhibition of $K_{ATP}$ channels, with values for half-maximal inhibition falling over three orders of magnitude for tolbutamide ($K_I$ = 10–17 $\mu$M), meglitinide (2.1 $\mu$M), glipizide (6.4 $\mu$M), and glibenclamide (4–20 nM).[73–75]. Sulphonylureas are effective from both sides of the plasma membrane; inhibition of channels is not dependent upon the availability of $Mg^{2+}$, but quite interestingly the magnitude of channel inhibition is enhanced by the addition of cytosolic ADP.[73]

From a functional perspective, the first indications of how important the sulphonylureas were to understanding the activity of $K_{ATP}$ channels, came from ligand-binding observations that the specificity and density of sulphonylureas binding sites correlates very well with: (i) estimates of the number of $K_{ATP}$ channels per β-cell; (ii) the efficacy of glibenclamide inhibition of $K_{ATP}$ channels; and (iii) the ability of the sulphonylureas to elicit insulin secretion. The affinity of glibenclamide for the ATP-sensitive $K^+$ channel is extraordinarily high, and therefore the drug was first used in the partial purification and isolation of a putative $K_{ATP}$ channel protein from pig brain. This protein had a molecular weight of approximately 150 kDa and was subsequently shown not to possess the iontophoretic activity.[76] A biologically active derivative of glibenclamide – 5-indo-2-hydroxyglyburide – was then used to partially purify a similar size protein (140 kDa) from the membranes of HIT insulinoma cells,[77] and this was subsequently followed by the cloning of the high-affinity receptor for sulphonylurea, SUR1.[78] The determination of the structure of SUR1 signalled a major advance, not only in our understanding of the pharmacology of sulphonylureas, but also in understanding the architecture of $K_{ATP}$ channels. First, the cDNA of SUR1 encodes 1582 amino acides, and when COSm6 cells are transfected with SUR1 cDNA this results in the expression of specific glibenclamide binding activity with a $K_d$ similar to that observed in native $\beta$-cells, and the rank order of potency in displacing [$^3$H]glibenclamide by glibenclamide, iodoglibenclamide, and tolbutamide was also similar to that recorded in $\beta$-cells. Secondly, sequence analysis has produced a 'predicted' structure of SUR1 that has several discrete features: (i) multiple membrane-spanning domains arranged into two groups of nine and four; (ii) an N-terminus that is extracellularily located and glycosylated; (iii) a cytoplasmic C-terminal; (iv) consensus sequences for three protein kinase A and 20 protein kinase C phosphorylation sites; and finally (v) two large intracellular domains that contain Walker A and B motifs, which are consensus recognition sites for the binding of nucleotides. The predicted primary structure of SUR1 therefore displays many features common to ATP-binding cassette (ABC) proteins; a superfamily of membrane proteins that also include the multi-drug resistance protein (MDR) and the cystic fibrosis (CF) transmembrane conductance regulator (CFTR).

What is now clear is that SUR1 alone cannot conduct ions, in particular $K^+$. However, when coexpressed in recombinant studies with the inward rectifier $K^+$ channel clone $K_{IR}6.2$ [also termed BIR ('β-cell inward rectifier')], these two proteins will recombine and form functional $K^+$ channels that have many of the characteristic properties of the native $K_{ATP}$ channel in $\beta$-cells; ATP inhibition, ADP-dependent gating, inward current rectification, high sensitivity to sulphonylurea, and modulation by diazoxide.[60] $K_{IR}6.2$ belongs to the ATP/

inward rectifier family of $K^+$ channel clones. These proteins have a predicated structure of two putative transmembrane domains that are linked by a P-region, which by analogy with the predictions made for other $K^+$ channel is involved in the pore formation that will allow $K^+$ to flux across the membrane. $K_{IR}6.2$ is novel among this family of proteins in one key respect: that in the absence of SUR1, $K_{IR}6.2$ is silent and therefore unable to act as a $K^+$ channel. Future studies will no doubt reveal more about the detail of the coalescence of these two heterogeneous proteins and define those key components/amino acid sequences of $K_{IR}6.2$ and SUR1 that are involved in the recognition and the formation of functional channels. These types of studies will also be important to determine the site of sulphonylurea binding to the complex and the precise location of the nucleotide sensitivity of the complex; no simple task when one considers the complexity of the positive and negative ionic effects of 'free'- and Mg-bound ATP and ADP on the native channel. Progress in this area is also coming from studies of PHHI and the identification of mutations within both the sulphonylurea receptor and the inward rectifier $K^+$ channel components of the channel. Two mutations are noteworthy at this stage, since they come from patients with the different degrees of severity of the disease:

1. The exon 35 mutation, which has been reported in PHHI patients arising from consanguineous relations among Ashkenazi Jews, Saudi, and Palestinian Arabs causes a complete loss of $K_{ATP}$ channel function in $\beta$-cells from these patients. This mutation is a guanine to adenine single base mutation that results in loss of a restriction endonuclease recognition site. This will cause abnormal RNA splicing, leading to the generation of a premature stop codon in the SUR1 mRNA. The net effect of this mutation is a truncation of the gene product with the loss of 200 amino acids that essentially comprise the majority of the second nucleotide binding fold. In order to assess directly whether this mutation will cause loss of $K_{ATP}$ channels in recombinant cells, a parallel mutation was engineered which, when coexpressed with native $K_{IR}6.2$, led to the abolition of functional $K_{ATP}$ channels when assayed by. $^{86}Rb^+$ efflux and electrophysiology.[79] These data suggest that the second nucleotide-binding fold (NBF) of SUR1 is not involved with nucleotide binding, but also with the assembly of recombinant complex. These findings correlate very well with the clinical phenotype of patients with this particular form of PHHI; an aggressive, early-onset type of disease that was completely unresponsive to medical therapy with diazoxide and somatostatin – data that we can now explain upon the basis of the absence of functional $K_{ATP}$ channels.
2. A similar correlation between clinical disease and the genetics of $K_{ATP}$ channels was also observed in another patient with PHHI. In this second case, a non-conserved substitution of glycine to arginine at residue 1479 (the G1479R mutation) was detected in a Sephardic Jewish proband who originally presented with a mild form of the disease phenotype. Transient hypoglycaemia diagnosed at birth in the patient, was treated by dietary manipulation without the intervention of medical therapy or surgery. The amino acid residue 1479 resides in the cytoplasmic linker region between the Walker A and B in the second NBF, and because of a series of recombinant experiments we now believe that this site plays a key role in the nucleotide regulation of $K_{ATP}$ channels. Using site-directed mutagenesis, the engineering of the G1479R mutation in SUR1 cDNA did not prevent

the generation of ATP-sensitive $K^+$ currents in COS cells when G1479R-SUR1 cDNA was coexpressed with $K_{IR}6.2$. However, the recombinant channels were no longer sensitive to the stimulatory effects of ADP. The consequence of this for β-cells is that they will be intolerant to glucose metabolism and that this would result in loss of regulated glucose-dependent insulin secretion; consistent with PHHI disease phenotype.[80]

Persistent hyperinsulinaemic hypoglycaemia of infancy is a devastating condition and while it is providing important clues to the molecular identity of the subunits of $K_{ATP}$ channels on the one hand, basic and applied medical studies of this rare condition are already providing opportunities for novel lines of therapy in the management of the disease,[81] and will continue to do so. In the light of molecular advances in the structure–function of $K_{ATP}$ channels, several studies have now begun to address the possibility that decreases in the rate of insulin secretion from non-insulin-dependent diabetics might also result from mutations in either $K_{IR}6.2$ or SUR1. Initial studies have so far failed to identify linkage between the coding region of $K_{IR}6.2$ and NIDDM;[82] however, in one study it was revealed that there was a highly significant increase in the prevalence of two silent polymorphisms in SUR1 from diabetics when compared with the normal population,[83] and the possibility exists that these polymorphims could be linked to other mutations with or near to the SUR1 locus and thus confer an increased risk of developing NIDDM.

Finally, while the principal action of sulphonylureas *in vivo* is the inhibition of $K_{ATP}$ channels leading to a depolarization of the cell membrane potential, it should be noted that these types of compounds also exert other effects on β-cells that might augment Ca-dependent release of insulin. First, sulphonylureas will inhibit $Cl^-$ channels in β-cells.[84] For these studies, β-cells are placed into hypotonic solutions and this leads to the activation of a cell-swelling $Cl^-$ current. This current is inhibited by stilbene derivatives such as DIDS, but can also be blocked by sulphonylureas. The current is far less sensitive to sulphonylureas than $K_{ATP}$ channels, but nevertheless blockade of these channels would contribute to a depolarization along with $K_{ATP}$ channel closure. Secondly, it has also been elegantly demonstrated that sulphonylureas will also interact with β-cells at a site downstream from the closure of $K_{ATP}$ channels and potentiate the effects of $Ca^{2+}$-dependent exocytosis.[85] The precise mechanism of action of this distal signalling phenomenon has yet to be determined, but the involvement of protein kinase C activation seems likely since the actions were abolished in the presence of inhibitors of the PKC. Furthermore, since these effects were observed at therapeutic concentrations of sulphonylureas, the data suggest that this augmentation of the exocytosis machinery might contribute to the hypoglycaemic actions of these compounds *in vivo*.

## Inhibition of $K_{ATP}$ channels by imidazolines

The concept that imidazoline compound might posses medicinally important secretagogue activity was developed a number of years ago. Much of the early work in this field surrounded observations made from studies involving phentolamine – an antagonist of α-adrenoceptors that was found to elicit an increase in the circulating concentrations of insulin when administered to experimental animal and man (for a recent comprehensive review, see Morgan *et al.*).[86] At that stage, the effects were attributed to the actions of the compound on the tonic inhibitory control of insulin secretion by catecholamines which act through $\alpha_2$-adrenoceptors. However, these effects were hard to reconcile with the results of experiments using the more potent and more selective $\alpha_2$-adrenoceptor antagonist idazoxan, which does not elevate serum insulin levels, and with the fact that *in vitro* imidazoline ring-containing α-adrenoceptor antagonists promoted insulin secretion from isolated islets of Langerhans. These types of studies argue strongly for a direct insulinotropic effect of these types of compounds on insulin-secreting cells, and the field has advanced to the stage now, where we currently believe that imidazolines are binding to discrete imidazoline receptor sites (IRs).

Imidazoline receptors can be broadly classified as $I_1$ and $I_2$ receptor subtypes. The principal pharmacological distinction between these sites relates to the fact that $I_1$ will bind clonidine with a high affinity and can be tagged with $^3$H-labelled derivatives, whereas $I_2$ sites are preferentially labelled with [$^3$H]idazoxan. Sites labelled by [$^3$H]idazoxan are found on membrane preparations from β-cells, but these appear *not* to contribute to the secretion of insulin, since idazoxan is only weakly effective and it will not readily antagonize the responsiveness of insulin-secreting cells to other imidazolines. Efaroxan and moxonidine, by contrast, are potent and selective $I_1$ agonists. Yet, while efaroxan will promote secretion from insulin-secreting cells, reverse the actions of diazoxide, decrease $^{86}Rb^+$, and inhibit $K_{ATP}$ channels, moxonidine fails to alter the effects of diazoxide on β-cells, or to modify the actions of efaroxan in anyway.[86] Part of the discrepancy and confusion surrounding the effectiveness of these compounds may ultimately be resolved by the fact that the β-cells might possess a novel type of IR, or alternatively that the potency of these compounds for the receptor sites in insulin-secreting cells may be somewhat different to the effects of the compounds on the same receptors in other tissues. Indeed, in β-cells it was recently demonstrated that whereas, in many tissues, radioisotope derivatives of clonidine preferentially labelled $I_1$ receptor sites, [$^{125}$I]clonidine sites were more closely related to $I_2$ than $I_1$ in β-cells and that the number of $I_1$ sites present on pancreatic β-cells was extremely small.

From an ion channel perspective, imidazolines remain one of the most effective groups of agents to inhibit $K_{ATP}$ channels. Several imidazoline receptor agonists will lead to the closure of $K_{ATP}$ channels; including phentolamine ($IC_{50} = 0.7\ \mu M$), efaroxan ($12\ \mu M$), antazoline ($<25\ \mu M$), tolazoline ($\sim 25\ \mu M$), and, surprisingly, idazoxan ($>100\ \mu M$). These agents are effective from either the inside of the cell membrane or the outside of the cell, and with the exception of idazoxan, $K_{ATP}$ channel closure is associated with a depolarization of the β-cell membrane, the initiation of action potentials generation, and the release of insulin (for review, see Dunne *et al.*).[87]

The subcellular mechanisms of action of imidazolines on β-cell signalling have only recently been elucidated, and involve, not changes in the accumulation of intracellular cAMP,[88] but increases in the free intracellular calcium ion concentration ($[Ca^{2+}]_i$). In a detailed study,[89] we have recently reported that both phentolamine and efaroxan are able to evoke a marked elevation of $[Ca^{2+}]_i$ in intact islets of Langerhans, and that the subcellular events are critically dependent upon $Ca^{2+}$ influx and not mobilization of stored $Ca^{2+}$. The evidence for this was that: (i) in the absence of outside $Ca^{2+}$, there is no increase in intracellular $Ca^{2+}$ by imidazolines; (ii) that acute removal of external $Ca^{2+}$ attenuated agonist-evoked increases in $[Ca^{2+}]$; (iii) that the hyperpolarizing $K_{ATP}$ channel agonist diazoxide attenuated both efaroxan- and phentolamine-evoked increases in $[Ca^{2+}]_i$ and caused a marked delay in the onset times of the responses; and finally that (iv) imidazoline-mediated rises in $[Ca^{2+}]_i$ were unaffected by the intracellular $Ca^{2+}$ ATPase inhibitor thapsigargin. These actions are entirely consistent with the predicted effects of imidazolines on $[Ca^{2+}]_i$, since $K_{ATP}$ channel closure initiates the generation of $Ca^{2+}$-dependent action potentials. Interactions between glucose and efaroxan and phentolamine, also suggested that the imidazolines are able not only to initiate $Ca^{2+}$ signalling events in $\beta$-cells, but that they will also potentiate the actions of glucose.[89] These experiments also enabled us to resolve the discrepancy surrounding the actions of idazoxan on ion channels and insulin secretion in $\beta$-cells. The substituted imidazoline idazoxan, at low concentrations, has little effect upon insulin secretion in the absence of an adrenoceptor agonist,[90,91] but at higher concentrations ($< 100\ \mu M$) will partially reverse the diazoxide-induced inhibition of insulin secretion.[91] This effect can be explained by weak block of $K_{ATP}$ channels,[88] but it is still somewhat surprising that a $K_{ATP}$ channel antagonist cannot promote insulin secretion. We now believe that this inconsistency is due to the fact that idazoxan will also inhibit voltage-gated $Ca^{2+}$ influx in $\beta$-cells.[89]

One of the most intriguing aspects of imidazolines and IRs is that, *in vivo*, these actions could be mediated by an endogenous ligand. For several years it has been known that an extract could be made from the brain that would displace labelled clonidine from the CNS. This endogenous agent was termed CDS – or 'clonidine displacing substance'.[92] Li and colleagues[93] recently reported that one a component of the 'endogenous ligand' for imidazoline receptors was agmatine. Agmatine is a relatively simple compound – decarboxylated arginine, – that will act as an insulin secretagogue.[94] Like efaroxan and phentolamine, agmatine will also block $K_{ATP}$ channels in $\beta$-cells;[90] however, unlike CDS – which will promote insulin secretion at sub-nanomolar concentrations[95], – agmatine is a relatively weak blocker of $K_{ATP}$ channels[89] and only effective as an insulin secretagogue at concentrations $> 100\ \mu M$.

The identity of the putative imidazoline receptor in $\beta$-cells remains elusive. However, it seems very likely that these compounds will interact directly with the $K_{ATP}$ channel complex and not through a separate protein subunit that couples to the SUR1/$K_{IR}$6.2 complex; functional studies would tend to be supportive of this idea. However, this does not preclude non-plasma membrane-related effects of imidazolines on insulin-secreting cells, and by analogy with the recently reported effects of sulphonylureas on distal signal events in insulin-secreting cells[85], three pieces of evidence might support a similar mechanism of action for the imidazolines. First, insulin secretion studies have recently demonstrated that the imidazoline RX871024 promotes insulin secretion by both $Ca^{2+}$-dependent and $Ca^{2+}$-independent modes of action.[96] Secondly, while the action of imidazolines on insulin secretion can be down-regulatable in a specify manner by overnight pretreatment of isolated islets with imidazolines,[97], there is no desensitization of the stimulatory actions of imidazoline receptor agonists on $Ca^{2+}$ signalling in intact glucose-responsive islets.[89] Thirdly, data from our laboratory has shown that imidazolines will promote insulin release from isolated PHHI islets, despite the absence of $K_{ATP}$ channels and imidazoline-induced elevation of $[Ca^{2+}]_i$ (M. Dunne and N. Morgan, unpublished results).

### Inhibition of $K_{ATP}$ channels by other agents

Finally, although the sulphonylureas and imidazoline derivatives are important groups of agents that will bring about the inhibition of $K_{ATP}$ channels and lead to insulin secretion *in vitro* and *in vivo*, it must also be pointed out that these channels are extraordinarily sensitive to a number of pharmacological agents. On the one hand, they are inhibited by well-recognized $K^+$ channel blockers such as quinine ($K_d = 40$ μM), aminoacridine (100 μM), 4-aminopyridine (100 μM), and tetraethylammonium ($TEA^+$).[14] However, on the other hand there are increasing numbers of agents that have a wide-ranging of biological activities, from anti-arrhythmic agents and anti-depressants to herbal plant extracts and commonly administered local anaesthetics, that will inhibit $K_{ATP}$ channels (Table 9.3).

**Table 9.3** Reported modulators of $K_{ATP}$ channels in pancreatic β-cells.*

| **Agonists** |
|---|
| Diazoxide, BPDZ-44/-62, cromakalim, pinacidil, nicorandil, RP49356, long-chain acyl-CoA esters, 5-hydroxydecanoate, lysophospholipids, oleic acid, linoleic acid, docosahexaenoic acid |

| **Antagonists** | | |
|---|---|---|
| *Sulphonylureas* | *Other antagonists* (contd) | *Other antagonists* (contd) |
| glibenclamide (4–20 nM) | 4-aminopyridine | mastopran |
| tolbutamide (10–17 μM) | agmatine (~20 μM) | lysophospholipids |
| meglitinide (2.1 μM) | yohimbine | arachidonic acid |
| glipizide (6.4 μM) | disopyrimide (3.6 μM) | ciclazindol |
| | cibenzoline (0.4 μM) | A-4166[e] |
| *Imidazolines* | ligustrazine | sparteine (170 μM) |
| phentolamine (0.7 μM) | pentobarbitone (360 μM) | amantadine (120 μM) |
| efaroxan (12 μM) | thiopentane (62 μM) | phloxine B |
| antazoline (<25 μM) | secobarbitone (250 μM) | eosin-5-maleimide |
| tolazoline (~25 μM) | linogliride (10–30 μM) | alloxan |
| idazoxan (>100 μM) | BTS 67 5 82[c] (~5 μM) | 5-hydroxydecanoate |
| RX811059 (>50 μM) | interleukin-1β | caffeine (10 mM) |
| RX871024 | activin A | verapamil |
| SL 84.0418[a] | AZ-DF-265 (1 nM) | TMB-8[f] |
| clonidine (4 μM) | DIP[d] (100 μM) | chlorpromazine (1 μM) |
| | KAD-1229[b] (~1 μM) | triflupromazine (4 μM) |
| *Other antagonists* | 8-methoxypsoralen | fluphenazine (6 μM) |
| $TEA^+$ (22 mM) | polymyxin B | trifluopromazine (20 μM) |
| quinine (40 μM) | SDZ PCO-400 | |
| aminoacridine | 2,3-butanedione monoxime (15 mM) | |

* All compounds have described as antagonizing the opening of ATP-sensitive $K^+$ channels when added to intact cells, whole cells or isolated patches of membrane. Where available, $IC_{50}$ values have been included.

a 2-(4,5-dihydro-1H-imidazol-2-yl)-1,2,4,5-tetrahydro-2-propyl-pyrrolo(3,2,1-hi)-indole hydrochloride

b calcium (2S)-2-benzyl-3-(CIS-hexahydro-2-isoindolinyl-carbonyl) propionate dihydrate

c 1,1-dimethyl-2[2-morpholinophenyl]guanidine monofumarate

d diazene dicarboxylic acid bis-(*N*'-methylpiperazide)

e *N*-((trans-4-isopropylcyclohexyl)-carbonyl)-D-phenylalanine)

f 8-(*N,N*-diethylamino)octyl-3,4,5-trimethoxybenzoate

Finally, '$K_{ATP}$ channel antagonists' are still a fruitful area for the strategic development of agents with which to treat diabetes. Despite the fact that sulphonylureas are very effective therapeutic adjuncts in the management of this condition, there are disadvantages to their widespread use in the clinical environment. Thus, some patients are known to acquire 'resistance' to sulphonylureas, in some groups of diabetics treated with sulphonylureas, there is an increased risk of cardiovascular complications, not all NIDDM patients will respond to sulphonylureas and, finally, because of their pharmacokinetic profile, some diabetics are susceptible to post-prandial hypoglycaemic episodes, particularly the elderly.[98] Additionally, in a recent review by the UK Prospective Diabetes Study Group,[99] one of the major conclusions drawn was that the most appropriate treatment for NIDDM patients was not known.[99] With this in mind, non-sulphonylurea-based agents are being examined for their effectiveness. On such agent is BTS 67 5 82, a novel oral hypoglycaemia-inducing agent that promotes insulin secretion both *in vivo*[100] and *in vitro*.[101] BTS 67 5 82 has been shown to be effective in alleviating the incidence of post-prandial hypoglyglycaemic episodes in diabetics, and the subcellular mechanisms that are responsible for this are associated with inhibition of $\beta$-cell $K_{ATP}$ channels, the generation of $Ca^{2+}$-dependent action potentials, and a marked increase in the cytosolic $Ca^{2+}$ concentration.[101] Linogliride is another non-sulphonylurea drug that lowers blood glucose levels in normal and diabetic donors. *In vitro*, linogliride also stimulates insulin release and affects the activity of ATP-sensitive $K^+$ channels, inhibiting $K^+$-currents with a half-maximal inhibition between 6–25 $\mu$M.[102]

## References


1. Wollheim, C.B. and Sharp, G.W.G. (1981). *Physiol. Rev.*, **61**, 914–973.
2. Hoenig, M. and Sharp, G.W.G. (1986). *Endocrinology*, **119**, 2502–2507.
3. Dean, P.M. and Matthews, E.K. (1970). *J. Physiol.*, **210**, 265–275.
4. Matthews, E.K. and Sakamoto, Y. (1975). *J. Physiol.*, **246**, 421–437.
5. Sehlin, J. and Taljedal, I.-P. (1975). *Nature*, **253**, 635–636.
6. Henquin, J.C. (1978). *Nature*, **271**, 271–272.
7. Henquin, J.C. (1980). *J. Biol. Chem.*, **186**, 541–550.
8. Ashcroft, F.M., Harrison, D.E., and Ashcroft, S.J.H. (1984). *Nature*, **312**, 446–448.
9. Misler, S., Falke, L.C., Gillis., K., and McDaniel, M.L. (1986). *Proc. Natl Acad. Sci. USA*, **83**, 7119–7123.
10. Cook, D.L. and Hales, C.N. (1984). *Nature*, **311**, 271–273.
11. Findlay, I., Dunne, M.J., and Petersen, O.H. (1985). *J. Membr. Biol.*, **88**, 165–172.
12. Dunne, M.J., Findlay, I., Petersen, O.H., and Wollheim, C.B. (1986). *J. Membr. Biol.*, **93**, 271–275.
13. Findlay, I. and Dunne, M.J. (1986). *Pflügers Archiv*. **407**, 238–240.
14. Dunne, M.J. and Petersen, O.H. (1991). *Biochim. Biophys. Acta*, **1071**, 67–82.
15. Rorsman, P. and Trube, G. (1990) In *Potassium channels*, (ed. N.S. Cook), pp. 96–116. J. Wiley & Sons, New York.

16. Dunne, M.J., Findlay, I., and Petersen, O.H. (1988). *J. Membr. Biol.*, **102**, 205–216.
17. Bokvist, K., Eliasson, L., Ämmälä, C., Renström, E., and Rorsman, P. (1995). *EMBO J.*, **14**, 50–57.
18. Ashcroft, F.M. and Kakei, M. (1989). *J. Physiol.*, **416**, 349–367.
19. Takano, M. and Ashcroft, F.M. (1996). *Pflügers Archiv.*, **431**, 625–631.
20. Arkhammar, P., Nilsson, T., Rorsman, P., and Berggren, P.-O. (1987). *J. Biol. Chem.*, **262**, 5448–5454.
21. Findlay, I. (1987). *J. Physiol.*, **391**, 611–629.
22. Larsson, O., Kindmark, H., Branstrom, R., Fredholm, B., and Berggren, P.-O. (1996). *Proc. Natl Acad. Sci. USA*, **93**, 5161–5165.
23. Proks, P., Takano, M., and Ashcroft, F.M. (1994). *J. Physiol.*, **475**, 33–44.
24. Lederer, W.J. and Nicholls, C.G. (1989). *J. Physiol.*, **419**, 193–211.
25. Davies, N.W. (1990). *Nature*, **343**, 375–377.
26. Ohno-Shosaku, T., Zunkler, B.J., and Trube, G. (1987). *Pflügers Archiv.*, **408**, 133–138.
27. Ribalet, B. and Ciani, S. (1987). *Proc. Natl Acad. Sci. USA*, **84**, 1721–1725.
28. Findlay, I., Dunne, M.J., Ullrich, S., Wollheim, C.B., and Petersen, O.H. (1985). *FEBS Lett.*, **185**, 4–8.
29. Dunne, M.J., West-Jordan, J., Abraham, R.J., Edwards, R.T.H., and Petersen, O.H. (1988). *J. Membr. Biol.*, **104**, 165–172.
30. Dunne, M.J., Illot, M.C., and Petersen, O.H. (1987). *J. Membr. Biol.*, **99**, 215–224.
31. Kozlowski, R.Z. and Ashford, M.L.J. (1990) *Proc. R. Soc., Lond. B.*, **240**, 397–410.
32. Ashcroft, F.M., Kakei, M., and Kelly, R.P. (1989). *J. Physiol.*, **408**, 413–430.
33. Dunne, M.J. and Petersen, O.H. (1986). *FEBS Lett.*, **208**, 59–62.
34. Kakei, M., Kelly, R.P., Ashcroft, F.M., and Ashcroft, S.H. (1986). *FEBS Lett.*, **208**, 63–66.
35. Petersen, O.H. and Findlay, I. (1987). *Physiol. Rev.*, **67**, 1054–1116.
36. Cook, D.L., Satin, L.S., Ashford, M.L.J., and Hales, C.N. (1988). *Diabetes*, **37**, 495–498.
37. Detimary, P., Jonas, J.-C., and Henquin, J.-C. (1995). *J. Clin. Invest.*, **96**, 1738–1745
38. Nilsson, T., Schultz, V., Berggren, P.-O., Corkey, B.E., and Tornheim, K. (1996). *Biochem. J.*, **314**, 91–94.
39. Civelek, V.N., Deeney, J.T., Kubik, K., Schultz, V., Tornheim, K., and Corkey, B.E. (1996). *Biochem. J.*, **315**, 1015–1019.
40. Dunne, M.J. and Petersen, O.H. (1986). *Pflügers Archiv.*, **407**, 564–565.
41. Dunne, M.J., Bullett, M.J., Li, G., Wollheim, C.B., and Petersen, O.H. (1989). *EMBO J.*, **8**, 412–420.
42. De Weille, J., Schmid-Antomarchi, H., Fosset, M., and Lazdunski, M. (1988). *Proc. Natl Acad. Sci. USA*, **85**, 1312–1316.

43. Ribalet, B. and Ciani, S. (1994). *J. Membr. Biol.*, **142**, 395–408.
44. Ribalet, B. and Eddlestone, G.T. (1995). *J. Physiol.*, **485**, 73–86.
45. Bozem, M., Nenquin, M., and Henquin, J.C. (1987). *Endocrinology*, **121**, 1025–1033.
46. Virji, M.A.G., Staffs, M.W., and Estensen, R.D. (1978). *Endocrinology*, **102**, 706–711.
47. Wollheim, C.B., Dunne, M.J., Peter-Reisch, B., Bruzzone, R., Pozzan, T., and Petersen, O.H. (1988). *EMBO J.*, **7**, 2443–2449.
48. Ribalet, B., Eddlestone, G.T., and Ciani, S. (1988). *J. Gen. Physiol.*, **92**, 219–237.
49. De Weille, J.R., Schmid-Antomarchi, H., Fosset, M., and Lazdunski, M. (1989). *Proc. Natl Acad. Sci. USA*, **86**, 2971–2975.
50. Dunne, M.J. (1994). *Am J. Physiol.*, **267**, C501–C506.
51. Rovira, J.M., Ripoll, C., and Soria, B. (1996). *Diabetologia*, **39**, A115.
52. Martin, S.C., Yule, D.I., Dunne, M.J., Gallacher, D.V., and Petersen, O.H. (1989). *EMBO J.*, **8**, 3595–3599.
53. Li, G.-D., Milani, D., Dunne, M.J., Pralong, W.-F., Petersen, O.H., and Wollheim, C.B. (1990). *J. Biol. Chem.*, **266**, 3449–3457.
54. Ashcroft, F.M. and Rorsman, P. (1989). *Prog. Biophys. Mol. Biol.*, **54**, 87–143.
55. Dunne, M.J. (1992) In *Potassium channel modulators, pharmacological, molecular and clinical aspects*, (ed. A.H. Weston and T.C. Hamilton), pp. 110–142. Blackwell, Oxford.
56. Petersen, O.H. and Dunne, M.J. (1989). *Pflügers Archiv.*, **414**, S115–S120.
57. Faivre, J.-J. and Findlay, I. (1989). *Biochim. Biophys. Acta*, **984**, 1–5.
58. Pirotte., B., De Tullio, P., Lebrun, P., Antoine, M., Fontaine, J., Masereel, B., Schynts, M., Dupont, L., Herchuelz, A., and Delarge, J. (1993). *J. Med. Chem.*, **36**, 3211–3213.
59. Lebrun, P., Antoine, M.-H., Ouedraogo, R., Dunne, M.J., Kane, C.M., Hermann, A., *et al.* (1996). *J. Pharmacol. Exp. Ther.*, **277**, 156–162.
60. Inagaki, N., Gonoi, T., Clement, J.P. IV, Namba, N., Inazawa, J., Gonzalez, G., Aguilar-Bryan, L., Seino, S., and Bryan, J. (1995). *Science*, **270**, 1166–1170.
61. Ämmälä, C., Moorhouse, A., Gribble, F., Ashfield, R., Proks, P., Smith, P.A., *et al.* (1996). *Nature*, **379**, 545–548.
62. Dunne, M.J., Aspinall, R.J., and Petersen, O.H. (1990). *Br. J. Pharmacol.*, **99**, 169–175.
63. Dunne, M.J. (1989), *FEBS Lett.*, **250**, 262–266.
64. Kozlowski, R.Z., Hales, C.N., and Ashford, M.L.J. (1989). *Br. J. Pharmacol.* **97**, 1039–1050.
65. Dunne, M.J. (1990). *Br. J. Pharmacol.*, **99**, 487–492.
66. Inagaki, N., Ganoi, T., Clement, J.P. IV, Wang, C.Z., Aguilar-Bryan, L., Bryan, J., and Seino, S. (1996). *Neuron*, **16**, 1011–1017.
67. Isomoto, S., Nondo, C., Yamada, M., Matsumoto, S., Higashiguchi, O., Horio, Y., Matsuzawa, Y., and Kurachi, Y. (1996). *J. Biol. Chem.*, **271**, 1789–1794.
68. Aynsley-Green, A. (1981). *Dev. Med. Child. Neurol.*, **23**, 372–379.

69. Grant, D.B., Dunger, D.B., and Burns, E.C. (1986). *Acta Endocrinol. Suppl.*, **279**, 340–345.

70. Kane, C, Shepherd, R.M., Squires, P.E., Johnson, P.R.V., James, R.F.L., Milla, P.J., Aynsley-Green, A., Lindley, K.J., and Dunne, M.J. (1996). *Nature Medicine*, **2**, 1344–1347.

71. Hellman, B. and Taljedal, I.-B. (1975). In *Handbook of experimental physiology*, Vol. XXXII, (ed. A. Hasselblatt and F. Bruchhausen). pp. 175–194. Springer-Verlag, Berlin.

72. Sturgess, N., Ashford, M.L.J., Cook, D.L., and Hales, C.N. (1985). *Lancet*, **2**, 474–475.

73. Zunkler, B.J., Lenzen, S., Manner, K., Panten, U., and Trube, G. (1988). *Naunyn-Schmiedebergs Arch. Pharmacol.*, **337**, 225–230.

74. Ashcroft, F.M., Kakei, M., Gibson, J.S., Gray, D.W., and Sutton, R. (1989). *Diabetologia*, **32**, 591–598.

75. Sturgess, N.C., Kozlowoski, R.Z., Carrington, C.A., Hales, C.N., and Ashford, M.L.J. (1988). *Br. J. Pharmacol.*, **95**, 83–94.

76. Bernadi, H., Fosset, M., and Lazdunski, M. (1988). *Proc. Natl. Acad. Sci. USA*, **85**, 9816–9820.

77. Aguilar-Bryan, L., Nelson, D.A., Vu, Q.A., Humphrey, M.B., and Boyd, A.E. III (1990). *J. Biol. Chem.*, **265**, 8218–8224.

78. Aguilar-Bryan, L., Nichols, C.G., Wechsler, S.W., Clement, J.P. IV, Boyd, A.E. III, Gonzalez, G., Herrera-Sosa, H., Nguy, K., Bryan, J., and Nelson, D.A. (1995). *Science*, **268**, 423–426.

79. Dunne, M.J., Kane, C., Shepherd, R.M., Sanchez, J., James, R.F.L., Johnson, P.R.V., Aynsley-Green, A., LU, S., Clement III, J.P., Lindley, K.J., Seino, S., and Aguilar-Bryan, L. (1997). *New Engl. J. Med.*, **336**, 703–706.

80. Nichols, C.G., Shyng, S.-L., Nestorowicz, A., Glaser, B., Clement, J.P., Gonzalez, C., *et al.* (1996). *Science*, **272**, 1785–1787.

81. Lindley, K.J., Dunne, M.J., Kane, C., Squires, P.E., James, R.F.L., *et al.* (1996) *Arch. Dis. Child.*, **74**, 373–378.

82. Seino, S., Inayaki, N., Namba, N., and Gonoi, T. (1996). *Diabetes Reviews*, **4**, 177–190.

83. Moller, A.M., Hansen, T., Almind, K., Echwald, S.M., Clausen, O., Urhammer, S.A., *et al.* (1996). *Diabetologia*, **39**, A5.

84. Kinard, T.A. and Satin, L.S. (1995). *Diabetes*, **44**, 1461–1466.

85. Eliasson, L., Renstrom, E., Ämmälä, C., Berggren, P.-O., Bertorello, A.M., Bokvist, K., *et al.* (1996). *Science*, **271**, 813–815.

86. Morgan, N.G., Chan, S.L.F., Brown, C.A., and Tsoli, E. (1995). *Ann. N.Y. Acad. Sci.*, **763**, 361–366.

87. Dunne, M.J., Harding, E.A., Jaggar, J.H., Squires, P.E., Liang, R., Kane, C., *et al.* (1995). *Ann. N.Y. Acad. Sci.*, **763**, 243–262.

88. Chan, S.L.F., Dunne, M.J., Stillings, M.R., and Morgan, N.G. (1991). *Eur. J. Pharmacol.*, **204**, 41–48.

89. Shepherd, R.M., Hashmi, M.N., Kane, C., Squires, P.E., and Dunne, M.J. (1996). *Br. J. Pharmacol.*, **119**, 911–916.

90. Ostenson, C.G., Pigon, J., Doxley J.C., and Efendic, S. (1988). *J. Clin. Endocrinol. Metab.*, **67**, 1054–1059.

91. Chan, S.L.F. and Morgan, N.G. (1990). *Eur. J. Pharmacol.* **176**, 97–101.

92. Atlas, D. and Burnstein, Y. (1984). *FEBS Lett.*, **170**, 887–891.

93. Li, G., Regunathan, S., Barrow, C.J., Eshraghi, J., Cooper, R., and Reis, D.J. (1994). *Science*, **263**, 966–969.

94. Sener, A., Lebrun, P., Blachier, F., and Malaisse, W.J. (1989). *Biochem. Pharmacol.*, **38**, 327–330.

95. Chan, S.L.F. and Morgan, N.G. (1996). *Diabetologia*, **39**,107.

96. Zairsev, S.V., Efanov, A.M., Efanov, I.B., Larsson, O., Ostenson, C.-G., Gold, G., Berggren, P.-O. and Efandic, S. (1996). *Diabetes*, **45**, 1610–1618.

97. Chan, S.L.F., Brown, C.A., and Morgan, N.G. (1993). *Eur. J. Pharmacol*, **230**, 375–378.

98. Leibowitz, G. and Cerasi, E. (1996). *Diabetologia*, **39**, 503–514.

99. UK Prospective Diabetes Study Group (1995). *Br. Med. J.*, **310**, 83–88.

100. Byrom, W.D., Weil, A., Brown, T.J., and Bratty, J.R. (1996). *Diabetologia*, **39**, A44.

101. Jones, R.B., Shepherd, R.M., Kane, C., Hashmi, M.N., Moores, A., Harding, E.A., and Dunne, M.J. (1996) *Diabetologia*, **39**, A43.

102. Ronner, P., Higgins, T.J., and Kimmich, G.A. (1991). *Diabetes*, **40**, 885–892.

# 10 Potential role of the pharmacology of $K^+$ channels in therapeutics

B. Soria

## Introduction

$K^+$ channels are a diverse and ubiquitous class of proteins that regulate a number of biological functions. Ligands for the study of a variety of $K^+$ channels are available. These include, for example, 'openers' and antagonists for the ATP-sensitive $K^+$ channel ($K_{ATP}$), or peptide toxins such as apamin and charybdotoxin that block other subtypes. Antagonists of the $K_{ATP}$ are useful in the treatment of type II diabetes, while openers of this channel are being tested in asthma and cardiovascular disease. Intracerebroventricular administration of $K^+$ channel openers block experimentally induced seizures in rodents through a hyperpolarization of neurones. The therapeutic applications for $K^+$ channel modulators include: hypertension, cardiac ischaemia and arrhythmia, protection against reperfusion damage, non-insulin-dependent diabetes mellitus, asthma, urinary incontinence, and hair growth.

The past decade has seen the emergence of a novel group of smooth muscle relaxant drugs known as $K^+$ channel openers. This group includes benzopyran derivatives (such as cromakalim, BRL34915, SDZ PCO 400, and Ro-31–6930), cyanoguanidine derivatives (pinacidil) and tetrahydrothiopyrans (RP 40356). $K^+$ channel openers may also be useful in the treatment of neurodegenerative diseases, pain, and cerebral ischaemia. Central nervous system selectivity is the key to the development of psychopharmacological agents to modify brain $K^+$ channel function. The ATP-sensitive $K^+$ channel openers which promise a bright future for this mechanism[1] comprise a large number of molecules that can be classified into three basic groups:[2] (1) agents such as levcromakalim that open a small-conductance (10–30 pS) glibenclamide-sensitive $K^+$ channel, currently known as the ATP-sensitive $K^+$ channel $K_{ATP}$; (2) hybrid molecules, such as nicorandil, that open $K_{ATP}$ channels and also activate the enzyme soluble guanylate cyclase; and (3) molecules such as dehydrosaponin 1 that open the large-conductance (100–150 pS) calcium-dependent $K^+$ channel, $BK_{Ca}$. $K^+$ channel openers in groups 1 and 2 are most potent in smooth muscle, but $K_{ATP}$ channels in cardiac muscle, neurones, and the pancreatic $\beta$-cell are also affected. *In vivo*, moderate to high doses produce a fall in diastolic pressure with reflex tachycardia; low doses may exert selective dilatory effects on specific vascular beds with little effect on systemic pressure. *In vitro*, all smooth muscles are relaxed with loss of spontaneous electrical and mechanical activity; hyperpolarization to values close to the $K^+$ equilibrium potential are often observed. These effects can be antagonized by glibenclamide and also by imidazolines and guanidines, such as phentolamine, guanethidine, and antazoline – agents that are also known to inhibit the smooth muscle delayed rectifier channel, $K_{DR}$. The mode and site of action of the $K^+$ channel openers is the subject of intense study. Irrespective of their specific mode of action, these openers – especially the hybrid molecules such as nicorandil – constitute a novel and promising approach to the treatment of cardiovascular disease.

## Effects of $K^+$ channel modulators on the cardiovascular system

### Effects on the heart.

Occlusion of the coronary artery results in myocardial ischaemia (insuficient blood supply to the myocardium to attend the myocardial metabolic demands). Restricted coronary blood flow generates 'angina pectoris', whereas complete occlusion of the coronary artery results in myocardial infarction. Arrhythmia, the major incidence event, is directly related to infarct size. Thus, pharmacological intervention to diminish the rise of the infarct or to control the arrhythmias should be beneficial in reducing mortality.

*$K^+$ channels in the heart*

Heart muscle possesses a wide variety of $K^+$ channels. It should be noted that agents used to isolate specific $K^+$ currents in other tissues fail to discriminate between different $K^+$ channels in the heart; hence, the number and type of $K^+$ channels actually present in the myocardial membrane, and their potential physiological role remain uncertain. The cardiac action potential results from a complex and precisely regulated movement of ions across the sarcolemmal membrane. Potassium channels represent the most diverse class of ion channels in heart and are the targets of several antiarrhythmic drugs. Potassium currents in the myocardium can be classified into one of two general categories: (i) inward rectifying currents such as $IK_{IR}$, $IK_{ACh}$, and $IK_{ATP}$; and (ii) primarily voltage-gated currents such as $IK_{DR}$ (slow and fast), $IK_A$ and others. The inward rectifier currents regulate the resting membrane potential, whereas the voltage-activated currents control action potential duration. The presence of these multiple, often overlapping, outward currents in native cardiac myocytes has complicated the study of individual $K^+$ channels; however, the application of molecular cloning technology to these cardiovascular $K^+$ channels has identified the primary protein structure and heterologous expression systems have allowed a detailed analysis of the function and pharmacology of a single-channel type. An important challenge for the future is to determine the relative contribution that each of these cloned channels makes to cardiac function.[3]

#### $K^+$ Channel blockers as antiarrhythmics

It is currently accepted that prolonging repolarization of the heart produces an antiarrhythmic effect. However, only recently have concerted efforts been made to take advantage of this in drug development. Following a period in which the antiarrhythmic effort was focused on $Na^+$ channel blocking (class I antiarrhythmics) drugs, the validity of the $Na^+$ channel approach has been recently brought into question by the troubling results of the Cardiac Arrhythmia Suppression Trial (CAST), a multicentre study sponsored by the National Institutes of Health. This study, devised to test the hypothesis that suppression of ventricular arrhythmias by antiarrhythmic drugs improves postinfarction survival, showed that the class I agents encainide and flecainide led to a 3.6-fold greater risk of arrhythmic death and/or cardiac arrest when compared with their respective placebo control groups. Although the mechanism of drug-related increases in mortality are unclear, proarrhythmia due to excessive conduction slowing has been suggested as a probable cause.[4]

Critical re-evaluation of the existing antiarrhythmic therapy has led to an increased interest in alternative approaches, particularly those based on a selective prolongation of repolarization and refractoriness. Studies which correlate antiarrhythmic efficacy with increases in the refractory period (rather than conduction slowing) support this approach. Antiarrhythmic agents can be classified into several classes.[5,6]

*Class I antiarrhythmic agents*

These have been classified into three subclasses based on their ability to prolong, shorten, or have no effect on repolarization and refractoriness[7].

1. Class Ia agents (quinidine, disopyramide, and procainamide) tend to increase the action potential duration. This results primarily from the blocking of the delayed rectifier current. Nevertheless, these agents are not specific for the delayed rectifier but also affect other $K^+$ currents ($K_{ACh}$, $K_{IR}$) and $Na^+$- and $Ca^{2+}$ currents underlying the plateau phase of the action potential, and as such these agents fail to induce prolonged repolarization.
2. Class Ib agents (lidocaine) have little or no $K^+$ channel-blocking activity.
3. Class $I_c$ agents (flecainimide) do not affect repolarization, although they may be effective blockers of the delayed rectifier, which is balanced by their effect on the $Na^+$-plateau current.

$K^+$-channel blocking activity of class I antiarrhythmic agents may be clinically important during premature stimulation. Action potentials elicited at an early phase of the diastole show a much clearer drug-induced prolongation. It therefore seems logical that the efficacy of these agents against re-entrant arrhythmias may be related to changes in the repolarization time course rather than to conduction velocity.

*Class III antiarrhythmic agents*

These lengthen repolarization (refractoriness) without slowing intracardiac conduction. These compounds, which represent a logical evolution from the class I antiarrhyhmics discussed above, were discovered serendipitously during the clinical use of drugs which were initially developed to treat indications other than arrhythmias (e.g. amiodarone, sotalol). As a result, the first generation of class III compounds often had ancillary pharmacological activities that diminished their usefulness as antiarrhythmic agents.[8] For example, amiodarone, initially developed as an antianginal agent, was soon recognized as possessing class III properties. Vaughan-Williams class III antiarrhythmic agents act mainly by prolonging the duration of the cardiac action potential and, thus, the refractory period. This effect may be obtained either by (i) increasing the inward sodium or calcium currents, which may lead to an intracellular calcium overload and induce a very proarrhythmic situation, or (ii) decreasing the outward potassium currents – the objective of the new class III antiarrhythmic drugs under development[9]. These new drugs selectively block one or several potassium channels regulated by membrane potential (transient outward current $IK_A$, delayed rectifying current $IK_{DR}$ and rectifying inward current $IK_{IR}$). Under physiological conditions, the blockade of potassium channels regulated by a ligand (for example, ATP-dependent) does not lead to a class III effect. Prolongation of ventricular repolarization is accompanied by a slowing of the heart rate and a positive inotropic effect. It is attenuated by rapid rhythms and amplified by slow rhythms: this is the reverse frequency-dependent phenomenon. However, normal frequency dependence (or 'use-dependence') has been reported within the ionic channel, this paradox apparently being due to the complexity of the relationship between the relative contributions of the ionic currents of repolarization and their modulation by the heart rate. The class III effect confers a proarrhythmic potential and may lead to torsades de pointes, favoured by bradycardia, hypokalaemia, and hypomagnesaemia. Experimentally, it favours early after-depolarizations which are presumed to be the cellular trigger event. The greater understanding of factors influencing the antiarrhythmic and proarrhythmic class III effects has led to the establishment of a pharmacological profile for the 'ideal' drug conferring the least proarrhythmic risk and the greatest efficacy[9]

The rational design of antiarrhythmic agents has led to several compounds which are now in different stages of clinical and preclinical development. The non-selective compounds amiodarone and sotalol were the first to be marketed. Whereas sematilide is in the preregistration stage, others are in phase II (MK-499, artilide, etc.), phase III (E-4031, dotelilide, almokalant UK68.798), or in preclinical (terikalant) studies.

As stated earlier, the interplay between ionic conductances in the generation of cardiac membrane potentials leads to situations in which low doses of $K^+$ channel openers may result in antiarrhythmic action or in which $K^+$ channel blockers attenuate the cardioprotective effects of isofluorane in the stunned myocardium.[10] For example, the contribution of $K_{ATP}$ channels and adenosine to the coronary vasodilation that occurs in the normal heart during exercise and in the presence of a coronary artery stenosis is poorly understood. Not all $K_{ATP}$ channels are activated during exercise at pressures in the autoregulatory range; however, most $K_{ATP}$ channels are recruited as pressures approach the lower end of the autoregulatory plateau. Thus, $K_{ATP}$ channels and endogenous adenosine play a synergistic role in maintaining vasodilation during exercise in normal hearts, and distal to a coronary artery stenosis that results in myocardial hypoperfusion during exercise.

Claims in the past that treatment with sulphonylureas increases cardiovascular mortality are not supported by sound evidence. Sulphonylureas may even reduce cardiovascular mortality by protecting against ventricular arrhythmias during cardiac ischaemia. $K_{ATP}$-opening drugs are under investigation for the treatment of essential hypertension and angina pectoris. They are at least as effective as calcium channel blockers in achieving adequate blood pressure control. The recently introduced coronary vasodilating drug, nicorandil, exerts its effect by two mechanisms of action: (i) opening $K_{ATP}$ in vascular smooth muscle cells of coronary arteries; and (ii) activation of guanidyl cyclase by its nitro group in these cells. A proarrhythmic effect of $K_{ATP}$ channel openers has only been observed at very high doses, but at the low doses used in angina pectoris and hypertension. *In vivo*, no negative effect is found of $K_{ATP}$-opening drugs on insulin secretion.[11]

Establishing the dose and therapeutic protocols must take into account not only the pharmacological properties and interactions but also the physiopathological context which will determine therapeutic success.

#### Antischaemic activity of $K^+$-channel openers

Potassium channel activators have the ability to open potassium channels in a variety of cells. Since most of their effects are antagonized by antidiabetic sulphonylureas, the ATP-sensitive potassium channel is their most likely target. The opening of $K^+$ channels leads to hyperpolarization of the surface membrane and the consequent closure of voltage-dependent ion channels and reduction of free intracellular calcium ions. Currently available $K^+$ channel activators, including aprikalim, bimakalim, cromakalim, emakalim, nicorandil, pinacidil, etc., display a high affinity for the $K^+$ channels of vascular smooth muscle. Vasodilation and a reduction in systemic vascular resistance are their most prominent pharmacological effects. Both coronary and cerebral arteries are highly sensitive to $K^+$ channel activator-induced dilation.

During myocardial ischaemia, extracellular potassium rises rapidly, the increase coinciding with the onset of ventricular arrhythmias. The extracellular accumulation of potassium can induce abnormalities in both impulse conduction and impulse generation. For example, uneven potassium conductances will elicit regional differences in action potential duration and repolarization. Ventricular fibrillation may result from regional differences in the refractory period. In a similar manner, the spread of injury current from the ischaemic tissue to surrounding normal tissue can trigger extrasystoles (depolarization-induced automaticity). It has been hypothesized that, during myocardial ischaemia, the activation of the ATP-sensitive $K^+$ channel may contribute significantly to the reduction of action potential duration, and an increase in extracellular potassium accumulation during myocardial ischaemia. ATP-sensitive $K^+$ channel antagonists prevent ischaemically induced reductions in action potential duration and heterogeneity of the refractory period, and may also induce oscillatory after-potentials under certain conditions (for example, calcium overload). In contrast, $K^+$ channel agonists enhance the dispersion of refractory period ischaemia, which promotes the formation of re-entrant arrhythmias. The pharmacological modulation of ATP-sensitive $K^+$ channels could therefore offer a novel approach for the management of cardiac arrhythmias in patients with ischaemic heart disease. In general, channel antagonists prevent ventricular fibrillation, whereas high (hypotensive) doses of channel agonists can induce malignant arrhythmias during ischaemia in animal models. Recent evidence however, suggests that $K^+$ channel agonists may promote a better preservation of myocardial mechanical performance during reperfusion, while ATP-sensitive $K^+$ channel antagonists exacerbate mechanical depression during ischaemia in experimental models.[12]

$K^+$ channel openers have been shown to exert direct anti-ischaemic effects on the heart and to reduce infarct size. The $K^+$channel opener nicorandil reduced infarct size after coronary artery occlusion by 30% in anaesthetized open-chest dogs.[13] Unfortunately, the cardioprotective effect coincided with large falls in arterial blood pressure, total peripheral resistance, and rate–pressure product, suggesting that reduction in the infarct size may have resulted from a reduction in myocardial oxygen demand. Other $K^+$ channel openers such as cromakalim and pinacidil produce the same beneficial effects on the ischaemic heart, but without the haemodynamic side effects of nicorandil, and with a reduction in ventricular fibrillation (nicorandil did not protect against ventricular fibrillation). Interestingly, nicorandil inhibited superoxide production by neutrophils, which may have contributed to the preservation of the ischaemic reperfused myocardium.

#### Effects on coronary blood flow

Apart from treatment of hypertension, $K^+$ channel activators appear to have therapeutic potential in coronary heart disease. They reduce cardiac afterload, increase native and collateral coronary blood flow, and reduce the size of experimental myocardial infarcts. This last effect cannot be satisfactorily explained by the haemodynamic or coronary vascular actions of $K^+$ channel activators; a cardioprotective mechanism is therefore likely postulated. For these drugs, ischaemia-induced and activator-induced opening of cardiac ATP-sensitive $K^+$ channels appear to work in concert.[14] $K^+$ channel openers may act as selective coronary vasodilators. An increase in transmural pressure causes depolarization of coronary arterioles, which in turn increases smooth muscle tone. Under these conditions, the opening of $K_{ATP}$ channels can induce a much larger change in membrane potential than in relaxed arteries. Although patch-clamp studies suggest direct and potent relaxant effects on coronary blood flow, care should be taken when extrapolating patch-clamp results from isolated coronary smooth muscle cells to the *in vivo* function of $K_{ATP}$. Furthermore, the rate of ATP hydrolysis by contractile proteins, and thus the submembrane nucleotide concentrations, might also be different in the presence of myogenic tone. The opening of $K_{ATP}$ channels is increased in situations related to energy imbalance, such as hypoxia, adenosine release, intracellular acidification, and lactate accumulation. However, there is increasing evidence that $K_{ATP}$ channels also contribute to the setting of the membrane potentials of coronary smooth muscle cells under normoxic conditions. Thus, the modulation of $K_{ATP}$ channels by intracellular metabolites and by vasoactive autacoids may play an important role in the regulation of coronary blood flow, even in the presence of normal intracellular ATP concentrations. The smooth muscle cells of coronary terminal arterioles form an electrical syncytium. The opening of new $K_{ATP}$ channels in smooth muscle cells might induce a spatially homogeneous hyperpolarization of the entire arteriole. The resulting homogeneous decrease in tone of the coronary smooth muscle cells of the arteriole may induce a considerable change in vascular resistance. Furthermore, the selective inhibitor glibenclamide attenuates metabolic vasodilation induced by tachycardia in dogs.[15]

The $K^+$ channel opener RP 52891 is a selective coronary vasodilator. It produces a large increase in the coronary blood flow at doses that have no effect on either blood pressure or heart rate. Other $K^+$ channel openers produce the same effect, suggesting that these compounds may be useful in the treatment of myocardial ischaemia. Nicorandil combines the pharmacological properties of an organic nitrate with those of $K^+$ channel activators. Experimental and clinical results characterize nicorandil as a unique and promising drug for the treatment of coronary heart disease. *In vitro* studies have shown that its effects are in part mediated by the endothelium.[16] Large arteries respond to nicorandil with an increase in the blood flow. However, the ability of $K^+$ channel openers to cause an increased flow in thinner arteries is reduced by flow-limiting stenosis.

A slowly forming occlusion of coronaries, e.g. by arteriosclerosis, allows collateral circulation to develop that becomes determinant in order to maintain an adequate blood flow and then protects against ischaemia. Therapeutic compounds should be effective in relaxing collateral blood vessels. There is evidence that the $K^+$ channel opener cromakalim increases the collateral coronary blood flow.

## Effects on blood vessel

### $K^+$ channel openers

#### *Systemic haemodynamic effects*

The effects of cromakalim, pinacidil, minoxidil sulphate, nicorandil, and diazoxide on blood pressure have been reported in a variety of species including rat, rabbit, cat, and dog. They produce a fall in peripheral resistance that may last for hours. In addition to the opening of $K^+$ channels, these agents may have some ancillary pharmacological properties (such as the activation of guanylate cyclase in the case of nicorandil). Their *in vitro* and *in vivo* profiles derive from a combination of these two properties.

#### *Regional haemodynamic changes*

There is a considerable degree of interest in finding drugs with a pharmacological profile which shows territorial selectivity. $K^+$ channel openers reduce vascular resistance in a multiplicity of regional beds. The most common areas include renal, mesesenteric, and femoral vascular beds. Regional differences have been described, but there still exist profound discrepancies in the results obtained which most probably reflect the different protocols used. For example, it seems to be accepted that the hindquarter (iliac or femoral) vasculature is subject to marked falls in resistance. Cromakalim, SR-44866, and Rp31-6939 cause a 40% reduction in resistance, but little change is observed in the actual blood flow. Effects reported on the renal blood flow are also variable. In the anaesthetized cat, cromakalim increases flow,[17] while in the rat no such changes have been detected.

#### *Effects of $K^+$ channel openers on plasma humoral levels*

Pinacidil, cromakalim, and SR 44866 increase plasma renin activity.[17] It is well established that an increase in renal blood flow opposes the secretion of renin; thus, the distinct ability of $K^+$ channel openers to induce this effect probably reflects their selectivity in modifying renal blood flow. However, the direct potentiating effect of cromakalim on the release of renin in cultured juxtaglomerular cells has been demonstrated. $K^+$ channel openers are known to open $K_{ATP}$ channels in the pancreatic $\beta$-cell *in vitro*, leading to reduced insulin release.

However, the concentration needed to produce this effect is much higher than that needed to cause a fall in peripheral resistances. Thus, with the exception of diazoxide, hypotensive doses of $K^+$ channel openers have a negligible effect on insulin release.[18]

*Effects of K-channel openers on isolated blood vessels*

Contractile responses may be either induced or spontaneous (such as in the portal vein). Hamilton *et al.*[19] first reported that cromakalim reduced the amplitude and frequency of spontaneous activity in the rat portal vein. $K^+$ channel openers affect either the spontaneous activity or that induced by KCl depolarization. Intracellular recording of the membrane potential has shown that the effect of $K^+$ channel openers coincides with a reduction in the input conductance (as expected by the opening of the channels responsible for the maintenance of the resting conductance). Measurement of $^{86}Rb^+$ efflux also showed that $K^+$ permeability is increased. All these effects were counteracted by specific $K^+$ channel blockers.

#### Endogenous openers

It was recognized at an early stage that the endothelium plays a key role in acetylcholine-induced vasodilation through the release of an endogenous agent (the endothelium-derived relaxing factor) from the vascular endothelium. Work by Moncada and colleagues[20] established nitric oxide (NO) as the probable agent. NO is a potent vasodilator of vascular smooth muscle in several territories, and the underlying mechanism of its action is the opening of $K^+$ channels. The identity of the $K^+$ channel type involved has not yet been clearly established. Other endogenous vasodilators include calcitonin gene-related peptide (CGRP) and prostacyclin. CGRP is a polypeptide found in neurones closely associated with central and peripheral blood vessels. Its vasodilatory effects can hardly be dissociated from those of the endothelium; furthermore, CGRP directly affects the rate and force of heart contraction by increasing $Ca^{2+}$-currents and cAMP, and as such has an effect more similar to that of $\beta$-adrenoceptor agonists.

Inhibition of nitric oxide synthase by L-NAME (nitro-arginine methyl ester) produces systemic and pulmonary vasoconstrictor effects. The subsequent administration of a nitric oxide donor (sodium nitroprusside) causes significant reduction in systemic vascular resistance and mean arterial pressure, the effects being greater in animals pretreated with saline, rather than L-NAME. Pulmonary vascular resistance and mean pulmonary arterial pressure were also reduced, but to a lesser degree. Cardiac output was partially, but significantly, restored, the degree to which cardiac output is restored being limited by a concomitant reduction in preload.[21]

## Effects of $K^+$ channel openers in airways diseases

### In vivo bronchodilator activity

Cromakalim and Ro-31-6930 prolong the duration of respiratory distress in conscious guinea pigs challenged with an aerosol of histamine or ovoalbumin.[22] This effect was also evident in ovoalbumin-sensitized animals, both conscious and anaesthetized. To avoid systemic effects on vascular peripheral resistance, $K^+$ channel openers were administered by inhalation.

When $K^+$ channel openers were assayed in animal models they were found to inhibit the development of airways hyperreactivity, an effect shared by cromakalim (SDZ PCO 400). The latter has been shown to reduce airways hyperreactivity already established in guinea pigs following allergic reaction. SDZ PCO 400 has an effect on airways hyperreactivity at doses that do not induce relaxation of airways smooth muscle. $K^+$ channel openers have been tested in healthy human volunteers, and in patients with nocturnal asthma, with clear and promising beneficial effects in both situations.

The $K^+$ channel openers produced marked effects on airways secretory processes, suggesting that these compounds may have an antisecretory effect. The relevant use of $K^+$ channel openers in airways disease has yet to be determined.[23]

### In vitro effects of $K^+$ channel openers

Concentration-dependent suppression of the spontaneous tone of guinea pig trachealis has been described for cromakalim, pinacidil, RP 49356, and Ro 31-6930.[24] This observation has been extended to bovine trachealis (which is devoid of spontaneous tone) and to human bronchial smooth muscle *in vitro*. A large number of electrophysiological studies, together with the measurement of $^{86}Rb^+$ efflux, support the view that all these drugs act by opening $K^+$ channels in the plasmalemma. For example, the $K_{ATP}$ channel blocker glibenclamide antagonizes the relaxant effects of $K^+$ channel openers such as cromakalim, pinocidil, RP 493456, and celikalim.[25] The effect is selective in that the sulphonylureas do not antagonize $\beta$-adrenoceptor- or verapamil-associated relaxation. The experimental data support the hypothesis that an increase in resting $K^+$ conductance by $K^+$ channel openers may account for their relaxing effect on airways smooth muscles.

## $K^+$ channel modulators and nervous system diseases

The potential for pharmacological intervention on the central nervous system is extensive. The field of ion channel pharmacology has developed relatively recently, and is based mostly on drugs which interact with $Ca^+$ and $K^+$ channels. In this sense; as neural systems are rich in receptor-operated $K^+$ channels,[26,27] $K^+$ channel modulators which interact with synaptic transmission or $K^+$ channels are the target for neuroactive drugs such as ethanol,[28] or tricyclic antidepressants.[29]

The possible sites for therapeutic intervention are many in number. For example, $K_{ATP}$ channels are known to modulate insulin release from pancreatic $\beta$-cells, while, it has also been proposed that $K_{ATP}$ channels in the nervous system might similarly modulate neurotransmitter release. The effects of $K_{ATP}$-opening agents on GABA release in the globus pallidus has been investigated. Diazoxide and cromakalim decreased the KCl-evoked release of [$^3$H] GABA from pallidal slices. The effects of both cromakalim and diazoxide were significantly antagonized by the concurrent application of the sulphonylurea, glibenclamide (100 $\mu$M). Intrapallidal injections of diazoxide in the reserpine-treated rat model of Parkinson's disease reduced akinesia in a dose-dependent manner. These data suggest that manipulation of neuronal potassium channels with pharmacological responses similar to $K_{ATP}$ may prove to be useful in the treatment of Parkinson's disease.[30]

Ion channels are under the regulation of neurotransmitters. The muscarinic receptor inhibits the inward rectifier $IRK_1$ (widely distributed throughout the central nervous system), presumably via stimulation of protein kinase C. Such an action on $IRK_1$ underlies the inhibitory effects of muscarinic receptor stimulation on inwardly rectifying potassium conductances observed in the brain.[26]

Ion channels may also be preferential targets for pharmaceutical compounds. For example, tricyclic antidepressants (imipramine, amitriptyline) and related compounds (chlorpromazine, tacrine, carbamazepine) block $K^+$ currents in rat sympathetic neurones. Concentration–response relationships for imipramine and tacrine show that imipramine was about seven times more potent than tacrine, although maximum inhibition and the Hill coefficient were the same for both compounds. Amitriptyline, chlorpromazine and imipramine (at 10 $\mu$M) were two to three times more potent as inhibitors of sustained $K^+$-current (mostly $K_{DR}$) than transient $K^+$-current (mostly $K_A$). Tacrine, however, was equally effective in blocking both components.[29]

Furthermore, the brain contains endosulphines, an endogenous ligand for the sulphonylurea receptor.[31] Anti-diabetic sulphonylureas act via high-affinity binding sites coupled to $K_{ATP}$ channels. Endosulphine exists in alpha- and beta-forms, both in the pancreas and central nervous system. Porcine alpha-endosulphine is a protein with a molecular mass of 13 kDa. Comparison of these sequences with that present in the National Biomedical Research Foundation protein data bank indicated a 82% match with a 112-amino acid protein with a molecular mass of 12 kDa called 'cyclic AMP-regulated phosphoprotein-19', isolated from the bovine brain as a substrate for protein kinase A.

## Epilepsy

The epilepsy gene map has been re-examined with new information concerning benign familial neonatal convulsions, benign familial infantile convulsions, Unverrichf–Lundborg disease, epilepsy with progressive mental retardation, and juvenile myoclonic epilepsy. Understanding of the molecular basis of paroxysmal disorders affecting the central nervous system has been revolutionalized since mutations in genes for the neurotransmitter receptors, GLRA1 and CHRNA4, and a voltage-gated potassium channel, KCNA1, have been identified as causes of inherited neurological disease.[32] Accidental poisoning of humans with-amino-pyridine (4-AP), during the treatment of botulinum toxin poisoning, has been reported to induce 'grand mal'-type seizures. Other potent $K^+$ channel blockers also provoke seizures. In addition to all these studies there are also reports showing that anti-epileptic drugs can increase $K^+$ currents. Based on these results it was logical to propose that $K^+$ channel openers be used in the treatment of epilepsy.

## Alzheimer's disease and cognition

$K^+$ channels play a critical role in learning and memory. Sanchez-Andrés and Alkon[33] reported that classical conditioning reduced the slope values of $I_{AHP}$ (responsible for after-hyperpolarization) without affecting other $K^+$ currents. Etcheberrigaray *et al.*[34] reported that fibroblasts from Alzheimer's disease patients lacked a 113 pS tetraethylammonium

(TEA)-sensitive $K^+$ channel. It has been recently shown that $K^+$ channel openers (pinacidil and minoxidil) injected immediately after a training session caused a 30–70% reduction in memory retention, whereas the $K^+$ channel blockers apamin and charybdotoxin are endowed with antiamnesic properties. Intraseptal injection of the $K^+$ channel blocker glibenclamide enhances spontaneous alternation scores in the rat, whereas glucose has a dose-dependent facilitation action on long-term memory, suggesting that $K^+$ channel blockers may be used to enhance cognition in Alzheimer's disease.

### Multiple sclerosis

Because the symptomatic treatments for multiple sclerosis (MS) are limited, new approaches have been sought. Anatomical studies of MS lesions show a relative preservation of axons, and clinical studies suggest that some of the neurological impairment in patients with MS is physiological. Electrophysiological studies suggest that demyelination exposes axonal $K^+$ channels which decrease action potential duration and amplitude, thereby hindering action potential propagation. Potassium channel blockers, including aminopyridines, have been shown to improve nerve conduction in experimentally demyelinated nerves. Two $K^+$ channel blockers, 4-aminopyridine (4-AP) and 3,4 diaminopyridine (DAP) have been tested in patients with MS. Preliminary studies with 4-AP demonstrated some benefit in many temperature-sensitive multiple sclerosis patients, and improvement of function was found in a large randomized double-blind, placebo-controlled crossover trial over 3 months of oral treatment in 68 patients with multiple sclerosis. An open-label trial of DAP showed improvement in some deficits, and a double-blind, placebo-controlled trial showed significant improvements in defined neurological deficits. A crossover comparison of the two agents suggested that 4-AP produces more central nervous system side effects (dizziness and confusion), whereas DAP produces more peripheral side effects (paraesthesias and abdominal pain). Both agents have rarely caused seizures. These studies suggest that aminopyridines may provide a new approach to the symptomatic treatment of MS. In Bulgaria, 4-AP has been used to reverse the actions of curare.[35] Although 4-AP is not better than conventional anticholinesterase activity, there is a synergistic effect between the two that would allow the use of the combined drugs at lower concentrations. 4-AP is also active in the treatment of defects in neurotransmitter release (e.g. Eaton–Lambert syndrome) with limited success, and in the treatment of botulinum toxin poisoning. Other uses include the treatment of demyelinating disease, Huntington's chorea, etc.

## $K^+$-channels and neuromuscular diseases

Ion channel disfunctions associated with neuromuscular disease have been described in episodic ataxia amyotrophic lateral sclerosis, acquired myotonia, and Isaac's syndrome (Isaac's syndrome is an uncommon, but distressing, condition of spontaneous abnormal muscle activity caused by neuronal hyperexcitability, possibly due to damaged slow potassium channels). It is noteworthy to point out that the sensitivity to $K^+$ channel drugs may change under pathological circumstances. In denervated skeletal muscle, delayed outward current became resistant to 4-AP, whereas TEA became more effective in blocking the fast, and less effective in blocking the slow, components of this current. Even though the primary course of skeletal muscle disease is not an alteration in $K^+$ channels, it is possible to use $K^+$ channel modulators to counteract the effect of other ionic conductances. For example, $K^+$ channel openers cromakalim and EMD 52692 completely suppressed after-contractions in fibres from myotonia congenita patients in which a reduced conductance to $Cl^-$ has been described.

## $K^+$ channel openers and urogenital diseases

The therapeutic potential of $K^+$ channel openers in treating genitourinary tissues was appreciated at an early stage. The treatment of bladder dysfunction, preterm labour, impotence, and ureteric spasm are the main focuses of its applications.

### Urinary incontinence

Involuntary contraction of the detrusor smooth muscle leads to urinary incontinence which, together with urgent and frequent urination, is a common problem in the elderly. The International Continence Society Committee on Standarization and Terminology (1988) has classified two main forms of urinary incontinence: detrusor instability and detrusor hyperreflexia. Whereas the former is often found in patients with partial outflow obstruction (e.g. due to benign prostatic hyperreflexia), the latter is associated with perturbation of the nervous control mechanisms (spinal lesions, multiple sclerosis, or diabetes). Ideal treatment would abolish the involuntary contractions but not affect the normal emptying of the bladder. The currently available treatment (antimuscarinic and smooth muscle relaxants) are not devoid of uncomfortable side effects (dry mouth, constipation, blurred vision, tachycardia, etc.).

#### Treatment of urinary incontinence

*In vitro effects*

Cromakalim and pinacidil at concentrations greater than 100 nM inhibit in a dose-dependent manner the spontaneous activity and basal tone in strips of guinea pig detrusor muscle.[36] They also (i) abolish 20 mM KCl-induced contraction of rat and human detrusor,[36] but have no effect on contractions induced by KCl at concentrations $>40$ mM, and (ii) induce a rightward shift in the concentration–response curve of carbachol-induced contractions.[36] These results were supported both by the effects on $^{86}Rb^+$ efflux and by electrophysiological studies.

When $K^+$ channel openers were assayed from strips of unstable bladder (which show many of the phenomena seen in patients with bladder dysfunction) they were effective in abolishing spontaneous activity. Furthermore, some studies indicate that hypertrophic smooth muscle displays an increased sensitivity to $K^+$ channel openers, a phenomenon also detected in spontaneously hypertensive rats.

*In vivo effects*

Cromakalim and pinacidil do not affect bladder function in normal animals. However, when assayed in animal models of detrusor instability, they lowered micturition pressure in rats with bladder hypertrophy. Unfortunately, these effects occurred at doses that also affect the cardiovascular system. Development of $K^+$ channel openers selective for the bladder would be of great interest in the treatment of bladder dysfunction. However, pilot studies in patients have concluded that no effect on detrusor activity can be achieved without cardiovascular side effects.

## Premature labour

Conversion of the electrically silent pregnant uterus to highly excitable at term represents a dramatic physiological event which is poorly understood. Khan *et al.*[37] have described a large conductance (212 pS) calcium-activated potassium channel ($BK_{Ca}$) in pregnant human myometrium which, in labour tissue, is either absent or has been considerably altered in its physiological and pharmacological properties. In labour tissue, the $K^+$ channels have an identical conductance (221 pS) and $K^+$ selectivity to $BK_{Ca}$ channels, but exhibit no $Ca^{2+}$ or voltage sensitivity. The activity of the $BK_{Ca}$ channel in pregnant tissue is inhibited by internal application of $Ba^{2+}$, but not TEA, whereas the activity of the $BK_{Ca}$ channel is sensitive to internal TEA but not $Ba^{2+}$. The role of the $BK_{Ca}$ channel may be to suppress myometrial activity during gestation, whereas remnant K-channel activity may be important in providing a $Ca^{2+}$-independent $K^+$ conductance which would allow cytoplasmic $Ca^{2+}$ levels to rise without activating a counteracting $Ca^{2+}$-dependent outward current, normally provided by the $BK_{Ca}$ channels which, by their very nature, would tend to oppose depolarization. These findings suggest that $K^+$ channels may have an important role in determining the functional activity of the myometrium.[37] Premature labour is the main indication for smooth muscle relaxants. So far, $\beta$-adrenoceptor agonists are used for this indication; however their side effects on the heart, vascular system and skeletal muscle preclude their use at doses that may cause tachycardia, hypotension, or tremor. The need for a new therapy is reinforced by other relaxants such as $Ca^{2+}$-channel antagonists which can produce less tachycardia and greater vasodilation than $\beta$-adrenoceptor agonists. For example, lidocaine can selectively block myometrial $Na^+$ channels, thereby reducing excitability. It can prolong gestation in the pregnant rat. These observations suggest that lidocaine may be useful in the clinical management of preterm labour. On the other hand, diazoxide, pinacidil, cromakalim, RP 49356, aprikalim, and BRL 38227 exert a relaxant effect on uterine smooth muscle,[38] while the sulphonylurea, glibenclamide, appears to be a competitive antagonist of this inhibition. No antagonism was observed with the sulphonylurea, tolbutamide. The relaxant effect of $K^+$ channel openers seems to be initiated by the preferential action on the myometrium pacemaker cells. Unfortunately, there are no tissue-selective $K^+$ channel openers that could be used in premature labour.

## Erectile tissues, vas deferens and the ureter

### Erectile tissues

Penile erection results from a decrease in arterial resistance, leading to an increased arterial inflow to the penis with subsequent filling of erectile tissues. Cavernous smooth muscle relaxation is effected through a complex physiological pathway; therefore, a defect in any step of this pathway may result in erectile dysfunction. Administration of pharmacological agents which cause relaxation of the cavernous smooth muscle through a different mechanism may serve as an effective therapeutic alternative for impotent patients. Smooth muscle relaxants, when injected into the penis, induce erection in humans. Cromakalim and pinacidil relax contractions of the isolated corpus cavernosum induced by noradrenaline or histamine in rabbits, pig, monkey, and humans. Nitric oxide (NO) released from non-adrenergic–non-cholinergic (NANC) nerves seems to be a principal mediator for the relaxation of penile erectile tissue leading to erection, and drugs acting by NO release have been shown to produce erection when injected intracorporeally into impotent patients. As a result, $K^+$ channel openers may be useful in the diagnosis and treatment of erectile dysfunction. On the other hand, the $K_{Ca}$ channel plays an important role in corporal smooth muscle physiology and alterations in the function/regulation of $K_{Ca}$ channels may also be an important feature of organic erectile dysfunction. Altered $K_{Ca}$ channel behaviour may contribute to an impaired ability to hyperpolarize corporal smooth muscle, possibly by altering intracellular calcium homoeostasis and, perhaps, corporal smooth muscle reactivity and tone.[39] Nicorandil is classified as a $K^+$ channel opener, but it also acts as a donor of NO. Nicorandil is effective in relaxing human corpus cavernosum, chiefly by its $K^+$ channel-opening action, and to some extent by its ability to release NO. ATP-dependent $K^+$ channels seem to be of limited importance in nicorandil's relaxing effect. If effective in impotent patients, the drug may represent a new and interesting approach to the treatment of erectile dysfunction.

### Ureter

$K^+$ channel openers such as diazoxide, pinacidil, and cromakalim inhibit the contractile response of the ureter to electrical field stimulation,[40] ureteric peristalsis in dogs, and the spontaneous myogenic activity of the guinea pig isolated renal pelvis. The effects of cromakalim were reversed by BAY K 8644, suggesting that the relaxation involves reduction of $Ca^{2+}$ influx through voltage-activated $Ca^{2+}$ channels. Thus, the use of diazoxide and other $K^+$-channel openers has been proposed in the treatment of ureteric colic.

The effects of the $K^+$ channel modulators cromakalim, TEA, and glibenclamide have been assessed in segments of human ureter rings obtained from kidney donors 'leftovers'. In this preparation, the amplitude of the contraction induced by electrical stimulation was not changed by glibenclamide, but was enhanced by TEA. The resting tension of the ureter was not changed by either of these $K^+$ channel inhibitors. Cromakalim did not change the resting tension of the human ureter *per se*, but induced a concentration-dependent inhibition of the contractions induced by electrical stimulation. Phasic contractions were abolished by cromakalim, whereas tonic contractions were unaffected. These results suggest that $K^+$ channels are important in the control of human ureter contractility, and that $K^+$ channel openers may be an alternative therapeutic indication in the treatment of human ureteric colic.

### Vas deferens

The vas deferens forms part of the male reproductive tract and extends from the cauda epididymis to the prostate. Rat vas deferens contractions can be evoked by the $K^+$ channel blockers $BaCl_2$ and 4-AP.[41] Both $BaCl_2$ and 4-AP evoked phasic contractions dose-dependently. High extracellular potassium (35–40 mM) caused a tonic contraction, but abolished the $BaCl_2$- and 4-AP-induced phasic activity. It also reduced the $BaCl_2$-induced sustained component of contraction, but increased the 4-AP-induced tonic contraction. Subsequent calcium entry through the depolarized plasma membrane is needed to trigger the generation of phasic contractions. Omission of calcium from the extracellular medium totally abolished the 4-AP-induced response, but only reduced the mean amplitude of phasic contractions induced by $BaCl_2$. $BaCl_2$-evoked activity in the vas deferens was mainly due to a blockade of $Ba^{2+}$-sensitive $K^+$ channels in the smooth muscle plasma membrane. It has also been suggested that atrial natriuretic factor (ANF) exerts a neuromodulatory effect on the vas deferens.[42] ANF effects could be mediated by an activation of either calcium-activated or outward rectifying $K^+$ channels.

## $K^+$ channels and immunosuppression

Immunosuppresive therapy is dominated by cycloscoporin and related drugs which possess substantial renal toxicity. The role of ion channels in the lymphocyte are currently under study. Ion channel modulation may constitute a safe and effective means of immunosuppression.

Voltage-gated potassium channels regulate calcium-dependent pathways involved in human T-lymphocyte activation.[43] A voltage-gated $K^+$ channel present in lymphocytes, Kv1.3, plays an important role in modulating lymphocyte function. Blockers of Kv1.3, including several naturally occurring scorpion peptide-toxins (charybdotoxin, kaliotaxin, margatoxin, and noxiustoxin), are effective inhibitors of T-cell function, and might be used to prevent graft rejection and to treat autoimmune disorders.

Patch-clamp studies have shown that immunosuppressors inhibit voltage-gated $K^+$ channels in human peripheral blood lymphocytes.[44] Cyclosporin A, rapamycin, and FK-506 reduced the peak $K^+$-current without changing its reversal potential. The current inhibition was similar at all membrane potentials studied and was accompanied by an increase in the rate of $K^+$-current inactivation. Membrane potential measurements in current-clamp showed a marked depolarization of the membrane ($> 10$ mV) on adding either of the immunosuppressors to the cells. Voltage-dependent $K^+$-current in human peripheral blood lymphocytes is inhibited by cyclosporin A and other immunosuppressors, resulting in a depolarized membrane potential.

## Other clinical aspects

The ubiquitous presence of $K^+$ channels and their prominent physiological role in many cell functions opens up many therapeutic possibilities. $K^+$ channel modulation modifies sperm motility, regional blood flow, and secretion. As a result, some applications have been discovered as side effects, for example the effect of minoxidil sulphate on hair growth. A very promising field of research can be anticipated from the modulation of distinct $K^+$ channel subtypes and locations.

## Acknowledgements

The author has been supported during this period by Grants from the Ministry of Health (FIS 94–0014–01 and FIS96–1994–01), Generalitat Valenciana (GV-3117–95) and European Commission (Contract ERBSC1-CT 920833).

## References

1. Gehlert, D.R. and Robertson, D.W. (1994) *Prog. Neuropsychopharmacol. Biol. Psychiatry*, **18**(7), 1093–102.
2. Edwards, G. and Weston, A.H. (1995). *Cardiovasc. Drugs Ther.*, **9** (*suppl. 2*); 185–193.
3. Deal, K.K., England, S.K., and Tamkun, M.M. (1996). *Physiol. Rev.*, **76**(1): 49–67.
4. Task Force of the Working Group of Arrhythmias of the European Society of Cardiology (1990). *Circulation*, **81**, 1123–1127.
5. Vaughan-Williams, E.M. (1984). *J. Clin. Pharmacol.*, **24**, 129–147.
6. Colatsky, T.J. and Angentieri, T.M. (1994). *Drug Development Research*, **33**, 235–249.
7. Harrison, D.C. (1985). *Am. J. Cardiol.*, **56**, 185–187.
8. Colatsky, T.J. (1992). In: *Potassium Channel Modulators*, (ed. A.H. Weston and T.C. Hamilton) pp. 304–340. Blackwell Scientific Publications, London.
9. Adamantidis, M.M. (1995). *Arch. Mal. Coeur. Vaiss.*, **88**, 33–40.
10. Kersten, J.R., Lowe, D., Hettrick, D.A., Pagel, P.S., Gross, G.J., Wartier, D.C. (1996). *Anesth-Analg.*, **83**(1), 27–33.
11. Ligtenberg, J.J., van Haeften, T.W., Links, T.P., Smit, A.J., and Reitsma, W.D. (1995). *Neth. J. Med.*, **47**(5), 241–251.
12. Billman, G.E. (1994). *Cardiovasc. Res.*, **28**(6), 762–769.
13. Endo, T., Nejima, J., Kiuchi, K., Fujita, S., Kikuchi, K., Hayakawa, H., and Okumura, H. (1988). *J. Cardiovasc. Pharmacol.*, **12**, 587–592.
14. Haeusler, G. and Lues, I. (1994). *Eur. Heart. J.*, **15**(suppl. C), 82–88.
15. Daut, J., Klieber, H.G., Cyrys, S., and Noack, T. (1994). *Cardiovasc. Res.*, **28**(6), 811–817.
16. Cook, N.S. (1989). *J. Cardiovasc. Pharmacol.*, **13**, 299–306.
17. Clapham, J.C., and Longman, S.D. (1989). *Eur. J. Pharmacol.*, **171**, 109–117.
18. Quast, U. and Cook, N.S. (1989). *J. Pharmacol. Exp. Ther.*, **250**, 239–249.
19. Hamilton, T.C., Weir, S.W., and Weston, A.H. (1986). *Br. J. Pharmacol.*, **88**, 103–111.
20. Moncada, S., Palmer, R.M., and Higgs, E.A. (1989). *Biochem. Pharmacol.*, **38**, 1709–1715.
21. Herity, N.A., Allen, J.D., Silke, B. and Adgey, A.A. (1994). *Cardiovasc. Res.*, **28**(6), 894–900.
22. Arch, J.R.S., Buckle, D.R., Bumstead, J., Clarke, G.D., Taylor, J.F., and Taylor, S.G. (1988). *Br. J. Pharmacol.*, **95**, 763–770.

23. Griffin, A. (1995). *Eur. J. Pharmacol.*, **280**(3), 317–325.
24. Raeburn, D.M. and Brown, T.J. (1991). *J. Pharmacol. Exp. Ther.*, **256**, 492–499.
25. de Lorenzi, F.G. (1994). *Pulm. Pharmacol.*, **7**(2), 129–135.
26. Jones, S.V. (1996). *Mol. Pharmacol.*, **49**(4), 662–667.
27. Koumi, S., Wasserstrom, J.A., and Ten Eick, R.E. (1995). *J. Physiol. (Lond.)*, **486**, 647–659.
28. Covarrubias, M. and Rubin, E. (1993). *Proc. Natl Acad. Sci. USA*, **90**(15), 6957–6960.
29. Wooltorton, J.R. and Mathie, A. (1994). *Br. Pharmacol.*, **110**(3), 1126–1132.
30. Maneuf, Y.P., Duty, S., Hille, C.J., Crossman, A.R., and Brotchie, J.M. (1996). *Exp. Neurol.*, **139**(1), 12–16.
31. Virsolvy Vergine, A., Salazar, G., Sillard, R., Denoroy, L., Mutt, V., and Bataille, D. (1996). *Diabetologia*, **39**(2), 135–141.
32. Elmslie, F., and Gardiner, M. (1995). *Curr. Opin. Neurol.*, **8**(2), 126–129.
33. Sanchez-Andrés, J.V. and Alkon, D.L. (1991). *J. Neurophysiol.*, **65**, 796–807.
34. Etcheberrigaray, R., Ito, E., Oka, K., Tofel Grehl, B., Gibson, G.E., and Alkon, D.L. (1993) *Proc. Natl Acad. Sci USA*, **90**, 8209–8213.
35. Paskov, D.S., Staenov, E., and Mirov, V.V. (1973). *Esperimentalnaja Chirurgija i Anaesthesiologija*, **18**, 48–52.
36. Malmgren, A., Sjögren, C., Yvelius, B., Mattiason, A., Andersson, K.E., and Anderson, P.O. (1990). *J. Urol.*, **143**, 828–834.
37. Khan, R.N., Smith, S.K., Morrison, J.J., and Ashford, M.L. (1993). *Proc. R. Soc. Lond. Biol. Sci.*, **251**(1330), 9–15.
38. Morrison, J.J., Ashford, M.L., Khan, R.N., and Smith, S.K. (1993). *Am. J. Obstet. Gynecol.*, **169**(5), 1277–1285.
39. Fan, S.F., Brink, P.R., Melman, A., and Christ, G.J. (1995). *J. Urol.*, **153**, 818–825.
40. Maggi, C.A., Giuliani, S., and Santicioli, P. (1994). *Br. J. Pharmacol.*, **111**(3), 687–694.
41. Huang, Y. (1995). *Br. J. Pharmacol.*, **115**, 845–851.
42. Kanwal, S. and Trachte, G.J. (1994). *J. Pharmacol. Exp. Ther.*, **268**, 117–123.
43. Lin, C.S., Boltz, R.C., Blake, J.T., Nguyen, M., Talento, A., Fischer, P.A., *et al.* (1993). *J. Exp. Med.*, **177**(3), 637–645.
44. Panyi, G., Gaspar, R., Krasznai, Z., ter Horst, J.J., Ameloot, M., Aszalos, A., Steels, P., and Damjanovich, S. (1996). *Biochem. Biophys. Res. Commun.*, **221**(2), 254–258.

# 5 Chloride Channels

# 11 Chloride channel pharmacology

M. A. Valverde, P. Hardy, and A.S. Monaghan

## Introduction

It has become clear that chloride channels play a role in a number of different cellular functions. Until recently the γ-amino-butyric acid (GABA) and glycine-activated – receptor anion – selective channels were the only well-characterized anion channels but, following the introduction of the patch-clamp technique, it became possible to identify $Cl^-$ channels underlying anionic conductances and their different cellular functions in a wide variety of cells. Indeed, the combination of electrophysiological techniques and molecular biology has facilitated the identification of the chloride channel defects in diseases such as cystic fibrosis (CF) (Welsh and Smith 1993) and myotonia (*et al.* 1991).

We will discuss the interaction of different pharmacological agents with $Cl^-$ channels in an attempt to help the reader assess their value in differentiating $Cl^-$ channels and their functions. Considerable information exists on the neuronal ligand-gated chloride channels and the reader is referred elsewhere (Eldefrawi and Eldefrawi 1987) and the discussion will be limited to compounds reported to interact with $Cl^-$ channels at the macroscopic or single-channel levels.

$Cl^-$ channels have been found in many different cells, including circulating cells (refs 40 and 42 in Table 11.1), osteoblasts (ref. 55 in Table 11.1), and even intracellular organelles (Mannella 1992) but often their physiological function remains speculative. Anionic conductances are recognized to contribute to the regulation of cardiac action potentials (Ackerman and Clapham 1993), the control of muscle cell resting conductance (Bretag 1987), cell volume regulation (Hoffman and Simonseu 1989) and neuronal excitability (Mayer *et al.* 1990). Furthermore, chloride channels are recognized for their role in transcellular transport in polarized epithelia and their involvement in intestinal electrolyte movement serves as a good example of their importance in the functioning of a tissue. Polarized epithelia are characterized by their ability to secrete and absorb electrolytes and the subsequent obliged movement of water. Depending on whether the tissue is absorptive (e.g. distal colon, and kidney ascending Henle's loop and collecting tubule) or secretory (respiratory and small intestinal tract), the accumulation of ions across the epithelia is carried out by the strategic combination of different ion transporters and channels located to apical and basolateral membranes. Electrolyte and fluid secretion can be initiated by neurohumoral agents and certain bacterial toxins known to alter intracellular levels of cAMP, cGMP or

$Ca^{2+}$, and all these agents have been found to regulate the activity of $Cl^-$ channels in various secretory epithelia (for more detailed reviews, see Liedtke 1989; Gogelein 1988; Frizzell and Halm 1990; Brian and Andreoli 1992. At least three distinct chloride channels have been identified in intestinal cells *in vitro*, but little is known about their distribution and function along the crypt-villus because of the difficulties of access to the apical membranes. Selective chloride channel blockers would be of value in dissecting the contribution of different channels to secretory stimuli and studies carried out on isolated intestinal crypt preparations indicate the usefulness of such an approach (Fig. 11.1). The stilbene inhibitor 4,4′-diisothicyanostilbene-2,2′-disulphonic acid (DIDS) was able selectively to inhibit the crypt shrinkage (as a result of chloride secretion) in response to carbachol but not in response to vasoactive intestinal polypeptide (VIP) (Fig. 11.1).

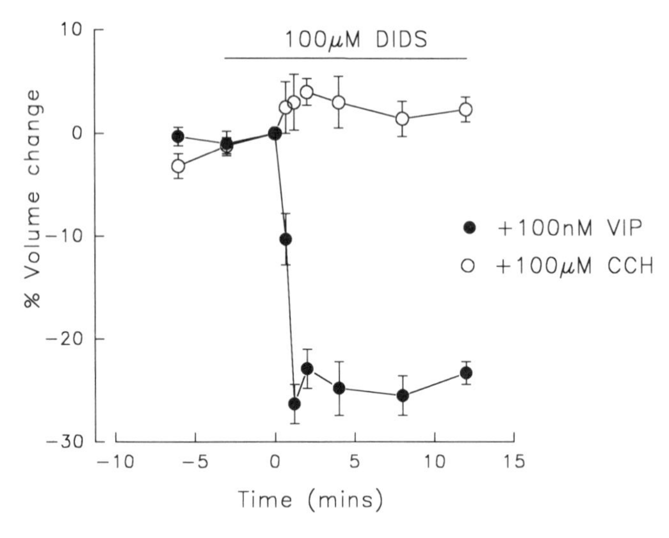


**Fig. 11.1** Relative changes in cell volume in isolated small intestinal crypts after the addition of the secretagogues vasoactive intestinal peptide (VIP) and carbachol (CCH). Activation of $Cl^-$ and $K^+$ conductances by the secretagogues determine the loss of $Cl^-$ and $K^+$ ions and the obliged water leading to shrinkage. That the secretagogue effects depend upon separate $Cl^-$ channels, activated by cAMP (VIP) and $Ca^{2+}$ (CCH) respectively, is suggested by the different sensitivity to DIDS. DIDS abolished the CCH-induced shrinkage (○) but not the response to VIP (●). (Modified from Valverde *et al.* 1993.)

## Chloride channel pharmacology

Ion channel blockers have been used extensively as tools for the identification, classification, and purification of cation-selective channels. The first blockers used to characterize selective ionic currents were natural toxins. Tetrodotoxin and saxitoxin were invaluable in the identification of $Na^+$ channels through their high specificity and potency (blocking at nanomolar concentrations), and natural poisons have proved to be a useful source of inhibitory compounds of both $Na^+$ and $K^+$ currents. The use of pharmacological agents as $Cl^-$ blockers has been less successful, with only a limited number of potent compounds identified to date and this has greatly hindered the study of anion channels in comparison. Table 11.1 includes a list of compounds which have been described to interact with chloride channels as a blocker, as well as some characteristics of the channels such as conductance and rectification.

**TABLE 11.1** Chloride channel blockers

| Cell type | Channel | Mode | Activation | Blocker | $IC_{50}$ | Reference |
|---|---|---|---|---|---|---|
| *Airways epithelia* | | | | | | |
| Trachea (dog) | 60 pS,OR | WCR | e+d | DPC(1)0.1–0.2 mM | | 1 |
| | | EIOP | e+d | DPC(1)0.1–1 mM | | 2 |
| | 50 pS,OR | EIOP | e+d | FA(6)5–400 μM | | 3,4 |
| | 50 pS,OR | EIOP | e+d | AA(6)10–400 μM | | 3 |
| Airways (human) | OR | WCR | s | DPC(1)2 mM | | 5 |
| | | | | NPPB(1) | 3.1 μM | 5,6 |
| | | EIOP | | FA(6) | | 3 |
| Human nasal | CFTR | WCR | | DPC(1)0.5 mM | | 7 |
| | | | | DIDS(1)0.1 mM | | |
| Alveoli | 375 pS | EIOP | e | SITS(1)0.1 mM | | 8 |
| Trachea (bovine) | 25–30 pS | PLB | | DIDS(1)0.1 mM | | 9 |
| Trachea (sheep) | 107 pS,IR | PLB | | NPPB(1)10–100 μM | | 10 |
| *Renal epithelia* | | | | | | |
| MDCK | 63 pS,OR | CAP | s | quinine(2)1 mM | | 11 |
| | | WCR | | ochratoxin A(7) | 30 nM | 12 |
| | | | | NPPB(1) | 600 nM | |
| A6 (*Xenopus*) | 360 pS | EOOP | e | SITS(1)1 mM | | 13 |
| GBK cortex (bovine) | 300 pS | EIOP | e | SITS(1)1 mM | | 14 |
| CCD (rabbit) | 303 pS | EIOP | e | DPC(1)0.1 mM | | 15 |
| | | | | NPPB(1)30 μM | | |
| | | | | DIDS(1)0.5 mM | | |
| | 96 pS,OR | EIOP | e | DIDS(1)0.1 mM | | 16 |
| | | | | NPPB(1)2,20 μM | | |
| TAL (rabbit) | | WCR | cAMP | DIDS(1)0.2 mM | | 17 |
| | | | | 9-AC(1)0.2 mM | | |
| | | | | DPC(1) | 0.2 mM | |
| Endosomal vesicle (rabbit cortex) | 73 pS | | | DIDS(1)1 mM | | 18 |
| | | | | NPPB(1)0.1 mM | | |
| *Gastrointestinal epithelia* | | | | | | |
| HT29 | 15 pS | EIOP | e+d | NPPB(1) | | 19 |
| | 50 pS,OR | EIOP | e+d | NPPB(1)1–50 μM | | |
| | 20–50 pS.OR | | e+d | NPPB(1) | 0.9 μM | 20 |
| | | | | cromalkin(7) | 19 μM | |
| | | | | RV31156(7) | 2.5 μM | |
| | 79 pS,OR | CAP/EIOP | e+d | NPPB(1)10,100 μM | | 21 |
| | | | | DNDS(1)20 μM | | |
| HT29 | 60 pS,OR | EOOP | e+d | NPPB(1) | 0.5 μM | 22 |
| | | | | verapamil(2) | 100 μM | |
| | | | | D888(2) | 60 μM | |
| | | | | nitrendipine(2)(100 μM) | | |
| | | | | Bay K(7)50 μM | | |
| | | | | diltazem(7)100 μM | | |

**TABLE 11.1** (contd)

| Cell type | Channel | Mode | Activation | Blocker | $IC_{50}$ | Reference |
|---|---|---|---|---|---|---|
| *Gastrointestinal epithelia* (contd) | | | | | | |
| T84 | OR | WCR | s | tamoxifen(3)10 μM | | 23 |
| | 210 pS | EIOP | e+d | bumetanide(1)10 μM | | 24 |
| Colon (rat) | | PLB | | DPC(1) | 565 μM cis | 25 |
| | | | | NPPB(1) | 24 μM cis | |
| | | | | IAA94(1) | 17 μM cis | |
| | | | | IAA95(1) | 16 μM cis | |
| | | | | DNDS(1) | 3 μM cis | |
| Colon (rat) | 80 pS | PLB | | TNP-ATP(5) | 0.3 μM cis | 26 |
| | | | | chlorotoxin(7) | | 27 |
| | 80 pS | EIOP | e | 9-AC(1)1 mM | | 28 |
| | 47 pS | EIOP | | NPPB(1)10 μM | | 29 |
| | | | | quinine(7) | | |
| Int407 | OR | WCR | s | SITS(1) | 1.5–6 μM | 30 |
| | | | | NPPB(1) | 25 μM | |
| | | | | DPC(1) | 350 μM | |
| | | | | AA(6) | 8 μM | |
| Shark rectal gland | 40–50 pS | EIOP | | DPC(1)100 μM | | 31 |
| | | | | NPPB(1)10 μM | | |
| Parietal cell (rat) | 25 pS,OR | EIOP | histamine | DPC(1)12 μM | | 32 |
| CAPAN-1 | 70 pS,OR | EIOP | e | DIDS(1)20–200 μM | | 33 |
| (pancreatic duct) | 350 pS | EIOP | e | DIDS(1)20 μM | | |
| PANC-1 (pancreatic) | 70 pS | EIOP | e+d | HEPES(7)10–67 mM | | 34 |
| *Other epithelia* | | | | | | |
| Choroid plexus (rat) | 28 pS,OR | EIOP | | SITS(1)1 mM | | 35 |
| | 375 pS | EIOP | e | furosemide(1)1 mM | | |
| | | | | SITS(1)1 mM | | |
| Epididymis (rat) | OR | WCR | s | NPPB(1) | 0.12 mM | 36 |
| | | | | DPC(1) | 0.5 mM | |
| | | | | DIDS(1) | 5 mM | |
| Placental microvillus (human) | 313 pS | EIOP | e | DIDS(1)0.1–1 mM | | 37 |
| | | | | DPC(1)1 mM | | |
| *Haematopoeitic cells* | | | | | | |
| Neutrophils (human) | OR | WCR | s | MK447A(1) | 37 μM | 38 |
| | | | | SITS(1)0.5 mM | | |
| Lymphocytes (murine) | OR | WCR | h | DIDS(1) | 17 μM | 39 |
| | | | | SITS(1) | 89 μM | |
| Lymphocytes (human) | 60 pS,OR | EIOP | e+d | IAA(1)20–200 μM | | 40 |
| | | | | SITS(1)1 mM | | |
| | 365 pS | EIOP | e | zinc(7)1 mM | | 41 |
| | | | | nickel(7)1 mM | | |
| Monocytes (human) | 28 pS,OR | EIOP | e | DIDS(1)10–100 μM | | 42 |
| | 328 pS | EIOP | e | DIDS(1)0.1–1 mM | | |

**TABLE 11.1** (contd)

| Cell type | Channel | Mode | Activation | Blocker | $IC_{50}$ | Reference |
|---|---|---|---|---|---|---|
| *Smooth muscle* | | | | | | |
| Portal vein (rabbit) | ICl(Ca) | WCR | $Ca^{2+}$ | SITS(1)<br>DIDS(1)<br>9-AC(1)0.1 mM | 640 μM<br>210 μM | 43 |
| Ventricular myocytes (rabbit) | ICl(Ca) | WCR | $Ca^{2+}$ | SITS(1)2 mM<br>DIDS(1)0.1 mM | | 44 |
| | IATP | WCR | ATP | DIDS(1)0.2 mM | | 45 |
| Ventricular myocytes (canine) | OR | WCR | s | 9-AC(1)1 mM<br>NPPB(1)40 μM | | 46 |
| Ventricular myocytes (rat) | 400 pS | EIOP | e | SITS(1)10 μM | | 47 |
| Ventricular myocytes (chick) | OR | WCR | s | DPC(1)0.2 mM | | 48 |
| *Skeletal muscle* | | | | | | |
| Frog | 260 pS | EIOP | e | tannic acid(7) 50 μM<br>gallic acid(7) 10 mM<br>zinc(7)0.1,1 mM | | 49 |
| Rat | | DEVC | | clofibric acid(7)1–100 μM | | 50 |
| (expressed in *Xenopus* oocytes) | IR(ClC-1) | DEVC | | 9-AC(1)0.1 mM | | 51 |
| Rabbit | 280 pS | EIOP | | IAA(1)10 μM | | 52 |
| *Neuronal cells* | | | | | | |
| NIE115 (neuroblastoma) | 343 pS | EIOP | e+d | 9-AC(1)1 mM<br>deltamethrin(7)2 μM | | 53 |
| Hippocampus (rat) | 30 pS | EIOP | e+d | 9-AC(1)1 mM<br>zinc(7)1 mM | | 54 |
| *Miscellaneous cells* | | | | | | |
| Osteoblast (rat) | OR | WCR | cAMP,PTH | DIDS(1)0.1 mM<br>DPC(1)0.1 mM | | 55 |
| Osteoclast (rat) | 19 pS OR | CAP/WCR | s | SITS(1)0.5,1 mM<br>DIDS(1)0.5 mM<br>DNDS(1)0.5 mM<br>niflumic acid(1)0.2 mM | | 56 |
| Keratinocyte (human) | OR | WCR | s | DDFSK(2)100 μM | | 57 |
| Fibroblast (human) | 300 pS | EIOP | e | DPC(1)0.7 mM | | 58 |
| *Xenopus* oocyte | IR(PLM) | DEVC | | 9-AC(1)200 μM<br>barium(7) | | 59 |
| | IR | DEVC | h | SITS(1)<br>DIDS(1)<br>niflumic acid(1)<br>9-AC(1)<br>bumetanide(1) | | 60 |

**TABLE 11.1** (contd)

| Cell type | Channel | Mode | Activation | Blocker | $IC_{50}$ | Reference |
|---|---|---|---|---|---|---|
| *Miscellaneous cells* (contd) | | | | | | |
| (oocytes) | (CFTR) | CAP/EIOP<br>DEVC | | DPC(1)200 μM<br>flufenamic acid(1)200 μM | | 61 |
| (oocytes) | OR | DEVC | s | NPPB(1)<br>DIDS(1)<br>SITS(1)<br>lanthanum(7)<br>gadolinium(7) | 10 μM<br>30 μM<br>60 μM<br>60 μM<br>250 μM | 62 |
| | ICl(Ca) | DEVC | $Ca^{2+}$ | NPPB(1) | 50 μM | 63 |
| Mitochondria | 1.3nS | PLB | | cyclosporin(2)20 nM | | 64 |
| 3T3 fibroblast | (CFTR) | WCR | | gliblenclamide(4)<br>tolbutamide(3)<br>diazoxide(7)<br>BRL38227(7)<br>minoxidil(7) | 22 μM<br>150 μM<br>260 μM<br>48 μM<br>38 μM | 65 |
| 3T3 fibroblast | OR(PgP) | WCR | s | DDFSK(2)<br>verapamil(2)<br>quinidine(2) | 40 μM<br>50 μM<br>60 μM | 66 |
| | | | | tamoxifen(3) | 0.4 μM | 67 |
| Lens fibre cells (bovine) | 17 pS | EIOP | e | NPPB(1)100 μM<br>DIDS(1)100 μM<br>verapamil(2)100 μM<br>quinidine(2)100 μM<br>DDFSK(2)100 μM<br>tamoxifen(3)100 μM | | 68 |
| Endothelium (human umbilical) | OR | WCR | s | DIDS(1)<br>DCDPC(1)20 μM<br>DDFSK(2)60 μM<br>verapamil(2)100 μM | 120 μM | 69 |
| HeLa (human) | OR | WCR | s | NPPB(1)100 μM<br>DIDS(1)100 μM<br>verapamil(2)10,100 μM<br>DDFSK(2)100 μM | | 70 |
| Chromaffin cells (bovine) | OR | WCR | GTP | DIDS(1)10 μM<br>chlorpromazine(7)5 μM<br>tolbutamide(3)0.5–5 mM | | 71 |

**Headings and abbreviations**: chloride channels are listed according to the tissue type and species in which they are found. CCD: cortical collecting duct; TAL: thick ascending limb.

**Channel**: channels are identified by channel conductance and/or rectification. When the channel rectifies the largest conductance is given. OR: outwardly rectifying; IR: inwardly rectifying; Specific channel proteins are given when identifiable and are shown in brackets when studied in an expression system e.g. (CFTR) in oocytes. PLM: phospholemman; PgP: P-glycoprotein.

**Mode**: technique used. WCR: whole-cell recording; DEVC: dual electrode voltage clamp; EIOP: excised inside-out patch recording; PLB: planar lipid bilayers; CAP: cell-attached patch recording; EOOP: excised outside-out patch recording.

**Activation**: e: excision of patch membrane; d: depolarization; h: hyperpolarization; s: anisosmotic swelling; PTH: parathyroid hormone; cAMP: agents increasing cAMP levels; GTP: G-protein mediated.

**Blocker**: the class in which the blocker has been classified is given in brackets along with the concentrations used and when available the $IC_{50}$. Abbreviations are given in text.

## References


1. Schoppa, N., *et al.* (1989). *J. Membr. Biol*, **108**, 73.
2. Welsh, M.J. (1986). *Pflügers Arch.*, **407**, S116.
3. Anderson, M.P., *et al.* (1990). Proc. Natl Acad. Sci. USA, **87**, 7334.
4. Hwang, T.C., *et al.* (1990). *Proc. Natl Acad. Sci. USA*, **87**, 5706.
5. McCann, J.D., *et al.* (1989). *J. Gen. Physiol*, **94**, 1015.
6. Li, M., *et al.* (1990). *Am. J. Physiol.*, **259**, C295.
7. Schwiebert, E.M., *et al.* (1992). *Proc. Natl Acad. Sci. USA*, **89**, 10623.
8. Kemp, P.J., *et al.* (1993). *Am. J. Physiol*, **265**, L323.
9. Ran, S., *et al.* (1992). *J. Biol. Chem.*, **267**, 20630.
10. Alton, E.W.F.W., *et al.* (1991). *J Physiol. (Lond.)*, **443**, 137.
11. Banderali, U., *et al.* (1992). *J. Membr. Biol.*, **126**, 219.
12. Gekle, M., *et al.* (1993). *Pflügers Arch*, **425**, 401.
13. Nelson, D.J., *et al.* (1984). *J. Membr. Biol.*, **80**, 81.
14. Velasco, G., *et al.* (1989). *Pflügers Arch.*, **414**, 304.
15. Light, D.B. *et al.* (1990). *Am. J. Physiol.*, **258**, F273.
16. Dietl, P. *et al.* (1992). *Am. J. Physiol.* **263**, F243.
17. Lu. L., *et al.* (1993). *Pflügers Arch.*, **135**, 181.
18. Shcmid, A., *et al.* (1989). *J. Membr. Biol.* **111**, 265.
19. Hayslett, J.P., *et al.* (1987). *Pflügers Arch.*, **410**, 487.
20. Reinsprecht, M., *et al.* (1992). *Biochem. Biophys. Res. Commun.*, **188**, 957.
21. Fischer, H., *et al.* (1992). *Pflügers Arch.*, **422**, 159.
22. Champigny, G., *et al.* (1990). *Biochem. Biophys. Res. Commun.*, **171**, 102.
23. Valverde, M.A., *et al.* (1993). *Pflügers Arch.*, **425**, 552.
24. Vaca, L., *et al.* (1992). *J. Membr. Biol.* **130**, 241.
25. Singh, A.K., *et al.* (1991). *Am. J. Physiol.*, **260**, C51.
26. Venglarik, C.J., *et al.* (1993). *J. Gen. Physiol.*, **101**, 545.
27. DeBin, J.A., *et al.* (1993). *Am. J. Physiol.*, **264**, C361.
28. Diener, M., *et al.* (1989). *J. Membr. Biol.*, **108**, 21.
29. Gogelein, H., *et al.* (1990). *Biochem. Biophys. Acta*, **1027**, 191–30.
30. Kubo, M., *et al.* (1992). *J. Physiol.(Lond.)*, **456**, 351.
31. Greger, R., *et al.* (1987). *Pflügers Arch.*, **409**, 114.
32. Sacchomani, G, *et al.* (1991). *Am. J. Physiol.*, **260**, C1000.
33. Becq, F., *et al.* (1992). *Pflügers Arch.*, **420**, 46.
34. Hanrahan, J.W., *et al.* (1990). *J. Membr. Biol.*, **116**, 65.
35. Christensen, O., *et al.* (1989). *Pflügers Arch.*, **415**, **36**.
36. Gunter-Smith, P.J. (1988). *Am. J. Physiol.*, **255**, C808.
37. Brown, P.D., *et al.* (1993). *Placenta*, **14**, 103.
38. Stoddard, J.S., *et al.* (1993) *Am. J. Physiol*, **265**, C156.
39. Lewis, R.S., *et al.* (1993). *J. Gen. Physiol*, **101**, 801.
40. Garber, S.S. (1992). *J. Membr. Biol.*, **127**, 49.
41. Schlichter, L.C., *et al.* (1990). *Pflügers Arch.*, **416**, 413.
42. Kanno, T., *et al.* (1990). *J. Membr. Biol.*, **116**, 149.
43. Hogg, R.C., *et al.* (1994). *Br. J. Pharmacol.*, **111**, 1333.
44. Zygmunt, A.C., *et al.* (1991). *Circ. Res.*, **68**, 424.
45. Kaneda, M., *et al.* (1994). *Br. J. Pharmacol*, **111**, 1355.
46. Tseng, G-N. (1992). *Am. J. Physiol*, **262**, C105.
47. Coulombe, A., *et al.* (1992). *Pflügers Arch.*, **422**, 143.
48. Zhang, J., *et al.* (1993). *J. Physiol. (Lond.)*, **474**, 80.
49. Woll, K.H., *et al.* (1987). *Pflügers Arch* **410**, 632.
50. Conte-Camerino, D., *et al.* (1988). *Pflügers Arch*, **413**, 105.
51. Steinmeyer, K., *et al.* (1991). *Nature*, **354**, 301.
52. Weber-Schurholz, S., *et al.* (1993). *J. Biol. Chem.*, **268**, 547.
53. Forshaw, P.J., *et al.* (1993). *Neuropharmacology*, **32**, 105.
54. Francolini, F., *et al.* (1987). *J. Gen Physiol*, **90**, 453.
55. Chesnoy-Marchais, D., *et al.* (1989). *Pflügers Arch.*, **415**, 104.
56. Kelly, M.E. *et al.* (1994) *J. Physiol.(Lond.)* **475**, 377.
57. Rugolo, M., *et al.* (1992). *Biochem. Biophys. Acta*, **1112**, 39.
58. Nobile, M., *et al.* (1988). *Biochem. Biophys. Res. Commun.*, **154**, 719.
59. Moorman, R. J., *et al.* (1992). *J. Biol. Chem.*, **267**, 14551.
60. Kowdley, G.C., *et al.* (1994). *J. Gen. Physiol.*, **103**, 217.
61. McCarty, N.A., *et al.* (1993). *J. Gen. Physiol*, **102**, 1.
62. Ackerman, M.J., *et al.* (1994). *J. Gen. Physiol.*, **103**, 153.
63. Wu, G., *et al.* (1992). *Pflügers Arch.*, **420**, 227.
64. Szabo, I., *et al.* (1992). *J. Biol. Chem.* **267**, 2940.
65. Sheppard, D.N., *et al.* (1992). *J. Gen. Physiol.*, **100**, 573.
66. Valverde, M.A., *et al.* (1992) *Nature*, **355**, 830.
67. Zhang, J.J. *et al.* (1994). *J. Clin Invest.*, **94**, 1690.
68. Zhang, J.J., *et al.* (1994) *Am. J. Physiol.*, **267**, C1095.
69. Nillius, B., *et al.* (1994). *J. Gen. Physiol.*, **103** 787.
70. Diaz, M., *et al.* (1993). *Pflügers Arch.*, **422**, 347.
71. Doroshenko, P., *et al.* (1991) *J. Physiol. (Lond.)*, **436**, 711.


Classifications of the various $Cl^-$ channel blockers have been attempted before (Greger 1990; Cabantchik and Greger 1992). We will try to include most agents which had been described to interact with $Cl^-$ channels tested at the level of either macroscopic or single-channel currents. We have defined seven groups of compounds: (1) anion transporter inhibitors, including stilbene disulphonates, aromatic carboxylic acids, and indanyl oxyacetic acids; (2) multidrug resistance reversers; (3) non-steroidal antioestrogens; (4) sulphonylureas; (5) ATP and derivatives; (6) fatty acids; and (7) a miscellaneous group. Although many of the blockers, even within the same family, do not share structural characteristics, they generally possess a net anionic charge.

Descriptions of the blocking characteristics in single-channel recordings is of value in analysing pharmacological interactions between ligands and receptors, and three types of blockade can be distinguished. The type of blockade seen depends on the residence time of a blocker molecule in/on the channel (Hille 1992) and at single-channel resolution three types can be distinguished:

1. Slow. The blocker binds to the channel for a long time (many milliseconds or longer). The channel is seen to close completely for long time periods. No flickering or changes in current amplitude are seen.
2. Intermediate. The blocker molecule binds to and dissociates from the channel in the space of a few milliseconds; thus, the channel is seen to open and close rapidly and for this reason is often referred to as a flickery-type blockade.
3. Fast. The blocker molecule binds to and dissociates from the channel very rapidly ($< 1$ ms). Thus, the flickering is too fast to be detected and the current is seen to have a lower amplitude and a noisier open level.

It is obvious that the type of blockade seen will depend on the resolution of the recording. For example, a recording acquired at 500 Hz may show a blockade to be of the 'fast' type, whereas when acquired at 5 KHz, the blockade may appear to be of the 'intermediate' type.

## Anion transporter inhibitors

The common feature of these compounds is their inhibitory effect on different anion transporters (Cabantchik and Greger 1992). The first group of chloride channel blockers to be studied extensively were the aromatic carboxylic acids. Bryant and Morales-Aguilera (1971) first used these compounds to inhibit the chloride conductance of muscle cells. The most potent inhibitor of this group was anthracene-9-carboxylic acid (9–AC). 9-AC and a derivative called diphenylamine-2-carboxylate (DPC) were subsequently found to block epithelial and non-epithelial chloride channels. However, DPC at millimolar concentrations also has other non-specific effects such as inhibition of cyclooxygenase activity (Greger 1990). Di Stefano *et al.* (1985) reported that DPC blocked the chloride conductance of rabbit cortical thick ascending loop of Henle with an $IC_{50}$ of 30 $\mu$M. Blockade only occurred when DPC was added to the basolateral membrane. In the same paper the authors also show that DPC blocked the chloride conductance of shark rectal gland when added to the apical membrane. DPC has also been shown to block the outwardly rectifying chloride channel (ORCC) of rat colonic enterocytes (ref. 25 in Table 11.1) and the volume-activated chloride currents of small intestinal (Intestine 407) cells (ref. 30 in Table 11.1). The high concentrations of 9-AC and DPC required to block chloride channels and lack of specificity at these concentrations led Wangemann *et al.* (1986) to carry out a study of over 200 derivatives of these compounds in order to find a more potent and specific blocker. 5-Nitro-2-(3-phenylpropylamino) benzoic acid (NPPB) was found to be the most potent. It was evident that a common feature of the structures was that lipophilic derivatives were more potent blockers of chloride conductance, while hydrophilic compounds tended to inhibit the $Na^{+}K^{+}2Cl^{-}$ cotransporter. NPPB blockade has been studied on many epithelial cells (Fig. 11.2) but its potency varies greatly between tissue preparations (Table 11.1). The potency is generally considered greater when applied to the extracellular face but exceptions exist (refs. 29 and 68 in Table 11.1). In addition, NPPB has been shown to block $K^{+}$ channels (Illek *et al.* 1992).

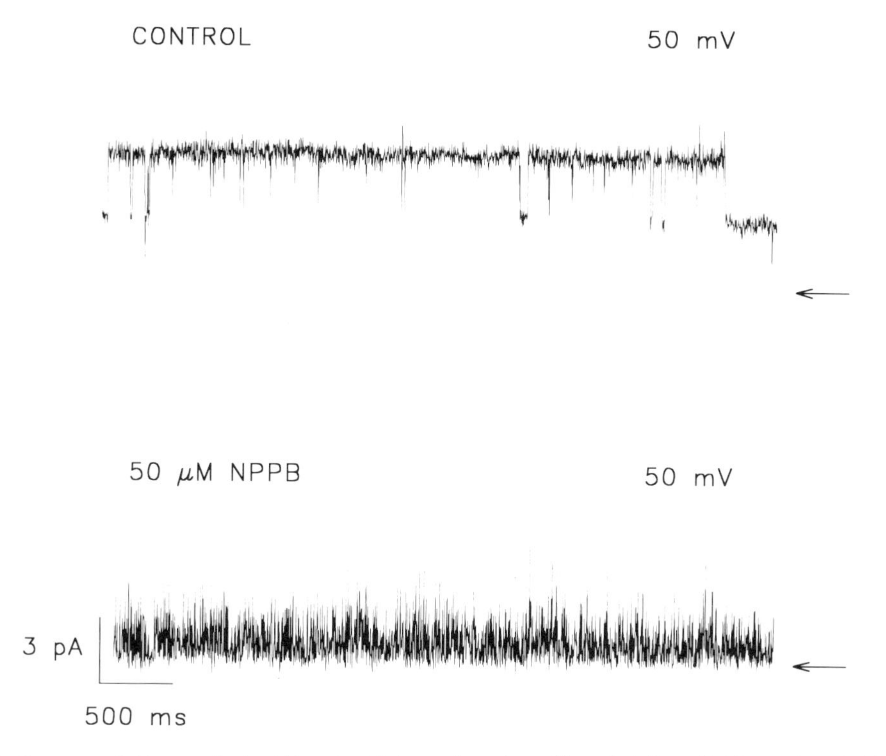


**Fig. 11.2** Traces recorded in an excised outside-out patch from guinea pig small intestinal villus cells. The patch was held at +50 mV in symmetrical NaCl solutions. In the upper trace, the bath solution was physiological saline and in the lower trace 50 $\mu$M NPPB was added to the bath solution. Arrows indicate the closed level. NPPB induced a flickery type of blockade (see text). (From Monaghan 1994.)

Stilbene disulphonates have been used as inhibitors of anion exchange transporters (for review, see Cabantchik and Greger 1992), but also act as chloride channel blockers. One such compound, 4,4′-dinitrostilbene-2,2′-disulphonic acid (DNDS) blocks an ORCC from rat colonic enterocytes with a greater potency when applied to the extracellular aspect of the membrane and caused a flickery-type block with an $IC_{50}$ of around 3 $\mu$M (ref. 25 in Table 11.1). Tilmann *et al.* (1991) tested the effects of stilbene derivatives on ORCCs of HT-29 cells and respiratory epithelial cells (REC) and found that DNDS was the most potent. They also saw a reversible flickery-type blockade with an $IC_{50}$ of 9 $\mu$M. Other members of this family of blockers such as DIDS and SITS (4-acetamido-4′-isothiocyanostilbene-2,2′-disulphonic acid) have also been shown to inhibit both anion exchangers (Cabantchik and Greger 1992) and $Cl^{-}$ channels (see Table 11.1) at concentrations generally above 100 $\mu$M.

Indanyl-oxyacetic acid (IAA) derivatives, structurally related to the diuretic ethacrynic acid, have been shown to be potent chloride channel blockers. One such derivative (IAA-94) has been used to purify a chloride channel from proximal renal tubule cells (Landry *et al.* 1987, 1989). IAA-94/95 and amidine block ORCCs in epithelial cell lines with an $IC_{50}$ of around 10 $\mu$M for both (Tilmann *et al.* 1991). The molecular structures of these compounds are unrelated to the other blockers in this group. Amidine actually carries a net positive charge which conflicts with the findings of Wangemann *et al.* (1986) which indicated that potency correlated with net negative charge. This suggests that the IAA derivatives have completely different interaction sites on the chloride channel. IAA derivatives have also been shown to block the ORCC from rat colonic enterocytes with a higher potency than NPPB (ref. 25 in Table 11.1).

Inhibition of muscle (Bretag 1987) and choroid plexus $Cl^-$ channels (ref. 35 in Table 11.1) by furosemide has been reported, although no such inhibition was confirmed at the single-channel level in intestinal epithelial cells (Greger 1990).

It is worth noting that almost all of these compounds have little effect in blocking the CFTR $Cl^-$ channel (Tabcharani *et al.* 1990; Anderson *et al.* 1992) apart from DPC and its derivative, flufenamic acid, which induce a voltage-dependent blockade when used at micromolar concentrations (ref. 61 in Table 11.1).

## Multidrug resistance reversers

It has been known for a long time that certain drugs, as diverse as $Ca^{2+}$ channel blockers, cationic compounds, and steroids are able to reverse a multidrug resistance (MDR) phenotype through their capacity to inhibit the transport of chemotherapeutic drugs by P-glycoprotein (P-gp) (for a review, see Ford and Hait 1990). Champigny *et al.* (ref. 22 in Table 11.1) first described the use of verapamil and other $Ca^{2+}$ channel blockers in blocking the ORCC of HT-29 intestinal cells. Verapamil is a phenalkylamine compound which is lipid-soluble and carries a net positive charge. It had been reported that verapamil partially reverses the MDR phenotype by direct interaction with P-gp (Ford and Hait 1990). Indeed, it was subsequently found that volume-activated $Cl^-$ currents associated with the expression of P-gp were also blocked by verapamil (ref. 66 in Table 11.1). Forskolin and 1,9-dideoxyforskolin (DDFSK, a derivative devoid of adenylate cyclase stimulatory effect), have been shown to bind P-gp and also to block the $Cl^-$ currents (ref. 66 in Table 11.1) along with other MDR reversers such as quinine, quinidine, and nifedipine, although the $IC_{50}$ values were always $>50$ $\mu$M (ref. 66 in Table 11.1). The blockade induced by these compounds is generally fast ($<1$ min) and reversible, suggesting an interaction with the channel molecule from the extracellular face of the membrane. Indeed, some were ineffective when added to the intracellular face of the membrane via the pipette solution. A more detailed description of the effect of these compounds, and P-gp substrates (colchicine, daunomycin, and vincristine) upon the $Cl^-$ currents can be found in the study by Mintenig *et al.* (1993) in which substrates of P-gp but not MDR reversers interfered with channel activation when added to the cytosolic side of the membrane, but only in the presence of hydrolysable analogues of ATP, suggesting that the action of the substrates occurs by inhibiting channel activation rather than a direct blockade of the channel. A more recent publication (Bear 1994) reported that several drugs transported by P-gp (colchicine, daunomycin) block a 40 pS outwardly rectifying $Cl^-$ channel in P-gp-expressing cells, but in this case the blockade was characterized by a reduction in the open time of the channel and took place in the absence of ATP. A simple explanation for this apparent discrepancy could lie in the fact that the channels activated by volume (ref. 66 in Table 11.1) and depolarization (Bear 1994) represent the activity of different proteins. The ability of MDR reversers to block volume-activated chloride currents is not dependent on the expression of P-gp, as similar blockade of $Cl^-$ channels has been reported in HeLa cells and lens fibre cells where P-gp has not been detected (refs. 67, 68, and 70 in Table 11.1).

### Non-steroidal antioestrogens

Tamoxifen and other non-steroidal antioestrogen derivatives such as toremifene and 4-hydroxytamoxifen, reverse multidrug resistance in P-glycoprotein-expressing cell lines (Kirk *et al.* 1993). When applied to the extracellular side of the membrane, tamoxifen at 10 $\mu$M can block volume-activated outwardly rectifying chloride currents in T84 cells (ref. 23 in Table 11.1), but has no effect on the linear cAMP-activated chloride conductance or the $Ca^{2+}$-activated chloride conductance in the same cells. In NIH3T3 fibroblasts, tamoxifen has an $IC_{50}$ of 0.3 $\mu$M for volume-activated chloride currents (ref. 67 in Table 11.1). This makes tamoxifen the most potent blocker of these currents identified to date, although it is ineffective against volume-activated currents in *Xenopus* oocytes (D. Clapham, personal communication). Tamoxifen appears to be a potentially useful agent in its ability to discriminate between chloride currents activated by different stimuli (cAMP, $Ca^{2+}$, cell swelling; Fig. 11.3), although data on its mode of action in channel blocking are needed. Tamoxifen, which is widely used in the treatment of breast cancer, has also been implicated in the production of cataracts, as a side effect of its chronic administration (Zhang *et al.* 1994). It has been suggested that the mechanism by which tamoxifen may induce cataract formation is through the blockade of a lens $Cl^-$ channel, involved in maintaining lens volume and hydration and, hence, transparency (refs 67 and 68 in Table 11.1).

### Sulphonylureas and $K^+$ channel openers

Sulphonylureas are a group of compounds used in the treatment of diabetes mellitus due to their effect on insulin release through inhibition of the ATP-sensitive $K^+$ channel (Sturgess *et al.* 1985). It has been reported that the sulphonylureas tolbutamide and glibenclamide are effective blockers of the CFTR $Cl^-$ channel (ref. 65 in Table 11.1). Moreover, in the same study it was found that certain $K^+$ channel openers like lemakalim (BRL 38227), minoxidil, and diazoxide when added extracellularly also blocked the $Cl^-$ channel, albeit slowly. The effect of glibenclamide on other $Cl^-$ channels was not studied and, subsequently, reports showing a lack of effect of the blocker upon cAMP-activated $Cl^-$ channels in gallbladder epithelium have appeared (Copello *et al.* 1993).

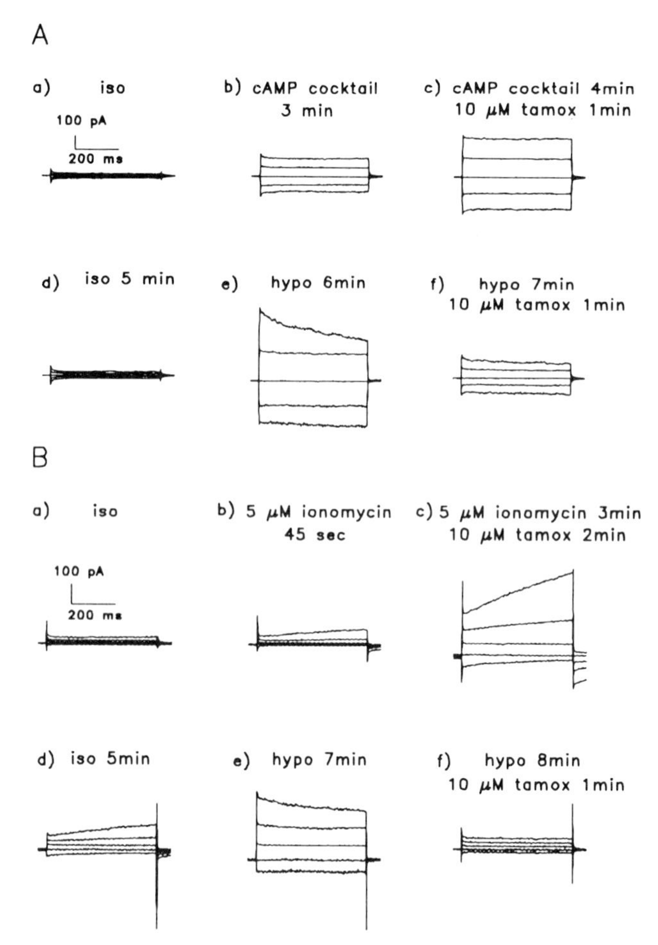


**Fig. 11.3** Differential effects of tamoxifen on epithelial $Cl^-$ currents recorded in two $T_{84}$ cells (panels A and B). cAMP – and $Ca^{2+}$-activated $Cl^-$ currents were not blocked by 10 $\mu$M tamoxifen (**Ac** and **Bc**). Hypotonicity-activated $Cl^-$ currents were rapidly blocked by 10 $\mu$M tamoxifen (**Af** and **Bf**). (Reproduced with permission from ref 23 in Table 11.1.)

## ATP nucleotides

Several reports have indicated that ATP, when added extracellularly, can inhibit the activity of outwardly rectifying $Cl^-$ channels. Manning and Williams (1989) found that ATP when applied to the extracellular (and not the intracellular) face of an outward rectifying $Cl^-$ channel reconstituted from human platelets in planar lipid bilayers caused a flickery-type blockade. A similar type of $Cl^-$ channel, reconstituted an ORCC from sheep tracheal epithelium into planar lipid bilayers, was found to be blocked by the external addition of ATP, also with a flickery-type blockade (ref. 10 in Table 11.1). The blockade was dose-dependent, with an $IC_{50}$ between 10 and 100 $\mu$M, and only occurred when the 'extracellular surface' was held at a negative potential. Ackerman *et al.* (ref. 62 in Table 11.1) have shown that *Xenopus* oocytes possess volume-activated outwardly rectifying $Cl^-$ currents which are reduced by the extracellular application of nucleotides. Application of the nucleotides cAMP,

ATP, and GTP at concentrations between 1 and 5 mM all inhibited these currents to a similar extent.

Stutts *et al.* (1992) have reported that extracellular ATP in human airways epithelial cells stimulates $Cl^-$ secretion with a half-maximal stimulation obtained at 40 $\mu$M ATP compatible with purinergic receptor activation. Further investigations on this effect on the outwardly rectifying $Cl^-$ channel in these cells found that ATP had no effect on the channel when added to the intracellular surface of membrane patches, but when added to the extracellular surface of patches the $P_o$ of the channel increased by 50% at 40 mV and by 40% at −40 mV. However, the increase in channel activity was also accompanied by increased flickery channel openings. This type of blockade by ATP has been described in other chloride channels (Manning and Williams 1989; Monaghan 1994) and could indicate that certain rectifying chloride channels possess extracellular binding sites for nucleotides. The ATP derivative 2′0-(2,4,6-trinitrocyclohexadienylidene) adenosine 5′-trisphosphate (abbreviated to trinitrophenyl-ATP or TNP-ATP), was first synthesized for use as a reporter-labelled substrate of heavy meromyosin ATPase and has been used to characterize nucleotide binding sites of many other ATPases. Recently, Venglarik *et al.* (ref. 26 in Table 11.1) studied the effect of TNP-ATP on rat colonic crypt enterocyte $Cl^-$ channel and found it to be a high-affinity blocker of the channel when reconstituted into planar lipid bilayers. TNP-ATP blocked with an $IC_{50}$ of 270 nM from the extracellular face of the channel and with an $IC_{50}$ of 20 $\mu$M from the intracellular face. ATP can compete with TNP-ATP for blockade of the channel, although concentrations greater than 1 mM are required. From these results the authors postulate that the channel has two nucleotide binding sites: one at the extracellular surface, and one at the intracellular face.

Blocking at nanomolar concentrations indicates that TNP-ATP is a useful compound and hopefully, specific. Indeed, TNP-ATP does not block volume-activated $Cl^-$ currents in HeLa cells at 1 $\mu$M (M.A. Valverde, unpublished data). Work also carried out in our laboratory found that in small intestinal villus cells outwardly rectifying $Cl^-$ channels recorded both in EIOP and WCR configurations were blocked by TNP-ATP ($K_D$ 18 nM). However, TNT-ATP was found to inhibit the $Cl^-$ currents only when cells were held a positive potentials (Fig. 11.4). At negative potentials, very little or no inhibition was seen and, moreover, even brief pulses to negative potentials caused a rapid reversal of blockade (Monaghan 1994). This voltage dependency is not fully understood, although some explanations have been attempted. TNP-ATP blockade of the ORCC has similar characteristics to that of charybdotoxin toxin (obtained from *Leirus quinquestriatus* venom) blockade of the maxi-$K^+$ channel (MacKinnon and Miller 1988). It is possible that the highly negatively charged TNP-ATP molecule binds to fixed positive charges on the extracellular surface of the ORCC very near to the pore. Hence, this relatively large molecule would effectively 'plug' the pore, thus preventing $Cl^-$ ions moving from the extracellular side to the intracellular side. In this case, membrane hyperpolarization would result in $Cl^-$ ions from the intracellular solution 'pushing' TNP-ATP off its binding site.

It is clear that a high potency as determined by a single $K_i$ or $IC_{50}$ alone is not necessarily sufficient in determining the usefulness of particular blocker. Voltage dependence of the blockade and type of single channel blocking may dramatically alter its usefulness for a particular tissue.

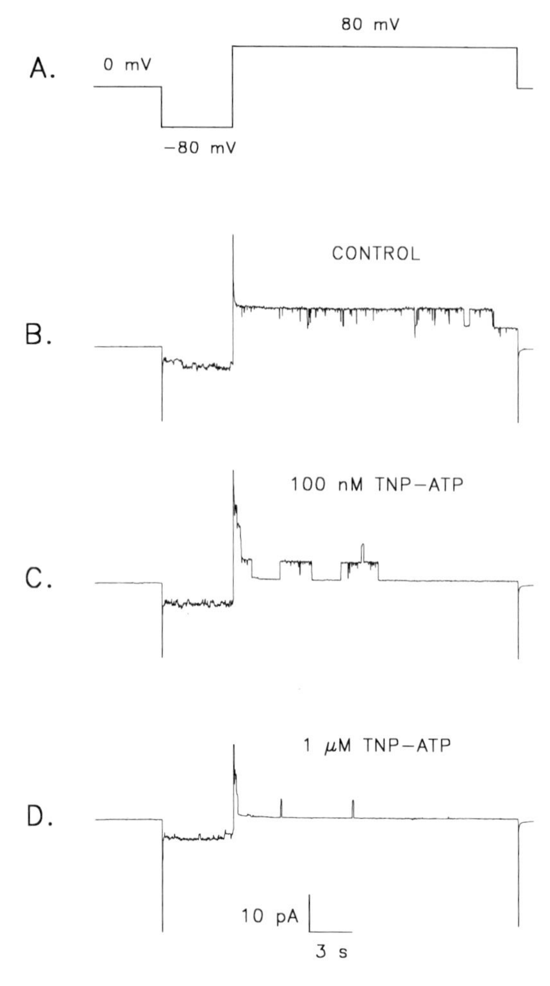


**Fig. 11.4** Voltage-dependent blockade of $Cl^-$ channels by TNP-ATP. $Cl^-$ channels recorded in the guinea pig enterocytes (as in Fig. 11.2). Bath and pipette solutions containing *N*-methyl-D-glucamine chloride. A. Pulse protocol used in this experiment. B–D. Current traces recorded under the indicated conditions. All current traces were digitized at 67 Hz with a low-pass filter at 30 Hz. Under control conditions (**B**), channel transitions were seen at both −80 and +80 mV. When TNP-ATP was added to the bath, the currents at 80 mV were markedly blocked, unlike the currents at −80 mV, which were not affected. (Data from Monaghan 1994.)

## Fatty acids

The direct effect of fatty acids upon the activity of ion channels has been discussed previously (Ordway *et al.* 1991). These effects include activation of $K^+$ channels in cardiac and smooth muscle cells, and inhibition of $Na^+$- and $Ca^{2+}$-currents. $Cl^-$ channels also are modulated by a direct effect of fatty acids, independent of the enzymatic pathways that convert arachidonic acid into its oxygenated metabolites. In the first study of fatty acids on $Cl^-$ channel activity, Hwang *et al.* (ref. 4 in Table 11.1) showed a blocking effect of fatty acids on $Cl^-$ channels of tracheal and colonic origin. This inhibition was characterized by a marked reduction in the mean open time of the channel without altering the closed time. Such an effect was observed when arachidonic acid (5–25 $\mu$M) and other unsaturated acids (linoleic and oleic acids) were added from the cytoplasmic side of the membrane. These effects were observed in the presence of inhibitors of the metabolic pathways for arachidonic acid, suggesting a direct blockade of the channel by arachidonic acid rather than a product of its metabolism. Similar results were obtained in cells from airway epithelia (ref. 3 in Table 11.1) and intestinal HT29 epithelial cells (Kunzelman *et al.* 1991) with a reported $IC_{50}$ for arachidonic acid of 5 $\mu$M in the later study. Kubo and Okada (ref. 30 in Table 11.1), using the WCR mode, recorded volume-activated $Cl^-$ currents in a cell line of human intestinal origin, which were blocked by arachidonic acid, albeit slowly, when added to the external side of the membrane. The fact that the blockade by fatty acids has a slow onset as well as a delayed, and incomplete reversibility, has been interpreted as an effect of the compounds incorporating into the membrane bilayers and altering the gating properties of the channels (Cabantchik and Greger 1992).

## Miscellaneous $Cl^-$ channel blockers

Blockade of $Cl^-$ channels by compounds with no apparent structural relationship will be covered in this section. Outwardly rectifying $Cl^-$ channels present in epithelial and mast cells are blocked by anti-allergic drugs (ref. 20 in Table 11.1). These compounds decrease the $P_o$ of the channel when added to the cytosolic side of the membrane, with an $IC_{50}$ of around 15 $\mu$M for cromolkyn and 1 $\mu$m for RU 31156. Similar channels recorded in a cell line from epithelial origin were blocked with some pH buffers (ref. 34 in Table 11.1). HEPES, a representative pH buffer, inhibited the channel activity by reducing the single-channel conductance at millimolar concentrations, an effect observed when the compound was added to either side of the membrane.

Pyrethroid insecticides (deltamethrin) decrease the $P_o$ of a large-conductance $Cl^-$ channel in a neuroblastoma cell line (ref. 53 in Table 11.1) using 2 $\mu$M deltamethrin in the bath solution, although full inhibition occurred after a delay of about 20 minutes.

Two different naturally occurring toxins have been associated with $Cl^-$ channel inhibition. Clorotoxin, a 35-amino acid peptide, purified from the scorpion, *Leiurus quinquestriatus*, blocks a colonic $Cl^-$ channel reconstituted in artificial lipid bilayers (ref. 27 in Table 11.1), with a $K_D$ of 1.15 $\mu$M, as a result of a decrease in the $P_o$. Ochratoxin A, a fungal mycotoxin associated with porcine nephropathy, also blocks $Cl^-$ currents in a collecting duct epithelium cell line (MDCK) (ref. 12 in Table 11.1). This toxin, structurally similar to the $Cl^-$ channel blocker NPPB, presents a $IC_{50}$ of about 30 nM.

A cytosolic inhibitor has also been postulated for the 40–70 pS $Cl^-$ channel found in intestinal (Kunzelman *et al.* 1991) and kidney epithelia (Krick *et al.* 1991). These experiments, initiated in a search for the postulated 'intracellular regulator' that might explain the lack of $Cl^-$ channel activity in cystic fibrosis cells, led to the identification of a heat-stable inhibitory component of cytosol with a relative molecular mass between 700 Da and 1.5 kDa (Kunzelman *et al.* 1991).

Finally, an inhibitory effect of $I^-$ upon the CFTR $Cl^-$ channel has been well described (Anderson *et al.* 1991; Tabcharani *et al.* 1992; also ref. 23 in Table 11.1). $I^-$, a halide normally more permeable than $Cl^-$ in many $Cl^-$ channels (Frizzel and Halm 1990), was found to induced a voltage-dependent blockade of the CFTR $Cl^-$ channel (Tabcharani *et al.* 1992; also ref. 23 in Table 11.1) in addition to its reduced permeability through the CFTR channel compared with that of $Cl^-$ (Anderson *et al.* 1991; also ref. 23 in Table 11.1).

## Clinical relevance

Although loop diuretics were found to have weak $Cl^-$ channel-blocking ability, most of the blockers described above have not been used therapeutically for this purpose. More recently, the CFTR $Cl^-$ channel blockers like gliburide (gliblenclamide) have been considered for the treatment of congenital diarrhoea and other related hypersecretory alteration, the objective being to block the apically located $Cl^-$ channels in secretory epithelia, thus preventing $Cl^-$ and water exit into the lumen. Their efficacy has yet to be established. The current interest in the treatment of breast cancer with tamoxifen offers an important opportunity to monitor the possible outcomes of therapeutic doses of a chloride channel. An increased incidence of cataracts has been reported in these patients and in studying the effects of tamoxifen upon the $Cl^-$ channels present in lens fibres, we have hypothesized that blockade of such channels might be the cause of cataract development in the patients on long-term treatment with tamoxifen (Fig. 11.5).

## Conclusions

The search for a potent, highly specific blocker of $Cl^-$ channels is ongoing. Such a blocker should help in future studies on channel characterization, location, and structure. The best blocker, described so far, appears to be TNP-ATP. However, the finding that this compound can block outwardly rectifying $Cl^-$ channels only at depolarizing potentials limits the conditions in which it may be used. To date, the pharmacological agents used to identify chloride channels have not been very discriminatory. Consequently, it is hoped that the more recent descriptions of new blockers will help to identify more potent compounds.

## Acknowledgements

We thank F.V. Sepúlveda for his support and advice. Work in the author's labs is funded by the Cystic Fibrosis Trust, The Royal Society, The Physidogical Society and The Wellcome Trust.

(A)

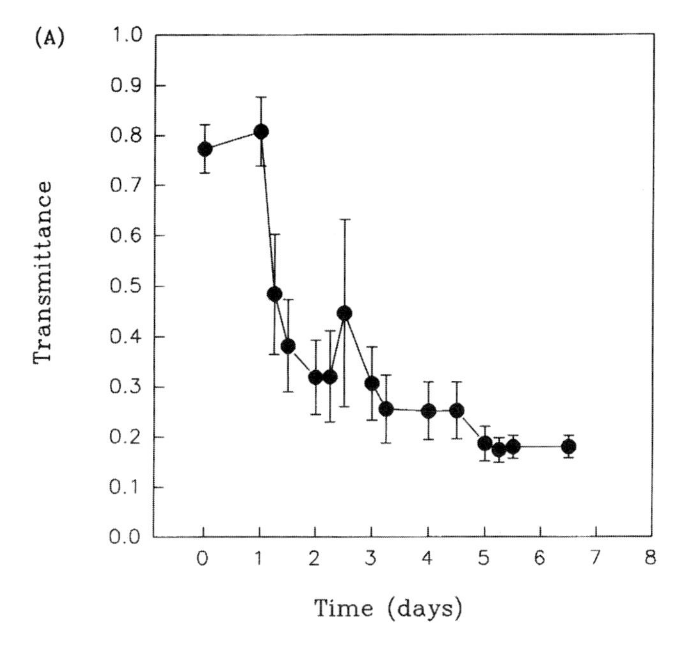


(B)

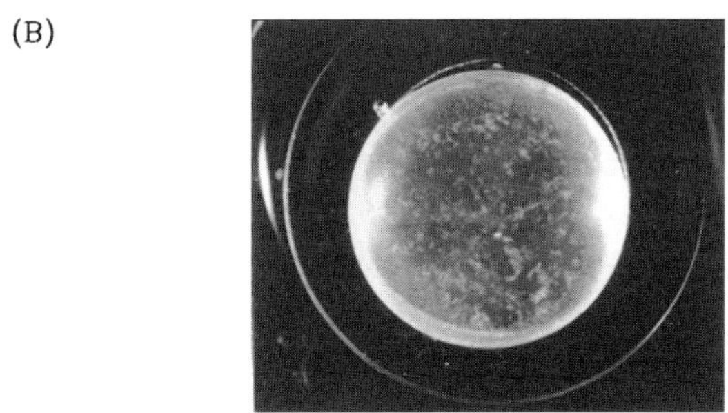

TAMOXIFEN 100$\mu$M

**Fig. 11.5** Effect of tamoxifen on lens opacity. Lens were cultured in a buffer containing tamoxifen (100 $\mu$M) and light transmittance measured as an indication of the lens opacification. A. The presence of the drug caused a severe reduction in the transmittance compared with the untreated lenses, where no changes were observed (results not shown). B. Lens opacity is also illustrated photographically (Lens viewed from above). (Reproduced from *The Journal of Clinical Investigation* (1994), **94**, 1690–1697, by copyright permission of The Society for Clinical Investigation.)

## References

Ackerman, M. J. and Clapham, D. E. (1993). Cardiac chloride channels. *Trends in Cardiac Medicine*, 23–28.

Anderson, M. P., Gregory, R. J., Thompson, S., Souza, D. W., Sucharita, P., Mulligan, R. C., *et al.* (1991). Demonstration that CFTR is a chloride channel by alteration of its anion selectivity. *Science*, **253**, 202–205.

Anderson, M. P., Sheppard, D. N., Berger, H. A., and Welsh, M. J. (1992). Chloride channels in the apical membrane of normal and cystic fibrosis airway and intestinal epithelia. *American Journal of Physiology*, **263**, L1–L14.

Bear, C. E. (1994). Drugs transported by P-glycoprotein inhibit a 40pS outwardly rectifying chloride channel. *Biochemical and Biophysical Research Communications*, **200**, 513–521.

Bretag, A. H. (1987). Muscle chloride channels. *Physiological Reviews*, **67**, 618–700.

Brian R. W. and Andreoli, T. E. (1992). Renal epithelial chloride channels. *Annual Review of Physiology*, **54** 29–50.

Bryant, S. H. and Morales-Aguilera, A. (1971). Chloride conductance in normal and myotonic muscle fibres and the action of monocarboxylic aromatic acids. *Journal of Physiology (London)*, **219**, 367–383.

Cabantchik, Z. I. and Greger, R. (1992). Chemical probes for anion transporters of mammalian cell membranes. *American Journal of Physiology*, **262**, C803–C827.

Copello, J., Heming, T. A., Segal, Y. and Reuss, L. (1993). cAMP-activated apical membrane chloride channels in *Necturus* gallbladder epithelium. *Journal of General Physiology*, **102**, 177–199.

DiStefano, A., Wittner, M., Schlatter, E., Lang, H. J., Englert, H., and Greger, R. (1985). Diphenylamine-2-carboxylate, a blocker of the $Cl^-$-conductive pathway in $Cl^-$-transporting epithelia. *Pflügers Archiv*, **405**, S95–S100.

Eldefrawi, A. T. and Eldefrawi, M. E. (1987). Receptors for $\gamma$-aminobutyric acid and voltage-dependent chloride channels as targets for drugs and toxicants. *The FASEB Journal*, **1**, 262–271.

Ford, J. M. and Hait, W. N. (1990). Pharmacology of drugs that alter multidrug resistance in cancer. *Pharmacological Reviews*, **42**, 155–199.

Frizzell, R.A. and Halm, D. (1990) Chloride channels in epithelial cells. In: *Current topics in membranes and transport*, (ed. S.I. Helman and W. Van Driessche), pp. 247–282. Academic Press, New York.

Gogelein, H. (1988). Chloride channels in epithelia. *Biochimica et Biophysica Acta*, **947**, 521–547.

Greger, R. (1990). Chloride channel blockers. *Methods in Enzymology*, **191**, 793–809.

Halm, D.R. and Frizzell, R.A. (1992) Ion transport across the large intestine. In: *Handbook of physiology – the gastrointestinal system, Vol. IV*, pp. 257–274. American Physiological Society Bethesda, Maryland.

Hille, B. (1992) *Ionic channels of excitable membranes*, 2nd edn. Sinauer Associates Inc., Sunderland, Massachusetts.

Hoffmann, E. K. and Simonsen, L. O. (1989). Membrane mechanisms in volume and pH regulation in vertebrate cells. *Physiological Reviews*, **69**, 315–373.

Illek, B., Fischer, H., Kreusel, K. M., Hegel, U., and Clauss, W. (1992). Volume sensitive basolateral $K^+$ channels in HT-29/B6 cells: block by lidocaine, quinidine, NPPB and $Ba^{2+}$. *American Journal of Physiology*, **263**, C674–C683.

Kirk, J., Houlbrook, S., Stuart, N. S. A., Stratford, I., Harris, A. L., and Carmichael, J. (1993). Selective reversal of vinblastine resistance in multidrug resistant cell lines by tamoxifen, toremifene and their metabolites. *British Journal of Cancer*, **29A**, 1152–1157.

Krick, W., Disser, J., Hazama, A., Burckhardt, G., and Fromter, E. (1991). Evidence for a cytosolic inhibitor of epithelial chloride channels. *Pflügers Archiv*, **418**, 491–499.

Kunzelman, K., Tilmann, M., Hansen, Ch. P., and Greger, R. (1991). Inhibition of epithelial chloride channels by cytosol. *Pflügers Archiv*, **418**, 479–490.

Landry, D. W., Akabas, M.H., Redhead, C., Edelman, A., Cragoe, E.J., and Al-Awqati, Q. (1989). Purification and reconstitution of chloride channels from kidney and trachea. *Science*, **244**, 1469–1472.

Landry, D.W., Reitman, M., Cragoe, E.J., Jr and Al-Awqati, Q. (1987). Epithelial chloride channel. Development of inhibitory ligands. *Journal of General Physiology*, **90**, 779–798.

Liedtke, C.M. (1989). Regulation of chloride transport in epithelia. *Annual Review of Physiology*, **51**, 143–160.

MacKinnon, R. and Miller, C. (1988). Mechanism of charybdotoxin block of $Ca^{2+}$-activated $K^{+}$ channels. *Journal of General Physiology*, **91**, 335–349.

Mannella, C.A. (1992). The 'ins' and 'outs' of mitochondrial membrane channels. *Trends in Biochemical Sciences*, **17**(8), 315–320.

Manning, S.D. and Williams, A.J. (1989). Conduction and blocking properties of a predominantly anion-selective channel from human platelet surface membrane reconstituted into planar phospholipid bilayers. *Journal of Membrane Biology*, **109**, 113–122.

Mayer, M.L., Owen, D.G., and Baker, J.L. (1990). Calcium-dependent chloride currents in vertebrate central neurons. In: *Chloride channels and carriers in nerve, muscle and glial cells*, (ed. F.J. Alvarez-Leefmans and J.M. Russell), pp. 355–364. Plenum Press, New York, London.

Mintenig, G.M., Valverde, M.A., Sepúlveda, F.V., Gill, D.R., Hyde, S.C., Kirk, J., and Higgins, C.F. (1993). Specific inhibitors distinguish the chloride channel and drug transporter functions associated with the human multidrug resistance P-glycoprotein. *Receptors and Channels*, **1**, 305–313.

Monaghan, A.S. (1994). Chloride and potassium channels of enterocytes isolated from guinea-pig small intestinal villi. Unpublished PhD Thesis. University of Cambridge.

Ordway, R.W., Singer, J.J., and Walsh, J.V., Jr (1991). Direct regulation of ion channels by fatty acids. *Trends in Neuroscience*, **14**, 96–100.

Steinmeyer, K., Klocke, R., Ortland, C., Gronemeier, M., Jockusch, H., Grunder, S., and Jentsch, T.J. (1991). Inactivation of muscle chloride channel by transposon insertion in myotonic mice. *Nature*, **354**, 304–308.

Sturgess, N.C., Ashford, M.L., Cook, D.L., and Hales, C.N. (1985). The sulphonylurea receptor may be an ATP-sensitive potassium channel. *Lancet*, **8453**, 474–475.

Stutts, M.J., Chinet, T.C., Mason, S.J., Fulton, J.M, Clarke, L.L., and Boucher, R.C. (1992). Regulation of chloride channels in normal and cystic fibrosis airway epithelial cells by extracellular ATP. *Proceedings of the National Academy of Sciences USA*, **89**, 1621–1625.

Tabcharani, J.A., Chang, X.-B., Riordan, J.R., and Hanrahan, J.W. (1992). The cystic fibrosis transmembrane conductance regulator chloride channel. Iodine block and permeation. *Biophysical Journal*, **62**, 1–4.

Tabcharani, J.A., Low, W., Elie, D., and Hanrahan, J.W. (1990). Low-conductance chloride channel activated by cAMP in the epithelial cell line T84. *FEBS Letters*, **270**, 157–164.

Tilmann, M., Kunzelmann, K., Frobe, U., Cabantchik, I., Lang, H.J., Englert, H.C., and Greger, R. (1991). Different types of blockers of the intermediate-conductance outwardly rectifying chloride channel in epithelia. *Plfügers Archiv*, **418**, 556–563.

Valverde, M.A., O'Brien, J.A., Sepúlveda, F.V., Ratcliff, R., Evans, M.J., and Colledge, W.H. (1993). Inactivation of the murine cftr gene abolishes cAMP-mediated but not $Ca^{2+}$-mediated secretagogue-induced volume decrease in small intestinal crypts. *Pflügers Archiv*, **425**, 434–438.

Wangemann, P., Wittner, M., DiStefano, A., Englert, H.C., Lang, H.J., Schlatter, E., and Greger, R. (1986). $Cl^-$ channel blockers in the thick ascending limb of the loop of Henle. Structure activity relationship. *Pflügers Archiv*, **407**, S128–S141.

Welsh, M.J. and Smith, A.E. (1993). Molecular mechanisms of CFTR chloride channel dysfunction in cystic fibrosis. *Cell*, **73**, 1251–1254.

Zhang, J.J., Jacob, T.J.C., Valverde, M.A., Hardy, S.P., Mintenig, G.M., Sepúlveda, F.V., *et al.* (1994). Tamoxifen blocks chloride channels: a possible mechanism for cataract formation. *Journal of Clinical Investigation*, **94**, 1690–1697.

# 6 Mechano-gated Channels

# 12 Drug effects on mechano-gated channels

O. P. Hamill and D. W. McBride Jr

## Introduction

Mechano-gated (MG) membrane ion channels provide a means of transducing membrane deformation or stretch into an electrical or ionic signal. They are characterized by the ability to change their open probability (i.e. gating) in response to mechanical stimulation. Together with voltage- and receptor-gated channels they form the three major functional classes of gated membrane ion channels. However, MG channels are a heterogeneous class, with members displaying a variety of ion selectivities (cation, $K^+$, $Cl^-$, non-selective), a wide range of conductances (10–3000 pS), and differences in their response to mechanical stimulation [i.e. stretch-activated (SA) and stretch-inactivated (SI). MG channels also vary considerably in their mechanosensitivities. At one extreme are MG channels that are sensitive to random thermal fluctuations or Brownian motion (e.g. MG channels in the hair cells of the ear), while at the other extreme are channels that only respond to strong mechanical stimulation, sometimes even exceeding the threshold for tissue or cell damage. The mechanisms that confer mechanosenitivity on a channel can be categorized into two main groups: (i) indirect mechanisms involving second messengers controlled by mechanosensitive (MS) enzymes; and (ii) direct mechanisms involving tension exerted directly on the channel proteins (for recent reviews see refs. 1–5).

Given the above heterogeneities in MG channels, it would be highly useful to have pharmacological agents that could discriminate between the various subclasses, and help in determining their role, if any, in specific MS processes. In contrast to the extensive literature on the pharmacology of voltage- and receptor-gated channels, much less information exists on MG channels. Nevertheless, over the past 10 years a number of chemicals have been shown to act on MG channels. Although no one compound has been discovered that is ideal in terms of its specificity and affinity for these channels, there do exist useful blockers and activators of MG channels (Table 12.1). Below we review a few of the better-characterized compounds.

## Mechano-gated channel blockers

### Gadolinium

Initial interest in testing the lanthanide gadolinium ($Gd^{3+}$) as a possible blocker of MG channels arose from the observation that $Gd^{3+}$ blocked geotropism and thigmotropism in

**Table 12.1** Some drugs and chemicals that act on mechano-gated channels.

| Channel | Compound | Effect | Reference |
|---|---|---|---|
| MG cation channels in hair cells and oocytes | Amiloride and various structural analogues; dimethylamiloride; phenamil; benzamil; hexamethyleneiminoamiloride; methoxynitriodbenzamil; bromohexamethyleneamiloride | Blockade | 42<br>43<br>44 |
| MG cation channels in hair cells and chick skeletal muscle | Aminoglycoside antibiotics (gentamicin, streptomycin, neomycin, ribostamycin dibekacin kanamycin) | Blockade | 28<br>32 |
| MG channels of diverse conductance and gating properties (e.g. cation, anion, $K^+$ and non-selective and both SA and SI channels) | Gadolinium | Blockade | 11<br>12 |
| MG anion channels in *E. coli* and cation channels in chick skeletal muscle | Amphipaths (chlorpromazine, trinitrophenol, procaine, tetracaine, dipyridamole, lysolecthin) | Activate | 49<br>32 |
| MG $K^+$ channels in gastric and vascular smooth muscle | Amphiphiles (e.g. fatty acids and lipids) | Negatives ones activate and positive ones suppress | 51<br>52 |

plants.[6] Subsequent patch-clamp studies indicated that $Gd^{3+}$ did indeed block MG channel activity in plants,[7] as well as in fungi,[8] bacteria,[9] and a variety of animal cells.[10–13] Such studies indicate that $Gd^{3+}$ blocks MG channels which display quite different ion selectivities and conductances, and can also block both SA and SI channels. However, the potency of $Gd^{3+}$ block varies between a few $\mu$M to several 100 $\mu$M or even mM concentrations.[10,15,16] While it would be hoped that $Gd^{3+}$ affected a structure common to all MG channels, this is not the case as there are at least two notable exceptions: (i) a $K^+$-selective MG channel in astrocytes; and (ii) a cation MG channel in cardiac muscle.[11,17]

Detailed analysis of $Gd^{3+}$ block has been carried out for the SA cation channel in *Xenopus* oocytes[11] and the SI cation channel in dystrophic mouse (*mdx*) muscle.[12] For the SA channel, extracellular $Gd^{3+}$ has at least three distinguishable effects – in the concentration range 1–5 $\mu$M there is: (i) a shift of ~25 mV of the single-channel current–voltage (I–V) curve along the voltage axis with no change in shape, which was suggested to be due to neutralization of negative charges in or near the vestibule of the channel; (ii) a voltage-independent reduction in open channel lifetime proposed to arise from $Gd^{3+}$ binding to an allosteric site outside the membrane field; and (iii) at concentrations of 10 $\mu$M and above, a complete inhibition of channel opening apparently caused by a highly cooperative transition or binding of $Gd^{3+}$ to the MG channel.[11] In contrast to these multiple effects on the SA channel, $Gd^{3+}$ block of the SI cation channel can be explained by a simple voltage-dependent open channel 'flickery' block in which $Gd^{3+}$ is driven into the channel by strong hyperpolarization, with no effect on open probability and the open channel conductance.[12] In neither SA nor SI channels does $Gd^{3+}$ alter the mechanosensitivity of the channel (i.e. the relationship between channel open probability and stimulus).

$Gd^{3+}$ sensitivity has been used to implicate MG channels in a variety of MS processes, including stretch-induced elevation of intracellular $Ca^{2+}$ in vascular smooth muscle.[18] skeletal muscle,[19] and endothelial cells[20] and stretch-induced arrhythmias in cardiac muscle.[21] However, the major drawback in the interpretation of such studies is that $Gd^{3+}$ also blocks other classes of channels. For example, it is a potent ($<0.5\ \mu M$) blocker of the voltage-gated high-threshold (L-type) $Ca^{2+}$ channels and the low-threshold (T-type) $Ca^{2+}$ channels[22,23] and blocks in higher concentrations ($\sim$mM) both voltage-gated $Na^+$ channels and $K^+$ channels,[24] as well as nucleotide-gated cation channels.[25]

## Aminoglycoside antibiotics

Antibiotics of the aminoglycoside family (e.g streptomycin, gentamicin, and neomycin) have been shown to block MG channels. Most studies have focused on their blocking action of MG channels in audio vestibular hair cells, where chronic exposure to these antibiotics is known to cause a clinically significant, and usually irreversible, hearing impairment.[26] The first acute studies were carried out on hair cells of the amphibian lateral line organ, where a reversible, dose-dependent block of extracellularly recorded microphonic potentials and afferent fibre activity was demonstrated.[27] Subsequent voltage-clamp studies indicated a rapid ($<100\ \mu s$), voltage- and concentration (1–10 $\mu M$)-dependent block of the vestibular hair cell MG channel.[28]

Aminoglycosides also block a variety of other MS processes, including mechanotransduction in slowly adapting cutaneous mechanoreceptors in cats,[29] in hair cells of the squid statocyst,[30] nematocyte discharge in hydrozoans,[31] and stretch-activated channels in chick skeletal muscle.[32] As with $Gd^{3+}$, aminoglycosides are not selective to MG channels and have been shown to block voltage-gated $Ca^{2+}$ channels,[33] $Ca^{2+}$-activated $K^+$ channels,[34] and ATP-sensitive channels,[35] as well as increasing desensitization of acetylcholine receptor channels.[36]

Perhaps the most elegant experimental use of aminoglycosides has been to demonstrate, using focal iontophoresis of gentamicin to specific regions of the hair bundle, the localization of the MG channels at the very tips of the stereocilia.[37]

## Amiloride and structural analogues

Amiloride is a potent diuretic and belongs to a group of more than 1000 compounds collectively called pyrazinecarboxyamides.[38] Amiloride is best known for its ability to block the epithelial $Na^+$ channel in submicromolar concentrations, but it has also been shown to block in higher concentrations (i.e. $>10\ \mu M$) a number of basic cellular physiological processes, including fertilization, volume regulation, and cellular proliferation[39]. Amiloride

(50–500 μM) also blocks mechanosensitivity in a number of different sensory and non-sensory systems including hair cells of the lateral line organ of *Necturus*[40], the inner ear of the chick[41] and rat,[42] and in *Xenopus* oocytes.[43–46]

In oocytes and hair cells, extracellular amiloride causes a highly voltage-dependent block of the MG channel in which inward current is reduced, but outward current is almost unaffected.[41–43] This voltage-dependent blocking behaviour might arise because the amiloride molecule, which is positively charged at pH 7.2, is driven into and 'plugs' the open channel. However, a number of observations argue against this simple model.[43] The model that appears best in explaining the block is one involving a voltage-dependent conformational change of the MG channel with subsequent voltage-independent and cooperative binding of two amiloride molecules. This model initially developed for channel block of the MG channel in *Xenopus* oocytes[43] has also been shown to be successful in explaining the MG channel block in mammalian hair cells.[42]

Structural analogues of amiloride that display nanomolar affinity in blocking the epithelial $Na^+$ channel proved critical in the labelling and purification studies of the channel proteins.[39] Therefore, attempts have been made to screen for higher-affinity amiloride analogues that may block MG channels in both oocytes[44] and hair cells.[42] Although more potent analogues have been found, none has shown submicromolar blocking affinities. For the analogues tested on hair cells, the order of increasing potency was amiloride $<$ dimethylamiloride $<$ phenamil $<$ benzamil $<$ hexamethyleneiminoamiloride $<$ methoxynitriodbenzamil, with half-blocking concentrations of 53, 40, 12, 5.5, 4.3, and 1.8 μÌM respectively[42]. For the oocyte, a similar order of potency was found with amiloride $<$ dimethylamiloride $<$ benzamil $<$ bromohexamethyleneamiloride and half-blocking concentrations of 500, 357, 94, and 34 μÌM respectively.[44] Although amiloride and some of its analogues show a similar sequence in blocking MG channels in hair cells and oocytes, there was 10-fold difference in potency. However, this difference was shown not to be due to the different ionic recording conditions, in particular $Ca^{2+}$ concentrations, used in the hair cell and oocyte studies.[46]

Amiloride sensitivity of the MG cation channel provided the initial clue that this channel may be structurally related to the amiloride-sensitive epithelial $Na^+$ channel. This idea was reinforced by two additional findings: (i) antibodies against the epithelial $Na^+$ channel label the tips of stereocilia where the MG channels are believed to be localized[47] and (ii) the sequence of the cloned epithelial $Na^+$ channel shows close homology to genes encoding putative MG channels identified using touch-insensitive mutants of *Caenorhabditis elegans*.[48] It remains to be determined whether MG channels in hair cells and oocytes also belong to this family.

## Mechano-gated channel activators

A diverse group of compounds, including lipid metabolites, free fatty acids, lipids, and amphipathic molecules are able to activate MG channels. Rather than acting 'indirectly' through signal transduction pathways, these compounds appear to act by 'direct' mechanisms, either by partitioning into the bilayer to alter membrane tension,[49] changing membrane deformation energy[50] or by interacting with allosteric sites on the channel protein.[51,52]

### Amphipathic molecules

Amphipathic molecules have both hydrophilic and hydrophobic groups and may be either positive, negative, or no net electric charge (e.g. chlorpromazine, trinitrophenol, and lysolecithin, respectively). Martinac *et al.*[49] demonstrated that amphipathic molecules, when introduced into the bathing solution, could reversibly increase the open probability of MG anion channels recorded in *E. coli* spheroblasts. When each was used alone, the effects were always positive (i.e. open probability increased) yet, when used in succession, cationic and anionic amphipaths were able to first neutralize each other's stimulatory effects and then had their own stimulatory effect. The effects were typically slow, sometimes taking up to an hour to increase open probability to unity. Amphipaths appeared to act by shifting the sigmoidal stimulus–response relation to the left, so that lower pressures activated the channel without affecting the slope of the Boltzmann curve. These effects were interpreted in terms of the bilayer couple hypothesis [53] in which cationic amphipaths insert into the positive inner monolayer and anionic species partition into the negative outer leaflet of the membrane. The introduction of molecules into one leaflet increases the membrane tension in the adjacent leaflet, and thereby activates the MG channel. Since the bilayer couple hypothesis depends on an asymmetry in the two leaflets, a feature of biological membranes including *E. coli*, one might predict that amphipaths would be inert when applied to symmetrical artificial lipid bilayers in which MG channel proteins had been reconstituted.

### (Fatty acids and lipids (amphiphilic molecules)

Amphiphilic molecules have polar heads attached to a long hydrophobic tail, and include such compounds as fatty acids and lipid molecules. Recent studies indicate that fatty acids activate various MG channels independent of cyclooxygenase or lipoxygenase pathways.[51, 52] Kirber *et al.* [51] have demonstrated that fatty acids, including arachidonic acid, increase or modulate the activity of a large-conductance, stretch-activated, $Ca^{2+}$-sensitive $K^+$ channel in rabbit arterial smooth muscle. In contrast to the slow effects of the amphipaths, fatty acid activation occurs in seconds rather than in tens of minutes. The fatty acid effects do not depend on the generation of arachidonic acid metabolites, since saturated fatty acids such as myristic acid and linoelaidic acid – which are not substrates for the enzymes that convert arachidonic acid to active metabolites – also activate the channel.[51] Further evidence for a direct effect was indicated by cell-free patch experiments in which activation by fatty acids was retained in the absence of soluble cytoplasmic enzymes.[51] A number of other features distinguish the amphiphilic from amphipathic activation. For example, fatty acids were not capable of activating the $K^+$ channel in the absence of basal activity (i.e. in zero $Ca^{2+}$ or at very negative potentials), indicating a modulatory role rather than direct activation. Furthermore, although this $Ca^{2+}$-sensitive $K^+$ channel is stretch-sensitive, it is not clear whether stretch and fatty acids act independently of one another. For example, it may be that membrane stretch activates a membrane-bound phospholipase, producing free fatty acids which could in turn activate the channel. Relaxation measurements should be able to distinguish such indirect effects from direct stretch activation. More recent studies have examined the structural requirements for fatty acid modulation of another mechanosensitive

$K^+$ channel in toad gastric muscle and demonstrated that, while negatively charged fatty acids activate the channel, positively charged fatty acids suppress activity and neutral amphiphiles are without effect. In this case the dependence of charge polarity differs from the polarity-independent stimulatory effect of amphipaths seen for the *E. coli* MG channel. One possible mechanism for the fatty acid effects may involve allosteric alteration of the channel protein.[52] Alternatively, the amphiphiles may, by partitioning into the bilayer, alter the bilayer deformation energy and in this way the probability of specific protein conformations.[50]

## Conclusions

In this article we have discussed a few of the better-studied ions and drugs that interact with MG channels. At present, no compound has been found that displays the high specificity and affinity exhibited by tetrodotoxin or α-bungarotoxin which proved so useful in the functional and structural characterization of the voltage-gated $Na^+$ channel and the acetylcholine receptor channel, respectively. Hopefully, continued screening of drugs and toxins on the different types of MG channels will identify more ideal compounds.

## Acknowledgements

We thank the Muscular Dystrophy Association and the National Institutes of Health for their support in these investigations.

## References

1. Hamill, O.P. and McBride, D.W. Jr. (1996). *Pharmacol. Rev.*, **48**, 231–252
2. Petrov, A.G. and Usherwood, P.N.R. (1994). *Eur. Biophys. J.*, **23**, 1–19.
3. Hudspeth, A.J. and Gillespie, P.G. (1994). *Neuron*, **12**, 1–9.
4. Martinac, B. (1993) In *Thermodynamics of membrane receptors and channels*, (ed. M.B. Jackson), pp. 327–352. CRC Press, Florida.
5. Sokabe, M. and Sachs, F. (1992). *In Advances in comparative and environmental physiology*, (ed. F. Ito Vol. 10), pp. 55–77. Springer-Verlag, Berlin.
6. Millet, B. and Pickard, B.G. (1988). *Biophys. J.*, **53**, 155A.
7. Garrill, A., Jackson, S.L., Lew, R.R., and Heath, I.B. (1993). *Eurp. J. Cell. Biol.*, **60**, 358–365.
8. Gustin, M.C., Zhou, X.-L, Martinac, B., and Kung, C. (1988). *Science*, **242**, 762–763.
9. Zhou, X.-L., Stumpf, M.A., Hoch, H.C., and Kung, C.A. (1991. *Science*, **253**, 1415–1417.
10. Berrier, C., Coulombe, A., Szabo, I., Zoratti, M., and Ghazi, A. (1992). *Eur. J. Biochem.*, **206**, 559–565.

11. Yang, X.-C. and Sachs, F. (1989). *Science*, **243**, 1068–1071.
12. Franco, A. Jr, Winegar, B.D., and Lansman, J.B. (1991). *Biophys. J.*, **59**, 1164–1170.
13. Filipovic, D. and Sackin, H. (1991). *Am. J. Physiol.*, **260**, F119–129.
14. Kawahara, K. and Matsuzaki, K. (1993). *Jpn. J. Physiol.* **43**, 817–832.
15. Quasthoff, S. (1994). *Neurosci. Lett.*, **169**, 39–42.
16. Zhou, X.-L. and Kung, C.A (1992). *EMBO J.*, **11**, 2869–2875.
17. Kim, D. (1993). *Circ. Res.*, **72**, 225–231.
18. Davis, M.J., Meininger, G.A., and Zawieja, D.C. (1992). *Am. J. Physiol.*, **263**, H1292–H1299.
19. Pressmar, J., Brinkmeier, H., Seewald, M.J., Naumann, T., and Ruedel, R. (1994). *Pflügers Arch.*, **426**, 499–505.
20. Naruse, K. and Sokabe, M. (1993). *Am. J. Physiol.*, **264**, C1037–C1044.
21. Stacy, G.P., Jr, Jobe, R.L., Taylor, K., and Hansen, D.E. (1992). *Am. J., Physiol.*, **263**, H613–H621.
22. Biagi, B.A. and Enyeart, J.J. (1990). *Am. J. Physiol.*, **264**, C1037–C1044.
23. Lacampagne, A., Gannier, F, Argibay, J., Garnier, D., and Le Guennec, J.-L. (1994). *Biochim. Biophys. Acta*, **1191**, 205–208.
24. Elinder, F. and Arhem, P. (1994). *Biophys. J.*, **67**, 71–83.
25. Popp, R. and Gogelin, H. (1992). *Biochim. Biophys, Acta*, **1108**, 59–66.
26. Schacht, J. (1986). *Hearing Res.*, **22**, 297–304.
27. Kroese, A.B.A. and Van Den Bercken, J. (1980). *Nature*, **283**, 395–397.
28. Kroese, A.B.A., Das, A., and Hudspeth, A.J. (1989) *Hearing Res.*, **37**, 203–218.
29. Bauman, K.I., Hamann, W., and Leung, M.S. (1988) *Prog. Brain Res.* **74**, 43–9.
30. Williamson, R. (1990). *J. Comp. Physiol. A.* **167**, 655–664.
31. Gitter, A.H., Oliver, D., and Thurm, U. (1993). *Naturwissenschaften*, **80**, 273–276.
32. Sokabe, M., Hasegawa, N., and Yamamori. K. (1993). *Ann. N.Y. Acad. Sci.* **707**, 417–421.
33. Nagawa, T., Kakehata, S., Akaike, N., Komune, S. Takasaka, T., and Uemura, T. (1992). *Brain Res.* **580**, 345–347.
34. Nomura, K., Naruse, K., Watanabe, K., and Sokabe, M. (1990). *J. Membr. Biol.*, **115**, 241–251.
35. Lin, X., Hume, R.I., and Nutall, A.L. (1993). *J. Neurophysiol.*, **70**, 1593–1605.
36. Okamoto, T. and Sumikawa, K. (1991). *Molec. Brain Res.* **9**, 165–168.
37. Jaramillo, F. and Hudspeth, A.J. (1991). *Neuron*, **7**, 409–420.
38. Kleyman, T.R. and Cragoe, E.J., Jr. (1988). *J. Membr. Biol.*, **105**, 1–21.
39. Benos, D. (1982). *Am. J. Physiol.*, **242**, C131–145.
40. Jorgensen, F.O. (1984). *Acta Physiol. Scand.*, **120**, 481–488.

41. Jorgensen, F. and Ohmori, H. (1988). *J. Physiol.*, **403**, 577–588.

42. Ruesch, A., Kros, C.J., and Richardson, G.P. (1994). *J. Physiol.*, **474** (1), 75–86.

43. Lane, J.W., McBride, D.W., Jr, and Hamill, O.P. (1991). *J. Physiol.*, **441**, 347–366.

44. Lane, J.W., McBride, D.W., Jr, and Hamill, O.P. (1992). *Br. J. Pharmacol.*, **106**, 283–286.

45. Hamill, O.P., Lane, J.W., and McBride, D.W. (1992). *Trends in Pharmacol. Sci.*, **13**, 372–376.

46. Lane, J.W., McBride, D.W., Jr, and Hamill, O.P. (1993). *Br. J. Pharmacol.*, **108**, 116–119.

47. Hackney, C.M., Furness, D.N., Benos, D.J., Woodley, J.F., and Barratt, J. (1992). *Proc. R. Soc. B.*, **248**, 215–221.

48. Canessa, C., Horisberger, J.D., and Rossier, B.C. (1993). *Nature*, **361**, 467–470.

49. Martinac, B., Adler, J., and Kung, C. (1990). *Nature*, **348**, 261–263.

50. Lundbaek, J.A. and Andersen, O.S. (1994). *J. Gen. Physiol.* **104**, 645–673.

51. Kirber, M.T., Ordway, R.W., Clapp, L.H., Walsh, J.V., Jr, and Singer, J.J. (1992). *FEBS Lett.*, **297**, 24–28.

52. Petrou, S., Ordway, R.W., Hamiliton, J.A., Walsh, J.V., Jr, and Singer, J.J. (1994). *J. Gen. Physiol.* **103**, 471–486.

53. Sheetz, M.P. and Singer, S.J. (1974). *Proc. Natl Acad. Sci. USA*, **71**, 4457–4461.

# 7 Intracellular Channels

# 13 Intracellular channels in muscle

E. Jaimovich

## Intracellular channels

It is now widely accepted that distinct channel activities are present on intracellular organelles, such as the endoplasmic reticulum, the Golgi apparatus, mitochondria, the nuclear envelope, and synaptic vesicles. However, we know very little about these channels because, as a result of their intracellular localization, they are not readily accessible to conventional electrophysiological techniques. Nonetheless, recent molecular and biochemical studies have begun to shed light on this unexplored field. The intracellular $Ca^{2+}$-release channels have been cloned and were found to include two gene families, the inositol 1,4,5-trisphosphate receptor ($IP_3R$) family,[1,2] and the ryanodine receptor (RyR) family.[2–5] Both the $IP_3R$[6–11] and RyR[11–17] families exist as at least three distinct types in mammals.

Several intracellular anion channels have also been cloned, including the voltage-dependent anion channel (VDAC),[18,19] which is permeable to $Cl^-$, and the newly discovered $Cl^-$ channel.[20] Three types of VDACs have been cloned in mammals.[18,19] Members of the VDAC family are considered to be associated with the mitochondrial benzodiazepine receptor,[18,19] whereas $Cl^-$ channels are found in intracellular organelles, such as the Golgi apparatus and endocytic vesicles. This chapter will focus exclusively, however, on the recently cloned $IP_3R$ and RyR families in general, and on their presence and roles in muscle cells in particular.

## Role of calcium in muscle cells

In all muscles, calcium ions represent the trigger for contraction which is called into action by raising their intracellular concentration from about 0.1 $\mu$M to 10 $\mu$M. Thus, calcium ions are, with a few exceptions, the main intracellular messengers carrying information from the excited cell membrane to the myofilaments in the interior of the fibres.[21] The modes and mechanisms of this information transfer are, however, quite different in various types of muscle. In fast skeletal muscle for instance, an electric signal – the action potential – travels along the membrane invaginations, the transverse tubules, and hits the triads where this extracellular message is transformed into an intracellular one: calcium ions are released from the terminal cisternae of the sarcoplasmic reticulum and diffuse into the myoplasm. This calcium signal is then recognized by troponin-C, the specific calcium-receptor protein, that 'turns on' the contractile mechanism. In vertebrate smooth muscle, on the other hand, calcium ions may also be released from the cell membrane to affect a different calcium-sensing protein, calmodulin (CaM), which, in conjunction with myosin light-chain kinase, catalyses the phosphorylation of the regulatory myosin light chain, thereby inducing contraction.

Depolarization of the muscle cell, therefore, does not imply changes in membrane potential between cisternae and myoplasm, though it does induce opening of calcium release channels in the cisternal membrane. Here, the closing and opening of the calcium channel is obviously determined by the potential of a membrane different from that in which the channels reside. The nature of this coupling process between the T-tubular depolarization and the change in calcium conductivity in the sarcoplasmic reticulum, the so-called T-SR coupling, is however, still a mystery.[22–24]

## Intracellular messages

Activation processes that precede and are responsible for the generation of a calcium transient may include calcium-induced calcium release, as well as inositol trisphosphate-induced calcium release (which may occur without membrane depolarization in smooth muscle). In cardiac muscle the gating of the membrane calcium channels is enhanced by cAMP. Calcium transients, therefore, may originate from quite different kinds of intracellular messages that will finally target the myofilaments.

Inositol trisphosphate is an important intracellular messenger that transmits signals from membrane receptors of smooth muscle (its role in other types of muscle remains controversial[25]) to the calcium-release sites of the sarcoplasmic reticulum.[26] For example, inositol trisphosphate is involved in the $\alpha$-adrenergic receptor contraction coupling: when the $\alpha$-receptors of the cell membrane are occupied with agonists, such as noradrenaline, the phospholipases of the cell membrane become activated and they degrade membrane lipids, in particular phosphoinositides to diacylglycerol and inositol trisphosphate. While the latter mobilizes calcium from the sarcoplasmic reticulum of smooth muscle, even after membrane skinning, the former bypasses the calcium signalling channel. Diacylglycerol may activate protein kinase-C, even at very low concentrations of myoplasmic free calcium, thus causing a phosphorylation of myosin light chains which, in turn, may be followed by the activation of the contractile mechanism. Diacylglycerol also serves as a substrate for the calcium-dependent formation of arachidonic acid which is the precursor of prostaglandins, leukotrienes, and thromboxanes. The role of all these intracellular and possibly intercellular messengers in the regulation of motile and secretory activity of cells can hardly be overemphasized. Further research is urgently needed to understand their mode of action.

## Skeletal muscle: the calcium release channels of the sarcoplasmic reticulum

The control of the contractile activity is exerted by two membrane systems in skeletal muscle, the T-tubules and the membrane of the sarcoplasmic reticulum. During excitation–contraction coupling (EC coupling), electrical depolarization of the T-tubular membranes induces the rapid release of calcium ions from the sarcoplasmic reticulum. However, it is not yet quite clear how the voltage is sensed in the T-tubules and transformed into a signal, which is then transmitted to the calcium-release sites of the sarcoplasmic reticulum. Recent work has established that this unknown step involves essentially the communication between two types of calcium channels, one which resides in the T-tubular membrane, and the other in the membrane of the sarcoplasmic reticulum.

It has long been suspected that a calcium release channel of the sarcoplasmic reticulum may be located in the junctional feet that span the gap between the membrane system of the T-tubules and the sarcoplasmic reticulum. The proof came with the discovery that the calcium channels of the lateral cisternae could be isolated and were shown to be identical in size and shape with the feet structure of the triads, as will be discussed below.

Calcium is released from the terminal cisternae of the sarcoplasmic reticulum at high rate[27,28], which is mediated via a highly conducting ion channel rather than by a carrier-coupled mechanism. Indeed, Meissner and his colleagues succeeded in reconstituting isolated vesicles of lateral cisternae containing the putative $Ca^{2+}$ channels into a planar lipid bilayer separating two compartments of a plastic chamber filled with saline.[29,30] By means of a current amplifier connected to two electrodes inserted into the compartments and suitably applied potentials, an ionic current could be measured across the bilayer due to the opening of pore proteins. The channels conducted preferentially $Ca^{2+}$ and $Ba^{2+}$ when these ions were present in the experimental chambers. Interestingly, the opening or 'gating' of the channels could be modulated by various ligands, including ATP and other adenine nucleotides, caffeine, and calcium itself. The plant alkaloid ryanodine also modulated channel activity and turned out to be particularly useful, as it is tightly bound to the channel protein.[31]

## Identity of isolated calcium release channels and 'feet'

The capacity of the putative channels of the lateral cisternae to bind ryanodine greatly facilitated its isolation.[32,33] Thus, during the purification procedure, the protein fractions containing the enriched ryanodine receptor (RyR) could be identified simply by monitoring ryanodine binding. The isolated protein, having a molecular weight of approximately 500 kDa,[34] was in its functional properties similar to the native sarcoplasmic reticulum channel.[32,35] As shown by electron microscopy,[36,37] it may form a tetramer resembling in its size and shape the 'feet' spanning the gap between T-tubular and sarcoplasmic reticulum membrane.[38,39] Thus, the feet, the RyR and the calcium channel were all shown to be the same protein.

To study its calcium conductance, the RyR protein was fused into liposomes which were then incorporated into an artificial lipid membrane separating the compartments of a double chamber, as described above. In this way, the ion-conductive properties of the reconstituted channel could be compared with that of the native sarcoplasmic reticulum membrane containing Ca-channels. These studies have been extensively reviewed.[40]

Interestingly, the values for (unit) $Ca^{2+}$ conductivity of both native or reconstituted channels in planar bilayers were similar (ranging from about 100 to 240 pS depending on the conditions) to those found for calcium channels of the sarcoplasmic reticulum from frog skeletal muscle.[41]

Furthermore, the open probability of the SR-calcium channels depended on the voltage across the sarcoplasmic reticulum membrane, or in the case of the purified RyR, across the planar bilayer.[42] In both cases, the open probability of the channels increased with the free calcium ion concentration,[43] suggesting that calcium itself was a modulator of the channel.

## The RyR family

As a result of molecular cloning, we now know that mammals have at least three distinct types of RyRs (for reviews see refs. 1, 3, 4): two types of muscular RyRs, a skeletal type or type 1 (sRyR or RyR1) [12,13] and a cardiac type or type 2 (cRyR or RyR2),[14,15] and a novel brain type RyR or type 3 (bRyR or RyR3).[16,17] The type names are based on their dominant tissue distribution (sRyR and cRyR) or the source of molecular cloning (bRyR). Recently, the RyR family has been shown also to be widely localized in non-muscle cells.[4]

### RyR1 (sRyR)

The complete sequence of RyR1 has been reported by two groups, one being 5037 amino acids long (565 kDa), found in the rabbit[12] and another being the short type (5032 amino acids; 564 kDa) lacking five amino acids (residues 3481–3485) in both the rabbit and human.[13] RyR1 has been postulated to have four (M1–M4)[12] or twelve (M′ M″ and M1–M10)[13] membrane-spanning domains clustered near the carboxy terminus. The last two of such domains (e.g. M3 and M4 in the four membrane-spanning domains model) are fairly homologous with the corresponding M5 and M6 of the $IP_3R$ family, as described below. There are the characteristic repeated segments occurring four times in two doublets (R1–R2 and R3–R4) in this large cytoplasmic region, which may form a foot structure.[13] An RyR1 complex observed by electron microscopy was seen to have a four-leaf clover structure (quatrefoil, $\sim 27 \times 27$ nm square) with a central hole 1–2 nm in diameter.[44]

RyR1 channel activities are affected by various modulators and compounds[5,45,46] (see below), and the functional sites for some of these factors have been proposed.[12,13] RyR1 expression in cDNA-transfected cells showed elevation of [$^3$H]ryanodine-binding activity with a $K_d$ of 19 nM[12] and a large-conductance channel activated by ATP and $Ca^{2+}$, and inhibited by $Mg^{2+}$ and ruthenium red.[47] Although a possible interaction between the dihydropyridine receptor (DHPR) and RyR1 for opening the RyR1-linked $Ca^{2+}$ channel remains to be investigated,[4] recent studies have suggested that functional parts of the RyR1 receptor molecule described below are involved in the channel opening mechanism. Three $Ca^{2+}$-binding sites were found in the amino-terminal vicinity of the channel domain in the rabbit RyR1.[12,48] The antibody against residues 4478–4512 containing the third $Ca^{2+}$-binding site increased the $Ca^{2+}$ sensitivity of RyR1 channels.[48] Recently, it was suggested that the Pro-Glu (PE) repeat, which matches the third high-affinity $Ca^{2+}$-binding site,[12] forms a site involved in the $Ca^{2+}$ activation mechanism.[49] In addition, the antibody against residues 4425–4621 of human RyR1 decreased $Ca^{2+}$-induced[45] $Ca^{2+}$ efflux from isolated terminal cisternae, suggesting the existence of a $Ca^{2+}$-dependent gating domain lying near this epitope region.[50]

ATP is known to activate RyR1 channel. Photo-affinity labelling by Bz2ATP (3'-O-(4-benzoylbenzoyl-adenosine-5'trisphosphate) showed the presence of a single ATP-binding site per RYR tetramer, but two or more sites, with different affinities, were also observed.[51] It was shown that one serine residue (Serine-2843) of rabbit RyR1 is phosphorylated by PKA, CaMKII and cGMP-dependent protein kinase (PKG), and one threonine residue by CaMKII.[52] Recordings from excised patches of sarcoplasmic reticulum membrane have suggested that RyR1-mediated $Ca^{2+}$-release channel activity is inactivated by CaMKII phosphorylation,[53] but RyR1 was shown not to be as good a substrate for CaMKII phosphorylation as RyR2.[54] The mutation was identified in the RyR1 of porcine malignant hyperthermia in which arginine-615 is replaced by cysteine (reviewed by Wang and Best[55]). The porcine malignant hyperthermia RyR1 channels were shown to be hypersensitive to various modulators. The region around arginine-15 of RyR1 has fragmentary sequence homology with the corresponding region of $IP_3R1$.[6] As this region of $IP_3RI$ is necessary for ligand binding, Arginine-615 of RyR1 may be involved in the binding of regulators of $Ca^{2+}$-channel gating.

## RyR2 (cRyR)

The primary structure of RyR2 was determined by cloning cDNA from rabbit cardiac muscle.[14,15] RyR2 consists of 4969 amino acids [14], and 4968 amino acids (or 4976 amino acids with an insertion of eight amino acids between residues 3715 and 3716).[15] The amino acid sequence of RyR2 was 66% identical to that of RyR1. The functional unit of the RyR2 channel contains only one high-affinity ryanodine-binding site/tetramer (with a four-leaf clover appearance on negative-stain electron microscopy[56] RyR2 is thought to function as a $Ca^{2+}$-induced $Ca^{2+}$ release (CICR) channel, thereby amplifying and/or regenerating $Ca^{2+}$ that is locally increased by $Ca^{2+}$ influx through cardiac DHPR activated by depolarization. The RyR2 channel is activated by micromolar $Ca^{2+}$ and millimolar ATP and inhibited by millimolar $Mg^{2+}$ and CaM.[4,54] CaM inhibits RyR2 channel activity, reducing the open probability, shortening the mean open time, and producing prolonged closures.[54] RyR2 is phosphorylated by PKA, PKG, PKC, and CaMKII.[57,58] The phosphorylation site for CaMKII was determined at Serine-2809 by sequencing phosphopeptides, and this CaMKII phosphorylation of RyR2 reverses the inhibitory effect of CaM on the open probability and restores prolonged channel openings.[54]

## RyR3 (bRyR)

Hakamata *et al.*[17] cloned RyR3 cDNA from a rabbit brain cDNA library by screening with RyR1 cDNA probes. Giannini *et al.*[16] independently isolated a partial cDNA clone of RyR3 as a gene β4 induced by transforming growth factor-β1 (TGF-β1) in mink lung epithelial cells (Mv1Lu). RyR3 consists of 4872 amino acids.[17] The amino acid sequence of RyR3 is 67% and 70% identical to that of RyR1 and RyR2, respectively. It has been well documented that the opening of both RyR1 and RyR2 channels is stimulated by either ryanodine or caffeine. In contrast, TGF-β1-treated Mv1Lu cells increase the intracellular concentration of free $Ca^{2+}$($[Ca^{2+}]_i$) in response to ryanodine but not caffeine.[16] The insensitivity of RyR3 to caffeine has not yet been confirmed at the molecular level.

### Other types of RyR

Recently, it was shown that the short RyR1 mRNA, which possibly encodes for the carboxy-terminal 656 amino acids of RyR1 (13% of the receptor protein), is expressed only in the brain of rabbit.[59] This short, brain-specific receptor is thought to contain a part of the putative modulatory domain (two $Ca^{2+}$-binding sites and one ATP-binding site) and a complete set of putative channel domains of RyR1. Therefore, this truncated RyR1 might function as a CICR channel that is more simply regulated than authentic RyR1, without forming a foot structure. Recently, the RyR gene has been cloned from *Drosophila*.[60] This putative *Drosophila* RyR has 45%, 47%, and 46% identity with RyR1, RyR2 and RyR3, respectively, indicating that this fly receptor does not belong to any mammalian RyR type.

## Modulation and pharmacology

Cyclic adenosine diphosphate ribose (cADPR) can induce $Ca^{2+}$ release from intracellular stores distinct from $IP_3$-sensitive calcium pools.[61–64] Brain extract contain the synthesizing enzymes for cADPR,[65] and brain microsomes have cADPR-induced $Ca^{2+}$-release activity that is sensitive to inhibition by ryanodine,[66,67] suggesting that cADPR is a major candidate(s) for the endogenous ligand of RyRs. A cADPR-induced $Ca^{2+}$-release activity was also observed in dorsal root ganglion neurones.[68] On the other hand, it has been reported that $IP_3$ can modulate a ryanodine-sensitive channel from skeletal muscle[69] although not all channels and/or species seem to be equally sensitive.[25,70]

RyR1 channel activities are affected by various modulators and compounds, including $Ca^{2+}$, $Mg^{2+}$, ATP, CaM, caffeine, ryanodine, ruthenium red, procaine, dantrolene, spermidine, and polyanions.[5,45,46] Besides the physiological modulators of the channel (calcium, magnesium, ATP, and other adenine nucleotides), Table 13.1 summarizes the long list of pharmacological compounds that have been reported to activate either calcium release or single-channel activity and, in some cases, ryanodine binding. Table 13.2 lists compounds reported to block channel activity or to inhibit calcium release. It is interesting to notice that, in analogy to the effects of calcium concentration, several compounds behave both as agonists and antagonists of calcium release depending on the system and the concentration used.

## The $IP_3R$ family

The $IP_3R$ channel release $Ca^{2+}$ from internal stores, such as the endoplasmic reticulum, in response to binding of the second messenger $IP_3$, which is produced by two phosphoinositide signal transduction cascades:[129] firstly, by the cascade starting at a seven membrane-spanning-type receptor (e.g. subtypes of metabotropic glutamate, muscarinic acetylcholine, and serotonin receptors), which is linked to a G protein and then to certain type(s) of phospholipase C (PLC), such as $\beta$-1 type (PLC $\beta$-1); and secondly, by the cascade from a receptor tyrosine kinase [e.g. receptors for PDGF (platelet-derived growth factor), EGF (epidermal growth factor), and NGF (nerve growth factor) to PLC$\gamma$-1. So far, the complete structure of three different receptor types in the $IP_3R$ gene family have been determined;[1,3,4,6–11] however, one more type identified by the PCR (polymerase chain reaction) technique[9] remains to be studied.

Although widely distributed in the central nervous system,[1] $IP_3Rs$ (measured as $[^3H]IP_3$ binding) have been studied in smooth muscle[130] and recently shown to be present in skeletal muscle[131] in cultured skeletal muscle cells.[132,133]

## $IP_3R1$

$IP_3R1$ cloned from either mouse cerebellum or rat brain consists of 2749 amino acids;[6,7] however, it comprises a heterogeneous group of receptor subtypes arising from alternative splicing in two regions:[134,135] SI (15 amino acids, $SI^+$ and $SI^-$) and SII (40 amino acids that are further subdivided into A, B, and C subregions, $SI^+$, $S11^+$, $S11^-$, $S11B^-$ and $S11BC^-$).[134,135] Recently, cDNAs of $IP^\times R1$ from *Xenopus laevis*[136] and humans have been cloned. $IP_3R1$ is functionally divided into three parts, an $IP_3$-binding (or ligand-binding) domain, a modulatory and transducing domain, and a channel domain.

The $IP_3$-binding domain comprises the 650 amino-terminal amino acids of $IP_3R1$, and, within this domain, residues 476–501 have been labelled by photo-affinity ligands.[137] The binding of (1,4,5)-$IP_3$ to this receptor ($K_D \sim 100$ nM; Hill coefficient 1.0)[138] is stoichiometric and affected by intracellular pH and the concentrations of $Ca^{2+}$ and $Mg^{2+}$;139 specific binding is inhibited by heparin. Newly identified $IP_3R$ agonists, referred to as adenophostin A and B, are 100-fold more potent than $IP_3$ in terms of $Ca^{2+}$ release.[140]

The modulatory and transducing domain is located between the $IP_3$-binding and channel domains, and appears to be involved in transducing ligand-binding signals into channel opening. $IP_3$-induced $Ca^{2+}$ release is modulated by other second messenger signalling ('cross-talk regulation'), such as cAMP-dependent protein kinase (protein kinase A: PKA), 141–143 and ATP,[144,145] as well as various modulators associated with the $Ca^{2+}$ signal itself ('feedback regulation'),[146–148] e.g. calmodulin, $Ca^{2+}$/CaM-dependent protein kinase II (CaMKII),[141,149] and protein kinase C (PKC),[149,150] and it has been proposed that functional sites for these modulators are localized within this modulatory domain. Moreover, $IP_3R1$ seems to autophosphorylate, and displays protein kinase-like activity.[151]

The channel domain contains six putative membrane-spanning domains (MSDs), M1 to M6. The monoclonal antibody mAb18A10, which recognizes the carboxy-terminus of $IP_3R1$, blocks $IP_3$-induced calcium release but not $IP_3$-binding.[152,153] Between domains M5 and M6, there are two N-linked glycosylation sites and a proposed 'pore'-forming sequence[153] similar to domains H5 or SS1–SS2 in the ion channel superfamily, which includes voltage-sensitive $Ca^{2+}$, $Na^+$ and $K^+$ channels and nucleotide-gated ion channels. Interestingly, the last two membrane-spanning domains (M5 and M6), the putative 'pore' sequence, and the carboxy-terminal cytoplasmic tail following M6 are homologous to those of RyRs.[6,7] This suggests that these regions of $IP_3Rs$ and RyRs play an indispensable role in common channel function.

The $IP_3R$ protein forms a tetramer structure that is square in shape, approximately 25 nm wide,[130,138] and has a pinwheel appearance with four-fold symmetry.[130] The functional $IP_3R$-channel complex, therefore, comprises four subunits, each of which has one ligand-binding site at the tip of the large amino-terminal cytoplasmic arm, making it look like a tentacle.

**Table 13.1** Known agonists of RyR $Ca^{2+}$ channels

| Agent | Effective concentration | System tested* | Species and tissue | References |
|---|---|---|---|---|
| *Anthraquinones* | | | | |
| Dounorubicine | 1–100 $\mu$M | a,b | Rabbit skeletal, rat cardiac and brain | 71–73 |
| Digoxin | 1–20 nM | a,c | Sheep cardiac | 74 |
| Doxorubicin | 1–300 $\mu$M | a,b,c | Rabbit skeletal, rat cardiac, dog cardiac | 71, 75–77 |
| Mitoxantrone | 5–10 $\mu$M | a | Rabbit skeletal | 71 |
| Rubidazone | 15–150 $\mu$M | a | Rabbit skeletal | 71 |
| *Polyamines* | | | | |
| Polylysine | 1–10 $\mu$g/ml | a | Rabbit skeletal | 78,79 |
| Protamine | 1 $\mu$g/ml | a | Rabbit skeletal | 78 |
| Putrescine | 1–100 mM | b | Rabbit skeletal | 80 |
| Spermidine | 1–100 mM | b | Rabbit skeletal | 80 |
| Spermine | 0.4–20 mM | a,b | Rabbit skeletal | 78,80 |
| *Local anaesthetics* | | | | |
| Lidocaine | | | | |
| Tetracaine | 0.1–15 mM | b | Rabbit skeletal, sheep brain | 81,82 |
| | 0.1–2 mM | a,c | Frog skeletal, cultured rat skeletal | 83,84 |
| *Volatile anaesthetics* | | | | |
| Enflurane | 2 % vol. | b | Pig cardiac | 85 |
| Halothane | 1.5%, 2% vol.<br>0.1–0.5 mM | a,b,c | Dog and pig cardiac, frog skeletal | 85–87 |
| Isoflurane | 2%, 2.5% vol. | a | Dog and pig skeletal | 85,86 |
| *Fatty acid derivatives* | | | | |
| Long-chain acyl CoA | 50 $\mu$M | a | Rabbit skeletal | 88 |
| Arachidonic acid | 16–50 $\mu$M | a | Rabbit skeletal, dog cardiac | 88–90 |
| Acyl carnitines | 50 $\mu$M | a | Rabbit skeletal | 88 |
| Palmitoyl carnitine | 1–100 $\mu$M | a,b,c | Rabbit skeletal | 88 |
| Sphingosine | 30–50 $\mu$M | a | Rabbit skeletal | 91 |
| Stearic acid | 16–32 $\mu$M | a | Rabbit skeletal | 89 |

**Table 13.1** (contd)

| Agent | Effective concentration | System tested* | Species and tissue | References |
|---|---|---|---|---|
| *Scorpion toxins* | | | | |
| *Buthotus* venom | 0.1–500 μg/ml | b,c | Rabbit skeletal, bovine cardiac, rat brain | 92 |
| Imperatoxin A | 1–1,000 nM | b,c | Rabbit skeletal | 93 |
| *Ryanodine analogues* | | | | |
| Ester E | | | | |
| Ester F | 1–1000 nM | a | Rabbit skeletal, dog cardiac | 94 |
| | 1–1000 nM | a | Dog cardiac, rabbit skeletal | 94 |
| *Others* | | | | |
| 4-Alkylphenol | 10–25 nmol/mg | a | Rat skeletal | 95 |
| Caffeine | 1–100 mM | a,b,c | Rabbit skeletal, dog and rat cardiac | 72,75,78,96–102 |
| Chlorocresol | 0.1–100 mM | a | Rabbit skeletal | 103 |
| Cyclic ADP ribose | 1–17 μM | a | Rabbit skeletal | 104 |
| | 1–2 μM | a,b | Dog cardiac | 105 |
| Dithiothreitol | 0.5–1 mM | b | Rabbit skeletal | 106 |
| α-HCH | 6–100 μM | a | Rat cardiac | 107 |
| MBED | 0.3–10 | a | Rabbit skeletal | 108 |
| Perchlorate | 8–100 mM | a,b,c | Rabbit skeletal, rabbit cardiac | 109 |
| Porphyrin | 1–60 μM | a,b,c | Rabbit skeletal | 110 |
| Rose bengal | 1–200 nM | a,c | Rabbit skeletal | 111,112 |

* a: calcium-release studies; b: ryanodine-binding studies; c: single-channel studies.

**Table 13.2** Known antagonists of RyR $Ca^{2+}$ channels

| Agent | Effective concentration | System used* | Species and tissue | References |
|---|---|---|---|---|
| *Polyamines* | | | | |
| Gentamicin | 1–20 μM | a | Rabbit skeletal | 78 |
| Neomycin | 0.01–20 μM | a,b | Rabbit skeletal, dog and rat cardiac | 78,79,97,101,112,113 |
| *Local anaesthetics* | | | | |
| Benzocaine | | | | |
| Chlorpromazine | 1–10 mM | b | Rabbit skeletal | 82 |
| Dibucaine | 0.1–1.5 mM | b | Rabbit skeletal | 82 |
| Procaine | 0.08–1.8 mM | b | Rabbit skeletal, sheep brain | 81,82,115 |
| Tetracaine | 1–20 mM | a,b,c | Rabbit skeletal | 82,116–119 |
| | 0.01–2 mM | a,b,c | Rabbit skeletal, dog cardiac, sheep brain, rat liver | 75,81,82,97,118 |
| *Fatty acid derivatives* | | | | |
| Sphingosine | 0.1–10 μM | a,b | Rabbit skeletal | 91 |
| *Scorpion toxins* | | | | |
| Imperatoxin I | 1–1000 nM | b,c | Rabbit skeletal, bovine | 93 |
| *Ryanodine analogues* | | | | |
| Dihydroryanodine | | | | |
| Ester E | 1–1000 nM | b | Rabbit skeletal, dog cardiac | 94 |
| | 1–1000 nM | b | Rabbit skeletal, dog cardiac | 94 |
| Ester F | 1–12 μM | a | Rabbit skeletal | 94 |
| | 1–1000 nM | b | Dog cardiac, rabbit skeletal | 94 |
| | 1–12 μM | a | Rabbit skeletal | 94 |
| *Others* | | | | |
| Dantrolene | 23 nM | b | Rabbit skeletal | 131 |
| DCCD | 25–200 μM | a,b | Rabbit skeletal, sheep cardiac | 121,122 |
| FLA365 | 0.01–20 μM | a,b | Rabbit skeletal, dog and rat cardiac | 106,113,114,123 |
| Ruthenium red | 0.001–20μM | a,b,c | Rabbit skeletal, dog and rat cardiac | 35,73,101,112,114,116,123–128 |
| Verapamil | 1–10 000 μM | b,c | Rabbit skeletal | 120 |

* a: calcium-release studies; b: ryanodine-binding studies; c: single-channel studies.

### $IP_3R2$

$IP_3R2$ cDNAs have been cloned from rat brain[8] and human cell lines.[1] $IP_3R2$ comprises 2701 amino acids and shares 68–69% sequence identity with mammalian $IP_3R1$. The $IP_3$-binding and channel domains are well conserved (72–74% and 69–75%, respectively) among the $IP_3R$ family.[1] Expression of the amino-terminal 1078 amino acids of $IP_3R2$ demonstrates high-affinity binding for $IP_3$, with similar specifity, but higher affinity than observed for the ligand-binding domain of $IP_3R1$.[8] Considering the sequence similarity, the $IP_3R$ cloned from *Drosophila*[154] seems, if anything, to belong to this type 2 category of $IP_3Rs$.

### $IP_3R3$

$IP_3R3$ cDNAS have been isolated from rat cell line RINm5F (2670 amino acids)[10] and human cell lines (2671 amino acids).[1,10] $IP_3R3$ has 62% and 65% identity with the entire amino acid sequence of mammalian $IP_3R1$ and $IP_3R2$, respectively. The conserved amino acid sequences are, for the most part, concentrated in the $IP_3$-binding (72–77%) and channel (66–69%) domains.[1] Interestingly, truncated rat $IP_3R3$ binds $IP_6$ (inositol hexakisphosphate) at a relatively high level in addition to binding to $IP_3$ and $IP_4$ (inositol 1,3,4,5-tetrakisphosphate) at levels similar to those of $IP_3R1$.[10] $IP_3$-binding activity of the expressed full-length human $IP_3R3$ was also suppressed to 30% by $IP_6$.[1] The $IP_3R3$ amino-terminal (750 amino acids) was expressed in cDNA-transfected cells and showed $[^3H]IP_3$-binding activity with a $K_D$ of 151 nM.[11]

### Other types of $IP_3R$

In addition to being present on the ER, $IP_3R$ activity (i.e. the presence of a $IP_3$-binding and $IP_3$-activated channel) and $IP_3R$-like immunoreactivity (using anti $IP_3R1$ polyclonal antibody) have been observed on the plasma membrane in some cell types.[155–160] ER-$IP_3Rs$ function as intracellular calcium release channels, whereas plasma membrane $IP_3Rs$ may mediate $Ca^{2+}$ influx in response to increased intracellular $IP_3$. We do not know, however, whether the currently known types of $IP_3Rs$ are incorporated into the plasma membrane, but we expect to find a new member of the $IP_3R$ family, the plasma membrane $IP_3R$ type. Evidence for the existence of $IP_3R$ activity in nuclei has accumulated.[150] It was reported that $IP_3R$ function is involved in the fusion of post-mitotic nuclear vesicles of *Xenopus* eggs, and $IP_3R1$ immunoreactivity has been localized on the outer membrane of the nucleus of Purkinje cells.[1,129] We do not know whether the other types are also localized on the outer membrane of the nucleus, or whether $IP_3R$ is also localized on the inner membrane. The finding of $IP_3R$ associated to cell nuclei as well as both nucleoplasmic $Ca^{2+}$ and $IP_3$ transients[132,162] supports a role for these receptors in skeletal muscle.

## Acknowledgements

The author's work has been supported by MDA, FONDECYT and the European Economic Community.

## References

1. Furuichi, T., Kohda, K., Miyawaki, A., and Mikoshiba, K. (1994). *Current Opinions in Neurobiology*, **4**, 294–303.
2. Pozzan, T., Rizzuto, R., Volpe, P., and Meldolesi, J. (1994). *Physiological Reviews*, **74** (3), 595–636.
3. Sorrentino, V. and Volpe, P. (1993). *Trends Pharmacol. Sci.*, **14**, 98–103.
4. McPherson, P.S. and Campbell, K.P. (1993). *J. Biol. Chem.*, **268**, 13765–13768.
5. Coronado, R., Morrissette, J., Sukhareva, M., and Vaughan, D.M. (1994). *Am. J. Physiol.*, **266**, C1485–C1504.
6. Furuichi, T., Yoshikawa, S., Miyawaki, A., Wada, K., Maeda, N., and Mikoshiba, K. (1989). *Nature*, **342**, 32–38.
7. Mignery, G.A., Newton, C.L., Archer, B.T., III, and Südhof, T.C. (1990). *J. Biol. Chem.*, **265**, 12679–12685.
8. Südhof, T.C., Newton, C.L., Archer B.T., III, Ushkaryov, Y.A., and Mignery, G.A. (1991). *EMBO J.*, **10**, 3199–3206.
9. Ross, C.A., Danoff, S.K., Schell, M.J., and Snyder, S.H. (1992). *Proc. Natl Acad. Sci USA*, **89**, 4265–4269.
10. Blondel, O., Takeda, J., Janssen, H., Seino, S., and Bell, G.I. (1993). *J. Biol. Chem.*, **268**, 11356–11363.
11. Maranto, A.R. (1994). *J. Biol. Chem.*, **269**, 1222–1230.
12. Takeshima, H., Nishimura, S., Matsumoto, T., Ishida, H., Kangawa, K., Minamino, N., *et al.* (1989). *Nature*, **339**, 439–445.
13. Zorzato, F., Fujii, J., Otsu, K., Phillips M., Green, N.M., Lai, F.A., Meissner, G. and MacLennan, D.H. (1990). *J. Biol. Chem.*, **265**, 2244–2256.
14. Otsu, K., Willard, H.F., Khanna, V.K., Zorzato, F., Green N.M., and MacLennan, D.H. (1990). *J. Biol. Chem.*, **265**, 13472–13483.
15. Nakai, J., Imagawa, T., Hakamata, Y., Shigekawa, M., Takeshima, H., and Numa, S. (1990). *FEBS Lett.*, **271**, 169–177.
16. Giannini, G., Clementi, E., Ceci, R. Marziali, G., and Sorrentino, V. (1992). *Science*, **257**, 91–94.
17. Hakamata, Y., Nakai, J., Takeshima, H., and Imoto, K. (1992). *FEBS Lett.*, **312**, 229–235.
18. Gureau, M.H., Khrestchatisky, M., Heeren, M.A., Zambrowicz, E.B., Kim, H., Grisar, T.M., *et al.* (1992). *J. Biol. Chem.*, **267**, 8679–8684.
19. Blachly-Dyson, E., Zambronicz, E.B., Yu, W.H., Adams, V., Mc-Cabe, E.R.B., Adelman, J., *et al.* (1993). *J. Biol. Chem.*, **268**, 1835–1841.
20. Landry, D., Sullivan, S., Nicolaides, M., Redhead, C., Edelman, A., Field, M., *et al.* (1993). *J. Biol. Chem.*, **268**, 14948–14955.

21. Rüegg, J.C. (1992). *Calcium in muscle contraction*. Springer-Verlag, Berlin.
22. Martonosi, A.N. (1984). *Physiol. Rev.*, **64**, 1240–1320.
23. Somlyo, A.P. (1985). *Nature*, **316**, 298–299.
24. Ebashi, S. (1991). *Annu. Rev. Physiol.*, **53**, 1–16.
25. Jaimovich, E. (1991). *J. Muscle Res. Cell. Motility*, **12**, 316–320.
26. Berridge, M.J. and Irvine, R.F. (1984). *Nature*, **312**, 315–321.
27. Melzer, W., Rios, E., and Schneider, M.F. (1984). *Biophys. J.*, **45**, 637–641.
28. Melzer, W., Rios E., and Schneider, M.F. (1987). *Biophys. J.*, **51**, 849–863.
29. Smith, J.S., Coronado, R., and Meissner, G. (1985). *Nature*, **316**, 446–449.
30. Smith, J.S., Coronado, R., and Meissner, G. (1986). *J. Gen. Physiol.*, **88**, 573–588.
31. Fleischer, S., Ogunbunmi, E., Dixon, M., and Fleer, E. (1985). *Proc. Natl. Acad. Sci. USA*, **82**, 7256–7259.
32. Imagawa, T., Smith, J.S., Coronado, R., and Campbell, K.P. (1987). *J. Biol. Chem.*, **262**, 16636–16643.
33. Inui, M., Saito, A., and Fleischer, S. (1987). *J. Biol. Chem.*, **262**, 1740–1747.
34. Zorzato, F., Fujii, J., Otsu, K., Phillips, M., Green, N.M., Lai, F.A., Messner, G., and MacLennan, D.H. (1990). *J. Biol. Chem.*, **265**, 2244–2256.
35. Lai, F.A., Erickson, H.P., Rousseau, E., Liu, Q.Y., and Meissner, G. (1988). *Nature*, **331**, 315–319.
36. Saito, A., Inui, M., Radermacher, M., Frank, J., and Fleischer, S. (1988). *J. Cell Biol.*, **107**, 211–219.
37. Wagenknecht, T., Grassucci, R., Frank, J., Saito, A., Inui, M., and Fleischer, S. (1989). *Nature*, **338**, 167–170.
38. Block, B.A., Imagawa, T., Campbell, K.P., and Franzini-Armstrong, C. (1988). *J. Cell Biol.*, **107**, 2587–2600.
39. Ferguson, D.G., Schwartz, H.W., and Franzini-Armstrong, C. (1984). *J. Cell Biol.*, **99**, 1735–1742.
40. Williams, A.J. (1992). *J. Muscle Res. Cell. Mot.*, **13**, 7–26.
41. Stein, P. and Palade, P. (1988). *Biophys. J.*, **54**, 357–363.
42. Ma, J., Fill, M., Knudson, M.C., Campbell, K.P., and Coronado, R. (1988). *Science*, **242**, 99–102.
43. Fill, M., Coronado, R., Mickelson, J.R., Vilven, J., Ma, J., Jacobson, B.A., and Louis, C.F. (1990). *Biophys. J.*, **57**, 471–475.
44. Radermacher, M., Wagenknecht, T., Grassucci, R., Frank, J., Inui, M., Chadwick, C., and Fleischer, S. (1993). *Biophys. J.*, **61**, 936–940.
45. Bezprozvanny, I.B., Ondrias, K., Kaftan, E., Stoyanovsky, D.A., and Ehrlich, B.E. (1993). *Mol. Biol. Cell*, **4**, 347–352.
46. Ehrlich, B.E., Kaftan, E., Bezprozvannaya, S., and Bezprozvanny, I. (1994). *Trends Pharmacol. Sci.*, **15**, 145–149.

47. Chen, S.R.W., Vaughan, D.M., Airey, J.A., Coronado, R., and MacLennan, D.H. (1993). *Biochemistry*, **32**, 3743–3753.

48. Chen, S.R.W., Zhang, L., and MacLennan, D.H. (1992). *J. Biol. Chem.*, **267**, 23318–23326.

49. Chen, S.R.W., Zhang, L., and MacLennan, D.H. (1993). *J. Biol. Chem.*, **268**, 13414–13421.

50. Treves, S., Chiozzi, P., and Zorzato, F. (1993). *Biochem. J.*, **2291**, 757–763.

51. Zarka, A., and Shoshan-Barmatz, V. (1993). *Eur. J. Biochem.*, **213**, 147–154.

52. Suko, J., Maurer-Fogy, I., Plank, B., Bertel, O., Wyskovsky, W., Hohenegger, M., and Hellmenn, G. (1993). *Biochim. Biophys. Acta*, **1175**, 193–206.

53. Wang, J. and Best, P.M. (1992). *Nature*, **359**, 739–741.

54. Witcher, D.R., Kovacs, R.J., Schulman, H., Cefali, D.C., and Jones, L.R. (1991). *J. Biol. Chem.*, **266**, 11144–11152.

55. MacLennan, D.H. and Phillips, M.S. (1992). *Science*, **256**, 789–794.

56. Anderson, K., Lai, F.A., Liu, Q.-Y., Rousseau, E., Erickson, H.P., and Meissner, G. (1989). *J. Biol. Chem.*, **264**, 1329–1335.

57. Takasago, T., Imagawa, T., and Shigekawa, M. (1989). *J. Biochem. (Tokyo)*, **106**, 872–877.

58. Takasago, T., Imagawa, T., Furukawa, K.–I., Ogurusu, T., and Shigekawa, M. (1991). *J. Biochem. (Tokyo)*, **109**, 163–170.

59. Takeshima, H., Nishimura, S., Nishi, M., Ikeda, M., and Sugimoto, T. (1993). *FEBS Lett.*, **322**, 105–110.

60. Takeshima, H., Nishi, M., Iwabe, N., MIyata, T., Hosoya, T., Masai, I., and Hotta, Y. (1994). *FEBS Lett.*, **337**, 81–87.

61. Lee, H.C. (1993). *J. Biol. Chem.*, **268**, 293–299.

62. Morrissette, J., Heisermann, G., Cleary, J., Ruoho, A., and Coronado, R. (1993). *FEBS Lett.*, **330**, 270–274.

63. Mészáros, L.G., Bak, J., and Chu, A. (1993). *Nature*, **364**, 76–79.

64. Galione, A. (1993). *Science*, **259**, 325–326.

65. Rusino, N. and Lee, H.C. (1989). *J. Biol. Chem.*, **264**, 11725–11731.

66. Takasawa, S., Nata, K., Yonekura, H., and Okamoto, H. (1993). *Science*, **259**, 370–373.

67. Suárez-Isla, B.A., Irribarra, V., Oberhauser, A., Larralde, L., Bull, R., Hidalgo, C. and Jaimovich, E. (1988). *Biophys. J.*, **54**, 737–741.

68. Suárez-Isla, B.A., Alcayaga, C., Marengo, J.J., and Bull, R. (1991). *J. Physiol. (Lond.)*, **441**, 575–592.

69. White, A.M., Watson, S.P., and Galine, A. (1993). *FEBS Lett.*, **318**, 259–263.

70. Currie, K.P.M., Swann, K., Galione, A., and Scott, R.H. (1992). *Mol. Biol. Cell*, **3**, 1415–1525.

71. Abramson, J.J., Buck, E., Salama, G., Casida, J.E., and Pessah, I.N. (1988). *J. Biol. Chem.*, **263**, 18750–18758.
72. Zimanyi, I. and Pessah, I. (1991). *J. Pharmacol. Exp. Ther.*, **256**, 938–946.
73. Zimanyi, I. and Pessah, I.N. (1992). *Brain Res.*, **561**, 181–191.
74. McGarry, S.J. and Williams, A.J. (1993). *J. Pharmacol.*, **108**, 1043–1050.
75. Lee, Y.S., Ondrias, K., Duhl, A.J., Ehrlich, B.E., and Kim, H. (1991). *J. Membr. Biol.*, **122**, 155–163.
76. Ondrias, K., Borgatta, L., Kim, D.H., and Ehrlich, B.E. (1990). *Circ. Res.*, **67**, 1167–1174.
77. Pessah, I., Durie, E., Schiedt, M., and Zimanyi, I. (1990). *Mol. Pharmacol.*, **37**, 503–514.
78. Cifuentes, M.E., Ronjat, M., and Ikemoto, N. (1989). *Arch. Biochem. Biophys.*, **273**, 554–561.
79. Kang, J.J., Tarcsafalvi, A., Carlos, A.D., Fujimoto, E., Shahrokh, Z., Thevenin, B.J.M., Shohet, S.B., and Ikemoto, N. (1992). *Biochemistry*, **31**, 3288–3293.
80. Zarka, A. and Shoshan-Barmatz, V. (1992). *Biochim. Biophys. Acta*, **1108**, 13–20.
81. Martin, C., Ashley, R., and Shoshan-Barmatz, V. (1993). *FEBS Lett.*, **328**, 77–81.
82. Shoshan-Barmatz, V. and Zchut, S. (1993). *J. Membr. Biol.*, **133**, 171–181.
83. Jaimovich, E. and Rojas, E. (1994). *Cell Calcium*, **15**, 356–368.
84. Bull, R. and Marengo, J.J. (1990). *Biophys J.*, **57**, 343A.
85. Connelly, T.J., El-Hayek, R., Rusy, B.F., and Coronado, R. (1992). *Biochem. Biophys. Res. Commun.*, **186**, 595–600.
86. Frazer, M.J. and Lynch, C. III. (1992). *Anesthesiology*, **77**, 316–323.
87. Bull, R. and Marengo, J.J. (1994). *Am. J. Physiol.*, **266**, C391–C396.
88. El-Hayek, R., Valdivia, C., Valdivia, H.H., Hogan, K., and Coronado, R. (1993). *Biophys. J.*, **65**, 779–789.
89. Cardoso, C. and De Meis, L. (1993). *Biochem. J.*, **296**, 49–52.
90. Dettbarn, C. and Palade, P. (1993). *Biochem. Pharmacol.*, **45**, 1301–1309.
91. Sabbadini, R.A., Betto, R., Teresi, A., Fachechi-Cassano, G., and Salvati., G. (1992). *J. Biol. Chem.*, **267**, 15475–15484.
92. Valdivia, H.H., Fuentes, O., El-Hayek, R., Morrissette, J. and Coronado, R. (1991). *J. Biol. Chem.*, **266**, 19135–19138.
93. Valdivia, H.H., Kirby, M.S., Lederer, W.J., and Coronado, R. (1992). *Proc. Natl. Acad. Sci. USA*, **89**, 12185–12189.
94. Humerickhouse, R.A., Besch, H.R., Gerzon, K., Ruest, L., Sutko, J.L. and Emmick, J.T., (1993). *Mol. Pharmacol.*, **44**, 412–421.
95. Beeler, T.J. and Gable, K.S., (1993). *Arch. Biochem. Biophys.*, **301**, 216–220.
96. Holmberg, S.R.M. and Williams, A.J. (1990). *Biochim. Biophys. Acta*, **1022**, 187–193.
97. Meissner, G. and Henderson, J.S. (1987). *J. Biol. Chem.*, **262**, 3065–3073.

98. Moutin, M. and Dupont, Y., (1988). *J. Biol. Chem.*, **263**, 4228–4235.
99. Pessah, I.N., Stambuk, R.A., and Casida, J.E., (1987). *Mol. Pharmacol.*, **31**, 232–238.
100. Su, J.Y. and Hasselbach, W. (1984). *Pflügers. Arch.*, **400**, 14–21.
101. Wyskovsky, W., Hohenegger, M., Plank, B., Hellmann, G., Klein, S. and Suko, J., (1990). *Eur. J. Biochem.*, **194**, 549–559.
102. Nagasaki, K. and Kasai, M. (1983). *J. Biochem. (Tokyo)*, **94**, 1101–1109.
103. Zorzato, F., Scutari, E., Tegazzin, V., Clementi, E., and Treves, S. (1993). *Mol. Pharmacol.*, **44**, 1192–1201.
104. Morrissette, J., Heisermann, G. Cleary, J., Ruoho, A., and Coronado, R. (1993). *FEBS Lett.*, **330**, 270–274.
105. Meszaros, L.G., Bak, J., and Chu, A. (1993). *Nature*, **364**, 76–79.
106. Zimanyi, I., Buck, E., Abramson, J., Mack, M., and Pessah, I. (1992). *Mol. Pharmacol.*, **42**, 1049–1057.
107. Pessah, I.N., Mohr, F.C., Schiedt, M. and Joy, M.R. (1992). *J. Pharmacol. Exp. Ther.*, **262**, 661–669.
108. Seino, A., Kobayashi, M. Kobayashi, J., Fang, Y., Ishibashi, M., Nakamura, H., Momose, K., and Ohizumi, Y. (1991). *J. Pharmacol. Exp. Ther.*, **256**, 861–867.
109. Ma, J., Anderson, K., Shirokov, R., Levis, R., Gonzalez, A., Karhaner, M., Hosey, M.M., Meissner, G. and Rios, E. (1993). *J. Gen. Physiol.*, **102**, 423–448.
110. Abramson, J.J., Milne, S., Buck, E. and Pessah, I.N. (1993). *Arch. Biochem. Biophys.*, **301**, 396–403.
111. Stuart, J., Pessah, I.N., Favero, T.G. and Abramson, J.J., (1992). *Arch. Biochem. Biophys.*, **292**, 512–521.
112. Xiong, H., Buck, E., Stuart, J., Pessah, I.N., Salama, G., and Abramson, J.J. (1992). *Arch. Biochem. Biophys.*, **292**, 522–528.
113. Calviello, G. and Chiesi, M. (1989). *Biochemistry*, **28**, 1301–1306.
114. Mack, M., Zimanyi, I., and Pessah, I. (1992). *J. Pharmacol. Exp. Ther.*, **262**, 1028–1037.
115. Nash-Adler, P., Louis, C.F., Fudyma, G., and Katz, A.M. (1980). *Mol. Pharmacol.*, **17**, 61–65.
116. Meissner, G. (1984). *J. Biol. Chem.*, **259**, 2365–2374.
117. Ogawa, Y. and Ebashi, S. (1976). *J. Biochem. (Tokyo)*, **80**, 1149–1157.
118. Xu, L., Jones, R., and Meissner, G. (1993). *J. Gen. Physiol.*, **101**, 207–233.
119. Zahradnikova, A., and Palade, P. (1993). *Biophys. J.*, **64**, 991–1003.
120. Valdivia, H., Valdivia, C., Ma, J., and Coronado, R. (1990). *Biophys. J.*, **58**, 471–481.
121. Michalak, M., Dupraz, P., and Soshan-Barmatz, V. (1988). *Biochim. Biophys. Acta.*, **939**, 587–594.
122. Yamamoto, N. and Kasai, M. (1982). *J. Biochem. (Tokyo)* **92**, 485–496.
123. Chiesi, M., Schwaller, R. and Calviello, G. (1988). *Biochem. Biophys. Res. Commun.*, **154**, 1–8.

124. Anderson, K., Lai, F.A., Liu, Q.-Y., Rousseau, E., Erickson, H.P., and Meissner, G. (1989). *J. Biol. Chem.*, **264**, 1329–1335.

125. Buck, E., Zimanyi, L. Abramson, J.J. and Pessah, I.N. (1992). *J. Biol. Chem.*, **267**, 23560–23567.

126. Hymel, L., Inui, M. Fleisher, S. and Schindler, H. (1988). *Proc. Natl Acad. Sci. USA*, **85**, 441–445.

127. Lai, F., Anderson, K. Rousseau, E. Liu, Q. and Meissner, G. (1988). *Biochem. Biophys. Res. Commun.*, **151**, 441–449.

128. Ma, J. (1993). *J. Gen. Physiol.*, **102**, 1031–1056.

129. Berridge, M.J. (1993). *Nature*, **361**, 315–325.

130. Chadwick CC, Saito A, Fleischer S. Proc. Natl. Acad. Sci USA. 87:2132–2136, 1990.

131. Moschella, M., Watras, J., Jayaraman, T., and Marks, A. (1995). *J. Muscle Res. Cell Motillity*, **16**, 390–400.

132. Jaimovich, E. and Liberona, J.L. (1997). *Biophys. J.*, **72**, A120.

133. Liberona, J.L., Caviedes, P., Tascón, S., Hidalgo, J., Giglio, J.R., Sampaio, S.V., Caviedes, R., and Jaimovich, E. (1997). *J. Muscle Res. Cell Motillity.*, (in press).

134. Danoff S.K., Ferris, C.D., Donath, C., Fischer, G., Munemitsu, S., Ullrich, A., and Snyder, S.H. (1991). *Proc. Natl Acad. Sci. USA*, **88**, 2951–2955.

135. Nakagawa, T., Okano, H., Furuichi, T., Aruga, J., and Mikoshiba, K. (1991). *Proc. Natl Acad. Sci. USA*, **88**, 6244–6248.

136. Kume, S., Muto, A., Aruga, J. Nakagawa, T., Michikawa, T., Furuichi, T., Nakade, S., Okano, H., and Mikoshiba, K. (1993). *Cell*, **873**, 555–570.

137. Mourey, R.J., Estevez, V.A., Marecek, J.F., Barrow, R.K., Prestwich, G.D., and Snyder, S.H. (1993). *Biochemistry*, **32**, 1719–1726.

138. Maeda, N., Niinobe, M., and Mikoshiba, K. (1990). *EMBO J.*, **9**, 61–67.

139. Van Delden, C., Foti, M., Lew, D.P., and Krause, K.-H. (1993). *J. Biol. Chem.*, **268**, 12443–12448.

140. Takahashi, M., Tanzawa, K., and Takahashi, S. (1993). *J. Biol. Chem.*, **268**, 369–372.

141. Yamamoto, H., Maeda, N., Niinobe, M., Miyamoto, E. and Mikoshiba, K. (1989). *J. Neurochem.*, **53**, 917–923.

142. Hajnóczky, G., Gao, E., Nomura, T., Hoek, J.B., and Thomas, A.P. (1993). *Biochem. J.*, **293**, 413–422.

143. Nakade, S., Rhee, S.K., Hamanaka, H., and Mikoshiba, K. (1994). *J. Biol. Chem.*, **269**, 6735–6742.

144. Ferris, C.D., Huganir, R.L., and Snyder, S.H. (1990). *Proc. Natl Acad. Sci. USA* **87**, 2147–2151.

145. Bezprozvanny, I. and Ehrlich, B.E. (1993). *Neuron*, **10**, 1175–1184.

146. Finch, E.A., Turner, T.J., and Goldin, S.M. (1991). *Science*, **252**, 443–446.

147. Iino, M. and Endo, M. (1992). *Nature*, **360**, 76–78.

148. Zhang, B.-X., Zhao, H., and Muallem, S. (1993). *J. Biol. Chem.*, **268**, 10997–11001.

149. Ferris, C.D., Huganir, R.L., Bredt, D.S., Cameron, A.M., and Snyder, S.H. (1990). *Proc. Natl Acad. Sci. USA*, **88**, 2232–2235.

150. Matter, N. Titz, M.-F., Freyernuth, S., Rogue, P., and Malviya, A.N. (1993). *J. Biol. Chem.*, **268**, 732–736.

151. Ferris, C.D., Cameron, A.M., Bredt, D.S., Huganir, R.L., and Snyder, S.H. (1992). *J. Biol. Chem*, **267**, 7036–7041.

152. Nakade, S., Maeda, N., and Mikoshiba, K. (1991). *Biochem. J.*, **277**, 125–131.

153. Michikawa, T., Hamanaka, H., Otsu, H., Yamamoto, A., Miyawaki, A., Furuichi, T., Tashiro, Y., and Mikoshiba, K. (1994). *J. Biol. Chem.*, **269**, 9184–9189.

154. Yoshikawa, S. Tanimura, T., Miyawaki, A., Nakamura, M., Yuzak, M., Furuichi, T., and Mikoshiba, K. (1992). *J. Biol. Chem.*, **267**, 16613–16619.

155. Khan, A.A., Steiner, J.P., and Snyder, S.H. (1992). *Proc. Natl Acad. Sci. USA*, **89**, 2849–2853.

156. Khan, A.A., Steiner, J.P., Klein, M.G., and Schneide M.F., and Snyder, S.H. (1992). *Science*, **257**, 815–818.

157. McDonald, T.V., Premack, B.A., and Gardner, P. (1993) *J. Biol. Chem.*, **268**, 3889–3896.

158. Sharp, A.H., Snyder, S.H., and Nigam, S.K. (1992) J. Biol. Chem. **267**, 7444–7449.

159. Fadool, D.A. and Ache, B.W. (1992). *Neuron*, **9**, 907–918.

160. Fujimoto, T., Nakade, S., Miyawaki, A., Mikoshiba, K., and Ogawa, K. (1992). *J. Cell Biol.*, **119**, 1507–1513.

161. Sullivan, K.M., Busa, W.B., and Wilson, K.L. (1993). *Cell*, **73**, 1411–1422.

162. Reyes, R. and Jaimovich, E. (1996). *Arch. Biochem. Biophys.*, **331**, 41–47.

# 14 Basic properties of calcium release channels in neural cells

P. Kostyuk and A. Verkhratsky

## Introduction

The resting cytoplasmic free calcium concentration ($[Ca^{2+}]i$) is maintained in all eukaryotic cells at a level not exceeding 0.1 $\mu$M (about four orders of magnitude lower than in the extracellular solution), but cellular excitation induces transient rise of $[Ca^{2+}]i$ up to several $\mu$M, or to even higher levels in tiny cellular compartments. These transient fluctuations of $[Ca^{2+}]i$ (attributed as 'calcium signals') trigger or regulate various intracellular events, including excitability, contraction, neurotransmitter release, synaptic plasticity, gene expression during development, differentiation, and ageing. The generation of calcium signal (Fig. 14.1) is determined by interaction of (i) two families of $Ca^{2+}$ permeable channels incorporated in plasmalemma and membrane of endo(sarco)plasmic reticulum (ER or SR) which form the pathways for $Ca^{2+}$ influx into the cytoplasm; (ii) cytosolic $Ca^{2+}$ buffering; (iii) $Ca^{2+}$ extrusion due to the activity of plasmalemmal $Ca^{2+}$ pumps; and (iv) $Ca^{2+}$ sequestration by intracellular calcium stores and mitochondria (for review, see refs. 1,2). The $Ca^{2+}$ influx into the cytoplasm which forms a rising phase of the calcium signal arrives from two sources, namely from extracellular space and internal stores through a particular subclass of $Ca^{2+}$ permeable channels incorporated in the membrane of the calcium stores, thus producing cytoplasmic $Ca^{2+}$ release. The properties of these channels were extensively studied in skeletal and cardiac muscle fibres, but much less in neural cells, although in the latter case they are also of great functional importance. The present review will be focused especially on the properties of neuronal calcium stores and the corresponding release channels.

## Calcium stores in neurones

Neural cells contain a number of $Ca^{2+}$-buffering systems which are able to accumulate and store $Ca^{2+}$ ions, thus contributing to cytoplasmic $Ca^{2+}$ homoeostasis. Cytoplasmic $Ca^{2+}$ can be buffered by cytoplasmic $Ca^{2+}$ binding proteins, sequestrated by the Golgi apparatus or cellular nucleus, pumped into the low-affinity mitochondrial $Ca^{2+}$ stores, or into rapidly

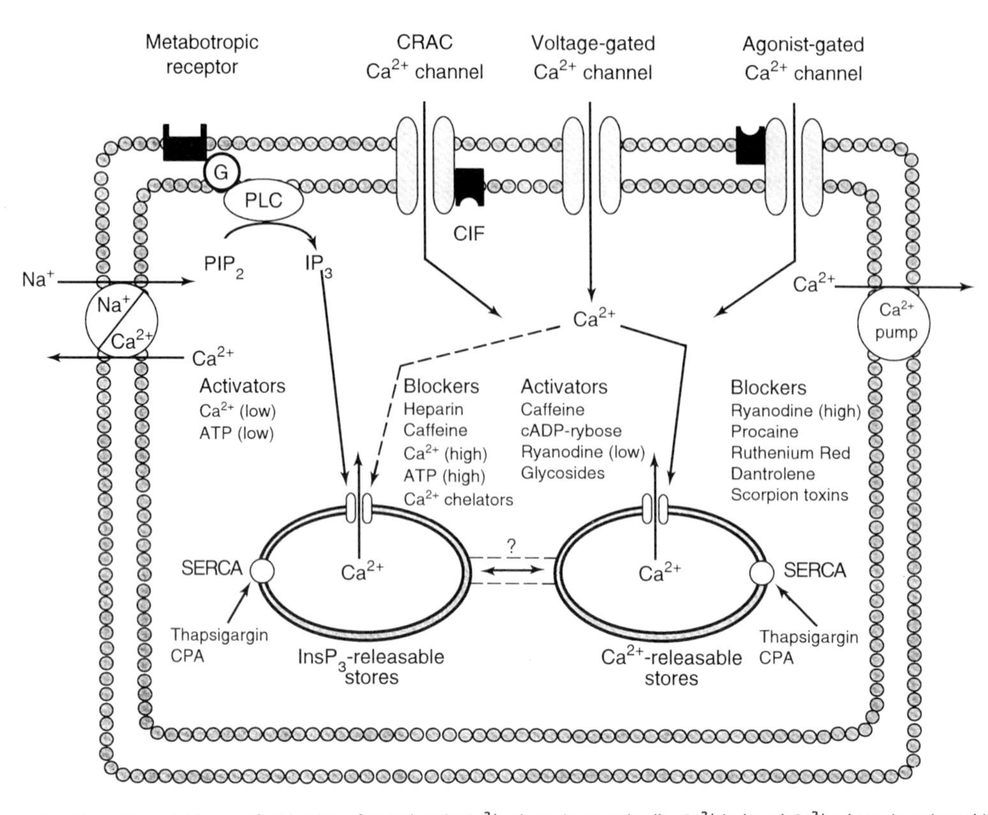


**Fig. 14.1** General scheme of activation of cytoplasmic $Ca^{2+}$ release in neural cells. $Ca^{2+}$-induced $Ca^{2+}$ release is activated in response to the increase of $[Ca^{2+}]_i$ due to the transmembrane $Ca^{2+}$ influx via voltage- and/or agonist-operated $Ca^{2+}$ channels. The $InsP_3$-activated release is activated by $InsP_3$ arriving from phosholipase C (PLC)-driven hydrolysis of the phosphatidylinositol 4,5-biphosphate ($PIP_2$). The filling of calcium stores by releasable $Ca^{2+}$ is achieved due to the activity of SERCA-pumps; the latter can be blocked by thapsigargin and cyclopyasonic acid (CPA). Excessive $Ca^{2+}$ ions can be eliminated from the cytoplasm by extrusion to the extracellular space through plasmalemmal $Ca^{2+}$ pumps and $Na^+/Ca^{2+}$ exchange mechanisms; in parallel, cytoplasmic $Ca^{2+}$ is buffered by proteins or (at some circumstances) pumped into mitochondria. Emptying of calcium stores can in turn trigger $Ca^{2+}$ influx into cytoplasm by activating CRAC membrane channels. CIF: calcium influx factor.

exchanging high-affinity stores associated with SR or ER. The latter contains $Ca^{2+}$ pumps known as sarcoplasmic–endoplasmic reticulum $Ca^{2+}$ ATPase-pumps (SERCA), which are effectively and selectively blocked by thapsigargin in nanomolar concentrations,[3] $Ca^{2+}$ release channels and $Ca^{2+}$ binding proteins such as calsequestrin and calreticulin. This type of $Ca^{2+}$ store is able not only to accumulate $Ca^{2+}$ ions, but also to release them in response to appropriate stimuli. It is still unknown whether $Ca^{2+}$ stores do coincide with the entire ER or they are restricted to its specific modified portions. Reticular structures are distributed quite evenly throughout the cell; in neurones, both the rough ER enriched by ribosomes and smooth ER form a continuous membrane network. Calcium stores are believed to be associated with smooth ER.[1,4]

Basically, ER calcium stores in neurones can be separated in at least two types which are controlled by two different second messengers, namely by inositol 1,4,5-trisphosphate ($InsP_3$) or by $Ca^{2+}$ ions (see refs. 1,4–6 and Fig. 14.1). Distinct spatial separation of both release mechanisms has been shown in many cases. In some *Aplysia* neurones a large response to

$InsP_3$ injection has been observed only at distances 120–160 μm below the surface.[7] In snail neurones, $Ca^{2+}$-dependent release was predominant in peripheral cytosol, whereas $InsP_3$-dependent release occurred around the nucleus.[8] Both stores are not overlapping or only partially overlapping in adrenal chromaffin cells.[9,10] It should be noted that, at least in chromaffin cells the presence of a third type of stores which is highly sensitive to both $InsP_3$ and caffeine–Ryanodine, has been postulated.[11]

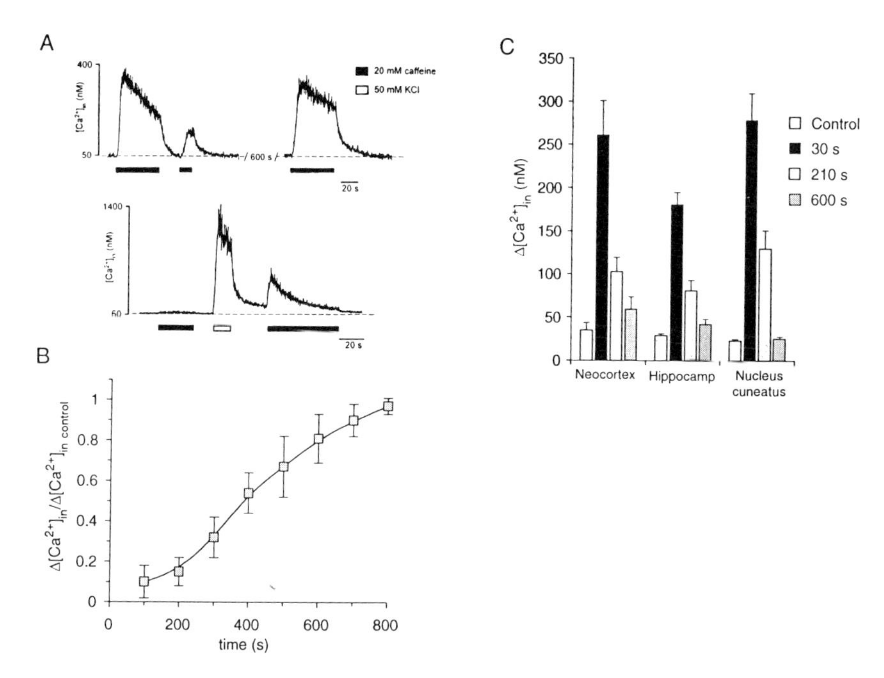


**Fig. 14.2** Caffeine-induced $[Ca^{2+}]_i$ transients in rat sensory and central neurones. A. Changes of $[Ca^{2+}]_i$ in response to caffeine applications measured from indo-1/AM loaded dorsal root gangia (top) and neocortical neurones. In all cases, caffeine was applied to a $Ca^{2+}$-free external solution in order to exclude possible transmembrane calcium influx. Times of application of caffeine (20 mM) and high-potassium (50 mM) are indicated below $[Ca^{2+}]_i$ trace. B. Time course of refilling of caffeine-sensitive $Ca^{2+}$ stores in dorsal root ganglion neurone. The amplitudes of caffeine-induced $[Ca^{2+}]_i$ transients recorded with different intervals between caffeine applications are normalized to the amplitude of the first (control) caffeine-mediated $[Ca^{2+}]_i$ response. Experimental points represent the average value collected from 16 cells; bars denote S.E.M. C. Spontaneous loss of releasable calcium from caffeine-sensitive pools in central neurones. Mean amplitudes (bars correspond to S.E.M.) of caffeine-mediated $[Ca^{2+}]_i$ transients recorded from various types of cultured central neurones under control conditions and at different times (30 s, 210 s and 600 s) after loading depolarization are shown. Experiments were performed on 23 neocortical, 49 hippocampal, and 29 nucleus cuneatus neurones. (Reproduced from ref. 25 with modifications.)

As in muscle fibres, the second messengers regulate the open probability of appropriate $Ca^{2+}$ release channels, thus changing the conductance of the ER membrane for $Ca^{2+}$. The availability of $Ca^{2+}$ stores for $Ca^{2+}$ release is controlled by both ER membrane $Ca^{2+}$ conductance and the amount of releasable $Ca^{2+}$ inside the ER lumen. The concentration of releasable $Ca^{2+}$ is determined by the activity of SERCA pumps (Fig. 14.2) and can also involve the activation of additional $Ca^{2+}$ influx from extracellular space via $Ca^{2+}$ release-activated $Ca^{2+}$ channels (CRAC),[12] activated by a special messenger – calcium influx factor (CIF)[13] (Fig. 14.1). The direct refilling of calcium stores from the extracellular space[14] or the preferential location of SERCA-rich parts of the ER in close proximity to plasmalemmal $Ca^{2+}$ channels[15] have also been suggested. As already mentioned, the SERCA pumps can be pharmacologically inhibited by thapsigargin and cyclopiasonic acid, thus preventing the replenishment of $Ca^{2+}$ stores. It is important that the $Ca^{2+}$ content of the stores can now be determined directly using the mag-fura-2 technique.[16] This dye accumulates in subcellular compartments; after cell permeabilization with digitonin and incubation in 'intracellular' buffer the cytoplasmic dye is released, and later the $Ca^{2+}$ release from the stores can be measured in a pure form.

Different morphological approaches revealed an uneven distribution of $Ca^{2+}$ and $InsP_3$ sensitive calcium stores in mammalian brain. At the cellular level, the ER structures containing $Ca^{2+}$ binding proteins are present in almost all parts of neuronal cells, including soma, axons, nerve terminals, dendrites, and dendritic spines. Functionally, both caffeine-mediated and $InsP_3$-mediated $[Ca^{2+}]_i$ release also has been found in all neuronal structural parts. However, the $Ca^{2+}$-sensitive and $InsP_3$-sensitive $Ca^{2+}$ release channels can be unevenly distributed within subcellular compartments. For instance, the ER localized in the dendritic spines of Purkinje neurones is primarily equipped with $InsP_3$-sensitive $Ca^{2+}$ release mechanism, but lacks both a $Ca^{2+}$-sensitive mechanism and the $Ca^{2+}$ storage protein (calsequestrin), whereas the juxtaspine regions of dendrites are rich in all these components.[17] Therefore within the dendritic spines, $[Ca^{2+}]_i$ changes of intracellular origin probably differ in the mechanism (exclusively $InsP_3$-dependent) from those in the dendritic stalk. Distinct separation of both release mechanisms is obvious from the studies of their presence in different brain structures by immunohistochemical methods: in cortical cells $InsP_3$-dependent mechanisms were found in cell bodies and proximal dendrites, $Ca^{2+}$-dependent mechanisms in apical dendrites.[18] Antibodies against the cardiac type of $Ca^{2+}$-sensitive $Ca^{2+}$ release channel stained mainly in the somata of Purkinje neurones.[17]

## Basic properties of $Ca^{2+}$-induced $Ca^{2+}$ release in neurones

### Caffeine-induced calcium release in neurones

Caffeine is one of the most popular and convenient tool for studying the properties of $Ca^{2+}$ sensitive $Ca^{2+}$ stores. It is generally accepted that caffeine interacts with $Ca^{2+}$-activated $Ca^{2+}$ release channels of the endoplasmic reticulum, making them more sensitive to $Ca^{2+}$, so that the $Ca^{2+}$ release mechanism can be activated even at resting levels of $[Ca^{2+}]_i$.[19] Within this framework, it can be argued that the responsiveness of caffeine-sensitive internal calcium stores is determined by: (i) relation between cytoplasmic calcium level and actual threshold of ER $Ca^{2+}$ channel, and (ii) the electro-driving force for $Ca^{2+}$ ions between the ER lumen and the cytoplasm, i.e. by releasable calcium content of internal stores. Caffeine-induced $Ca^{2+}$ release effectively deprives internal stores, which can be observed as a decrease of the store responsiveness to caffeine in response to its subsequent application.[20,21] The refilling of calcium stores is presumably a result of competition between the activity of SERCA-pumps located in the endoplasmic reticulum (which are responsible for $Ca^{2+}$ uptake by internal stores) and plasmalemmal calcium pumps (which underlie $Ca^{2+}$ extrusion from a cell). Blockade of $Ca^{2+}$ extrusion by $La^{3+}$ decreased both the rate of calcium store depletion and the velocity of the decay of the caffeine-induced $[Ca^{2+}]_i$ transients, reflecting an increased recirculation of $Ca^{2+}$ between the ER and the cytoplasm.[21] Moreover, it is believed that deprivation of internal stores of releasable $Ca^{2+}$ stimulates the $Ca^{2+}$ uptake by the ER. Quite often, following the wash-out of caffeine, a subresting drop of $[Ca^{2+}]_i$ is observed (the so-called 'post-caffeine undershoot').[20,21] These undershoots probably reflect the activity of SERCA-pumps which are thought to be responsible for the reaccumulation of $Ca^{2+}$ ions by internal stores, and to be a sign of the activation of SERCA-pumps following the depletion of internal stores.

The caffeine-evoked transient $[Ca^{2+}]_i$ elevation has been described for a variety of neuronal cells from both peripheral (sensory[21]; sympathetic[20,22,23]) and central[24–26] nervous systems. Though calcium stores have quite similar pharmacological properties, they differ in $Ca^{2+}$ ion handling in peripheral and central neurones. In peripheral sensory[21,25] and sympathetic[20] neurones under resting conditions, calcium stores are continuously filled with releasable $Ca^{2+}$ and after discharging they spontaneously refill. In contrast (Fig. 14.2), in central neurones, caffeine-sensitive stores contain minute amount of releasable $Ca^{2+}$ under resting conditions; nevertheless, they can be rapidly but transiently charged by depolarization-triggered $Ca^{2+}$ entry. In central neurones, in contrast to peripheral neurones, the calcium stores display a spontaneous depletion of releasable $Ca^{2+}$.[25]

In another class of central neurones – cerebellar Purkinje neurones – the situation seems to be even more complicated. It appeared that the amplitude of the caffeine-induced $Ca^{2+}$ release is controlled by the resting $[Ca^{2+}]_i$ level, which precedes the caffeine application. In resting conditions, application of 20 mM caffeine to Purkinje neurones located in a cerebellar slice had either no effect on $[Ca^{2+}]_i$ or induced a tiny $[Ca^{2+}]_i$ response. However, increase in the resting $[Ca^{2+}]_i$ led to an almost linear increase in the amplitude of the caffeine-mediated $[Ca^{2+}]_i$ response.[27] It seems that, in Purkinje neurones, resting $[Ca^{2+}]_i$ controls both the store filling state and the availability of $Ca^{2+}$-sensitive $Ca^{2+}$ release channel to be open in response to appropriate stimuli.

The ability of caffeine to induce $Ca^{2+}$ liberation from intracellular stores appeared to be concentration-dependent. The threshold caffeine concentration which induces caffeine-mediated $[Ca^{2+}]_i$ transients in sensory neurones was found to be in the range of 1 mM;[21] at higher concentrations, caffeine generated $[Ca^{2+}]_i$ transients with larger amplitude and a higher rate of rise. It is believed that, at concentrations below 5 mM, caffeine action is modulated by $[Ca^{2+}]_i$; at higher concentration caffeine releases $Ca^{2+}$ independently of the presence of $Ca^{2+}$ ions in the cytoplasm.[19]

## $Ca^{2+}$-induced $Ca^{2+}$ release in neurones

It is well established that, under physiological conditions, the calcium-induced calcium release (CICR) from internal stores plays an important role in generation of $Ca^{2+}$ signals in muscle cells.[28] In particular, in myocytes CICR acts as one of the main sources of $Ca^{2+}$ ions triggering contraction. In neuronal cells, the participation of CICR in modulation of depolarization-induced $[Ca^{2+}]_i$, rise is less obvious. Several reports have been published showing that a part of the cytoplasmic calcium signal during membrane depolarization is due to $Ca^{2+}$-triggered $Ca^{2+}$ release from caffeine-and ryanodine-sensitive internal $Ca^{2+}$ stores.

Such observations have been made on PC12 chromaffin cells,[29] bullfrog sympathetic neurones,[20] and *Xenopus* sensory neurones;[30] however, in some other observations small or no CICR was seen in neuronal cells.[22] The experimental design applied for discovery of CICR in neurones is based on pharmacological modification of the state of $Ca^{2+}$-sensitive $Ca^{2+}$ release channel or the filling of calcium stores with subsequent monitoring of the parameters of depolarization-triggered $[Ca^{2+}]_i$ transients (like their amplitudes and speed of rise). It has been reported for bullfrog sympathetic neurones[20] and rat sensory neurones,[21] that sensitization of ER $Ca^{2+}$ channels in the presence of low caffeine concentrations significantly increased the rate of rise of the depolarization-induced $[Ca^{2+}]_i$, transients, indicating an additional $Ca^{2+}$ release from intracellular stores. The treatment of neurones by 1 mM caffeine increased the sensitivity of the CICR mechanism, and $Ca^{2+}$ entry during depolarization evoked a larger release compared with control conditions. The depletion of calcium stores by prolonged exposure to high-caffeine-containing external solution had the opposite effect: when internal stores were emptied of releasable $Ca^{2+}$ both the rate of rise and the amplitude of depolarization-induced $[Ca^{2+}]_i$, elevation were decreased. Blockade of the CICR by ryanodine or dantrolene also reduced the amplitude of depolarization-activated $[Ca^{2+}]_i$ transients.[20,21] The involvement of CICR in the generation of $[Ca^{2+}]_i$ transients evoked by cell depolarization was clearly shown for bullfrog sympathetic neurones,[31] where ryanodine and dantrolene significantly reduced (up to 50%) the amplitudes of $[Ca^{2+}]_i$ transients in response to cell depolarization. Moreover, when two or three $[Ca^{2+}]_i$ transients were induced by a sequence of pulses from $-60$ mV to 0 mV, the second and third transients were facilitated, and this facilitation was completely blocked by dantrolene. It seems that, in contrast to muscle cells where CICR mechanisms develop as a regenerative (all or none) process, in neurones CICR has a rather gradual character.[31] In sympathetic neurones, the amount of $Ca^{2+}$ release increased with the increase of $Ca^{2+}$ entry. This property of CICR mechanism can be meaningful for $Ca^{2+}$-dependent cytoplasmic signaling in neuronal cells. It should be mentioned that CICR was much more prominent in cultured sympathetic neurones, as compared with freshly isolated cells,[31,32] suggesting the involvement of unknown metabolic mechanisms in regulation of calcium stores functional availability.

Interestingly, in isolated chromaffine cells in divalent cation-deficient external solutions, the caffeine-induced rise of $[Ca^{2+}]_i$ is accompanied by a large elevation of $[Na^+]_i$; manipulations with $[Na^+]_i$ modulate the caffeine-mediated $[Ca^{2+}]_i$ transients, suggesting a possible sodium dependence of cytoplasmic $Ca^{2+}$ release in these cells.[33]

For central neurones, the existence of CICR in normal physiological conditions (i.e. in the absence of exogenous activators) remains unclear. So far, the involvement of CICR mechanism in the generation of physiologically evoked $[Ca^{2+}]_i$ transients has been postulated for hippocampal[34] and cerebellar granule cells.[35] In both types of cells inhibition of CICR by ryanodine or dantrolene, as well as calcium store depletion by thapsigargin treatment, significantly attenuated the NMDA-activated $[Ca^{2+}]_i$ transients.

## $InsP_3$-Induced $Ca^{2+}$ release in neurones

$InsP_3$ is generated in response to the activation of phospholipase C (PLC) which is coupled to a variety of plasmalemmal receptors including those for glutamate, acetylcholine, bradykinin, substance P, tyrotropine-releasing factor, and many other neurotransmitters and hormones (for review see, ref. 5). The signal transduction between plasmalemmal receptors and PLC involves G proteins. Activated PLC hydrolyses the membrane-bound phospholipid phosphatidylinositol 4,5-biphosphate ($PIP_2$), forming diacylglycerol (DAG, an activator of protein kinase C) and $InsP_3$. A number of receptors coupled with $InsP_3$ turnover, which are termed as 'metabotropic' receptors, have been discovered in the nervous system.[36]

The glutamate-activated metabotropic receptors have been found to increase the cytoplasmic level of $InsP_3$ in a variety of mammalian neurones, including striatum neurones,[37] cerebellar granule cells,[38] and hippocampal neurones.[39] Subsequently, the glutamate-triggered $InsP_3$- mediated cytoplasmic $[Ca^{2+}]_i$ responses were shown in cultured hippocampal,[40] cerebellar granule neurones,[41] and cerebellar Purkinje neurones in slices.[42] The $InsP_3$-mediated mechanisms seem also to be involved in acetylcholine-,[15, 41] bradykinin-[15, 43] and platelet-activating factor[44] mediated elevation of $[Ca^{2+}]_i$ in both peripheral and central neurones.

A down-regulation of the $InsP_3$-mediated $[Ca^{2+}]_i$ release mechanism obviously also exists. It has been observed that in chick sensory neurones, carbachol and bradykinin may produce only a weak response in some cells, while direct injection of $InsP_3$ or GTP$\gamma$S induce large $[Ca^{2+}]_i$ transients. Pretreatment with protein kinase C inhibitors staurosporine or H-7 increased the agonist-induced responses, while application of phorbol esters abolished it completely.[45] Thus, the down-regulation might be induced by activation of protein kinase C; the activity of latter could be controlled by DAG co-produced with $InsP_3$.

## Molecular architecture of calcium release channels

$Ca^{2+}$ release channels are oligomeric proteins which span the membrane of SR or ER. These channels can be relatively easily withdrawn from the cell for ultrastructure analysis due to the fact that channel proteins bind specifically and with high affinity to $InsP_3$ (for $InsP_3$-sensitive channels) and ryanodine (for $Ca^{2+}$-sensitive channels); the latter are also known as ryanodine receptors. Using $InsP_3$ and ryanodine as highly specific tools, both types of $Ca^{2+}$ release channels have been purified and characterized; moreover, the cDNAs for these channel proteins have been cloned, sequenced, and expressed.

### $Ca^{2+}$-sensitive $Ca^{2+}$ release channels

The $Ca^{2+}$ release channels were first purified from skeletal and cardiac muscles. It appeared that $Ca^{2+}$ release channels belong to the family of channel proteins, and at least two types of channel have been described, namely the skeletal muscle type and the cardiac muscle type. The skeletal muscle $Ca^{2+}$ release channel is coupled directly to the plasmalemmal high-voltage-activated (L) $Ca^{2+}$ channels through the so-called 'foot' proteins, thus forming a functionally active structure – the 'triade'. In skeletal muscle, the plasmalemmal depolarization directly activates $Ca^{2+}$ release from sarcoplasmic reticulum, due to the fact that plasmalemmal $Ca^{2+}$ channels serve as a voltage sensor, transducing the activating signal directly to $Ca^{2+}$ release channel via the foot protein.[46] Thus, the skeletal muscle $Ca^{2+}$ stores possess depolarization-induced mechanisms of $Ca^{2+}$ release. In contrast, in cardiac muscle, the sarcoplasmic $Ca^{2+}$ release channels are not connected with the plasmalemma, and require an increase in $[Ca^{2+}]_i$ as an appropriate stimulus. The cDNAs for both types of $Ca^{2+}$ release channels have been cloned from rabbit and human[47–49] preparations. Being incorporated in the planar bilayer, the corresponding protein forms channels activated by $Ca^{2+}$ (50 nM) in the presence of ATP and is sensitive to ryanodine.[50] Cryoelectron microscopy of native channels from skeletal muscle with resolution of 3 nm has shown that they have a central cavity with radial channels and a peripheral vestibulus.[51] This is a preferred substrate for Ca/calmodulin-dependent phosphorylation occuring at a single site (serine-2809). Phosphorylation at this site activates the channel.[52]

The extension of these experiments to nervous tissue demonstrated that both types of $Ca^{2+}$ sensitive $Ca^{2+}$ release channel are expressed in mammalian brain, and with some species specificity. In the mouse brain, the skeletal muscle type of $Ca^{2+}$ release channel appeared to be expressed preferentially in the cerebellum, while the cardiac type occurred both in the cerebellum and in other brain regions. However, data concerning the expression of subtypes of $Ca^{2+}$-sensitive $Ca^{2+}$ release channels seem to be controversial – some reports shows that in chicken[53] and rabbit[54] brain the skeletal muscle type of $Ca^{2+}$ release channels dominates, while others state the existence in brain of only the cardiac subtype of $Ca^{2+}$-sensitive $Ca^{2+}$ release channels.[55,56] They are present in dendrites, cell-bodies, and terminals of rat forebrain, and are highly enriched in the hippocampus.[57] Northern blot analysis also revealed in the rabbit brain only the mRNA encoding the cardiac type of $Ca^{2+}$-sensitive release channel.[49] Data from physiological experiments also favour the existence of only the $Ca^{2+}$-sensitive type of $Ca^{2+}$ release in neurones, whereas nobody has yet shown the existence of depolarization-induce $Ca^{2+}$ release in nerve cells.

From the structural point of view, the $Ca^{2+}$-sensitive $Ca^{2+}$ release channel in skeletal muscle SR is a homotetramer composed of monomers with a molecular weight of ~500 kDa.[58] The terminal domains of these monomers protrude into the cytoplasm, and four transmembrane domains near the C-terminal probably form the channel itself. The primary structure of the ryanodine receptor/$Ca^{2+}$ release channel monomer in the brain is similar to that in cardiomyocytes,[59] comprising 4872 amino acids;[60] it is also a tetramer that is distinct from the $InsP_3$-receptor.[55]

## $InsP_3$-sensitive $Ca^{2+}$ release channels

$InsP_3$-sensitive release channels were first detected in the SR from skeletal muscle fibres. They were half-maximum activated by 15 $\mu$M $InsP_3$ and blocked by $La^{3+}$ ions.[61] Later on, they were also purified from the brain of mammals and their amino acid sequence has been determined. It has been shown that $InsP_3$-sensitive channels are derived from the family of three or four distinct genes, and are characterized by their different sensitivity to $InsP_3$.[62] Although $InsP_3$- and $Ca^{2+}$ sensitive release channels show little homology, their structure is to a certain extent similar. The $InsP_3$-sensitive release channel is also thought to be a homotetramer (the molecular weight of the single monomer is in the range of ~-260 KDa) with four transmembrane regions. In Purkinje neurones, they are maximally activated by 5–10 $\mu$M $InsP_3$ released by flash-photolysis from caged compound. In astrocytes, higher concentrations of $InsP_3$ are needed for activation.[63]

## Single-channel properties of $Ca^{2+}$ release channels

Both types of purified $Ca^{2+}$ release channels have been characterized at the single-channel level using brain microsomes or purified channels reconstituted in planar lipid bilayers.

The $Ca^{2+}$-sensitive $Ca^{2+}$ release channel from brain microsomal membranes has multiple conductance states with the most frequently observed single-channel conductance of 100 pS (at 54 mM $Ca^{2+}$) and $P_{Ca2+}/P_{Tris+}$ about 11.[64] Other divalent cations pass the channel in the order $Ba^{2+} > Ca^{2+} > Sr^{2+} > Mg^{2+}$. They retain high-affinity binding sites for ryanodine ($K_d$ ~ 3 nM) and can be activated by caffeine, ATP, and AMP-PCP.[64,65] The open state probability of $Ca^{2+}$-sensitive release channel increases substantially at $[Ca^{2+}]_i$ levels between 0.1 and 1 $\mu$M, while higher levels of $[Ca^{2+}]_i$, (10 to 1000 $\mu$M) inhibit channel openings.[66]

The single-channel properties of the neuronal $InsP_3$-sensitive release channel have been also studied on purified cerebellar proteins incorporated into lipid membranes. The conductance of $InsP_3$-sensitive release channels has been estimated at the level of 10–26 pS, this being four-fold smaller compared with the $Ca^{2+}$-sensitive $Ca^{2+}$ release channels, which also show multiple conductance states (about four).[66,67] Based on the conductance and kinetics of single cerebellar $InsP_3$-gated channels, the amount of $Ca^{2+}$ ions released during each opening could be estimated: each single opening of the $InsP_3$-sensitive $Ca^{2+}$ release channel liberated ~4000 $Ca^{2+}$ ions from the internal store.[68] Binding of the $InsP_3$ to the channel causes its opening; however, cytosolic $Ca^{2+}$ ions act as a coagonist together with $InsP_3$. The $[Ca^{2+}]_i$ dose–response curve for $InsP_3$ receptor has a bell-shaped form, with a maximum channel open probability at $[Ca^{2+}]_i$ levels of 0.2–0.3 $\mu$M. The increase in $[Ca^{2+}]_i$ to levels exceeding 1 $\mu$M causes a sharp decrease in the open probability of the $InsP_3$-sensitive $Ca^{2+}$ release channel.[66]

An important novel finding is a blocking effect of caffeine on $InsP_3$-sensitive channels from cerebellar microsomes. At 5 mM concentration, caffeine produced a 3.3-fold increase in the duration of closed time of the channel, with little effect on the mean open time. Increase in $InsP_3$ to 20 $\mu$M partially reversed the effect. Thus, the action of caffeine on $Ca^{2+}$ release must be considered with caution.[69] The similar warning has been appeared recently concerning the validity of fura-2-based measurements of $InsP_3$-mediated $Ca^{2+}$ release: it was demonstrated[70] that $InsP_3$ binding to the cerebellar $Ca^{2+}$ release channel can be competitively inhibited by fura-2 (with $IC_{50}$ ~120 $\mu$M) as well as BAPTA ($IC_{50}$ ~340 $\mu$M) and EDTA ($IC_{50}$ ~8.7 mM).

In general, biophysical properties of both types of release channels in brain tissue seem to be quite similar to those of SR channels. In cardiac fibres, the $Ca^{2+}$-sensitive channels have unitary conductance of 70 pS at 50 mM $Ca^{2}$,[71] whereas in vascular smooth muscle fibres it is 110 pS at 100 mM $Ca^{2+}$.[72] The open probability is increased by caffeine (0.5–2 mM), the effect is observed at 0.1–10 mM $Ca^{2+}$, but absent at 80 nM; at higher caffeine concentrations (> 5 mM), channel openings become $Ca^{2+}$-insensitive.[19] Probably, an identical model may explain these properties of release channels in different tissues. The channel has several open and closed states, and $Ca^{2+}$ interact with at least one closed state.[73] A single-ion occupancy is suggested with four energy barriers and three binding sites in the channel. Discrimination between ions may be due to a high-affinity central binding site with $K_d = 150$ $\mu$M.[74,75]

## Pharmacological modulators of $Ca^{2+}$-sensitive $Ca^{2+}$ release channels

The $Ca^{2+}$-sensitive $Ca^{2+}$ release channels are modulated by numerous substances. Cations, like $Mg^{2+}$, $H^{+}$, and $Ba^{2+}$, inhibit openings of purified $Ca^{2+}$ release channels isolated from muscle SR.[6] It has also been found that $Ba^{2+}$ ions in submillimolar concentrations could effectively block the $Ca^{2+}$ and caffeine-induced calcium release from isolated SR and from ER of snail[76] and sensory[21] neurones. In contrast, ATP promotes openings of $Ca^{2+}$-sensitive $Ca^{2+}$ release channels.[6] The activity of these channels is also regulated by a number of pharmacological agents described below (see also Table 14.1).

**Table 14.1** Pharmacological modulators of $Ca^{2+}$ release channels.

| Channel | Modulator | Effect | Reference |
|---|---|---|---|
| $Ca^{2+}$ sensitive $Ca^{2+}$ release channel | Caffeine | Activation | 19 |
| | Ryanodine | Activation (1 $\mu$M) Blockade (50–100$\mu$M) | 55 |
| | Procaine | Blockade | 21 |
| | Dantrolene | Blockade | 86 |
| | Ruthenium red | Blockade | 66 |
| | Digitoxin | Activation | 90 |
| | Imperatoxin I | Blockade | 89 |
| | Imperatoxin II | Activation | 89 |
| | *Buthotus hottentota* toxin | Activation | 88 |
| | Doxorubicine (Adriamicine) | Blockade | 91 |
| | cADP-ribose | Activation | 92 |
| | Palmitoyl carnitine | Activation | 94 |
| $InsP_3$-sensitive $Ca^{2+}$ release channel | Heparin | Blockade | 66 |
| | Caffeine | Blockade | 98 |
| | $Ca^{2+}$ chelators | Blockade | 70 |

### Ryanodine

Ryanodine is a plant alkaloid[77] which binds specifically to the $Ca^{2+}$-activated $Ca^{2+}$ channels of endoplasmic reticulum and substantially modulates its function. The binding of ryanodine to the channel is reported to be use-dependent: ryanodine preferentially binds to the open channel.[78] Following binding, ryanodine at 1–10 $\mu$M concentrations locks the channel in a subconductance state (with a preferential conductance of around 40 pS[79] that causes the inhibition of both $Ca^{2+}$- and caffeine-mediated calcium release due to the prevention of $Ca^{2+}$ accumulation by internal stores or, at higher concentrations (around 100 $\mu$M) induces blockade of the $Ca^{2+}$-activated $Ca^{2+}$ release channel.[55] For skeletal and heart muscles, positive cooperativity of ryanodine binding to the $Ca^{2+}$ release channel has been reported; the $Ca^{2+}$ release channel appears to have two ryanodine binding sites with $K_d$ 5–10 nM and ~3 $\mu$M, respectively. Ryanodine binding to the high-affinity site stabilizes the channel in the open state, while occupation of the low-affinity site locks the channels in the closed state.[80] Recently, basic ryanodine derivatives (amino- and guanidinoacrylryanodines) with enhanced affinity to SR $Ca^{2+}$ release channels have been also described.[81]

In both peripheral and central neurones, ryanodine treatment completely and irreversibly blocked the caffeine-induced $[Ca^{2+}]_i$ transients. In cultured bovine chromaffin cells, ryanodine at concentrations of 0.4–50 $\mu$M suppressed caffeine-mediated $[Ca^{2+}]_i$ elevation and catecholamine secretion.[82] At similar concentrations (0.1–100 $\mu$M), ryanodine inhibited caffeine-evoked $[Ca^{2+}]_i$ transients in cultured rat sensory,[21] hippocampal, neocortical, and nucleus cuneatus[25] neurones, as well as in mouse cortical neurones.[83] Examples of such an inhibitory action of ryanodine on the caffeine-induced $[Ca^{2+}]_i$ transients in rat sensory and hippocampal neurones are shown in Fig. 14.3. In these experiments[21], we used a protocol designed to obtain a maximal release in response to each caffeine application. For this purpose we alternatively exposed the cell to high-potassium (30 or 50 mM $K^+{}_o$) solution (in order to load internal stores by releasable $Ca^{2+}$) and caffeine (10 or 20 mM)-containing solutions. In this case, control applications of caffeine after depolarization-induced loading of internal stores evoked $[Ca^{2+}]_i$ transients with similar properties. After the addition of 10 $\mu$M ryanodine to the extracellular solution, only the first caffeine application was able to induce a $[Ca^{2+}]_i$ transient, while the following ones failed to release calcium. Obviously, this property reflects the use-dependent interaction of $Ca^{2+}$ release channels with ryanodine.

In peripheral neurones, ryanodine also inhibits several mechanisms thought to be coupled with $Ca^{2+}$ release from internal stores, such as after-hyperpolarization (AHP) in sympathetic neurones.

It has also to be noted that the treatment of sensory neurones by ryanodine caused a slow-down of the kinetics of recovery of depolarization-evoked $[Ca^{2+}]_i$ transients, probably reflecting the decrease of total buffer capacity due to the blockade of $Ca^{2+}$ accumulation in internal stores.[21]

## Ruthenium red

Ruthenium red has been reported to block opening of the purified SR $Ca^{2+}$-sensitive $Ca^{2+}$ release channels as well as cerebellar $Ca^{2+}$-sensitive $Ca^{2+}$ release channels incorporated into lipid bilayers.[66] Ruthenium red effectively inhibited caffeine-induced $[Ca^{2+}]_i$ elevation in Purkinje neurones in slices;[27] however, in bullfrog sympathetic neurones, ruthenium red failed to inhibit caffeine-induced $Ca^{2+}$ release.[84]

## Procaine

The local anaesthetic, procaine, is known to inhibit reversibly the $Ca^{2+}$-activated $Ca^{2+}$ release in muscle fibres[85] and in bullfrog sympathetic neurones.[84] In DRG and hippocampal neurones, similarly to the above mentioned cells, procaine also caused an inhibition of

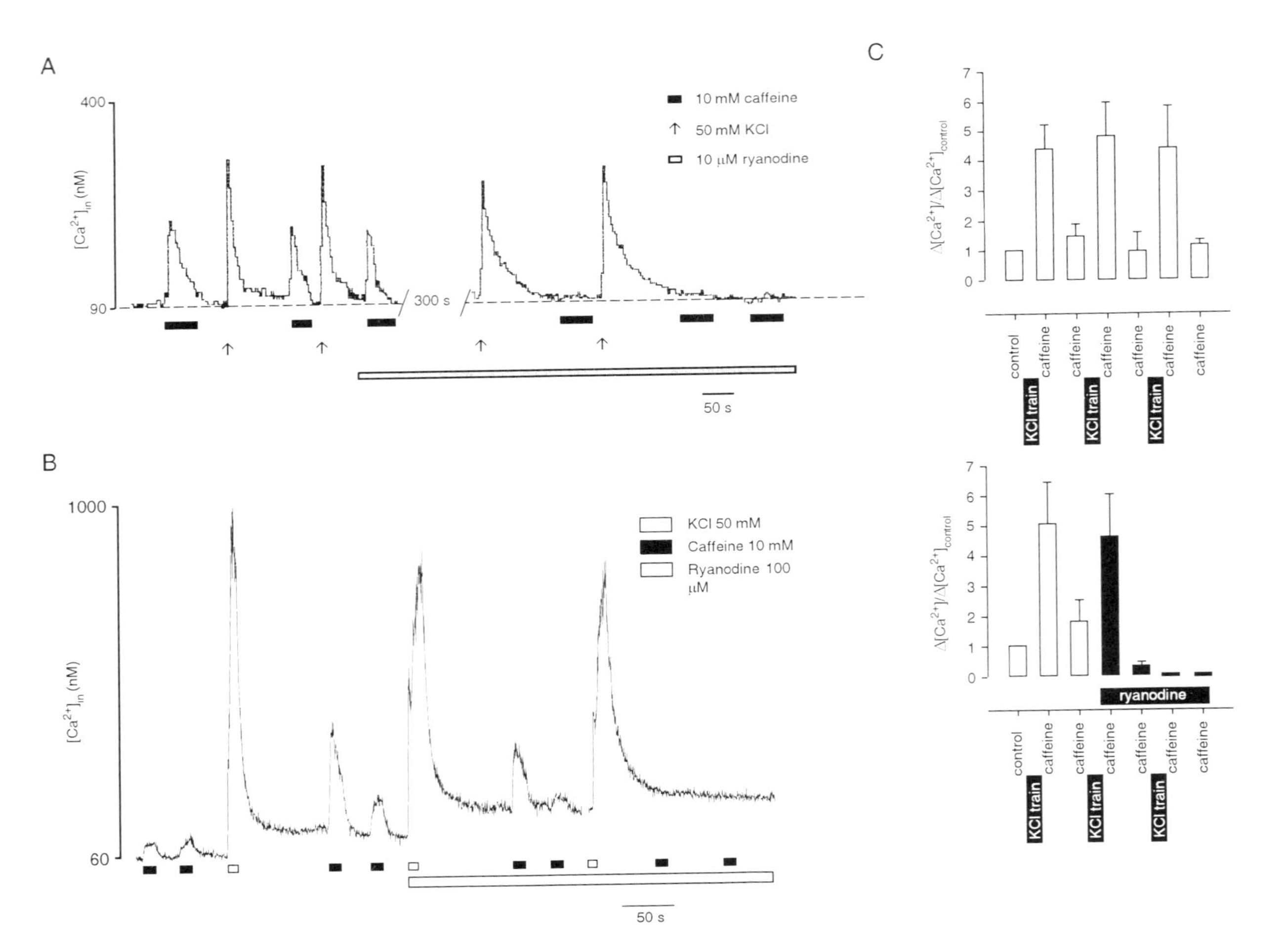


**Fig. 14.3** Ryanodine inhibition of the caffeine-induced $[Ca^{2+}]_i$ transients in rat dorsal root ganglia (A) and hippocampal (B) neurones. Times of application of caffeine (10 mM), high-potassium (50 mM) and ryanodine (10 $\mu$M) are indicated below $[Ca^{2+}]_i$ trace. Ryanodine had no significant effect on the first $[Ca^{2+}]_i$ transient evoked by caffeine in the presence of ryanodine, but the second and following responses were completely blocked. Wash-out of ryanodine did not restore the caffeine-induced $[Ca^{2+}]_i$ transients. (C) Normalized amplitudes of caffeine-induced $[Ca^{2+}]_i$ transients recorded from hippocampal neurone in control (top) and after external ryanodine (10 $\mu$M) administration (bottom). This graph clearly shows the use-dependent action of ryanodine. (Reproduced from ref. 21, with modifications.)

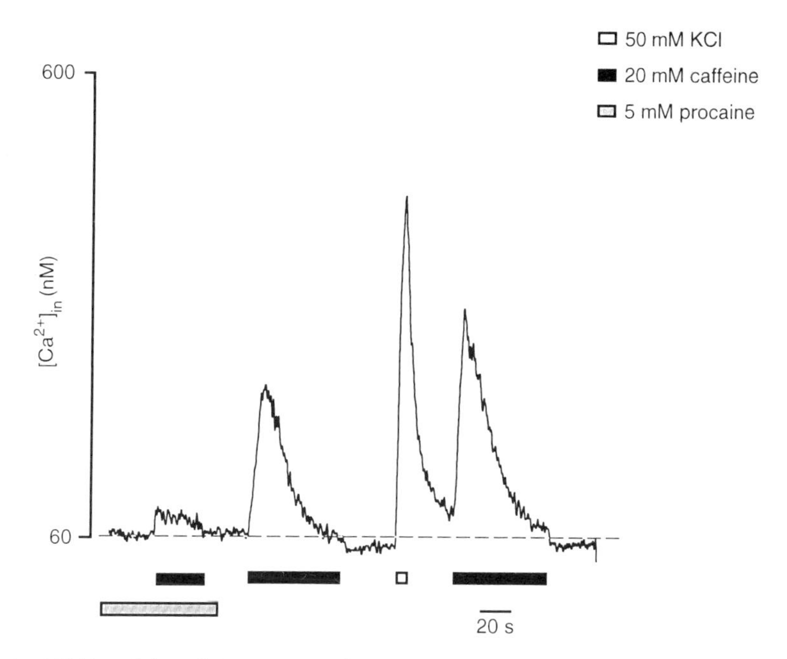


**Fig. 14.4** Procaine inhibition of the caffeine-induced $[Ca^{2+}]_i$ transients in cultured rat dorsal root ganglion neurone. The duration of caffeine and procaine applications are indicated below the $[Ca^{2+}]_i$ traces. The cell was bathed in 5 mM procaine-containing external solution 120 s before the start of the recording. Note that procaine blocked the undershoot following caffeine wash-out. (Reproduced from ref. 21.)

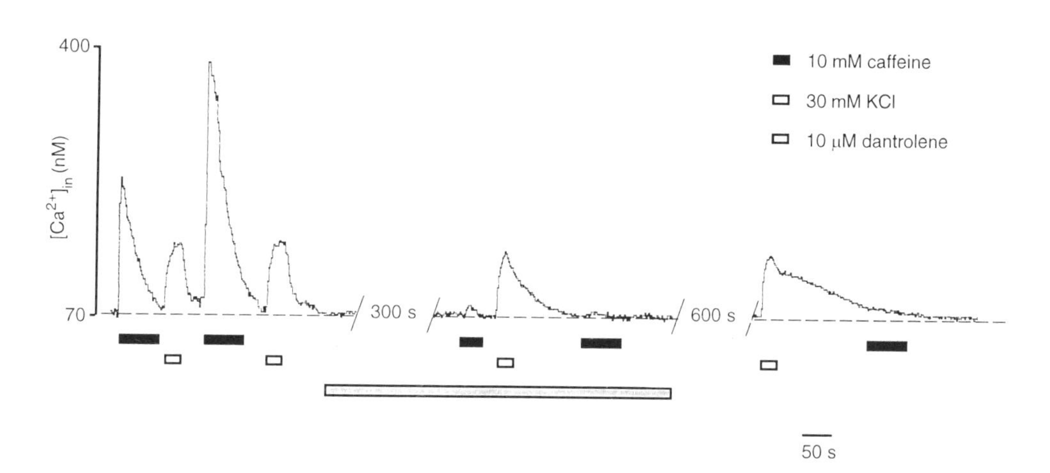


**Fig. 14.5** The effect of dantrolene sodium on the caffeine-induced $[Ca^{2+}]_i$ transients in rat dorsal root ganglion neurone. The duration of caffeine and dantrolene applications are indicated below the $[Ca^{2+}]_i$ traces. Note that blockade of internal calcium stores by dantrolene slightly diminished the amplitude and slowed down the recovery of the depolarization-induced $[Ca^{2+}]_i$ transients. The time constants of the recovery of the depolarization-induced $[Ca^{2+}]_i$ transients were 6.3 s under control conditions, and 28.9 s after treatment with dantrolene. (Reproduced from ref. 21.)

caffeine-induced $Ca^{2+}$ release from internal stores.[21] Figure 14.4 demonstrates an example of such an inhibition. In contrast to the ryanodine inhibition, procaine actually showed not a use-dependent, but rather a time-dependent blockade of caffeine-induced $Ca^{2+}$ release. Usually it was necessary to incubate the cell in procaine-containing solution for 30–40 s to achieve an effective inhibition of caffeine-induced $[Ca^{2+}]_i$ transients. The wash-out of procaine by normal physiological solution completely restored the ability of caffeine to release $Ca^{2+}$ from internal stores.

### Dantrolene

It has been reported that dantrolene sodium (Dantrium) blocked $Ca^{2+}$-induced $Ca^{2+}$ release from sarcoplasmic reticulum vesicles and inhibited $Ca^{2+}$ liberation from internal calcium stores in muscle fibres[86] and nerve cells.[21,87] Like procaine, dantrolene exhibited a clear time-dependent inhibition of caffeine-induced $[Ca^{2+}]_i$ release: usually the blockade of $[Ca^{2+}]_i$ transients developed within 2–3 minutes after addition of dantrolene into the external solution (Fig. 14.5). Simultaneously with inhibition of caffeine-induced $[Ca^{2+}]_i$ transients, dantrolene slightly reduced the amplitude of depolarization-induced $[Ca^{2+}]_i$ transients and significantly slowed down its recovery.[21]

### Neurotoxins

Recently, several new toxins reported to interact with $Ca^{2+}$-sensitive $Ca^{2+}$ release channels have been purified and their activity tested on SR $Ca^{2+}$ release channels. The toxin with a molecular weight of 5–8 kDa purified from the venom of the scorpion *Buthotus hottentota* opens $Ca^{2+}$-sensitive $Ca^{2+}$ release channels with a $K_d$ of 20–30 nM,[88] while two other toxins obtained from the venom of the scorpion *Pandinus imperator* (Imperatoxins) with molecular weights 10.5 and 8.7 kDa exert a dual action on these channels: the first blocks them with $ED_{50} = 10$ nM, while the second activates with $ED_{50} = 6$ nM.[89]

### Cardiac glycosides

Surprisingly, one of the oldest known cardiac glycosides, digitoxin, believed to act via inhibition of plasmalemmal Na/K-ATPase, appears to bind with $K_d$ 12.5 nM to cardiac $Ca^{2+}$-sensitive $Ca^{2+}$ release channels; moreover, digitoxin caused $Ca^{2+}$ release from cardiac $Ca^{2+}$ stores.[90]

### Doxorubicine

On cardiac SR $Ca^{2+}$ release channels the action of another modulator, a doxorubicine (Adriamicine), a widely used chemotherapeutic agent, has been described.[91] Doxorubicine caused an initial activation of $Ca^{2+}$-sensitive $Ca^{2+}$ release channels which turned to irreversible inhibition after several minutes incubation with doxorubicine.

Although the effects of neurotoxins and doxorubicine have been tested only on muscle preparations, it would be very interesting to monitor the effects of all these substances on brain $Ca^{2+}$ release channels.

## Endogenous modulators of $Ca^{2+}$-sensitive $Ca^{2+}$ release channels

### cADP-ribose

Although for some time caffeine was the only known activator of the $Ca^{2+}$-sensitive release channel, an endogenous substance which may function as a physiological modulator of $Ca^{2+}$-induced $Ca^{2+}$ release has recently been found. The candidate for such a role is a $NAD^+$ metabolite, cyclic ADP-ribose, which derives from NAD as a result of the activity of the ADP-ribosilyl cyclase. This enzyme has been described in a variety of tissues, including brain.[92] The cADP-ribose has a molecular weight of 541 Da and appeared to be the most potent $Ca^{2+}$-mobilizing agent; it has been reported that cADP-ribose releases $Ca^{2+}$ from sea urchin egg microsomes with a $EC_{50}$ of 18 nM.[92] The effects of cADP-ribose could be abolished by pretreatment with drugs known to interact with $Ca^{2+}$-sensitive $Ca^{2+}$ release channels, such as caffeine, ryanodine, procaine, and ruthenium red,[92] suggesting that $Ca^{2+}$-sensitive $Ca^{2+}$ release channels are a target for cADP-ribose. Up to now, the ability of cADP-ribose to release $Ca^{2+}$ in neurones remained untested. However, it has been recently shown to release $Ca^{2+}$ from brain microsomes at a concentration of $\sim$150 nM; this release was suppressed by ryanodine.[93]

### Fatty acid metabolites

Palmitoyl carnitine, an endogenous fatty acid metabolite, has been reported to directly activate $Ca^{2+}$-sensitive $Ca^{2+}$ release channels from skeletal muscle SR at micromolar concentrations.[94] Palmitoyl carnitine increased the open time of $Ca^{2+}$ release channels incorporated in planar bilayers by a factor of seven. It was suggested that in this way the metabolism of fatty acids can modulate $[Ca^{2+}]_i$ in myocytes; moreover, this pathway may be involved in the development of muscle disorders in palmitoyl transferase II-deficient patients.

## Pharmacological modulators of InsP3-sensitive $Ca^{2+}$ release channels

The functional activity of $InsP_3$-sensitive $Ca^{2+}$ release channel can be inhibited by $H^+$ [95] and $Mg^{2+}$ [96] ions and enhanced by ATP in submillimolar concentrations. ATP alone is not sufficient to open the cerebellar $InsP_3$-sensitive channel incorporated in lipid bilayers, but addition of ATP (0.1–2 mM) in the presence of $InsP_3$ led to a several-fold increase in the channel open probability.[97] The dose–response curve for ATP action on $InsP_3$-sensitive $Ca^{2+}$ release channels was bell-shaped: at high ATP concentrations (>4 mM), a decrease of the $InsP_3$-gated channel open probability was observed. Moreover, it seems that the $InsP_3$-sensitive $Ca^{2+}$ release channel can be modulated by phosphorylation, and also appeared sensitive to caffeine, with 0.1–10 mM caffeine inhibiting $Ca^{2+}$ release activated by photo-released $InsP_3$ in *Xenopus* oocytes.[98]

The single pharmacological tool which interacts directly with $InsP_3$-sensitive $Ca^{2+}$ channel is heparin, which is believed to compete with $InsP_3$ for the binding site. Heparin effectively blocks the opening of $InsP_3$-gated channels at micromolar concentrations[97] (see also Table 14.1).

## Interactions between $Ca^{2+}$ stores

It seems that, in different types of neurones, $InsP_3$- and $Ca^{2+}$-sensitive $Ca^{2+}$ release mechanisms can either operate on separate calcium stores or that a single functional ER compartment can posess both release mechanisms. In cultured hippocampal neurones, activation of glutamate metabotropic receptors induced $Ca^{2+}$ release from $InsP_3$-sensitive stores, but did not deplete caffeine- and $Ca^{2+}$-sensitive pools.[43] In contrast, in PC12 cells[15] and in cerebellar granule cells,[41] caffeine and metabotropic agonists demonstrated cross-depletion of the presumably single $Ca^{2+}$-storing ER compartment. Another way for the interaction between the stores can be mediated via $Ca^{2+}$ ions: these, when released from one ER compartment, may activate additional $Ca^{2+}$ release from other compartments. Such a mechanism has been described for bovine adrenal chromaffin cells: in this preparation, histamine-triggered $InsP_3$-mediated $Ca^{2+}$ release supplies $Ca^{2+}$-sensitive $Ca^{2+}$ release channels with $Ca^{2+}$, producing a subsequent $Ca^{2+}$ release from caffeine – ryanodine-sensitive stores.[11,99]

## Functional role of intracellular $Ca^{2+}$ release

### Regulation of membrane excitability

Fluctuations of $[Ca^{2+}]_i$ can effectively control membrane excitability via modulation of a variety of $Ca^{2+}$-dependent conductances. Among them, the $Ca^{2+}$-dependent potassium conductance plays an important role in regulating the resting membrane potential and excitation threshold. A broad family of $Ca^{2+}$-dependent $K^+$ channels has been discovered in neuronal cells; these channels are characterized by different conductance and $Ca^{2+}$ dependence. The activity of these channels mostly determines the after-hyperpolarization (AHP) observed in many neurones; the latter is believed to suppress firing activity due to increasing of threshold for subsequent action potential. Several lines of evidence favour the idea that AHP may be driven not only by $Ca^{2+}$ entering the cell via plasmalemmal $Ca^{2+}$ channels during action potentials, but also by $Ca^{2+}$ released from internal calcium stores. Caffeine application enhanced AHP in frog and rat sympathetic neurones,[100,101] while ryanodine displayed an opposite action, shortening the AHP.[100] Extended caffeine applications led to the appearance of hyperpolarization in frog sympathetic[102] and mouse sensory[103] neurones. A similar hyperpolarization induced by bradykinin-triggered $InsP_3$-mediated $Ca^{2+}$ release has been reported for neuronal hybrid NG108–15 cells.[104]

A new aspect of interrelations between $Ca^{2+}$ stores and membrane permeability is obvious from the discovery of plasmalemmal $Ca^{2+}$-permeable channels controlled by the filling state of calcium stores.[12,105] It has been suggested that the activation of these $Ca^{2+}$ channels is mediated by a diffusable messenger released by emptying of intracellular $Ca^{2+}$ stores. This diffusable messenger (molecular weight $\sim 500$ Da) called CIF has been postulated to activate CRAC channels in macrophages, fibroblasts, and astrocytes.[13,106]

### Neurotransmitter release

It is generally accepted that $[Ca^{2+}]_i$ determines the probability of fusion of synaptic vesicles to the presynaptic membrane, thus controlling the efficacy of neurotransmitter release.[107]

Increasingly, data indicate that $Ca^{2+}$ liberation from internal stores may play a role in the regulation of neurotransmitter release. Presynaptic nerve terminals contain ER enriched by $Ca^{2+}$ storing systems, as well as by $Ca^{2+}$ release channels, and caffeine can release $Ca^{2+}$ in close proximity with presynaptic active zones.[108,109]

### Control of synaptic plasticity

$[Ca^{2+}]_i$ could play an important role in the regulation of synaptic plasticity by modulating both the properties of the pos'synnaptic membrane and the efficacy of presynaptic neurotransmitter release. One form of synaptic plasticity is related to long-lasting changes in synaptic function [long-term potentiation (LTP) or depression (LTD)] which are thought to be the cellular substrates for learning and memory. It seems that $[Ca^{2+}]_i$ plays an important role for generation of both LTP and LTD. For instance, in CA1 hippocampal neurones, the injection of $Ca^{2+}$-chelating agents in post-synaptic neurones effectively inhibited the induction of LTP.[110] Similarly, the elevation of $[Ca^{2+}]_i$ in the dendrites of Purkinje neurones appeared to be a key point for initiation of LTD; the inhibition of this $[Ca^{2+}]_i$ elevation prevented the development of LTD.[111] The elevation of $[Ca^{2+}]_i$ involved in generation of long-lasting synaptic plasticity may be achieved either due to transmembrane $Ca^{2+}$ influx (presumably via NMDA-operated channels) or to $Ca^{2+}$ liberation from internal stores. The latter consideration is favoured by observations that blockade of CICR mechanisms by dantrolene,[112] as well as depletion of $Ca^{2+}$ stores by thapsigargin-mediated SERCA inhibition,[113] inhibited the development of LTP in hippocampal slices.

### Cell proliferation and growth

It is well known that $Ca^{2+}$ ions activate gene transcription, thus regulating cell growth and development, and may modulate cytoplasmic protein synthesis. Intracellular $Ca^2$ release coupled with metabotropic receptor activation is suggested to be involved in the control of neuronal growth and differentiation. This suggestion is based on the correlation between the density of glutamatergic metabotropic receptors and periods of extensive developmental changes. It has been proposed, for instance, that metabotropic-triggered $Ca^{2+}$ release is involved in the morphogenesis of Purkinje neurones, as well as in the control of astrocytic differentiation.[114,115]

### Neuronal ageing

Increasingly, data also indicate that brain ageing is accompanied by the alteration of mechanisms of calcium homoeostasis. Functional properties of the calcium stores also seem to change during neuronal ageing. It appeared that the amount of $Ca^{2+}$ stored in the caffeine-sensitive compartment of ER is significantly higher in old central neurones, presumably due to overload of these stores by releasable calcium.[116,117] In contrast with data obtained from neurones in young and adult animals, the releasable calcium content of caffeine-sensitive stores in the neurones of old rats was initially high enough to produce $[Ca^{2+}]_i$ transient in response to caffeine, and the charging depolarization did not change the amplitude of these transients. However, the overloaded calcium stores in these neurones cannot effectively participate in $Ca^{2+}$ removal after periods of neuronal activity, and the $[Ca^{2+}]_i$ transients became extensively prolonged. It is possible to suggest that the reduction of $Ca^{2+}$ sequestration by internal stores, as well as previously reported age-dependent decreases of cytoplasmic $Ca^{2+}$ buffering, mitochondrial $Ca^{2+}$ accumulation, and plasmalemmal calcium extrusion by $Ca^{2+}$ pumps and $Ca^{2+}$ exchangers, are responsible for age-dependent alteration of $[Ca^{2+}]_i$ homoestasis in neuronal cells.[117]

## References

1. Tsien, R.W. and Tsien, R.Y. (1990). *Annu. Rev. Cell. Biol.*, **6**, 715–760.
2. Kostyuk, P.G. (1992). *Calcium ions in nerve cell function*, Oxford University Press, Oxford, New York, Tokyo.
3. Lytton, J., Westlin, M., and Hanley, M.R. (1991). *J. Biol. Chem.*, **266**, 17067–17071.
4. Meldolesi, J., Madeddu, L., and Pozzan, T. (1990). *Biochem. Biophys. Acta*, **1055**, 130–140.
5. Berridge, M.J. (1993). *Nature*, **361**, 315–325.
6. Henzi, V. and McDermott, A.B. (1992). *Neuroscience*, **46**, 251–274.
7. Levy, S. (1992). *J. Neurosci.*, **12**, 2120–2129.
8. Kostyuk, P.G. and Kirischuk, S.I. (1993). *Neuroscience*, **53**, 943–947.
9. Liu, P.S., Lin, Y.J., and Kao, L.S. (1991). *J. Neurochem.*, **56**, 172–177.
10. Robinson, I.M. and Burgoyne, R.D. (1991). *J. Neurochem.*, **56**, 1587–1593.
11. Stauderman, K.A. and Murawsky, M.M. (1991). *J. Biol. Chem.*, **266**, 19150–19153.
12. Hoth, M. and Penner, R. (1992). *Nature*, **355**, 353–356.
13. Randriamampita, C. and Tsien, R.Y. (1993). *J. Neurochem.*, **61**, **(Suppl.)**, S122–C.
14. Putney, J.W. (1986). *Cell Calcium*, **7**, 1–12.
15. Reber. B.F.X., Stucki, J.W., and Reuter, H. (1993). *J. Physiol. (Lond)*, **468**, 711–727.
16. Hofer, A.M. and Machen, T.E. (1993). *Proc. Natl. Acad. Sci. USA* **90**, 2598–2602.
17. Takei, K., Stukenbrok, H., Metcalf, A., Mignery, G.A., Sudhof, T.C., Volpe, P., and De Camilli, P. (1992). *J. Neurosci*, **12**, 489–505.
18. Sharp, A.H., McPherson, P.S., Dawson, T.M., Aoki, C., Campbell, K.P., and Snyder, S.H. (1993). *J. Neurosci.*, **13**, 3051–3063.
19. Sitsapesan, R. and Williams, A.J. (1990). *J. Physiol. (Lond.)*, **423**, 425–439.
20. Friel, D.D. and Tsien, R.W. (1992). *J. Physiol. (Lond.)*, **450**, 217–246.
21. Usachev, Y., Shmigol, A., Pronchuk, N., Kostynk, P., and Verkhrtasky, A. (1993). *Neuroscience*, **57**, 845–859.
22. Lipscombe, D., Madison, D.V., Poenie, M., Reuter, H. Tsien, R.W., and Tsien, R.Y. (1988) *Neuron*, **1**, 355–365.
23. Nohmi, M., Hua, S.Y., and Kuba, K. (1992). *J. Physiol. (Lond.)*, **450**, 513–528.

24. Brorson, J.R., Bleakman, D., Gibbons, S.J., and Miller, R.J. (1991). *J. Neurosci.* **11**, 4024–4043.
25. Shmigol, A., Kirischuk, S., Kostyuk, P., and Verkhratsky, A. (1994). *Pflügers Arch.*, **426**, 174–176.
26. Bleakman, D., Roback, J.D., Wainer, B.H., Miller, R.J., and Harrison, N.L. (1993). *Brain Res.*, **600**, 257–267.
27. Verkhratsky, A., Kano, M., and Konnerth, A. (1993). Abstracts of the XXXII IUPS Congress, Glasgow, August 1–6, 1993, 200/50.
28. Endo, M. (1985). *Curr.* Top. *Membr. Transport*, **25**, 181–230.
29. Reber, B.F.X. and Reuter, H. (1992). *J. Physiol. (Lond.)*, **435**, 145–162.
30. Barish, M. (1991). *J. Physiol. (Lond.)*, **444**, 545–565.
31. Hua, S.Y., Nohmi, M., and Kuba, K. (1993). *J. Physiol. (Lond.)*, **464**, 245–272.
32. Nohmi, M., Hua, S.Y., and Kuba, K. (1993). *J. Physiol. (Lond.)*, **458**, 171–190.
33. Sorimachi, M., Yamagami, K., Nishimura, S., and Kuramoto, K. (1992). *J. Neurochem.* **59**, 2271–2277.
34. Alford, S., Frenguelli, B.G. and Collingridge, G.L. (1992). *J. Physiol. (Lond.)*, **452**, 178P.
35. Simpson, P.B., Chaliss, R.A.J., and Nahorski, S.R. (1993). *J. Neurochem.*, **61**, 760–763.
36. Fisher, S.K. and Agranof, B.W. (1987). *J. Neurochem.*, **48**, 999–1017.
37. Sladeczek, F., Pin, J.P., Recasens, M., Bockaert, J., and Weiss, S. (1985). *Nature*, **317**, 717–719.
38. Nicoletti, F., Wroblewski, J.T., Novelli, A., Alho, H., Guidotti, A., and Costa, E. (1986). *J. Neurosci.* **6**, 1905–1911.
39. Nicoletti, F., Meek, L.J., Iadarola, M.J., Chuang, D.M., Roth, B.L., and Costa, E. (1986). *J. Neurochem.*, **46**, 40–46.
40. Murphy, S.N. and Miller, R. (1988). *Proc. Natl Acad. Sci. USA* **85**, 8737–8741.
41. Irwing, A.J., Collingridge, G.L., and Schofield, J.G. (1992). *J. Physiol. (Lond.)*, **456**, 667–680.
42. Llano, I., Dreessen, J., Kano, M., and Konnerth, A. (1991). *Neuron*, **7**, 577–583.
43. Thayer, S.A., Perney, T.M., and Miller, R.J. (1988). *J. Neurosci*, **8**, 4089–4097.
44. Bito, H., Nakamura, M., Honda, Z., Izumi, T., Iwatsubo, T., Seyama, Y., Ogura, A., Kudo, Y., and Shimizu, T. (1992). *Neuron*, **9**, 285–294.
45. Mironov, S.L. (1994). *Neuroreport*, **5**, 445–448.
46. Rios, E. and Brum, G. (1987). *Nature*, **325**, 717–720.
47. Takeshima, H., Nishimura, S., Matsumoto, T., Ishida, H., Kanagawa, K. Miniamino, N., Matsuo, H., Ueda, M., Hanaoka, M., Hirose, T., and Numa, S. (1989). *Nature*, **339**, 439–445.
48. Zorzato, F., Fujii, J., Otsu, K., Philips, M., Green, N.M., Lai, F.A., Meissner, G., and MacLennan, D.H. (1990). *J. Biol. Chem.*, **265**, 2244–2256.

49. Otsu, K., Willard, H.F., Khanna, V.K., Zorzato, F., Green, N.M., and MacLennan, D.H. (1990). *J. Biol. Chem.*, **265**, 13472–13483.

50. Hymel, L., Inui, M., Fleischer, S., and Schindler, H. (1988). *Proc. Natl Acad. Sci.* USA, **85**, 441–445.

51. Radermacher, M., Wagenknecht, T., Grassucci, R., Frank, J., Inui, M., Chadwick, C., and Fleischner, S. (1992). *Biophys. J.*, **61**, 936–940.

52. Witcher, D.R., Kovacs, R.J., Schulman, H., Cefali, D.C., and Jones, L.R. (1991). *J. Biol. Chem.* **266**, 11144–11152.

53. Ellisman, M.H., Deerinck, T.J., Ouyang, Y., Beck, C.F., Tanksley, S.J., Walton, P.D., Airey, J.A., and Sutko, J.L. (1990). *Neuron*, **5**, 135–146.

54. McPherson, P.S. and Campbell, K.P. (1990). *J. Biol. Chem.*, **265**, 18454–18460.

55. McPherson, P.S., Kim, Y.K., Valdivia, H., Knudson, C.M., Takekura, H., Franzini-Armstrong, C., Coronado, R., and Campbell, K.P. (1991). *Neuron*, **7**, 17–25.

56. Kawai, T., Ishii, Y., Imaizumi, Y., and Watanabe, M. (1991). *Brain Res.*, **540**, 331–334.

57. Lai, F.A., Dent, M., Wickenden, C., Xu, L., Kumari, G., Misura, M., Lee, H.B., Sar, M., and Meissner, G. (1992). *Biochem. J.*, **288**, 553–564.

58. Rubtsov. A.M. and Murphy, A.J. (1988). *Biochem. Biophys. Res. Commun*, **154**, 462–468.

59. Nakai, J., Imagawa, T., Hakamata, Y., Shigekawa, M., Takeshima, H., and Numa, S. (1990). *FEBS Lett.*, **271**, 169–177.

60. Hakamata, Y., Nakai, J., Takeshima, H., and Imoto, K. (1992). *FEBS Lett.*, **312**, 229–235.

61. Suarez-Isla, B.A., Alcayaga, C., Marengo, J.J., and Bull, R. (1991). *J. Physiol. (Lond.)*, **441**, 575–591.

62. Mackgrill, J.J. and Lai, F.A. (1994). *Biophys. J.*, **66**, A147.

63. Khodakhah, K. and Ogden, D. (1993). *Proc. Natl Acad. Sci. USA*, **90**, 4976–4980.

64. Ashley, R.H. (1989) *J. Memb. Biol.*, **111**, 179–189.

65. Sierralta, J., Suarez-Isla, B., and Fill, M. (1994). *Biophys. J.*, **66(2)**, A147.

66. Bezprozvanny, I., Watras, J., and Ehrlich, B.E. (1991). *Nature*, **351**, 751–754.

67. Maeda, N., Kawasaki, T., Nakade, S., Yokota, N., Taguchi, T., Kasai, M., and Mikoshiba, K. (1991). *J. Biol. Chem.*, **266**, 1109–1116.

68. Bezprozvanny, I. and Ehrlich, B.E. (1994). *Biophys. J.*, **66**, A146.

69. Bezprozvannaya, S., Bezprozvanny, I., and Ehrlich, B.E. (1994). *Biophys. J.*, **66(2)** A146.

70. Richardson, A. and Taylor, C.W. (1993) *J. Biol. Chem.*, **268**, 11528–11533.

71. Anderson, K., Lai, F.A., Liu, Q.Y., Rousseau, E., Erickson, H.P., and Meissner, G. (1989). *J. Biol. Chem.*, **264**, 1329–1335.

72. Herrman-Frank, A., Darling, E., and Meissner, G. (1991). *Pflügers Arch.*, **418**, 353–359.

73. Ashley, R.H. and Williams, A.J. (1990). *J. Gen. Physiol.*, **95**, 981–1005.
74. Tinker, A. and Williams, A.J. (1992). *J. Gen. Physiol.*, **100**, 479–493.
75. Tinker, A., Lindsay, A.R.C., and Williams, A.J. (1992). *J. Gen. Physiol.*, **100**, 495–517.
76. Mironov, S.L. and Usachev, J.M. (1990). *Neurosci. Lett.*, **112**, 184–189.
77. Jenden, D.J. and Fairhurst, A.S. (1969). *Pharmacol. Rev.*, **21**, 1–25.
78. Nagasaki, K. and Fleishner, S. (1988). *Cell Calcium*, **9**, 1–7.
79. Rousseau, E., Smith, J.S., and Meissner, G. (1987). *Am. J. Physiol.*, **253**, C364–C368.
80. McGrew, S.G., Wolleben, C., Siegl, P., Inui, M., and Fleischer, S. (1989). *Biochemistry*, **28**, 1686–1691.
81. Gerzon, K., Humerickhouse, R.A., Besch, H.K., Bidasee, K.R., Emmick, J.T., Roeske, R.W., Tian, Z., Ruest, L., and Sutko, J.L. (1993). *J. Med. Chem.*, **36**, 1319–1323.
82. Teraoka, H., Nakazato, Y., and Ohga, A. (1991). *J. Neurochem.*, **57**, 1884–1890.
83. Tsai, T.D. and Barish, M.E. (1991). *Biophys. J.*, **59(2)** 599A.
84. Marrion, N.V. and Adams, P.R. (1992). *J. Physiol. (Lond.)*, **445**, 515–535.
85. Konishi, M. and Kurihara, S. (1987). *J. Physiol. (Lond.)*, **383**, 269–283.
86. Ohta, T. and Ohga, A. (1990). *Eur. J. Pharmacol.*, **178**, 11–19.
87. Thayer, S.A., Hirning, L.D., and Miller, R.J. (1988). *Mol. Pharmacol.*, **34**, 664–673.
88. Valdivia, H.H., Fuentes, O., El-Hayek, R., Morrissette, J., and Coronado, R. (1991). *J. Biol. Chem.*, **266**, 19135–19138.
89. Valdivia, H.H., Kirby, M.S., Lederer, W.J., and Coronado, R. (1992). *Proc. Natl Acad. Sci. USA.*, **89**, 12185–12189.
90. Hymel, L.J., Whitworth, A., Wang, S.N., and Clarcson, C.W. (1994). *Biophys. J.*, **66**, A19.
91. Ondrias, K., Borgatta, L., Kim, D.H., and Ehrlich, B.E. (1990). *Circ. Res.*, **67**, 1167–1174.
92. Galione, A. (1992). *Trends Pharmacol. Sci.*, **13**, 304–306.
92A. Galione, A., Lee, H.C., and Busa, W.B. (1991). *Science*, **253**, 1143–1146.
93. Meszaros, L.G., Bak, J., and Chu, A. (1993). *Nature*, **364**, 76–79.
94. El-Hayek, R., Valdivia, C., Valdivia, H.H., and Coronado, R. (1993). *Biophys. J.*, **65**, 779–789.
95. Joseph, S.K., Rice, H.L., and Williamson, J.R. (1989). *Biochem. J.*, **258**, 261–265.
96. Volpe, P., Alderson-Lang, B.H., and Nickols, G.A. (1990). *Am. J. Physiol.*, **258**, C1077–C1085.
97. Bezprozvanny, I. and Ehrlich, B.E. (1993). *Neuron*, **10**, 1175–1184.
98. Parker, I. and Ivorra, I. (1991). *J. Physiol. (Lond.)*, **433**, 229–240.
99. Stauderman, K.A., McKinney, A., and Murawski, M.M. (1991). *Biophys. J.*, **278**, 643–650.

100. Kawai, T. and Watanabe, M. (1989). *Pflügers Arch.*, **413**, 470–475.
101. Kuba, K., Morita, K., and Nohmi, M. (1983). *Pflügers Arch.*, **399**, 194–202.
102. Kuba, K. (1980). *J. Physiol. (Lond.)*, **298**, 251–269.
103. Mathers, D.A. and Barker, L.J. (1984). *Brain Res.*, **293**, 35–47.
104. Higashida, H. and Brown, D.A. (1986). *Nature*, **323**, 333–335.
105. Penner, R., Fasolato, C., and Hoth, M. (1993). *Curr. Opin. Neurobiol.*, **3**, 368–374.
106. Lückhoff, A. and Clapham, D.E. (1994). *Biophys. J.*, **66(2)**, A153.
107. Smith, S.J. and Augustine, G.J. (1988). *Trends Neurosci.*, **11**, 458–464.
108. Martinez Serrano, A. and Satrustegui, J. (1989). *Biochem. Biophys. Res Commun.*, **161**, 965–971.
109. Fujimoto, S., Yamamoto, K., Kuba, K., Morita, K., and Kato, E. (1980). *Brain Res.*, **202**, 21–32.
110. Lynch, G., Larson, J., Kelso, S., Barrionuevo, G., and Schottler, F. (1983). *Nature*, **305**, 719–721.
111. Konnerth, A., Dreessen, J., and Augustine, G.J. (1992). *Proc. Natl Acad. Sci. USA*, **89**, 7051–7055.
112. Obenaus, A., Mody, I., and Baimbridge, K.G. (1989). *Neurosci. Lett.*, **98**, 172–178.
113. Harvey, J. and Collingridge, G.L. (1992). *Neurosci. Lett.*, **139**, 197–200.
114. Maeda, N., Niinobe, M., and Mikoshiba, K. (1990). *Eur. Molec. Biol. Org.*, **9**, 61–67.
115. Nicolettu, F., Magri, G., Ingrao, F., Bruno, V., Catania, M.V., Dell Albani, P., Condorelli, D.F., and Avola, R. (1990). *J. Neurochem.*, **54**, 771–777.
116. Kirischuk, S., Pronchuk, N., and Verkhratsky, A. (1992). *Neuroscience*, **50**, 947–951.
117. Verkhratsky, A., Shmigol, A.V., Kirischuk, S., Pronchuk, N., and Kostyuk, P.G. (1994). *Ann. N.Y. Acad. Sci.*, **747**, 365–381.

# 15 Mitochondrial membrane channels and their pharmacology

M.-L. Campo, H. Tedeschi, C. Muro, and K. W. Kinnally

## Introduction

Mitochondria are the primary energy providers of most eukaryotic cells because of their role in oxidative phosphorylation. It is therefore not surprising that mitochondria have been the focus of important studies for the past 40 years. These organelles are enclosed by an outer and an inner membrane. The outer membrane has important permeability properties and a probable regulatory role that is just beginning to be recognized. Aside from its role as a permeability barrier and in the translocation of solutes, the inner membrane containing the components responsible for electron transport and the ATP-synthase is closely associated with energy transduction.

Mitochondrial membrane channels have been the subject of several excellent recent reviews.[1–4] The present review will examine the current state of knowledge on this topic and the pharmacology of the channels.

The study of mitochondrial channels has taken three distinct pathways. Some isolated proteins, when reconstituted into bilayers, were found to have electrophysiological properties of ion channels. They display discrete current transitions compatible with the opening and closing of channels and exhibit voltage gating. In addition, studies of permeability using mitochondrial suspensions postulated a variety of channels in the inner membrane from data indicating permeability transitions. More recently, channels were revealed from patch-clamping of the native membranes or from reconstitution of membrane fragments into liposomes or planar bilayers. The inner mitochondrial membrane has generally been studied by stripping the outer membrane by a variety of means in the preparation of mitoplasts. It is important, however, to recognize that the areas of contact between inner and outer membranes (the so-called contact sites[5] are likely to remain after most of these procedures. Contact sites are thought to have special relevance in a number of physiological processes. The sections which follow will discuss outer membrane and inner membrane separately, and in addition will address the evidence for interactions between inner and outer membrane components.

## Channels of the outer mitochondrial membrane

### Voltage-dependent anion-selective channels

A protein behaving as a channel when incorporated in planar bilayers was first discovered by Schein *et al.*[6] The channel was shown to be associated with the mitochondrial outer membrane and was referred to as voltage-dependent anion-selective channel (VDAC, or mitochondrial porin). The VDAC protein of approximately 30 kDa was found to be the major component of the outer membrane. The VDACs of a variety of organisms have been sequenced. Despite the remarkable similarities in electrophysiological characteristics,[7] the various VDACs from different organisms differ significantly in their amino acid sequences.

The amino acid sequences of VDAC provide alternating polar and hydrophobic stretches which can be folded into a $\beta$-sheet, or barrel with a hydrophobic exterior and a hydrophilic channel[8] similar to that shown for the porin of Gram-positive bacteria with electron and X-ray crystallography.[9,10] Circular dichroism supports the notion that the VDAC's secondary structure is predominantly a $\beta$-sheet.[11]

VDAC forms two-dimensional crystals when the outer membrane phospholipid of *Neurospora* mitochondria is depleted using phospholipase $A_2$.[12,13] Electron microscopic studies of these crystals reveal VDAC channels arranged in a group of six. Crystal density and space-filling considerations suggest that each VDAC channel is composed of a single 31 kDa polypeptide[14] a conclusion also supported by scanning transmission electron microscopy mass measurements on freeze-dried membrane crystals[15] and by experiments with yeast expressing both wild-type and mutant VDACs.[16]

Reconstruction of the membrane crystals embedded in negative stain suggest that each pore traverses the outer membrane with an inner diameter of 3 nm,[17] as confirmed with cryoelectronmicroscopy of unstained frozen-hydrated VDAC arrays.[18,19] The studies with cryo-imaging and embedding in aurothioglucose suggest that the lumen wall is not uniform in height. In addition, there are dense arms of protein extending from the channels. These arms are labelled by antibodies for the first 20 residues of *Neurospora* VDAC[20] and each is thought to correspond to an amphipathic $\alpha$-helix which are not part of the lumen.

Several models of VDAC structure are now under consideration (see ref. 4 for a review).

The insertion of the channels into planar bilayers provide 650 pS increments in 150 mM KCl with subconductance levels. The channel is voltage-gated, with a predominant conductance level of 300 pS when partially closed by either positive or negative potentials. The shift from open to partially closed is accompanied by a shift from slight anion-selectivity, and in some cases virtually no selectivity,[7,21] to cation-selectivity.[22] At least in planar bilayer systems,[23] closing takes several seconds, whereas the opening is much more rapid, the absolute values depending on the potential difference across the bilayer.

Results obtained with liposomes containing VDAC or outer membrane fragments are somewhat different. Most results were obtained using outer membrane fragments. At least some of the differences may result from the presence of other channels, demonstrated to be present in VDAC-less mutants.[24] The lack of symmetry in the voltage dependence, however, may result from the presence of modulators which in some cases have been shown to produce such asymmetry.[25]

The parameter $n$ is related to voltage sensitivity and can be obtained from the steepness of the current versus voltage curves. In conventional models of voltage gating, $n$ corresponds to the number of charges that move across the electric field during gating events. In the studies of VDAC using planar bilayers, $n$ ranges between 2 and 4,[7,22] regardless of the source of VDAC and despite differences in amino acid sequence. The interspecies variation is systematic and $n$ is inversely related to $V_o$ so that the amount of energy required for the closing is approximately constant (see discussion of ref. 1).

Zimmerberg and Parsegian[26] showed that VDAC could be partially closed by the osmotic pressure exerted by polymers. These results suggest that the gating results from large-scale conformational changes. The use of appropriate VDAC mutants allowed identification of the transmembrane strands involved in gating. This was done by determining which charged residues affect $V_o$ (the voltage at which the open probability is 0.5).[27] The findings support a role of domains at both the N- and C-terminals of VDAC. These studies postulate a mechanism for gating in which part of the protein is removed from the wall of the lumen, thereby reducing the pore diameter. However, other models are also possible such as one initially involving the insertion of the N-terminal region into the lumen after detaching from the bilayer.[28]

Voltage gating of VDAC is markedly pH-dependent. Voltage-induced closure of the VDAC channel[29] is inhibited by alkaline pH. The midpoint of inhibition was found to be near the $pK_a$ of lysine (10.6), implicating positively charged amino acids. Reversing the charge of exposed lysines by succinylation[21] produces a loss of voltage dependence. In addition, acid conditions (pH 3–4) enhance voltage-dependent closure.[30] Reversing the charge on carboxyl groups of acidic residues of the protein by amidation also enhances the closure. These results indicate that both acidic and basic amino acids play a role in gating. Neither elevated pH nor succinylation cause large-scale changes in the structure of crystalline VDAC,[31,32] suggesting that their effects on gating are not due to gross conformational changes in the protein.

In nanomolar concentrations, a copolymer of methacrylate, styrene, and maleate (Konig's polyanion of 10 kDa) alters the VDAC's voltage sensitivity.[33] This polyanion's effects are somewhat complex and depend on the polarity of the applied potential with respect to the side on which the polyanion is added. Konig's polyanion reduces $V_o$ from $-30$ mV to $-5$ mV when added to the negative-potential side, while increasing $V_o$ for positive potentials to 80–100 mV.[34] Another polyanion, dextran sulphate (8 kDa), exerts a similar effect at higher (micromolar) levels on VDAC at negative potentials, increasing $n$ by up to 14-fold, but maintaining $nV_o$ constant.[35] Unlike Konig's polyanion, dextran sulphate does not stabilize the open state when added to the positive side of the bilayer.

A soluble protein fraction released from lysed mitochondria has been found to affect VDAC's voltage dependence analogously to polyanions.[36] The modulator activity appears to be associated with a protein of 54 kDA and pI around 5.[37] $V_o$ is decreased about one-half, while $n$ is increased two-to three-fold, the net effect being a slight increase in $nV_o$.[38]

There is evidence that diffusion of ATP across the outer membrane is greatly restricted by preincubation of intact mitochondria with the polyanion, presumably due to the switching of VDAC to its less-conducting, cation-selective substate.[38,39]

NADH has a specific effect on VDAC from several species,[40] increasing $n$ from 2–3 to 4–6 without altering $V_o$ at concentrations of 100 μM. A sequence near the C-terminus of the polypeptide is similar to the dinucleotide binding site of the adenine nucleotide transporter.[28,41] Regulation of outer membrane permeability by interaction of the VDAC channel with NADH might have important physiological implications. For example, as Zizi *et al.*[40] point out, NADH is a byproduct of glycolysis and glucose is known to suppress mitochondrial respiration in several tissues (the Crabtree effect).

### Other outer membrane channels

Evidence for other channels stems in part from studies using VDAC-less mutants and incorporating extracts into planar bilayers.[24] Studies with mitochondrial extracts using a bilayer technique[42] have also implicated a slightly cation-selective channel other than VDAC. Both sets of experiments displayed conductance steps of 300 pS. The latter study, however, also revealed 100 to 540 pS transitions. The cation-selective channel is transiently blocked by signal peptides,[43] suggesting a role in proteins imported into mitochondria, and has been referred to as the peptide-sensitive channel (PSC).

There is at least one report of an outer membrane channel with conductance steps ranging from 10 to 50 pS.[44]

### Patch-clamping results with the native membranes

The high density of VDAC (several hundreds or even 1000 in each patch, calculated for example from published data)[7,45,46] in the outer membrane predict a very low resistance. For this reason, early studies of the native outer membrane (see ref. 47 for a review), were carried out at very low ionic strength. A very low resistance was in fact observed even at these low ionic strengths. No discrete channel behaviour was observed as expected from the high conductance, implying that several hundred channels were involved in each patch. VDAC was also implicated by the response at positive voltage to polyanion addition and the effect of succinic anhydride treatment. At negative voltages, the conductance decreased only rarely and most frequently increased. These latter results were consistent with a biphasic effect, an initial activation of channels followed by an assembly into still larger channels. The involvement of protein channels in these effects is supported by experiments in which the patch is exposed to proteases which decreased the rectification and increased the resistance (B. Popp, H. Tedeschi, and K. W. Kinnally, unpublished results).

Many uncertainties have been introduced by the observation that patch-clamping intact mitochondria has resulted in the occasional observation of patch resistances in the range of 0.5 GΩ and at times as high as 10 GΩ.[2,48] These observations suggest that, at least in these mitochondria, the majority of the VDAC are closed completely. The previous reports of much lower patch resistances[49,50] may indicate that completely closed VDAC can be activated.

## Channels of the inner mitochondrial membrane

### Channels deduced from permeability studies

Study of mitochondrial permeability to ions have observed permeability transitions which are most readily interpreted on the basis of activation of channels in the inner mitochondrial membrane.

Uncoupling protein (UCP) – also called thermogenin – is present in the inner membrane of brown adipose tissue where it constitutes as much as 15% of the protein. UCP may also correspond to a channel. Studies of UCP have been reviewed broadly by Nicholls *et al.*[51] and more recently by Ricquier *et al.*[52] and Klaus *et al.*[53] UCP appears to confer a high degree of permeability to $H^+$ (or perhaps $OH^-$) and possibly halides to the inner mitochondrial membrane.[51] Therefore, its uncoupling effect is attributed to its ability to collapse the electrochemical gradient needed for energy coupling.

Permeability transitions have been observed in mitochondrial suspensions. Alkaline pH or lowering matrix $Mg^{2+}$ (for review, see refs. 54 and 55) induce a change attributed to the opening of the inner membrane anion channel (IMAC). The transition produced by the presence of $Ca^{2+}$ (see ref. 56 for a review) is attributed to the opening of the permeability transition pore (PTP). A $K^+$-uniporter has been found to be activated by depletion of $Mg^{2+}$,[57,58] and appears to be relatively unselective, transporting $Rb^+$ $Na^+$ and $Li^+$.[58] This uniporter is most likely to correspond to a channel.

#### Uncoupling protein

UCP activity is very significant in mammals at birth, during cold adaptation, or in arousal from hibernation. The protein uncouples oxidation from phosphorylation, thereby generating heat. The uncoupling effect can be blocked by purine di- and trinucleotides and the block can be overcome by fatty acids.[51]

UCP can be induced by cold acclimation *in vivo* or by exposure of brown fat cultured cells to noradrenaline.[59] Removal of the noradrenaline results in rapid specific degradation of UCP and UCP-mRNA.[60, 61] The production of UCP was found to be transcriptionally and post-transcriptionally controlled by noradrenaline and thyroid hormones.

When UCP is reconstituted into liposomes, the system accumulates $H^+$ in the presence of valinomycin when the liposomes are preloaded with $K^+$.[62] This system, however, is unable to transport $Cl^-$. Garlid[63] re-examined the question and found a GDP-inhibitable anion transport in proteoliposomes containing UCP. However, in the intact mitochondrial system only the permeability of the $H^+$ is affected by fatty acids and therefore it is entirely possible that two closely related proteins may be present.

Nicholls *et al.*[51] consider the UCP pathway to correspond to a channel. Klingenberg and Winkler[62] have argued against this alternative because the turnover of UCP in the reconstituted liposome system is low. However, although a high turnover is prima facie evidence of a channel, the opposite is not true.

Mirzabekov and Akhemerov[64] have observed a high-conductance channel activity in planar bilayers after the addition of UCP extracted from brown fat. The channel was found to be slightly anion-selective and to resemble VDAC. However, the report is not sufficiently detailed for critical evaluation and it should be noted that even a slight contamination by VDAC could account for these results.

Nicholls *et al.*[51] calculate a flow of protons of 16 nmol $min^{-1}$ (per mg protein) $mV^{-1}$ and 0.8 nmol of channels per mg protein. The conductance of an individual channel would be 10 ions $min^{-1}$ $mV^{-1}$, or a conductance of 0.03 fS per channel. The halide conductance would be of the same order of magnitude. This would preclude detecting individual conductances. It should be noted however that, according to Klitsch and Siemen,[65] the conductance per mitochondrion would correspond to 2.7 pS and would therefore be detectable by patch-clamping.

UCP is a protein of 32 kDa which has been isolated from brown fat after affinity-labelling with 8-azido ATP.[66] The protein is distinct from the adenine nucleotide translocator as shown immunologically[67] and from the amino acid sequences.[68, 69] Immunological procedures were also used to demonstrate its absence from a variety of mammalian tissues other than brown fat.[70]

UCP has been sequenced[68, 69] and found closely related to the adenine nucleotide and the $P_i$ translocators.[71, 72] All three contain a similar three-fold repeat of approximately 100 amino acids. The physical properties of UCP (e.g. its interaction with detergent)[73] and its amino acid sequence suggest a molecule which may be entirely embedded in the membrane. Hydropathy and circular dichroism studies[68] suggest six α-helical segments spanning the membrane as well as one β-strand. Since the helices and the β-strand contain extensive hydrophilic residues, it would not be difficult to hypothesize convincingly a possible channel spanning the membrane.

### The inner membrane anion channel

IMAC are activated by alkalinization of the matrix side of the membrane.[74–76] The activation is reversible and inhibited by endogenous $Mg^{2+}$. IMAC lacks specificity and can transport many anions, although it favours smaller size and multivalence.

IMAC can be found in many of tissues, including plants, and has been studied pharmacologically. Inhibitors range from local anaesthetics to benzodiazepines and cationic amphiphiles.[55]

### The permeability transition pore (PTP)

The permeability transition of the inner mitochondrial membrane induced by $Ca^{2+}$ requires $Ca^{2+}$ accumulation in the matrix and is facilitated by the presence of one of many possible inducing agents. These agents are chemically diverse (ref. 56 lists more than 60), suggesting that the activation is a complex process. The increase in permeability may result from the activation of a channel or pore,[77] although there are dissenting voices. The transition permits the passage of rather large molecules such as sucrose,[78] pyridine nucleotides[79] and, in extreme cases, may include matrix proteins.[80] Since the activation of PTP may be progressive, the increases in permeability may be graded depending on a variety of conditions. The process can be inhibited by a variety of agents (e.g. protons, $Mg^{2+}$, ADP, bongkrekic acid, and $Ca^{2+}$-chelation – again suggesting a complex effect). The transition can be reversed by the chelation of $Ca^{2+}$ or the presence of cyclosporin A, ADP, $Mg^{2+}$, or reduced pyridine nucleotides. There is evidence that the channels act cooperatively[81] and there are suggestions that the channel may be formed by assembly via crosslinking by disulphide bridges.[82] The possible physiological role of this pore is still not clear. The permeability transitions attributed to the activation of PTP has been linked to such pathological events as ischaemia–reperfusion injury.[56] As discussed below, PTP has many features in common with MCC and the two might be equivalent.

## Electrophysiological studies

Mironova *et al.*[83] were the first to demonstrate channel behaviour of an inner mitochondrial membrane component. The study incorporated the partially purified $K^+$ translocator in planar bilayers. However, demonstration of the presence of ion channels in the native inner mitochondrial membrane required the application of patch-clamp techniques to mitoplasts.[84, 85]

Patch-clamping of mitoplasts indicates that the mitochondrial inner membrane has a high resistance and is usually electrically silent.[85–87] At least five distinct classes of channel activities, which are summarized in Table 15.1, can be activated by a variety of conditions (divalent ions, pH, transmembrane voltage). These channel activities display a wide range of ion selectivities and voltage dependence. Generally, each activity responds to at least one physiological effector, e.g. ATP, $Ca^{2+}$, or pH. So far, despite these discoveries none of these activities has been clearly ascribed to a particular protein.

**Table 15.1** Summary of mitoplast channel activity

| Channel activity | Size (pS)* | Voltage dependence | Selectivity | Effectors | Probable equivalent from permeability studies |
|---|---|---|---|---|---|
| MCC | 10 to >1000 | Yes | Slight cation to none | $Ca^{2+}$, $Mg^{2+}$, ADP, pH, voltage | PTP |
| mCS | ≈100 | Yes | Slight anion | $Ca^{2+}$, voltage | – |
| $K^+$ | 9 | No | $K^+$ | ATP | Selective |
| ACA | 15 | No | Slight cation | pH, $Mg^{2+}$ | $K^+$ – uniporter – unselective |
| AAA | 45 | No | Slight anion | pH, $Mg^{2+}$ | IMAC |

* All measurements made in symmetrical 0.15 M KCl except $K^+$ channel.

### The 100-pS voltage-gated channel (mCS)

The first activity detected in single-channel recordings from the native inner mitochondrial membrane is a voltage-gated, 100-pS conductance[84] now called mCS (mitochondrial centum picosiemen activity). This activity has since been recorded in mitoplasts from a variety of tissues including rat and mouse liver, heart, rat brown adipose,[65, 84–87] and kidney (K.W. Kinnally, unpublished results), but not yeast.[129] There is still some question about the activation of mCS. Sorgato *et al.*[84] find the activity without submitting the mitoplasts to a specific activation procedure. However, other experiments suggest that they are activated by exposure of the cytoplasmic side of the inner membrane to $Ca^{2+}$-chelators.[88] Once activated, the activity is insensitive to $Ca^{2+}$.

The conductance of mCS in 150 mM KCl is generally about 110 pS but may vary between 90–140 pS. Substates on the order of 0.25, 0.5, 0.7, and 1.3 times the primary conductance value also have been reported.[2,65] mCS is slightly anion-selective ($P_{Cl^-}/P_{K^+} = 4.5$) in its main 110 pS conductance state and in at least one (50 pS) substate.[2] This channel is strongly voltage-dependent, closing with matrix negative potentials. Kinetic analysis of this bursting activity indicates that mCS has multiple open and closed states, as well as intermediate subconductance states.[65,89,90] Typical traces for mCS are shown in Fig. 15.1, while Fig. 15.2 illustrates the voltage dependence (the control curve).

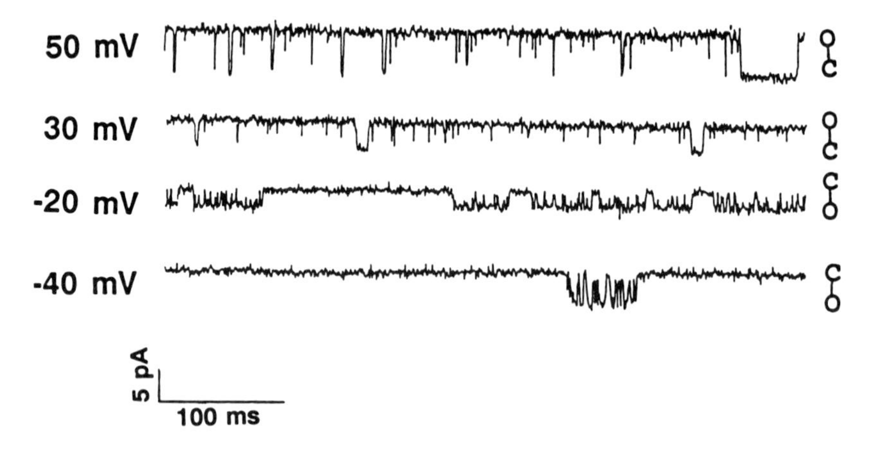


**Fig. 15.1** Sample current traces of mCS activity show the strong voltage dependence of the mCS activity. Current traces were recorded from an attached patch from a mouse liver mitoplast in 150 mM KCl, 5 mM HEPES, 1 mM EGTA, 2 mM $MgCl_2$ 2.5 $\mu$M Rotenone, pH 7.4. (Reproduced from Tedeschi and Kinnally 3 with permission.)

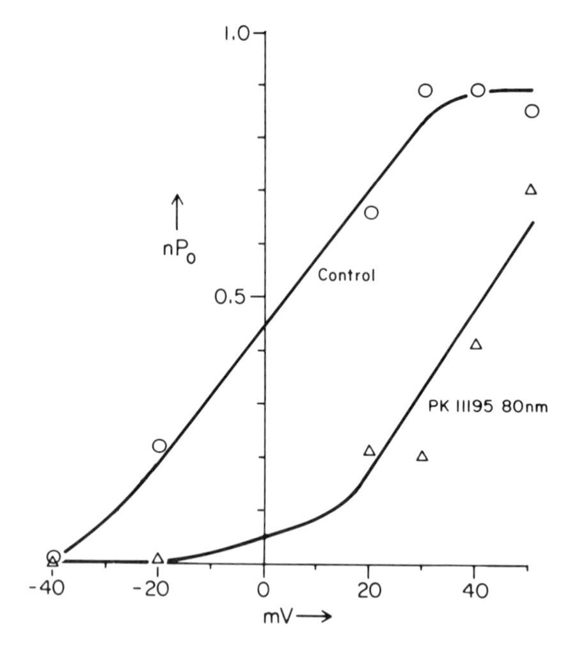


**Fig. 15.2** Open probability as a function of voltage in the absence (○) and in the presence (△) PK11195. The medium contained 150 mM KCl, 5 mM HEPES, 1 mM EGTA and 0.75 mM $CaCl_2$, pH 7.4. (Reproduced from Kinnally *et al.* 87, with permission.)

Early efforts to partially purify the mCS channel indicate that it co-isolates with the $F_o$ region of the ATP-synthase.[91] It is *not*, however, thought to be a component of the ATP-synthase, since its activity is insensitive to oligomycin or *N*, *N*′-dicydocarbodiimide (DCCD).

mCS is insensitive to changes in matrix pH from 6–9[84] and is not affected by millimolar $Mg^{2+}$ or micromolar ATP and ADP if applied on the matrix side of excised patches.[92] Klitsch and Seimen[65] have reported that mCS is inhibited by submillimolar levels of not only di- and trinucleotide phosphates, but also of GMP when added to the outside of patched mitoplasts, providing evidence that this channel activity is not related to the uncoupling protein, thermogenin, which is insensitive to GMP.

While a variety of compounds have no effect on mCS activity,[89] Antonenko *et al.*,[93] found this channel activity to be inhibited by the amphiphilic cationic drugs amiodarone and propranolol. Campo *et al.*[90] have described an inhibition of mCS by antimycin A binding at a site distinct from that involved with respiratory inhibition, as indicated by differences in $IC_{50}$. Kinnally *et al.*[87] reported that ligands of the mitochondrial benzodiazepine receptor (mBzR), such as R05-4864, PK11195, and protoporphyrin IX (PPIX), are high-affinity effectors but that clonazepam, a central benzodiazepine receptor ligand, is without effect. Figure 15.2 shows the inhibition of mCS by PK 11195.

Some postulated roles for this channel include volume regulation[65] and protein import.[1]

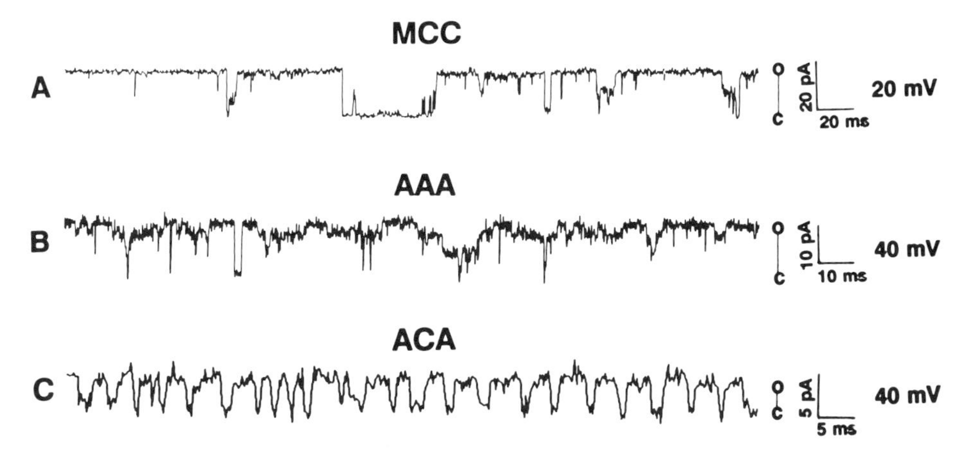


**Fig. 15.3** Sample current traces of inner membrane channel activities. **A**. MCC activity was recorded from an excised patch from rat heart mitoplast in symmetrical 150 mM KCl, 5 mM HEPES, 1 mM EGTA, 0.75 mM $CaCl_2$ ($\sim 10^{-7}$M free Ca) pH 7.4. **B**. AAA was recorded from an excised patch from mouse liver mitoplast in symmetrical 1 M KCl, 5 mM HEPES, 10 $\mu$M $CaCl_2$ with a pipette medium at pH 6.8 and a bath of pH 8.2. **C**. ACA was recorded from an excised patch from a mouse liver mitoplast in symmetrical 1 M KCl, 5 mM HEPES, 10 $\mu$M $CaCl_2$ with the pipette medium at pH 6.8 and the bath at pH 8.2. (Reproduced from Tedeschi and Kinnally 3, with permission.)

### The multiple conductance channel (MCC)

A variety of channel transition sizes of mitoplasts (10 to 1000 pS, in 150 mM KCl) and with a peak conductance of 1000–1500 pS have been grouped in the same class because they: (i) frequently occur in a single patch; (ii) can be induced together by elevated $Ca^{2+}$ levels or voltages greater than ±60 mV; (iii) are activated by voltage progressively, from low-conductance to high-conductance states, and the converse is true when inhibitors are introduced;[94] and (iv) have the same pharmacology.[89] Typical traces are shown in Fig. 15.3A. Although more than one channel may be involved, this activity has been ascribed to the multiple conductance channel or MCC (also called mitochondrial megachannel by Petronilli *et al*).[95] The largest (nS) transitions are typically observed in about 25% of the patches from mammalian and yeast mitoplasts.[94,96] The conductance levels between 300 and 1300 pS are probably substates of one channel, since multiple small openings can close simultaneously with a single large transition.[95] MCC shows little or no ion selectivity, except in some of its lower-conductance substates in which it is slightly cation-selective.

MCC is an inner mitochondrial membrane channel activity since it is recorded from patches which also contain mCS activity. Kinnally *et al.*[89] have proposed that MCC is associated with mitochondrial contact sites at which the inner and outer membranes closely adhere. In this model, inner and outer membrane components would act in series (like gap–junction conexons) and allow communication directly between the matrix and cytoplasm. Support for such a model is provided by the reports of MCC-like transitions in reconstituted contact site preparations[97] as well as by the fact that levels of $Ca^{2+}$ which activate MCC also increase the density of mitochondrial contact sites.[98]

Once activated (by $Ca^{2+}$ or large positive potentials) MCC is voltage-dependent, opening with negative matrix potentials and closing to lower conductance substates for small positive potentials. Increasing the transmembrane voltage above 50 mV usually reopens the channel (Fig. 15.4; control curve). The highest-conductance states often have extremely long open times, of the order of seconds to minutes.

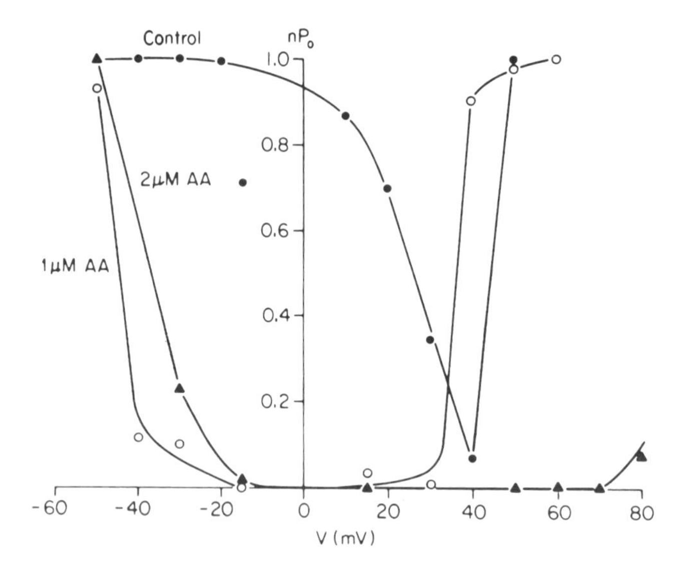


**Fig. 15.4** The open probability ($nP_o$) of MCC as a function of voltage and antimycin A. Amplitude diagrams were used to estimate the open probability in the absence (●) or presence of 1 μM (○) or 2 μM (▲) antimycin. (Reproduced from Campo *et al.* 90, with permission.)

Recently, Szabó and Zoratti[99] reported closure of the large-conductance mitoplast channel at both negative and positive potentials, a voltage dependence reminiscent of VDAC. However, a direct involvement of VDAC in MCC activity is not likely, since MCC activity was found in mitoplasts of yeast strains lacking the VDAC gene.[96] However, there is evidence for the interaction of VDAC with mitoplast channels and several indications of functional interactions between the outer membrane and mitoplast channels, including the inhibition by mBzR ligands of MCC (as well as mCS), and a difference in the behaviour at large negative or positive potentials of the MCC activity of wild-type and VDAC-less yeast strains.[96] The interaction between VDAC and MCC is most likely to occur at intermembrane contact sites.

A variety of physiologically significant effectors for MCC activity have been defined. MCC are activated by exposure of mitoplasts to micromolar or higher levels of $Ca^{2+}$, and subsequent treatment with the chelator EGTA will, under some conditions, reverse this activation.[100] This inner membrane channel activity is also regulated by such physiological effectors as $Mg^{2+}$, acidic pH, and ADP.[100]

Several pharmacological agents have been shown to affect MCC activity. The high-affinity ligands of mBzR (e.g. R05-4864 and PPIX) affect MCC activity in a biphasic manner, inhibiting at nanomolar levels and activating at higher concentrations.[87] However, the structurally similar compounds clonazepam and R015-1788 have no effects on MCC in the same range of concentrations. Several other MCC effectors, including antimycin A, cyclosporin A, and the uncoupler CCCP, are also mBzR ligands.[101] In addition, the list of compounds affecting MCC activity include the amphiphilic cations amiodarone, propranolol, and quinine, as well as dibucaine (see Table 15.2 and ref. 89 for summary). The inhibition and activation by antimycin is shown in Fig. 15.4.

There are several interesting correlations between MCC activity observed in patch-clamp experiments and the so-called PTP (see above). For example, both MCC and PTP are activated by $Ca^{2+}$ and inhibited by $Mg^{2+}$, ADP, $H^+$, amiodarone, and cyclosporin A, with most effects occurring at comparable levels for the two channels. Thus, it has been suggested that MCC and PTP are the same entity.[100]

MCC has several functional similarities to PSC, the peptide-sensitive outer-membrane cation channel (i.e. sizes of conductance transitions, range of voltage activation), implicating a possible role for MCC in protein import (which probably occurs at the sites of intermembrane contact.[103,104]

#### Channels induced by alkaline pH

Two pH-sensitive mitoplast channel activities have been described, both displaying greater open probabilities at alkaline pH and both requiring depletion of $Mg^{2+}$.[105,106] One of these channel activities is cation-selective (ACA, alkaline-induced cation activity), while the other is slightly anion-selective (AAA, alkaline-induced anion activity).[106] The two pH-sensitive channel activities were originally inferred from whole-patch currents which were later resolved into discrete transitions by high ionic strength (0.5–2 M salt) and high voltage (around $\pm 100$ mV). Typical traces in 1 M KCl for AAA and ACA are shown in Fig. 15.3B and C respectively.

**Table 15.2** Pharmacological survey for mitoplast channel inhibition.

| | Effector (μM) | Channel activity | | | | | Reference |
|---|---|---|---|---|---|---|---|
| | | MCC | mCS | AAA | ACA | $K^+$ | |
| Aluminium | 10–100 | + | + | nt | nt | nt | 127 |
| Amiodarone | 0.4–8 | + | + | + | + | nt | 93 |
| Antimycin A | 0.5–2 | + | + | nt | nt | nt | 90 |
| 4–Aminopyrine | 5000 | nt | nt | – | – | + | 125 |
| CCCP | 0.1–10 | + | + | nt | nt | nt | 92 |
| Clonazepam | >15–20 | – | – | nt | nt | nt | 87 |
| Cyclosporin | 0.25 | + | – | nt | nt | nt | 128 |
| Dibucaine | 100–500 | + | + | nt | nt | nt | * |
| FCCP | 0.1–10 | + | + | nt | nt | nt | 125 |
| Glibenclamide | 5 | nt | nt | – | – | + | 92 |
| Propranolol | 70–700 | + | + | + | + | nt | 93 |
| Protoporphyrin IX | 0.02–0.04 | + | + | nt | nt | nt | 87 |
| PK11195 | <0.08 | nt | + | nt | nt | nt | 87 |
| Quinine | 150–700 | + | + | + | + | nt | 89, 93 |
| Rotenone | 2–10 | – | – | nt | nt | nt | + |
| Ro5–4864 | 0.02–0.07 | + | + | nt | nt | nt | 87 |
| Ro15–1788 | 16 | + | nt | nt | nt | nt | 87 |
| Tributyl tin | 100 | nt | nt | + | + | nt | 89 |

* E. Holmunamedov and K. W. Kinnally (1993), unpublished results
+ M. L. Campo, K. W. Kinnally, and H. Tedesch: (1992) unpublished results
nt: not tested.

The cation channel ACA has a conductance of about 15 pS in 0.15 M KCl and is relatively voltage insensitive. Its voltage dependence and unit conductance are similar to those of the ATP-sensitive $K^+$ channel. However, unlike the latter channel, ACA is relatively non-selective for cations and is not affected by 4-aminopyridine and glibenclamide plus ATP. Instead, like mCS and MCC, ACA is inhibited by amiodarone and propranolol. ACA's cation selectivity and $Mg^{2+}$ sensitivity suggest that it may correspond to one of the cation uniporters whose presence has been inferred from solution studies and which are implicated in volume homoeostasis.[107]

The anionic channel AAA has an open-state conductance of about 45 pS and, like ACA, is relatively voltage insensitive. AAA has two conductance substates of one-third and two-thirds the fully open state, and is only slightly selective for different anions.

AAA may correspond to a low-conductance anion channel activity recently reconstituted in bilayers by Hayman *et al.*[108] There are also many similarities between AAA and IMAC, the inner membrane anion channel inferred from mitochondrial suspension studies (see above). This includes pH and inhibitor sensitivity. However, AAA and IMAC differ in estimated pore size and degree of anion selectivity – a question that requires further analysis.

### ATP-sensitive $K^+$ channels

Inoue *et al.*[92] described a $K^+$ channel in fused mitoplasts. This activity has a conductance of about 10 pS in 0.1 M KCl and is voltage independent. This channel is localized in the inner mitochondrial membrane, as shown by its presence in patches also containing mCS activity.

The channel activity is inhibited by known effectors of the plasma membrane ATP-sensitive $K^+$ channels, such as ATP, 4-aminopyridine, and glibenclamide plus ATP.

It has been suggested by Inoue *et al.*[92] and Mironova *et al.*[83] that the ATP-sensitive $K^+$ channel might form multimeric complexes capable of coordinated gating, which would explain the range of step conductances observed in reconstituted systems (see below). A possible role for the ATP-dependent $K^+$ channel is the regulation or fine-tuning of the mitochondrial metabolism, where metabolism could be channelled to either oxidative phosphorylation or ion transport. The channel could also have a role in thermal regulation (G. Mironova, personal communication).

Paucek *et al.*[109] partially purified a protein from rat liver or rat heart which catalyses electrophoretic $K^+$ uptake in liposomes and, when reconstituted in bilayers, shows a unit conductance of 30 pS. The activity is inhibited by ATP and ADP in the presence of divalent cations, and by glibenclamide in the absence of $Mg^{2+}$. The channel resembles the $K^+$ of the plasma membrane[110] and the channel of Inoue *et al.*,[92] although the properties are somewhat different, possibly because of the different conditions of the assay (see discussion in ref. 109).

### Channel activity from reconstitution of transport activity

Proteins extracted from mitochondria have been tested for channel activity by incorporation into planar bilayers and liposomes. Whether they function as channels in the native system remains an open question because of the paradoxical behaviour of translocators. For example, the $Na^+$, $K^+$-ATPase of the plasma membrane can behave as a channel in planar bilayers.[111,112] This channel activity is relevant to the physiology of the molecule since it can be inhibited by ouabain. However, in other reconstitution experiments the ATPase behaves as expected for a translocase.[113]

In addition to the $K^+$-transport protein discussed above, Mironova *et al.*[114] found channel behaviour after incorporating mitochondrial transport proteins into planar bilayers either the 40 kDa glycoprotein or the 2 kDa polypeptide associated with the $Ca^{2+}$-transporting system. The activity was blocked by ruthenium red and thiol reagents.

Voltage-dependent cation-selective channel activities with step conductances in the range 24–175 pS have been reported to be associated with mitochondrial membrane protein, such as those isolated by ethanol extraction,[83] quinine affinity chromatography,[115] and ion exchange chromatography.[109]

It is difficult to attempt to correlate results from these reconstitution experiments with those obtained with native membranes. The conductances in the bilayer systems are generally much higher. However, as suggested by Inoue *et al.*,[92] Mironova *et al.*,[83] Zorov *et al.*[94] and Tedeschi and Kinnally,[3] many of these proteins may be able to form multimeric complexes capable of coordinated gating, which would explain higher conductances of reconstituted systems or the range of step conductances observed.

## Possible interactions at contact sites

As originally described by Hackenbrock,[5] the inner and the outer membranes make contact at specific sites, the so-called contact sites. A variety of experiments have suggested a special organization and role of these sites.

The suggestion has been made that, at contact sites, outer and inner mitochondrial channels are in register to provide patent channels between the medium and the patching pipette.[89] This suggestion is based on results of experiments in which membrane preparations from brain mitochondria enriched in contact sites[116] showed channel activity that we consider to correspond to the MCC.[117] This model is represented in Fig. 15.5.

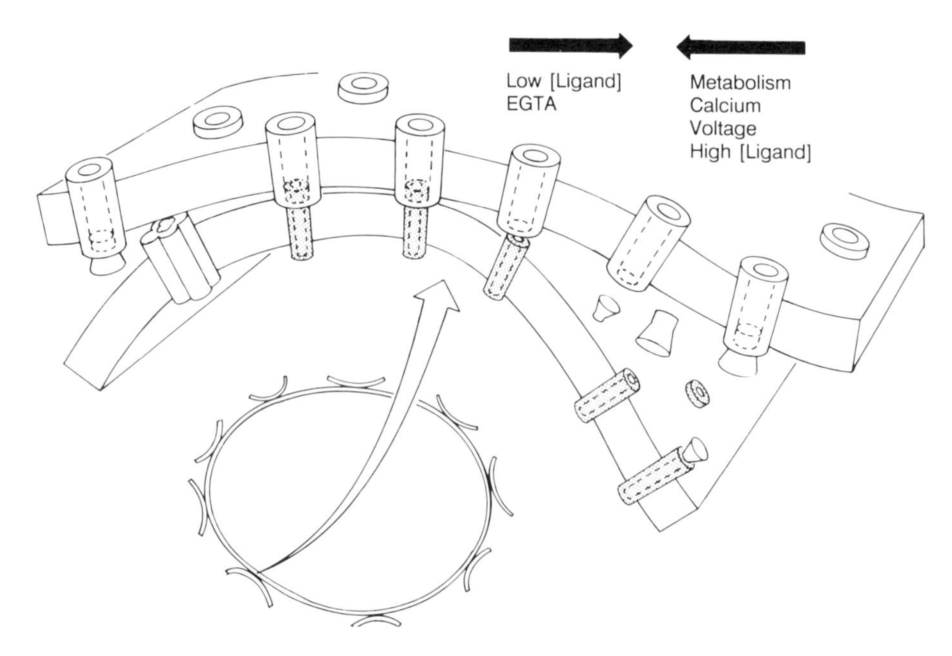


**Fig. 15.5** Surrealistic model involving the interaction of the inner and outer mitochondrial membrane channels to provide a patent ion pathway. Regulation of the apposition of the two membranes could control activity. Channels outside of the contact site are thought generally to be inactive as indicated by 'corks'. (Reproduced from Kinally and Tedeschi 2, with permission.)

As already mentioned, the results from many laboratories seem to implicate complexes between various inner and outer mitochondrial membrane components, including the adenine nucleotide translocator (ANT), VDAC, hexokinase, creatine kinase, and other components possibly at contact sites. These observations have led to the suggestion of a functional complex of VDAC and ANT. In the appropriate tissues, hexokinase (in the cytoplasmic side)[102] or creatine kinase (at the surface of the inner membrane)[118] are also thought to form part of this complex. The effect of a synthetic polyanion (for review, see ref. 25) or an endogenous protein regulator[38] on membrane function in intact mitochondria support these views. The polyanion has been shown to interact with VDAC and to affect mitochondrial functions associated with ATP utilization. The modulator has been shown to reduce state 3 respiration and respiratory control ratios (RCR) in intact mitochondria. However, the fact that both polyanions and modulators are effective when placed on the outside of the mitochondrion remains a puzzle.

The finding that NADH increases the sensitivity of VDAC to voltage point to a possible role in the Crabtee effect,[40] again suggesting an interaction between outer and inner membrane functions.

In agreement with the proposed models, a functioning mitochondrial benzodiazapine receptor (mBzR) complex has been recently isolated and was found to comprise an 18 kDa protein as well as ANT and VDAC, the latter two identified by Western blot analysis.[119] These findings suggest the possibility of investigating the relationship between outer and inner mitochondrial membrane channels and the mBzR through an examination of the sensitivity of the channel activity to benzodiazepines. In accordance with these expectations, these drugs were found to block inner mitochondrial membrane channels.[87]

## Discussion

The number of compounds which have an effect on channel activity is extensive and some are discussed earlier. We have summarized the salient inhibitors in Table 15.2 from studies using patch-clamping. An extensive list of drugs which block IMAC (which may correspond to AAA) has been published[55] and will not be repeated here. They include cationic amphiphiles such as the $\beta$-blocker propranolol, the antidepressant amitriptyline, quinine, and local anaesthetic. Mercurials and *N*-ethylmaleimide, triorganotins, and DCCD are also inhibitory.

The pharmacology of mitochondrial channels has several distinct features which deserve further discussion.

1. Few of the channels have been found to be inhibited with specific compounds. Only some drugs have been found to have specific effects on mitochondrial channels. Since the testing has not been extensive, the possibility remains that some of these drugs may not be completely specific. Cyclosporin is probably specific for MCC (which we believe corresponds to PTP, which is also inhibited). The ATP-sensitive $K^+$-channel is inhibited by glibenclamide and 4-aminopyridine.
2. The effect of many drugs is pleiotropic. Many of the drugs tested are not specific for any one channel. For example, amiodarone blocks all channels that were tested (Table 15.2). There are several possible explanations: (i) the effect is not on the channels but on the membrane; (ii) many channels share the same receptor; (iii) some channel activities (e.g. mCS and MCC) might be different functional states of the same channel; and (iv) there are homologous domains in different channel proteins.

A general effect of all these drugs on the membrane is possible. Most of the drugs are hydrophobic and the effect may result from binding to the hydrocarbon chains of the bilayers. Common effects of so-called stabilizers, which include tranquillizers and anaesthetics has been recognized.[120] However, there are several arguments that can be raised against this alternative. Some of the channel inhibitors have opposite effects on membranes. For example, quinidine, an isomer of quinine, has been shown to decrease the fluidity of liver plasma membranes.[121] Propranolol, on the other hand, disrupts the order of the hydrocarbon chains in platelet membranes[122] but both inhibit MCC activity. In addition, closely related compounds have very different effects, suggesting the involvement of specific receptors. For example, the benzodiazepine Ro5–4864 is extremely effective in blocking both MCC and mCS, while the closely related clonazepam is almost completely ineffective.[87] However, dynamic membrane states could explain some of the results. In this model, open channels could be restricted to specialized structures such as contact sites.[2] There is some indication of dynamic regulatory effects on contact sites, e.g. presence of $Ca^{2+}$[98,116] and metabolism.[123] This model (represented in Fig. 15.5) would provide an explanation, albeit incomplete, of why the activation of PTP (which probably corresponds to MCC) is affected by so many divergent agents.[56]

The presence of a single receptor capable of interacting with many channels is worth considering. The mBzR binds a broad spectrum of drugs,[101,124] including many affecting IMAC[55] and channels observed with patch-clamping.[2]

The alternative – that the various channel activities are manifestation of different states of the same channel – cannot be easily dismissed with presently available data. Some of the inhibitors already mentioned are specific, or act differentially on the various channel activities. For example, cyclosporin does not affect mCS, and micromolar levels of the unclouplers FCCP or CCCP that increase the open probability of MCC inhibit mCS at similar concentrations.[125] However, the sensitivity to chemicals could be a function of various conformations of the same protein. Lohret and Kinnally[96] report MCC in yeast mitochondria, but were unable to detect mCS in more than 200 patches. However, this may simply mean that the putative single-channel protein is unable to assume the mCS conformation in yeast.

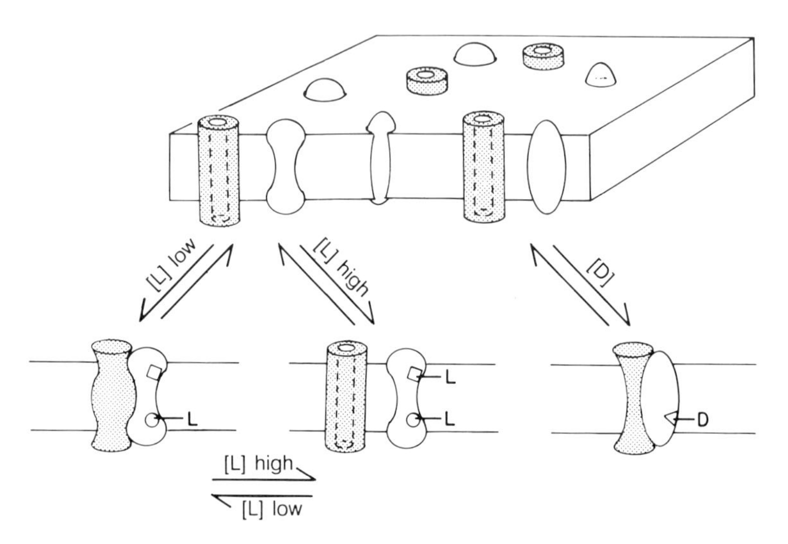


**Fig. 15.6** Hypothetical model of generic channels in a membrane to illustrate the inhibition by low ligand (L) or drug (D) levels by binding to high-affinity sites (△, ○) and the activation by high ligand levels due to binding of ligands to a lower-affinity site (□). (Reproduced from Kinnally and Tedeschi 2, with permission.)

3. Some drugs have a biphasic effect. We have found that some chemicals decrease the open probability of inner membrane channels at low concentrations, but increase it at higher concentrations. This is the case, for example for the effect of antimycin A[90] and protoporphyrin IX[87] on MCC activity. Gudz *et al.*[126] report similar effects of DCCD and butylated hydroxytoluene on PTP.

These observations suggest the presence of two different binding sites with different affinities either on the channel or a separate receptor. One of these would be a high-affinity inhibitory site. The other would be a lower-affinity activating site. A model which would conform with this postulate is represented in Fig. 15.6.

## Acknowledgements

The authors'research is supported in part by National Science Foundation grant MCB9117658 and by Spanish grant DGICYT PB92-0720.

## References

1. Sorgato, M.C. and Moran, O. (1993). *Crit. Rev. Biochem. Molec. Biol.*, **18**, 127–171.
2. Kinnally, K.W. and Tedeschi, H. (1994). Mitochondrial channels: an integrated view. In *Molecular biology of mitochondrial transport systems*, (ed. M. Forte and M. Colombini), pp. 169–198. Springer Verlag, Berlin, New York.
3. Tedeschi, H. and Kinnally, K.W. (1994). Mitochondrial membrane channels. In *Membrane channels – molecular and cellular physiology* (ed. C. Peracchia), pp. 529–548. Academic Press, New York, London.
4. Mannella, C.A. and Kinnally, K.W. (1996). Ion channels of mitochondrial membranes. In *Treatise on biomembranes*, Vol. 6, pp. 377–410. JAI Press.
5. Hackenbrock, C.R. (1968). *Proc. Natl Acad. Sci. USA*, **61**, 598–605.
6. Schein, S.J., Colombini, M., and Finkelstein, A. (1976). *J. Membr. Biol.*, **30**, 99–120.
7. De Pinto, V., Ludwig, O., Krause, J., Benz, R., and Palmieri, F. (1987). *Biochim. Biophys. Acta*, **894**, 109–119.
8. Forte, M., Guy, H.R. and Mannella, C.A. (1987). *J. Bioenerg. Biomembr.*, **19**, 341–350.
9. Weiss, M.S., Wacker, T., Weckesser, J., Welte, W., and Schulz, G.E. (1990). *FEBS Lett.*, **267**, 268–272.
10. Jap, B.K., Walian, P., Gehring, K., and Earnest, T. (1991). *Biophys. J.*, **59**, 15A.
11. Shao, L., Van Roey, P., Kinnally, K.W., and Mannella, C.A. (1994). *Biophys. J.*, **66**, A21.
12. Mannella, C.A. (1984). *Science*, **224**, 165–166.
13. Mannella, C.A. (1986). In: *Methods in enzymology* (ed. S. Fleischer and B. Fleischer), Vol. **125**, pp. 595–610. Academic Press, Inc., London.
14. Mannella, C.A. (1987). *J. Bioenerg. Biomembr.*, **19**, 329–340.
15. Thomas, L., Kocsis, E., Colombini, M., Erbe, E., Trus, B.L., and Steven, A.C. (1991). *J. Struct. Biol.*, **106**, 161–171.
16. Peng, S., Blachly-Dyson, E., Colombini, M., and Forte, M. (1992). *J. Bioenerg. Biomembr.*, **24**, 27–31.
17. Mannella, C.A. (1989). *J. Bioenerg. Biomembr.* **21**, 427–437.

18. Mannella, C.A., Guo, X.W., and Cognon, B. (1989). *FEBS Lett.*, **253**, 231–234.
19. Guo, X.W. and Mannella, C.A. (1993). *Biophys. J.*, **64**, 545–549.
20. Mannella, C.A., Forte, M., and Colombini, M. (1992). *J. Bioenerg. Biomembr.*, **24**, 7–19.
21. Doring, C. and Colombini, M. (1985). *J. Membr. Biol.*, **83**, 87–94.
22. Colombini, M. (1989). *J. Membr. Biol.*, **111**, 103–111.
23. Ludwing, O., Benz, R., and Schulz, J.E. (1989). *Biochim. Biophys. Acta*, **978**, 319–327.
24. Dihanich, M., Schmid, A., Oppliger, W., and Benz, R. (1989). *Eur. J. Biochem.*, **181**, 703–708.
25. Benz, R. and Brdiczka, D. (1992). *J. Bioenerg. Biomembr.*, **24**, 33–46.
26. Zimmerberg, J. and Parsegian, V.A. (1986). *Nature*, **323**, 36–39.
27. Thomas, L., Blachly-Dyson, E., Colombini, M., and Forte, M. (1993). *Proc. Natl Acad. Sci. USA*, **90**, 5446–5449.
28. Mannella, C.A. (1990), *Experientia*, **46**, 137–145.
29. Bowen, K.A., Tam, K., and Colombini, M. (1985). *J. Membr. Biol.*, **86**, 51–59.
30. Mirzabekov, T.A. and Ermishkin, L.N. (1989). *FEBS Lett.*, **249**, 375–378.
31. Mannella, C.A. and Frank, J. (1986). *J. Ultrastruct. Molec. Struct. Res.*, **96**, 31–40.
32. Guo, X.W. and Mannella, C.A. (1992). *Biophys. J.*, **63**, 418–427.
33. Colombini, M. (1987). *J. Bioenerg. Biomembr.*, **19**, 309–320.
34. Benz, R., Kottke, M., and Brdiczka, D. (1990). *Biochim. Biophys. Acta*, **1022**, 311–318.
35. Mangan, P.S. and Colombini, M. (1987). *Proc. Natl Acad. Sci. USA*, **84**, 4896–4900.
36. Holden, M.J and Colombini, M. (1988). *FEBS Lett.*, **241**, 105–109.
37. Liu, M.Y., Togrimson, A., and Colombini, M. (1994). *Biochim. Biophys. Acta*, **1185**, 203–212.
38. Liu, M.-Y. and Colombini, M. (1992). *J. Bioenerg. Biomembr.*, **24**, 41–46.
39. Benz, R., Wojtczak, L., Bosch, W., and Brdiczka, D. (1988). *FEBS Lett.*, **231**, 75–80.
40. Zizi, M., Forte, M., Blachly-Dyson, E., and Colombini, M. (1994). *J. Biol. Chem.*, **269**, 1614–1616.
41. Mannella, C.A. and Auger, I. (1986). *Biophys. J.*, **49**, 272A.
42. Chich, J.-F., Goldshmidt, D., Thieffry, M., and Henry, J.-P. (1991). *Eur. J. Biochem.*, **196**, 29–35.
43. Henry, J.-P., Chich, J.-F., Goldschmidt, D., and Thieffry, M. (1989). *J. Membr. Biol.*, **112**, 139–147.
44. Moran, O. and Sorgato, M.C. (1992). *J. Bioenerg. Biomembr.*, **24**, 91–98.
45. Freitag, H., Neupert, W., and Benz, R. (1982). *Eur. J. Biochem*; **123**, 619–636.
46. Lindén, M., Andersson, G., Gellefors, P., and Nelson, B.D. (1984) (1984). *Biochim. Biophys. Acta*, **770**, 93–96.

47. Tedeschi, H., Kinnally, K.W., and Mannella, C.A. (1989). *J. Bioenerg. Biomembr.*, **21**, 451–459.

48. Moran, O., Sciancalepore, M., Sandri, G., Panfili, E., Bassi, R., Ballarin, C., and Sorgato, M.C. (1992). *Eur. Biophys. J.*, **21**, 311–319.

49. Tedeschi, H., Mannella, C.A., and Bowman, C.L. (1987). *J. Membr. Biol.*, **97**, 21–29.

50. Kinnally, K.W., Tedeschi, H., and Mannella, C.A. (1987). *FEBS Lett.*, **226**, 83–87.

51. Nicholls, D.G., Cunningham, S.A., and Rial, E. (1986) In *Brown adipose tissue*, (ed. P. Trayhurn and D.G. Nicholls), pp. 52–85. Edward Arnold (Publishers) Ltd., London.

52. Ricquier, D., Casteilla, L., and Bouillaud, F. (1991). *FASEB J.*, **5**, 2237–2242.

53. Klaus, S., Casteilla, L., Bouillaud, F., and Ricquier, D. (1991). *Int. J. Biochem.*, **23**, 791–801.

54. Garlid, K.D. and Beavis, A.D. (1986). *Biochim. Biophys. Acta*, **853**, 187–204.

55. Beavis, A.D. (1992). *J. Bioenerg. Biomemb.*, **24**, 77–90.

56. Gunter, T.E. and Pfeiffer, D.R. (1990). *Am. J. Physiol. Cell. Physiol.*, **27**, C755–C786.

57. Bernardi, P., Angrilli, A., Ambrosin, V., and Azzone, G.F. (1989). *J. Biol. Chem.*, **264**, 18902–18906.

58. Nicolli, A., Redetti, A., and Bernardi, P. (1991). *J. Biol. Chem.*, **266**, 9465–9470.

59. Puigserver, O., Herron, D., Gianotti, M., Palou, A., Cannon, B., and Nedergaard, J. (1992). *Biochem. J.*, **284**, 393–398.

60. Rehnmark, S., Bianco, A.C., Kiefer, J.D., and Silva, J.E. (1992). *Am. J. Physiol.*, **262**, E58–E67.

61. Bianco, A.C., Kieffer, J.D., and Silva, J.E. (1992). *Endocrinology*, **130**, 2625–2633.

62. Klingenberg, M. and Winkler, E. (1985). *EMBO J.*, **4**, 3087–3092.

63. Garlid, K.D. (1990). *Biochim. Biophys. Acta*, **1018**, 151–154.

64. Mirzabekov, T.A., and Akhemerov, R.N. (1987). *Biofizika*, **32**, 345–346.

65. Klitsch, T. and Siemen, D. (1991). *J. Membr. Biol.*, **122**, 69–75.

66. Heaton, G.M., Wagenvoord, R.J., Kemp, A., and Nicholls, D.G. (1978). *Eur. J. Biochem.*, **82**, 515–521.

67. Ricquier, D., Barlet, J.P., Garel, J.M., Combes-George, M., and Dubois, M.P. (1983). *Biochem. J.*, **210**, 859–866.

68. Aquila, H., Link, T.A., and Klingenberg, M. (1987). *FEBS Lett.*, **212**, 1–9.

69. Bouillaud, F., Weissenbach, J., and Ricquier, D. (1986). *J. Biol. Chem.*, **261**, 1487–1490.

70. Ricquier, D. and Bouillaud, F. (1986). The brown adipose tissue mitochondrial uncoupling protein. In *Brown adipose tissue*, (ed. P. Trayhurn and D.G. Nicholls), pp. 86–104. Edward Arnold (Publishers) Ltd., London.

71. Aquila, H., Link, T.A., and Klingenberg, M. (1985). *EMBO J.*, **4**, 2369–2376.

72. Runswick, M.J., Powell, J.T., Nyren, P., and Walker, J.E. (1987). *EMBO J.*, **6**, 1367–1373.

73. Lin, C.S. and Klingenberg, M. (1980) *FEBS Lett.*, **113**, 299–303.
74. Brierley, G.P. (1970). *Biochemistry*, **9**, 697–707.
75. Selwyn, M.J., Dawson, A.P., and Fulton, D.V. (1987). *Biochem. Soc. Trans.*, **7**, 216–219.
76. Beavis, A.D. and Garlid, K.D. (1987). *J. Biol. Chem.*, **262**, 15085–15093.
77. Haworth, R. A. and Hunter, D.R. (1980). *J. Membr. Biol.*, **54**, 231–236.
78. Al-Nasser, I. and Crompton, M. (1986). *Biochem. J.*, **239**, 19–29.
79. Gazzotti, P. (1975). *Biochem. Biophys. Res. Commun.*, **67**, 634–638.
80. Igbavboa, U., Zwizinski, C.W., and Pfeiffer, D.R. (1989). *Biochem. Biophys. Res. Commun.*, **161**, 619–625.
81. Crompton, M. and Costi, A. (1988). *Eur. J. Biochem.*, **178**, 488–501.
82. Fagian, M.M., Pereira da Silva, L., Martins, I.S., and Vercesi, A.E. (1990). *J. Biol. Chem.*, **265**, 19955–19960.
83. Mironova, G.D., Fedotcheva, N.I., Makarov, P.R., Pronevich, L.A., and Mironov, G.P. (1981). *Biofizika*, **26**, 451–457.
84. Sorgato, M.C., Keller, B.U., and Stühmer, W. (1987). *Nature*, **330**, 498–500.
85. Kinnally, K.W., Campo, M.L. and Tedeschi, H. (1988). Abstracts, 14th International Congress of Biochemistry, p. 168.
86. Kinnally, K.W., Campo, M.L., and Tedeschi, H. (1989). *J. Bioenerg. Biomembr.*, **21**, 497–506.
87. Kinnally, K.W., Zorov, D.B., Antonenko, Yu., Snyder, S., McEnery, M.W., and Tedeschi, H. (1993). *Proc. Natl Acad. Sci. USA*, **90**, 1374–1378.
88. Kinnally, K.W., Zorov, D.B., Antonenko, Y.U., and Perini, S. (1991). *Biochem. Biophys. Res. Commun.*, **176**, 1183–1188.
89. Kinnally, K.W., Antonenko, Y., and Zorov, D.B. (1992). *J. Bioenerg. Biomembr.*, **24**, 99–110.
90. Campo, M.L., Kinnally, K.W., Perini, S., and Tedeschi, H. (1992). *J. Biol. Chem.*, **267**, 8123–8127.
91. Sorgato, M.C., Moran, O., De Pinto, V., Keller, B.U., and Stühmer, W. (1989). *J. Bioenerg. Biomembr.*, **21**, 485–496.
92. Inoue, I., Nagase, H., Kishi, K., and Higuti, T. (1991). *Nature*, **352**, 244–247.
93. Antonenko, Y.N., Kinnally, K.W., Perini, S., and Tedeschi, H. (1991). *FEBS Lett.*, **285**, 89–93.
94. Zorov, D., Kinnally, K.W., and Tedeschi, H. (1992). *J. Bioenerg. Biomembr.*, **24**, 119–124.
95. Petronilli, V., Szabó, I., and Zoratti, M. (1989). *FEBS Lett.*, **259**, 137–143.
96. Lohret, T. and Kinnally, K.W. (1994). *Biophys. J.*, **66**, A22.

97. Moran, O., Sandri, G., Panfili, G., Stühmer, W., and Sorgato, M.C. (1990). *J. Biol. Chem.*, **265**, 908–913.

98. Bakker, A., De Bie, I., Bernaert, T., Ravingerova, A., Ziegelhoffer, H., Van Belle, H., and Jacob, W. (1993). *Eur. J. Morphol.*, **31**, 46–50.

99. Szabó, I. and Zoratti, M. (1993). *FEBS Lett.*, **330**, 201–205.

100. Szabó, I. and Zoratti, M. (1992). *J. Bioenerg. Biomembr.*, **24**, 111–117.

101. Hirsch, J.D., Beyer, C.F., Malkowitz, L. Beer, B., and Blume, A.J. (189). *J. Pharmacol.*, **34**, 157–163.

102. McEnery, M.W. (1992). *J. Bioenerg. Biomembr.*, **24**, 63–69.

103. Hartl, F.U., Pfanner, N., Nicholson, D.W., and Neupert, W. (1989). *Biochim. Biophys. Acta*, **988**, 1–45.

104. Brdiczka, D. (1991). *Biochim. Biophys. Acta*, **1071**, 291–312.

105. Antonenko, Y.N., Kinnally, K.W., and Tedeschi, H. (1991). *J. Membr. Biol.*, **124**, 151–158.

106. Antonenko, Y.N., Kinnally, K.W., and Tedeschi, H. (1994). *Biophys. J.*, **64**, 343A.

107. Bernardi, P., Zoratti, M., and Azzone, G.F. (1992). In *NATO ASI Series, Vol. H 64, Mechanics of swelling*, (ed. T.K. Karalis), pp. 357–377. Springer-Verlag, Berlin.

108. Hayman, K.A., Spurway, T.D., and Ashley, R.H. (1993). *J. Membr. Biol.*, **136**, 181–190.

109. Paucek, P., Mironova, G., Mahdi, F., Beavis, A., Woldegiorgis, G., and Garlid, K.D. (1992). *J. Biol. Chem.*, **267**, 26062–26060.

110. Davis, N.W., Standen, N.B., and Stanfield, P.R. (1991). *J. Bioenerg. Biomembr.*, **23**, 509–535.

111. Last, T.A., Gantzer, M.L., and Tyler, C.D. (1983). *J. Biol. Chem.*, **258**, 2399–2404.

112. Mironova, G.D., Bocharnikova, N.I., Mirsalikhova, N.M., and Mironov, G.P. (1986). *Biochim. Biophys. Acta*, **861**, 224–236.

113. Eisenrauch, A., Grell, E., and Bamberg, E. In *The sodium pump: structure, mechanism, and regulation*, (ed. J.H. Kaplan and P. De Weer), pp. 317–326. The Rockefeller University Press, New York.

114. Mironova, G.D., Sirota, T.V., Pronevich, L.A.,. Trofimenko, N., Mironov, G.P., Gregorjev, P.A., and Kondrashova, M.N. (1982). *J. Bioenerg. Biomembr.*, **14**, 213–225.

115. Diwan, J.J. and Costa, G. (1994). Purification and patch clamp analysis of a 40 pS mitochondrial channel. In *Molecular biology of mitochondrial transport systems*, (ed. M. Forte and M. Colombini), pp. 199–208. Springer Verlag, Berlin, New York.

116. Sandri, G., Siagri, M., and Panfili, E. (1988). *Cell Calcium*, **9**, 159–165.

117. Moran, O., Sandri, G., Panfili, E., Stühmer, W., and Sorgato, M.C. (1990). *J. Biol. Chem.*, **265**, 908–913.

118. Schlegel, J., Wyss, M., Eppenberger, H.M., and Walliman, T. (1990). *J. Biol. Chem.*, **265**, 9221–9227.

119. McEnery, M.W., Snowman, A.M., Trifiletti, R.R., and Snyder, S.H. (1992). *Proc. Natl Acad. Sci. USA*, **89**, 3170–3174.

120. Seeman, P.M. (1966). *Int. Rev. Neurobiol.*, **9**, 146–221.

121. Needham, L., Dodd, N.J.F., and Houslay, M.D. (1987). *Biochim. Biophys. Acta*, **899**, 44–50.

122. Dash, D. and Rao, G.R. (1990). *Arch. Biochem. Biophys.*, **276**, 343–347.

123. Knoll, G. and Brdiczka, D. (1983). *Biochim. Biophys. Acta*, **733**, 102–110.

124. Verma, A. and Snyder, S.H. (1988). *Mol. Pharmacol.*, **34**, 800–805.

125. Campo, M.L., Kinnally, K.W., and Tedeschi, H. (1994). *Biophys. J.*, **66**, A22.

126. Gudz, T.I., Novrogodov, S.A., and Pfeiffer, D.R. (1992). *Seventh European Bioenergetics Conference*, **7**, 125.

127. Zorov, D., *et al.* (1991). *J. Cell Biol.*, **111**, 60A.

128. Szabó, I. and Zoratti, M. (1991). *J. Biol. Chem.*, **266**, 3376–3379.

129. Lohret, T.A. and Kinnally, K.W. (1995). *Biophys. J.*, **68**, 2299–2309.

# 16 Modulation of a large-conductance $K^+$ channel from chromaffin granule membranes by adrenergic agonists and G proteins

N. Arispe, P. De Mazancourt, H. B. Pollard, and E. Rojas

## Introduction

Ion channels occur in the membrane of secretory granules from adrenal chromaffin cells (Picaud *et al.* 1984), neurohypophyseal granules Lemos *et al.* 1989; Stanley *et al.* 1988, and synaptic vesicles from *Torpedo* electric organ (Rahamimoff *et al.* 1988, 1989). Anion ($Cl^-$ channels were found to occur in rat neurohypophysis (Stanley *et al.* 1988), and fused *Torpedo* secretory vesicles (De Riemer *et al.* 1987; Rahamimoff *et al.* 1988). Cationic ($K^+$), $Ca^{2+}$-dependent channels have been reported in rat hypophyseal granules (Stanley *et al.* 1988; Lemos *et al.* 1989). In addition, Rahamimoff *et al.* (1990) reported two types of cation-selective channels in *Torpedo* vesicles which were characterized as being of large-conductance, low-selectivity, calcium-activated and voltage-regulated. However, of the increasing number of proteins found to be localized in small synaptic vesicles (Knaus *et al.* 1990) and, to some extent, in adrenal chromaffin granules (Picaud *et al.* 1984), only synaptophysin has been shown to exhibit cation-selective channel activity (Thomas *et al.* 1988). Elemental imaging studies of secretory granules, both inside adrenal chromaffin cells and following isolation, indicate that concentrations of $K^+$ and $Na^+$ can change substantially during subcellular fractionation (Ornberg *et al.* 1988). Yet, other constituents such as nitrogen (ATP and catecholamines), phosphorus (ATP), calcium, and magnesium remained constant (Ornberg *et al.* 1988). Taken together, these results suggest that cation channels occur within the vesicle membranes, and that their function is compatible with sustained stability of the granules. Chromaffin granule membranes have been studied from this perspective only recently and now we definitely know they contain such channels (Arispe *et al.* 1992; Ashley *et al.* 1994). Based on these data we concluded that this channel may be important in granule assembly (Arispe *et al.* 1992) and in catecholamine equilibrium distribution. In this regard, cytosolic levels of catecholamines (CA) depend on a delicate balance of electrochemical driving forces across the membrane of the chromaffin granules (Pollard *et al.* 1976: Johnson and Scarpa, 1979; Phillips 1978; Phillips and Apps 1980; Ashley *et al.* 1994). Since the $K^+$ gradient across the granule membrane changes during maturation and recycling of the vesicles after exocytosis, we considered the possibility that the difference in $K^+$ across the granular membrane during the different phases of the secretory cycle may give rise to a $K^+$ diffusion electrical potential. We hypothesized that the expression of this potential by activation of the $K^+$ channel might provide a highly efficient mechanism to modulate the uptake of CA by the secretory vesicles. To test the hypothesis that the regulation of the $K^+$-selective channel is intimately linked with the intracellular trafficking of CA, we studied the effects of various adrenergic agonists on the activity of the $K^+$-channel and searched for possible regulatory mechanisms. In support for this model we found and report here that adrenergic receptor agonists, including classical $\beta$-adrenergic agonists epinephrine (adrenaline), norepinephrine (noradrenaline), and isoproterenol, and $\alpha_1$-adrenergic agonist phenylephrine, induced a dramatic increase in the open-probability $p_o$ of the channel. Furthermore, $K^+$ channel activation by CA could be reverted by adrenergic receptor antagonists such as propranolol (Arispe and Rojas 1995). From these studies we conclude that $K^+$ channel $p_o$ is modulated by a mechanism involving the activation of adrenergic receptors known to be present in chromaffin granule membranes. Finally, we recently discovered that this $K^+$ channel is tightly controlled by inhibitory as well as stimulatory heterotrimeric GTP-binding proteins. Using antibodies against specific $\alpha$ subunits for immunoblot analysis, we were able to identify the presence of the inhibitory $G_{i2}$ and $G_{i3}$ subtypes, as well as the stimulatory $G_o$ and $G_s$ subtypes, but not $G_{i1}$ in adrenal chromaffin granules. Furthermore, functional analysis of the $K^+$ channel incorporated into planar lipid bilayers showed that GDP$\beta$S and GTP$\gamma$S have opposite effects on channel activity inducing interconversions between a low and a high open-probability states. Consistent with these findings, the same antibodies antagonized the effects of the non-hydrolysable analogues on the open-probability of the $K^+$ channel (Arispe *et al.* 1995).

The activity of the $K^+$ channel incorporated into planar bilayers directly from intact chromaffin granules exhibits two distinct modalities characterized by the open-probability $p_o$. We observed that all the adrenergic agonists tested induced a dramatic activation of the channel from the usual low $p_o$ ($<0.3$) to the less frequently observed high $p_o$ ($>0.7$). We concluded that the two modalities of $K^+$ channel activity present in our records were generated by the unique large-conductance $K^+$-selective channel of the chromaffin granule membrane. Since in the majority of the $K^+$ channel incorporations the open-probability $p_o$ was found to be rather low, we concluded that, in our chromaffin granule preparation, the $K^+$ channel was under tonic inhibition by an undetermined mechanism. Since these studies showed that the open-probability $p_o$ of the $K^+$ channel can be dramatically changed by adrenergic agonists from the cytosolic aspect of the incorporated vesicle, we hypothesized that the $K^+$ channel might well be directly controlled by specific G proteins (Arispe *et al.* 1995). We found that different antibodies, each one raised against a specific $\alpha$ subunit of a trimeric G protein, identified the presence of $G_o$, $G_s$, $G_{i2}$ and $G_{i3}$ but not $G_{i1}$ in our highly purified preparation of adrenal chromaffin granules (McKenzie *et al.* 1988; Simonds *et al.* 1989*b*; Spiegel *et al.*, 1990*a*, 1992; Spiegel, 1991) Furthermore, functional analysis of the $K^+$ channel incorporated into a planar lipid bilayer membrane showed that the non-hydrolyzable analogues guanosine 5′-O-(2-thiodiphosphate) (GDP$\beta$S) and guanosine 5′-O-(3-thiotriphosphate) (GTP$_\gamma$S) had opposite effects on the open-probability of the channel. Finally, we found that the same antibodies used in the immunoblot analysis could either activate or inhibit the channel in much the same way as GDP and GTP analogues. We conclude that the activity of the chromaffin granule large-conductance $K^+$ channel is kept under tight control by direct action of both inhibitory and stimulatory G proteins on the channel.

These data led us to propose that the regulation of the $K^+$ channel involve adrenergic receptors, known to be present in the chromaffin granules (Nikodejevic *et al.* 1976; Hoffman *et al.* 1976; Zinder *et al.* 1977), directly coupled via $G\alpha_i$ proteins to the channel.

## Investigational methods

### Preparation of intact chromaffin granules and chromaffin granule ghosts

Intact chromaffin granules were prepared from bovine adrenal medulla tissue by homogenization in 0.3 M sucrose and purified on a metrizamide gradient, as previously described (Pollard *et al.* 1979*a*). Following equilibrium centrifugation, the metrizamide was diluted by three volumes of 0.3 M sucrose and centrifuged at 49 000 *g*. The granules were then resuspended in 0.3 M sucrose and stored at 4°C.

Chromaffin granule ghosts were prepared from bovine adrenal medulla tissue by homogenization in 0.3 M sucrose and purification over a 1.6 M sucrose step, as previously described (Brocklehurst and Pollard 1990). Thereafter, the pellet from the 1.6 M sucrose step was resuspended by homogenization in 50 volumes of ice-cold 5 mM Na-HEPES, pH 7.2, and allowed to remain on ice for 30 min. The solution was centrifuged at 49 000 *g*, and the resulting pellet then resuspended by homogenization in 5 ml of 0.3 M sucrose to effect a second osmotic shock. Final purification was carried out by centrifugation over a 1 M sucrose step (density = 1.10 g $cm^{-3}$). The ghosts remaining at the top of the gradient after centrifugation at 36 000 r.p.m. were subjected to a final osmotic shock by a 10-fold dilution in buffer. These membranes were then centrifuged at 49 000 *g*, and resuspended in a minimal volume of buffer before being stored at −70°C.

### Bilayer chamber and membrane formation

The plexiglass bilayer chamber consisted of two compartments separated by a thin Teflon film (Fig. 16.1). During experiments, the solutions were simultaneously stirred by two Teflon-coated magnets placed in a restricted space at the bottom of each compartment. Ag/AgCl pellet electrodes were immersed in a small pool containing 0.5 M KCI and were electrically connected to the solutions in each compartment via agar bridges (2% agar in 0.5 M KCI). Single-channel currents were recorded using a patch-clamp amplifier (EPC-7, List Medical Electronics or Axopatch-1D, equipped with a CV-48 bilayer headstage, Axon Instruments) and were stored on magnetic tape using a PCM/VCR digital system (Digital-4, Toshiba) with a frequency response in the range from DC to 25 000 Hz. Records were made from playbacks through a low-pass filter (8-pole Bessel 902 LPF, Frequency devices) set in the range from 200 to 500 Hz.

Planar bilayers were formed by applying a suspension of synthetic palmitoyl – oleolyl – phosphatidyl - ethanolamine (POPE) and phosphatidylserine (PS, 50 mg $ml^{+1}$) in decane.

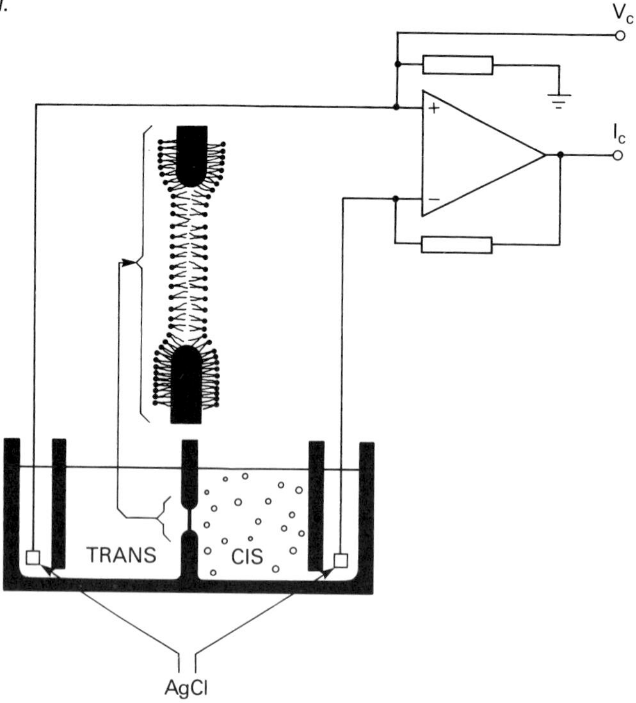


**Fig. 16.1** Simplified diagram of the bilayer membrane chamber and amplifier. For explanations, see text.

A small glass rod was used to deliver the lipids to a hole of ca. 100–150 $\mu$m in diameter in a Teflon film separating two compartments that contained the required salt solutions.

Ion channels, presumably present in secretory granules from adrenal chromaffin cells, were incorporated into a bilayer by adding a small volume (5–10 $\mu$l) of a suspension of intact secretory vesicles (or purified ghosts) made in a K-HEPES solution. Vesicles were added on only the *cis* side of the chamber, and incorporation occurred directly from the experimental solutions. In the majority of the experiments, to fuse chromaffin granules with a lipid bilayer we prepared a suspension of the intact vesicles in a K-HEPES solution (in mM: 200 K-HEPES, pH 7.4). Ion channels present in secretory granules from adrenal chromaffin cells were incorporated into a bilayer by adding a small volume (5–10 $\mu$l) of this suspension of intact vesicles to a different K-HEPES solution (either 200 or 400 mM). $CaCl_2$ was added (free $Ca^{2+}$ ca 200 $\mu$M) to facilitate incorporation of the vesicles to the bilayer. Granules were added on only the *cis* side of the chamber. Solutions containing adrenergic agonists and antagonists (Research Biochemicals International, MA) were always prepared a few minutes before the addition to the solutions in the compartments.

## Biochemical procedures

The C-terminal decapeptides of $\alpha_o$, $\alpha_{i3}$, $\alpha_s$, yeast GP1$\alpha$ (the product of GPA1), the internal decapeptides of $\alpha_{i1}$, $\alpha_{i2}$ and $\alpha_{i3}$ (see Table 16.1) and the common C-terminal decapeptide of $\alpha_{i1}$ and $\alpha_{i2}$ were synthesized (Doherty *et al.* 1991), conjugated to keyhole limpet haemocyanin with glutaraldehyde, and injected into rabbits. Some antisera were affinity-purified on Affi-Gel 15 columns (Bio-Rad) containing the corresponding immobilized peptide (McKenzie *et al.* 1988; Simonds *et al.* 1989*a,b*; Spiegel *et al.* 1990*a,b*). The protein concentration was determined by the Bradford method. For immunoblot, P2 membranes and chromaffin granules (150–300 $\mu$g protein) were diluted 1 : 1 in 2× Laemmli buffer and resolved using a SDS – polyacrylamide gel (10% acrylamide, 0.13% bisacrylamide), transferred to PDVF membranes (Millipore) and immunoblotted (20 h at room temperature) with the anti-sera. The antibody – antigen complex was detected by $^{125}$l-labelled protein.

**Table 16.1** Summary of the effects of NaF, non-hydrolysable guanosine analogues (GTP$\gamma$S and GDP$\beta$S), and antibodies on $p_o$.

| Treatment | Channel state | |
|---|---|---|
| | hop | lop |
| NaF ($\alpha_{all}$) | $p_o$ ↑ (Fig. 16.15) | None* |
| GTP$\gamma$S | $p_o$ ↑ (Figs. 16(a,b)) | $p_o$ ↓ (Fig. 16.17) |
| GDP$\beta$S | None* | $p_o$ ↓ (Figs. 16.18, 16.19) |
| EC ($\alpha_{i3}$, $\alpha_o$) | None* | $p_o$ ↓ (Fig. 16.21) |
| AS7 ($\alpha_{i2}$) | None* | $p_o$ ↓ (Fig. 16.22(a)) |
| GO ($\alpha_o$) | None* | $p_o$ ↓ (Figs. 16.22(b,c)) |
| RM ($\alpha_s$ | None* | $p_o$ ↑ (Fig. 16.23) |

* Tested; data not shown.

# Results of investigations

## Cation channels from intact chromaffin granules

Intact chromaffin granules (and highly purified chromaffin granule membrane ghosts) possess a $K^+$ channel of large-conductance, i.e. 140–380 pS in symmetrical KCl (or K-HEPES) solutions of increasing concentrations from 140 to 400 mM.

In symmetrical KCl (200 mM) solutions, exposure of the *cis* side of the bilayer to intact granules caused the otherwise electrically silent bilayer to acquire ion channel activity. In the specific example shown in Fig. 16.2A, one channel with two levels of conductance was expressed. Portions of the traces in Fig. 16.2A (denoted by the numbers) are shown in Fig. 16.2B using expanded (12-fold) time scale. In segment 1, to the right of the record, there is evidence of two levels. On average, the first level was estimated as 160 pS and the second level as 380 pS. Under these recording conditions (symmetrical KCl), transitions among these levels occurred very frequently (Fig. 16.2B, segments 2 and 3).

In KCl solutions the $K^+$ channel kinetics always depended on transmembrane potential (Fig. 16.3A). As illustrated in Fig. 16.3B, the open-probability varied from ca. 0.2 to 1 as the transmembrane potential was increased from −30 to 30 mV. Furthermore, changing to asymmetrical KCl solutions (Fig. 16.3C) shifted the reversal potential of the currents from 0 mV (measured in symmetrical 200 mM KCI; ○,□) to −24 mV (measured in asymmetrical KCl: 200 mM *cis* 50 mM *trans*; ● while the open-probability $p_o$ values at each potential remained unchanged. Assuming that the Goldman – Hodgkin – Katz equation,

$$V_{rev} = RT/F\ In\{P_K[K^+]_{trans} + P_{Cl}[Cl^-]_{cis}\}/\{P_K[K]_{cis} + P_{Cl}[Cl^-]_{trans.}\}$$

is applicable to this system, from the reversal potential of −24 mV (Fig. 16.3C, ●) we estimate the permeability ratio $P_K/P_{Cl}$ in the channel to be 6.7. Thus, in all the channel records acquired in KCl solutions, $K^+$ is the main ion carrying the current. The other cations ($Ca^{2+}$) contributed little, if at all.

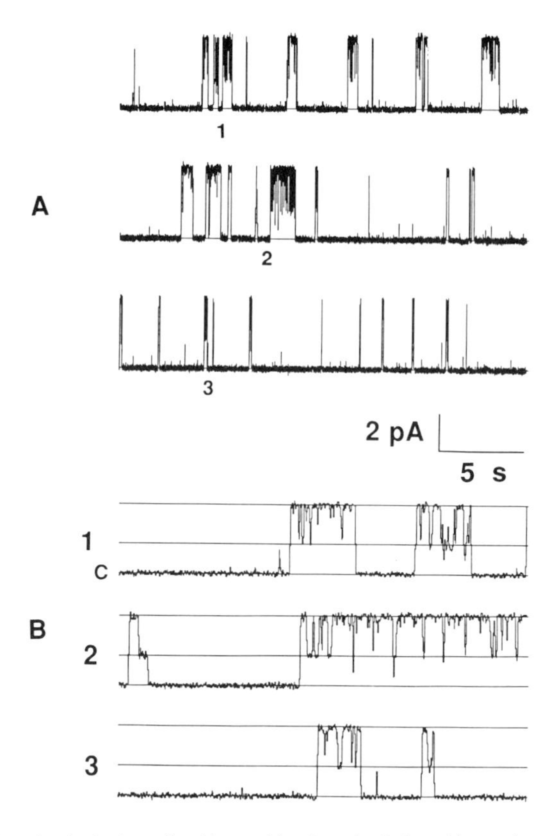


**Fig. 16.2** Incorporation of cationic channel activity resulting from the fusion of intact chromaffin granules with the bilayer. Symmetrical $K^+$ solutions (in mM: 200 KCl, 5 K-HEPES, pH 7.4) in both compartments of the bilayer chamber. A. Three consecutive segments of a continuous record at a transmembrane potential of −10 mV. B. Segments 1–3 were taken from the records shown in panel A, as indicated by the numbers beneath the records in A, and are shown on expanded time scale.

## Potassium channels from purified chromaffin granule membrane ghosts

Although the method used to prepare the adrenal chromaffin granules employed in the preceding experiments yielded highly purified intact vesicles (Pollard *et al.* 1979*a*,*b*), we now included additional purification steps to exclude possible contaminating membranes. Furthermore, we wanted to study any possible modulation of the $K^+$ channel activity by intracellular molecules, we decided to minimize the concentration of non-membrane granule constituents. We therefore lysed adrenal chromaffin cell granules and prepared highly purified granule ghosts by a procedure employing two separate gradients and three osmotic lysis steps.

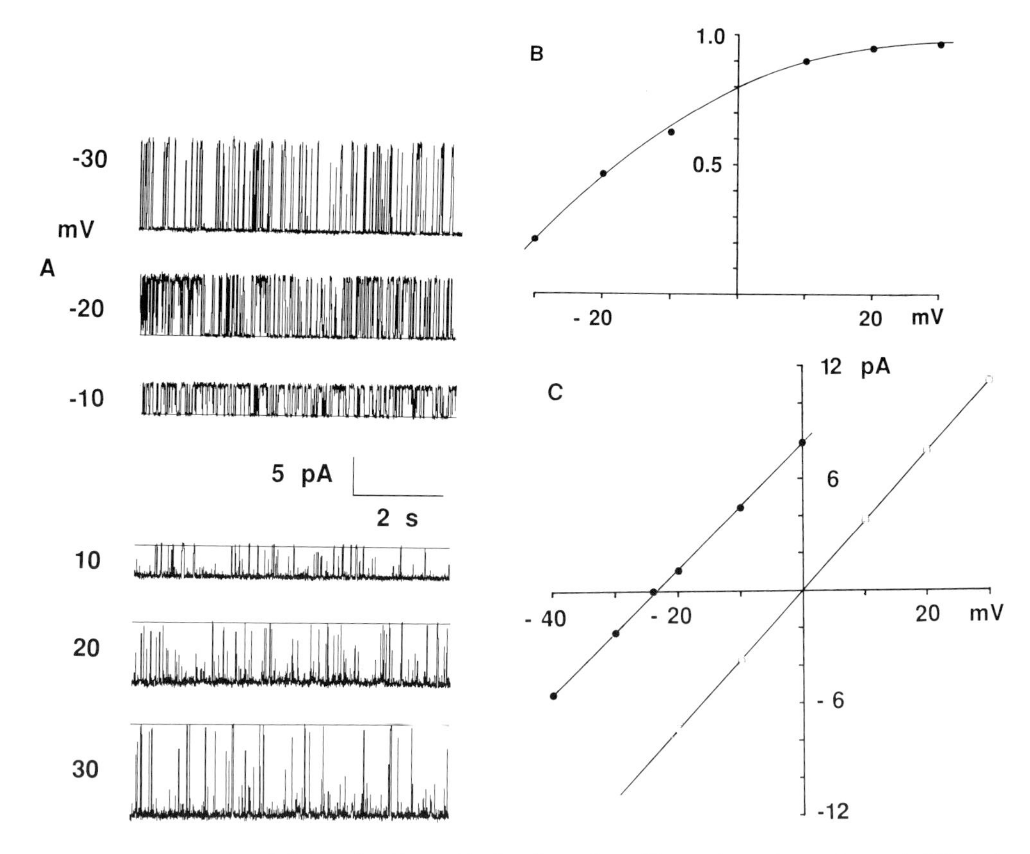


**Fig. 16.3** Incorporation of $K^+$ channel activity resulting from intact chromaffin granule membranes. Symmetrical $K^+$ solutions (in mM: 200 KCl, 5 K-HEPES, pH 7.4) in both compartments of the bilayer chamber. A. Six consecutive segments of a continuous record at different transmembrane potentials indicated in mV on the left side. Continuous horizontal line represents zero current level or closed channel. B. Open channel probability $p_o$ as a function of transmembrane potential. These were calculated from the ratio between the open time and the total time. C. I-V curves in symmetrical 200 mM KCl (○, □) and in asymmetrical 200 mM KCl (*cis* side) 50 mM KCl (trans side) (•). $V_{rev}$ was measured as −24 mV, resulting in a $P_K/P_{Cl}$ value of 6.7.

As expected, we were able to detect only the same type of $K^+$ channel with chromaffin granule ghosts as the donor membrane. This conclusion is based on a large number of incorporations using three different preparations of granule ghosts. From one preparation to the next, the conductance in symmetrical $K^+$ solutions was rather similar in the two preparations. The only measurable difference was the channel kinetics, i.e. the open probability. Records from two different incorporations are shown in Fig. 16.4 (symmetrical 400 mM K-HEPES). As illustrated in Fig. 16.4A the frequency of channel openings is smaller than that shown in Fig. 16.4B (exhibiting long-lasting silent intervals). As expected, the current – voltage relationship depicted in Fig. 16.4C is relatively linear from −40 to 40 mV, with a slope conductance of ca. 160 pS.

Low selectivity for cations has been a hallmark of the secretory vesicle membrane cationic channels hitherto reported (De Riemer *et al.* 1987; Rahamimoff *et al.* 1988; Lemos *et al.* 1989; Lee *et al.* 1992). We therefore estimated the selectivity of the present chromaffin

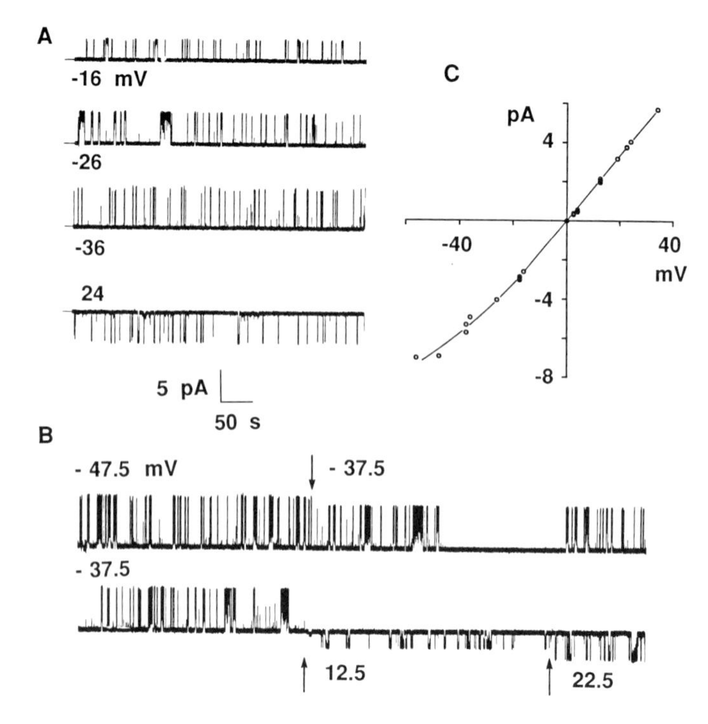


**Fig. 16.4** Incorporation of potassium-channel activity resulting from the fusion of highly purified chromaffin granule membrane ghosts with the bilayer. A, B. Single $K^+$ channel activity from two different preparations of granule ghosts. Symmetrical K-HEPES solution (400 mM) in both compartments. C. Values for the I–V curve were obtained from amplitude histograms constructed from databases of at least 2000 events.

granule ghosts channel by measuring the reversal potential of the single channel current ($V_{rev}$) under bi-ionic conditions. First, we measured the size of the events (channel activity depicted in Fig. 16.5, inset on the right) and constructed amplitude distribution histograms at various transmembrane potentials. Next, we plotted the current – voltage curve, i.e. mean amplitude of the single $K^+$ channel current as a function of transmembrane potential Fig. 16.5). For the asymmetrical bi-ionic system (200 mM K-HEPES on the *cis* side and 100 mM K-HEPES plus 100 Na-HEPES on the *trans* side), a $V_{rev}$ of $-15$ mV was determined from the I–V curve in Fig. 16.5. From the Goldman – Hodgkin – Katz equation

$$V_{rev} = RT/F \ln \{P_{Na}[Na]_{trans} + P_K[K]_{trans}/P_K[K]_{cis}\}$$

we calculate the permeability ratio $P_K/P_{Na}$ to be ca. 10. From these data it is clear that the cationic channels incorporated together with the membranes of the chromaffin granule ghosts were also highly selective for $K^+$. We therefore feel it appropriate to refer to this chromaffin granule ghost channel as a potassium channel (now on referred to as the CG-type $K^+$ channel), and thus distinct from the previously described cation-selective channels in other secretory vesicle membrane preparations.

The CG-type $K^+$ channel is unique not only in its high selectivity for $K^+$ over $Na^+$, but

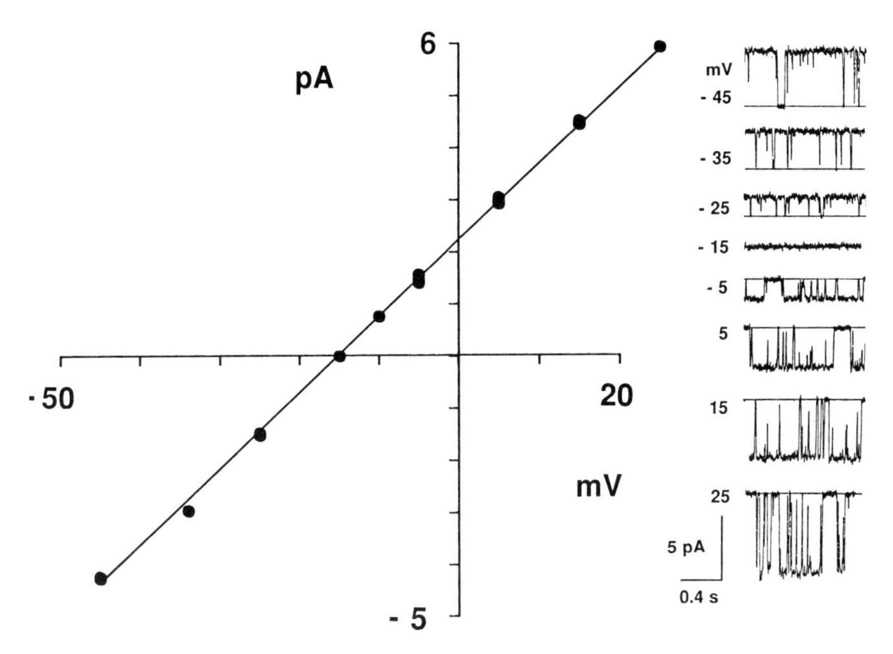


**Fig. 16.5** Potassium selectivity of the cation channel from granule ghosts. Vertical axis represents the mean value from the amplitude histogram of single-channel events at the indicated transmembrane potential (abscissa). Inset: channel activity with $K^+$ as charge carrier.

also, as previously shown (Arispe *et al.* 1992), in its insensitivity to the presence of $Ca^{2+}$ or charybdotoxin, a specific blocker of the $[Ca^{2+}]$-activated $K^+$ channel of large-conductance (Anderson *et al.* 1988; MacKinnon and Miller 1988).

## Kinetic behaviour of the CG-type K⁺ channel

The CG-type $K^+$ channel exhibits a very characteristic type of bursting behaviour. As illustrated in Fig. 16.6, we found the channel to go through brief (ca 1 ms) and frequent transitions from the open to the closed state. We also observed single-channel current levels of smaller amplitude which we interpreted as incomplete closures (Fig. 16.2). From these records we measured the time intervals between channel openings (distribution of closed times) and between channel closures (distribution of open times). The database used for the open time and closed time histograms corresponds to recordings with a signal-to-noise ratio of $>10:1$ (Fig. 16.6). All events of duration $<0.2$ ms have been discarded, although omitting such events does introduce some distortion of the observed channel kinetics. The number of events for the database associated with each histogram was always $>2000$. The minimum number of open states entered by the CG-type $K^+$ channel can be estimated determining the number of exponential components necessary to fit the observed open time distribution (Sakmann and Neher 1983). As shown in Fig. 16.7, open time and closed time histograms clearly required more than one time constant to fit the points. At a transmembrane potential of $-50$ mV the time constants were 10.5 and 40.5 ms for the open distribution (Fig. 16.7, upper right side) and 1.8 and 7.3 ms for the closed distribution

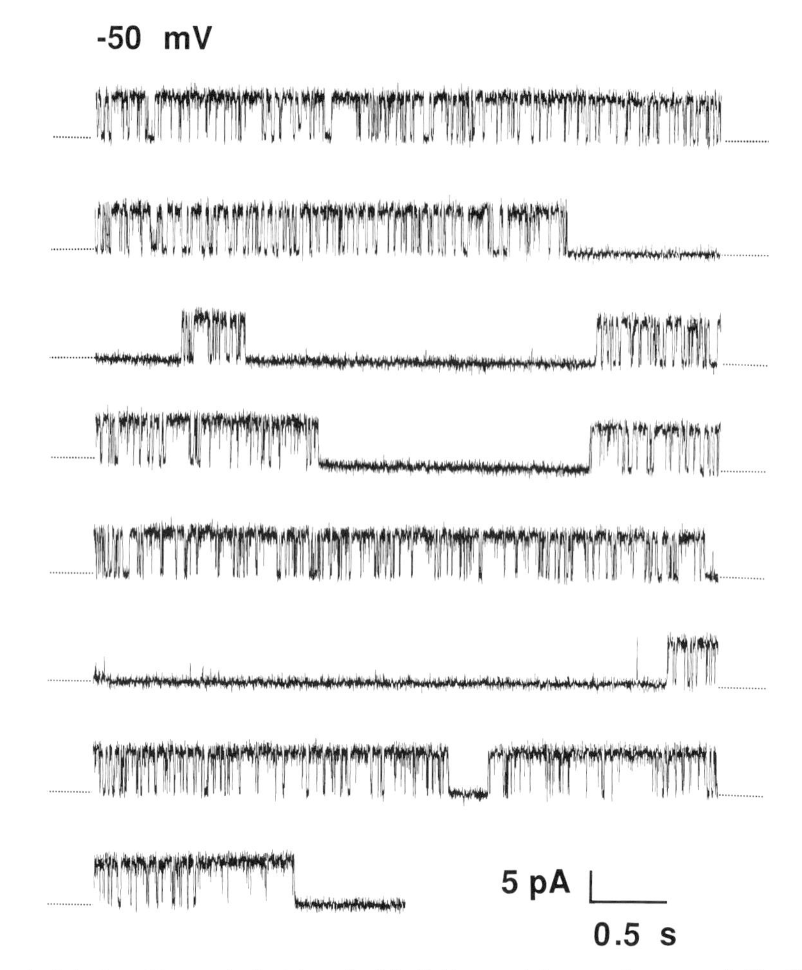


**Fig. 16.6** Bursting behaviour of the granule ghost channel activity. Eight segments from a continuous record of the $K^+$ channel activity. Records made in the presence of symmetrical K-HEPES (400 mM). The closed state of the channel is indicated by the dotted lines on the left and right side of each record.

(Fig. 16.7, lower right side). These kinetic parameters were not affected by $Ca^{2+}$ (see next section). We concluded that the flickering of the channel current is presumably the consequence of the passage through more than one state during the transition from closed to fully open configuration. The presence of more than one closed configuration for the CG-type $K^+$ channel might explain the interburst behaviour.

### The large-conductance $K^+$ channel chromaffin granule membranes exhibits two functional states

The CG-type $K^+$ channel exhibits two distinct patterns of activity. As described in a previous publication (Arispe *et al.* 1995), in a large proportion of the incorporations (70%; $n = 108$ incorporations), $K^+$ channel openings were characterized by low open-probability (lop) values ($p_o < 0.3$). Figure 16.8 shows that $K^+$ channel openings occurred in brief bursts

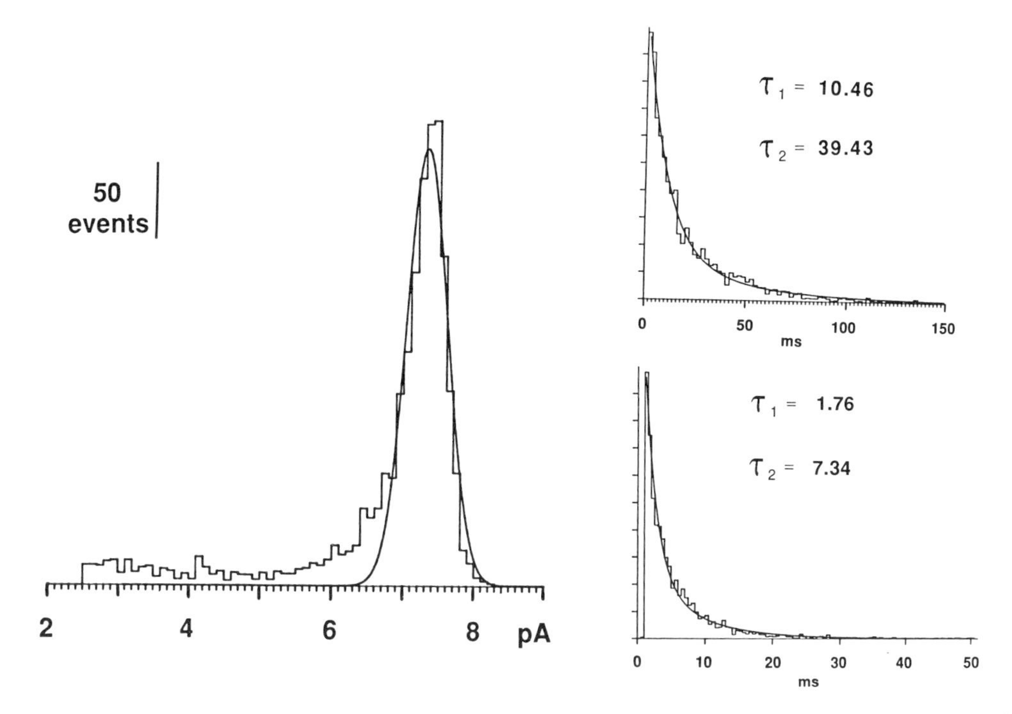


**Fig. 16.7** Amplitude, open- and closed-time histograms of the granule ghost channel. Transmembrane potential equal to −50 mV and symmetrical K-HEPES (400 mM). Distribution of channel current amplitudes constructed with 2616 events (left side). Fitted curve was calculated as,

$$dp/di = 1/(2\pi)^{0.5} \exp[-(1 - \langle i \rangle)^2/2\,\sigma^2]$$

with the following values for the parameters:

$$\langle i \rangle = 7.3 \text{ pA and } \sigma^2 = 0.3 \text{ pA}.$$

Only values <8.4 pA and >6.5 pA were included. The bars from 2.2 to 6.5 represent different open states of the channel. Right side: open (upper) and closed time histograms (lower). For each histogram the fitted line corresponds to

$$NE_o = \Sigma NE_i.\exp[-t/\tau_i]$$

where i represents the number of states and $NE_o$ open state event frequency. $\tau_i$ values are given next to the corresponding fitted curve.

(Fig. 16.8A: upward deflections of the trace at −10 mV). Long-lasting inter-burst intervals (closed conformation of the channel) were often observed. Open probability $p_o$ values estimated from the 10-min segments of a continuous record in Fig. 16.8A were 0.041, 0.042, 0.043 and 0.038 from top to bottom record, respectively. In contrast, during the other type of activity, present in a smaller fraction of the successful incorporations (22%), the channel exhibited high open-probability (hop) values ($p_o > 0.7$), remaining in the open conformation during long-lasting periods, with frequent rapid transitions to the closed state (Fig. 16.8B; brief downward deflections of the trace at −10 mV; $p_o = 0.88$). Occasionally (8%) two or more channels exhibiting the hop and lop patterns of activity were incorporated in the bilayer (Fig. 16.8C).

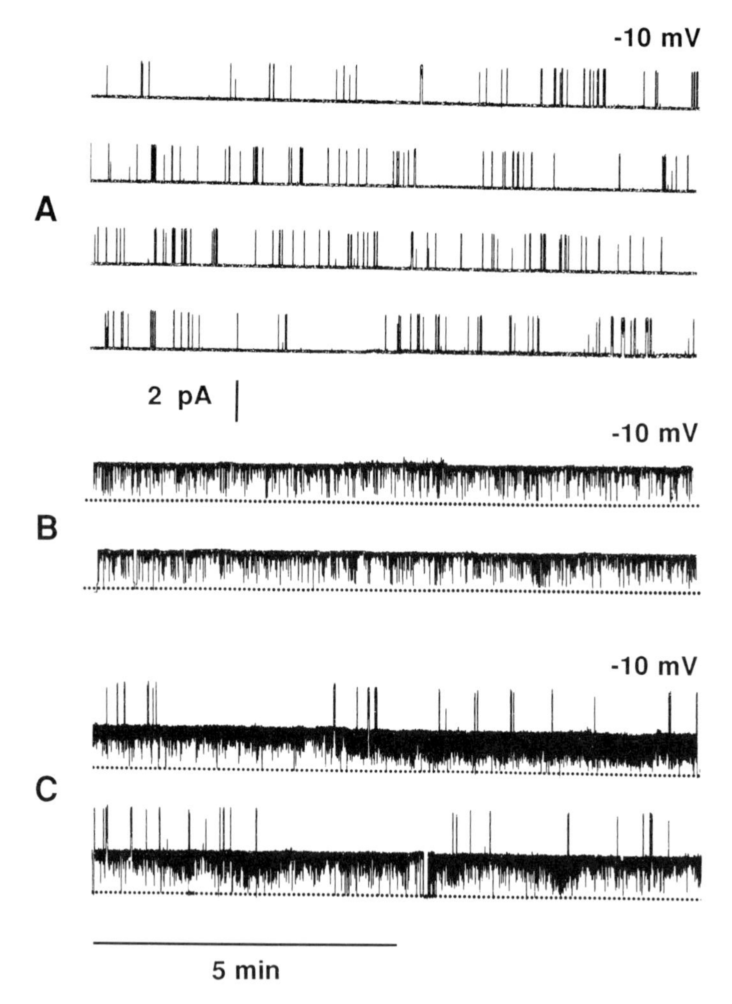


**Fig. 16.8** The chromaffin granule membrane $K^+$ channel exhibits two modes of activity. Symmetrical K-HEPES (200 mM) throughout. A. Channel in the low open probability mode. Single-channel conductance at $-10$ mV estimated from the amplitude histogram was 190 $\pm$ 6 pS; $p_o = 0.041 \pm 0.008$. B. Channel in the high open probability mode. Conductance at $-10$ mV was 195 $\pm$ 7 pS; $p_o = 0.78$. C. Conductances were estimated as 202 $\pm$ 4 pS and 198 $\pm$ 8 pS for first and second level, respectively.

As anticipated from the data in Fig. 16.8, the current – voltage relationships (I–V curves) calculated from both lop and hop modalities of $K^+$ channel activity were found to be linear, the single-channel conductance being equal to ca. 195 pS (symmetrical 200 mM K-HEPES) for both patterns of activity. In all the experiments in which the channel events occurred in the lop modality ($p_o > 0.3$), the frequency of the $K^+$ channel openings seemed insensitive to the transmembrane potential (Fig. 16.9A). By contrast, in the hop modality ($p_o > 0.7$), the frequency of closures depended on the transmembrane potential (Fig. 16.9B: compare frequency of closures at $-20$ and $+20$ mV). Indeed, $p_o$ decreased as the transmembrane potential was made more negative.

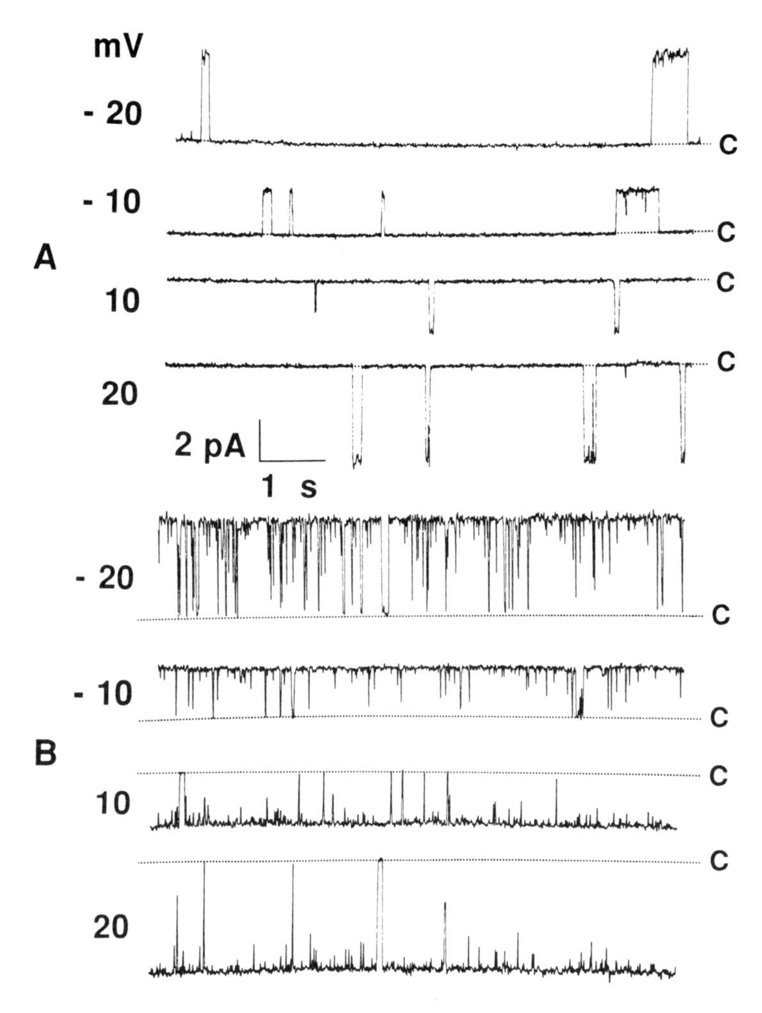


**Fig. 16.9** The chromaffin granule membrane $K^+$ channel in the hop modality exhibits voltage-dependent gating. Symmetrical K-HEPES (200 mM) throughout. Transmembrane potential is given in mV on the left side of each record, and closed state of the $K^+$ channel is indicated by the dotted line in each trace. A. Pattern of channel activity defined as lop. At −20 mV, $K^+$ channel openings correspond to the upward deflections of the trace. B. Pattern of channel activity defined as hop. Transitions to the closed state are less frequent at more positive transmembrane potentials.

## The CG-type $K^+$ channel is insensitive to charybdotoxin and $Ca^{2+}$

Throughout these experiments we maintained the $Ca^{2+}$ concentration at either 1 mM (in KCI solutions) or <0.2 mM (in K-HEPES solutions). However, early in our studies we noted that addition of EGTA had no demonstrable effect on the CG-type $K^+$ channel activity, once it was incorporated and functional in the bilayer. In addition, to further allay our continuous concerns regarding possible trace contamination of our ghosts preparation by non-chromaffin granule membranes, we also tested the CG-type $K^+$ channels for sensitivity to charybdotoxin. This toxin is a highly selective inhibitor of the $Ca^{2+}$-sensitive $K^+$ channel of large conductance known to be present in the plasma membrane of chromaffin cells (Fenwick *et al.* 1982). The record in Fig. 16.10 illustrates that the activity of the CG-type $K^+$ channel remained unchanged after the consecutive additions of charybdotoxin (0.1 $\mu$M) and EGTA (1.5 mM) to both *cis* and *trans* compartments. From these results, we conclude that the CG-type $K^+$ channel is insensitive to Ca, and that it is very unlikely derived from the chromaffin cell membrane.

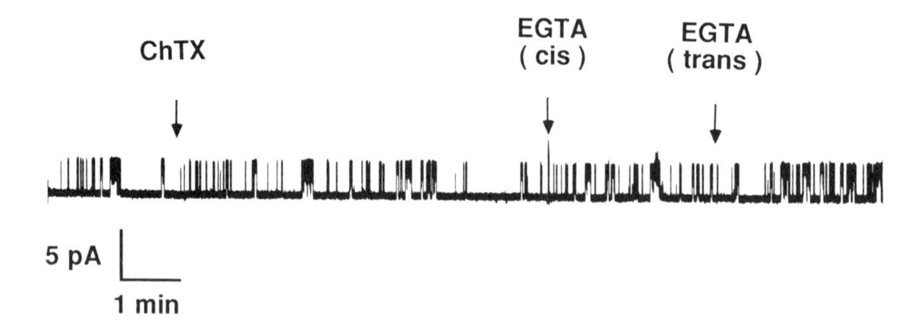


**Fig. 16.10** The granule ghost $K^+$ channel is insensitive to charybdotoxin (ChTX) and $Ca^{2+}$. Transmembrane potential equal to −25 mV and symmetrical K-HEPES (400 mM). ChTX was added to both sides (100 nM) as indicated (left arrow). K-EGTA (1.5 mM) was added as indicated to the *cis* side (centre arrow) and to the *trans* side (right arrow). During each addition at the time indicated by the arrows, the recording was interrupted and the solutions on both sides were stirred.

## The CG-type $K^+$ channel is regulated by adrenergic receptors

As shown in Fig. 16.11, the physiological agonist for adrenergic receptors norepinephrine (0.01–1 mM) can evoke a dramatic switch in the pattern of $K^+$ channel activity from lop to hop. In the control records at −10 and 10 mV (upper two traces), the CG-type $K^+$ channel was found in the lop mode and, therefore $p_o$ was insensitive to transmembrane potential and remained unchanged at ca. 0.05. A few minutes after the addition of a high dose of norepinephrine (1 mM) to the *cis* side, $p_o$ increased to ca. 0.8 at −10 mV and 0.95 at 10 mV. Several minutes after the application of adrenergic agonist, the conductance of the channel, estimated from I–V curves, decreased from its control value of 190 to 145 pS. As a rule, in the presence of the agonist, the overall channel activity increased with time and multiple current levels were often observed, even at low doses of norepinephrine. Thus, a substantial fraction of the observed increase in channel activity ($p_o$) evoked by the adrenergic agonist could result from the activation of several $K^+$ channels. However, statistical analysis of records (bottom records) revealed that application of norepinephrine to the *cis* augments the open-probability of the $K^+$ channel. Thus, the adrenergic agonist norepinephrine induced a profound change not only in the $K^+$ channel gating, but also affected the conduction mechanisms of the channel. Furthermore, the data support the idea that the lop and hop types of activity originate from two different states of the same CG-type $K^+$ selective channel. Addition of norepinephrine on the *trans* side was without effect in all the experiments carried out different granule preparations. In contrast, the addition of norepinephrine (10–1000 $\mu$M) to the *cis* side dramatically activated the CG-type $K^+$ channel after a time delay, from seconds to minutes depending on agonist concentration.

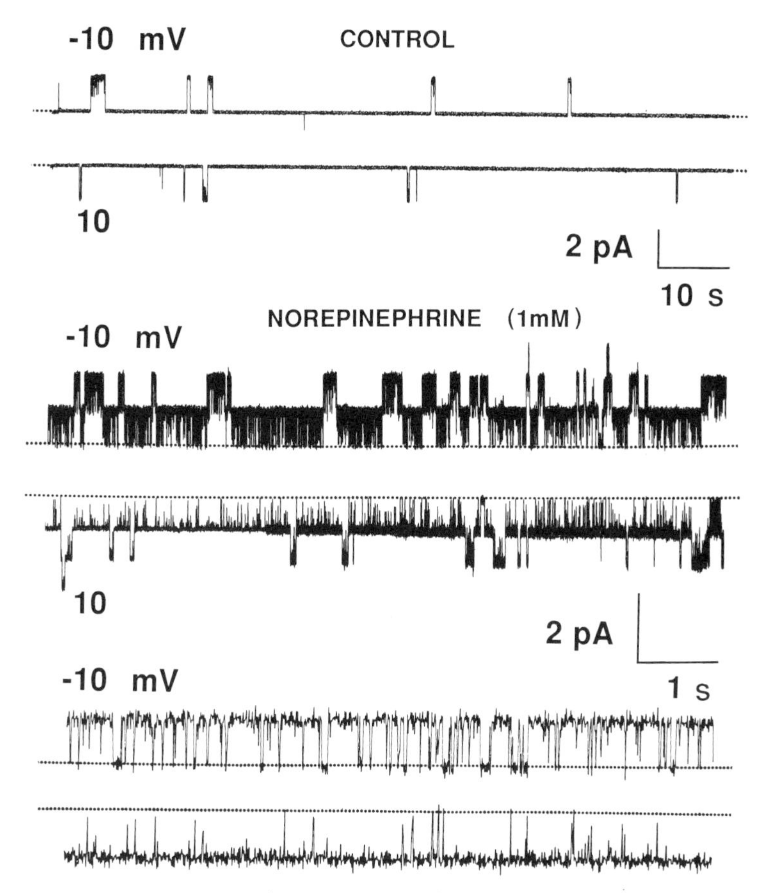


**Fig. 16.11** Norepinephrine modifies both the $K^+$ channel gating and the conductance. Symmetrical K-HEPES (200 mM; $[Ca^{2+}]$ ca. 80 $\mu$M). Single $K^+$ channel activity at $-10$ and 10 mV during the control period in the absence of agonist (upper records). Upwards deflections at $-10$ mV and downwards at 10 mV represent open-channel currents. The records in the middle were made 10 min after the addition of the adrenergic receptor agonist norepinephrine to the *cis* side of the chamber. Shown in the lower part of the figure are two segments of the channel activity depicted in the middle on an expanded time base.

These results suggest that the activation of the $K^+$ channel involves adrenergic receptors on the cytosolic aspect of the vesicles. In an attempt to identify the receptor type involved, we used specific different adrenergic agonists and antagonists. Figure 16.12 shows that the CG-type $K^+$ channel can be activated by a physiological dose of the $\beta$-adrenergic agonist isoproterenol (10 $\mu$M). $K^+$ channel activity prior to the application of isoproterenol to the *cis* side was recorded and remained unchanged during a control period of 13 min (upper two records at $-14$ mV; 170 pS in symmetrical 200 mM K-HEPES). Application of a small dose of isoproterenol (10 $\mu$M) after a long delay (8–10 min) stimulated channel activity. Although at this time only one $K^+$ channel current level was observed (third record from the top), 5 min later the record exhibited three channel current levels (Fig. 16.12, bottom record), suggesting the recruitment of other channels. Activation of the $K^+$ channel by isoproterenol at higher doses ca. 1 mM) occurred with a brief delay ($<2$ min) after the addition of the agonist. Thus, the $K^+$ channel can be activated switching from lop to hop mode by a relatively low dose of the $\beta$-adrenergic agonist isoproterenol.

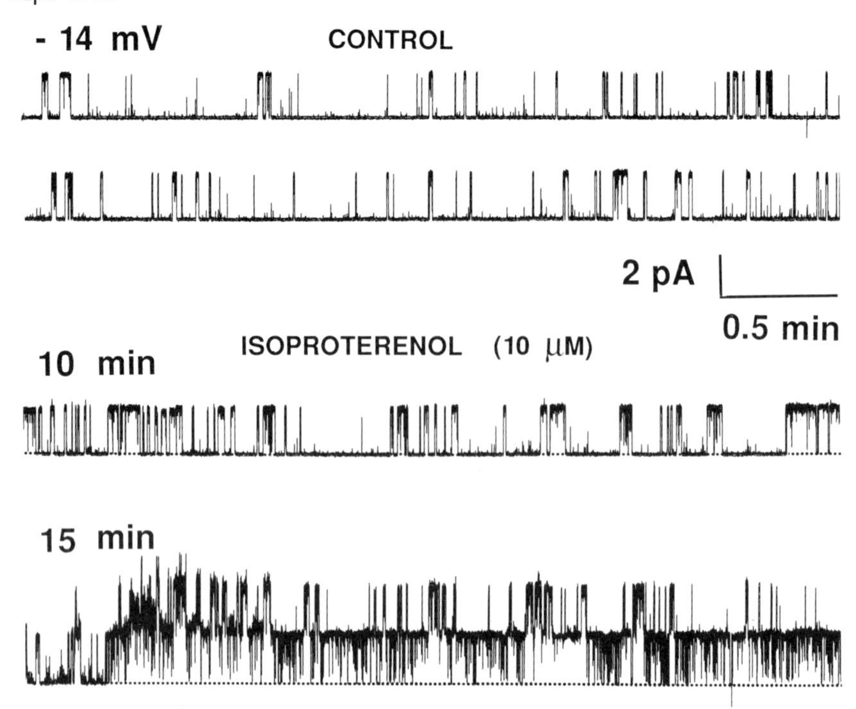


**Fig. 16.12** The $\beta$-adrenergic receptor agonist isoproterenol activates the potassium channel. Conditions as for Fig. 16.11. Single $K^+$ channel activity at −14 mV during the control period in the absence of agonist (upper records). Upward deflections represent open-channel currents. The traces shown below the control records were made in the presence of the agonist 10 and 15 min after the addition of isoproterenol to the *cis* side of the bilayer chamber.

We also tested the $\alpha$-adrenergic agonist, phenylephrine (0.01–1 mM), and found that relatively high doses ca. 1 mM) were needed to achieve activation of the $K^+$ channel (Fig. 16.13). During the control period (11 min) prior to the application of phenylephrine, the $K^+$ channel exhibited infrequent brief bursts of openings (Fig. 16.13). Despite the fact that the concentration of phenylephrine was two orders of magnitude higher than the concentration of isoproterenol needed to activate the channel (compare Figs. 16.11 and 16.12), it took ca. 18 min to produce a noticeable phenylephrine-induced activation of the channel. At lower doses (< 1 mM), phenylephrine failed to activate the channel.

Our results clearly show that adrenergic receptors modulate the activity of the $K^+$ selective channel present in the secretory granule membranes. Furthermore, the results also show that the specific $\beta$-agonist isoproterenol interacts with the receptor from the cytosolic side. We also tested the ability of the $\beta$-adrenergic antagonist, propanolol, to reverse the activation of the $K^+$ channel activation by isoproterenol. As illustrated in Fig. 16.14, $p_o$ after stimulation of the $K^+$ channel by isoproterenol (250 $\mu$M) from the *cis* side reached a value of ca. 0.6 (control). Propranolol (100 $\mu$M) applied to the *trans* side induced a moderate decrease to ca. 0.4. In contrast, in less than 5 min, the same dose of propanolol added to the *cis* side was able to completely block the $K^+$ channel (Fig. 16.14; third and fourth records from the top). The record at the bottom illustrates that the further addition to the *cis* side of isoproterenol (250 µM) was without effect.

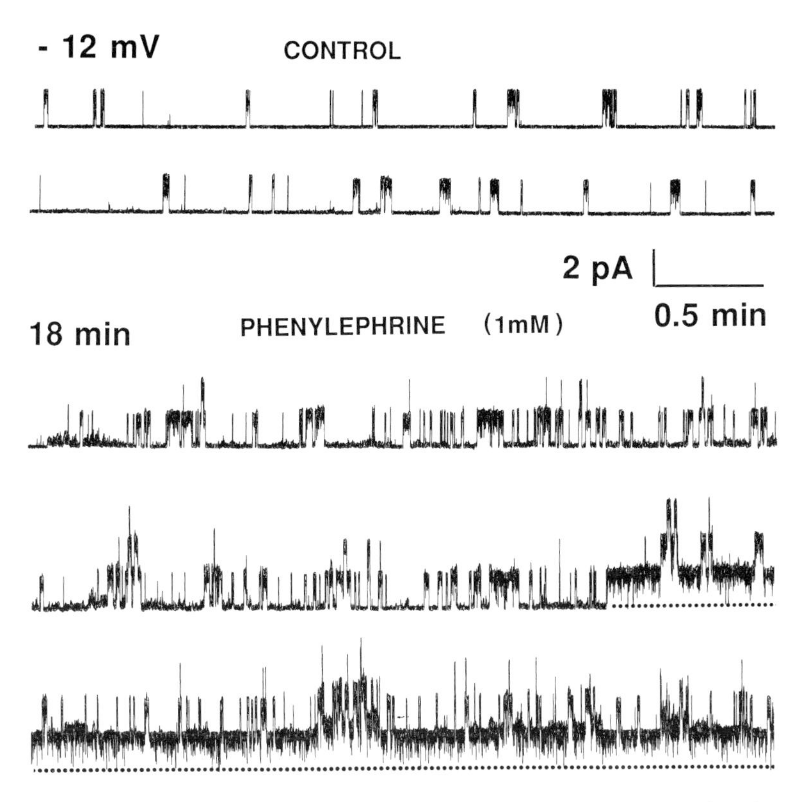


**Fig. 16.13** The $\alpha_1$-adrenergic receptor agonist phenylephrine also activates the potassium channel. Conditions as for Fig. 16.11. Two consecutive segments of a continuous record of the $K^+$ channel activity during the control period are shown in the upper part of the figure. Upward deflections represent open-channel currents. The three consecutive records shown below were made 18 min after addition of phenylephrine (1 mM) to the *cis* side.

## G protein coupling between a $\beta$-adrenergic receptor and the CG-type $K^+$ channel

G proteins are responsible for the coupling of cell membrane receptors to membrane resident ion channels or enzymes (Cerione *et al.* 1986; Yatani *et al.* 1988; Brown and Birnbaumer 1990; Spiegel *et al.* 1992; Brown 1993), and there is now growing evidence that the distribution of G protein is not limited to plasma membranes (Carlson *et al.* 1986; Barr *et al.* 1991; Donaldson *et al.* 1991; Carrasco *et al.* 1994; De Mazancourt *et al.* 1994). In addition, $G_o$ immunoreactivity has been detected in granule membranes from adrenal chromaffin cells (Toutant *et al.* 1987). Since no specific function of this protein was described, we examined the possibility of $G_o$ involvement in $K^+$ channel regulation and we asked whether the receptor-channel mechanism, present in the plasma membrane, might also be operative in the chromaffin granule membrane.

## Effects of NaF and non-hydrolysable guanosine nucleotide analogues on $K^+$ channel activity

We also resorted to the classical protocol of testing the effects of NaF, GTP$\gamma$S and GDP$\beta$S on chromaffin granule $K^+$ channel activity. The results of a typical experiment with $F^-$ (20 mM NaF) are depicted in Fig. 16.15 and the effects of GTP$\gamma$S (40 and 50 μM) are shown in

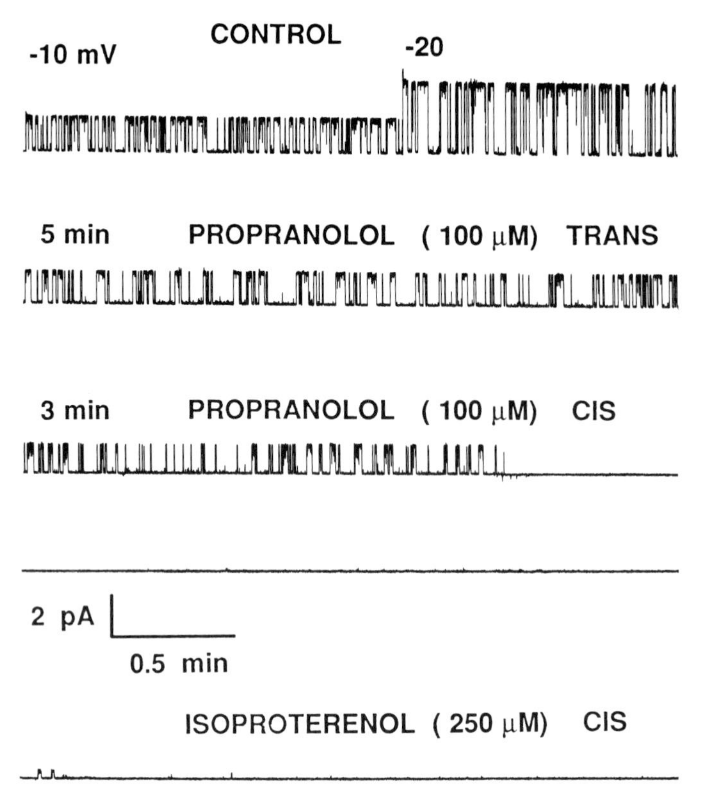


**Fig. 16.14** The $\beta$-adrenergic receptor antagonist propranolol reverses $K^+$ channel activation by isoproterenol. Conditions as for Fig. 16.11. Single $K^+$ channel activity at $-10$ and $-20$ mV during the control period in the presence of agonist (top record). Upward deflections represent open-channel currents. The records shown below the control record were made in the presence of the antagonist propranolol (100 $\mu$M), 5 min after its application from the *trans* side and 3 min after the addition of propranolol to the *cis* side of the bilayer chamber. The fourth record from the top shows complete blockade. Further addition of isoproterenol (250 $\mu$M) on both sides in the presence of propranolol had no effect (bottom record).

Fig. 16.16. We started the experiments with the $K^+$ channel in the hop modality ($p_o \geqslant 0.7$). Both, NaF amd GTP$\gamma$S evoked a transition in the pattern of $K^+$ channel activity from hop to lop. As shown in Fig. 16.16A, three levels of channel current are apparent, suggesting the presence of at least three channels; a low dose of GTP$\gamma$S (40 $\mu$M) applied to the *trans* but not to the *cis* side was sufficient to induce the transition; $p_o$ decreased from 0.98 (upper record) to 0.43 (lower record) in ca. 4 min. The result exemplified in Fig. 16.16B with only one channel in the bilayer clearly demonstrates the inhibitory effect of GTP$\gamma$S (50 $\mu$M). As a rule, application of higher doses of GTP$\gamma$S to both sides of the bilayer reduced the time required to observe a discernible decrease in channel activity (data not shown). These experiments were repeated with similar results using five different preparations of intact chromaffin granules.

Although totally unexpected, we also observed occasionally that GTP$\gamma$S (50 $\mu$M) from the *trans* side can activate the channel. Figure 16.17 depicts one of the few instances (three of eight experiments) in which we observed activation instead of inhibition of the $K^+$ channel induced by application of GTP$_\gamma$S. During the control period (upper record) only one channel

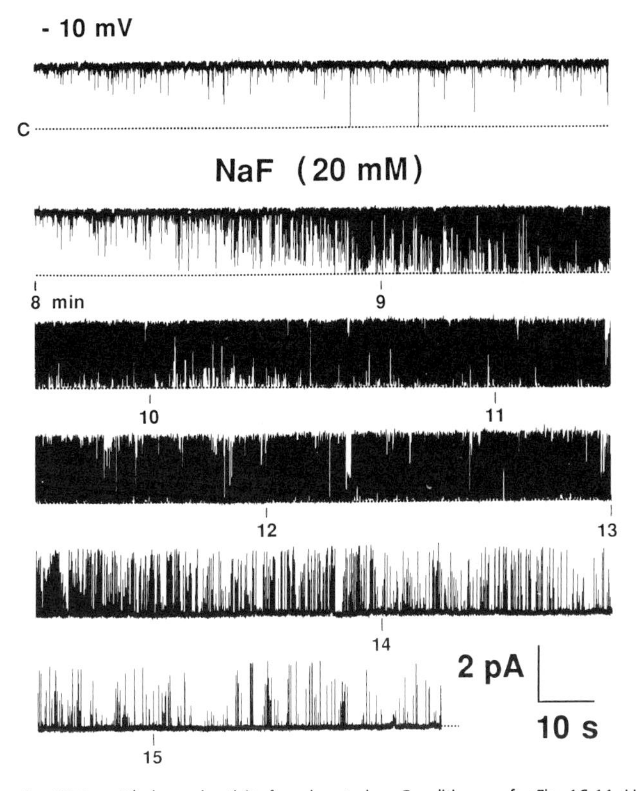


**Fig. 16.15** $F^-$ switches the CG-type $K^+$ channel activity from hop to lop. Conditions as for Fig. 16.11. Upper record: control, before application of 20 mM NaF. Horizontal dotted line: closed channel.

current level was observed ($p_o$ ca. 0.5). Three minutes after the addition of GTPγS to the *trans* side, channel activity was dramatically increased with at least five channel current levels (Fig. 16.17, lower record).

Application of the non-hydrolysable GDP analogue GDPβS (500 μM) to the *cis* side was without effect (Fig. 16.18, second record from the top), even at the end of a 8-min period of exposure to GDPβS. In contrast, the same dose of GDPβS applied to the *trans* side activated the CG-type $K^+$-selective channel increasing $p_o$ from ca. 0.4 (third record from the top) to 0.9 after 26 min (bottom record). A substantial fraction of the observed increase in $p_o$ evoked by GDPβS could result from the activation of several $K^+$ channels (Fig. 16.18, bottom record). However, statistical analysis of records which exhibited only one or two channels revealed that application of GDPβS increases the mean-open time of the CG-type $K^+$ channel.

One distinctive property of the hop mode of $K^+$ channel activity is the voltage sensitivity of the open-probability $p_o$. Figure 16.19 shows that GDPβS can induce a switch in the mode of $K^+$ channel activity from lop (Fig. 16.19, control record at −10 mV) to hop (Fig. 16.19, records at −40 through 30 mV). Furthermore, while $p_o$ during the control period remained constant (ca. 0.1) at different transmembrane potentials (not shown), after activation by GDPβS the $K^+$ channel acquired voltage sensitivity (Fig. 16.19, right side, lower panel). I–V curves from measurements obtained prior to (□) and after (■) the application of GDPβS

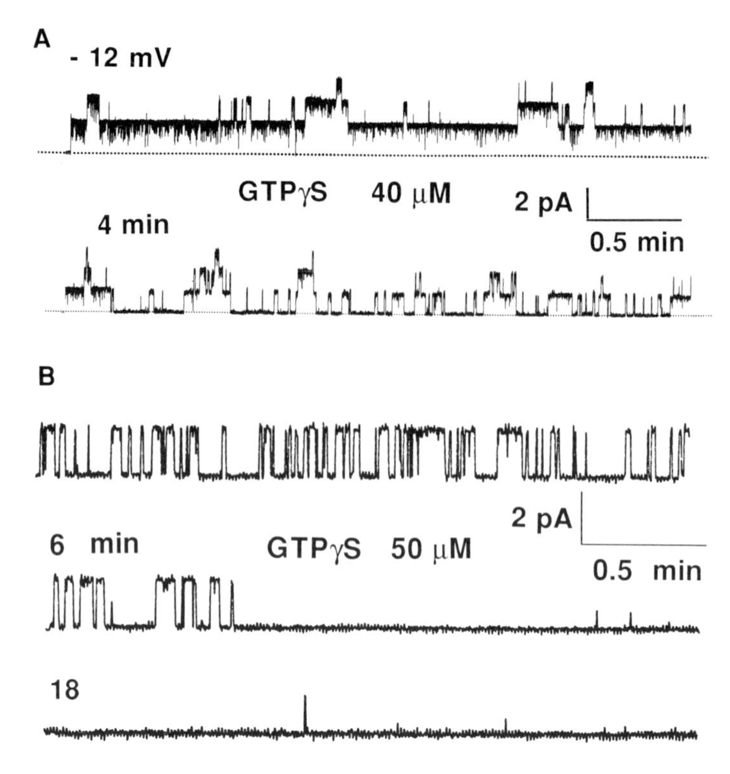


**Fig. 16.16** GTPγS converts the activity of the $K^+$ channel from hop to lop mode. Conditions as for Fig. 16.11. A. $K^+$ channel activity (at least two channels in the bilayer) before addition of GTPγS was in the hop mode (top record). Lower record was made 4 min after addition of GTPγS (40 μM) to the *trans* side. B. The control record was made with the channel in the hop mode (top record); then GTPγS (50 μM) was added to the *trans* side.

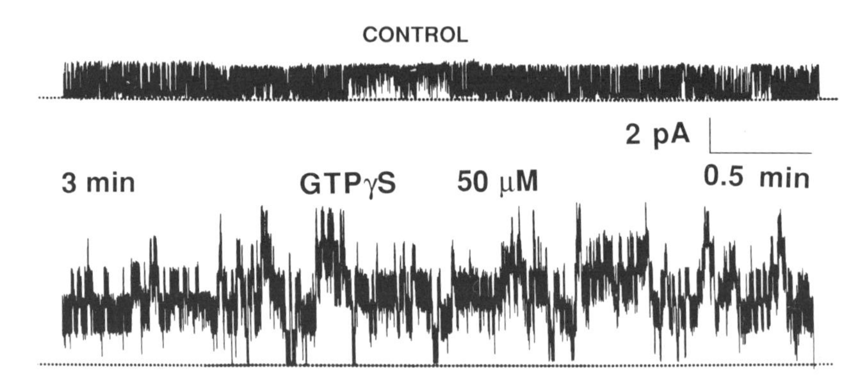


**Fig. 16.17** GTPγS can activate the CG-type $K^+$ channel into the hop mode. Conditions as for Fig. 16.11. Before addition of GTPγS, channel activity exhibited a single level (upper record). At 3 min after the addition of GTPγS (50 μM) to the *trans* side, several levels were apparent in the lower record, suggesting the switch to the **hop** modality.

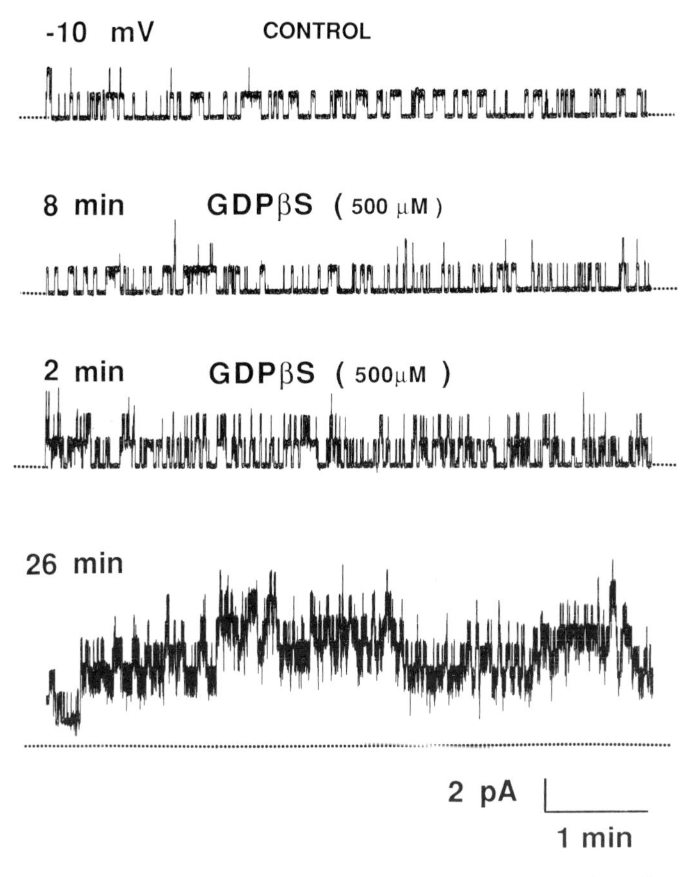


**Fig. 16.18** GDP$\beta$S activates the CG-type $K^+$ channel. Conditions as for Fig. 16.11. Before addition of GDP$\beta$S, channel activity from two channels was the lop modality (top record). After the addition of GDP$\beta$S (500 $\mu$M) to the *cis* side, channel activity remained unaltered for ca. 8 min (second record from top). Addition of the same dose of GDP$\beta$S (500 $\mu$M) to the *trans* side (third record from top) evoked a clear increase in channel activity. After ca. 26 min, several identical levels were apparent, suggesting that other CG-type $K^+$ channels had been converted to hop (bottom record).

were found to be linear, with similar slopes of ca. 185 pS (Fig. 16.19, right side, upper panel). From these experiments with GTP and GDP non-hydrolysable analogues we concluded that both GTP$\gamma$S and GDPβS acted on sites accessible from the *trans* side. Since the GTP$\gamma$S molecules can switch the $K^+$ channel activity from the hop ($p_o > 0.7$) to the lop ($p_o < 0.3$) mode, and GDP$\beta$S can induce a change in the opposite direction, we further hypothesized that the $K^+$ channel may be under direct control by more than one G-protein.

To determine the molecular components accounting for the observed inhibitory and stimulatory effects evoked by the non-hydrolysable GTP and GDP analogues, we screened our highly purified preparation of adrenal chromaffin granules for different types of $\alpha$ subunits of trimeric G proteins using highly selective antibodies against specific subunits.

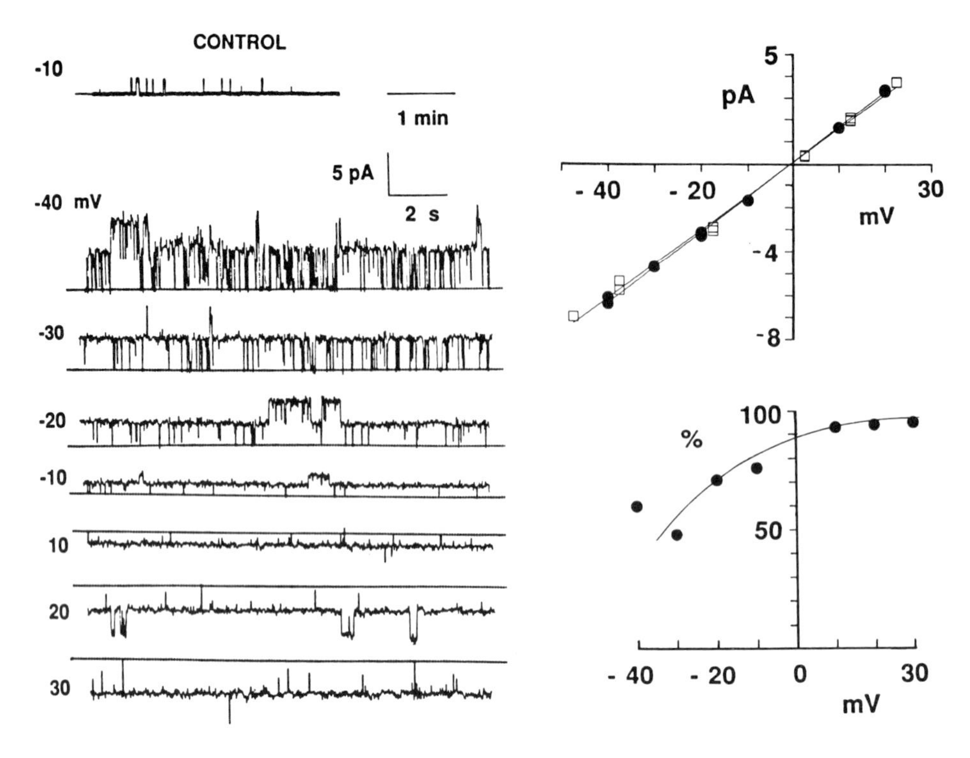


**Fig. 16.19** Current–voltage relationship and fractional-open time after activation of the $K^+$ channel by GDP$\beta$S. Left side: Representative segment of a continuous record of the $K^+$ channel activity before application of GDP$\beta$S (control, top record). Records below were made at different transmembrane potentials (indicated in mV next to each record) after the activation by GDP$\beta$S. Right side: upper panel represents the I–V curve, before application of GDP$\beta$S (□) and after activation by GDP$\beta$S (■) Lower panel: fractional open-time as a function of membrane potential after the activation by GDP$\beta$S (●).

## Identification of $G_o$, $G_{i2}$, $G_{i3}$, and $G_s$, but not $G_{i1}$ in adrenal chromaffin granules

Figure 16.20 depicts seven representative autoradiograms of proteins from highly purified granule membranes (CG, lanes a) and P2 crude membrane fractions (P2MF, lanes b). The antibody AS7, against $\alpha_t$, $\alpha_{i1}$ and $\alpha_{i2}$ (with weak $\alpha_{i3}$ cross-reactivity), was used in autoradiogram labelled AS7. Only one thick band (39–41 kDa) is apparent in lane a corresponding to CG and two bands in lane b corresponding to P2MF. The upper band in lane b is a 67 kDa protein only present in P2MF, a result that confirms the purity of the CG fraction tested. The $\alpha_{i2}$ protein is in excess compared with $\alpha_{i1}$ in lane b and, for this reason, the $\alpha_{i1}$ band is absent from lane b of autoradiogram AS7. In another experiment, the lower band labelled $\alpha_{i2}$ in lane a and the lower band in lane b comigrated with the band in brain cholate extract identified as $\alpha_{i2}$ (not shown). Furthermore, both CG and P2MF reacted with another antibody (given name LE) raised against an internal domain of $\alpha_{i2}$ (not shown). Taken together, these data indicate that $\alpha_{i2}$ is expressed in roughly similar amounts in both CG and P2MF.

The antibody LD (autoradiogram labelled LD) raised against an internal domain of the protein $\alpha_{i1}$ revealed a faint band only in the P2MF lane, suggesting that $\alpha_{i1}$ is not present in CG. Autoradiograms labelled GO and GC2 are antibodies raised against the C- and

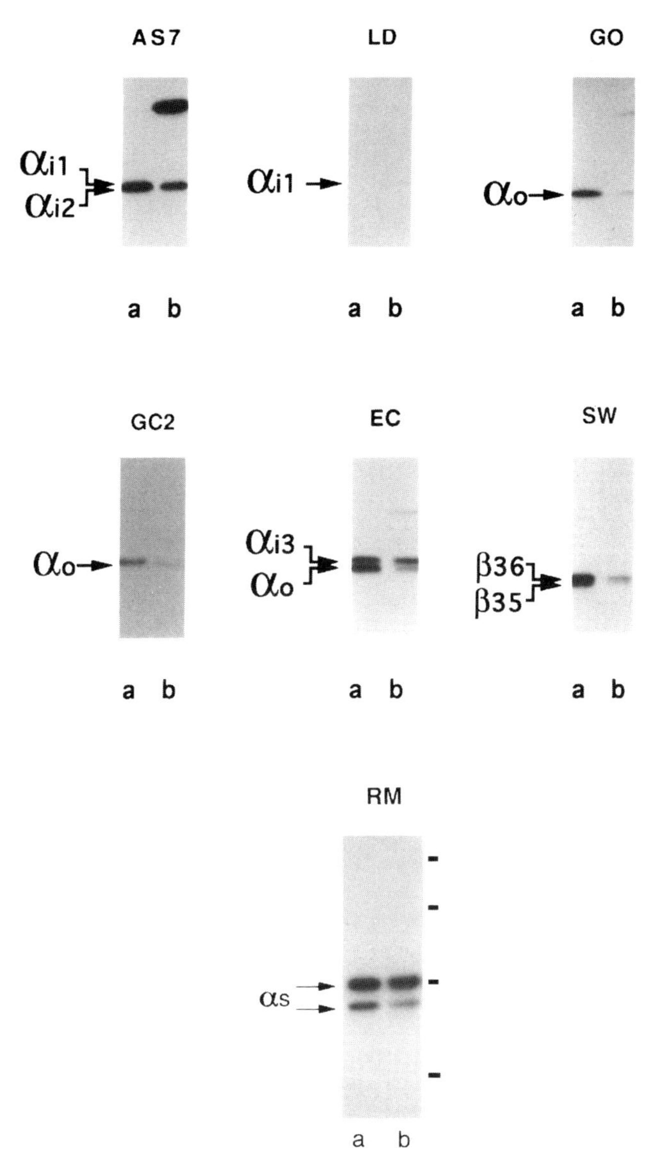


**Fig. 16.20** Immunoblot identification of $G_o$, $G_{i2}$, $G_{i3}$ and $G_s$ but not $G_{i1}$ in a highly purified preparation of adrenal chromaffin granules. Each autoradiogram can be identified by the letter code given above. Lane **a** corresponds to the highly purified adrenal chromaffin granule preparation; lane **b** corresponds to a P2 crude membrane fraction prepared from bovine adrenal chromaffin cells.

N-terminal domains of $\alpha_o$, respectively. Both antibodies gave consistent labelling, i.e. strong in CG (lane a) and faint in P2MF (lane b). These results indicate a high content of $\alpha_o$ in CG. Also shown in autoradiogram EC is the labelling with EC, an antibody raised against the C-terminal domain of $\alpha_{i3}$ which cross-reacts with $\alpha_o$. As shown on autoradiogram EC, there is a clear doublet at 39–41 kDa in each lane, the lower band being presumably $\alpha_o$. Consistently with the data displayed on the other autoradiograms GO and GC2, the lower band corresponding to $\alpha_o$ was faint in the P2MF (lane b) (McKenzie *et al.* 1988; Spiegel *et al.* 1990*a*,*b*, 1992; Spiegel 1991).

Autoradiogram SW shows the labelling with an antibody raised against the C-terminal undecapeptide common to $\beta_{1-4}$. It is apparent that although SW detected a doublet in both fractions, the labelling in lane a is stronger. This result demonstrates that the $\beta$ subunit and, probably the $\gamma$ subunit responsible for membrane attachment of the G protein (Simonds *et al.* 1991), are also present in adrenal chromaffin granules. Finally, autoradiogram RM depicts the labelling obtained with antibody against $\alpha_s$. Autoradiogram RM shows clear labelling of a doublet in the P2MF lane b, corresponding to the short and the long $\alpha_s$ isoforms. The antibody RM also indicated the presence of two $\alpha_s$ isoforms (doublet in lane a) in adrenal chromaffin granules.

In summary, $\alpha_o$ and $\alpha_s$ are by far the predominant G protein subunits in adrenal chromaffin granules. Of the other subunit types, we detected the presence of equivalent amounts of $\alpha_{i2}$ and $\alpha_{i3}$ in both CG and P2MF.

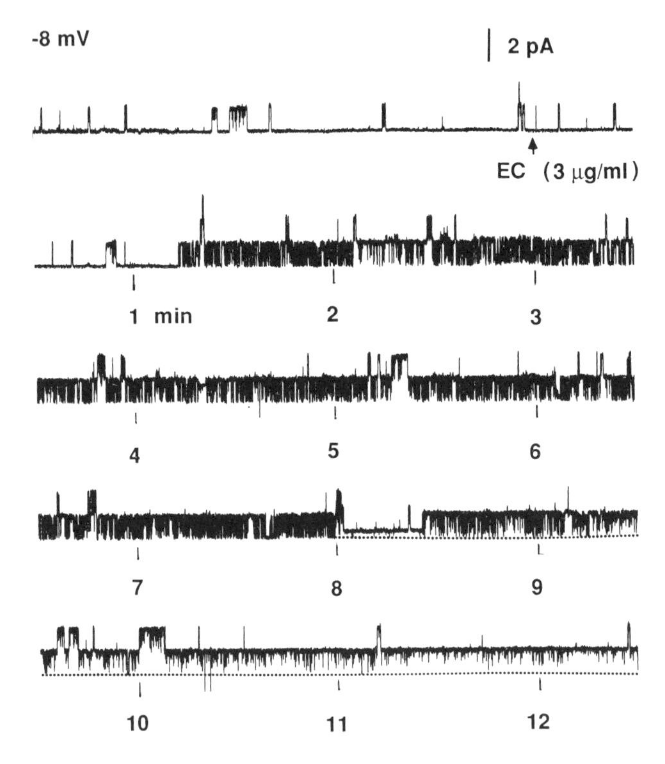


**Fig. 16.21** The antibody EC to $\alpha_{i3}$ converts the activity from lop to hop mode. Conditions as for Fig. 16.11. EC was added to the *trans* side as indicated by the arrow. The numbers underneath the records indicate the time of exposure in minutes. The control record exhibits two levels in the lop modality.

## Effects of antibodies against specific $\alpha$ subunits of trimeric G proteins on $K^+$ channel activity

The same specific antibodies used in the immunoblot analysis were examined for identification of the G protein(s) regulating the large-conductance $K^+$ channel activity. In perfect agreement with the results from the immunoblot analysis, we also found that each one of the antibodies that gave positive labelling had a profound effect on the activity of the $K^+$ channel. The records shown in Fig. 16.21 were made during experiments in which the $K^+$ channel exhibited a $p_o < 0.3$ (top records). As illustrated in Fig. 16.21 (second record from the top), antibody EC at a concentration of 3 $\mu$g ml$^{-1}$ evoked a dramatic conversion of the pattern of $K^+$ channel activity from lop to hop ($p_o$ ca. 0.9). EC is not specific for $G\alpha_{i3}$ and cross-reacts with $G\alpha_o$ (see Fig. 16.20). However, it is safe to assume that prior to the application of EC, the $K^+$ channel was under tonic inhibition by activated $G\alpha_{i3}$ molecules and the antibody EC removed this inhibition by complexation with the $\alpha_{13}$subunits.

Similar effects were also observed with more specific antibodies raised against other G proteins. This was the case for AS7 (Fig. 16.22A; 5.4 $\mu$g/ml$^{-1}$) which exhibits a reasonable specificity for $G\alpha_{i2}$, although it cross-reacts with $G\alpha_{i2}$. Like the antibody EC, AS7 also switched the $K^+$ activity from lop to hop. Taken together, these data suggest that the large-conductance $K^+$ channel is under inhibitory control by $G_i$ proteins. However, GO raised against $\alpha_o$, a classical activator of $K^+$ channels in other cell types (Madison and Nicoll 1986), also induced the transition from the lop to the hop activity of the $K^+$ channel. As shown in Fig. 16.22B, GO (8.3 $\mu$g/ml$^{-1}$ added to the *trans* side activated the channel and, as expected, a further addition of GTP$\gamma$S on the *trans* side was unable to inhibit the activity (Fig. 16.22 B, bottom record). Immunoblots using GO exhibited two clear bands corresponding to $\alpha_o$ and $\alpha_{i3}$, so cross-reaction of GO with $\alpha_{i3}$ provides an alternative explanation for the result. In this case, complexation of the $\alpha_{i3}$ subunit by GO prevented either the inhibitory interaction between the GTP$\gamma$S-activated $\alpha_{i3}$ subunit and the $K^+$ channel, or prevented the activation of $\alpha_{i3}$ subunit by GTP$\gamma$S, or both. Consistent with this alternative mechanism we found that GO (8.3 $\mu$g ml$^{-1}$ applied to the *trans* side can revert the inhibition of the $K^+$channel activity induced by NaF (Fig. 16.22C). The unavoidable conclusion is that fluoride set the $K^+$ channel under inhibitory control by a G space protein and, the antibody GO, by binding to the fluoride activated G space protein subunit, released the blockade.

The immunoblot analysis using the RM antibody against $\alpha_s$ also revealed the presence of this subunit in our preparation of adrenal chromaffin granules (Fig. 16.20). The $G_s$ system is a classical example of a plasma membrane trimeric G space protein that couples receptor activation with specific ion channels in different cell types (Scamps *et al.*, 1992). We found that RM is a potent blocker of the $K^+$ channel activity and is only effective from the *trans* side (Fig. 16.23). The experiment was carried out with only one channel in the bilayer (Fig. 16.23, upper record). Ten minutes after the addition of RM to the *cis* side, no noticeable effects were apparent, $p_o$ remaining unchanged at ca. 0.4. By contrast, 6 min after the application of the antibody to the *trans* side, the $K^+$ channel activity was completely inhibited.

From the functional analysis of the effects of specific antibodies used here we conclude that both inhibitory as well as stimulatory G proteins are acting on the $K^+$-channel.

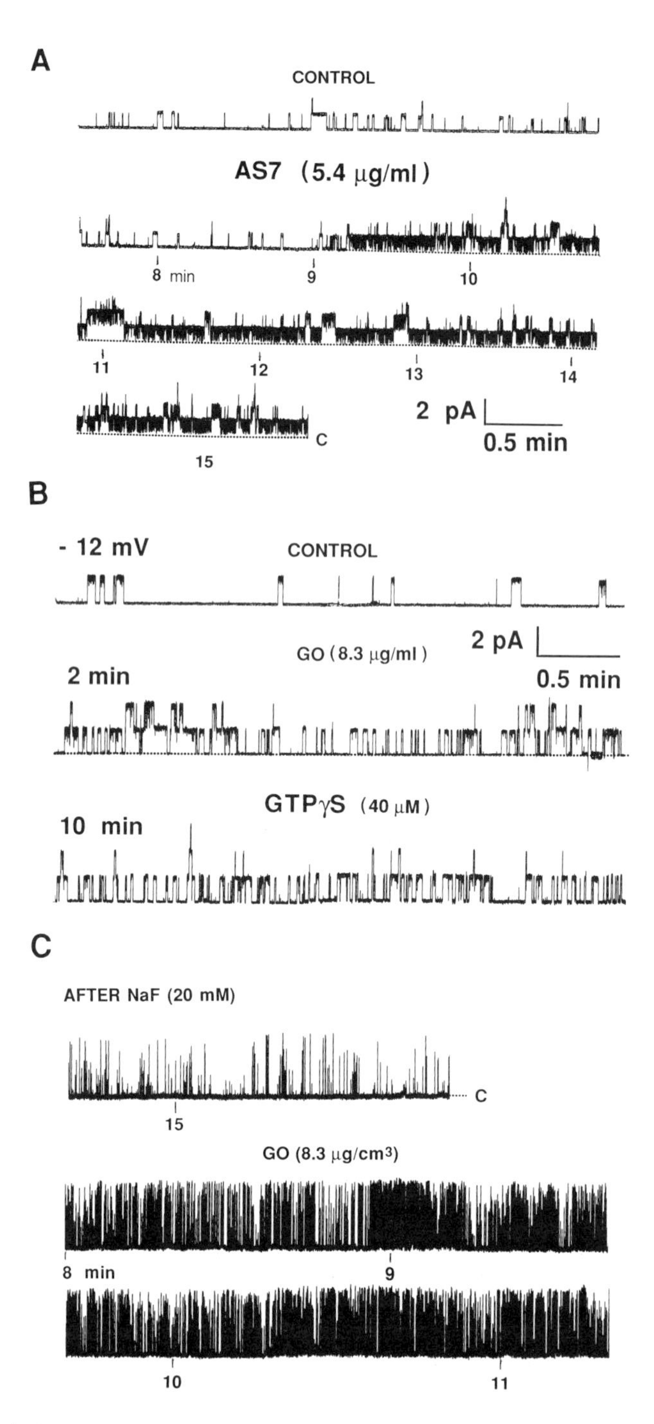


**Fig. 16.22** The antibodies AS7 to $\alpha_{i2}$ and GO to $\alpha_o$ convert the activity from lop to hop mode. Conditions as for Fig. 16.11. Channel activity before addition of antibodies was in the lop mode (control top records). A. Records made after the addition of AS7 (5.4 $\mu$g ml$^{-1}$) to the *trans* side (second record from the top). The numbers below the records indicate time in minutes. B. At 2 min after addition of GO (8.3 $\mu$g ml$^{-1}$ to the *trans* side the channel activity has changed to hop mode (middle record). Further addition of GTPγS (40 $\mu$M) to the *trans* side was without effect on the channel activity (bottom record). C. Channel activity converted to the lop mode by ca. 15 min exposure to 20 mM NaF (top record). Application of GO (8.3 $\mu$g ml$^{-1}$) to the trans side activated the CG-type $K^+$ channel (bottom records). The numbers below the records indicate the time of exposure in minutes.

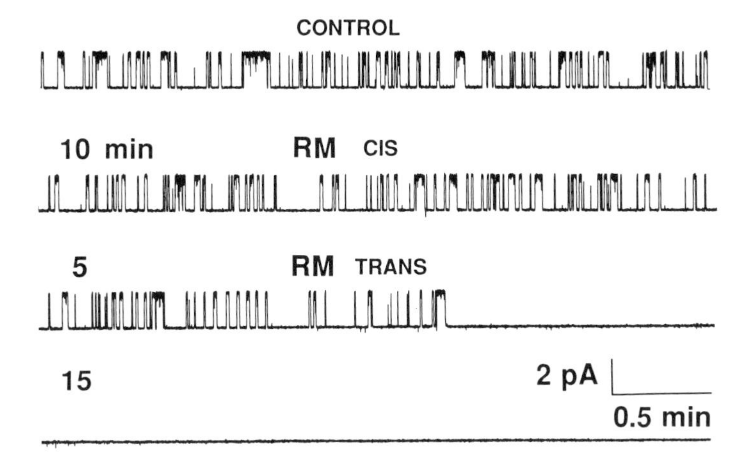


**Fig. 16.23** The antibody RM to $\alpha_s$ converts the activity from hop to lop mode. The upper record was made before addition of RM. Nearly 10 min after the addition of RM to the *cis* side, no noticeable effects were detected. At 7 min after the addition of RM to the *trans* side, the $K^+$ channel was blocked (lower record).

## Discussion

### A $K^+$-selective channel of high-conductance is present in the adrenal chromaffin granule membrane

The main finding in our study is that the chromaffin granule membrane contains a $K^+$-selective channel with large conductance $K^+{}_{CG}$ which lacks sensitivity to either $Ca^{2+}$ or charybdotoxin, a selective blocker of the high-conductance, $Ca^{2+}$-activated $K^+$ channel (MacKinnon and Miller 1988). These characteristics clearly distinguish the $K^+{}_{CG}$ channel from other $K^+$ channels previously reported for rat hypophyseal granules (Lemos *et al.*, 1989; Lee *et al.*, 1992) and secretory cholinergic vesicles from *Torpedo* (Rahamimoff *et al.*, 1989). Furthermore, the high selectivity for potassium ions of the CG-type $K^+$ channel are clearly distinct from those briefly reported for the non-selective synaptophysin cation channel (Thomas *et al.*, 1988). Thus, synaptophysin, known to be present in chromaffin secretory vesicles (Obendorf *et al.*, 1988), cannot be responsible for the channel activity described here.

### Possible physiological role of the CG-type $K^+$ channel

The properties of the chromaffin granule $K^+$ channel do not necessarily exclude a possible role in secretion, but they do indicate that it might also be reasonable to consider the channel in terms of other granule functions. Some of these other functions impact on establishment of ion gradients. The granules are assembled in the Golgi, where they begin to assume the composition of the mature granule. We lack information about the composition for this nascent state, and so must leave it blank for the present. The mature granule then moves into the cytosol, where it is poised ready to fuse during exocytosis. The concentrations of different ions are known with accuracy and precision for granules in this state (in mM): $Na^+ = 0$, $K^+ = 83$, $Ca^{2+} = 9$ and $Cl^- = 49$ (Ornberg *et al.*, 1988). Following fusion and exocytosis, the loss of granule contents is paralleled by the acquisition of ions and other bulk phase components from the extracellular medium. Substantial experimental data exist to substantiate this claim. For example, the exposure of dopamine $\beta$-hydroxylase and other glycoproteins on the inner surface of the chromaffin granule during exocytosis allows antibodies to these proteins to be bound and internalized upon retrieval of expended granule membranes (Lingg *et al.*, 1983). The extent of this replacement of granule contents by bulk phase solution can also be appreciated by the fact that retrieval is even accompanied by uptake of macromolecules such as horseradish peroxidase, which are then available for resecretion (Baker and Knight 1981; Von Grafenstein *et al.*, 1986). It is therefore likely that the retrieved ghost is equilibrated with an ionic composition that is similar to that of the medium, and thus entirely different from that of the cytosol or of the mature granule.

The recycling process in the chromaffin cell is also well established on both morphological and biochemical grounds (for review see Trifaro and Poisner 1982) and newer membrane capacitance data are also consistent with this concept (Neher and Marty 1982; Penner and Neher 1989; Lindau 1991). Of course, we need not assume that the chromaffin granule must have a specific ionic composition in order to be competent to accumulate new transmitter. However, it is likely that sometime during the recycling process the intragranular [$Na^+$] must change from the extracellular concentration, roughly 152 mM, to the concentration of the mature granule (ca.0 mM). Coincidentally, the intragranular [$K^+$] must increase from 6 mM (extracellular concentration) to 83 mM. The time scale for resecretion of acquired contents from the medium is known to be ca. 30 minutes (Lingg *et al.*, 1983), but no information is available regarding the rate for resetting the intragranular milieu with ions or other granular contents.

We are now in a position to ask in what manner the CG-type $K^+$ channel might contribute to any of the above-described processes in the chromaffin cell. An obvious possibility is the regulation of chemo-osmotic events across the granule membrane (for review, see Brocklehurst and Pollard 1988*a,b*). It is well-established that the inwardly directed proton pump present in the chromaffin granule membrane creates a proton gradient with the granule interior having a lower pH than the outside. The activity of this proton pump is also expected to make the granule transmembrane electrical potential positive inside. Indeed, while in the absence of MgATP the transmembrane potential of isolated chromaffin granules was estimated as close to $-10$ mV, in the presence of MgATP the proton pump in the granule membrane generates a potential of 50 to 70 mV inside positive (Pollard *et al.*, 1976; Holz 1979; Salama *et al.*, 1980). Since transport and accumulation of catecholamines depends on the electrochemical force across granule membrane, one might then ask whether the CG-type $K^+$ channel might be involved in setting the electrical potential component of the electrochemical gradient across the granule membrane. It should be kept in mind that for retrieved ghosts the concentration gradient for $K^+$ might be capable of sustaining a large negative potential similar to that of the plasma membrane of the chromaffin cell, i.e. $-65$ mV (Fenwick *et al.*, 1982). At the very least, this potential gradient will favour the catecholamine loading of the granule.

With regard to the granule assembly and recycling processes, it is also certainly possible that the ghost $K^+$ channel could mediate potassium ion movement to establish the 83 mM [$K^+$] level. During the assembly process, the nascent granule could begin with an intragranular potassium concentration of 123 mM, similar to that of the cytosol. In this case, progress to the mature 83 mM state would have to entail loss of $K^+$. Previous studies have anticipated that the granule ATPase might be involved in coincident acidification of the nascent granule and, if this occurred, entering $H^+$ could exchange for exiting $K^+$. To lose ca. 40 mEq $L^{-1}$ of $K^+$, the granule would have to acquire 40 mE/l of $H^+$. Indeed, since the buffering capacity of the granule contents is close to 300 mE/l (Salama *et al.* 1980), the added protons could be accommodated easily. By contrast, the exchange of endogenous, buffered protons for exogenous potassium with the antibiotic nigericin is known to destabilize mature granules (Pazoles and Pollard 1978). Therefore, we cannot exclude the possibility that the opposite action, of replacing potassium with buffered protons during assembly, could be responsible for the acquired stability of the otherwise 'hypertonic' granule contents in an isotonic (300 mOsm) cytosolic medium. Clearly, any required potassium movements could be mediated by the new $K^+$ channel.

The case for invoking a role for the CG-type $K^+$ channel in the assembly or recycling process is only compelling in the sense that the chemical data requires a mechanism for moving large amounts of $K^+$ through the granule membrane. Until the present discovery, there was no basis for understanding how this could happen. We can conclude now that the CG-type $K^+$ channel is thus a candidate for this transfer mechanism. We do not exclude this or other membrane resident channels from involvement in the exocytosis process, except to emphasize that the properties of the new $K^+$ channel do not make an obvious case for this possibility.

We have also shown here, for the first time, that the CG-type $K^+$ channel is regulated by a $\beta$-adrenergic receptor, probably from the cytosolic aspect of the chromaffin secretory granule membrane. The underlying mechanism might include activation of G proteins resident in the membrane of the secretory vesicles (Arispe *et al.* 1995). In the present work we have provided, for the first time, evidence for the existence of a large-conductance $K^+$ channel in adrenal chromaffin granule membranes that is directly controlled by inhibitory as well as stimulatory G-protein coupled mechanisms. This conclusion is based on three lines of evidence. First, functional analysis of the $K^+$ channel incorporated into planar lipid bilayer membranes showed two preferred patterns of activity which define the lop and hop states of the channel. We found that non-hydrolysable analogues GDP$\beta$S and GTP$\gamma$S have opposite effects on $K^+$-channel activity. While GDP$\beta$S, thought to compete with endogenous GDP for a binding site on the $\alpha$ subunit, stimulated the $K^+$ channel in the lop mode of activity, GTP$\gamma$S, mimicking the activation of the $\alpha$ subunit by endogenous GTP, inhibited the $K^+$ channel in the hop mode of activity. Second, NaF blocked the large-conductance $K^+$ channel in much the same way as GTP$\gamma$S. Third, using antibodies against C-terminal decapeptides or internal decapeptides of the $\alpha$ subunit of specific trimeric G proteins (McKenzie *et al.* 1988; Simonds *et al.* 1989*a*,*b*; Spiegel *et al.* 1990*a*,*b*, 1992; Spiegel 1991, we were able to identify the presence of $G_o$, $G_s$, $G_{i2}$, and $G_{i3}$, but not $G_{i1}$, in a highly purified preparation of adrenal chromaffin granules. Since the application of the same specific antibodies used in the immunoblot analysis reverted the effects of the non-hydrolysable analogues on the activity of $K^+$ channel, we concluded that the activity of the chromaffin granule large-conductance $K^+$ channel is kept under tight control by a mechanism involving G proteins (Arispe *et al.* 1995). Taken together, our results strongly support the idea that the activity of the CG-type $K^+$ channel is tightly regulated and that this regulation might depend on the cytosolic concentration of CA (Arispe *et al.* 1995).

Previous works have identified the presence of $G_o$ immunoreactivity in chromaffin cell secretory granule membranes (Toutant *et al.* 1987). Furthermore, in the hippocampus, $G_o$ couples β-adrenergic receptors with adenylate cyclase molecules and thereby regulates cyclic adenosine monophosphate (cAMP) production. This messenger is a crucial component of the mechanism of modulation of the $[Ca^{2+}]_i$-activated $K^+$ channel present in the plasma membrane of pyramidal cells (Madison and Nicoll 1986). Although it is well established that antibodies against specific G proteins disrupt receptor/G protein coupling, there is growing evidence that G proteins interact directly with ion channels. Indeed, a recent report provides some evidence for direct interactions between $G\alpha_s$ and a $Ca^{2+}$ channel incorporated in a lipid bilayer (Hamilton *et al.* 1991). Using antibodies against specific C-terminal decapeptides of $\alpha_o$, $\alpha_{i3}$, $\alpha_s$ the internal decapeptides of $\alpha_{i1}$ and $\alpha_{i2}$, and $\alpha_{i3}$ and the common C-terminal decapeptide of $\alpha_{i1}$ and $\alpha_{i2}$, and the N-terminal domain of $\alpha_o$ (antibody GC2), our immunoblot analysis not only revealed the presence of $\alpha_o$, but also the presence of $\alpha_s$, $\alpha_{i2}$, and $\alpha_{i3}$ in the purified granules. Consistent with these data, $\beta$ subunits were also found in our preparation of chromaffin granules. Originally, two of these $G\alpha$ proteins were found to couple adrenergic receptors to adenylate cyclase (stimulatory $G_s$ and inhibitory $G_{i3}$) with either a $Ca^{2+}$ channel ($G_s$) or a $K^+$ channel ($\alpha_{i2}$) at the end of the chain of events, all components within the plasma membrane (Brown 1993).

The immunoblot analysis used here also identified the dimer $\beta\gamma$ in our preparation of chromaffin granule membranes (Simonds *et al.* 1991). This structural component, common to many trimeric GTP-binding proteins, plays a crucial role in the functional cycle, allowing the release of the activated $\alpha$ subunit after the exchange of GDP for GTP which occurs while the $\alpha$ subunit and the $\beta\gamma$ dimer are tightly associated. This finding lends strong support to the idea that the trimeric form of GTP-binding proteins also operate within the secretory vesicle membranes. Of the other trimeric G proteins found in the secretory vesicles from chromaffin cells, $\alpha_o$ couples an adrenergic receptor to adenylate cyclase, providing a mechanism to regulate a ca. 55 pS $K^+$ channel (Madison and Nicoll 1986); $\alpha_{i2}$ couples a $M_2$ subtype muscarinic receptor to a ca. 40 pS $K^+$ channel in the plasma membrane (Brown 1993). Although we cannot completely rule out the possibility of contamination of our secretory granules with membranes from other intracellular organelles or even plasma membrane, we have compelling reasons to believe that these G proteins labelled by our set of highly specific antibodies are part of the granule membrane.

1. The method used to prepare the adrenal chromaffin granules (Pollard *et al.* 1979*a*,*b*) yielded highly purified intact vesicles.
2. For the present experiments, we included additional purification steps to exclude possible contaminating membranes.
3. The $\alpha_{i1}$ G protein subunit commonly found in the plasma membrane was completely absent from all the autoradiograms examined with the specific antibody against its internal decapeptide (Spiegel *et al.* 1990*a*,*b*).
4. The large-conductance $K^+$ channel activated by $[Ca^{2+}]_i$ and voltage, abundant in the plasma membrane of chromaffin cells (Fenwick *et al.* 1982; Glavinovic and Trifaro 1988) was never co-reconstituted into our bilayer membranes.

These data serve to ascertain the purity of the preparation, and thus, we conclude that the set of G proteins detected are most likely to be part of the granule membrane.

## Role of the channel in vesicle function

Regulation of the transport of vesicles from the endoplasmic reticulum to the Golgi apparatus involves heterotrimeric guanidine nucleotide-binding proteins (Stow *et al.* 1991; Colombo *et al.* 1992; Aridor *et al.* 1993). Thus, it is possible that G proteins remain attached to the chromaffin granule membrane after the vesicles are assembled by the Golgi system. On the other hand, modulation of the $K^+$ channel activity by adrenergic agonists (Arispe and Rojas 1995) strongly suggests a mechanism whereby only the stimulatory $G_s$ serves to couple adrenergic receptors facing the cytosolic side of the chromaffin granule membrane to the channel (Kume *et al.* 1992). The vectorial character of this model arose from the striking polarity found in terms of sites of action for both non-hydrolysable GTP and GDP analogues as well as the antibodies. One last argument in support of the present model is the observation that adrenergic agonists activate the channel only from the *cis* side of the bilayer. After fusion of the secretory vesicle with the bilayer membrane, the cytosolic side of the vesicle faces the *cis* side of the bilayer chamber.

## Functional adrenergic receptors are present in the membrane of chromaffin granules

Early studies of adenylate cyclase activity in chromaffin cell secretory vesicle membranes showed that the enzyme is controlled by $\beta$-adrenergic agents. Furthermore, granule adenylate cyclase is activated by fluoride and non-hydrolysable GTP analogues such as GMP-PNP, indicating the functional existence of heterotrimeric G proteins in this membrane system (Hoffman *et al.* 1976; Nikodejevic *et al.* 1976; Zinder *et al.* 1977; Winkler and Westhead 1980). Although the specific receptors which might be present in the chromaffin granule remains to be elucidated, we do not rule out the possibility that, in addition to the adenylate cyclase described by Nikodejevic *et al.* (1976), the chain of events leading to the expression of the diffusion potential may also be initiated by the binding of epinephrine to $\beta$-adrenergic receptors on the cytosolic aspect of the granules. If this were the case, $\alpha_s$ or $\alpha_o$ subunits might couple the receptor to the channel (Spiegel *et al.* 1990*b*; Brown 1993).

## Dual stimulatory and inhibitory control of the K+ channel by G proteins

As summarized in Table 16.1, we have shown here that GTP$\gamma$S can induce the transition from hop to lop or lop to hop. In addition, antibody GO raised against $\alpha_o$ (Morris *et al.* 1990) is able to revert the effects of $F^-$ and neither NaF nor GTP$\gamma$S can revert the stimulation of the $K^+$ channel by each of the antibodies tested, including GO. Furthermore, we have observed, although less frequently, that the antibody RM against $\alpha_s$ can also block the $K^+$ channel, suggesting that both stimulatory as well as inhibitory G proteins can interact with the channel. Dual stimulatory and inhibitory regulation has been described before for a $[Ca^{2+}]_i$ activated $K^+$ channel present in the airways smooth muscle plasma membrane (Kume *et al.* 1992). However, unlike the situation found in smooth muscle, where the stimulatory control occurs via adenylate cyclase, in chromaffin granules regulation of the $K^+$ channel is mediated directly by both stimulatory and inhibitory $\alpha$ subunits. Thus, $K^+$ channel activation (or inactivation) by either GTP$\gamma$S or NaF may indeed be due to activation of either $G\alpha_o$ or $G\alpha_{i2}$ and/or $G\alpha_{i3}$ subunits by GTP$\gamma$S or NaF, respectively. Appropriately, both events can be prevented if the $\alpha$ subunit is removed by complexation with the corresponding antibody.

The requirements for a minimal model to explain the interaction of G proteins and the chromaffin granule $K^+$ channel can be summarized as follows (see Table 16.1). GTP$\gamma$S activates both $\alpha_s$ and $\alpha_i$. Interaction of GTP$\gamma$S-activated $\alpha_s$ or $\alpha_i$ with the $K^+$ channel will increase or decrease $p_o$, respectively. GDP$\beta$S will bind to both $\alpha_s$, and $\alpha_i$. GDP$\beta$S binding will deactivate both types of subunits and, release the $K^+$ channel from the corresponding effect. The highly specific antibodies EC, AS7, GO and RM used to identify the presence of specific $G\alpha$ proteins were also tested in the functional analysis of the $K^+$ channel. In all instances we obtained results consistent with the inhibition of the specific $G\alpha$ protein owing to complexation with the antibody (Table 16.1).

In conclusion, the $Ca^{2+}$-insensitive $K^+$ channel of large-conductance present in adrenal chromaffin granule membranes is directly gated by inhibitory as well as stimulatory G proteins. Although the physiological role of the $K^+$ channel remains to be elucidated, we do not rule out the possibility that this regulation is directed towards catecholamine trafficking across the granule membrane during the secretion cycle and membrane fusion at the site of exocytosis.

## References


Anderson, C.S., MacKinnon, R., Smith, C., and Miller, C. (1988). Charybdotoxin block of single $Ca^{2+}$-activated $K^{2+}$-channels. Effects of channel gating, voltage, and ionic strength. *J. Gen. Physiol.* **91**, 317–333.

Aridor, M., Rajmilevich, G., Beaven, M.A., and Sagi-Eisenberg, R. (1993). Activation of exocytosis by the heterotrimeric G protein $G_{i3}$. *Science*, **262**, 1569–1572.

Arispe, N., Pollard, H.B, and Rojas, E. (1992). Calcium-independent $K^+$-selective channel from chromaffin granule membranes. *J. Membr. Biol.*, **130**, 191–202.

Arispe, N., Rojas, E. (1995). Adrenergic agonist modulation of the large conductance $K^+$-channel from chromaffin granule membranes. (submitted.)

Arispe, N., De Mazancourt, P., and Rojas, E. (1995). Direct control of a large conductance $K^+$-selective channel by G-proteins in adrenal chromaffin granule membranes. *J. Membr. Biol.*, **147**, 109–119.

Ashley, R.H., Brown, D.M., Apps, D.K., and Phillips, J.H. (1994). Evidence for a $K^+$ channel in bovine chromaffin granule membranes: single-channel properties and possible bioenergetic significance. *Eur. Biophys. J.*, **23**, 263–275.

Barr, F.A., Leyte, A., Mollner, S., Pfeuffer, T., Tooze, S.A., and Huttner, W.B. (1991). Trimeric G-proteins of the *trans*-Golgi network are involved in the formation of constitutive secretory vesicles and immature secretory granules. *FEBS Lett.*, **294**, 239–243.

Baker, P.F. and Knight, D. E. (1981). Calcium control of exocytosis and endocytosis in bovine adrenal medullary cells. *Phil. Trans. R. Soc. Ser. B.*, **296**, 83–103.

Brocklehurst, K. W. and Pollard, H.B. (1988*a*). Osmotic effects in membrane fusion during exocytosis. *Curr. Topics in Membranes and Transport*, **32**, 203–225.

Brocklehurst, K.W. and Pollard, H.B. (1988*b*). Chemosmotic events. In *Energetics of secretion responses*, (ed. J-.W.N. Akkerman), Vol. 2, pp. 47–61. CRC Press, Inc.

Brocklehurst, K.W. and Pollard, H.B. (1990). Cell biology of secretion. In *Peptide hormone secretion. A practical approach*, (ed. J.C. Hutton and K. Siddle), pp. 233–255. IRL/Oxford, Oxford, New York, Tokyo.

Brown, A.M. and Birnbaumer, L. (1990). Ionic channels and their regulation by G-protein subunits. *Annu. Rev. Physiol.*, **52**, 197–213.

Brown, A.M. (1993). Membrane-delimited cell signaling complexes: direct ion channel regulation by G proteins. *J. Membr. Biol.*, **131**, 93–104.

Carlson, K.E., Woolkalis, M.J., Newhouse, M.G., and Manning, D.R. (1986). Fractionation of the $\beta$ subunit common to guanine nucleotide-binding regulatory proteins with the cytoskeleton. *Mol. Pharmacol.*, **30**, 463–468.

Carrasco, M.A., Sierralta, J., and De Mazancourt, P. (1994). Characterization and subcellular distribution of G-proteins in highly purified skeletal muscle fractions from rabbit and frog. *Arch. Biochem. Biophys.*, **310**, 76–81.

Cerione, R.A., Regan, J.W., Nakata, H., Codina, J., and Spiegel, A.M. (1986). Functional reconstitution of the alpha 2-adrenergic receptor with guanine nucleotide regulatory proteins in phospholipid vesicles. *J. Biol. Chem.*, **261**, 3901–3909.

Colombo, M.I., Mayorga, L.S., Casey, P.J., and Stahl, P.D. (1992). Evidence of a role for heterotrimeric GTP-binding proteins in endosome fusion. *Science*, **255**, 1695–1697.

De Riemer, S.A., Rahamimoff, R., Sakmann, B., and Stadler, H. (1987). Conductances and channels in fused synaptic vesicles from Torpedo electric organ. *J. Physiol.*, **368**, 84P

De Mazancourt, P., Goldsmith, P.K., and Weinstein, L.S. (1994). The inhibition of adenylyl cyclase by galanin in rat insulinoma cells is mediated by the G protein $G_{i3}$. *Biochem. J.*, **303**, 369–375.

Doherty, P., Ashton, S.V., Moore, S.E, and Walsh, F.S. (1991). Morphoregulatory activities of NCAM and N-cadherin can be accounted for by G protein-dependent activation of L- and N-type neuronal calcium channels. *Cell*, **67**, 21–33.

Donaldson, J.G., Kahn, R.A., Lippincott-Schwartz, J., and Klausner, R.D. (1991). Binding of ARF and $\beta$-COP to Golgi membranes. Possible regulation by a trimeric G protein. *Science*, **254**, 1197–1199.

Fenwick, E.M., Marty, A., and Neher, E. (1982). A patch-clamp study of bovine chromaffin cells and of their sensitivity to acetylcholine. *J. Physiol.*, **331**, 557–597.

Glavinovic, M.I. and Trifaro, J.M. (1988). Quinine blockade of currents through $Ca^{2+}$-activated $K^{+}$-channels in bovine chromaffin cells. *J. Physiol. (Lond.)*, **399**, 139–152.

Hamilton, S.L., Codina, J., Hawkes, M.J., Yatani, A., Sawada, T., Strickland, F.M., *et al.* (1991). Evidence for direct interaction of $G_s\alpha$ with the $Ca^{2+}$ channel of skeletal muscle. *J. Biol. Chem.*, **266**, 19528–19535.

Hoffman, P.G., Zinder, O., Nikodejevic, O., and Pollard, H.B. (1976). ATP-stimulated transmitter release and cyclic AMP-synthesis in isolated chromaffin granules. *J. Supramol. Structure*, **4**, 181–184.

Holz, R.W. (1979). Measurement of membrane potential of chromaffin granules by accumulation of triphenylmethylphosphonium cation. *J. Biol. Chem.*, **254**, 6703–6709.

Johnson, R.G. and Scarpa, A. (1979). Protonmotive force and catecholamine transport in isolated chromaffin granules. *J. Biol. Chem.*, 3750–3760.

Kume, H., Graziano, M.P. and Kotlikoff, M.I. (1992). Stimulatory and inhibitory regulation of calcium-activated potassium channels by guanine nucleotide-binding proteins. *Proc. Natl Acad. Sci. USA*, **89**, 11051–11055.

Knaus, P., Marqueze-Pouey, B. Scherer, H. and Betz, H. (1990). Synaptoporin, a novel putative channel protein of synaptic vesicles. *Neuron*, **5**, 453–462.

Lee, C.J., Dayanithi, Nordmann, J.J. and Lemos, J.R. (1992). Possible role during exocytosis of a $Ca^{2+}$-activated channel in neurohypophysial granules. *Neuron*, **8**, 335–342.

Lemos, J.R., Ocorr, K., and Nordmann, J.J. (1989). Possible role for ionic channels in neurosecretory granules of the rat neurohypophysis. *Soc. Gen. Physiol. Ser.*, **44**, 333–347.

Lingg, G., Fischer-Colbrie, R., Schmidt, W. and Winkler, H. (1983). Exposure of an antigen of chromaffin granules on cell surface during exocytosis. *Nature*, **301**, 610–611.

Lindau, M. (1991). Time-resolved capacitance measurements: monitoring exocytosis in single cells. *Q. Rev. Biophys.*, **24**, 75–101.

Madison, D.V. and Nicoll, R.A. (1986). Cyclic adenosine 3′,5′-monophosphate mediates $\beta$-receptor actions of noradrenaline in rat hippocampal pyramidal cells. *J. Physiol.*, **372**, 245–259.

MacKinnon, R. and Miller, C. (1988). Mechanism of charybdotoxin block of the high-conductance $Ca^{2+}$-activated $K^+$-channel. *J. Gen. Physiol.*, **91**, 335–349.

McKenzie, F.R., Kelly, E.C., Unson, C.G., Spiegel, A.M. and Milligan, G. (1988). Antibodies which recognize the C-terminus of the inhibitory guanine-nucleotide-binding protein ($G_i$) demonstrate that opioid peptides and foetal-calf serum stimulate the high-affinity GTPase activity of two separate pertussis-toxin substrates. *Biochem. J.*, **249**, 653–659.

Morris, D., McHugh-Sutkowski, Moos, M., Simonds, W.F. and Spiegel, A.M. (1990). Immunoprecipitation of adenylate cyclase with an antibody to a carboxyl-terminal peptide from G. alpha. *Biochemistry*, **29**, 9079–9084.

Neher E. and Marty, A. (1982). Discrete changes of cell membrane capacitance observed under conditions of enhanced secretion in bovine adrenal chromaffin cells. *Proc. Natl Acad. Sci. USA*, **79**, 6712–6716.

Nikodejevic, O., Nikodejevic, B., Zinder, O., Guroff, G., Yu, M.-Y., and Pollard, H.B. (1976). Control of adenylate cyclase from secretory vesicles membranes by beta-adrenergic agents and nerve growth factor. *Proc. Natl Acad. Sci. USA*, **73**, 771–774.

Obendorf, D., Schwarzenbrunner, U., Fischer-Colbrie, R., Laslop, A. and Winkler, H. (1988). In adrenal medulla synaptophysin (protein p38) is present in chromaffin granules and in a special vesicle population. *J. Neurochem.*, **51**, 1573–1580.

Ornberg, R.L., Kuijpers, G.A., and Leapman, R.D. (1988). Electron probe microanalysis of the subcellular compartments of bovine adrenal chromaffin cells. *J. Biol. Chem.*, **263**, 1488–1493.

Pazoles, C.J. and Pollard, H.B. (1978). Evidence for activation of anion transport in ATP-evoked transmitter release from isolated secretory vesicles. *J. Biol. Chem.*, **253**, 3962–3969.

Penner, R. and Neher, E. (1989). The patch-clamp technique in the study of secretion. *Trends Neurosci.*, **12**, 159–163.

Phillips, J.P. and Apps, D.K. (1980). Stoichiometry of catecholamine/proton exchange across the chromaffin granule membrane. *Biochem. J.*, **192**, 273–278.

Phillips, J.H. (1978). 5-hydroxytryptamine transport by Bovine Chromaffin-Granule Membrane. *Biochem. J.*, **170**, 673–679.

Picaud, S., Marty, A., Trautmann, A., Grynszpan-Winograd, O., and Henry, J.-P. (1984). Incorporation of chromaffin granule membranes into large-size vesicles suitable for patch-clamp recording. *FEBS Lett.*, **178**, 20–24.

Pollard, H.B., Zinder, O., Hoffman, P.G., and Nikodijevik, O. (1976). Regulation of the transmembrane potential of isolated chromaffin granules by ATP, ATP analogs, and external pH. *J. Biol. Chem.*, **251**, 4544–4551.

Pollard, H.B., Pazoles, C.J., Creutz, C.E. and Zinder, O. (1979*a*). The chromaffin granule and possible mechanisms of exocytosis. *Int. Rev. Cytol.*, **58**, 159–197.

Pollard, H.B., Shindo, H., Creutz, C.E. and Zinder, O. (1979*b*). Internal pH and state of ATP in adrenergic chromaffin granules determined by 31P nuclear magnetic resonance spectroscopy. *J. Biol. Chem.*, **254**, 1170–1177.

Rahamimoff, R., De Riemer, S.A., Sakmann, B., Stadler, H. and Yakir, N. (1988). Ion channels in synaptic vesicles from *Torpedo* electric organ. *Proc. Natl Acad. Sci. USA*, **85**, 5310–5314.

Rahamimoff, R., De Riemer, S.A., Ginsburg, S., Kaiserman, I., Sakmann, B., Stadler, H. and Yakir, N. (1989). Ionic channels in synaptic vesicles: are they involved in transmitter release? *Q. J. Exp. Physiol.*, **74**, 1019–1031.

Rahamimoff, R., De Riemer, S.A., Ginsburg, S., Kaiserman, I., Sakmann, B., Stadler, H. and Yakir, N. (1990). Ionic channels and proteins in synaptic vesicles: facts and speculations. *J. Basic Clin. Physiol. Pharmacol.*, **1**, 7–17.

Sakmann, B. and Neher, E. (1983). *Single-channel recording*. Plenum Publishing Co., New York.

Salama, G., Johnson, R.G., and Scarpa, A. (1980). Spectrophotometric measurements of transmembrane potential and pH gradients in chromaffin granules. *J. Gen. Physiol.*, **75**, 109–140.

Scamps, F., Rybin, V., Puceat, M., Tkachuk, V. and Vassort, G. (1992). A Gs protein couples P2-purinergic stimulation to cardiac Ca-channels without cyclic AMP production. *J. Gen. Physiol.*, **100**, 675–701.

Simonds, W.F., Goldsmith, P.K., Woodard, C.J., Unson, C.G., and Spiegel, A.M. (1989*a*). Receptor and effector interactions of Gs. Functional studies with antibodies to the alpha s carboxyl-terminal decapeptide. *FEBS Lett.*, **249**, 189–194.

Simonds, W.F., Goldsmith, P.K., Codina, J., Unson, C.G., and Spiegel, A.M. (1989*b*). $G_{i2}$ mediates alpha 2-adrenergic inhibition of adenylyl cyclase in platelet membranes: *in situ* identification with G alpha C-terminal antibodies. *Proc. Natl Acad. Sci. USA*, **86**, 7809–7813.

Simonds, W.F., Butrynski, J.E., Gautam, N., Unson, C.G., and Spiegel, A.M. (1991). G-protein beta gamma dimers. Membrane targeting requires subunit coexpression and intact gamma C-A-A-X domain. *J. Biol. Chem.*, **266**, 5363–5366.

Spiegel, A.M., Simonds, W.F., Jones, T.L., Goldsmith, P.K. and Unson, C.G. (1990*a*). Antibodies as probes of G-protein receptor-effector coupling and of G-protein membrane attachment. *Biochem. Soc. Symp.*, **56**, 61–69.

Spiegel, A.M., Simonds, W.F., Jones, T.L., Goldsmith, P.K, and Unson, C.G. (1990*b*). Antibodies against synthetic peptides as probes of G protein structure and function. *Soc. Gen. Physiol.*, **45**, 185–195.

Spiegel, A.M. (1991). Receptor-effector coupling by G-proteins: implications for neuronal plasticity. *Prog. Brain. Res.*, **86**, 269–276.

Spiegel, A.M., Shenker, A., and Weinstein, L.S. (1992). Receptor-effector coupling by G-proteins: implications for normal and abnormal signal transduction. *Endocr. Rev.*, **13**, 1–30.

Stanley, E.F., Ehrenstein, G., and Russell, J.T. (1988). Evidence for anion channels in secretory vesicles. *Neuroscience*, **25**, 1035–1039.

Stow, J.L., De Almeida, J.B., Narula, N., Holtzman, E.J., Ercolani, E.J., and Ausiello, D.A. (1991). A heterotrimeric G protein, $G\alpha_{i3}$, on Golgi membrane regulates the secretion of a heparan sulfate proteoglycan in LLC-PK, epithelial cells. *J. Cell Biol.*, **114**, 1113–1124.

Thomas, L., Hartung, K., Langosch, D., Rehm, H., Bamberg, E., Franke, W., and Betz, H. (1988). Identification of synaptophysin as a hexameric channel protein of the synaptic vesicle membrane. *Science*, **242**, 1050–1053.

Toutant, M., Aunis, D., Bockaert, J., Homburger, V. and Rouot, B. (1987). Presence of three pertussis toxin substrates and GO$\alpha$ immunoreactivity in both plasma and granule membranes of chromaffin cells. *FEBS Lett.*, **215**, 339–343.

Trifaro, J.M. and Poisner, A.M. (1982). Common properties in the mechanisms of synthesis, processing and storage of secretory products. In *The secretory granule*, (ed. A.M. Poisner and J.M. Trifaro), Elsevier Biomedical Press.

Von Grafenstein, H., Roberts, C.S., and Baker, P.F. (1986). The kinetics of the exocytosis endocytosis secretory cycle in bovine adrenal medullary cells. *J. Cell Biol.*, **103**, 2343–2352.

Winkler, H. and Westhead, E.W. (1980). The molecular organization of adrenal chromaffin granules. *Neuroscience*, **5**, 1803–1823.

Yatani, A., Hamm, H., Codina, J., Mazzoni, M.R., Birnbaumer, L., and Brown, A.M. (1988). A monoclonal antibody to the alpha subunit of $G_k$ blocks muscarinic activation of atrial $K^+$ channels. *Science*, **241**, 828–831.

Zinder, O., Menard, R., Lovenberg, W., and Pollard, H.B. (1977). Direct evidence for co-localization of adenylate cyclase, dopamine-beta-hydroxylase and cytochrome $b_{562}$ to bovine chromaffin granule membranes. *Biochem. Biophys. Res. Commun.*, **79**, 707–712.

# 8 Cyclic Nucleotide-gated Channels

# 17 Cyclic nucleotide-gated ion channels: physiology, pharmacology, and molecular biology

G. Matthews

## Introduction

Ion channels that are directly gated by cyclic nucleotides represent a class of ligand-gated channel whose ligand binding site is on the intracellular side of the channel, rather than the extracellular side. Channels of this type were first described in the outer segments of vertebrate rod photoreceptors by Fesenko *et al.*[1] Subsequently, evidence for cyclic nucleotide-gated channels has been found in a variety of cells and tissues.[2–5] This chapter will review the physiological, pharmacological, and molecular properties of cyclic nucleotide-gated channels. The principal focus will be the cGMP-gated channel of rod photoreceptors, which is the most thoroughly studied example. However, comparisons with other types of cyclic nucleotide-gated channel, particularly the olfactory channel, will be made where possible.

## Physiological role of cGMP-gated channels in phototransduction

Vertebrate photoreceptors are depolarized in darkness and hyperpolarize in response to light. It is now clear that this occurs because cGMP is elevated inside the photoreceptor in the dark, keeping cGMP-gated channels open. Because the channels are non-specific cation channels with a reversal potential near zero, this maintains the photoreceptor in a partially depolarized state in darkness. Upon illumination, phosphodiesterase is activated and cGMP levels fall, causing the channels to close (see refs. 6–9 for reviews). The most direct evidence for this mechanism was obtained in experiments in which inside-out patches were excised from rod photoreceptor outer segments and cGMP was added to the solution bathing the intracellular face of the patch,[1] as illustrated in Fig. 17.1. cGMP reversibly increased membrane conductance to cations in a dose-dependent manner. This is a direct action of cGMP on membrane conductance, as evidenced by the fact that the effect is observed rapidly and repeatedly for the lifetime of the patch (up to 1 h) in the absence of ATP, thus ruling out indirect action via activation of protein kinase and phosphorylation of the channel. In addition, activation of the channels by cGMP is observed in purified channels reconstituted in artificial bilayer membranes,[10] and a region homologous to the cGMP binding site of cGMP-dependent protein kinase is part of the channel structure[11] (see below).

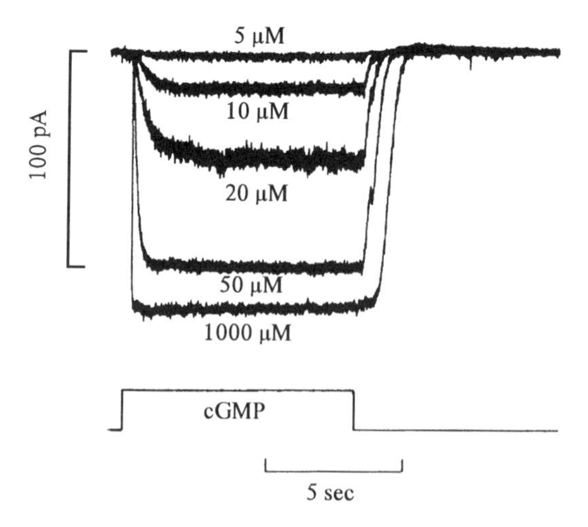


**Fig. 17.1** Dose – response relation of cGMP-activated conductance. An inside-out patch was excised from the outer segment of a rod photoreceptor from toad retina, and cGMP at the indicated concentrations was applied to the intracellular face of the patch at the time indicated by the lower trace. The superimposed traces show recordings made with symmetrical NaCl solutions on the two sides of the patch membrane and with a holding potential of − 61 mV. Under these conditions, activation of cGMP-gated conductance produces an inward membrane current. (Figure modified from ref. 22.)

Experiments with excised patches demonstrated that cGMP activates cation channels in photoreceptor membranes, but what is the evidence that the channel activated by cGMP is the same as the light-sensitive channel of the photoreceptor? To examine this point, the quantitative properties of the conductance activated by cGMP in excised patches were compared with the properties of the light-sensitive conductance in cell-attached patches. Noise analysis of both conductances[12,13] showed that the kinetics of the underlying channels were similar and that the single-channel currents were also similar. This similarity in noise properties suggested that the light-sensitive channel and the cGMP-gated channel are one and the same. To examine this directly, single light-sensitive channels were recorded in cell-attached patches of dark-adapted rods and compared with the channels subsequently opened by cGMP in the same membrane patch after it was excised in the inside-out configuration.[14,15] As shown in Fig. 17.2, the single-channel properties were identical. This provided direct demonstration that the channel closed by light in the intact cell is the same as the channel opened by cGMP in excised patches. The concentration of cGMP required in excised patches to match the amount of channel activity recorded in darkness in cell-attached patches was 5–7 $\mu$M, which provides an estimate for the free cytoplasmic concentration of cGMP in the rod outer segment in darkness.

Experiments have also been done using broken-open photoreceptor outer segments to demonstrate that the conductance opened by exogenously added internal cGMP is closed by illumination.[16] This demonstrates that the phototransduction machinery in the outer segment is able to counteract the effect of added cGMP on membrane conductance, presumably by activation of phosphodiesterase. Thus, not only is the channel opened by cGMP the same as the endogenous light-sensitive channel (Fig. 17.2), but also the cGMP-gated channels can be shut by activation of the phototransduction process. Coupled with the wealth of biochemical information about the linkage between photo-activated rhodopsin and phosphodiesterase

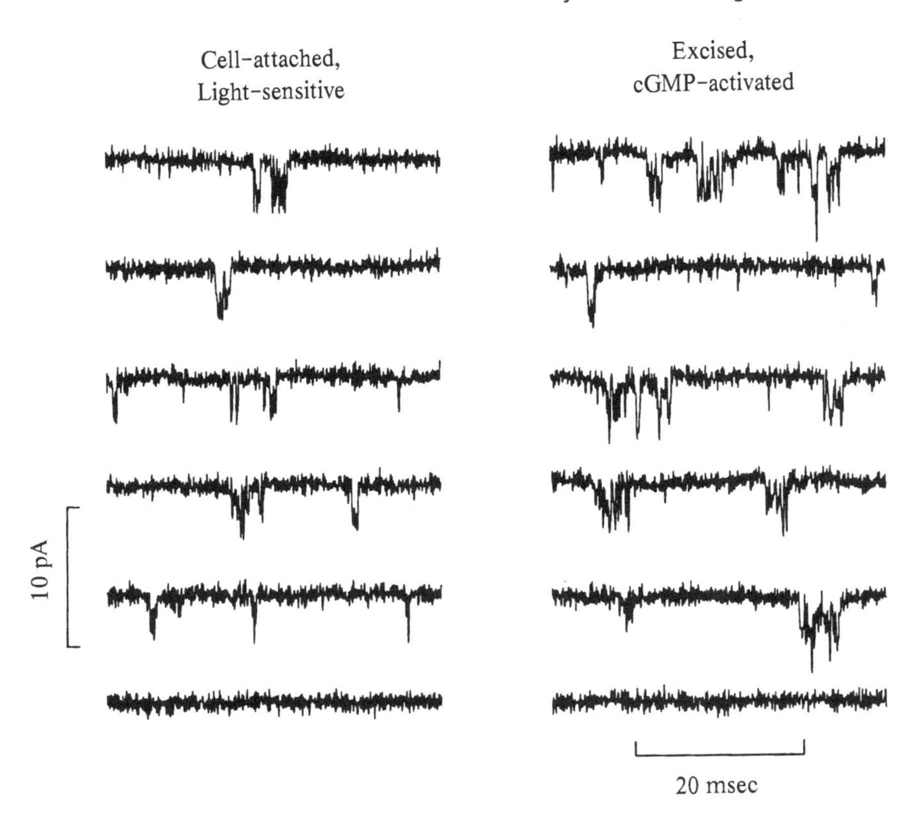


**Fig. 17.2** Comparison of light-sensitive (left column) and cGMP-gated (right column) channel activity within the same patch of membrane. Light-sensitive channel activity was recorded from a cell-attached patch on the outer segment of a dark-adapted rod photoreceptor from toad retina. The bottom trace on the left is in the presence of steady illumination, while all other traces are in darkness. Subsequently, the same patch of membrane was excised in inside-out configuration, and 10 $\mu$M cGMP was applied to the intracellular face to elicit the channel activity shown on the right. The bottom trace on the right is in the absence of cGMP. In order to reveal the brief openings and closings characteristic of the channel events, the recordings were made at wide bandwidth (0–5000 Hz). Recordings were made with divalent cation-free Ringer solution on both sides of the membrane. (Figure modified from ref. 13.)

activity,[6] experiments of this type demonstrate conclusively that the cGMP-gated channels of photoreceptor outer segments mediate the electrical response to light.

## Gating of the photoreceptor channel by cGMP

A fundamental question about the cGMP-gated channel is the relation between cGMP concentration and channel-open probability. The simplest way to obtain this relation over a wide range of cGMP concentration is to examine the gating of the channel in an excised patch containing only a single copy of the cGMP-gated channel. This proved possible in membrane patches obtained from rod inner segments,[17] where the density of cGMP-gated channels is much lower than in the outer segment.[17,18] Figure 17.3 shows the concentration dependence of channel gating in such a one-channel patch. Two features are readily apparent. First, even at saturating concentrations of cGMP, the maximal open probability was substantially less than 1 (0.3 on average); this is because of rapid flicker of the channel between open and closed

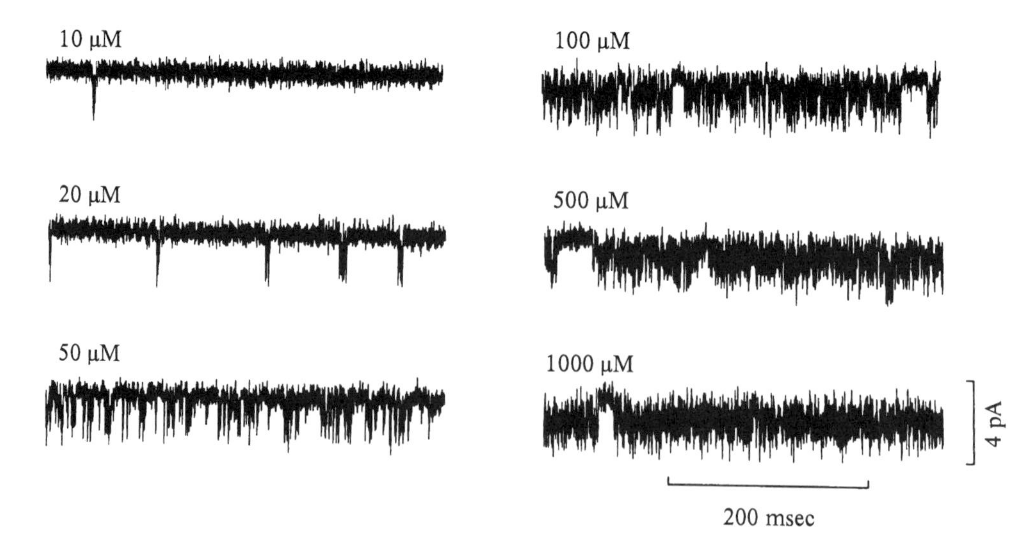


**Fig. 17.3** Concentration dependence of channel gating by cGMP in an excised, inside-out patch of membrane from the inner segment of a rod photoreceptor. The patch contained only one cGMP-gated ion channel. Recording bandwidth: 0–2000 Hz. (Figure reproduced from ref. 17.)

states, which was observed at all levels of cGMP. Second, the open probability at low concentrations of cGMP was steeply dependent on concentration, with a Hill coefficient of about 3 (Fig. 17.4). Given that molecular structure reveals only one cGMP binding site per channel unit (see below), this indicates that multiple units are required to make up a functional cGMP-activated channel.

In the experiments on the concentration dependence of channel gating in one-channel patches,[17] there was considerable variability in the sensitivity of individual channels to cGMP, with the half-saturating concentration ($K_{½}$) varying from 17 to 50 $\mu$M. This mirrors the variability in reported values for $K_{½}$ in experiments on macroscopic currents in outer-segment patches containing several hundred cGMP-gated channels.[7,9] In the one-channel patches, however, the relation between cGMP concentration and open probability for an individual channel was always well described by the Hill equation with a coefficient of 3, despite the variation in $K_{½}$ among channels, while in outer-segment patches there is considerable variation in the Hill coefficient, with reported values ranging from 1.7 to 3.1 (see ref. 7). Also, published dose–response relations for outer-segment patches containing large numbers of channels show consistent deviation from the Hill equation, unlike the one-channel, inner-segment patches. This consistent deviation in the multi-channel patches can be explained under the assumption that the population of channels in an outer-segment patch shows the same degree of variation in sensitivity to cGMP as observed for individual channels in experiments on inner-segment patches that contained a single channel.[17] The cGMP dose–response relation for individual channels from inner-segment patches containing only a single cGMP-activated channel is summarized in Fig. 17.4A To compare results across channels, the open probability is normalized with respect to the asymptotic open probability and [cGMP] is normalized with respect to $K_{½}$ for each channel. The curve through the data was drawn according to the Hill equation:

$$R = G^3/(1 + G^3)$$

where $R$ is the normalized open probability and $G$ is $[cGMP]/K_{½}$. Data from seven outer segment patches, similarly normalized, are shown in Fig. 17.4B; the curve was again drawn according to the Hill equation above. Note that for the outer-segment patches, there was consistent deviation from the curve at intermediate concentrations, in the region near $K_{½}$ (the 'knee' region of the dose–response curve). The transition occurred less abruptly than anticipated from the Hill equation; similar behaviour can also be seen in previously published data.[9] No such deviation in the knee region was seen for the one-channel patches from the inner segment (Fig. 17.4A)

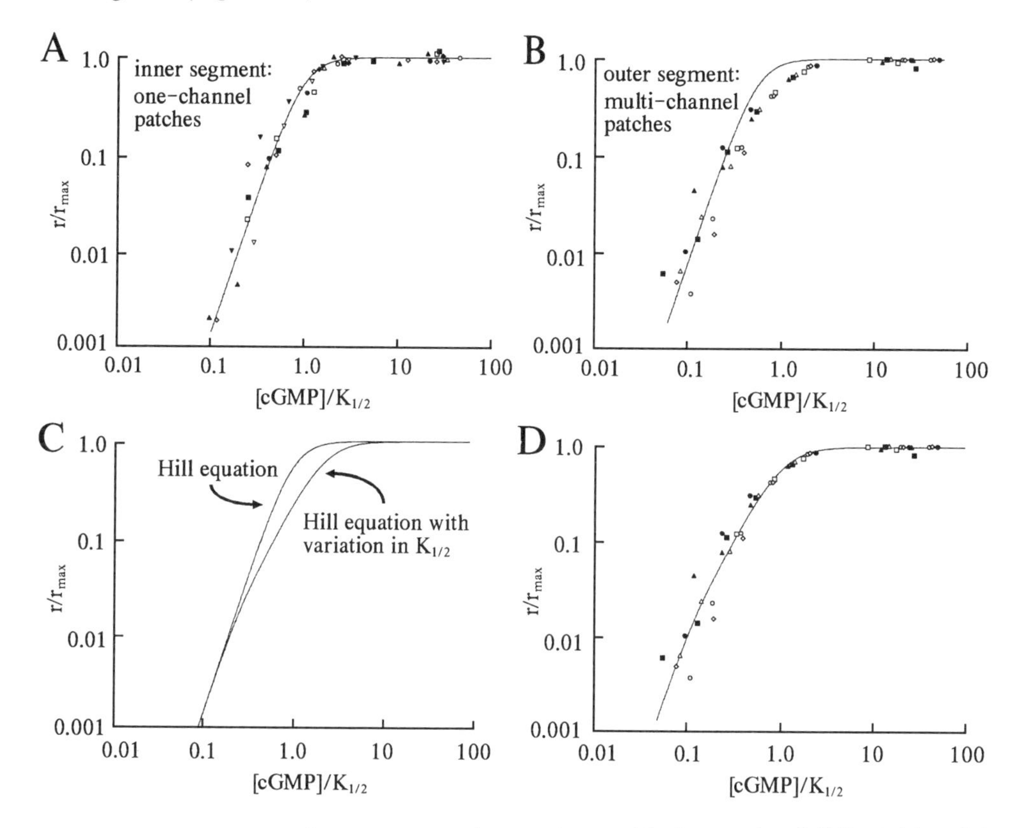


**Fig. 17.4** Dose – response curve of outer-segment membrane patches containing large number of cGMP-gated channels is described accurately if variability in $K_{1/2}$ across channels is allowed for. A. Summary of dose–response relations from nine inner-segment patches, each containing one cGMP-gated channel. The response is the open probability at each [cGMP], expressed relative to the saturating open probability at high [cGMP]. Although the individual values of $K_{1/2}$ for single channels ranged from 17–50 $\mu$M, the dose–response curve for each individual channel was well described throughout the range of [cGMP] by the Hill equation for cooperative cGMP binding with a Hill coefficient of 3 (see text). Each symbol type shows results from a different channel. B. Summary of dose–response relations for cGMP-gated channels in 10 outer-segment patches, each containing many channels. The response for these patches was the conductance increase elicited by cGMP, expressed relative to the conductance increase at saturating [cGMP]. Unlike the one-channel patches shown in A, the multi-channel patches deviated consistently from the Hill equation. C. Dose–response relation predicted from the standard Hill equation for cooperative binding with a Hill coefficient of 3, compared with the relation predicted if the population of channels in multi-channel patches has a Gaussian distribution of $K_{1/2}$, with mean and variance given by the population mean and variance observed in the sample of one-channel patches. D. Dose–response relation for outer-segment patches containing large numbers of patches (replotted from B) is well described by the modified Hill equation that allows for variation in $K_{1/2}$ in the population of channels (replotted from C).

Typically, a shallower transition in the dose–response curve would be allowed for by reducing the value of the Hill coefficient; however, this has the disadvantage that the asymptotic slope at low [cGMP] is also shallower, and the fit to the data at low [cGMP] would then be inadequate. Another alternative would be to use the non-cooperative binding form of the Hill equation, but this results in a transition in the knee region of the curve that is too gentle.[19] The observation that individual cGMP-activated channels show considerable variation in sensitivity to cGMP suggests a different approach to fitting the macroscopic dose–response curves. If the population of channels present in a single outer-segment patch shows variability in sensitivity similar to that observed across patches in one-channel patches, then deviation from the Hill equation in the knee region of the dose-response curve is to be expected. Qualitatively, as [cGMP] is progressively reduced from a saturating level, activity in channels with a high $K_{1/2}$ will begin to fall off first, while activity of more sensitive channels will be sustained at a high level. As [cGMP] is reduced further, activity falls off in more and more channels, until at sufficiently low [cGMP], all the channels are in the linear portion (on double-logarithmic coordinates) of the dose-response relation. Thus, the macroscopic dose–response relation at low [cGMP] should fall off with a slope of 3 on double-logarithmic coordinates, as the relation for individual channels does, but the transition from saturation should be less abrupt. The form that the dose–response relation takes under these conditions will depend on the assumed distribution of $K_{1/2}$ for the population of channels in the patch. The population dose–response relation expected if $K_{1/2}$ were normally distributed, with mean and standard deviation taken from the observed group of one-channel patches, is shown in Fig. 17.4C; for comparison, the relation observed for individual channels is also shown in Fig. 17.4C. As shown in Fig. 17.4D, the dose–response data for macroscopic conductance changes in multi-channel, outer-segment patches are well described by the relation predicted under the assumption that the distribution of sensitivity observed across patches in one-channel patches also applies to the group of channels present in a single multi-channel patch. Thus, the apparent variability in the reported values from different laboratories for $K_{1/2}$ and the Hill coefficient for cGMP-gated channels can be accounted for in a simple way by the variability in sensitivity of individual channels to cGMP. The variation in sensitivity may arise from differences in phosphorylation of the channel.[20]

The concentration dependence of channel gating also gives information about the possible mechanisms underlying the various kinetic components of channel gating by cGMP. In Fig. 17.3, one readily apparent feature of the cGMP-gated channel is the rapid flicker between open and closed states, which continued even at saturating concentrations of cGMP. As shown in Fig.17.5, the average durations of these rapid openings and closings were unaffected by cGMP concentration. This suggests that the flicker of the channel is not due to binding and unbinding of cGMP, but is instead accounted for by a non-cGMP-dependent mechanism. One such possibility is that the rapid openings and closings reflect opening and closing of an independent gate (perhaps analogous to a ball-and-chain inactivation gate of voltage-dependent channels), whose activity is only manifested when the cGMP-dependent gate is open. Recent evidence suggests that this non-cGMP-dependent gate might reside in the newly described β-subunit of the cGMP-activated channel[21] (also see below). By contrast, the principal way in which increasing concentrations of cGMP increase the open probability of the channel is by increasing the average duration of the bursts of rapid flicker and decreasing the average closed duration between bursts (Figs. 17.3 and 17.5). A simple interpretation is that the duration of a burst corresponds to the duration of the fully liganded state of the channel, during which the cGMP-dependent gate is open, while the duration of the closed state between bursts corresponds to the excursions into various partially liganded closed states.[17,22]

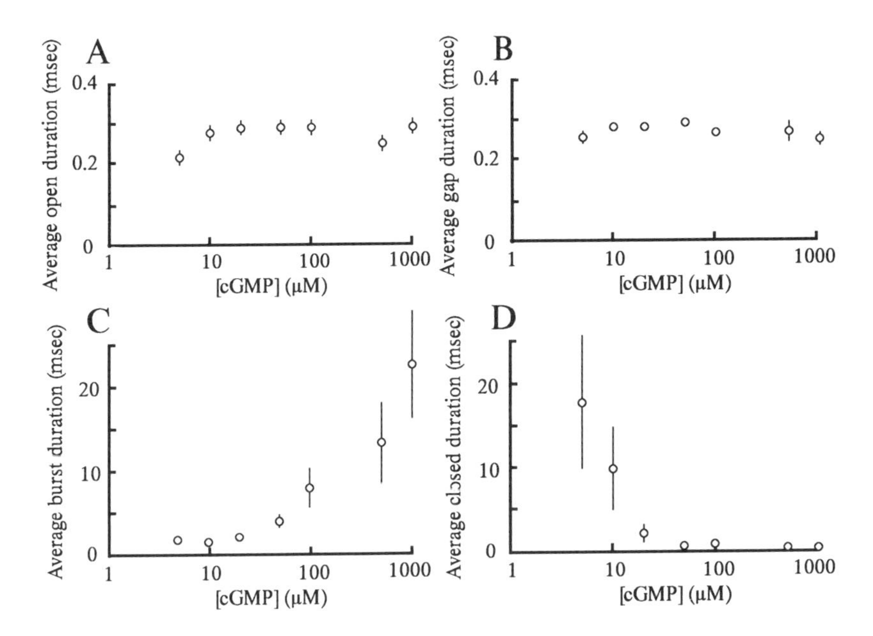


**Fig. 17.5** Kinetics of gating of single cGMP-activated channels at various concentrations of cGMP. Gating properties of single channels were measured in inner-segment patches containing one cGMP-gated channel. A. Average open duration versus [cGMP]. Open duration was defined as the duration of each individual excursion into the open state, defined by a half-amplitude criterion. B. Gap duration was defined as the duration of each individual brief excursion into the closed state that occurred within a burst of brief openings and closing. Both open duration and gap duration correspond to the brief flickers between open and closed states within a burst, and both were independent of cGMP concentration. C. Burst duration was defined as the total duration of a group of openings separated by closings briefer than 1 ms (i.e. a burst was terminated by any closing longer than 1 ms). Burst duration increased dramatically with [cGMP]. D. Closed duration decreased dramatically with [cGMP], as intervals between bursts became progressively shorter until activity fused into single long bouts of rapid flicker between open and closed states (called openings and gaps, and shown in A and B). (Reproduced from ref. 22.)

## Lack of desensitization of the cGMP-gated channel

Unlike other ligand-gated channels, the cGMP-gated channel shows no slow desensitization in the prolonged presence of agonist.[23] This is an important feature, since cGMP is present continuously in darkness. Experiments to test for both slow and rapid components of desensitization upon rapid exposure to cGMP are shown for a patch containing a single cGMP-gated channel in Fig.17.6, which shows channel activity elicited by a saturating concentration of cGMP. Activity continued unabated for tens of seconds, even in the maintained presence of agonist. From the very onset of channel activity after application of cGMP, the channel adopted the rapid flicker between open and closed states characteristic of the channel (see above); no change in gating behaviour was apparent either within the first few milliseconds of activity (Fig. 17.6B trace a,) or after many seconds of maintained activity (Fig. 17.6B trace b,). Thus, there was no change in single-channel gating either on the time scale of milliseconds or on the time scale of seconds after application of cGMP. The cGMP-gated channel differs in this regard from other ligand-gated channels and appears to be specialized to track changes in internal cGMP concentration rapidly and faithfully, without time-dependent changes in sensitivity to its agonist.

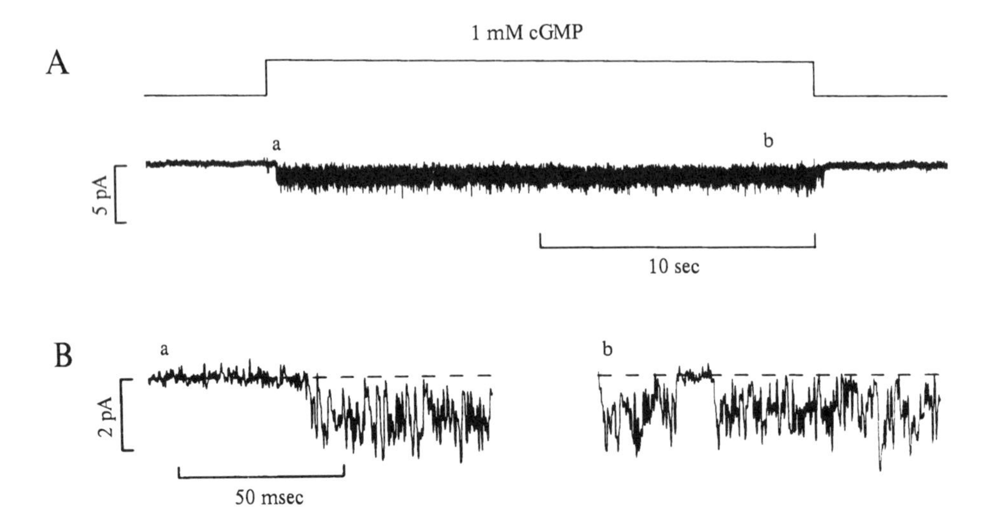


**Fig. 17.6** Lack of rapid and slow desensitization in cGMP-gated channels. Recordings show response of an inner-segment patch containing one cGMP-gated channel to application of a saturating concentration of cGMP (1 mM). A. An overview of the response at slow time scale. B. Higher-resolution views of activity at the onset of channel activity (position a in A) and after several seconds of continuous cGMP application (position b in A). (Reproduced from ref. 23.)

## Permeation properties of cyclic nucleotide-gated channels

Cyclic nucleotide-gated channels of both olfactory receptors and photoreceptors are non-specific cation channels, showing little permeation selectivity among alkali monovalent cations.[7] In this regard, cyclic nucleotide-gated channels are more like the nicotinic acetylcholine-activated channel and are different from the voltage-gated channels to which they otherwise show greater molecular similarity.[24,25] Also like the actylcholine-activated channel, the cGMP-gated channel of photoreceptors is permeable to a variety of organic monovalent cations, although the maximum size of permeant molecules is somewhat smaller than in the acetylcholine-activated channel,[26] suggesting a somewhat smaller pore size.

Given that the conductivity of the cGMP-gated channel of rod photoreceptors is about equal to $K^+$, the principal internal cation, and to $Na^+$, the principal external cation under physiological conditions, it would be expected that the light-sensitive conductance of intact rods would have an approximately ohmic current–voltage relation with a reversal potential near zero. Although the reversal potential is in fact near zero, the current–voltage relation in intact rods shows pronounced outward rectification,[27] with little inward current at negative potentials but rapidly increasing outward current at positive potentials. The resolution of the apparent paradox is in the effect of divalent cations on the cGMP-gated channel: under physiological ionic conditions, with mM levels of $Ca^{2+}$ and $Mg^{2+}$ in the external solution, inward current through the channel is substantially reduced because of partial blockade by external divalent cations. This effect of external divalent cations on the current–voltage relation of the cGMP-gated conductance is illustrated in Fig. 17.7A. In this experiment, the intracellular face of the patch was bathed in a 0 $Ca^{2+}$/0 $Mg^{2+}$ solution containing a saturating concentration of cGMP and the extracellular face was bathed either in Ringer solution with normal levels of $Ca^{2+}$ and $Mg^{2+}$, or in Ringer solution without $Ca^{2+}$ and $Mg^{2+}$. In the presence of external $Ca^{2+}$ and $Mg^{2+}$, the current-voltage relation showed pronounced outward rectification like that of the light-sensitive conductance in intact photoreceptors. When external $Ca^{2+}$ and $Mg^{2+}$ were removed, so that both faces of the excised patch were bathed in 0 $Ca^{2+}$/0 $Mg^{2+}$ solution, the rectification vanished. Figure 17.7B shows a similar experiment on an inside-out patch. When $Ca^{2+}$ and $Mg^{2+}$ were present on the intracellular face of the patch, but not the extracellular face, the current–voltage curve showed inward rectification (i.e. the opposite of that in Fig. 17.7A). Again, when 0 $Ca^{2+}$/0 $Mg^{2+}$ solution bathed both faces of the patch, the cGMP-activated conductance was ohmic. Similar results have been reported by others (see ref. 9 for a recent review). This behaviour suggests that $Ca^{2+}$ and $Mg^{2+}$ enter the channel and interact strongly with a binding site or sites within the channel,[28] so that the rate of divalent ion flux through the channel is low. While the site is occupied, the channel is plugged, so that the effective single-channel conductance is small whenever current is being driven from the side of the membrane (either intracellular or extracellular) containing divalent cations. Under physiological ionic conditions, approximately 15% of the inward dark current in rod photoreceptors is carried by $Ca^{2+}$ (ref.9) with most of the remainder being carried by $Na^{+}$.

## Pharmacology of cyclic nucleotide-gated channels

The natural intracellular ligand for cyclic nucleotide-gated channels is either cAMP (e.g. in olfactory cilia) or cGMP (e.g. in photoreceptors). However, both the relative and the absolute sensitivities to cAMP and cGMP vary considerably among the various types of cyclic nucleotide channels that have been described to date. For cGMP-gated channels of photoreceptors, cAMP is not a very effective agonist, with a $K_{1/2}$ of $\sim$1 mM and incomplete maximal activation,[29] compared with a $K_{1/2}$ of 20–50 $\mu$M for cGMP. Olfactory channels, however, are approximately equally sensitive to cGMP and cAMP,[2,30] and the absolute sensitivity is about 10-fold higher than in photoreceptor channels. There is some variability in the reported sensitivity of olfactory cyclic nucleotide-gated channels from different preparations to cAMP and cGMP, particularly for cloned channels expressed in artificial systems compared with native channels in excised patches.[30] Some of the variability might therefore reflect cell-specific modification of the channels or modulation of sensitivity.[20,29,31]

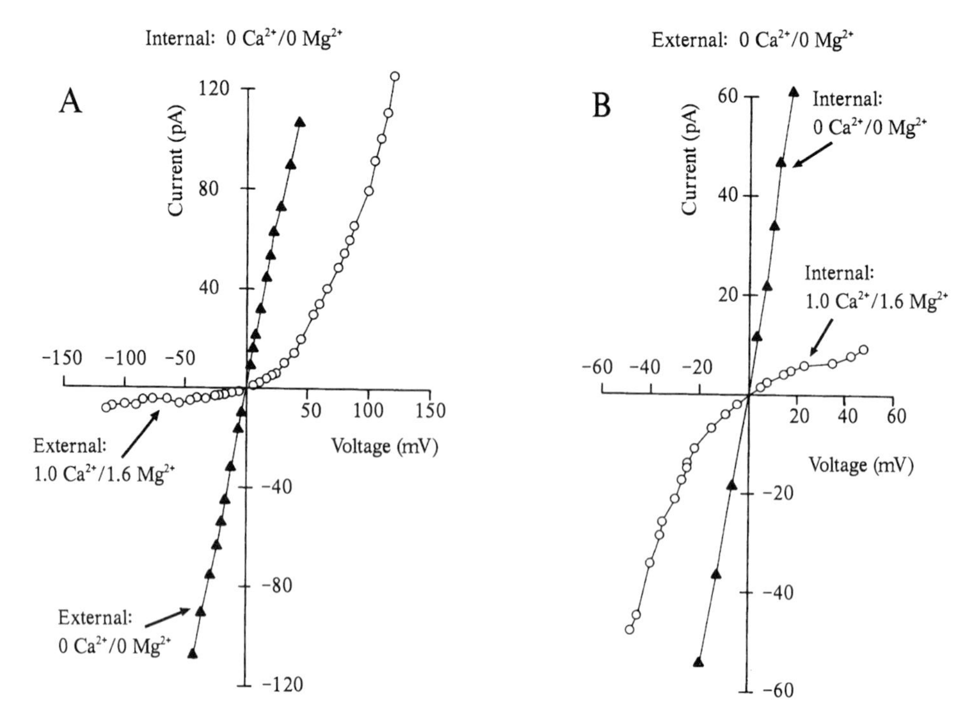


**Fig. 17.7** Current – voltage relations of cGMP-dependent conductance in outer-segment patches from rod photoreceptors. A. In this experiment, the internal face of the patch was exposed to divalent cation-free Ringer solution + 1.25 mM cGMP. The external face was then exposed to either normal Ringer solution, containing $Ca^{2+}$ and $Mg^{2+}$ (O), or to divalent cation-free Ringer solution (▲). B. In this experiment, the external face of the patch was exposed to divalent cation-free Ringer solution, and the internal face was exposed to normal Ringer solution, with $Ca^{2+}$ and $Mg^{2+}$ (O), or to divalent cation-free Ringer solution (▲). Both internal solutions contained 1.25 mM cGMP.

In addition to naturally occurring nucleotides, derivatives are also able to activate the channels. In particular, 8-substituted derivatives are frequently more effective than the natural cyclic nucleotide.[32] For example, in photoreceptor channels, the $K_{1/2}$ for 8-bromo-cGMP is approximately 10-fold lower than for cGMP itself.[33] This can be a useful experimental tool, because 8-bromo-cGMP is also resistant to phosphodiesterase, making it a good choice for activation of cGMP-gated channels in intact cells or excised native membrane patches, which might have endogenous phosphodiesterase activity. Comparison of the effectiveness of a variety of 8-substituted derivatives showed that uncharged substitutions were more effective than negatively charged substitutions, which were in turn more effective than positively charged moieties.[32] An 8-substituted derivative of cAMP (8-(4-chlorophenylthio)-cAMP) has also been reported to be about 10-fold more effective than cAMP in activating the olfactory cyclic nucleotide-gated channel.[30]

Although cAMP is not particularly effective in activating the cGMP-gated channel of photoreceptors, low levels of cAMP that do not directly activate the channel have been reported to enhance channel activation by cGMP.[29,34] This may reflect an allosteric effect of cAMP on channel gating, or an interaction with the cGMP-binding site.

Other than cyclic nucleotides, alternative agonists for the cyclic nucleotide-gated channels have not yet been reported. Blockers are also few, other than the channel-blocking actions of divalent cations and large organic cations mentioned above. The calcium-channel blocker diltiazem has been reported to block the cGMP-gated conductance of photoreceptors,[35] although the L-*cis*-isomer is more effective than the D-*cis*-isomer, which is usually the more effective isomer.[36] L-*cis*-diltiazem produces almost complete block of photoreceptor channels at 10 $\mu$M,[35] but olfactory channels are much less sensitive.[30] Sensitivity of cloned photoreceptor channels to L-*cis*-diltiazem is lower than in native channels, but high sensitivity is imparted to the cloned channel by coexpression of the newly described second subunit.[21] Thus, sensitivity of cyclic nucleotide-gated channels to diltiazem might depend on the properties of the β-subunit. In addition to diltiazem, the calcium-channel antagonist pimozide has also been reported to block photoreceptor cGMP-gated channels[37] with half-maximal inhibition at submicromolar concentration, and D 600, a verapamil derivative, has been reported to block olfactory cyclic nucleotide-gated channels,[30] although its potency is not great. Finally, a derivative of amiloride, 3′,4′-dichlorobenzamil, blocks the cGMP-gated channels of photoreceptor membranes.[38]

## Molecular properties of cyclic nucleotide-gated channels

Molecular characterization of the cyclic nucleotide-gated channels initiated with the biochemical isolation and characterization of the photoreceptor channel protein,[39] a 63 kDa protein that produced functional cGMP-gated channels when reconstituted into lipid bilayers.[10] Amino acid sequences of peptide fragments from this protein were then used to obtain cDNA for the entire protein, which was then functionally expressed in *Xenopus* oocytes.[11] Subsequently, this sequence for the cGMP-gated channel of rod photoreceptors was used to guide the characterization of cyclic nucleotide-gated channels in other cells and tissues, including olfactory neurones,[40,41] cone photoreceptors,[42] heart,[4,5] and kidney.[4] The members of the family of cyclic nucleotide-gated channels show high amino acid sequence homology, but are quite different from other ligand-gated channels, such as nicotinic acetylcholine receptors, $GABA_A$ receptors, and glycine receptors (see ref. 43 for comparisons). Instead, the cyclic nucleotide-gated channels are more similar to voltage-gated channels,[24] such as potassium channels, in the proposed number of transmembrane segments,[43] in the structure of the putative pore region,[25] and in possessing a region similar to the proposed voltage-sensor motif (S4 region).[43] Despite this similarity to voltage-gated channels, cyclic nucleotide-gated channels have little voltage sensitivity.[15] By analogy with voltage-gated channels, which are thought to form functional channels from aggregates of four subunits (in the case of potassium channels) or from arrangements of four repeated domains (in the case of sodium and calcium channels), functional cyclic nucleotide-gated channels are likely formed from the interaction of multiple subunits. This is consistent with the high cooperativity of cGMP binding in activation of native cGMP-gated channels (see Fig. 17.4).

Near the C-terminus of cyclic nucleotide-gated channels is a region that shows homology with cyclic nucleotide binding sites of cGMP-dependent and cAMP-dependent protein kinases,[11,43] which suggests that this region forms the ligand-binding site of the channel. Consistent with this view, substitutions of single amino acids within this region have been shown to alter the relative sensitivity of photoreceptor and olfactory channels to cGMP and cAMP.[44] However, the relative effectiveness of cGMP and cAMP in cloned olfactory channels expressed in cell lines or oocytes differs from that of native olfactory channels,[30] and thus it is likely that additional factors influence ligand specificity in cyclic nucleotide-gated channels. In this regard, the recent discovery of a new subunit of cGMP-gated channel[21] raises the possibility that native channels may be hetero-oligomers with properties different from the presumed homo-oligomers formed by artificially expressed channels. In addition, ligand sensitivity of cGMP-gated channels can be regulated by calmodulin[31] and by phosphatases.[20]

Another similarity between voltage-gated and cyclic nucleotide-gated channels is in the H5 domain that is thought to form the pore and to regulate the permeation properties of the channel. The olfactory and photoreceptor cyclic nucleotide-gated channels have been shown to differ somewhat in the apparent size of the pore and in permeation properties, and these differences have been demonstrated to be determined by differences in the H5 domains of the two types of channel.[45] Also, lack of discrimination among alkali cations and blockade by divalent cations – which are characteristics common to cyclic nucleotide-gated channels but not to potassium channels – can be produced in chimeric channels in which the H5 region of a normally potassium-specific channel is replace by the H5 region of cGMP-gated channels.[25] Thus, it is likely that the permeation properties of channels in the cyclic nucleotide-gated family are determined by the structure of the H5 portion of the channel.

## Summary

The family of ion channels directly gated by cyclic nucleotides represents a new class of channel with hybrid properties of voltage-gated channels and ligand-gated channels. Although channels of this family have been most thoroughly studied in the primary sensory neurones of the visual and olfactory systems, they will likely be found in any cellular signalling system in which changes in intracellular cyclic nucleotide levels must be coupled rapidly to changes in plasma membrane ionic conductance. Although cyclic nucleotide-gated channels described to date have been non-specific cation channels, ion-specific variants could conceivably arise via simple alterations in the pore-forming region. Because an appreciable portion of the inward ionic current through cyclic nucleotide-gated channels is carried by calcium ions, channels of this type might also offer a mechanism by which an increase in internal cGMP or cAMP could induce an increase in intracellular $Ca^{2+}$.

## References

1. Fesenko, E.E., Kolesnikov, S.S., and Lyubarsky, A.L. (1985). *Nature*, **313**, 310–313.
2. Nakamura, T. and Gold, G.H. (1987). Nature, **325**, 442–444.
3. Matthews, G. (1991). *Trends Pharmacol. Sci.*, **12**, 245–247.
4. Ahmad, I., Redmont, L.J., and Barnstable, C.J. (1990). *Biochem. Biophys. Res. Commun.* **173**, 463–470.
5. Biel, M., Altenhofen, W., Hullin, R., Ludwig, J., Freichel, M., Flockerzi, V., Dascal, N., Kaupp, U.B., and Hofmann, F. (1993). *FEBS Lett.*, **329**, 134–138.
6. Stryer, L. (1986). *Annu. Rev. Neurosci.* **9**, 87–119.

7. Yau, K.W. and Baylor, D.A. (1989). *Annu. Rev. Neurosci.* **12**, 289–327.
8. Baylor, D.A. (1987). *Invest. Ophthal. Vis. Sci.*, **28**, 34–49.
9. Yau, K.-W. (1994). *Invest. Ophthal. Vis. Sci.*, **35**, 9–32.
10. Hanke, W., Cook, N.J., and Kaupp, U.B. (1988). *Proc. Natl Acad. Sci.* USA, **85**, 94–98.
11. Kaupp, U.B., Niidome, T., Tanabe, T., Terada, S., Bönigk, W., Stühmer, W., *et al.* (1989) *Nature*, **342**, 762–766.
12. Matthews, G. (1986). *J. Neurosci.* **6**, 2521–2526.
13. Matthews, G. and Watanabe, S. (1987). *Neurosci. Res.*, **Suppl. 6**, S55–S66.
14. Matthews, G. (1987). *Proc. Natl Acad. Sci. USA*, **84**, 299–302.
15. Matthews, G. and Watanabe, S. (1987). *J. Physiol. (Lond.)* **389**, 691–715.
16. Yau, K.-W. and Nakatani, K. (1985). *Nature*, **317**, 252–255.
17. Matthews, G. and Watanabe, S. (1988). *J. Physiol. (Lond.)*, **403**, 389–405.
18. Watanabe, S. and Matthews, G. (1988). *J. Neurosci.* **8**, 2334–2337.
19. Zimmerman, A.L. and Baylor, D.A. (1986). *Nature*, **321**, 70–72.
20. Gordon, S.E., Brautigan, D.L., and Zimmerman, A.L. (1992). *Neuron*, **9**, 739–748.
21. Chen, T.-Y., Peng, Y.-W., Dhallan, R.S., Ahamed, B., Reed, R.R., and Yau, K.-W. (1993). *Nature*, **362**, 764–767.
22. Watanabe, S. and Matthews, G. (1989). *Neurosci. Res.*, **Suppl. 10**, S1–S8.
23. Watanabe, S. and Matthews, G. (1990). *Visual Neurosci.* **4**, 481–487.
24. Jan, L.Y. and Jan, Y.N. (1992). *Cell*, **69**, 715–718.
25. Heginbotham, L., Abramson, T., and MacKinnon, R. (1992). *Science*, **258**, 1152–1155.
26. Picco, C. and Menini, A. (1993). *J. Physiol. (Lond.)*, **460**, 741–758.
27. Bader, C.R., MacLeish, P.R., and Schwartz, E.A. (1979). *J. Physiol. (Lond.)*, **296**, 1–26.
28. Zimmerman, A.L. and Baylor, D.A. (1992). *J. Physiol. (Lond.)*, **449**, 759–783.
29. Ildefonse, M., Crouzy, S., and Bennett, N. (1992). *J. Membr. Biol.*, **130**, 91–104.
30. Frings, S., Lynch, J.W., and Lindemann, B. (1992). *J. Gen. Physiol.*, **100**, 45–67.
31. Hsu, Y.-T. and Molday, R.S. (1993). *Nature*, **361**, 76–79.
32. Brown, R.L., Bert, R.J., Evans, F.E., and Karpen, J.W. (1993). *Biochemistry*, **32**, 10089–10095.
33. Zimmerman, A.L., Yamanaka, G., Eckstein, F., Baylor, D.A., and Stryer, L. (1985). *Proc. Natl Acad. Sci. USA*, **82**, 8813–8817.
34. Furman, R.E. and Tanaka, J.C. (1989). *Biochemistry*, **28**, 2785–2788.
35. Stern, J.H., Kaupp, U.B., and MacLeish, P.R. (1986). *Proc. Natl Acad. Sci. USA*, **83**, 1163–1167.
36. Koch, K.-W. and Kaupp, U.B. (1985). *J. Biol. Chem.*, **260**, 6788–6800.
37. Nicol, G.D. (1993). *J. Pharmacol. Exp. Ther.*, **265**, 626–632.

38. Nicol, G.D., Schnetkamp, P.P.M., Saimi, Y., Cragoe, E.J., Jr, and Bownds, M.D. (1987). *J. Gen. Physiol.* **90**, 651–669.

39. Cook, N.J., Hanke, W., and Kaupp, U.B. (1987). *Proc. Natl Acad. Sci. USA*, **84**, 585–589.

40. Dhallan, R.S., Yau, K.-W., Schrader, K.A., and Reed, R.R. (1990). *Nature*, **347**, 184–187.

41. Ludwig, J., Margalit, T., Eismann, E., Lancet, D., and Kaupp, U.B. (1990). *FEBS Lett.*, **270**, 24–29.

42. Bönigk, W., Altenhofen, W., Müller, F., Dose, A., Illing, M., Molday, R.S., and Kaupp, U.B. (1993). *Neuron*, **10**, 865–877.

43. Kaupp, U.B. (1991). *Trends Neurosci.*, **14**, 150–157.

44. Altenhofen, W., Ludwig, J., Eismann, E., Kraus, W., Bönigk, W., and Kaupp, U.B. (1991). *Proc. Natl Acad. Sci. USA*, **88**, 9868–9872.

45. Goulding, E.H., Tibbs, G.R., Liu, D., and Siegelbaum, S.A. (1993). *Nature*, **364**, 61–64.

# 9 Receptor-operated Channels

# 18 Excitatory amino acid-activated channels

J. Lerma, A.V. Paternain, N. Salvador, F. Somohano, M. Morales, and M. Casado

## Introduction

Besides the widespread metabolic function, it is now well established that the dicarboxylic amino acid glutamate plays a major role in excitatory synaptic transmission. Glutamate interacts with receptors forming cationic channels with different ion permeabilities, conductances, and activation–deactivation kinetics, as well as demonstrating distinct pharmacological properties. Excitatory amino acid receptors not only mediate normal synaptic transmission, but also participate in a number of complex physiological phenomena, such as plasticity, and ultimately trigger neuronal degeneration and cell death in pathological conditions.[1,2] For these reasons, the study of excitatory amino acid receptors and the development of excitatory amino acid antagonists that would prevent excitotoxic damage has become one of the major areas of basic and clinical neurobiology. Knowledge of the pharmacological and biophysical properties and modulatory mechanisms of glutamate receptors is a prerequisite for the development of drugs directed specifically toward one of the multiple receptor sites in order to influence its activity, thus preventing or ameliorating overstimulation and eventually neurodegeneration.

A breakthrough took place when the cDNA coding for a subunit of the glutamate receptor was isolated by the first time in 1989 by expression cloning.[3] The subsequent isolation of cDNAs coding for different subunits of the ionotropic glutamate receptors demonstrated that they belong to a common gene family (see refs. 4 and 5 for reviews). The expression of recombinant receptors in oocytes or mammalian cells is providing some clues as to how glutamate receptors may function in native membranes, and determining the essential parts of the receptor molecule that are involved in ion selectivity, channel gating, and agonist interaction. Several recent reviews have considered the structure of genes coding for glutamate receptors, brain distribution, and protein sequence homology, as well as the structural and functional properties of cloned receptor subunits.[4,6–11] In this article, we review some of our recent data in this field as a step toward a better understanding of how the different types of glutamate receptors exist in native membranes, and also how multiple binding sites control channel activity.

## Fast perfusion system

In voltage- or patch-clamp conditions, rapid activation of voltage-dependent channels can be easily evoked by steepening the membrane potential. However, rapid activation of ligand-gated channels are greatly hampered by the difficulty of fast change in external solution around the cell. Rapid activation–deactivation properties (e.g. desensitization), as shown by many of the known neurotransmitter receptor channels, may remain unobserved if the solution change is made slowly. Modulation of such a process cannot be studied unless a rapid and uniform application of the agonist takes place. Rapid application of external solution allows not only examination of the rapid desensitization of neurotransmitter receptors, but also an investigation of kinetics concerning the interaction between the drug and its receptor.

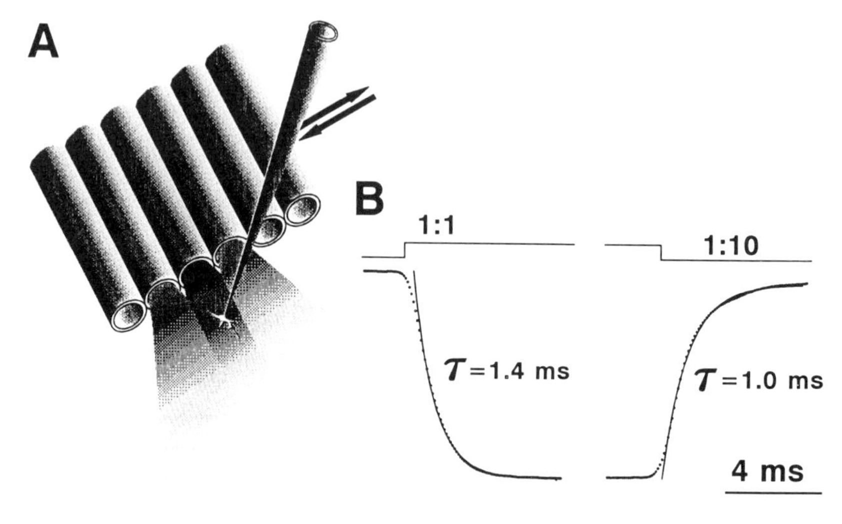


**Fig. 18.1** Perfusion apparatus used for fast solution changes. A. Schematic representation of an array of six glass tubes. Solutions flowing from adjacent barrels are quickly available by rapidly displacing the whole array to position the desired barrel above the cell under recording B. Current relaxations recorded by an open patch pipette during a jump from a barrel containing diluted extracellular solution (Ringer 1 : 10) to a normal-strength solution and back. The variation in tip current is due to the change in junction potential. The solid lines are single exponentials fitted to the current relaxations with the indicated time constants (τ).

Several systems have been developed to replace the external solution quickly around the cells (for detailed information, see ref. 12). In order to make rapid solution changes around neurones under voltage-clamp conditions, we have used a perfusion apparatus similar to that described by Johnson and Ascher.[13] A cartoon showing the main features of this perfusion system is illustrated in Fig. 18.1A. Basically, the apparatus consists of an array of several glass tubes (300–400 μm in outer diameter) which are glued together. The diameter of each barrel is enough to completely bathe a neurone, including its branches. Each tube contains a different solution, with solution changes being made by laterally displacing the whole array, using a motorized device, to position a given barrel above the cell. Solution is allowed to flow only from the barrel containing Ringer solution and the barrel to which the jump is to be made. To test the speed of solution changes with this system, the tip potential of an open patch pipette was measured while jumping from a barrel containing diluted Ringer solution to another containing normal strength solution. Tip currents relaxed and recovered consistently following single exponentials with time constants around 1 ms (Fig. 18.1B). The motorized device may be under the control of a personal computer or a pulse generator. Thus, reproducible solution changes suitable for response average could be obtained with this system. Since several solutions are available quickly, pharmacological studies and dose–response curves can be made (e.g. see Figs. 18.3 and 18.5).

## The NMDA receptor channel complex

The *N*-methyl-D-aspartate (NMDA) subtype of glutamate receptors is a ligand-gated ion channel which appears to be involved in complex physiological phenomena in the central nervous system (CNS). In addition to its involvement in pathological processes, activation of the NMDA receptor–channel complex is required at many CNS synapses for the generation of long-term potentiation, a model of learning at the cellular level.[14] NMDA receptor activity has also been implicated in developmental cell structuring.[15,16] The high $Ca^{2+}$ permeability of NMDA receptors underlies their role in all these phenomena, the increase in intracellular $Ca^{2+}$ being thought to trigger the biochemical processes responsible not only for plasticity, but also for excitotoxic cell death.

The NMDA receptor channel complex has a number of regulatory sites that are targets for modulation by endogenous as well as exogenous compounds. The regulatory sites (Fig. 18.2) include a binding site for the endogenous agonist glutamate (and the synthetic NMDA), a high-affinity binding site at which glycine acts to allow agonist-induced channel opening,[13,17,18] which is different from the strychnine-sensitive, chloride-permeable, glycine receptor; and sites within the channel lumen where $Mg^{2+}$[19,20] and phencyclidines (PCP, MK-801, TCP, ketamine) bind[21–23] to produce a voltage-dependent open channel block. There is an additional site or sites where $Zn^{2+}$ acts to inhibit allosterically the agonist-induced response in a voltage-independent manner.[24–26] The endogenous polyamines, spermine and spermidine, potentiate allosterically the response to NMDA acting on an additional site.[27–29] Moreover, it has been described that NMDA receptors are sensitive to high extracellular $H^+$ concentration[30,31] and to the redox state.[32]

Activation of NMDA receptors by glutamate produces a response with a similar shape to current induced by NMDA. Hill analysis reveals a value of about 3 $\mu$M as the concentration for half-maximal activation ($EC_{50}$) of NMDA receptors by glutamate (in the presence of glycine) (J. Lerma, unpublished observation; also ref. 33). The Hill coefficient is $>1$, suggesting the existence of at least two agonist binding sites on the receptor. The specific synthetic agonist, NMDA, activates the receptor less potently, with an $EC_{50}$ of 20–30 $\mu$M and a Hill coefficient of $>1$.[33–35]

### The glycine (coagonist) site

The action of glycine on NMDA receptors is dual: it reduces desensitization and is required for NMDA or glutamate to induce channel opening (see Fig. 18.2).[17,18,19] The glycine

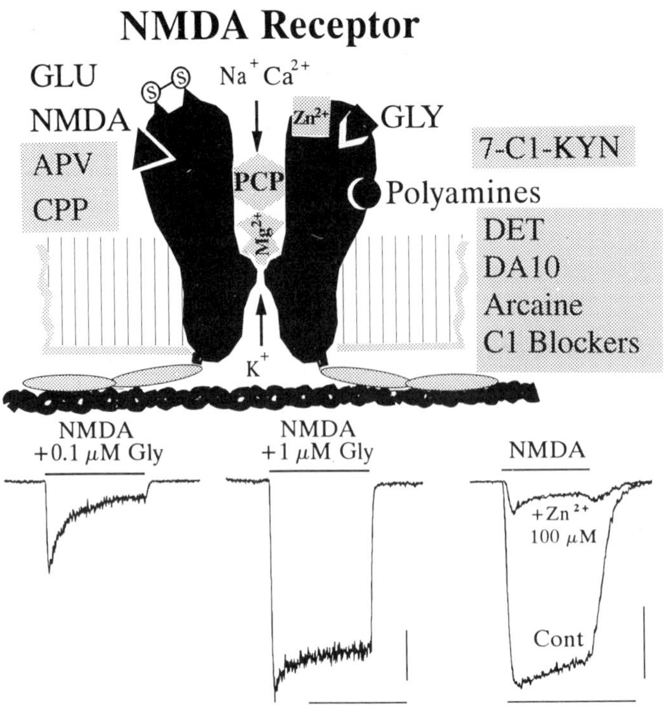

**Fig. 18.2** Sites of modulation of the NMDA receptor channel complex. At least six sites of modulation can be distinguished on the NMDA receptor molecule: the agonist site, where glutamate and NMDA bind to operate the channel; the glycine site, where glycine acts to allow channel opening and to decrease desensitization; the polyamine site, where spermine and spermidine act to regulate desensitization and glycine affinity; the channel, where $Mg^{2+}$ and phencyclidines block permeation in a voltage-dependent manner. In addition, there is a site (S-S) which confers receptor sensitivity to redox agents. Antagonists at each site are indicated within shaded boxes. Records on the bottom show the inward current recorded under voltage-clamp conditions upon a jump to NMDA with a low and a high concentration of glycine (left records, $V_m$=−60 mV). Note that glycine potentiates the response and largely decreases desensitization. The right record shows the blocking effect of $Zn^{2+}$ on the response induced by 100 μM NMDA + 10 μM glycine ($V_m$=−60 mV). Calibrations: 1 s and 200 pA and 1000 pA for the two left and right records, respectively.

potentiation of NMDA responses was first demonstrated by Johnson and Ascher[13] in cultured cortical cells. An interesting aspect of the glycine modulation is that the dose–effect curve for glycine potentiation of NMDA current shows no steady response at no glycine added, as reported by ourselves.[37] The proposal that glycine is required for steady activation of NMDA receptors, however, was stated more emphatically by Kleckner and Dingledine.[17] Nevertheless, some response to NMDA is normally detected with zero glycine added. Contaminating glycine is extremely difficult to avoid and we and others have reported background levels of glycine of 40–50 nM, as determined by HPLC. Thus, with the lowest possible background contamination of glycine, some response to NMDA ought to be been seen. The NMDA-induced current in the absence of added glycine is completely blocked by a low concentration of 7-chloro-kynurenic acid, a competitive antagonist of the glycine binding site.[18,38] Thus, the normal contamination level of glycine accounts for the small response obtained in the absence of added glycine, and this amino acid should be considered as a required coagonist of glutamate at the NMDA receptor complex. The physiological role of glycine still remains unclear since the high affinity of its binding site ($EC_{50} \approx 300$ nM)[17,18,36,37] implies that it has to be saturated at the resting extracellular levels of glycine in hippocampus (3 $\mu$M, as determined *in vivo* by microdialysis: J. Lerma and R. Martín del Río, unpublished results; also see ref. 39). One hypothesis is that there is a physiological antagonist or partial agonist. Whether this compound exists and what kind of regulation it undergoes are important and, as yet, unanswered questions.

Selective antagonists for the glycine site include the halogenated derivatives of kynurenate. Although the first identified antagonist, 7-chlorokynurenate, was moderately potent (0.5 $\mu$M of $IC_{50}$ in [$^3$H]glycine binding experiments; see ref. 38), 5-iodo, 7-chlorokynurenate showed a potency about an order of magnitude higher ($IC_{50}$ of 40 nM). The 4-thio analogues of these compounds have more recently been reported to be slightly more potent. Thus, thiokynurenate, 7-chlorothiokynurenate and 5,7-dichlorothiokynurenate are now commercially available (see the Tocris Neuramin catalogue).

## The polyamine site

It has been shown that the endogenous polyamines, spermine and spermidine, are able to enhance binding of the phencyclidine receptor ligands [$^3$H]MK-801 and [$^3$H]TCP to brain membranes.[29] Polyamines stimulated [$^3$H]MK-801 binding, even in the presence of maximally effective concentrations of glutamate and glycine,[29,40] suggesting that the effect of polyamines was mediated by an additional and unknown recognition site in the NMDA receptor–channel complex. Polyamines are ubiquitous compounds, present at high concentrations in the brain and other tissues,[41,42] where they may regulate cell growth and differentiation.[43] Although their concentration in interstitial fluid and their effect on the NMDA receptor were unknown, it was suggested some years ago that polyamines may influence neuronal activity.[44] An additional interest in the polyamine site comes from experiments showing that several compounds with protective action against excitotoxicity appear to be acting as antagonists at this site.[45,46] For instance, ifenprodil has been shown to be neuroprotective in models of focal ischaemia and neurotoxicity,[46] acting at the polyamine site, although its selectivity still is a matter of controversy. Thus, the polyamine receptor may be a potential target for drugs intended to prevent excitotoxic cell death.

The action of spermine on functional responses to NMDA (plus 1 $\mu$M glycine) in cultured spinal cord neurones under voltage-clamp and concentration-clamp conditions is shown in Fig. 18.3. Spermine potentiates both phases of the response in a dose-dependent manner. Hill analysis revealed a similar $EC_{50}$ for the steady state and peak currents (71.5 $\mu$M and 80.9 $\mu$M respectively). Hill coefficients were 1.69 for the steady state and 1.63 for the peak, suggesting the presence of multiple binding sites. These values are comparable with those reported from electrophysiological experiments with NMDA receptors expressed in oocytes[28,47] and more recently in hippocampal neurones,[48] but are slightly higher than the concentrations needed to produce a half-maximal increase in the binding of [$^3$H]MK-801.[49] Despite the lack of different apparent affinities for the polyamine, spermine potentiates the steady-state response more than the peak current, i.e. it reduces desensitization. The effect of spermine on desensitization can be clearly seen in Fig. 18.3A, where average responses to NMDA with and without added spermine have been superimposed. With 100 $\mu$M NMDA and 1 $\mu$M glycine, the time course of desensitization could be well described by a single exponential decay with a time constant of 234±13 ms. With 500 μM spermine, the time constant of desensitization onset ($\tau_{on}$) was 2.9-fold that of the control. Thus, it is apparent that spermine not only decreases the extent to which the NMDA-induced responses desensitize, but also slows the onset of desensitization. On the other hand, twin-pulse applications of NMDA in the absence and in the presence of several concentrations of spermine (Fig. 18.3B,C) revealed that the recovery rate from the desensitized state is not altered by spermine. In control conditions, the recovery process was fitted to a single exponential with a time constant ($\tau_{rec}$) of 462 ms, corresponding to a recovery rate of $2.16 \pm 0.11\ s^{-1}$. The $\tau_{rec}$ was minimally affected by spermine.

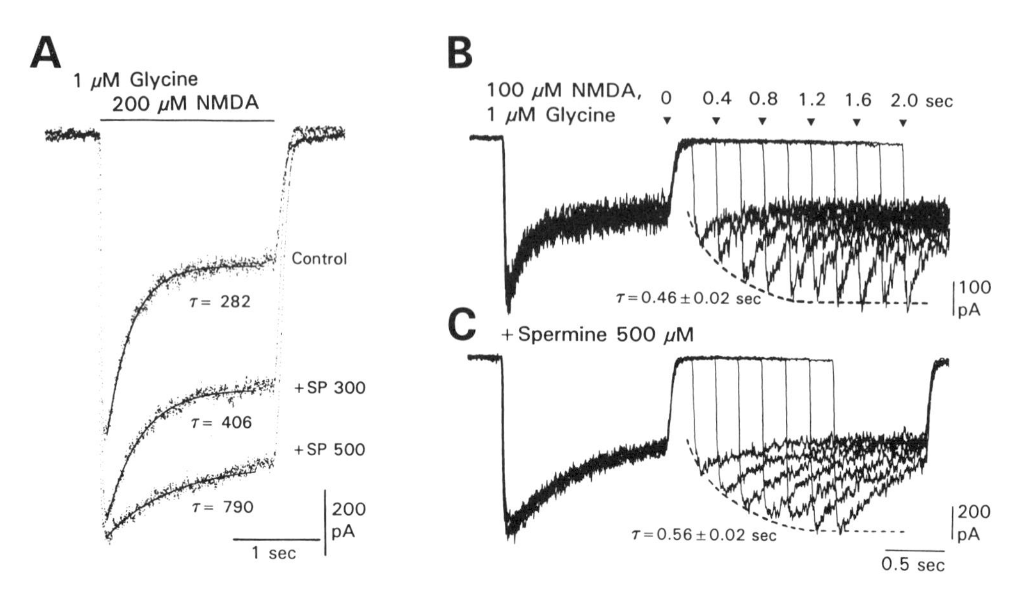


**Fig. 18.3** Spermine decreases desensitization of NMDA-induced currents and slows its onset rate while it has no effect on recovery from the desensitized state. A. Averaged inward currents (*n*=3) obtained with and without spermine (SP) are superimposed. Lines drawn through the data points are single exponential functions fitted to the onset of desensitization with the indicated time constants (τ). The cell was exposed to spermine for 0.5 s before the NMDA was applied. B and C. Spermine does not affect the rate of recovery from desensitization. The interval between first (conditioning) and second (test) pulses of NMDA was varied to measure the speed of recovery from desensitization. Trials were separated by 20 s to allow complete recovery from desensitization and are shown superimposed when no spermine was added (b) and when spermine (500 μM) was continuously present (c). The illustrated cases correspond to different cells and were chosen to facilitate comparison since they showed similar degrees of desensitization. (Modified from Lerma[27].)

The potentiating action of spermine on NMDA-induced currents can also be seen in internally washed cells,[27] a situation in which desensitization becomes independent of glycine.[50] Furthermore, in our previous study,[27] some cells showed almost complete desensitization, despite the fact that glycine was present. This desensitization should, by definition, be glycine-independent. However, spermine greatly increased steady-state currents and slowed desensitization onset in those cells. Thus, the modulation of NMDA receptor desensitization by spermine and glycine are mechanistically separated, although both mechanisms could cooperate to greatly reduce NMDA receptor desensitization. In fact, at subsaturating concentration of glycine, polyamines increase glycine affinity.[28,48]

Although it remains to be demonstrated, it is possible that, as in cultures, polyamines modulate *in vivo* the susceptibility of NMDA receptors to desensitize, in this respect cooperating with glycine. Some compounds with protective activity against neurotoxicity and cell death have been suggested to act as antagonists at the polyamine binding site.[45,46] Thus, the polyamine recognition site of the NMDA receptor may constitute an additional therapeutic target for the prevention of neurotoxicity-associated cell loss. A number of compounds acting as polyamine antagonists have been described (see Fig. 18.2), and some have been shown to act as inverse agonists, while many others may have multiple actions on the NMDA receptor (see below). The pharmacological characteristics of this allosteric site have not been examined in detail with electrophysiological approaches (but see ref. 51). Thus, the activity profile of polyamine analogues and other active drugs at the polyamine site is not well understood.

## Sites of antagonism of the NMDA receptor

Activation of NMDA receptors can be prevented by a number of endogenous and exogenous compounds, with different properties. D-(−)-2-amino-5-phosphonovalerate (D-APV), 3-((R)-2-carboxypeperazin-4-yl)-propyl-1-phosphonate (D-CPP), $Zn^{2+}$, and $Mg^{2+}$ all antagonize NMDA-induced currents.[19,20,25,26,37] While the inhibition by APV and CPP show the characteristic competitive profile, the endogenous divalent cations $Zn^{2+}$ and $Mg^{2+}$ act non-competitively. Whole-cell experiments and single-channel recordings have indicated that, unlike $Zn^{2+}$, the binding site for $Mg^{2+}$ is located deep within the membrane-spanning domain of the NMDA receptor channel[52]. The importance of the observation that physiological concentrations of $Mg^{2+}$ (1–2 mM; $IC_{50} \approx 5$ $\mu$M, at $-60$ mV of membrane potential) inhibit NMDA-induced currents is nowadays clear. Because $Mg^{2+}$ selectively blocks NMDA receptor activity at hyperpolarized potentials (Fig. 18.4),[19,20] this ion gates the NMDA channel and gives to the NMDA receptor the unique characteristic among the ligand-operated channels – its voltage dependence.

NMDA receptors are involved in the generation of long-term potentiation (LTP),[14] a model of learning at the cellular level. Properties of LTP support the hypothesis, proposed long ago,[53,54] that alterations of synaptic efficacy underly information storage in the brain. LTP is an activity-dependent synaptic plasticity which consists of an enhancement of synaptic strength, due to usage, that lasts at least several weeks in intact animals.[55] The voltage-dependent block of NMDA channels by $Mg^{2+}$ provides the foundation of molecular understanding for such an operation and explains why, for LTP to occur under physiological conditions, the post-synaptic membrane must be depolarized to get sufficient $Ca^{2+}$ influx through the NMDA-activated channels. The combined biophysical properties and the localization on dendritic spines of NMDA receptors, together, explain the main characteristics of LTP – associativity and specificity.

Like $Mg^{2+}$, phencyclidine (PCP) and related drugs also block NMDA receptors at a site within the channel. There is enough evidence indicating that this site is responsible for psychotomimetic and neuroprotective actions of PCP and PCP-like drugs. Among others, our studies have shown that in *Xenopus* oocytes injected with rat brain mRNA, block of NMDA-induced current by PCP site ligands (e.g. TCP, PCP, SKF-10,047, MK-801) requires

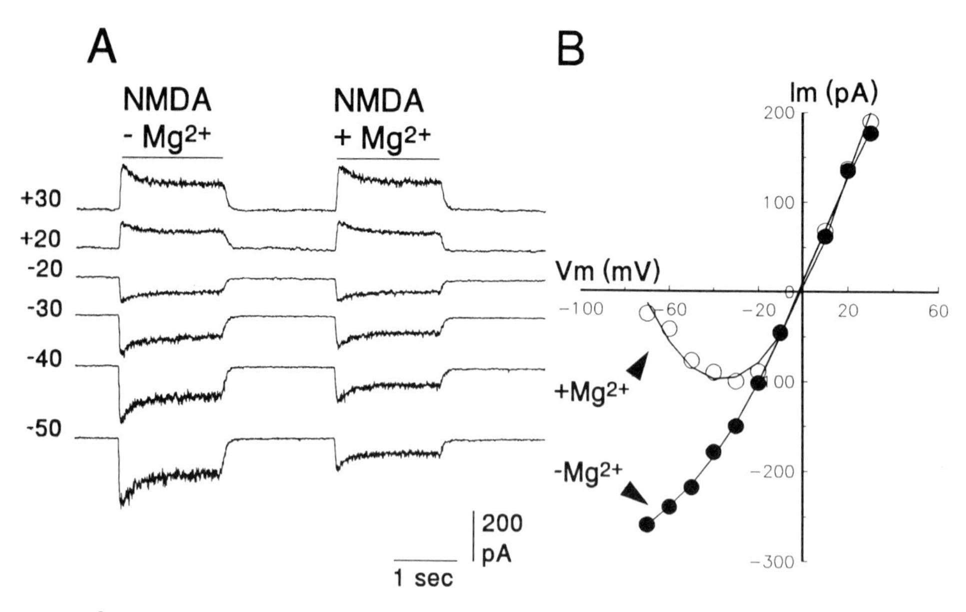


**Fig. 18.4** $Mg^{2+}$ inhibits the NMDA-induced current in a voltage-dependent manner. A. In a cultured hippocampal cell, the membrane potential was varied from −80 mV to +30 mV and NMDA (100 $\mu$M) alternatively applied with and without $Mg^{2+}$ (200 $\mu$M). B. Current–voltage relationship for NMDA-induced currents with and without $Mg^{2+}$. Note the presence of a region of negative slope in the conductance when $Mg^{2+}$ was present.

the presence of agonist (i.e. block is use-dependent).[22,37] PCP can be trapped inside the channel when agonist is removed and, consequently, recovery of the PCP block also is use-dependent, i.e. PCP leaves the channel only when it is open. Lerma *et al.* have used 'trapping' to investigate the interaction of PCP and $Mg^{2+}$ in blocking the NMDA channel.[22] They found that $Mg^{2+}$ partially prevented PCP block and shifted the PCP dose–inhibition curve to the right, without affecting the maximal effect, indicating that $Mg^{2+}$ competitively reduced PCP block. This observation suggests that $Mg^{2+}$ tends to prevent PCP from reaching its binding site in the channel and is consistent with the hypothesis that only one or the other blocker can occupy the channel at a time. An interesting possibility raised by these results is that the binding sites for $Mg^{2+}$ and PCP overlap to some extent. Recently, it has been shown by site-directed mutagenesis, that replacement by glutamine of an asparagine in the region that most probably forms the NMDA channel wall strongly reduces the sensitivity of recombinant NMDA receptors to block by both $Mg^{2+}$ and MK-801.[56] Although it would be difficult to demonstrate that PCP within the channel prevents $Mg^{2+}$ from entering it, these data strongly indicate that the *in vivo* action of PCP and PCP ligands will be affected by $Mg^{2+}$ levels in the synaptic cleft. Such an interaction may be more complex, as depolarization reduces $Mg^{2+}$ occupancy of the channel more than it reduces PCP occupancy,[57] and therefore $Mg^{2+}$ less effectively reduces PCP block at depolarized levels.

The mammalian brain contains large amounts of $Zn^{2+}$, much of it localized in synaptic vesicles of excitatory terminals. $Zn^{2+}$ is a very effective NMDA antagonist ($IC_{50}$ = 11 $\mu$M) showing in its action only a weak voltage dependence.[25,26] The primary effect of $Zn^{2+}$ on NMDA-activated channels is a reduction in both the frequency and the mean lifetime of channel openings.[24,58] This mechanism is sufficient to account for most of the inhibition of NMDA-induced currents. Thus, $Zn^{2+}$ should be considered an allosteric inhibitor of NMDA receptors. A second, less potent, inhibitory effect of $Zn^{2+}$ on NMDA channels is voltage-dependent,[24,58] suggesting that $Zn^{2+}$ may act at a site in the NMDA receptor channel. However, this effect should be small under physiological conditions, since it is likely to be obscured by the interaction of $Ca^{2+}$ and $Mg^{2+}$ with the NMDA channel[58] There is some evidence that $Zn^{2+}$ may be synaptically released upon depolarization[59] in hippocampal slices. However, the physiological role of $Zn^{2+}$ remains unclear.

In cultured spinal cord neurones we found that blockers of chloride transport prevented NMDA receptor activation in a dose-dependent manner and are specific for this class of glutamate receptor (Fig. 18.5). Furosemide is a widely used loop diuretic, while piretanide and bumetanide are related compounds. Niflumic and flufenamic acids are used as antiinflamatory agents. Antagonism of NMDA-mediated currents by chloride transport blockers is voltage-independent and shows fast on–off kinetics. The action was non-competitive with NMDA and did not arise from interaction with the $Zn^{2+}$ inhibitory site, since blockade of NMDA-induced responses by furosemide and $Zn^{2+}$ was additive. In contrast, the inhibition was attenuated by the polyamine spermine. Since the presence of spermine was not required for inhibition to develop, we concluded[60] that chloride transport blockers are non-competitive antagonists of the NMDA receptor, likely acting as inverse agonists at the polyamine site. However, another interesting possibility is that inhibitors of chloride transport are agonists of a second (inhibitory) polyamine recognition site associated with the NMDA receptor complex.[51] This action may explain the protective effect that has been shown for some of these drugs in neuronal degeneration and, since they also prevent neuronal swelling, they may be good starting compounds for synthesis of appropriate therapeutic agents to ameliorate excitotoxicity.

## Non-NMDA receptors

The so-called non-NMDA ionotropic receptors include AMPA receptors, which were initially characterized by rapid activation kinetics and fast desensitization, and kainate receptors which were differentiated from AMPA receptors in that they were non-desensitizing.[61] Despite the lack of specific pharmacology, responses to AMPA and kainate were postulated to occur at different molecular entities. Some evidence from electrophysiological experiments indicated that both responses could be due to the interaction of AMPA and kainate with the same receptor.[62] However, it has not been until the cloning of the AMPA receptor subunits that conclusive evidence was reported showing that both agonists activate the same molecular complex.[63,64] Thus, molecular biology studies have recently provided new insight into the functional and structural diversity of the non-NMDA type of glutamate receptor channels.[5] These studies have indicated that all glutamate receptor channels may be assembled from structurally homologous subunits, which can be grouped into subfamilies according to sequence characteristics. On the basis of ligand affinity, the GluR-A to D (GluR-1 to 4) subunits are constituents of AMPA receptor channels,[64–67] while the GluR-5, GluR-6, GluR-7 and KA subunits form recombinant receptors with high affinity for kainate.[68–70] However, as mentioned, kainate gates AMPA receptors and AMPA can

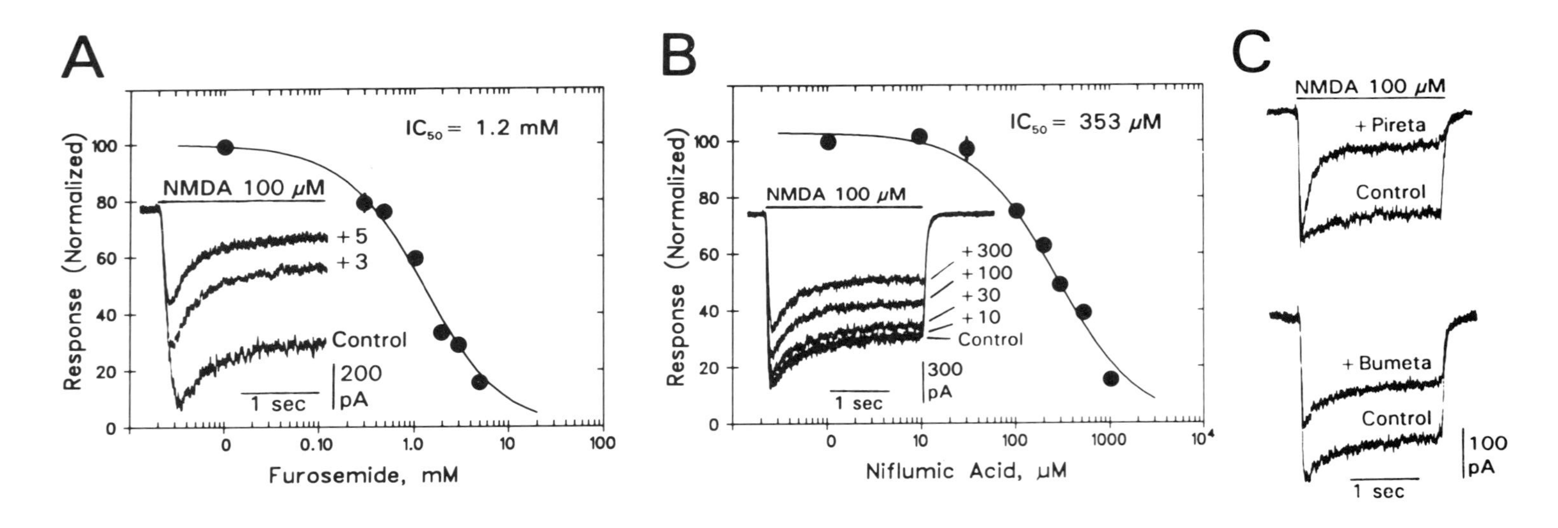


**Fig. 18.5** Inhibitors of chloride transport blocked the current induced by perfusion of NMDA in cultured spinal cord neurones. Complete dose–inhibition curves were constructed for furosemide (A) and niflumic acid (B). Points represent the mean ± S.E.M. of five and six neurones in A and B, respectively and were fitted to the Hill equation (solid curves). Insets show superimposed averaged ($n$=3) records. C. Inhibition of NMDA-induced currents by coapplication of piretanide (5 mM) and bumetanide (2 mM). In all cases, 10 μM glycine was present and the membrane potential clamped at −60 mV. (From Lerma and Martín del Río[60].)

activate receptors formed by particular combinations of the high-affinity kainate receptor subunits.[5] Furthermore, it is now clear that whereas AMPA, quisqualate, and glutamate activate AMPA receptors in CNS neurones, inducing responses which exhibit rapid and strong desensitization, domoate and kainate produce sustained currents by opening the same receptor channels (Fig. 18.6). All these responses are competitively antagonized by the quinoxalinediones CNQX, DNQX, and BNQX.[71] Given the lack of specific agonists and/or antagonists, the functional distinction between AMPA and kainate receptors in the brain is difficult at the present time (see below).

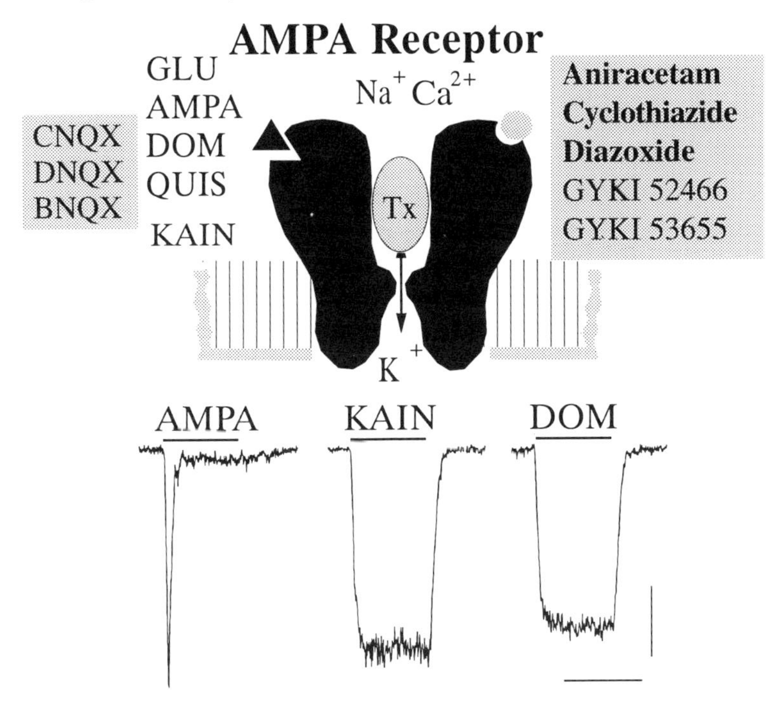


**Fig. 18.6** Sites of modulation of the AMPA receptor. Glutamate, AMPA and quisqualate all activate and inactivate the AMPA receptor as shown in the bottom left record. Kainate and domoate activate, but do not inactivate, the AMPA receptor (bottom middle and right records). The quinoxalinediones, CNQX, DNQX, and BNQX, antagonize competitively this receptor. There is an allosteric modulatory site (the bezodiazepine site), where aniracetam, cyclothiazide, and diazoxide act regulating desensitization, while other compounds, like GYKI 52466 and GYKI 53655, prevent activation of the AMPA receptor. Calibrations: 0.5 s and 50 pA.

Patch-clamp recordings from CNS neurones revealed that native non-NMDA receptors are similar to the AMPA-type recombinant glutamate receptors expressed from cDNA clones,[72,73] but failed so far to detect receptor channels of the kainate type. Kainate-induced responses with properties similar to recombinantly expressed kainate receptors have remained elusive for a long time. Functional receptors with higher affinity for kainate than previously observed were first described in peripheral neurones some time ago.[74] However, rapidly desensitizing responses to kainate have only very recently been demonstrated in hippocampal neurones[75] and glia,[76] indicating that functional receptors of the kainate type are also expressed by native brain cells.

**Table 18.1** Compounds modulating glutamate receptors channels.

| Receptor (site) | Substance(s) | Effect | Reference |
|---|---|---|---|
| NMDA (agonist) | Glutamate, aspartate | Endogenous agonist | |
| | NMDA | Specific agonist | |
| | APV | Competitive antagonists | 112 |
| | CPP | | 113 |
| NMDA (glycine, coagonist) | Glycine | Endogenous agonist | |
| | HA-966 | Partial agonist | 114 |
| | 7-Cl-KYN | Competitive on glycine | 38 |
| | 5, 7-DiCl-KYN | Non-competitive on NMDA | |
| | Thiokynurenines | Antagonists | |
| NMDA (Channel) | $Mg^{2+}$ | Endogenous open channel blocker | 52 |
| | Phencyclidine | Open channel blockers | 22 |
| | MK-801 (dizolcipine) | | |
| | Ketamine | | 23 |
| NMDA (Polyamine) | Spermine | Potentiation (slows desensitization) | 27 |
| | Spermidine | | |
| | DET | Antagonists/ | 49 |
| | DEA-10 | Inverse agonists | |
| | Arcaine | | 115 |
| | Ifenprodil | | 45 |
| | $Cl^{-}$ Transport blockers | | 60 |
| AMPA (Agonist) | Glutamate | Endogenous agonist | |
| | AMPA | Desensitizing agonist | 116 |
| | Kainate, domoate | Non-desensitizing agonists | |
| | CNQX | Competitive antagonists | 71 |
| | DNQX | | |
| | BNQX | | |
| AMPA (Benzodiazepine) | Diazoxide | Potentiation (decrease | 93 |
| | Cyclothiazide | desensitization) | 94 |
| | GYKI 52466 | Antagonists | 98, 99 |
| | GYKI 53655 | (Inverse agonists?) | 99 |
| Kainate (Agonist) | Glutamate, kainate | Desensitizing agonists | 75 |
| | Domoate | | |
| | NS-102 | Competitive antagonist | 105, 105a |

## AMPA receptors

Pharmacological studies indicate that AMPA receptors are responsible for most of the synaptic responses in many neuronal pathways. It appears, however, that fast synaptic transmission at excitatory synapsis is not mediated by a single receptor class, as molecular biology studies have demonstrated that AMPA receptors of different properties may be obtained by different combinations of the four AMPA subunits (GluR-A-D).

Expression of AMPA subunits has demonstrated that $Ca^{2+}$ can permeate through either homomeric or heteromeric AMPA receptors whenever the assembled receptor lacks edited GluR-B subunits.[77,78] Another feature that can be observed in the absence of the edited GluR-B subunit is a marked inward rectification of AMPA receptor-mediated currents, starting near a membrane potential of 0 mV. In contrast, homo-oligomeric GluR-B channels or heteromeric assemblies containing this subunit exhibit slight outward rectifying I–V relationships.[79] These results indicate that the presence of edited GluR-B determines the I–V behaviour of the whole assembly as well as the $Ca^{2+}$ permeability of the formed channel. It has been demonstrated that a single amino acid residue in the second transmembrane domain of the GluR-B subunit governs electrophysiological properties and also $Ca^{2+}$ permeability of formed channels.[79,80] In agreement with these findings, it has recently been reported that AMPA receptors in Bergmann glia, which do not express the GluR-B subunit, are permeable to $Ca^{2+}$ and also exhibit inwardly rectifying currents.[81,82]

### $Ca^{2+}$ permeability of AMPA receptor channels

Most reports have described linear or outwardly rectifying I–V relationships for AMPA receptors,[61] with negligible $Ca^{2+}$ permeability.[83] Indeed, for a long time, glutamate-triggered $Ca^{2+}$ increase was attributed to the opening of NMDA receptor channels, while non-NMDA-mediated $Ca^{2+}$ increase was considered a secondary event, caused by opening of voltage-gated $Ca^{2+}$ channels following either excitatory amino acid-induced depolarization or the activation of metabotropic glutamate receptors. Recently, some $Ca^{2+}$ influx through AMPA receptors has been detected in various nerve cell types.[81,82,84–87] We have evaluated $Ca^{2+}$ permeability in hippocampal cells in culture after completion of the I–V curve in $Na^+$ solution to determine the rectification properties of the native AMPA receptors. $Ca^{2+}$ permeability was evaluated in 10 mM $Ca^{2+}$, rather than completely substituting $Ca^{2+}$ for $Na^+$, to avoid a drastic change in negative surface potential,[52] which could cause greater deviation in the reversal potential than expected according to the constant-field equation. The impermeant cation *N*-methyl-D-glucamine substituted for $Na^+$ (see ref. 88 for complete details). Most cultured hippocampal neurones (52.2%) showed little or no inward current in $Ca^{2+}$ solution, at negative potentials up to −90 mV, in response to kainate or AMPA. These cells were characterized by an outwardly rectifying I–V relationship (Fig. 18.7A). Considering these cells as a whole population, the average ratio of permeability divalent to monovalent cations ($P_{Ca}/P_{Cs}$) was 0.10.

A clear $Ca^{2+}$ permeability was detected in 6.5% of cells, in which the reversal potential of the kainate-induced current was −42 mV with $Ca^{2+}$ as the only carrier of inward current. This equilibrium potential corresponded to an average $P_{Ca}/P_{Cs}$ of 0.9, ranging from 0.36 to

1.57. The reversal potential for kainate-induced current with $Na^+$ outside and $Cs^+$ inside the pipette was $-0.25$ mV in this group of cells, indicating a $P_{Na}/P_{Cs}$ of 0.78. The $P_{Ca}/P_{Na}$ was consequently 1.15, suggesting that these cells were equally permeable to $Na^+$ and $Ca^{2+}$ through AMPA receptor channels. An example of results obtained in this type of cells is shown in Fig. 18.7B.

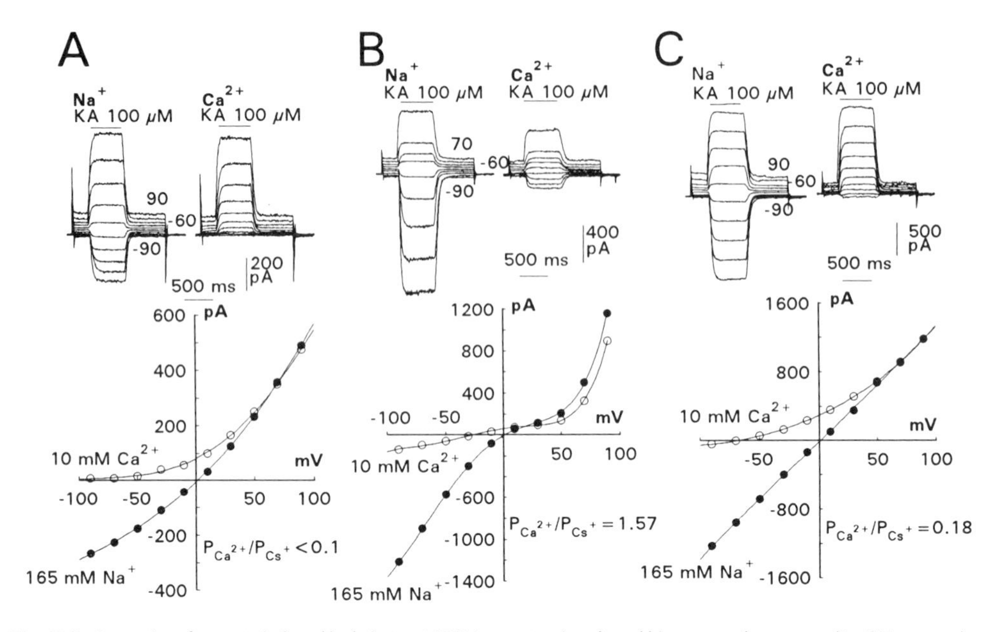


**Fig. 18.7** Properties of currents induced by kainate at AMPA receptors in cultured hippocampal neurones. (A–C) Top records, from a resting potential of $-60$ mV, the membrane potential was varied from $-90$ to $+90$ mV in 20-mV steps. A 0.6-s pulse of a known concentration of agonist (horizontal bar above the current records) was applied at each potential in a normal solution ($Na^+$) and in a solution containing no $Na^+$ but 10 mM $Ca^{2+}$ as the only charge carrier ($Ca^{2+}$). The I–V relationships were calculated and represented after leak substraction (bottom plots). The solid lines represent the 2nd or 4th order polynomials fitted to current amplitudes. Responses induced by kainate at AMPA receptors could be classified in hippocampal cells as outward rectifying and non-$Ca^{2+}$-permeable (A), inward rectifying and $Ca^{2+}$-permeable (B), and cells with linear and low, but not negligible, permeability to $Ca^{2+}$ (C). The ratio $P_{Ca}/P_{Cs}$ is indicated for each representative example. (From Lerma *et al.*[88].)

Some 41% of the cells had an almost linear I–V relationship (Fig. 18.7C) and a significant permeability to $Ca^{2+}$. The ratio of permeability $Ca^{2+}$ to $Cs^+$ ranged from 0.01 to 0.62 (average 0.18). This result indicates that some cells may show a substantial permeability to $Ca^{2+}$ whereas the I–V relationship does not indicate inward rectification. However, $Ca^{2+}$ permeability and degree of rectification are inversely correlated (see ref.88). Whole-cell $Ca^{2+}$ permeability was still lower for AMPA than for NMDA responses. Experiments performed with NMDA (100 μM) indicated a $P_{Ca}/P_{Cs}$ ratio of 5.8

#### The benzodiazepine site

In contrast to the NMDA receptor, where a number of modulatory sites have been found, AMPA receptors apparently lack such a complex pharmacology. However, in 1990 it was described that a nootropic drug, aniracetam (1-p-anisol-pyrrolidione), potentiated responses to quisqualate through AMPA receptors expressed in *Xenopus* oocytes after injection of whole-brain mRNA.[89] Similarly, this drug enhanced population EPSP responses in hippocampal slices. This result provided the first evidence for the existence of an allosteric modulatory site on the AMPA receptor molecule. It was shown that aniracetam enhances glutamate-evoked currents in whole-cell recordings and slows desensitization kinetics of AMPA-induced responses.[90–92] Subsequent investigation demonstrated that a family of compounds, the benzothiazides, are able to reduce desensitization of AMPA receptors. Although diazoxide, a clinically useful antihypertensive drug, is an order of magnitude more potent than aniracetam, cyclothiazide is the most potent member of this family for reducing fast desensitization of AMPA receptors and enhancing post-synaptic currents.[93–96]

Further evidence supporting the concept that the AMPA receptor channel protein contains a recognition site where receptor activity may be regulated, comes from recent experiments using novel compounds with non-competitive antagonistic properties of AMPA receptors. Tarnawa and colleagues[97] described a 2, 3-benzodiazepine, GYKI 52466, which non-competitively blocks the ion current induced by activation of AMPA receptors.[98,99] Although there is some evidence indicating a competitive interaction between cyclothiazide and GYKI 52466, whether or not all these drugs act on the same allosteric site or the AMPA receptor contain multiple binding loci for all these drugs, remains to be elucidated. Nevertheless, the non-competitive antagonism of AMPA receptor activity could offer some advantages over competitive blockers in studying receptor function. Similarly, the availability of a whole family of compounds affecting fast desensitization of AMPA receptors may help to clarify the mechanisms underlying desensitization and the physiological role played by this phenomenon.

Several toxins isolated from the venom of insects interact with glutamate receptors. As for different voltage-gated channels, these toxins may represent useful tools to analyse pharmacologically different types of glutamate receptors in the CNS.[100] Argiotoxin, initially isolated from the spider *Argiope lobata*[101] and the Joro toxin, isolated from *Nephila clavata*[102] block glutamate receptor channels in a use-dependent manner with different potencies. Interestingly, it has been found that sensitivity of AMPA receptors to blockade by these toxins correlates with the charge and size of the same amino acid that also controls $Ca^{2+}$ permeability: those formations lacking the GluR-B subunits in the structure have high-affinity binding for these toxins, whereas the presence of arginines (i.e. GluR-Bs), which determine low $Ca^{2+}$ permeability, confers insensitivity to toxin block.[103,104] The possibility to block selectively those $Ca^{2+}$-permeable AMPA receptors may be useful for treatment of degenerative diseases.

## Kainate-selective receptors

Probably, the most important characteristic of the kainate-selective receptors is that AMPA (up to 500 $\mu$M) fails to activate them, whereas a fast desensitizing inward current can be observed upon rapid kainate perfusion (Fig. 18.8). Glutamate and quisqualate also activate kainate receptors, inducing responses with similar shapes. Responses induced by glutamate and quisqualate are cross-desensitized by the previous application of kainate, and kainate induces no response after glutamate or quisqualate perfusion, indicating that all these agonists are able to activate and totally desensitize the same population of receptor channels. In contrast, responses to the potent neurotoxin domoic acid desensitizes only partially, and deactivate very slowly upon domoate wash-out (single exponential decay time constant 715 ms). Responses to domate are also cross-desensitized by the previous application of kainate. The onset of desensitization of kainate-activated currents in most cases follows a single exponential decay with a time constant of about 20 ms (300 $\mu$M kainate). Full recovery of kainate-induced peak current takes about 60 s, indicating that kainate-selective receptors recover from desensitization considerably more slowly than NMDA or AMPA receptors. Concentrations of kainate much lower than 1 $\mu$M do not induce any noticeable current, half-maximal activation being achieved with about 20 $\mu$M. A similar analysis in cells showing exclusively non-desensitizing responses to kainate (i.e. expressing only receptors of the AMPA type) revealed a lower sensitivity to kainate ($EC_{50} = 240$ $\mu$M). Hill coefficients were in both cases close to unity (1.1 and 0.9, respectively). Keinänen *et al.*[64] introduced the term 'AMPA-selective' glutamate receptors to designate those formations made by GluRA-D (GluR1–4). Similarly, the term 'kainate-selective' was initially used to describe those receptors made by GluR-6 subunits,[67] since these recombinant receptors were not sensitive to AMPA. Because, as for the GluR6 homomeric formations, AMPA is unable to gate responses in cells where kainate, glutamate, and quisqualate induce fully desensitizing currents, we have adopted the name 'kainate-selective receptors' to denote the glutamate receptors mediating such currents.

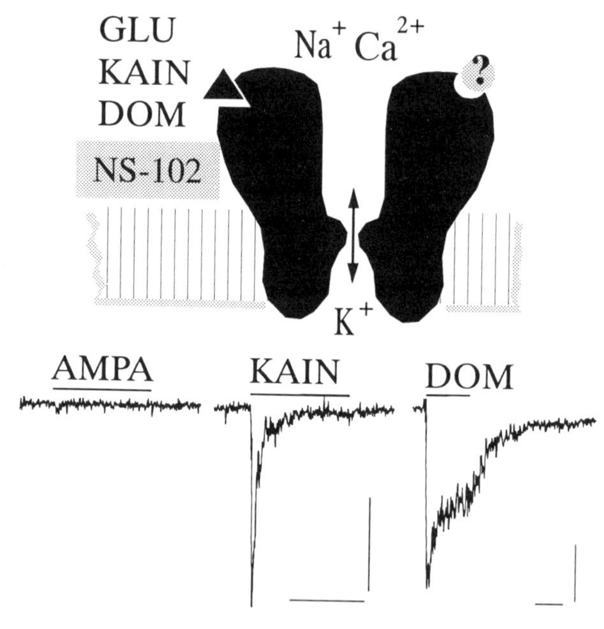


**Fig. 18.8** Sites of modulation of the kainate-selective receptor. In contrast to the AMPA receptor (see Fig. 18.6), AMPA fails to activate kainate receptors (bottom left record), while they show rapid activation–deactivation kinetics upon kainate binding (bottom middle record). Domoate activates and partially desensitizes kainate receptors (bottom right record). NS-102 acts as a competitive antagonist. No allosteric modulatory sites are known for these receptors. Calibrations: 400 ms and 20 pA. (From Lerma *et al.*[75].)

Kainate-selective receptors show different pharmacological properties compared with AMPA receptors. The competitive antagonist of AMPA receptors CNQX (10 μM)[71] slightly attenuates the peak current induced by kainate in cells where no AMPA responses are detected, or where this component is minor. In contrast, 10 μM CNQX largely blocked the induction of non-desensitizing currents by kainate in cells not expressing kainate-selective receptors (85% of inhibition with 300 μM kainate) (Fig. 18.9). In cells showing both responses, the introduction of 10 μM CNQX greatly reduces the steady component of the response, unmasking the transient current. Thus, CNQX is a less potent blocker of kainate-selective receptors than of AMPA receptors. Conversely, the newly developed glutamate antagonist, 5-nitro-6, 7, 8, 9-tetrahydrobenzo[G]indole-2, 3-dione-3-oxime (Ns–102)[105] is able to block reversibly the peak current more potently than the steady response induced by kainate (Fig 18.9). However, a more recent and detailed study has shown that competitive antagonists display weak selectivity between AMPA and kainate receptors in the hippocampus (see ref. 105a).

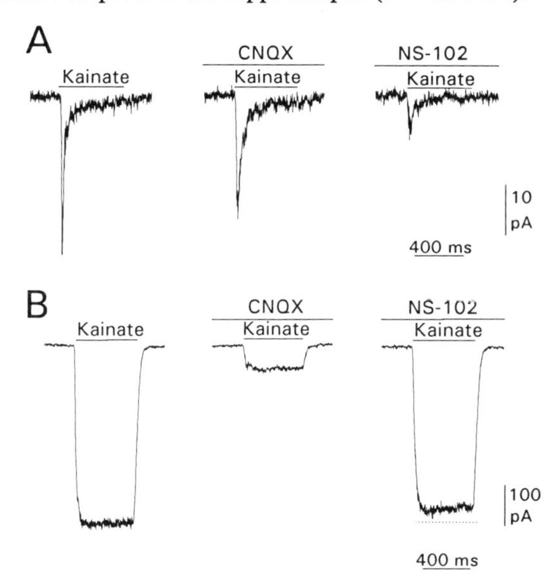


**Fig. 18.9** Different pharmacological properties for kainate-selective and AMPA receptors. The effect of CNQX (10 μM) and NS-102 (3 μM) on the response to kainate (300 μM) in cells expressing only kainate-selective receptors (A) and only AMPA receptors (B). (From Lerma *et al.*[75].)

In summary, hippocampal neurones express glutamate receptors that are selectively activated by kainate. Kainate-evoked responses undergo rapid and total desensitization, the recovery from which is rather slow. Other glutamate receptor agonists like glutamate, quisqualate, or domoate are also able to gate kainate-selective receptors. AMPA, even at high concentrations, does not activate these responses. In addition, the kainate-selective receptor has a much higher affinity for kainate than has the AMPA receptor, even considering that fast desensitization of kainate-selective receptors may lead to overestimation of the value for half-maximal activation in dose–response curves. Glutamate receptors with a similar affinity and with desensitizing properties when activated by kainate have been observed in dorsal root ganglion (DRG) cells in culture.[74] However, unlike DRG cell receptors, the type of high-affinity kainate receptors we found in hippocampal cells: (i) are not activated by AMPA; (ii) are totally desensitized in the presence of ligand; (iii) exhibit fast and monoexponential desensitizing kinetics; (iv) are less sensitive to CNQX than AMPA receptors; and (v) may coexist with AMPA receptors. More recently, receptors sharing many of the properties described here have been found in O-2A progenitors. The pharmacological profile, and the fast and the total desensitization of kainate-selective receptors, are consistent with properties observed in GluR-6 homomeric assemblies expressed in mammalian cells[69,106] (see also ref. 106a). Indeed GluR-6 subunits form homomeric channels which are the only receptor channels known that are gated by kainate but not by AMPA,[69,106–108]

## Therapeutic perspectives for excitatory amino acid pharmacology

Although aspartate and possibly other excitatory amino acids may contribute to neurotransmission, glutamic acid is thought to be the neurotransmitter of most synapses in the central nervous system. The different types of receptor activated by glutamate, mediate normal synaptic transmission along excitatory pathways. However, glutamate abnormally elevated in the extracellular space, even for fairly short periods, may also induce neuronal death. We know that certain classes of vulnerable neurones disappear after even brief hypoxic episodes, as induced for instance during stroke. Medical treatment of stroke patients is still uncertain in outcome and, as Choi[1] recently stated, we are no more able than Hippocrates to treat cerebral hypoxia itself. That is the reason why stroke is a major cause of death and especially of disability in the world. As key mediators of excitotoxicity, the receptors for excitatory amino acids have been extensively studied in recent years with pharmacological, biochemical, and electrophysiological approaches. Fortunately, the availability of specific agonists and antagonists for the NMDA receptor has made possible the acquisition of a detailed knowledge of several of the NMDA receptor mechanisms. Molecular knowledge is increasing, and the cloning of the NMDA and non-NMDA glutamate receptors is providing information on the structural domains of the receptor channel complexes. At the NMDA receptor, a reduction of ions (i.e. $Ca^{2+}$) flowing through the channel may be obtained by manipulating the regulatory sites with different pharmacological properties: competitive and non-competitive antagonism, use-dependent ion channel block, and desensitization. Modulation of desensitization might be one of the avenues to follow as a therapeutic strategy to reduce brain susceptibility to hypoxic or ischaemic insults. One would hope that a more desensitizing receptor would be only transiently activated (e.g. during fast transmission) while the prolonged exposure to the neurotransmitter would favour the receptor entering into the desensitized (i.e. inactive, non-conducting) state, therefore avoiding overload of the cytoplasm with $Ca^{2+}$. Recovery from desensitization would proceed only upon return of extracellular glutamate toward normal levels. Considering that the large desensitization that AMPA receptors undergo upon activation seems to be a very effective neuroprotective mechanism in non-NMDA receptor-induced neuronal damage,[109] modulation of NMDA receptor desensitization by antagonizing the spermine and/or glycine site might be a mechanism for controlling neuronal susceptibility to excitotoxic levels of glutamate. In fact, antagonists of the polyamine or the glycine sites (two desensitization-regulating sites) have been shown to protect against excitotoxicity in neuronal cultures.[110, 111] Although this possibility is attractive from a physiologist's point of view, a careful evaluation of desensitizing mechanisms and their modulation is mandatory. With the help of molecular and cellular neurobiology on the one hand, and the study of excitatory amino acid receptors in intact systems on the other, the future will elucidate many of these questions.

## Acknowledgements

This work was supported by grants from the DGICYT (PB89-0061; PB93-0150), FISSS (92/0266; 95/0869) and the European Community (BIO2-CT93-243 -DG 12-). F.S. was a bursary holder from the European Community and A.V.P. holds a fellowship from Glaxo-CSIC.

## References

1. Choi, D.W. (1990). *Trends in Neurosci.*, **11**, 465–469.
2. Meldrum, B. and Garthwaite J. (1990). *Trends in Neurosci.* **11**, 379–387.
3. Holmann, M., O'Shea-Greenfield, A., Rogers, S.W., and Heinemann, S. (1989). *Nature*, **342**, 643–648.
4. Gasic, G.P. and Heinemann, S. (1991). *Curr. Opin. Neurobiol.*, **1**, 20–26.
5. Sommer, B. and Seeburg, P.H. (1992). *Trends Pharmacol. Sci.*, **13**, 291–296.
6. Barnes, J.M. and Henley, J.M. (1992). *Prog. Neurobiol.*, **39**, 113–133.
7. Gasic, G.P. and Hollmann, M. (1992). *Annu. Rev. Physiol.*, **54**, 507–536.
8. Hollmann, M. and Heinemann, S. (1994). *Annu. Rev. Neurosci.*, **17**, 31–108.
9. Nakanishi, S. (1992). *Science*, **258**, 597–603.
10. Seeburg, P.H. (1993). *Trends Neurosci.*, **16**, 359–365.
11. Wisden, W. and Seeburg, P.H. (1993). *Curr. Opin. Neurobiol.*, **31**, 291–298.
12. Kettemann, H. and Grantyn, R. (1992). *Practical electrophysiological methods.* Wiley-Liss. New York.
13. Johnson, J.W. and Ascher, P. (1987). *Nature*, **325**, 529–531.
14. Nicoll, R.A., Kauer, J.A., and Malenka, R.C. (1988). *Neuron*, **1**, 97-103.
15. Brewer, G.J. and Cotman, C.W. (1989). *Neurosci. Lett.*, **99**, 268–273.
16. Ranschecker, J.P. and Hahn, S. (1987). *Nature*, **326**, 183–185.
17. Kleckner, N.W. and Dingledine, R. (1988). *Science*, **241**, 835–837.
18. Lerma, J., Zukin, R.S., and Bennett, M.V.L. (1990). *Proc. Natl Acad. Sci. USA*, **87**, 2354–2358.
19. Mayer, M.L., Westbrook, G.L., and Guthrie, P.B. (1984). *Nature*, **309**, 261–263.
20. Nowak, L., Bregestovski, P., Ascher, P., Herbert, A., and Prochiantz, A. (1984). *Nature*, **307**, 462–465.
21. Huettner, J.E. and Bean, B.P. (1988). *Proc. Natl Acad. Sci. USA*, **85**, 1307–1311.
22. Lerma, J., Zukin, R.S., and Bennett, M.V.L. (1991). *Neurosci. Lett.*, **123**, 187–191.

23. McDonald, J.F., Bartlett, M.C., Mody, I, Papahill, P., Reynolds, J.N., Salter, M.W., Schneiderman, J.H., and Pennefather, P.S. (1991). *J. Physiol. (Lond.)*, **432**, 483–508.
24. Christine, C.W. and Choi, D.W. (1990). *J. Neurosci.*, **10**, 108–116.
25. Peters, S., Koh, J., and Choi, D.W. (1987). *Science*, **236**, 589–593.
26. Westbrook, G.L. and Mayer, M.L. (1987). *Nature*, **328**, 640–643.
27. Lerma, J. (1992). *Neuron*, **8**, 343–352.
28. McGurk, J.F., Bennett, M.V.L., and Zukin, R.S. (1990). *Proc. Natl Acad. Sci. USA*, **87**, 9971–9974.
29. Ransom, R.W. and Stec, N.L. (1988). *J. Neurochem.*, **51**, 830–836.
30. Tang, C.M., Dichter, M., and Morad, M. (1990). *Proc. Natl Acad. Sci. USA*, **87**, 6445–6449.
31. Traynelis, S.F. and Cull-Candy, S.G. (1990). *Nature*, **345**, 347–350.
32. Aizenman, E., Lipton, S.A., and Loring, R.H. (1989). *Neuron*, **2**, 1257–1263.
33. Patneau, D.K. and Mayer M.L. (1990). *J. Neurosci.*, **10**, 2385–2399.
34. Lerma, J., Kushner, L., Zukin, R.S., and Bennett, M.V.L. (1989). *Proc. Natl Acad. Sci. USA*, **86**, 2083–2087.
35. Verdoorn, T.A. and Digledine, R. (1988). *Mol. Pharmacol.*, **34**, 298–307.
36. Vyklicky, Jr L., Benveniste, M., and Mayer, M.L. (1990). *J. Physiol. (Lond.)*, **428**, 313–331.
37. Kushner, L., Lerma, J., Zukin, R.S., and Bennett, M.V.L. (1988). *Proc. Natl Acad. Sci. USA*, **85**, 3250–3254.
38. Kemp, J.A. Foster, A.C., Leeson, P.D., Priestley, T., Tridgett, R., Ivensen, L.L. and Woodruff, E.N. (1988). *Proc. Natl Acad. Sci. USA*, **85**, 6547–6550.
39. Lerma J., Herranz AS., Herreras O., Abraira A., and Martín del Rio R. (1986). *Brain Res.*, **384**, 145–155.
40. Williams, K., Romano, C., and Molinoff, P.B. (1989). *Mol. Pharmacol.*, **37**, 575–581.
41. Seiler, N. and Schmidt-Glenewinkel, T. (1975). *J. Neurochem.*, **24**, 791–795.
42. Shaw, G.G. and Pateman, A.J. (1973). *J. Neurochem.*, **20**, 1225–1230.
43. Tabor, C.W. and Tabor, H. (1984). *Biochemistry*, **53**, 745–790.
44. Wedgwood, M.A. and Wolstencroft, J.H. (1977). *Mol. Pharmacol.*, **16**, 445–446.
45. Carter, C.J., Lloyd, K.J., Zivkoviz, B., and Scatton, B. (1990). *J. Pharmacol. Exp. Ther.*, **253**, 475–482.
46. Zeevalk, G.D. and Nicklas, W.J. (1990). *Brain Res.*, **522**, 135–139.
47. Brackley, P., Goodnow, R., Jr, Nakanishi, K., Sudan, H.L., and Usherwood, P.N.P. (1990). *Neurosci. Lett.*, **114**, 51–56.
48. Benveniste, M. and Mayer, M. (1993). *J. Physiol.*, **464**, 131–163.
49. Williams, K., Dawson, V.L., Romano, C., Dichter, M., and Molinoff, P.B. (1990). *Neuron*, **5**, 199–208.

50. Sather, W., Johnson, J.W., Henderson, G., and Ascher, P. (1990). *Neuron*, **4**, 725–731.
51. Subramanian, S., Donevan, S.D., and Rogawski, M.A. (1994). *Mol. Pharmacol.*, **45**, 117–124.
52. Ascher, P. and Nowak, L. (1988). *J. Physiol (Lond.)*, **399**, 247–266.
53. Cajal, S.R. (1911). *Histologie du Systeme Nerveux de L'homme et des vertébrés*, Vol. 2. Maloine: Paris.
54. Hebb, D.O. (1949). *The organization of behavior: a neuropsychological theory*. Wiley, New York.
55. Bliss, T.V.P. and Gardner-Medwin, A.R. (1973). *J. Physiol. (Lond.)*, **232**, 357–374.
56. Mori, H., Masaki, H., Yamakura, T., and Mishina, M. (1992). *Nature*, **358**, 673–675.
57. Bennett, M.V.L., García-Ballesteros, A., Zukin, R.S., and Lerma, J. (1991). *Soc. Neurosci. Abs.*, **17**, 1292.
58. Legendre, P. and Westbrook, G.L. (1990). *J. Physiol. (Lond.)*, **429**, 429–449.
59. Assaff, S.Y. and Choung, S.H. (1984). *Nature*, **308**, 734–736.
60. Lerma, J. and Martín del Río, R. (1992). *Mol. Pharmacol.*, **41**, 217–222.
61. Mayer, M.L. and Westbrook, G.L. (1987). *Prog. Neurobiol. (Oxford)*, **28**, 197–276.
62. Patneau, D.K. and Mayer, M.L. (1991). *Neuron*, **6**, 785–798.
63. Lambolez, B., Curutchet, P., Stinnakre, J., Bregestovski, P., Rossier, J., and Prado de Carvalho, L. (1990). *Nature*, **347**, 26.
64. Keinänen, K., Wisden, W., Sommer, B., Werner, P., Herb, A., Verdoorn, T.A., Sakmann, B., and Seeburg, P.H. (1990). *Science*, **249**, 556–560.
65. Boulter, J., Hollmann, M., O'Shea-Greenfield, A., Hartley, M., Deneris, E., Maron, C., and Heinemann, S. (1990). *Science*, **249**, 1033–1037.
66. Nakanishi, N., Shneider, N.A., and Axel, R. (1990). *Neuron*, **5**, 569–581.
67. Sakimura, K., Morita, T., Kushiya, E., and Mishina, M., (1992). *Neuron*, **8**, 267–274.
68. Bettler, B., Boulter, J., Hermans-Borgmeyer, I., O'Shea-Greenfield, A., Deneris, E.S., Moll., C., Borgmeyer U., Hollmann, M., and Heinemann, S. (1990). *Neuron*, **5**, 583–595.
69. Herb, A., Burnashev, N., Werner, P., Sakmann, B., Wisden, W., and Seeburg, P.H. (1992). *Neuron*, **8**, 775–785.
70. Sommer, B., Burnashev, N., Verdoorn, T.O., Keinänen, K., Sakmann, B., and Seeburg, P.H. (1992). *EMBO J.*, **11**, 16151–1656.
71. Honore, T. Davies, S.N., Drejer, J., Fletcher, E.J., Jacobsen, P., Lodge, D., and Nielsen, F.E. (1988). *Science*, **241**, 701–703.
72. Jonas, P. and Sakmann, B. (1992). *J. Physiol. (Lond.)*, **455**, 143–171.
73. McBain, C. and Dingledine., R. (1993) *J. Physiol. (Lond.)*, **462**, 373–392.
74. Huettner, J.E. (1990) *Neuron*, **5**, 255–266.
75. Lerma, J., Paternain, A.V., Naranjo, J.R. and Mellström, B. (1993). *Proc. Natl Acad. Sci. USA*, **90**, 11688–11692.

76. Patneau, D.K., Wright, P.W., Winters, C., Mayer, M.M. and Gallo, V. (1994). *Neuron*, **12**, 357–371.

77. Hollman, M., Hartley, M., and Heinemann, S.F. (1991) *Science*, **252**, 851–853.

78. Burnashev, N., Monyer, H., Seeburg, P.H., and Sakmann, B. (1992). *Neuron*, **8**, 189–198.

79. Verdoorn, T.A., Burnashev, N., Monyer, H., Seeburg, P. and Sakmann, B. (1991), *Science*, **252**, 1715–1718.

80. Hume, R.I., Dingledine, R., and Heinemann, S.F. (1991), *Science*, **253**, 1028–1031.

81. Burnashev, N., Khodorova, A., Jonas, P., Helm, J., Wisden, W. Monyer, H., Seeburg, P.H., and Sakmann, B. (1992) *Science*, **256**, 1566–1570.

82. Müller, T., Möller, T., Berger, T., Schnitzer, J., and Kettenmann, H. (1992), *Science*, **256**, 1563–1566.

83. McDermont, A.B., Mayer, M.L., Westbrook, G.L., Smith, S.L., and Banker, J.L. (1986). *Nature*, **321**, 519–522.

84. Iino, M., Ozakawa, S, and Tsuzuki, K. (1990). *J. Physiol. (Lond.)*, **424**, 151–165.

85. Murphy, S.N. and Miller, R.J. (1989) *J. Pharmacol. Exp. Ther*, **249**, 84–193.

86. Rörig, B. and Grantyn, R. (1993). *Neurosci. Lett.*, **153**, 32–36.

87. Williams, L.R., Pregenzer, J.F., and Oostveen, J.A. (1992) *Brain Res*, **581**, 181–189.

88. Lerma, J., Morales, M., Ibarz J.M., and Somohano, F. (1994), *Eur. J. Neurosci*, **6**, 1080–1088.

89. Ito, I., Tanabe, S., Kohda, A., and Sugiyama, H. (1990) *J. Physiol*, **424**, 533–543.

90. Isaacson, J.S. and Nicoll, R.A. (1991) *Proc. Natl Acad. Sci. USA* **88**, 10936–10940.

91. Vyklicky, L., Patneau, D.K., and Mayer, M. (1992) *Neuron* **7**, 971–984.

92. Tang, C.M., Shi, Q.Y., Katchman, A. and Lynch, G. (1991) *Science* **254**, 288–290.

93. Yamada, K.A. and Rothman, S.M. (1992). *J. Physiol.* **458**, 409–423.

94. Yamada, K.A. and Tang, C.M. (1993). *J. Neurosci.* **13**, 3904–3915.

95. Patneau, D.K., Vyklicky, L., Jr, and Mayer, M.L. (1993). *J. Neurosci.* **13**, 3496–3509.

96. Trussell, L.O., Zhang, S., and Raman, I.M. (1993) *Neuron* **10**, 1185–1196.

97. Tarnawa, I., Farkas, S., Berzsenyi, P., Pataki, A., and Andrasi, F. (1989). *Eur. J. Pharmacol.* **167**, 193–199.

98. Zorumski, C.F., Yammada, K.A., Price, M.T., and Olney, J.W. (1993) *Neuron*, **10**, 61–67.

99. Donevan, S.D. and Rogawski, M.A. (1993) *Neuron*, **10**, 51–59.

99a. Paternain, A.V., Morales, M., and Lerma, J. (1995) *Neuron*, **14**, 185–189.

100. Jackson, H. and Usherwood, P.N.R. (1987). *Trends Neurosci*, **11**, 278–282.

101. Grishing, E.V., Volkova, T.M., Arsenev, A.S., Reshetova, O.S., Onoprienko, V.V.,

Magazanik, L.G., Antonov, S.M., and Fedorova, I.M. (1986) *Bioorgam Khim.* **12**, 1121–1124.

102. Usherwood, P.N.R. (1991) In *Excitatory amino acids*, Fidia Research Foundation Symposium Series, vol. **5**, pp. 379–395, Raven Press, N.Y.

103. Herlitze S., Raditsch, M., Ruppersberg, J.P., Jahn, W., Monyer, H., Schoepfer, R., and Witzemann, V. (1993), *Neuron*, **10**, 1131–1140.

104. Blaschke, M., Keller, B.V., Rivosecchi, R. Hollmann, M., Heinemann, S. and Konnerth, A. (1993) *Proc. Natl Acad. Sci. USA*, **90**, 6528–6532.

105. Johansen, T.H., Drejer, J., Wätjen, F., and Nielsen, E.O. (1993), *Eur. J. Pharmacol.* **246**, 195–204.

105a. Paternain, A.V., Vicente, A., Nielsen, E., Ø., and Lerma, J. (1996) *Eur. J. Neurosci.*, **8**, 2129–2136.

106. Khöler, M., Burnashev, N., Sakmann, B. and Seeburg, T.H. (1993). *Neuron*, **10**, 491–500.

106a. Ruano, D., Lambolez, B., Rossier, J., Paternain, A.V. and Lerma, J. (1995) *Neuron*, **14**, 1009–1017.

107. Egebjerg, J., Bettler, B., Hermans-Borgmeyer, I and Heinemann, S. (1991). *Nature*, **351**, 745–748.

108. Morita, T., Sakimura, K., Kushiya, E., Yamazaki, M., Meguro, H., Araki, K., Abe, T., Mori, K.J. and Mishina, M. (1992). *Mol. Brain Res.* **14**, 143–146.

109. Zorumski, C.F., Thio, L.L., Clark, G.D. and Clifford D.B. (1990) *Neuron*, **5**, 61–66.

110. McNamara, D. and Dingledine, R. (1990) *J. Neurosci*, **10**, 3970–3976.

111. Priestly, T., Horne, A.L., McKernan R.M., and Kemp, J.A. (1990) *Brain Res.* **531**, 183–188.

112. Davies, J., Francis, A.A., Jones, A.W. and Watkins, J.C. (1981). *Neurosci. Lett.*, **21**, 77–81.

113. Davies, J., Evans, R.H., Herrling, P.L., Jones, A.W., Olverman, H.J., Pook, P., and Watkins, J.C. (1986) *Brain Res*, **382**, 169–173.

114. Henderson, G., Johnson, J.W. and Ascher, P. (1990) *J. Physiol* **430**, 189–212.

115. Reynolds, I.J. (1990) *J. Pharmacol. Exp. Ther* **255**, 1001–1007.

116. Krogsgaard-Larsen, P., Honoré, T., Hansen, J.J., Curtis, D.R., and Lodge, D. (1980). *Nature*, **284**, 64–66.

# 19 Comparison of native and recombinant NMDA receptor channels

P. Stern and D. Colquhoun

## Introduction

One of the most important problems in synaptic physiology is the identification of which receptor subtypes are responsible for mediating synaptic transmission. The advent of the methods of molecular genetics has revealed that all of the fast neurotransmitter-activated receptor channels that mediate synaptic transmission are made up of several subunits. These studies have also shown that there are many sorts of subunits. The question is, which particular subunit combination mediates the signal in each sort of synapse? This may, of course, be an overoptimistic way of putting the question. There may not be just one answer. It is known, for example, that there is often more than one receptor type present in the same cell (and, indeed, often in the same patch of membrane). Another major problem is that most cells appear to have not only subsynaptic receptors (those located directly under the sites of transmitter release), but also many extrasynaptic receptors (the function of which, if any, is not known). Membrane patches are almost always taken from the somatic membrane, and will almost always contain extrasynaptic receptors. If we are lucky, it may turn out that synaptic receptors are homogeneous (even though extrasynaptic receptors are not always homogeneous). In this case there will at least be a unique answer to the question that was posed above, though of course the answer will be more difficult to find if synaptic and extrasynaptic receptors differ. Recent work has shown a close similarity between somatic and dendritic glutamate receptors (Spruston *et al.* 1995); this is encouraging, though the dendritic receptors are not necessarily subsynaptic.

Comparison of native and recombinant channels is one of the main methods used to elucidate the subunit composition of native receptors. The now classical studies of Mishina *et al.* (1986) showed clearly that the two forms of muscle nicotinic acetylcholine receptors (the embryonic and the adult form) were due to a switch in the subunit composition of these receptors. Since then it has been the aim of a number of groups to repeat this type of experiment in the central nervous system. Unfortunately, the diversity of subunits for every known agonist-activated receptor has turned out to be much greater than could have been anticipated. Furthermore, it has become apparent that the systems used for heterologous expression of cloned receptor subunits may not always produce receptors that behave in exactly the same way as the 'same' receptors do when in their native environment. These differences appear to result, in some case, from production of receptors from which a subunit has been omitted, but in other cases the reason for differences from native receptors remains unexplained (see review by Edmonds *et al.* 1995). Here, we review some of the attempts that have been made to answer these questions in the case of the NMDA type of glutamate receptor.

## Studies with NMDA receptors

It has become increasingly clear that glutamate is the most important excitatory transmitter in the central nervous system (see reviews by Streit 1985; Headley and Grillner 1990; Orrego and Villanueva 1993). It has been suggested that the *N*-methyl-D-aspartate (NMDA) type of glutamate receptors is important for the understanding of a large range of physiological phenomena. For example, the NMDA receptor is thought to play an important role in cortical development (Kleinschmidt *et al.* 1987; Komuro and Rakic 1993; Piggott *et al.* 1993) and in synaptic plasticity (for reviews, see Bliss and Collingridge 1993; Malenka and Nicoll 1993), as well as in pathological events such as neurotoxicity (Choi 1988; Meldrum and Garthwaite 1990). However, the evidence for these roles is not always as unequivocal as often seems to be assumed (though their contribution to hyperbole in scientific writing is undeniable). For example, Fox and Daw (1993) give a critical evaluation concerning the role of NMDA receptors in cortical plasticity. And, although there is evidence that NMDA receptors are involved in some (but not all) sorts of long-term potentiation, there is still a good deal of ambiguity concerning their precise role relative to other postulated mediators. In a recent study in which the NMDA (NR1 subunit) gene was knocked out, mice died within 2 days of birth (Li *et al.* 1994); it will require more subtle experiments of this sort, with different subunits (e.g. Ito *et al.* 1996), and with inducible promoters, to give a more solid idea as to the physiological role of NMDA receptors in normal developing and adult animals.

The first studies using single-channels of NMDA receptors were done by Ascher and coworkers (Nowak *et al.* 1984; Ascher and Nowak 1988). They showed that NMDA receptors are non-specific cation channels with a single-channel conductance of about 50 pS and with a rather high calcium permeability. Furthermore, they clearly established that the unusual voltage dependence, with a region of negative slope conductance in the negative voltage range, is due to a rapid block of the ion channels caused by magnesium ions. A far more unusual characteristic of NMDA receptors is their requirement for glycine to be present before the channel can be opened by glutamate (Johnson and Ascher 1987; Kleckner and Dingledine 1988). This adds another potential complication to studies of NMDA channel function.

To some extent it is possible to compare published results on native and recombinant receptors. However, it is remarkably rare for experiments from different laboratories to be done under comparable conditions. Almost always there are differences of cell preparation, temperature, ionic composition, and so on. To cite one example, the single-channel conductance is an important criterion for comparison of native and recombinant channels, but it is quite sensitive, not only to the monovalent cation concentration (which is not the same in all bath solutions), but also to the extracellular free calcium concentration (which varies even more from lab to lab, and also depends on whether bicarbonate buffer is used or not). It was shown by Gibb and Colquhoun (1992) that the main single-channel conductance level for the most common sort of NMDA receptor channel was 50 pS in 1 mM calcium, but 42 pS in 2.5 mM calcium and 66 pS in EDTA-buffered solution, and a similar dependence is also found in a low-conductance' channel (recombinant NR1/NR2D; Wyllie *et al.* 1996). Thus, it is particularly valuable to work, in parallel, on both native and recombinant receptors in the same laboratory so that comparable results can be ensured.

## Characteristics of native NMDA receptor channels

In order to compare the behaviour of native and recombinant receptors it is desirable to have detailed knowledge of both of them. Several publications described the details of the single-channel behaviour of native NMDA receptors before the first receptor clones were found. This detailed information was, at the time, of limited usefulness in improving our understanding of synaptic transmission, but it has turned out to be of great value since, because it can be compared with the results obtained with different combinations of recombinant subunits.

Howe *et al.* (1991) described two conductance levels which could account for the large majority (more than 90%) of glutamate, aspartate, and NMDA-induced single-channel openings in cultured cerebellar granule cells. The main conductance level was again about 50 pS (in 1 mM calcium), while a second, briefer and rarer, 40 pS subconductance level was also found. They also analysed the number and frequency of transitions between the different conductance levels (including the shut state). In addition, they investigated the distribution of open times conditional on amplitude level. They found that the largest conductance level (50 pS) could be best fitted by two exponential components. The distribution of shut times was fitted with four exponentials. The discrepancy between these results and subsequent studies (see below) can best be explained by the different agonist concentrations (which may have led to a certain degree of receptor desensitization, especially in the higher concentration range). These experiments suffered, however, from two disadvantages. Firstly they were performed on cultured cells, and it cannot be guaranteed that the receptor type(s) produced in culture are the same as *in vivo* (see for example Bekkers *et al.* 1996). Secondly, the experiments were done before the importance of glycine was appreciated, so no glycine was added to the solutions and high agonist concentrations were used; the effective agonist concentration was therefore not known, and this is a great disadvantage if it is wished to investigate, for example, the structure of individual channel activations by agonist.

Since that time many authors have used either freshly dissociated cells, or cells in brain slices, to overcome the possible hazards of culturing. Gibb and Colquhoun (1991, 1992) used both methods on hippocampal (CA1) cells. By using very low glutamate concentrations (20–100 nM), in the presence of glycine, they were able to separate individual receptor activations (see Edmonds *et al.* 1995 for a precise definition of this term). They observed 40 and 50 pS conductance levels (in 1 mM calcium) which were very similar to those seen by Howe *et al.* (1991). They fitted shut time distributions with five exponentials, and found that the three fastest components, and probably also the fourth, were not dependent on the glutamate concentration. This allowed them to describe quantitatively the complex patterns of channel openings which can be grouped into bursts, clusters, and super-clusters. It was clear that individual channel activations were often very long, and that this was one reason for the remarkable slowness of the NMDA receptor-mediated synaptic current.

### 'Low-conductance' NMDA channels

The majority of native NMDA channels that have been reported seem to be similar to those mentioned above, having a main conductance level of 50 pS (in 1 mM $Ca^{2+}$) with a brief 40 pS sublevel. There have also been several reports of lower conductance channels too. These have a main conductance level of 33–38 pS and a subconductance level of 16–20 pS, about half of the main level (Cull-Candy and Usowicz 1987; Ascher *et al.* 1988; Farrant *et al.* 1994; Momiyama *et al.* 1996). In these channels the mean lifetime of the sublevel is usually as long, or longer than that of the main level. Ascher *et al.* (1988) reported 35/16 pS channels in cultured embryonic spinal neurones (mouse). Farrant *et al.* (1994) reported similar channels in granule cells after postnatal day 13 (P13). Momiyama *et al.* (1996) report channels with similar conductance appearing by themselves in young (before P12) Purkinje cells, and mixed with the more common 50/40 'high-conductance' channels in neurones from deep cerebellar nuclei and from the dorsal horn of the spinal cord. However, these channels differ from those in granule cells in two ways; they show temporal asymmetry of sublevel transitions, and the mean lifetime of the subconductance level is *longer* than that of the main level. These studies are discussed further below.

## Recombinant NMDA receptor channels

In 1991, the first subunit of the NMDA receptor was found (Moriyoshi *et al.* 1991). In two electrode voltage-clamp experiments with RNA-injected *Xenopus* oocytes, this group showed that this subunit (now termed NMDAR1, or NR1 for short) could reproduce, qualitatively at least, many of the pharmacological properties of native NMDA receptors. Soon afterwards a second family of NMDA receptor subunits was found. They are termed the NMDAR2 (NR2) family, which, at present, consists of four members (NR2A to NR2D) (Monyer *et al.* 1992; Ishii *et al.* 1993). This nomenclature applies to the NMDA receptor clones found in the rat. The homologues of these subunits in the mouse, named $\xi1$ (NR1) and $\varepsilon1$ to $\varepsilon4$ (NR2A to NR2D), have also been found and characterized on the whole-cell level (Yamazaki *et al.* 1992; Meguro *et al.* 1992; Kutsuwada *et al.* 1992, Ikeda *et al.* 1992). The first human NMDA receptor subunits followed very soon (Karp *et al.* 1993; Planells-Cases *et al.* 1993). The situation became more complicated when several alternatively spliced variants of the NR1 and NR2D subunits were discovered (Anantharam *et al.* 1992; Sugihara *et al.* 1992; Nakanishi *et al.* 1992; Durand *et al.* 1992; Durand *et al.* 1993; Kusiak and Norton 1993; Ishii *et al.* 1993). Differences in zinc, spermine, and pH sensitivity have been found between one splice variant and another, but effects on single-channel conductance are not known (Durand *et al.* 1992, 1993; Hollmann *et al.* 1993; Zhang *et al.* 1994; Zheng *et al.* 1994; Traynelis *et al.* 1995).

## Stoichiometry of NMDA receptors

When subunits are expressed in pairs (NR1 with one of the four NR2 subunits) it has been found that there are two NR1 subunits per receptor, because coexpression of wild-type and mutant NR1 receptors (together with NR2A) generates one new sort of channel (which presumably contains one of each type of NR1) with properties intermediate between those found when only mutant, or only wild-type NR1 messages are injected (Béhé *et al.* 1995). The number of NR2 subunits in the complex is still uncertain.

There is, of course, no guarantee that native receptors contain only two sorts of subunit; there could be three or even more. Wafford *et al.* (1993) showed that the NR1/NR2A/NR2C combination has a glycine sensitivity which is distinct from either NR1/NR2A or NR1/NR2C alone, or from any mixture of these pairs. They thus conclude that the combined injection of RNA coding for three different subunits produces a new type of channel. Immunoprecipitation studies have also suggested the formation of channels with three sorts of subunit (Chazot *et al.* 1994; Sheng *et al.* 1994). At the single-channel level, we have found that such triplets often give rise to channels which behave unstably, as shown by stability plots for both the channel amplitude and for the probability of being open (P. Stern *et al.*, unpublished results; see also Colquhoun and Sigworth 1995; Fig.14C and D). This question is clearly not yet settled.

## High-conductance channels NR1/NR2A and NR1/NR2B

As mentioned above, when mRNA for NR1 alone is injected into *Xenopus* oocytes it can produce functioning channels. However, the whole-cell responses, measured with a two-electrode voltage-clamp, are very small, so it is most unlikely that outside-out patches will contain even one active channel. The subunits of the NR2 family do not produce any channels at all as homomers. On the contrary, the pairwise combination of NR1 with any one of the NR2 subunits gives large responses in the oocyte expression system. Stern *et al.* (1992) therefore compared single-channel records of several pairs of heteromeric NMDA receptor clones. In outside-out patches from *Xenopus* oocytes, the combination of NR1/NR2A and NR1/NR2B gave single channels with a main conductance level of 50 pS and a subconductance level of about 40 pS which resemble closely the native receptors that have been described in cerebellar granule cells as well as the very similar channels in hippocampal neurones (Gibb and Colquhoun 1991, 1992; Howe *et al.* 1991; also see above). This similarity was not restricted to the two conductance levels (40 and 50 pS in 1 mM calcium), but it was found that the frequency of transitions between different conductance levels was in close quantitative agreement with the measurements of Howe *et al.* (1991) in cultured cerebellar granule cells. Furthermore, the quantitative similarity extended to the distributions of open times, shut times, burst lengths, etc. (P. Béhé *et al.* in preparation). The conclusion of this work was that low-concentration equilibrium single-channel records are indistinguishable for; (i) NR1/NR2A channels expressed in oocytes; (ii) NR1/NR2B channels expressed in oocytes; and (iii) the '50 pS' channels observed in various cultured, dissociated or brain-slice neurones. However, recombinant NR1/NR2A and NR1/NR2B channels are clearly distinguishable on the basis of their glycine sensitivity (which is ten times lower for the NR1/NR2A combination than for NR1/NR2B; Kutsuwada *et al.* 1992; Stern *et al.* 1992). Results of

Monyer *et al.* (1994) suggest that they may also be distinguishable on the basis of the offset decay rate of whole-cell currents following a concentration jump.

The conductances observed for normal and native high-conductance channels are summarized in Table 19.1.

**Table 19.1** Summary of single-channel conductances* for 'high-conductance' channels, recombinant (upper part) and native (lower part).

| | Main level (pS) | Sublevel (pS) | Area (%) | Temporal asymmetry | Reference |
|---|---|---|---|---|---|
| NR1a-NR2A (oocyte) | 50.1 | 38.3 | 20.1 | No | a |
| NR1a-NR2B (oocyte) | 50.9 | 38.7 | 16.9 | No | a |
| NR1a-NR2A (HEK293 cell) | 51.4 | 38.1 | 23.1 | No | b |
| Hippocampus (CA1) | 51 | 37 | 23 | No | c |
| Cerebellar granule cell (cultured) | 51.0 | 42.1 | 15.6 | No | d |
| Cerebellar granule cell (brain slice) | 49.9 | 40 | 17 | No | e |

* Measurements in 1 mM $Ca^{2+}$. It is also noted whether or not there is asymmetry in the number of main level ⟷ sublevel transitions.
[a] Stern *et al.* 1992
[b] Stern *et al.* 1994
[c] Gibb and Colquhoun 1992
[d] Howe *et al.* 1991
[e] Farrant *et al.* 1994

## Low-conductance channels NR1/NR2C and NR1/NR2D

The NR1/NR2C and NR1/NR2D subunit combinations, on the other hand, give single channels which differ in nearly all the parameters analysed, when expressed in *Xenopus* oocytes (Stern *et al.* 1992; Wyllie *et al.* 1996).

The NR1/NR2C channels have a main conductance level of 36 pS with a sublevel of 19 pS, about half the amplitude of the main level, with a fairly brief open time (Stern *et al.* 1992). These channels resemble closely the low-conductance channels observed in brain slices by Farrant *et al.* (1994). They described the development of NMDA receptors in rat cerebellum, and found that in 14-day-old rat cerebellar slices (the usual age for brain slice experiments), the channels all appeared to be of the '50 pS' type, described above. However, in older animals (in which brain slices are harder to work with) they found that mature post-migratory granule cells contain a subtype of receptor which shows conductance levels of 33 and 20 pS in 1 mM calcium; these values are similar to those found with the NR1/NR2C combination. This result is in beautiful agreement with *in situ*-hybridization studies (Watanabe *et al.* 1992; Monyer *et al.* 1994) which show that there is little messenger RNA for NR2C present in the cerebellum at 7 days after birth, but after 14 days the amount of NR2C message has increased to something near to the adult level.

The NR1/NR2D channels have very similar conductance levels (35 and 17 pS), but differ from NR1/NR2C in two ways. Firstly, the mean duration of the subconductance level is *longer* than that of the main level, and secondly, the records show temporal asymmetry. This asymmetry is manifested by the fact that direct transitions from the 35 pS level to the 17 pS level are more common than transitions in the opposite direction (Wyllie *et al.* 1996). Such asymmetry is not observed with any of the other subunit combinations, or with native 'high-conductance' channels. A similar asymmetry was, however, reported by Cull-Candy and Usowicz (1987, 1989). They reported single channels in cultured 'large cerebellar neurones' (it is not certain what sort of cell these correspond to in the cerebellum – they could originate from Purkinje cell or from neurones in the deep cerebellar nuclei). Their records apparently contained two sorts of channels one of which was the usual 50/40 pS type, but the other had conductance levels of 18 pS and 38 pS which are similar to those found in the NR1/NR2C and NR1/NR2D combinations. But the lower-conductance type showed asymmetry which resembles only the NR1/NR2D combination. More recently, Momiyama *et al.* (1996) investigated Purkinje cells in cerebellar slices, and found, up to P12, very similar low-conductance channels (only); after this time NMDA channels are absent from Purkinje cells. The suggestion that these channels are the NR1/NR2D type agrees very well with the results of *in situ*-hybridization studies which show that only NR1 and NR2D mRNA are detectable in Purkinje cells at this age (Akazawa *et al.* 1994). Very similar low-conductance channels, along with typical high-conductance channels, were detected in deep cerebellar nuclei and in spinal cord dorsal horn neurones by Momiyama *et al.* (1996), and these too may be NR1/NR2D channels (see Ishii *et al.* 1993; Tölle *et al.* 1993; Monyer *et al.* 1994).

The conductances observed for normal and native low conductance channels are summarized in Table 19.2.

**Table 19.2** Summary of single-channel conductances* for 'low-conductance' channels, recombinant (upper part) and native (lower part).

| | Main level (pS) | Sublevel (pS) | Area (%) | Temporal asymmetry | Reference |
|---|---|---|---|---|---|
| NR1a-NR2C (oocyte) | 35.9 | 19.2 | 25.1 | No | a |
| NR1a-NR2D (oocyte) | 35 | 17 | 39 | Yes | b |
| Cerebellar granule cell (after P13) | 33 | 20 | 14 | No | c |
| 'Large cerebellar neurones' (cultured) | 38 | 18 | ? | Yes | d |
| Cerebellar Purkinje cell (before P12) | 38 | 18 | 17 | Yes | e |

* Measurements in 1 mM $Ca^{2+}$ (or 0.85 mM $Ca^{2+}$ in the absence of bicarbonate: ref b). It is also noted whether or not there is asymmetry in the number of main level ⟷ sublevel transitions.
[a] Stern *et al.* 1992
[b] Wyllie *et al.* 1996
[c] Farrant *et al.* 1994
[d] Cull-Candy and Usowicz 1987
[e] Momiyama *et al.* 1996

## Other differences between high- and low-conductance channels

One feature shared by NR1/NR2C and NR1/NR2D channels is that they both show lower sensitivity to block by magnesium than NR1/NR2A and NR1/NR2B channels (Monyer *et al.*

1994). In view of the importance of this magnesium block, and its voltage dependence, for the physiological function of NMDA receptors, this difference in magnesium sensitivity could well be reflected in different physiological functions for high- and low-conductance channels, though such differences have not so far been demonstrated. Investigation of block by magnesium is hindered by the fact that it is not a simple open channel blocker, and its various effects are complex and ill-understood (Nowak *et al.* 1984; Ascher and Nowak 1988; Johnson and Ascher 1990; Paoletti *et al.* 1995). This means that block must be characterized by relatively crude methods such as $IC_{50}$ measurements (which are likely to vary with agonist concentration as well as membrane potential). Nevertheless, useful conclusions have been drawn.

The question naturally arises of what is the structural basis for the similarity between NR2A and NR2B on one hand, and NR2C and NR2D on the other hand. When the NMDA subunits were first cloned it was supposed that they resembled topologically the nicotinic and GABA receptors which are generally believed to have four membrane-spanning regions. More recently, it has been suggested that glutamate receptors (both NMDA and non-NMDA) may belong to a different subfamily, and have only three membrane-spanning regions, with the very long C-terminal region being intracellular rather than extracellular (Hollmann *et al.* 1994; Bennett and Dingledine 1995; Wo and Oswald 1995*a*, *b*). On these schemes the 'M2' region is seen as a loop that dips into the membrane rather than crossing it. Regardless of the precise topology, it is nevertheless certainly true that mutations in the M2 region have strong effects on ion permeation, so it is interesting to note that the M2 domain is identical in the NR2A and the NR2B subunits, whereas that of NR2C differs in three positions. However, NR2C and NR2D differ in only one position in the M2 domain. The overall similarity in amino acid sequence, and the similarity in the M1 – M4 region is also greatest between NR2A and the NR2B on one hand, and between NR2C and the NR2D on the other (Ishii *et al.* 1993). The differences in magnesium sensitivity (discussed above) do not depend on a single site, but can be transferred by making chimeric receptors with the whole M1 – M4 region transplanted, or by transplanting three small separate domains within this region (Kuner and Schoepfer 1996).

## Validity of the expression system

The close similarity between the characteristics of native NMDA receptor single channels, and those expressed in *Xenopus* oocytes is gratifying, but it is somewhat circular to use this observation as an answer to the question if the same channel type has the same behaviour, whether it is in oocytes or in its native environment. It was, therefore, very reassuring to find that the single-channel behaviour of channels produced by transient transfection of HEK 293 cells was indistinguishable from their behaviour when produced by RNA injection into oocytes (Stern *et al.* 1994; see Table 19.1). In this study the following single-channel properties were compared: channel conductances and transition frequencies between the different levels, shut time distributions, and open time distributions conditional on the conductance level. All were similar for the two different expression systems, and this adds weight to the view that both expression systems work well for NMDA receptors (probably better than they do for nicotinic receptors; see Edmonds *et al.* 1995).

### Concentration jump experiments

A different and complementary approach to investigation of the function of channels is the analysis of channels measured in outside-out patches after brief agonist concentration jumps. Experiments of this sort have already given valuable information about native NMDA channels (e.g. Lester *et al.* 1990; Edmonds and Colquhoun 1992). Very brief agonist pulses (down to about 200 $\mu$s in duration) can be applied to mimic synaptic currents, which has the advantage that one can not only compare the results of concentration jumps on native receptors with those on recombinant receptors, but one can also compare the results with synaptic responses evoked by stimulation of presynaptic inputs in brain slices (see review by Edmonds *et al.* 1995). This may allow inferences about the subunit composition of the subsynaptic receptors. Experiments of this sort have yet to be published in detail on recombinant NMDA receptors, though they have already given valuable information about non-NMDA receptors (Colquhoun *et al.* 1992; Hestrin 1992, 1993; Raman and Trussell 1992).

## Conclusions

Transfection of cloned subunits into heterologous expression systems like the *Xenopus* oocyte does not invariably reproduce accurately the behaviour of receptors in their native environment (Edmonds *et al.* 1995). However, it seems clear from the work described above that transfection of NMDA receptors into both oocytes and into HEK 293 cells can reproduce the behaviour of native NMDA receptors rather accurately. This is a great convenience. In particular, the single-channel records at equilibrium are quantitatively indistinguishable in the two expression systems, and in various neurones. It is also clear that such records may provide a very clear distinction between different subunit combinations in some cases (e.g. NR1/NR2A versus NR1/NR2C), whereas in other cases they are of no help (e.g. NR1/NR2A versus NR1/NR2B). In general, we need *both* single channel records and other information (e.g. concentration jump responses, glycine and magnesium sensitivity, and pharmacological profile) in order to demonstrate similarity between an expressed subunit combination and a native receptor. The evidence is still quite incomplete; there are, for example, few good estimates (e.g. from Schild plots) of antagonist equilibrium constants, and more quantitative information about glycine and magnesium sensitivity would be valuable.

## Outlook

The electrophysiological experiments discussed above are, of course, only half the story. Parallel work on the molecular biology and biochemistry of the receptors is equally necessary, though it is electrophysiological work which must, ultimately, be the criterion for relevance to synaptic transmission.

Recently, several papers have appeared which describe *in situ*-hybridization studies using some or all of the published NMDA receptor subunits (Watanabe *et al.* 1992, 1993; Mikkelsen *et al.* 1993; Pujic *et al.* 1993; Standaert *et al.* 1993; DellaVedova *et al.* 1994; Ishida *et al.* 1994; Monyer *et al.* 1994). Some of them deal with only a specific region of the brain or with only one developmental stage. But taken together they show the heterogeneity in time and space of CNS NMDA receptor subunit distribution, and give a hint as to which subunit combinations one might expect in different types of cell. It is nevertheless essential to confirm the data on the level of functioning channels, i.e. on an electrophysiological basis, because *in situ* tells us only that messenger RNA coding for a certain subunit can be found in a cell. This does not tell anything about the active receptor finally incorporated into the cell membrane.

It should soon be possible to select a specific brain area (during a specific developmental stage), or even a certain cell type in a nucleus, which shows only a clearly defined subunit composition and do electrophysiological recordings from these membranes. The 'single-cell PCR' approach, which allows amplification of the subunits that are coded for in one individual cell (possibly one from which a recording has already been made) should be a valuable approach too; it has already been used successfully on non-NMDA-type glutamate receptors (Lambolez *et al.* 1992; Bochet *et al.* 1994). The method of patch-clamping in brain slices (Edwards *et al.* 1989) whereby one can see, and choose, the cells of interest is the ideal technique to pursue these goals. The first results confirming *in situ*-hybridization studies in this way have recently been published (Farrant *et al.* 1994; see above).

A different, and valuable, approach which gives related information is the use of specific antibodies directed against particular subunits (Petralia *et al.* 1994). This method allows one to deal directly with the translated protein (perhaps already incorporated into the cell membrane; see for example Aoki *et al.* 1994 and Ehlers *et al.* 1995).

To account for the high level of organization in the nervous system one should also consider the possibility that even in one and the same neurone there might be differences in receptor subunit structure according to their location. It is easily possible to imagine that dendritic receptors are different from somatic ones, or even that different receptors might be expressed under different synapses. The advent of new methods, which make it possible to see fine structures like dendrites in brain slices, have already started to contribute towards the solution of these problems (Stuart *et al.* 1993; Stuart and Sakmann 1994; Spruston *et al.* 1995).

## Acknowledgements

P.S. was a recipient of a long-term fellowship of the Human frontier science programme organization.

## References

Akazawa, C., Shigemoto, R., Bessho, Y., Nakanishi, S., and Mizuno, N. (1994). Differential expression of five *N*-Methyl-D-Aspartate receptor subunit mRNAs in the cerebellum of developing and adult rats. *J. Comp. Neurol.*, **347**, 150–160.

Anantharam, V., Panchal, R.G., Wilson, A., Kolchine, V.V., Treistman, S.N., and Bayley, H. (1992). Combinatorial RNA splicing alters the surface charge on the NMDA receptor. *FEBS Lett.*, **305**, 27–30.

Aoki, C., Venkatesan, C., Go, C.-G., Mong, J.A., and Dawson, T.M. (1994). Cellular and subcellular localization of NMDA-R1 subunit immunoreactivity in the visual cortex of adult and neonatal rats. *J. Neurosci.*, **14**, 5202–5222.

Ascher, P. and Nowak, L. (1988). The role of divalent cations in the *N*-methyl-D-aspartate responses of mouse central neurones in culture. *J. Physiol.*, **399**, 247–266.

Ascher, P., Bregestovsky, P., and Nowak, L. (1988). *N*-methyl-D-aspartate-activated channels of mouse central neurones in magnesium-free solutions. *J. Physiol.*, **399**, 207–266.

Béhé, P., Stern, P., Wyllie, D.J.A., Nassar, M., Schoepfer, R., and Colquhoun, D. (1995). Determination of the NMDA NR1 subunit copy number in recombinant NMDA receptors. *Proc. R. Soc. Lond. B.*, **262**, 205–213.

Bekkers, J.M., Vidovic, M., and Ymer, S. (1996). Differential effects of histamine on *N*-methyl-D-aspartate channel in hippocampal slices and cultures. *Neuroscience*, **72**, 669–677.

Bennett, J.A. and Dingledine, R. (1995). Topology profile for a glutamate receptor: three transmembrane domains and a channel-lining reentrant membrane loop. *Neuron*, **14**, 373–384.

Bliss, T.V.P. and Collingridge, G.L. (1993). A synaptic model of memory: long-term potentiation in the hippocampus. *Nature*, **361**, 31–39.

Bochet, P., Audinat, E., Lambolez, B., Crepel, F., Rossier, J., Lino, M., Tsuzuki, K., and Ozawa, S. (1994). Subunit composition at the single-cell level explains functional properties of a glutamate-gated channel. *Neuron*, **12**, 383–388.

Chazot, P.L., Coleman, S.K., Cik, M., and Stephenson, F.A. (1994). Molecular characterization of *N*-methyl-D-aspartate receptors expressed in mammalian cells yields evidence for for the coexistence of three subunit types within a discrete receptor molecule. *J. Biol. Chem.*, **269**, 24403–24409.

Choi, D.W. (1988). Glutamate neurotoxicity and diseases of the nervous system. *Neuron*, **1**, 623–634.

Colquhoun, D., Jonas, P., and Sakmann, B. (1992). Action of brief pulses of glutamate on AMPA/kainate receptors in patches from different neurones of rat hippocampal slices. *J. Physiol.*, **458**, 261–287.

Colquhoun, D. and Sigworth, F.J. (1995). Analysis of single channel data. In *Single channel recording*, 2nd ed, (ed. B. Sakmann and E. Neher), pp. 483–587. Plenum Press, New York.

Cull-Candy, S.G. and Usowicz, M.M. (1987). Multiple-conductance channels activated by excitatory amino acids in cerebellar neurons. *Nature*, **325**, 525–528.

Cull-Candy, S.G. and Usowicz, M.M. (1989). On the multiple-conductance single channels activated by excitatory amino acids in large cerebellar neurones of the rat. *J. Physiol.*, **415**, 555–582.

DellaVedova, F., Bonecchi, L., Bianchetti, A., Fariello, R.G., and Speciale, C. (1994). Age-related changes in the relative abundance of NMDAR1 mRNA spliced variants in the rat brain. *Neuroreport*, **5**, 581–584.

Durand, G.M., Gregor, P., Zheng, X., Bennett, M.V.L., Uhl, G.R., and Zukin, R.S. (1992). Cloning of an apparent splice variant of the rat *N*-methyl-D-aspartate receptor NMDAR1 with altered sensitivity to polyamines and activators of protein kinase C. *Proc. Natl Acad. Sci. USA*, **89**, 9359–9363.

Durand, G.M., Bennett, M.V.L., and Zukin, R.S. (1993). Splice variants of the *N*-methyl-D-aspartate receptor NR1 identify domains involved in regulation by polyamines and protein kinase C. *Proc. Natl Acad. Sci. USA*, **90**, 6731–6735.

Edmonds, B., Gibb, A.J. and Colquhoun, D. (1995). Mechanisms of activation of acetylcholine and glutamate receptors and the time course of synaptic currents. *Annu. Rev. Physiol.*, **57**, 495–519.

Edmonds, B. and Colquhoun, D. (1992). Rapid decay of averaged single-channel NMDA receptor activations recorded at low agonist concentrations. *Proc. R. Soc. Lond. B.*, **250**, 279–286.

Edwards, F.A., Konnerth, A., Sakmann, B., and Takahashi, T. (1989). A thin slice preparation for patch clamp recordings from synaptically connected neurones of the mammalian central nervous system. *Pflügers Arch. Eur. J. Physiol.*, **414**, 600–612.

Ehlers, M.D., Tingley, W.G., and Huganir, R.L. (1995). Regulated subcellular distribution of the NR1 subunit of the NMDA receptor. *Science*, **269**, 1734–1737.

Farrant, M., Feldmeyer, D., Takahashi, T., and Cull-Candy, S.G. (1994). NMDA-receptor channel diversity in the developing cerebellum. *Nature*, **368**, 335–339.

Fox, K. and Daw, N.W. (1993). Do NMDA receptors have a critical function in visual cortical plasticity? *Trends Neurosci.*, **16**, 116–122.

Gibb, A.J. and Colquhoun, D. (1991). Glutamate activation of a single NMDA receptor-channel produces a cluster of channel openings. *Proc. R. Soc. Lond. B.*, **243**, 39–45.

Gibb, A.J. and Colquhoun, D. (1992). Activation of NMDA receptors by L-glutamate in cells dissociated from adult rat hippocampus. *J. Physiol.*, **456**, 143–179.

Headley, P.M and Grillner, S. (1990). Excitatory amino acids and synaptic transmission: the evidence for a physiological function. *Trends Pharmacol. Sci.*, **11**, 205–211.

Hestrin, S. (1992). Activation and desensitization of glutamate-activated channels mediating fast excitatory synaptic currents in the visual cortex. *Neuron*, **9**, 991–999.

Hestrin, S. (1993). Different glutamate receptor channels mediate fast excitatory synaptic currents in inhibitory and excitatory cortical neurons. *Neuron*, **11**, 1083–1091.

Hollmann, M., Boulter, J., MAron, C., Beasley, L., Sullivan, J., Pecht, G., and Heinemann, S. (1993). Zinc potentiates agonist-induced currents at certain splice variants of the NMDA receptor. *Neuron*, **10**, 943–954.

Hollmann, M., Maron, C., and Heinemann, S. (1994). N-Glycosylation site tagging suggests a three transmembrane domain topology for the glutamate receptor GluR1. *Neuron*, **13**, 1331–1343.

Howe, J.R., Cull-Candy, S.G. and Colquhoun, D. (1991). Currents through single glutamate receptor channels in outside-out patches from rat cerebellar granule cells. *J. Physiol.*, **432**, 143–202.

Ikeda, K., Nagasawa, M., Mori, H., Araki, K., Sakimura, K., Watanabe, M., Inoue, Y., and Mishina, M. (1992). Cloning and expression of the NMDA receptor channel. *FEBS Lett.*, **313**, 34–38.

Ishida, N., Matsui, M., Mitsui, Y., and Mishina, M. (1994). Circadian expression of NMDA receptor mRNAs, ε3 and ξ1, in the suprachiasmatic nucleus of rat brain. *Neurosci. Lett.*, **166**, 211–215.

Ishii, T., Moriyoshi, K., Sugihara, H., Sakurada, K., Kadotani, H., Yokoi, M., *et al.* (1993). Molecular characterization of the family of the *N*-Methyl-D-Aspartate receptor subunits. *J. Biol. Chem.*, **268**, 2836–2843.

Ito, I., Sakimura, K., Mishina, M., and Sugiyama, H. (1996). Age-dependent reduction of hippocampal LTP in mice lacking *N*-methyl-D-aspartate receptor ε1 subunit. *Neurosci. Lett.*, **203**, 69–71.

Johnson, J.W. and Ascher, P. (1987). Glycine potentiates the NMDA response in cultured mouse brain neurons. *Nature*, **325**, 529–531.

Johnson, J.W. and Ascher, P. (1990). Voltage-dependent block by intracellular $Mg^{2+}$ of *N*-methyl-D-aspartate-activated channels. *Biophys. J.*, **57**, 1085–1090.

Karp, S.J., Masu, M., Eki, T., Ozawa, K., and Nakanishi, S. (1993). Molecular cloning and chorosomal localization of the key subunit of the human *N*-methyl-D-aspartate receptor. *J. Biol. Chem.*, **268**, 3728–3733.

Kleckner, N.W. and Dingledine, R. (1988). Requirement for glycine in activation of NMDA-receptors expressed in *Xenopus* oocytes. *Science*, **241**, 835–837.

Kleinschmidt, A., Bear, M.F. and Singer, W. (1987). Blockade of NMDA receptors disrupts experience-dependent plasticity in kitten striate cortex. *Science*, **238**, 355–358.

Komuro, H. and Rakic, P. (1993). Modulation of neuronal migration by NMDA receptors. *Science*, **260**, 95–97.

Kuner, T. and Schoepfer, R. (1996). Multiple structural elements determine subunit specificity of $Mg^{2+}$ block in NMDA receptor channels. *J. Neurosci.*, **16**, 3549–3558.

Kusiak, J.W. and Norton, D.D. (1993). A splice variant of the *N*-methyl-D-aspartate (NMDAR1) receptor. *Mol. Brain Res.*, **20**, 64–70.

Kutsuwada, T., Kashiwabuchi, N., Mori, H., Sakimura, K., Kushiya, E., Araki, K., *et al.* (1992). Molecular diversity of the NMDA receptor channel. *Nature*, **358**, 36–41.

Lambolez, B., Audinat, E., Bochet, P., Crepel, F., and Rossier, J. (1992). AMPA receptor subunits expressed in single Purkinje cells. *Neuron*, **9**, 247–258.

Lester, R.A.J., Clements, J.D., Westbrook, G.L., and Jahr, C.E. (1990). Channel kinetics determine the time-course of NMDA receptor-mediated synaptic currents. *Nature*, **346**, 565–567.

Li, Y., Erzurumlu, R.S., Chen, C,. Jhaveri, S., and Tonegawa, S. (1994). Whisker-related neuronal patterns fail to develop in the trigeminal brainstem nuclei of NMDAR1 knockout mice. *Cell*, **76**, 427–437.

Malenka, R.C. and Nicoll, R.A. (1993). NMDA-receptor-dependent synaptic plasticity: multiple forms and mechanisms. *Trends Neurosci.*, **16**, 521–527.

Meguro, H., Mori, H., Araki, K., Kushiya, E., Kutsuwada, T., Yamazaki, M., *et al.* (1992). Functional characterization of a heteromeric NMDA receptor channel expressed from cloned cDNAs. *Nature*, **357**, 70–74.

Meldrum, B. and Garthwaite, J. (1990). Excitatory amino acid neurotoxicity and neurodegenerative disease. *Trends Pharmacol. Sci.*, **11**, 379–387.

Mikkelsen, J.D., Larsen, P.J., and Ebling, F.J.P. (1993). Distribution of *N*-methyl-D-aspartate (NMDA) receptor mRNAs in the rat suprachiasmatic nucleus. *Brain Res.*, **632**, 329–333.

Mishina, M., Takai, T., Imoto, K., Noda, M., Takahashi, T., Numa, S., Methfessel, C., and Sakmann, B. (1986). Molecular distinction between fetal and adult forms of muscle acetylcholine receptor. *Nature*, **321**, 406–411.

Momiyama, A., Feldmeyer, D., and Cull-Candy, S.G. (1996). Identification of a native low-conductance NMDA-channel with reduced sensitivity to $Mg^{2+}$ in rat central neurones. *J. Physiol.*, **494**, 479–492

Monyer, H., Sprengel, R., Schoepfer, R., Herb, A., Higuchi, M., Lomeli, H., *et al.* (1992). Heteromeric NMDA receptors: molecular and functional distinction of subtypes. *Science*, **256**, 1217–1221.

Monyer, H., Burnashev, N., Laurie, D.J., Sakmann, B., and Seeburg, P.H. (1994). Developmental and regional expression in the rat brain and functional properties of four NMDA receptors. *Neuron*, **12**, 529–540.

Moriyoshi, K., Masu, M., Ishii, T., Shigemoto, R., Mizuno, N., and Nakanishi, S. (1991). Molecular cloning and characterization of the rat NMDA receptor. *Nature*, **354**, 31–37.

Nakanishi, N., Axel, R., and Shneider, N.A. (1992). Alternative splicing generates functionally distinct *N*-methyl-D-aspartate receptors. *Proc. Natl Acad. Sci. USA*, **89**, 8552–8556.

Nowak, L., Bregestovski, P., Ascher, P., Herbet, A., and Prochiantz, A. (1984). Magnesium gates glutamate-activated channels in mouse central neurones. *Nature*, **307**, 462–463.

Orrego, F. and Villanueva, S. (1993). The chemical nature of the main central excitatory transmitter: a critical appraisal based upon release studies and synaptic vesicle localization. *Neuroscience*, **56**, 539–555.

Paoletti, P., Neyton, J., and Ascher, P. (1995). Glycine-independent and subunit-specific potentiation of NMDA responses by extracellular $Mg^{2+}$. *Neuron*, **15**, 1109–1120.

Petralia, R.S., Yokotani, N., and Wenthold, R.J. (1994). Light and electron microscope distribution of the NMDA receptor subunit NMDAR1 in the rat nervous system using a selective anti-peptide antibody. *J. Neurosci.*, **14**, 667–696.

Piggott, M.A., Perry, E.K., Perry, R.H., and Scott, D. (1993). *N*-methyl-D-aspartate (NMDA) and non-NMDA binding sites in developing human frontal cortex. *Neurosci. Res. Commun.*, **12**, 9–16.

Planells-Cases, R., Sun, W., Ferrer-Montiel, A.V., and Montal, M. (1993). Molecular cloning, functional expression, and pharmacological characterization of an *N*-methyl-D-aspartate receptor subunit from human brain. *Proc. Natl Acad. Sci. USA*, **90**, 5057–5061.

Pujic, Y., Matsumoto, I., and Wilce, P.A. (1993). Expression of the gene coding for the NR1 suunit of the NMDA receptor during rat brain development. *Neurosci. Lett.*, **162**, 67–70.

Raman, I.M. and Trussell, L.O. (1992). The kinetics of the response to glutamate and kainate in neurons of the avian cochlear nucleus. *Neuron*, **9**, 173–186.

Sheng, M., Cummings, J., Roldan, L.A., Jan, Y.N., and Jan, L.Y. (1994). Changing subunit composition of heteromeric NMDA receptors during development of rat cortex. *Nature*, **368**, 144–147.

Spruston, N., Jonas, P., and Sakmann, B. (1995). Dendritic glutamate receptor channels in rat hippocampal CA3 and CA1 pyramidal neurones. *J. Physiol.*, **482**, 325–352.

Standaert, D.G., Testa, C.M., Penney, J.B., and Young, A.B. (1993). Alternatively spliced isoforms of the NMDAR1 glutamate receptor subunit: differential expression in the basal ganglia of the rat. *Neurosci. Lett.*, **152**, 161–164.

Stern, P., Béhé, P., Schoepfer, R., and Colquhoun, D. (1992). Single-channel conductances of NMDA receptors expressed from cloned cDNAs: comparison with native receptors. *Proc. R. Soc. Lond. B.*, **250**, 271–277.

Stern, P., Cik, M., Colquhoun, D., and Stephenson, F.A. (1994). Single channel properties of cloned NMDA receptors in HEK 293 cells: comparison with results from *Xenopus* oocytes. *J. Physiol.*, **476**, 391–397.

Streit, P. (1985). Glutamate and aspartate as transmitter candidates for systems of the cerebral cortex. In *Cerebral cortex*, Vol. 2, (ed. A. Peters, and E.G. Jones), pp. 119–143. Plenum Press, New York.

Stuart, G.J., Dodt, H.U., and Sakmann, B. (1993). Patch-clamp recordings from the soma and dendrites of neurons in brain slices using infrared video microscopy. *Pflügers Arch. Eur. J. Physiol.*, **423**, 511–518.

Stuart, G.J. and Sakmann, B. (1994). Active propagation of somatic action potentials into neocortical pyramidal cell dendrites. *Nature*, **367**, 69–72.

Sugihara, H., Moriyoshi, K., Ishii, T., Masu, M., and Nakanishi, S. (1992). Structures and properties of seven isoforms of the NMDA receptor generated by alternative splicing. *Biochem. Biophys. Res. Commun.*, **185**, 826–832.

Tölle, T.R., Berthele, A., Zieglgansberger, W., Seeburg, P.H., and Wisden, W. (1993). The differential expression of 16 NMDA and non-NMDA receptor subunits in the rat spinal cord and in periaqueductal gray. *J. Neurosci.* **13**, 5009–5028.

Traynelis, S.F., Hartley, M., and Heinemann, S.F. (1995). Control of proton sensitivity of the NMDA receptor by RNA splicing and polyamines. *Science*, **268**, 873–876.

Wafford, K.A., Bain, C.J., Le Bourdelles, B., Whiting, P.J., and Kemp J.A. (1993). Preferential co-assembly of recombinant NMDA receptors composed of three different subunits. *Neuroreport*, **4**, 1347–1349.

Watanabe, M., Inoue, Y., Sakimura, K., and Mishina, M. (1992). Developmental changes in distribution of NMDA receptor channel subunit mRNAs. *Neuroreport*, **3**, 1138–1140.

Watanabe, M., Inoue, Y., Sakimura, K., and Mishina, M. (1993). Distinct distributions of five *N*-methyl-D-aspartate receptor channel subunit mRNAs in the forebrain. *J. Comp. Neurol.*, **338**, 377–390.

Wo, Z.G. and Oswald, R.E. (1995*a*). A topological analysis of goldfish kainate receptors predicts three transmembrane segments. *J. Biol. Chem.*, **270**, 2000–2009.

Wo, Z.G. and Oswald, R.E. (1995*b*). Unraveling the modular design of glutamate-gated ion channels. *Trends Neurosci.*, **18**, 161–168.

Wyllie, D.J.A., Béhé, P., Nassar, M., Schoepfer, R., and Colquhoun, D. (1996). Single-channel currents from recombinant NMDA NR1a/NR2D receptors expressed in *Xenopus* oocytes. *Proc. R. Soc. Lond. B.*, **263**, 1079–1086

Yamazaki, M., Mori, H., Araki, K., Mori, K.J., and Mishina, M. (1992). Cloning, expression and modulation of a mouse NMDA receptor subunit. *FEBS Lett.*, **300**, 39–45.

Zhang, L., Zheng, X., Paupard, M.C., Wang, A.P., Santchi, L., Friedmann, L.K., Zukin, R.S., and Bennett, M.V. (1994). Spermine potentiation of recombinant *N*-methyl-D-aspartate receptors is affected by subunit composition. *Proc. Natl Acad. Sci. USA*, **91**, 10883–10887.

Zheng, X., Zhang, L., Durand, G.M., Bennett, M.V., and Zukin, R.S. (1994). Mutagenesis rescues spermine and $Zn^{2+}$ potentiation of recombinant NMDA receptors. *Neuron*, **12**, 811–818.

# 20 A unified theory of antidepressant action: evidence for adaptation of the NMDA receptor following chronic antidepressant treatments

R. T. Layer, P. Popik, G. Nowak, I. A. Paul, R. Trullas, and P. Skolnick

## Introduction

Clinically effective antidepressants (AD) include drugs with a remarkable structural diversity as well as non-pharmacological interventions such as electronconvulsive shock (ECS).[1,2] These therapies produce a variety of *in vitro* effects (e.g. blockade of monoamine reuptake) that cannot be causally related to antidepressant efficacy.[3–5] Moreover, chronic AD treatment, which appears to be required for clinical improvement across therapies,[1,2] has not been demonstrated to effect consistent changes in any described transmitter system.[4–8] The failure to evince consistent changes in noradrenergic, dopaminergic, serotonergic, GABAergic, and peptidergic neurotransmitter systems across chronic AD treatments has led to the hypothesis that perturbation of any of these transmitter systems alone cannot account for either AD activity across therapies or provide a coherent biochemical basis of depressive symptomatology.[9,10]

Following our initial report demonstrating that functional NMDA antagonists are active in the forced swim and tail suspension tests,[11] several laboratories have independently demonstrated the ability of this group of compounds to mimic the effects of clinically active AD in a variety of preclinical models. More recent findings from our laboratory demonstrate that chronic treatment with AD can effect adaptive changes in the radioligand-binding properties of the NMDA receptor. The consistency of these neurochemical changes following AD drawn from every principal therapeutic class, when taken together with the slowly developing nature of these phenomena, strongly suggests the NMDA receptor may be a final common mediator of AD action. Based on this pharmacological evidence, it may be hypothesized that glutamatergic pathways also play an essential role in the pathophysiology of depression.

## Functional NMDA antagonists exhibit AD-like actions in preclinical models

In our initial studies, the actions of a competitive NMDA antagonist (2-amino-7-phosphonoheptanoic acid; AP-7), a non-competitive NMDA antagonist (dizocilpine; MK-801), and a partial agonist at strychnine-insensitive glycine receptors (1-aminocyclopropanecarboxylic acid; ACPC) were evaluated in a forced swim test (FST) adapted for mice.[11] Behavioural procedures such as the FST[12] and tail suspension test (TST)[13] are not generally considered animal models of depression. Nonetheless, both tests will detect potential AD and many variations of these procedures are commonly employed for this purpose.[14,15] AP-7 produced a significant, dose-dependent reduction in the duration of immobility in the FST at doses that did not affect ambulatory time in an open field (Table 20.1). This latter finding is particularly significant, since motor stimulants can produce false positives in the FST. Dizocilpine also significantly reduced the duration of immobility in this test at doses as low as 0.1 mg $kg^{-1}$). Nonetheless, this effect of dizocilpine was biphasic (Table 20.1), with the maximum reduction in immobility (92% at 0.5 mg $kg^{-1}$) produced by a dose which also significantly increased (34%) time spent ambulating in an open field. While the highest dose of dizocilpine used in this study (1 mg $kg^{-1}$) produced a significant reduction in immobility without an accompanying change in time spent ambulating in an open field, doses of dizocilpine in this range produce a variety of psychomotor effects in mice.[16] Panconi *et al.*[17] confirmed the efficacy of MK-801 (0.1 mg $kg^{-1}$) in the FST at doses below those stimulating motor activity (0.3 mg $kg^{-1}$), but this separation of activity was not manifested in the TST. These investigators concluded that MK-801 may be a false positive in these screening models.

**Table 20.1** Effects of AP-7 and MK-801 on swim-induced immobility and ambulatory activity in an open field.

| Drug | Dose (mg $Kg^{-1}$ | No. of mice | Immobility (s) | δ (%) | No. of mice | Activity (s) | δ (%) |
|---|---|---|---|---|---|---|---|
| Control | – | 28 | 144±6 | | 9 | 145±5 | |
| AP-7 | 40 | 7 | 141±21 | | 4 | 144±5 | |
| | 80 | 10 | 153±9 | | 3 | 167±16 | |
| | 100 | 20 | 93±8* | −35 | 11 | 153±5 | |
| | 200 | 11 | 55±13*# | −62 | 17 | 134±10 | |
| Control | – | 8 | 140±12 | | 8 | 150±10 | |
| MK-801 | 0.1 | 8 | 80±10* | −43 | 8 | 144±4 | |
| | 0.5 | 8 | 11±4*# | −92 | 8 | 201±3*# | +34 |
| | 1.0 | 8 | 64±23* | −54 | 8 | 170±14 | |

MK-801 and AP-7 were administered i.p. to male NIH/HSD mice 15 min and 30 min before testing, respectively. Controls received an equivalent volume of the corresponding vehicle. Immobility (s) during forced swim and ambulatory activity (s) in an open field were measured as described in the text. Values represent mean ± S.E.M.*, indicates significantly different from control group;#, significantly different from all other groups ($P<0.05$, Student–Newman–Keuls test). (Data from ref. 11.)

Parenteral administration of ACPC also reduced immobility in the FST in a dose-dependent fashion (Fig. 20.1), with a minimum effective dose of 100 mg $kg^{-1}$ and a maximum reduction in immobility (65%) manifested at 200 mg $kg^{-1}$ (Fig. 20.1). Parenteral administration of glycine completely abolished this effect, which is consistent with a functional antagonist action of ACPC mediated at strychnine-insensitive glycine receptors. The dose of glycine used to block this effect of ACPC in the FST elevated hippocampal glycine levels by ~60%, but was devoid of intrinsic activity in the FST and did not affect imipramine (IMI)-induced reductions in immobility.[11] These findings are also consonant with the ability of glycine to reverse other pharmacological actions of ACPC (including its neuroprotective actions) both *in vitro*[18] and *in vivo*.[19,20]

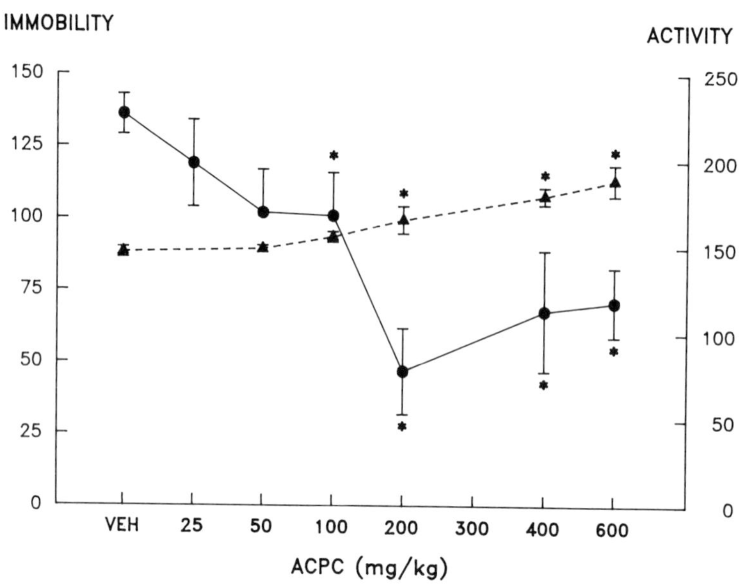


**Fig. 20.1** Effects of ACPC on duration of immobility in the forced swim test (FST) and ambulatory time in an open field. Values represent mean ±S.E.M. Symbols: ●, immobility (s); ▲, ambulatory time (s). * indicates significantly different from vehicle (VEH) group, $P<0.05$, Student–Newman–Keuls test. The number of animals examined in the FST test was: VEH, 40; ACPC (25 mg $kg^{-1}$) 11, and (50–600 mg $kg^{-1}$), 10. In the open field: VEH, 23; ACPC (50 and 600 mg $kg^{-1}$), 5; and ACPC (100–400 mg $kg^{-1}$) 10. Male NIH/HSD mice (25–30 g bodyweight) were injected i.p. with ACPC dissolved in saline. Controls received an equivalent volume of saline. Animals were tested 15 min after injection. Several putative antagonists at strychnine-insensitive glycine receptors (ACBC, I2CL, 7C-KYN, HA-966 and C-LEU) did not reduce immobility in the FST but some produced severe ataxia and some were lethal when combined with forced swim. (Data from Trullas and Skolnick[11].)

The maximum reductions in immobility observed following both AP-7 and ACPC were comparable with those produced by the prototypic tricyclic AD, imipramine.[11] ACPC also produced modest but statistically significant increases in ambulatory time that appear unrelated to its actions in the FST. Thus, while ACPC-mediated reductions in immobility reached a plateau at 200 mg $kg^{-1}$, ambulatory time continued to increase, albeit modestly, as a function of dose (Fig. 20.1). Motor stimulants detected as false positives in the FST can completely eliminate immobility, while the reductions in immobility produced by clinically effective AD generally reach a plateau, as was observed with both AP-7 and ACPC. Subsequent studies comparing the effects of orally and parenterally administered ACPC[21] also evinced a temporal dissociation between reductions in immobility in the FST and increased motor activity in an open field. Since the predictive validity of the FST may be lower in mice than rats,[22] the effects of ACPC were examined in another commonly used model to detect clinically effective AD, the TST.[13] ACPC produced a dose-dependent reduction in immobility in this measure with a minimum effective dose of 200 mg $kg^{-1}$.[11] Since the latter studies were performed in C57B1/6 mice (chosen for their high basal immobility scores in this test relative to other inbred mouse strains), it was concluded the ability of functional NMDA antagonists to mimic AD in these tests is not strain-dependent; subsequent studies by other laboratories have demonstrated this efficacy predictive of AD action is not species-dependent (see below). More recent studies from our laboratory, using the FST, have demonstrated the efficacy of another functional NMDA antagonist acting at the polyamine site. SL 82.0715 (eliprodil) produces a dose-dependent reduction in immobility in the FST at doses which have no significant effect on motor activity in the open field. Increasing doses of eliprodil depress motor activity, and at these doses, immobility in the FST increased above baseline. Thus, the efficacy of eliprodil in the FST provides an independent line of evidence that the ability of functional NMDA antagonists to mimic AD in these preclinical tests is not simply related to locomotor stimulation (Layer *et al.* 1994, in preparation).

Subsequent studies have demonstrated that functional NMDA antagonists have AD-like actions in other variations of the FST. For example, Maj *et al.*[23] demonstrated the efficacy of dizocilpine in a FST procedure using rats. Moreover, dizocilpine potentiated the effects of imipramine and citalopram (but not mianersin) in the FST. Both the direct and potentiating effects of dizocilpine in the FST were manifested at doses that did not increase motor activity. Maj *et al.*[24] also examined the effects of competitive NMDA antagonists CGP 37849 and CGP 39551 in the FST using rats. Both compounds were active in the FST at doses which either had no effect or decreased motor activity.

While the FST and TST are generally predictive of clinical AD activity, these screening procedures are not generally considered to model depression.[25] Willner and colleagues[26] have attempted to construct a behavioural paradigm in rats using chronic, mild stress-induced anhedonia to model aspects of depression.

Following 3 or more weeks of exposure to chronic mild stressors, rats typically decrease consumption of a 1% sucrose solution, which has been proposed as the development of anhedonia. While this procedure has not yet been as extensively validated as the FST and TST using pharmacological agents, chronic treatment with AD[27] can reverse stress-induced reductions in sucrose consumption. Papp and Moryl[28,29] have examined the effects of NMDA antagonists in this paradigm and demonstrated that both a non-competitive (dizocilpine) and competitive (CGP 37849) NMDA antagonist were as effective as imipramine in restoring sucrose consumption. The antidepressant-like actions of functional NMDA antagonists in several preclinical tests provides an indication of the therapeutic potential of this class of compounds as AD (Table 20.2). While the clinical application of some classes of NMDA antagonists may be problematic, particularly on a long-term basis, these preclinical studies provided the impetus for recent studies which led us to conclude that adaptive changes in the NMDA receptor complex may be the final common mechanism of AD action.

## Neurochemical studies

### Chronic treatment with functional NMDA antagonists down-regulates cortical $\beta$-adrenoceptors

Chronic, but not acute treatment of laboratory animals with many clinically effective AD therapies down-regulates (i.e., reduces the maximum number of) cortical $\beta$-adrenoceptors.[30,31] While the relationship of $\beta$-adrenoceptor down-regulation to both the mechanism

**Table 20.2** Summary of studies demonstrating functional NMDA antagonists mimick clinically active AD in preclinical models.

| **Functional NMDA antagonists are active in models that predict antidepressant action in humans** |
|---|
| 1. AP-7, dizocilpine, and ACPC are active in the forced swim test; ACPC is active in the tail suspension test.[11, 21, 53] |
| 2. Dizocilpine is active in forced swim test + potentiated 'classical' antidepressants in this paradigm.[23] |
| 3. CGP 37849 and CGP 39551 (competitive NMDA antagonists) are active in the forced swim test.[24] |
| 4. The polyamine antagonist SL 82.0715 (eliprodil) is active in forced swim test (Layer *et al.* in preparation). |
| **Like many clinically active antidepressants, functional NMDA antagonists are active in putative models of depression** |
| 1. CGP 37849, CGP 40116, and dizocilpine reduce the anhedonia (deceased consumption of sucrose solution) produced by chronic, mild stress (Willner's model).[28, 29] |
| 2. Dizocilpine blocks development of learned helplessness.[67] |

7-Chlorokynurenic acid (25–150 mg $kg^{-1}$), indole-2-carboxylic acid (25–200 mg $kg^{-1}$), HA-966 (2.5–10 mg kg and ACBC (50–100 mg $kg^{-1}$) were inactive in the forced swim test (mice). Numbers in parentheses refer to references. Several of these compounds were lethal when combined with forced swim.[11]

of antidepressant action and the pathophysiology of depression remains obscure, this phenomenon has been the among the most consistent findings across antidepressant therapies.[31] The activity of functional NMDA antagonists in preclinical models predictive of antidepressant action led us to examine the effects of chronic treatment with such compounds on radioligand binding to cortical $\beta$-adrenoceptors. In our initial studies,[32] mice were treated for 7 days with doses of ACPC (200 mg $kg^{-1}$) or dizocilpine (0.1 and 1 mg $kg^{-1}$) that reduced immobility in the FST (Table 20.1; Fig. 20.1) and [$^3$H]dihydroalprenolol (DHA) binding to $\beta$-adrenoceptors was measured in cerebral cortex. The prototypic antidepressant imipramine (15 mg $kg^{-1}$ × 7 days) reduced [$^3$H]DHA binding by 23% ($P < 0.05$) compared with 19% and 21% following chronic ACPC and dizocilpine, respectively (each $P < 0.05$ compared with vehicle-treated mice). The lower dose of dizocilpine (0.1 mg $kg^{-1}$) did not significantly affect [$^3$H]DHA binding. Intermediate doses of dizocilpine were not examined in these experiments. Subsequently, we found that chronic (14-day) administration of the polyamine antagonist eliprodil (which is also active in the FST) also produces a robust down-regulation of $\beta$-adrenoceptors (R. Layer *et al.* in preparation). *In toto*, these findings are consistent with the hypothesis that functional NMDA receptor antagonists are capable of producing an AD action and, by implication, that this family of ligand-gated ion channels may be involved in the pathophysiology of depression.[11] Both the behavioural and neurochemical effects of functional NMDA antagonists led to a series of experiments examining the effects of chronic AD treatment on the NMDA receptor.

## Chronic treatment with imipramine effects adaptive changes in the NMDA receptor

Our initial experimental design[33] consisted of injecting four groups daily for 2 weeks with: (i) imipramine (15 mg $kg^{-1}$ per day); (ii) ACPC (200 mg $kg^{-1}$ per day); (iii) dizocilpine

(0.1 mg $kg^{-1}$); or (iv) vehicle. Parallel groups of mice received one injection of drug or vehicle. Twenty-four hours after the last injection, brains were removed, dissected, and fast-frozen over solid $CO_2$. Radioligand binding to NMDA receptors, strychnine-insensitive glycine receptors, and NMDA receptor-coupled cation channels were evaluated using [$^3$H]CGP 39653 (a specific, high-affinity NMDA antagonist[34], [$^3$H]5,7-dichlorokynurenic acid (DCKA) (a specific high-affinity glycine antagonist[35]), and [$^3$H]dizocilpine[36], respectively. Chronic, but not acute treatment with imipramine significantly altered the radioligand-binding properties of the NMDA receptor which can be summarized as follows:

1. A ~2.5-fold reduction (compared with vehicle-treated mice) in the potency of glycine to inhibit [$^3$H]5,7-DCKA binding to strychnine-insensitive glycine receptors (Fig. 20.2). No significant difference in basal [$^3$H]5,7-DCKA binding was observed.
2. A 28% reduction in the proportion of high-affinity glycine sites inhibiting [$^3$H]CGP 39653 binding to NMDA receptors (Fig. 20.3). No significant difference in basal [$^3$H]CGP 39653 binding was observed.
3. A decrease (~36%) in basal [$^3$H]MK-801 binding to sites within NMDA receptor coupled cation channels which was reversible by the addition of glutamate.

These effects were observed in cerebral cortex but not in hippocampus, striatum, or basal forebrain (Figs. 20.2 and 20.3). Chronic treatment with ACPC resulted in changes which paralleled those of imipramine on ligand binding to the NMDA receptor complex (an approximately 2-fold reduction in the potency of glycine and an approximately 22% reduction in the proportion of high-affinity, glycine-displaceable [$^3$H]CGP 39653 binding sites, respectively), but the reduction in basal [$^3$H]dizocilpine binding (~25%) did not achieve statistical significance. In contrast, a regimen of 0.1 mg/ $kg^{-1}$ dizocilpine did not significantly alter these measures. These findings led us to hypothesize that chronic, but not acute, treatment with AD can effect adaptive changes in the NMDA receptor. If this hypothesis is correct, then such changes should: (i) be manifested across AD therapies independent of the chemical structure or *in vitro* effects; and (ii) develop slowly and persist for some time following cessation of treatment.

## Time course and dose-dependence studies

Manifestation of the therapeutic response to AD treatment typically requires 3–6 weeks,[37] and the remission of depressive symptomatologies can persist up to several months after discontinuation of treatment.[7,31] These clinical observations have focused preclinical studies of the mechanism of AD action on various adaptive responses to antidepressants.[3,4] For example, monoaminergic receptor adaptation (such as $\beta$ adrenoceptor, $5HT_2$, or dopamine $D_1$ receptor down-regulation) in the central nervous system has been observed following chronic administration of a wide range of AD therapies to rodents, including various pharmacotherapies, electroconvulsive shock (ECS), and rapid eye movement sleep deprivation (reviewed in ref. 7). While current theories underlying the therapeutic response to antidepressant treatments are generally based on adaptation of one or more monoaminergic receptor systems (reviewed in refs. 5, and 10) none of the

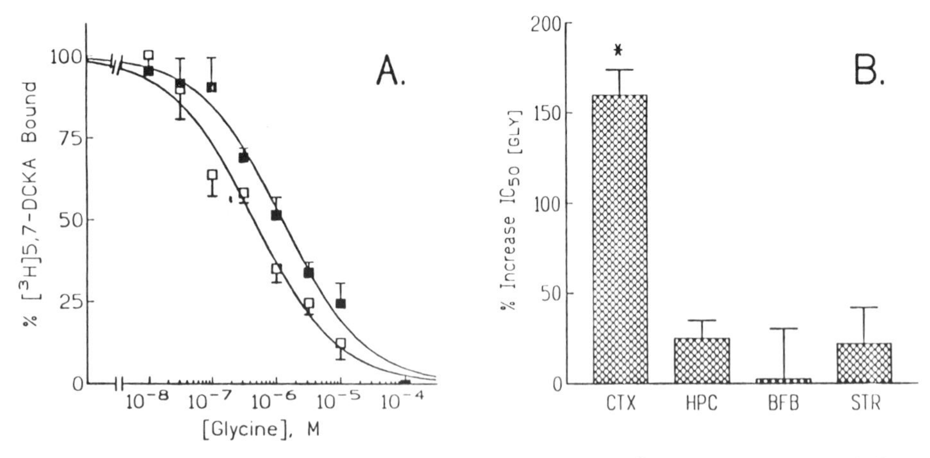


**Fig. 20.2** Effect of chronic imipramine (IMI) treatment (filled symbols, shaded bars) on [$^3$H]5,7-DCKA binding to cortical NMDA receptors. Data are presented as the mean ± S.E.M. of 5–12 animals/group. A. Glycine inhibition of [$^3$H]5,7-DCKA binding to cortical membranes. Data are expressed as percent specific binding of 20.2 ± 0.6 nM nM [$^3$H]5,7-DCKA. Untransformed values are presented in Table 20.1. Open symbols: saline injections; filled symbols, IMI injections. B. Effect of chronic IMI treatment on glycine inhibition of [$^3$H]5,7-DCKA binding in cortex (CTX), hippocampus (HPC), basal forebrain (BFB), and caudate nucleus (STR). Data are expressed as the percent increase in $IC_{50}$ versus. saline and represent the mean ± S.E.M. of 4–6 animals/group. *$P < 0.05$ versus. saline (Dunnett's *post hoc* comparison). (Data from Nowak *et al.*[33].)

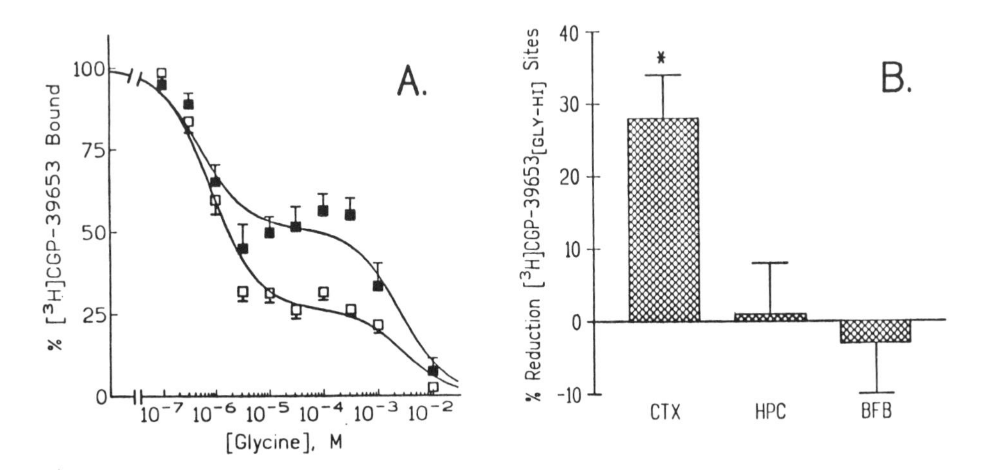


**Fig. 20.3** Effect of chronic imipramine (IMI) treatment on glycine inhibition of [$^3$H]CGP-39653 binding. A. Glycine inhibition of [$^3$H]CGP-39653 binding to cortical membranes. Data are expressed as percent specific binding of 5.3 ± 0.3 nM [$^3$H]CGP-39653. Untransformed values are presented in Table 20.2. Data were best fit to a two-site model with $F(1,8) = 37.7$, $P < 0.0005$ and $F(1,8) = 18.1$, $P < 0.004$ for saline (open) and IMI (filled) treated, respectively. B. Effect of chronic IMI treatment on glycine inhibition of [$^3$H]CGP-39653 binding in cortex (CTX), hippocampus (HPC), and basal forebrain (BFB). Data are expressed as the percent reduction in the apparent density of high-affinity, glycine-displaceable [$^3$H]CGP-39653 binding sites and represent the mean ± S.E.M. of 4–6 animals/group. *$P < 0.05$ versus. saline (Dunnett's *post hoc* comparison). (Data from Nowak *et al.*[33].)

adaptive phenomena examined to date is uniformly produced by all antidepressant treatments (see refs. 7 and 31 for reviews). Moreover, the adaptation of monoaminergic systems disappears rapidly, in some cases lasting only 24 hours after withdrawal of treatment.[1,31]

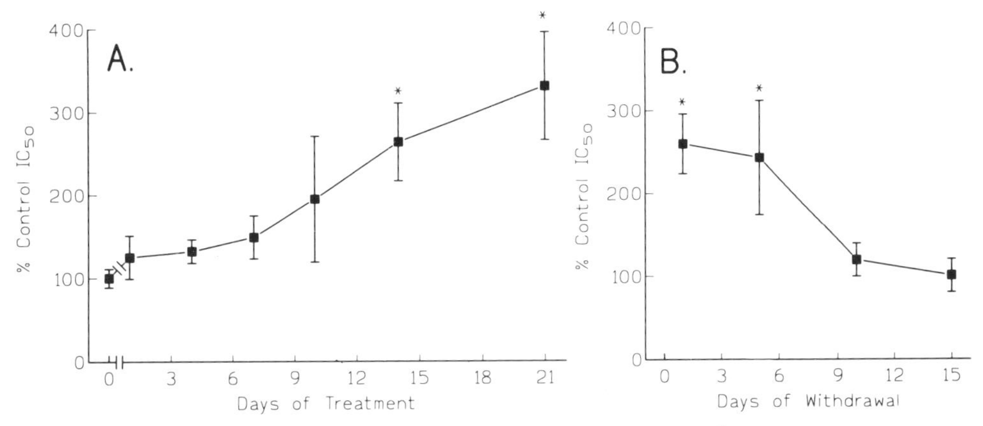


**Fig. 20.4** Time course of imipramine-induced increases in the $IC_{50}$ of glycine to inhibit [$^3$H]5,7-DCKA binding. Mice were treated with imipramine (15 mg $kg^{-1}$, i.p.) for 1–21 days and killed 24 h after the last treatment (A), or treated for 14 days and killed 1–15 days after cessation of treatment (B). Data are expressed as the percent of basal (control) $IC_{50}$ values and represent the mean ± S.E.M. of 4–15 animals/group. In cortices from saline-treated mice, basal [$^3$H]5,7-DCKA binding was 1229 ± 85 fmol $mg^{-1}$ protein and the $IC_{50}$ value of glycine was 0.39 ± 0.04 μM (*n* = 15 animals). *, $P < 0.05$ versus saline-treated animals. (Data from Paul *et al.*[38])

Since a 14-day regimen of IMI reduced the $IC_{50}$ of glycine to inhibit [$^3$H]5,7-DCKA binding in cortex, we examined the interval required for manifestation of this effect, the time required for this effect to dissipate following cessation of treatment, and the dose – response relationship for this action.[38] The first statistically significant increase was observed after 14 days (15 mg $kg^{-1}$ per day) of treatment (264% of control; Fig. 20.4). Continued treatment with imipramine for 21 days produced a further increase in the $IC_{50}$ of glycine to 330% of control (Fig. 20.4). The increased $IC_{50}$ of glycine produced by 14 days of treatment with imipramine was present at both 1 and 5 days after cessation of treatment, but was no longer apparent after 10 days (Fig. 20.4). When administered for 14 days, imipramine (2.5–30 mg $kg^{-1}$) significantly increased the $IC_{50}$ for glycine to displace [$^3$H]5,7-DCKA at 15 and 30 mg $kg^{-1}$ with an $ED_{50}$ of ~5 mg $kg^{-1}$. Chronic imipramine treatment (2.5–30 mg $kg^{-1}$) was without effect on basal [$^3$H]5,7-DCKA binding. Parallel studies were conducted with both the 5-HT uptake blocker citalopram (CIT) and ECS in order to determine the generality of these phenomena in response to other AD.[38] CIT (20 mg $kg^{-1}$) required at least 10 days of treatment to significantly increase in the $IC_{50}$ of glycine to inhibit [$^3$H]5,7-DCKA binding (200% of control). Continuing treatment for 14 or 21 days did not result in a further increase in the $IC_{50}$ of glycine. When administered for 14 days, CIT (5–40 mg $kg^{-1}$) produced significant increases in the $IC_{50}$ of glycine at doses ≥10 mg $kg^{-1}$ with an $ED_{50}$ value of ~5 mg $kg^{-1}$. Citalopram treatment did not significantly affect the basal [$^3$H]5,7-DCKA binding under any condition.[38] ECS produced at 370% increase in the $IC_{50}$ value of glycine to inhibit [$^3$H]5,7-DCKA binding by the 7th day of treatment. This effect persisted through the 10th day following cessation of treatment. ECS treatment did not significantly affect basal [$^3$H]5,7-DCKA binding.[38]

The prolonged drug treatment required to effect significant reductions in the potency of glycine in this measure is indicative of an adaptive response. While the response to a non-pharmacological AD treatment (ECS) also requires chronic administration to achieve significant effects, these effects appear to develop more rapidly (7 days versus 10–14 days) than those following either IMI or CIT. Moreover, these effects were remarkably persistent, requiring at least 7 days following cessation of treatment to return to control values. These findings are generally consistent with clinical observations that AD require chronic administration to achieve therapeutic results[3,37] and that the beneficial effects of these treatments can persist following discontinuation of treatment (reviewed in refs. 11 and 31). Furthermore, the rapid effects of ECS (compared with pharmacotherapies) which we observed are consistent with the rapid remission of depressive symptoms during ECS treatment.[39] In view of the apparent adaptive nature of these AD-induced changes in the ligand binding properties of the NMDA receptor complex, we examined the pharmacological specificity of these phenomena.

## Pharmacological specificity

If adaption of the NMDA receptor is a requisite for AD action, then such changes should be manifested after chronic treatment with a wide variety of AD, regardless of chemical structure or *in vitro* actions. In order to test this hypothesis, mice were chronically injected with 17 AD and the potency of glycine to inhibit [$^3$H]5,7-DCKA binding and/or its effects on [$^3$H]CGP 39653 binding examined in cortical tissues.[38] These findings are summarized in Tables 20.3 and 20.4. The 17 AD used to test the pharmacological specificity of NMDA receptor adaptation have demonstrated efficacy in controlled, double-blind studies [1,2]; (Table 20.3 legend). These compounds include tricyclic AD, monoamine oxidase inhibitors (MAOI), the so-called 'atypical' agents, and ECS. Among the AD tested, only chronic treatment with the MAOI pargyline failed to significantly reduce the potency of glycine to inhibit [$^3$H]5,7-DCKA binding. Nonetheless, the ability of chronic treatment with these AD to produce this adaptive change in the ligand-binding properties of the NMDA receptor complex is a slightly more robust predictor of clinical efficacy (16/17, 94%) than either changes in $\beta$-adrenoceptor density (12/17) or efficacy in the FST (12/17, 71%; $P < 0.08$, Fisher's exact test) (Table 20.3). These latter measures are perhaps the most widely used biochemical and behavioural predictors of AD activity, respectively.[8,12,22] A more limited number of substances considered 'false positives' or 'negatives' in either the FST or $\beta$-adrenoceptor down-regulation were also examined for their effects on ligand binding to the NMDA receptor complex. If this group of compounds is also considered, an increase in the $IC_{50}$ of glycine was a significantly better predictor of antidepressant action (22/23, 96%) than either the FST (13/23, 57%; $P < 0.002$, Fisher's exact test) or reductions in $\beta$-adrenoceptor density (15/23; 65%; $P < 0.01$, Fisher's exact test) (Table 20.3). The range of compounds tested also provides some indication of the specificity of this phenomenon. While only one dose of each isomer was studied, chronic administration of the inactive (D-) isomer of L-deprenyl did not produce a significant change in the potency of glycine to inhibit [$^3$H]5,7-DCKA binding, thus providing

**Table 20.3** Effect of chronic treatment with antidepressants and non-antidepressants on [$^3$H]5,7-DCKA binding to mouse cortex: comparison with biochemical and behavioural actions.

| Treatment | ↑ $IC_{50}$ (P <0.05) | FST[a] | ↓ β-AR[b] | Clinical Activity |
|---|---|---|---|---|
| *Tricyclics* | | | | |
| Imipramine (15) | Yes | Yes | Yes | Yes |
| Desipramine (10) | Yes | Yes | Yes | Yes |
| Amitriptyline (15) | Yes | Yes | Yes | Yes |
| Doxepin (15) | Yes | Yes | Yes | Yes |
| *MAOIs* | | | | |
| Pargyline (20) | No | Yes | Yes | Yes |
| Tranylcypromine (2) | Yes | Yes | Yes | Yes |
| Clorgyline (5) | Yes | No | Yes | Yes |
| L-Deprenyl (2) | Yes | Yes | Yes | Yes |
| D-Deprenyl (2) | No | ? | ? | No |
| *Atypicals* | | | | |
| Alaproclate (20) | Yes | No | No | Yes |
| Mianserin (15) | Yes | No | No | Yes |
| Fluoxetine (20) | Yes | No | No | Yes |
| Nisoxetine (30) | Yes | Yes | Yes | Yes† |
| Citalopram (20) | Yes | No | No | Yes |
| Bupropion (40) | Yes | Yes | Yes | Yes |
| Sertraline (25) | Yes | Yes | Yes | Yes |
| *Other* | | | | |
| Chlorprothixene (2) | Yes | No | No | Yes |
| Chlorpromazine (2) | No | No | No | No |
| Salbutamol (2) | No | No | Yes‡ | ? |
| Chlordiazepoxide (10) | No | No | No | No |
| Chlorpheniramine (30) | No | Yes? | ? | No |
| Scopolamine (0.5) | No | Yes | ? | No |
| Electroconvulsive shock | Yes | Yes | Yes | Yes |

The drug doses and electroshock regime employed were based on the reported abilities of these regimes to either reduce cortical β-adrenoceptor density ($B_{MAX}$) following chronic treatment [7,8] or exhibit efficacy in the forced swim test (FST) [12,22] which has predictive validity for clinically effective AD. For those drugs which have no effects on either β-adrenoceptor density or the FST (alaproclate, mianserin, fluoxetine, citalopram, chlorprothixene, and chlorpromazine) doses were chosen on the basis of characteristic effects on other neurotransmitter systems reported in the literature. The data in column two represent the presence of a statisically significant ($P$<0.05; Dunnett's $t$-test) increase in the $IC_{50}$ of glycine to inhibit [$^3$H]5,7-DCKA binding to mouse cortical membranes. Values in parentheses are the doses in mg/kg$^{-1}$ per day. Data represent the mean ± S.E.M. of 5–11 animals per group, except alaproclate ($n$ = 3). In saline and sham ECS mouse cortex, basal [$^3$H]5,7-DCKA binding was 1295 ± 48 fmol/mg$^{-1}$ protein ($n$=71) from ten experiments); the $IC_{50}$ and slope factor for glycine in these mice was 0.32 ± 0.02 μM (range of the mean values from ten experiments: 0.25–0.38 μM) and 0.74 ± 0.03 (range: 0.57–1.05), respectively ‡. In an independent series of experiments, we determined the ability of salbutamol (2 mg kg$^{-1}$ for 14 days) to down-regulate cortical β-adrenoceptors. This regimen reduced the $B_{max}$ of [$^3$H]dihydroalprenolol by ~20% (saline: 168±10 fmol/mg$^{-1}$ protein; salbutamol: 135±5 fmol/mg$^{-1}$ protein [$n$=9/group] $P$<0.05 compared with saline-treated animals), which is comparable with the reduction in $B_{max}$ observed in this mouse strain following chronic treatment with imipramine.[32] [a]FST: treatment significantly reduces immobility in the mouse FST. Data reviewed in ref. 22. [b]↓β-AR: treatment significantly reduces the density ($B_{MAX}$) of forebrain β-adrenoceptors. Data reviewed in refs. 7 and 9. [c]Clinical activity: treatment possesses antidepressant properties in clincial trials. Data reviewed in ref 1 and 2. Only data from double-blind, placebo-controlled studies was accepted as evidence of clinical activity and is presented here for comparison. †R. Fuller, personal communication. Although clinically active as an antidepressant in double-blind trials, nisoxetine was withdrawn due to a relatively high incidence of adverse side effects. (Data from refs. 33 and 38.)

**Table 20.4** Effect of chronic treatment with selected AD on glycine-displaced [$^3$H]CGP-39653 binding to mouse cortex.

| Treatment | $IC_{50}$ Site 1 ($\mu$M) | % Site 1 | $IC_{50}$ Site 2 ($\mu$M) |
|---|---|---|---|
| Saline | 0.95 ± 0.24 | 71 ± 5 | 4113 ± 1129 |
| Imipramine (15) | 0.64 ± 0.10 | 54 ± 3* | 3200 ± 900 |
| Amitriptyline (15) | 1.33 ± 0.35 | 44 ± 4* | 668 ± 197 |
| Citalopram (15) | Undetectable | 0* | 2098 ± 1866 |
| Pargyline (20) | 1.68 ± 0.84 | 33 ± 8* | 3412 ± 2271 |

Data represent the mean ± S.E.M. of 4–5 animals per group. Acute treatment was without effect on either basal [$^3$H]CGP-39653 binding or the displacement of this radioligand by glycine (data not shown). * $P < 0.05$ versus saline-treated animals. Portions of these data were published in Nowak *et al.*[33] and Paul *et al.*[38] Values in parentheses represent the daily dose of drug in mg kg$^{-1}$.

some indication that this effect is stereoselective. Moreover chlorpromazine, a non-antidepressant congener of chlorprothixene, was ineffective in this measure. Chronic treatment with salbutamol, which to date has not been demonstrated to produce antidepressant actions in double-blind clinical trials (but like other $\beta$-adrenoceptor agonists reduces the density of cortical $\beta$-adrenoceptors (Table 20.3 legend)[40] did not affect the potency of glycine to inhibit [$^3$H]5,7-DCKA binding. While the potential for obtaining false positives can never be completely excluded, these data, taken together with the effects of chronic treatment with selected AD on [$^3$H]CGP 39653 binding (see below) suggest a high degree of specificity for clinically active agents.

The AD doses used in these studies were primarily based on reports demonstrating either a reduction in cortical $\beta$-adrenoceptor density following chronic treatment (reviewed in refs. 7 and 8) or efficacy in the FST.[12,22] Since the reduction in potency of glycine to inhibit [$^3$H]5,7-DCKA binding is dose- and/or time-dependent following chronic treatment with imipramine, CIT, and ECS (see above), it is possible that either the dose of pargyline or the duration of treatment used was insufficient to significantly alter this measure. However, since chronic treatment with imipramine and ECS also reduces the proportion of high-affinity, glycine-displaceable [$^3$H]CGP-39653 binding sites to cerebral cortex,[33,34] (Fig. 20.3) the effects of chronic treatment with pargyline were also examined in this measure. In contrast to its modest effects on the $IC_{50}$ of glycine, pargyline produced a more robust reduction in the proportion of high-affinity, glycine-displaceable [$^3$H]CGP-39653 binding sites than imipramine (Table 20.4). All of the AD tested mimicked this action, and it is noteworthy that chronic treatment with CIT abolished high-affinity, glycine-displaceable binding of this NMDA antagonist (Table 20.4). While decreasing (or even abolishing) the proportion of high-affinity, glycine-displaceable [$^3$H]CGP 39653 binding sites, these AD treatments had no consistent effects on the potency of glycine, which is what might be predicted based on their ability to reduce the potency of glycine to inhibit [$^3$H]5,7-DCKA binding. Re-examination of the ability of adaptive changes in radioligand binding to the NMDA receptor to assess potential AD actions using both neurochemical measures (i.e. [$^3$H]5,7-DCKA and [$^3$H]CGP 39653) perfectly predicts (17/17) clinically active compounds, and is a significantly more robust predictor ($P < 0.02$, Fisher's exact test) than either changes in $\beta$-adrenoceptor density or efficacy in the FST.

These observations clearly dissociate AD-induced changes in glycine inhibition of [$^3$H]5,7-DCKA and [$^3$H]CGP-39653 binding. The instability of [$^3$H]CGP 39653 precluded a more thorough description (that is, additional AD and a detailed time course study) of this phenomenon, but the results obtained with two tricyclic AD, an 'atypical' agent, and an MAOI are consistent with the hypothesis that chronic AD treatment induces an adaptation of the NMDA receptor, manifested in the latter set of experiments as an effect on the coupling between glycine and glutamate recognition sites. Despite an incomplete understanding of the stoichiometric relationship between these two neurochemical measures, these findings are consistent with the hypothesis that antidepressant-induced changes in the NMDA receptor reflect an alteration in the strychnine-insensitive glycine receptor and its ability allosterically to modulate other sites on this ligand-gated ion channel. This hypothesis is also consistent with the small (~30%) but statistically significant reductions in basal [$^3$H]5,7-DCKA binding observed following chronic treatment with desipramine, amitriptyline, and chlorprothixene.[38] Whether the failure of the other antidepressants to produce similar reductions in basal [$^3$H]5,7-DCKA binding reflects pharmacokinetic or pharmacodynamic differences among these agents is unknown, since only one dose level of each agent was employed.

While the AD used in these studies are effective against major (sometimes called unipolar) depression, bipolar disorders are most effectively treated with a different set of drugs, most commonly lithium salts and carbamazepine.[42] In pilot experiments, rats were chronically treated (14 days) with lithium (as LiCl), carbamazepine, and a combination of LiCl and carbamazepine to determine whether these compounds would also effect adaptive changes in the NMDA receptor (Table 20.5). No statistically significant reductions in the $IC_{50}$ of glycine to inhibit [$^3$H]5,7-DCKA binding were observed following these treatments. While it must be noted that both the drug doses and time course were fixed, these experiments indicate that only treatments effective in major depression effect an adaptation of the NMDA receptor. A more detailed investigation varying both dose and length of administration must be performed to confirm or refute this hypothesis.

**Table 20.5** Effect of chronic treatment with carbamazepine and LiCl on the $IC_{50}$ of glycine to inhibit [$^3$H]5,7 DCKA binding to rat neocortical membranes.

| Treatment | $IC_{50}$(nM) | % |
|---|---|---|
| Vehicle | 332 ± 20 | 100 |
| Carbamazepine | 348 ± 38 | 105 |
| LiCl | 471 ± 82 | 142 |
| Carbamazepine + LiCl | 503 ± 63 | 152 |

Wistar rats were treated for 14 days with carbamazepine (20$^T$ mg kg$^{-1}$ p.o.) and/or LiCl (35 mg kg$^{-1}$ p.o). One day following the last drug treatment, the animals were sacrificed, neocortices dissected and [$^3$H]5,7 DCKA binding performed as described. Nowak *et al.*[33] Data represent the means ± S.E.M. of 6–8 animals/group. % represents increase relative to vehicle treated control rats. The authors thank Drs. Violetta Klimek and Marta Dziedzicka, Institute of Pharmacology, Polish Academy of Sciences (Krakow) for contributing to the injection and tissue harvesting phases of these experiments.

## What are the consequences of AD-induced alterations in the NMDA receptor?

While chronic AD treatments produce significant reductions in the potency of glycine to inhibit [$^3$H]5,7-DCKA binding[38] which appear to be region-specific,[33] the absolute reductions in potency are modest, ranging from ~2–4-fold (Table 20.3). If strychnine-insensitive glycine receptors were saturated under physiological conditions, the impact of these AD-induced changes on the operation of NMDA receptor-coupled cation channels would be marginal. Glycine concentrations in extracellular and cerebrospinal fluids are more than sufficient to saturate these receptors, and several studies have failed to demonstrate a modulation of NMDA receptors by application of glycine.[43] Nonetheless, several other studies have demonstrated that application of glycine or glycine mimetics augment the effects of NMDA *in situ*.[44] For example, glycine can potentiate the actions of NMDA in both neocortical slices[45] and thalamic neurones.[46] Furthermore, administration of glycine or the glycine mimetic D-serine elevates cerebellar cyclic GMP levels, an effect which can be blocked by NMDA antagonists.[47–49] Based on these findings, it appears that strychnine-insensitive glycine receptors are not saturated *in situ*, which suggests that AD-induced adaptation of the NMDA receptor could have a significant impact on the operation of NMDA receptor-coupled cation channels.

In view of the essential role of glycine in the operation of NMDA receptor-coupled cation channels [50], the reduced potency of glycine at strychnine-insensitive glycine receptor which is apparently produced by chronic AD treatment could result in impaired spontaneous channel opening. This hypothesis is also consistent with the ~50% reduction in basal (non-equilibrium) [$^3$H]dizocilpine binding to the NMDA receptor-coupled ionophore following chronic imipramine, with no change in the maximum response to either glycine or glutamate.[33] The effects of chronic treatment with other AD on basal [$^3$H]dizocilpine binding has not been examined. The reduction in the proportion of high-affinity, glycine-displaceable [$^3$H]CGP-39653 binding to the glutamate recognition site produced by all AD examined in this measure[33,38,41] (Table 20.4) is also consistent with this hypothesis. Since this appears restricted to cerebral cortex (and further confined to specific layers and/or regions of the cerebral cortex) it may be possible that this hypothesis may be more directly testable using electrophysiological techniques.

Several other related findings have emerged from these studies which have not yet been fully explored. For example, representatives from both the dihydropyridine (e.g. nifedipine, nimodipine) and benzothiazepine (e.g. diltiazem) classes of voltage-dependent calcium channel (VDCC) antagonists are active in the FST[51–53] and low doses of tricyclic AD augment the actions of subeffective doses of dihydropyridine VDCC antagonists.[51,52] Moreover, the dihydropyridine calcium channel agonist, BAY K 8644 enhances immobility in the FST[54] and antagonizes the effects of imipramine.[52] Based on these preclinical behavioural findings, we determined whether VDCC antagonists could mimic AD in producing adaptive changes in the radioligand-binding properties of the NMDA receptor.[55] Two-week treatment of mice with nimodipine resulted in a dose-dependent increase in the $IC_{50}$ of glycine to inhibit [$^3$H]DCKA binding (maximum change 1.9-fold), while diltiazem reduced the potency of glycine by ~2.8-fold (Fig. 20.5). The magnitude of these reductions is comparable with those (~2–4.3-fold) produced by chronic AD treatment. While it has not been rigorously established that VDCC antagonists are clinically effective AD, the ability of

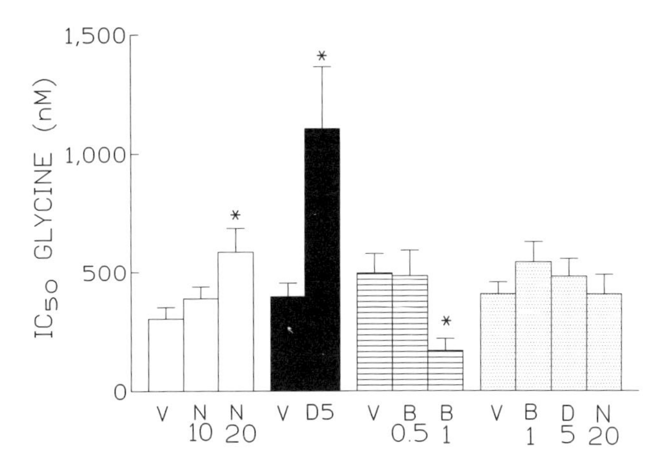


**Fig. 20.5** Effects of acute and chronic treatment with modulators of voltage-dependent calcium channels on glycine inhibition of [$^3$H]5,7-DCKA binding to mouse cerebral cortex. Mice were injected once (acute: stippled bars) or daily for 14 days. Animals were sacrificed 24 h later. [$^3$H]5,7-DCKA binding to well-washed membranes was performed as previously described[33]. No statistically significant differences in basal [$^3$H]5,7-DCKA binding were observed among treatment groups. Data represent the mean ±S.E.M. of 6–8 animals /group. V, vehicle N, nimodipine; D, diltiazem; B, BAY K 8644. Values are the doses in mg $kg^{-1}$ per day. *, $P < 0.05$ versus. the corresponding vehicle group (Dunnett's *post hoc* comparison). (Data from Nowak *et al.*[55].)

AD to induce adaptation of the NMDA receptor indicates such trials are warranted. One VDCC antagonist not examined in our pilot study was the diphenylalkylamine, verapamil. The effects of this compound (which acts at a distinct locus from either dihydropyridines or benzothiazepines on VDCC) have been examined in depressed subjects, with bipolar patients being the most responsive.[56] Verapamil was recently reported to be ineffective in a cohort of depressed patients that were also resistant to tricyclic AD.[57] If verapamil does produce an adaptation in the NMDA receptor, clinical trials in a cohort of patients responsive to tricyclic AD would clearly be warranted. Perhaps more intriguing is the observation that chronic treatment with BAY K 8644 significantly *increased* the potency of glycine to inhibit [$^3$H]5,7-DCKA binding (Fig.20.5), an action opposite to that produced by AD. These findings, together with the report that elevated calcium intake increased learned helplessness in rats,[58] prompted an examination of the sucrose preference of mice following chronic treatment with BAY K 8644 (Fig.20.6). BAY K 8644 produced a small, albeit statistically significant, reduction in sucrose preference (measured in a two-bottle choice situation) first manifested during the second trial block of 3 days. While the relative diminution in sucrose preference produced by BAY K 8644 is more modest than obtained in the chronic, mild stress-induced model of anhedonia,[27] these findings are consistent with the potential for developing a chemically induced model of 'depression'.

## A unified theory of antidepressant action

Adaptation of the NMDA receptor produced by the prototypic AD imipramine appears restricted to the cerebral cortex.[33] If a similar situation arises following chronic treatment with other AD, it will be critical to determine the regional and laminar distribution of these changes. Two recent reports underscore the potential importance of the frontal cortex to depression. Thus, Drevets *et al.*[59] have reported an increased blood flow in the left prefrontal

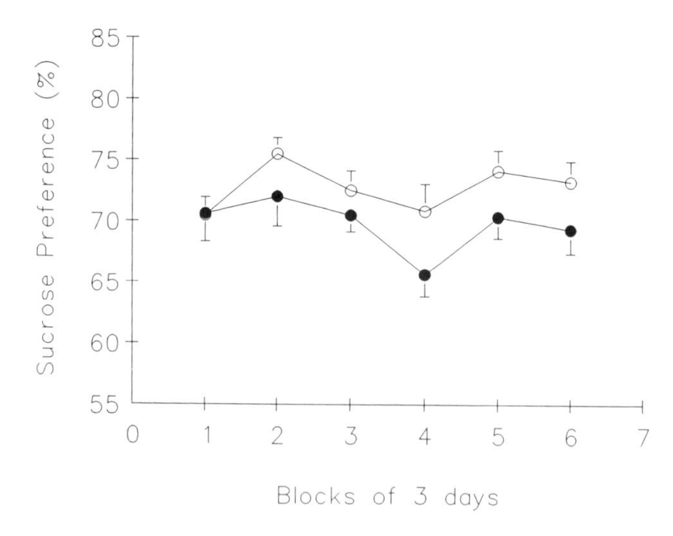


**Fig. 20.6** Effects of chronic treatment with Bay K 8644 on sucrose consumption: Mice were housed individually and adapted to the experimental procedure for 1 week before drug injections were started. During that time, mice were injected with vehicle and the sucrose (1%) or water preference (two-bottle choice) was measured (not shown). Injections were then initiated with (−)Bay K 8644 (1.0 mg $kg^{-1}$, i.p., ●) or vehicle (○). This figure presents the relative preference for sucrose solution, measured every day and pooled in blocks of 3 days for graphical purposes. Two-way ANOVA showed significant effect of drug $F(1,364) = 7.29$, $P < 0.01$ and block: $F(7,364) = 2.71$, $P < 0.01$, with no significant drug × block interaction. On the last day of experiment, the mice were killed and their forebrains used for [$^3$H]5,7-DCKA binding. $IC_{50}$ values for glycine (in nM) for placebo- and Bay K 8644-treated mice were: 609 + 102 and 375 ± 39 respectively (*t*-test = 2.04, df 17, $P < 0.05$). Ten mice were used in each group.

cortices of a relatively homogeneous subject group with familial pure depressive disease that was absent in asymptomatic subjects. These authors suggest that this abnormality represents a state marker of familial pure depressive disease. Moghaddam[60] reported a remarkable increase in glutamate and aspartate release in prefrontal cortex relative to other brain areas following swim and restraint stress. In view of the association between repeated stress and the development of depression,[26] these findings provide a framework for linking the results summarized in this review to the development of depressive disorders. An examination of the NMDA receptor following conditions of chronic mild stress which lead to anhedonia[26] will constitute an important first step in this direction. The contribution of monoaminergic pathways in AD-induced adaptation of the NMDA receptor is unknown, but in view of the vast literature linking monoamines to depression, a pivotal study will evaluate the importance of intact monoaminergic pathways for AD-induced changes in NMDA receptors.

The molecular mechanisms responsible for the adaptative changes in ligand binding to the NMDA receptor summarized in this review are unknown. While some AD such as desipramine act as channel blockers at NMDA receptors,[61] the relatively low affinities of other AD (e.g. mianserin) to inhibit [$^3$H]dizocilpine binding makes this an unlikely mechanism to account for the effects observed across treatments. Moreover, this explanation cannot be reconciled with the efficacy of a non-pharmacological treatment (ECS) in these studies. Like other ligand-gated ion channels, NMDA receptors possess multiple, allosterically coupled recognition sites which act in concert to modulate channel activity (i.e. the frequency and duration of channel opening). While neither the stoichiometry nor composition of native NMDA receptors is known, expression studies using recombinant receptors indicate both the potency and efficacy of ligands acting at these recognition sites is highly dependent upon subunit composition.[62] Thus, chronic AD treatment could induce changes in subunit composition or stoichiometry. Alternatively, AD may affect postranslational modification (e.g. phosphorylation) of the NMDA receptor. Current studies in our laboratory are focused on these latter issues.

A review of the clinical literature suggests a role for calcium homeostasis in the pathophysiology of human depression.[63] For example, alterations in CSF or serum calcium levels have been reported in depressed patients which was not directly associated with endocrine dysfunction.[64, 65] Conversely, reduced plasma, serum, or CSF calcium levels are observed with improvement of depressive symptomatology after either ECS or AD drug therapy.[64–66] In view of the role of NMDA receptors in regulating $[Ca^{2+}]_i$, AD-induced adaptation of this family of ligand-gated ion channels may be means of dampening/controlling elevated $[Ca^{2+}]_i$ within circumscribed areas of the central nervous system. While further neurochemical, electrophysiological, and behavioural studies will be required to appreciate fully the functional significance of changes in the radioligand-binding properties of the NMDA receptor, these studies represent the first description of a consistent, selective adaptation of a neurotransmitter-associated receptor produced by all major classes of AD. By comparison, 15 years after the demonstration that chronic antidepressant treatment results in down-regulation of forebrain $\beta$-adrenoceptors,[30] the physiological and behavioural consequences of this phenomenon remain unknown.[9, 10] In summary, both the neurochemical and behavioural evidence presented in this review provide compelling evidence that adaptation of the NMDA receptor is a common feature of AD drug action. Based on this pharmacological evidence, it is hypothesized that this family of ligand-gated ion channels also plays a pivotal role in depressive disorders.

---

**Substances interacting with NMDA receptors (cited in Layer *et al.*)**

*NMDA recognition site*

a) 2-Amino-7-phosphonoheptanoic acid (AP-7)/competitive antagonist/Evans *et al.* (1982). *Brit. J. Pharmacol.* **75**, 65–75.
b) CGP 39653/competitive antagonist/Sills *et al.* (1991). *Eur. J. Pharmacol.*, **192**, 19–24.
c) CGP 37849/competitive antagonist/Schmutz *et al.* (1990). *Naunyn-Schmiedeberg's Arch. Pharmacol.*, **342**, 61–66.
d) CGP 39551/competitive antagonist/Schmutz *et al.* (1990). *Naunyn-Schmiedeberg's Arch. Pharmacol.*, **342**, 61–66.

*Strychnine-insensitive glycine receptor*

a) 5,7-dichlorokynurenic acid (5,7-DCKA)/competitive antagonist/Baron *et al.* (1991). *Eur. J. Pharmacol.*, **206**, 149–154.
b) 1-aminocyclopropanecarboxylic acid (ACPC)/partial agonist/Marvizon *et al.* (1989). *J. Neurochem.*, **52**, 992–994.

*NMDA receptor-coupled cation channel*

a) Dizocilpine (MK-801)/antagonist/Wong *et al.* (1988). *J. Neurochem.*, **50**, 274–281.

4) Polyamine site

a) Eliprodil (SL 82.0715)/antagonist/Scatton *et al.* (1994). In *Direct and allosteric control of glutamate receptors*, (ed. M. Palfreyman, I. Reynolds, and P. Skolnick), pp. 139–154, Chapter 10. CRC Press, Boca Raton.

## References

1. Hollister, L.E. and Csernansky, J.G., (ed.) (1990) *Clinical pharmacology of psychotherapeutic drugs*, 3rd edn. Churchill Livingstone, New York.
2. Baldessarini, R.J. (ed.) (1983) *Biomedical aspects of depression and its treatment.* American Psychiatric Press, New York.
3. Oswald, I., Brezinova, V., and Dunleavy, D.L.F. (1972). *Br. J. Psychiatr.*, **120**, 673–677.
4. Segal, D.S., Kuczenski, R., and Mandell, A.J. (1974). *Biol. Psychiatry*, **9**, 19–24.
5. Maj, J., Przegalinski, E., and Moglnicka, E. (1984). *Rev. Physiol. Biochem. Pharmacol.*, **100**, 1–74.
6. Sugrue, M.F. (1985). *Psychopharmacology*, **21**, 619–622.
7. Hollister, L.E. (1986). *Annu. Rev. Pharmacol. Toxicol.*, **26**, 23–37.
8. Caldecott-Hazard, S., Morgan, D.G., DeLeon-Jones, F., Overstreet, D. H., and Janowsky, D. (1991). *Synapse*, **9**, 251–301.
9. Vetulani, J. (1991). *Pol. J. Pharmacol.*, **43**, 323–338.
10. Stone, E.A. (1983). *Behav. Brain Sci.*, **6**, 555–577.
11. Trullas, R. and Skolnick, P. (1990). *Eur. J. Pharmacol.*, **185**, 1–10.
12. Porsolt, R.D., Bertin, A., and Jalfre, M. (1977). *Arch. Int. Pharmacodyn. Ther.*, **229**, 327–336.
13. Steru, L., Chermat, R., Theirry, B., and Simon, P. (1985). *Psychopharmacology*, **85**, 367–370.
14. Porsolt, R.D. (1981) In *Antidepressants, neurochemical, behavioral and clinical perspectives*, (ed. S.J. Enna, J.B. Malik and E. Richelson), pp. 121–139. Raven Press, NY.
15. Steru, L., Chermat, R., Theirry, B., Mico, J.A., Lenegre, A., Steru, M., Simon, P., and Porsolt, R.D. (1987). *Prog. Neuro-Psycopharmacol. & Biol. Psychiatr.*, **11**, 659.
16. Evoniuk, G.E., Hertzman, R.P., and Skolnick, P. (1991). *Psychopharmacology*, **105**, 125–128.
17. Panconi, E., Roux, J., Altenbaumer, M., Hampe, S., and Porsolt, R.D. (1993). *Pharmacol. Biochem. Behav.*, **46**, 15–20.
18. Boje, K.M., Wong, G., and Skolnick, P. (1993). *Brain Res.*, **603**, 207–214.
19. Winslow, J., Insel, T., Trullas, R., and Skolnick, P. (1990). *Eur. J. Pharmacol.*, **190**, 11–21.
20. Long, J. and Skolnick, P. (1994). *Eur. J. Pharmacol.*, **261**, 295–301.
21. Trullas, R., Folio, T., Young, A., Miller, R., Boje, K., and Skolnick, P. (1991). *Eur. J. Pharmacol.*, **203**, 379–385.
22. Borsini, F. and Meli, A. (1988). *Psychopharmacology*, **94**, 147–160.
23. Maj, J., Rogoz, Z., Skuza, G., and Sowinska, H. (1992) *Eur. Neuropsychopharm.*, **2**, 37–41.

24. Maj, J., Rogoz, Z., Skuza, G., and Sowinska, H. (1992). *Pol. J. Pharmacol. Pharm.*, **44**, 337–346.
25. Willner, P. (1989). *Pharmacol. Ther.*, **45**, 425–455.
26. Willner, P., Muscat, R., and Papp, M. (1992). *Neurosci. Biobehav. Rev.*, **16**, 525–534.
27. Muscat, R., Papp, M., and Willner, P. (1992). *Biol. Psychiatry*, **31**, 937–946.
28. Papp, M. and Moryl, E. (1993). *Eur. Neuropsychopharm.*, **3**, 348–349.
29. Papp, M. and Moryl, E. (1994). *Eur. J. Pharmacol.*, **263**, 1–7.
30. Banerjee, S.P., Kung, L.S., Riggi, S.J., and Chanda, S.K. (1977). *Nature*, **268**, 455–456.
31. Heninger, G. and Charney, D. (1987) In *Psycopharmacology: the third generation of progress*, (ed. H. Meltzer), pp. 535–544, Chapter 54. Raven Press, NY.
32. Paul, I.A., Trullas, R., Skolnick, P., and Nowak, G. (1992). *Psychopharmacology*, **106**, 285–287.
33. Nowak, G., Trullas, R., Layer, R.T., Skolnick, P., and Paul, I.A. (1993). *J. Pharmacol. Exp. Ther.*, **265**, 1380–1386.
34. Sills, M.G., Fagg, G., Pozza, M., Angst, C., Brundish, D., Hurt, S.D., Wilusz, E.J., and Williams, M. (1991). *Eur. J. Pharmacol.*, **192**, 19–24.
35. Baron, B.M., Siegel, B.W., Sloan, A.L., Harrison, B.L., Palfreyman, M.G., and Hurt, S.D. (1991). *Eur. J. Pharmacol.*, **206**, 149–154.
36. Wong, E., Knight, A., and Woodruff, G. (1988). *J. Neurochem.*, **50**, 274–281.
37. Klein, D.F. and Davis, J.M. (1969) *Diagnosis and treatment of psychiatric disorders.* Williams and Wilkins, Baltimore.
38. Paul, I.A., Nowak, G., Layer, R.T., Popik, P., and Skolnick, P. (1994). *J. Pharmacol. Exp. Ther.*, **269**, 95–102.
39. Fink, M. (1984) In *Neurobiology of mood disorders*, (ed. R.M. Post and J. C. Ballenger). pp. 721–730. Williams and Wilkins, Baltimore.
40. Beer, M., Hacker, S., Poat, J., and Stahl, S.M. (1987). *Br. J. Pharmacol.*, **92**, 827–824.
41. Paul, I.A., Layer, R. T., Skolnick, P., and Nowak, G. (1993). *Eur. J. Pharmacol. – Mol. Pharmacol. Sec.*, **247**, 305–312.
42. Dunner, D. and Clayton, P.J. (1987) In *Psychopharmacology: the third generation of progress*, (ed. H.Y. Meltzer), pp. 1077–1083, Chapter 109. Raven Press, New York.
43. Kemp, J.A. and Leeson, P.D. (1993). *Trends Pharmacol. Sci.*, **14**, 20–25.
44. Thomson, A.M. (1990). *Prog. Neurobiol.*, **35**, 53–74.
45. Thomson, A.M., Walker, V.E., and Flynn, D.M. (1989). *Nature*, **338**, 422–424.
46. Salt, T.E. (1989). *Brain Res.*, **481**, 403–406.
47. Wood, P.L., Emmett, M.R., Rao, T.S., Mick, S., Cler, J., and Iyengar, S. (1989). *Neurochemistry*, **53**, 979–981.
48. Danysz, W., Wroblewski, J.T., Brooker, G., and Costa, E. (1989). *Brain Res.*, **479**, 270–276.

49. Rao, T., Cler, J., Emmet, M., Mick, S., Iyengar, S., and Wood, P. (1990). *Neuropharmacology*, **29**, 1075–1080.
50. Kleckner, N.W. and Dingledine, R. (1988). *Science*, **241**, 835–837.
51. Mogilnicka, E., Czyrak, A., and Maj, J. (1987). *Eur. J. Pharmacol.*, **138**, 413–416.
52. Martin, P., Laurent, S., Massol, J., Childs, M., and Puech, A.J. (1989). *Eur. J. Pharmacol.*, **162**, 185–188.
53. Skolnick, P., Miller, R., Young, A., Boje, K., and Trullas, R. (1992). *Psychopharmacology*, **107**, 489–496.
54. Mogilnicka, E., Czyrak, A., and Maj, J. (1988). *Eur. J. Pharmacol.*, **151**, 307–311.
55. Nowak, G., Paul, I.A., Popik, P., Young, A., and Skolnick, P. (1993). *Eur. J. Pharmacol.–Mol.Pharmacol. Sec.*, **247**, 101–102.
56. Hoschl, C. and Kozeny, J. (1989). *Biol. Psychiatry*, **25**, 128–140.
57. Adlersberg, S., Toren, P., Mester, R., Rehavi, M., Skolnick, P., and Weizman, A. (1994). *Clin. Neuropharmacol.*, **17**, 294–297.
58. Trulson, M.E., Arasteh, K., and Ray, D.W. (1986). *Pharmacol. Biochem. Behav.*, **24**, 445–448.
59. Drevets, W.C., Videen, T.O., Price, J.L., Preskorn, S.H., Carmichael, S.T., and Raichle, M.E. (1992). *J. Neurosci.*, **12**, 3628–3641.
60. Moghaddam, B. (1993). *J. Neurochem.*, **60**, 1650–1657.
61. Kitamura, Y., Zhoa, X.-H., Takei, M., Yonemitsu, O., and Nomura, Y. (1991). *Neurochem. Int.*, **3**, 247–253.
62. Meguro, H., Mori, H., Araki, K., Kushiya, E., Kutsuwada, T., Yamazaki, M., *et al.* (1992). *Nature*, **357**, 70–74.
63. Ortolano, G.A., Swonger, A.K., Kaiser, E.A., and Hammond, R.P. (1983). *Med. Hypoth.*, **10**, 207–221
64. Linder, J., Brismar, K., Beck-Friis, J., Saaf, J., and Wetterberg, L. (1989). *Acta. Psychiatr. Scand.*, **80**, 527–537
65. Carman, J.S., Post, R.M., Goodwin, F.K., and Bunney, W.E. (1977). *Biol. Psychiatry*, **12**, 5–17.
66. Mellerup, E.T., Bech, P., Sorenson, T., and Rafaelsen, O.J. (1979). *Biol. Psychiatry*, **14**, 711–714.
67. Meloni, D., Gambarana, C., De Montis, M.G., Dal Pra', P., Taddei, I., and Tagliamonte, A. (1993). *Pharmacol. Biochem. Behav.*, **46**, 423–426

# 21 Excitatory synaptic transmission in hippocampus: modulation by purinoceptors via $Ca^{2+}$ channels

O. Krishtal and A. Klishin

## Introduction

Existing paradigm indicates adenosine as being a major modulator of both metabolic and electrical events in the mammalian central nervous system (CNS). Its major action is inhibitory: in the presence of a sufficiently high concentration of Ado, the brain is predominantly silent. In micromolar concentration range, this substance is so effective that some investigators use it in the search for natural pacemakers – presuming that in the presence of sufficiently high adenosine concentration only those neurones that are supplied with intrinsic generator produce nerve impulses.

In general terms, there are two ways of depressing cell-to-cell signalling: (i) by diminishing the excitability of pre- and/or post-synaptic cells by activation of hyperpolarizing influences; and (ii) by decreasing neurotransmitter release probability. In various preparations, adenosine is found to be active in both ways.[1]

One of the most extensively examined regions of the CNS is the hippocampus, especially in its sliced form. The main reasons for this are: (i) the tissue's orderly organized pyramidal cell layers make it especially easy to apply the hippocampus to virtually all existing electrophysiological approaches; and (ii) the hippocampus is really a champion in expressing various synaptic plasticity properties. Famous long-term potentiation, the only known so far possible electrical manifestation of memory, has been first described and is most strongly expressed in the system of hippocampal excitatory synapses.[2]

In various preparations, it has been shown that adenosine is able to inhibit the release not only of excitatory amino acids (EAA), glutamate (Glu) and aspartate (Asp), but also the release of many other neurotransmitters, such as acetylcholine, noradrenaline, dopamine, serotonin, and $\gamma$-aminobutyric acid (GABA)[3]. It should be mentioned, however, that in hippocampus adenosine seems to leave the GABAergic inhibitory synaptic transmission unaffected.[4] This at least partially simplifies the situation with the most prominent action of adenosine in the hippocampus, which is the powerful inhibition of excitatory synaptic transmission.

## Sites of action

Is this action of adenosine post-synaptic or pre-synaptic? Intracellular microelectrode measurements indicated that, post-synaptically, adenosine evoked the outward current (single electrode voltage-clamp) carried by $K^+$ ions.[5] Haas and co-workers[1, 6] distinguish two different components of such conductance, namely voltage- and calcium-insensitive potassium conductance and voltage-insensitive calcium-dependent conductance. Activation of voltage- and $Ca^{2+}$-insensitive conductance occurs via G protein associated with $A_1$ adenosine receptor. The pharmacology of adenosine-mediated hyperpolarization is peculiar: it is unaffected by such common blockers of potassium conductances as tetraethylammonium or 4-aminopyridine, but is effectively inhibited by $Ba^{2+}$ ions.[7] However, the inhibition of synaptic transmission persists when the post-synaptic hyperpolarizing action of adenosine is blocked by $Ba^{2+}$.[8] Still in 1983, Dolphin and colleagues[9] demonstrated that adenosine blocks Glu and Asp, but not GABA release from hippocampal slices, and then showed that this inhibition can be removed by pertussis toxin.[10] However, the question still remained as to how relevant this observation was specifically for the synaptic transmission. Taken together, these data indicated that the post-synaptic effect of adenosine only accompanies something more powerful than that which occurs at the pre-synapse. The only alternative to be eliminated in this situation was a possibility that adenosine changes the reaction of the cell to the transmitter reaching it from the pre-synapse. Quite recently this question was elegantly solved almost simultaneously by the three groups of investigators. Between May and November 1992, Prince and Stevens,[11] Scholz and Miller,[12] and finally Thompson *et al.*[13] published the data on miniature post-synaptic current analysis. The possibility for this approach evolved with the development of an *in situ* whole-cell patch-clamp technique which was pioneered in the Sakmann laboratory.[14] In this case the cell is voltage-clamped and internally perfused while preserving its synaptic input. This allows low noise measurements to be performed, thus resolving the unitary events in synaptic transmission.

According to the existing paradigm which has survived much examination over several decades, such an event as 'single quantum of post-synaptic current' or 'miniature EPSC' (mEPSC) is a response to the amount of transmitter liberated by one synaptic vesicle. All three groups of investigators found that adenosine does not change the amplitude of mEPSCs, though their frequency of occurrence falls dramatically. Thus, the latter phenomenon accounts for the inhibition of amplitude of macroscopic EPSC under the action of adenosine: it is a decrease in the release of transmitter due to the decrease in the probability of a single event.

Although this concept of both macro- and microscopic events demonstrating mutually corresponding behaviour is attractive, it does not account for all the problems encountered. To illustrate this, let us examine how this paradigm accounts for the actual mechanism of adenosine action.

## Effector mechanisms

The evoked transmitter release is triggered by so-called '$Ca^{2+}$ signal'. This term is introduced now to substitute for the expression 'increase in intracellular $Ca^{2+}$ concentration', the latter probably being much too long for the ubiquitous use.

It was found that one of the types of $Ca^{2+}$ channels, namely N-type channels, is blocked by micromolar concentrations of adenosine. According to current opinion, the background adenosine concentration in the CNS is of the same range.[1] The pharmacology of excitatory synaptic transmission in the hippocampus appears to be remarkably similar to that of N-type $Ca^{2+}$ channels: both are blocked by adenosine, by EAA, and also by a specific N-type $Ca^{2+}$ channel blocker, $\omega$-conotoxin GVIA ($\omega$-CTX).[15,16] Thus, N-type channels are regulated not only by adenosine, but also by EAA *per se*. This seems to provide further circumstantial evidence indicating the possible ultimate role of N-type channels in the transmission. Furthermore, both adenosine and EAA receptors seem to be operationally interconnected in their modulatory influence on the N-type $Ca^{2+}$ channels[16] (Fig. 21.1).

Unfortunately, this seems at present to be only a part of the truth. One of the problems is as follows: $Ca^{2+}$ channel blockers neither eliminate the action of adenosine on the frequency of mEPSC, nor inhibit the spontaneously occuring mEPSC.[12] The $Ca^{2+}$ channel blocker used by these authors ($Cd^{2+}$) non-specifically blocks all types of $Ca^{2+}$ channels. So, at present, the minimum assumption is that the blocking action on $Ca^{2+}$ current is not the only negative feedback mechanism that allows adenosine to regulate the excitatory transmitter release. This difficulty itself seems to provide the inspiration for further studies.

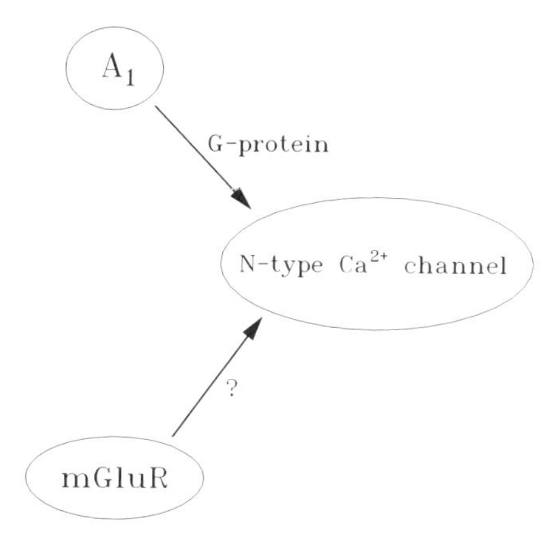


**Fig. 21.1** Interaction of N-type $Ca^{2+}$ channels with metabotropic receptors in the pre-synapse. $A_1$, $A_1$ adenosine receptor; mGluR, metabotropic glutamate receptor.

So far, we have discussed inhibition: both pre- and post-synaptically evoked decrease in the efficacy of synaptic transmission. However, recently it has been shown that the role of adenosine is in fact dual. While inhibiting excitatory synaptic transmission via $A_1$ receptors, adenosine also facilitates transmission via $A_2$ receptors. Several groups have recorded corresponding enhancement of field potentials and EPSC.[17–19] It should be noted, however, that in direct release measurements the selective $A_2$ agonist CGS-21680 enhanced EAA release only in ischaemic cerebral cortex.[20] Powerful potentiation of synaptic transmission has been observed in hippocampal slices under the combined action of adenosine and the selective $A_1$ antagonist cyclopenthyltheophylline (CPT)[19]. Recently Mogul, *et al.*[21] have demonstrated in acutely isolated hippocampal neurones the up-regulation of high-threshold

$Ca^{2+}$ current in similar conditions, namely under high concentrations of 2-chloroadenosine and CPT.[21] This effect appeared to be insensitive to ω-CTX, but could be blocked by ω-agatoxin-IVA (ω-Aga). The conclusion of the authors was that P-type $Ca^{2+}$ channels are up-regulated via activation of $A_{2b}$ adenosine receptors. This effect could well be on the background of the potentiation of synaptic transmission observed under adenosine and CPT, although the latter phenomenon is easily obtained under the combined action of the specific $A_1$ receptor agonist cyclopenthyladenosine (CPA) and antagonist (CPT) (O. Garaschuk, Y. Kovalchuk, and O. Krishtal, unpublished observations).

## Bicomponental hippocampal EPSC

It is well known that, in the hippocampus, EAA activate two major subclasses of post-synaptic ionotropic receptors, *N*-methyl-D-aspartate (NMDA) and non-NMDA,[2] leading to the appearance of bicomponental EPSC (NMDA- and non-NMDA-components respectively). Taking into account the different kinetic properties (rise-and decay-time) of both constituents[22] one can use the early and the late components of EPSC as respective measures of contribution of non-NMDA- and NMDA-components into total EPSC. In a number of cases, including the potentiation of EPSC by Ado + CPT, it was found that the ratio of late and early components of EPSC (NMDA/non-NMDA ratio) is not changed under certain experimental influences that involve considerable changes in the EPSC amplitude.[19] This observation includes the case of potentiation under CPA + CPT and has been used to support the idea of all-or-nothing mode of functioning of individual synapses.[23] We have re-examined the dependence of both EPSC components on the activity of Ado receptors at different concentrations of external $Ca^{2+}$ and $Mg^{2+}$.

## Effect of CPT administration on hippocampal EPSC

When measured at increased $[Ca^{2+}]_o$ and decreased $[Mg^{2+}]_o$ (2.5 mM $Ca^{2+}$ and 0.5 mM $Mg^{2+}$ as compared with 1.5 and 1.1 mM respectively in normally used artificial cerebrospinal fluid) the ratio NMDA/non-NMDA becomes dependent on the activity of $A_1$ receptors. Figure 21.2 demonstrates a prominent action of CPT resulting in a slow-down of the EPSC decay which accompanies an earlier-described increase in the peak amplitude of the current.[19] After reaching its peak value, the EPSC started slowly to diminish, but the NMDA/non-NMDA ratio remained increased, even after washing out CPT for as long as the neurone lasted. Repeated application of CPT led to a much smaller increase in EPSC amplitude, with no further change in the NMDA/non-NMDA ratio. We used various concentrations of CPT in the range of 60 nM to 8 μM, and found no differences in the described phenomena.

What is the cause of the change in EPSC kinetics under CPT? We have found that it is the rearrangement in the relative contribution of NMDA and non-NMDA components into total EPSC, rather than the change in the kinetics of separate EPSC components which results in the change of total EPSC kinetics. 1 μM CPT influenced only the amplitude of each EPSC components (data not shown). The mean values of EPSC potentiation are: $1.9 \pm 0.3$ ($n = 3$) in the presence of 100 μM DL-2-amino-5-phosphonovaleric acid (APV), an NMDA-receptor antagonist, and $3.6 \pm 0.6$ ($n = 3$) in the presence of 10 μM 6-cyano-7-nitroquinoxaline-2,3-dione (CNQX), a non-NMDA-receptor antagonist.

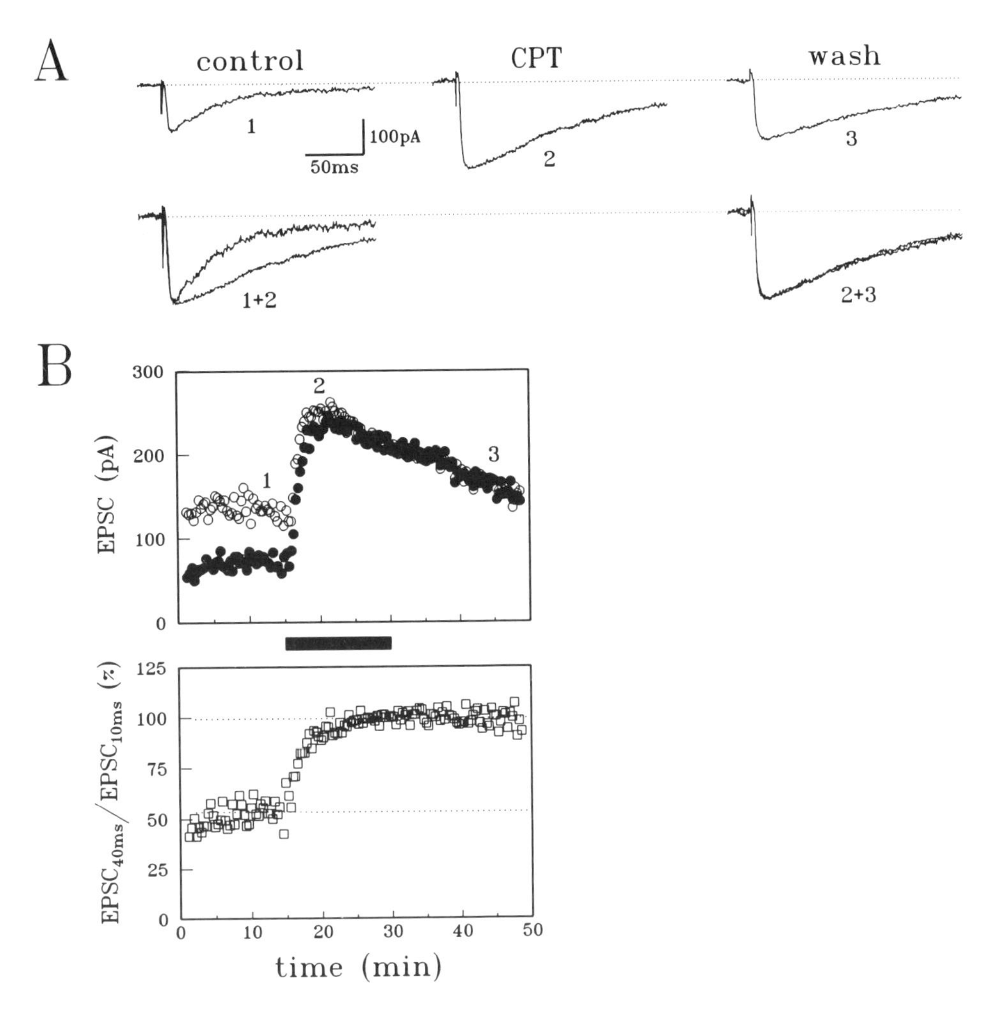


**Fig. 21.2** Effect of application of CPT on the EPSC. The CA1 pyramidal neurone was held at −50 mV. A. CPT (250 nM) potentiates EPSC amplitude (top curves) and dramatically slows down its decay (normalized currents on the bottom). Decay remains slower after washing out CPT. Each curve is the mean of four sequential traces. Digits indicate the moments (on B) when EPSC were measured. **B.** CPT induces irreversible change in the kinetics of EPSC. Top graph: time course of amplitude of total EPSC measured in 10 ms (○) and in 40 ms (●) after the stimulus. Bottom graph: ratio of amplitudes at 40 and 10 ms. Filled bar indicates when 250 nM CPT was applied.

Is this action of CPT (change of NMDA/non-NMDA ratio) post-synaptic or pre-synaptic? Possible effects of CPT on the sensitivity of isolated hippocampal neurones were studied in a separate series of experiments. NMDA receptor activation by rapid application of 10 mM Asp + 10 $\mu$M glycine was not altered by the preapplication of 2 $\mu$M CPT. The same result was obtained for non-NMDA receptors activated by quisqualate. These experiments indicate the lack of a direct action of CPT on EAA receptors, and argue for the pre-synaptic action of CPT.

### Dependence on the $Ca^{2+}$ and $Mg^{2+}$ concentration

High concentrations of extracellular $Mg^{2+}$ ions prevented preferential up-regulation of the NMDA-component by CPT (Fig 21.3). Moreover, it also led to a quicker return of EPSC to the control value after washing out CPT. The same behaviour of EPSC was in the 'usual' solution (used by most investigators: 1.5 mM $Ca^{2+}$ and 1.5 mM $Mg^{2+}$). These experiments indicate at a strong dependence of modulatory effect of CPT on extracellular concentrations of $Ca^{2+}$ and $Mg^{2+}$. They also explain why this effect has not been observed before.

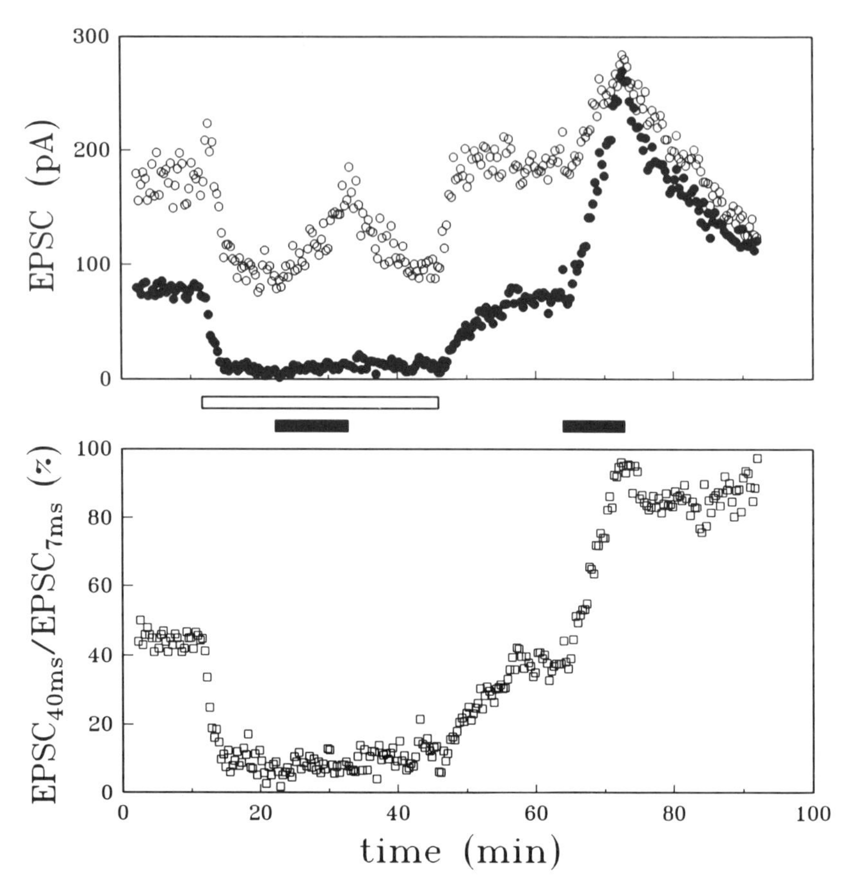


**Fig. 21.3** High $[Mg^{2+}]_0$ prevents the modulatory effect of CPT. The CA1 pyramidal neurone was held at −50 mV. Top graph: time course of amplitude of total EPSC measured in 7 ms (○) and in 40 ms (●) after the stimulus. Bottom graph: ratio of amplitudes at 40 and 7 ms. Open bar indicates when 5 mM $Mg^{2+}$ (instead of 0.5 mM) was applied; filled bars: 125 nM CPT.

### $\omega$-Conotoxin prevents the effect of CPT

As already mentioned, Ado is known to block N-type $Ca^{2+}$ channels in hippocampal neurones via $A_1$ adenosine receptors.[24,25] A specific blocker of N-type channels, $\omega$-CTX, strongly and irreversibly, but (contrary to Ado) non-completely inhibits the excitatory synaptic transmission in hippocampus.[26,27] We searched for a possible effect of ω-CTX on the NMDA/non-NMDA ratio.

In four experiments ($n = 10$) ω-CTX (1 μM) preferably inhibited the NMDA-component of EPSC. Subsequent application of CPT (125 nM) led to usual EPSC potentiation, but the change of NMDA/non-NMDA ratio occurred only in three experiments, and in two cases its effect was dramatically slowed down. In other cases (five experiments) CPT did not affect the kinetics of EPSC (Fig. 21.4).

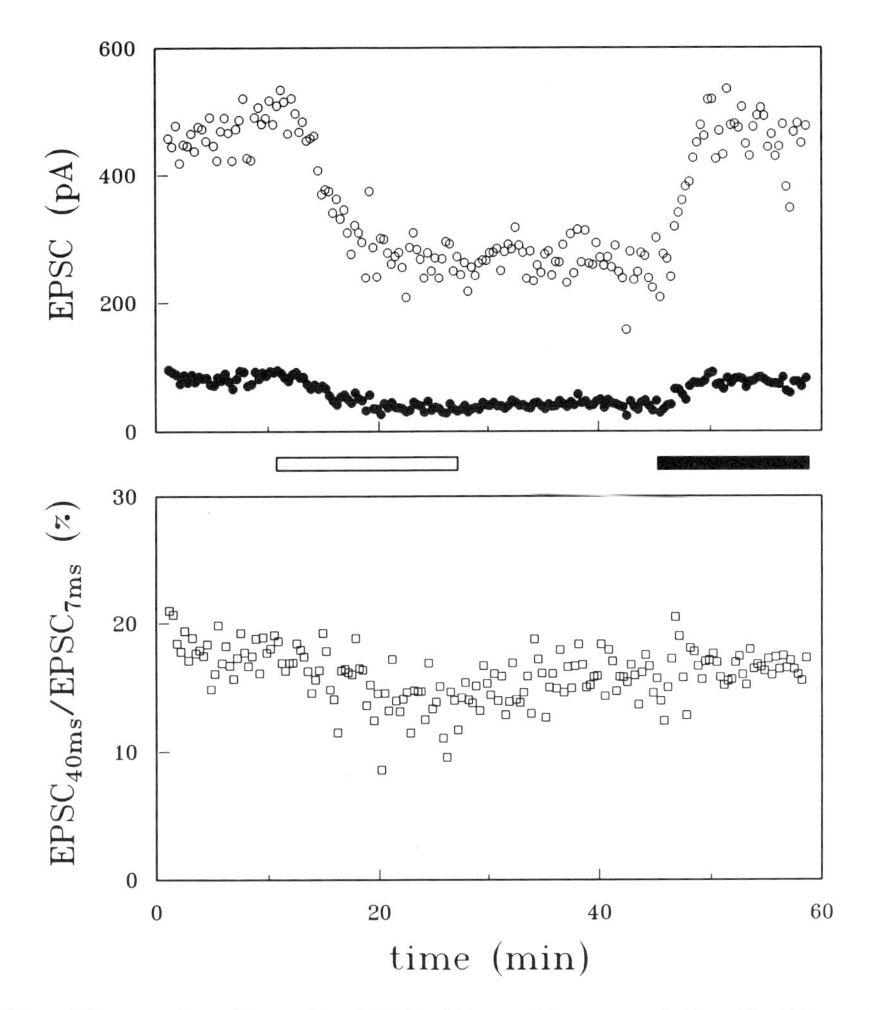


**Fig. 21.4** ω-CTX prevents the modulatory effect of CPT. The CA1 pyramidal neurone was held at −50 mV. Top graph: time course of amplitude of total EPSC measured in 7 ms (○) and in 40 ms (●) after the stimulus. Bottom graph: ratio of amplitudes at 40 and 7 ms. Open bar indicates when 1 μM ω-CTX was applied; filled bar: 125 nM CPT.

## Diadenosine polyphosphates as neuromodulators

Diadenosine polyphosphates ($Ap_4A$ and $Ap_5A$) are present in the secretory granules of chromaffin cells[28] as well as in the rat brain synaptic terminals.[29] Their contribution to the exocytosis of the total synaptosomal content is considerable, ranging from 7–12%. $Ap_4A$ and $Ap_5A$ are released from synaptosomes in a $Ca^{2+}$-dependent manner.[29] Recently, it has been reported that both substances induce a blocking action on the excitatory synaptic transmission in the rat hippocampus, an action which is elicited via $A_1$ receptors and differs in some respects from the action of Ado.[30]

$Ap_4A$ and $Ap_5A$ caused the inhibition of extracellular field potential (FP) and EPSC. This effect was fully reversible and antagonized by CPT (125 nM). In the majority of cases, the prolonged application of $Ap_5A$ to the slices revealed a gradual recovery of synaptic transmission from the inhibition. The recovery could be observed on FP, but surprisingly not on the EPSC. This process was practically absent in the case of Ado. In many cases, it was possible to inhibit the recovery by incubating the slices with protein kinase C inhibitors staurosporin (5 $\mu$M for 30–40 min) or sphingosine (0.1 $\mu$M for 30–40 min).[30]

Although the inhibitory action of $Ap_5A$ on synaptic transmission in the hippocampus is mediated by $A_1$ receptors like the effect of Ado, this action cannot be explained by the instability of the compound tested and subsequent usual action of Ado resulting from $Ap_5A$ decomposition: the stable agonist of $P_2$ receptor, $\beta,\gamma$-methylene-ATP, mimics the action of diadenosine polyphosphates in the hippocampus (Fig. 21.5). Moreover, in isolated pyramidal neurones where Ado blocks N-type $Ca^{2+}$ current (also via $A_1$ receptors) $Ap_5A$ is absolutely ineffective (A.Tsyndrenko, V.Panchenko and O. Krishtal, unpublished data). These observations indicate the heterogeneity of $A_1$ receptors associated with different effector mechanisms.

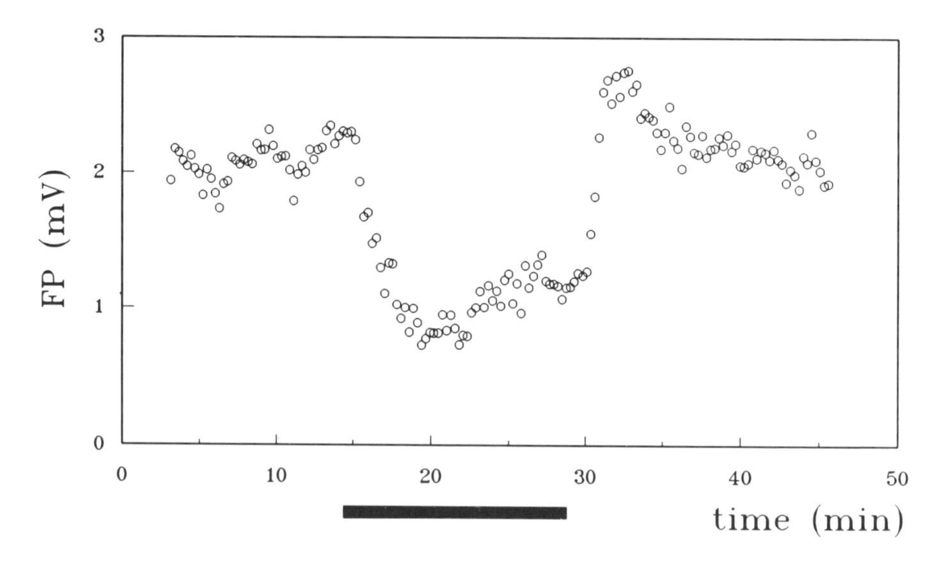


**Fig. 21.5** Effect of application of $\beta,\gamma$-methylene-ATP on the field potential. Time course of population spike amplitude in CA1 field. Filled bar indicates when 10 $\mu$M $\beta,\gamma$-methylene-ATP was applied.

## Contribution of different $Ca^{2+}$-currents to hippocampal synaptic transmission

The initial classification of $Ca^{2+}$ inward currents in mammalian neurones introduced by Nowycky *et al.*[31] indicates the existence in these cells of at least three types of $Ca^{2+}$-current, namely low-threshold-activated T-type, and high-threshold-activated N- and L-types. With the development of molecular biology (cloning of channel molecule subunits) and discovery of novel pharmacological tools, the classification of the channels including permeability for $Ca^{2+}$ ions, becomes more sophisticated. Thus, first the discovery of F-toxin and then of still more selective $\omega$-Aga allowed confirmation of the early suggestion of Llinas about the existence of at least one more type of $Ca^{2+}$ channel, namely the P-type.[32] Several very recent papers indicate the presence of $\omega$-Aga sensitivity in the hippocampus, although the conclusions of the authors are somewhat different. Thus, according to Takahashi and Momiyama[33] the major part of the EPSC is blocked by concentrations of $\omega$-Aga as low as 50 nM. This indicates the P-type channels to be the major source of $Ca^{2+}$ triggering the excitatory synaptic transmission. However, Tsien and colleagues[34] needed micromolar concentration of this toxin to achieve a similar effect. This brought them to the conclusion that, in fact, Q-type channels are supporting hippocampal synaptic transmission in addition to the N-type channels. Another argument favouring this conclusion was the sensitivity of synaptic transmission to the blocking action of $\omega$-conotoxin-MVIIC which specifically blocks Q-type $Ca^{2+}$ channels first described in cerebellar granule neurones.[34] In terms of sensitivity of hippocampal synaptic transmission to $\omega$-Aga, our observations well confirm those made by the group of Tsien: in concentrations up to 300 nM we saw minimal inhibition of EPSC.

Further studies are still needed to reveal possible interaction between newly identified hippocampal $Ca^{2+}$ conductance components with the system of adenosine receptors.

## Experimental procedures

Experiments were carried out on transverse 300–500 $\mu$m thick hippocampal slices of Wistar rats (21-day-old animals) at room temperature 22–25°C. During preincubation and recording, the slices were kept in oxygenated (95% $O_2$, 5% $CO_2$) solution of the following composition (in mM): NaCl 138, KCl 2.7, $CaCl_2$ 2.5, $MgSO_4$ 0.5, $NaHCO_3$ 26, $NaH_2PO_4$ 0.4, $K_2HPO_4$ 0.26, glucose 15 (pH 7.4).

To obtain access to the CA1 pyramidal neurones, surface tissue was removed by a saline jet flowing from a micropipette. It was then possible to obtain high-resistance electrical contact with the cell membrane. The experiments were carried out on completely submerged slices in a 0.7 ml experimental chamber. For electrical stimulation, a bipolar nichrom electrode (each wire 200 $\mu$m thick) was positioned on the surface of the slice. The patch pipette was filled with intracellular solution of the following composition (in mM): CsF 100, $NaH_2PO_4$ 40, HEPES – CsOH 10, Tris – Cl 10 (pH 7.3). The neurones demonstrated voltage-operated inward and outward currents. Stimulation (with frequency 0.2 Hz) of Schaffer collateral – commissural pathway resulted in the appearance of intracellularly recorded EPSC with a reversal potential of 5–10 mV. The current remained stable for up to 3 h of recording.

The removal of tissue in stratum oriens of field CA1 of hippocampus led to a loss of a considerable part of the inhibitory pathways. Addition of 100 $\mu$M bicuculline did not change the shape of EPSC in these slices. A still larger contribution of the NMDA component was achieved by using artificial cerebrospinal fluid with slightly elevated $Ca^{2+}$ and decreased $Mg^{2+}$.

## Conclusions and discussion

In addition to straightforward – although still far from being understood – inhibitory and excitatory actions, Ado has becomes strongly suspected of exerting modulatory effects via metabotropic actions. This would be no surprise, since $A_1$ and $A_2$ receptors are correspondingly known to decrease and increase intracellular cAMP levels.[35] However, the existence of still unidentified metabotropic pathways other than via protein kinase C is also highly probable.[21]

There is still a way to go in establishing a mutual correspondence between electrophysiologically and biochemically detected receptor functions. Probably, important news in this field will come from further studies of the changes in the sensitivity of excitatory synaptic transmission to the blocking action of Ado. Nonetheless, in 1988, our group demonstrated that the inhibitory action of Ado can be greatly reduced by prolonged incubation of slices with EAA,[36] data which were confirmed in 1990 by Bartrup and Stone.[37] NMDA receptor pharmacology is involved in exerting this action. In 1991, Phillis *et al.*,[38] when measuring release of EAA from brain, found that this process is escaping from Ado control in the conditions of ischaemia. A new aspect in these regulatory relationship is the long-term modulatory basis of some influences. At least two such observations can be mentioned. Nishimura *et al.*[39] have recently demonstrated a long-term post-inhibitory excitation of neurotransmission in the hippocampus after application of Ado. The enhancement of excitatory transmission by a combined action of Ado and $A_1$ receptor antagonist also lasts more than 40 min.[19] The increase in NMDA/non-NMDA ratio of the EPSC demonstrated in Fig. 21.2 lasts for many hours.

Diadenosine polyphosphates are stored in brain synaptosomes and are subjected to $Ca^{2+}$-dependent release. They appear to be at least as effective in exerting negative feedback onto excitatory synaptic transmission in the hippocampus as Ado. The apparent link of their blocking (and potentiating) action with the activity of PKC allows the assumption that they may have a specific neuromodulatory role.

Intrinsically plastic properties of synaptic transmission are possibly expressed in the case of hippocampus in their limiting form. Their further description will probably smooth away the existing discrepancies concerning the weights and roles of various $Ca^{2+}$ channels in synaptic transmission and regulation of the latter by Ado receptors and other mechanisms (Table 21.1). In 1989, our group demonstrated that, under prolonged application of EAA, hippocampal synaptic transmission loses sensitivity to the blocking action of $\omega$-CTX.[26] Interestingly, this effect was found to be reversible. It can be suggested that, depending on still non-identified factors, different populations of $Ca^{2+}$ channels shift their duties in delivering the major part of $Ca^{2+}$ used for transmitter release at different times. In this case it may be understood how and why one and the same system of synapses often exhibits so variable pharmacological properties.

**Table 21.1** Compounds affecting hippocampal ion channels.

| Channel | Compound | Effect | Reference |
|---|---|---|---|
| N-type $Ca^{2+}$ | ω-Conotoxin GVIA | Blockade | 21, 27 |
| | Adenosine | Blockade | 16, 24, 25 |
| | Glutamate | Blockade | 15, 16 |
| P-type $Ca^{2+}$ | ω-Agatoxin IVA | Blockade | 32 |
| Q-type $Ca^{2+}$ | ω-Conotoxin MVIIC | Blockade | 34 |
| Adenosine-activated $K^{+}$ | $Ba^{2+}$ | Blockade | 7 |

## References


1. Greene, R.W. and Haas, H.L. (1991). *Prog. Neurobiol.*, **36**, 329–341.
2. Collingridge, G.L. and Lester, R.A.J. (1989). *Pharmacol. Rev.*, **40**, 143–210.
3. Dunwiddie, T.V. (1985). *Int. Rev. Neurobiol.*, **27**, 63–139.
4. Lamber, N.A. and Teyler, T.J. (1991). *Neurosci. Lett.*, **122**, 50–52.
5. Okada, Y. and Ozawa, S. (1980). *Eur. J. Pharmacol.*, **68**, 483–492.
6. Gerber, U., Greene, R.W., Haas, H.L., and Stevens, D.R. (1989). *J. Physiol.*, **417**, 567–578.
7. Greene, R.W. and Haas, H.L. (1985). *J. Physiol.*, **366**, 119–127.
8. Thompson, S.M., Haas, H.L., and Gähwiler, B.H. (1992). *J. Physiol.*, **451**, 347–363.
9. Dolphin, A.C. and Archer, E.R. (1983). *Neurosci. Lett.*, **43**, 49–54.
10. Dolphin, A.C. and Prestwich, S.A. (1985). *Nature*, **316**, 148–150.
11. Prince, D.A. and Stevens, C.F. (1992). *Proc. Natl Acad. Sci. USA*, **89**, 8586–8590.
12. Scholz, K.P. and Miller, R.J. (1992). *Neuron*, **8**, 1139–1150.
13. Scanziani, M., Capogna, M., Gähwiler, B.H., and Thompson, S.M. (1992). *Neuron*, **9**, 919–927.
14. Sakmann, B., Edwards, F., Konnerth, A., and Takahashi, T. (1989). *Q. J. Exp. Physiol.*, **74**, 1107–1118.
15. Swartz, K.J. and Bean, B.P. (1992). *J. Neurosci.*, **12**, 4358–4371.
16. Chernevskaya, N. I., Obukhov, A. G., and Krishtal, O.A. (1991). *Nature*, **349**, 418–420.
17. Sebastiao, A.M. and Ribeiro, J.A. (1992). *Neurosci. Lett.*, **138**, 41–44.
18. Nishimura, S., Mohri, M., Okada, Y., and Mori, M. (1990). *Brain Res.*, **525**, 165–169.
19. Garaschuk, O., Kovalchuk, Y., and Krishtal, O. (1992). *Neurosci. Lett.*, **135**, 10–12.
20. O'Regan, M.H., Simpson, R.E., Perkins, L.M., and Phillis, J.W. (1992). *Neurosci. Lett.*, **138**, 169–172.
21. Mogul, D.J., Adams, M.E., and Fox, A.P. (1993). *Neuron*, **10**, 327–334.
22. Hestrin, S., Nicoll, R.A., Perkel, D.J., and Sah, P. (1990). *J. Physiol.*, **422**, 203–225.

23. Perkel, D.J. and Nicoll, R.A. (1993). *J. Physiol.*, **471**, 481–500.
24. Gross, R.A., MacDonald, R.L., and Ryan-Jastrow, T. (1989). *J. Physiol.*, **411**, 585–595.
25. MacDonald, R.L., Skerritt, J.H., and Werz, M.A. (1986). *J. Physiol.*, **370**, 75–90.
26. Krishtal, O.A., Petrov, A.V., Smirnov, S.V., and Nowycky, M.C. (1989). *Neurosci. Lett.*, **102**, 197–204.
27. Kamiya, H., Sawada, S., and Yamamoto, C. (1988). *Neurosci. Lett.*, **91**, 84–88.
28. Rodriguez del Castillo, A., Torres, M., Delicado, E.G., and Miras-Portugal, M.T. (1988). *J. Neurochem.*, **51**, 1696–1703.
29. Pintor, J., Diaz-Rey, M.A., Torres, M., and Miras-Portugal, M.T. (1992). *Neurosci. Lett.*, **136**, 141–144.
30. Klishin, A., Lozovaya, N., Pintor, J., Miras-Portugal, M.T., and Krishtal, O. (1994). *Neuroscience*, **58**, 235–236.
31. Nowycky, M.C., Fox, A.P., and Tsien, R.W. (1985). *Nature*, **316**, 440–443.
32. Llinás, R., Sugimori, M., Hillman, D.E., and Cherksey, B. (1992). *Trends Neurosci.*, **15**, 351–355.
33. Takahashi, T. and Momiyama, A. (1993). *Nature*, **366**, 156–158.
34. Wheeler, D.B., Randall, A., and Tsien, R.W. (1994). *Science*, **264**, 107–111.
35. Van Calker, D., Müller, M., and Hamprecht, B. (1979). *J. Neurochem.*, **33**, 999–1005.
36. Krishtal, O.A., Smirnov, S.V., and Osipchuk, Y.V. (1988). *Neurosci. Lett.*, **85**, 82–88.
37. Bartrup, J.T. and Stone, T.W. (1990). *Brain Res.*, **530**, 330–334.
38. Phillis, J.W., Walter, G.A., and Simpson, R.E. (1991). *J. Neurochem.*, **56**, 644–650.
39. Nishimura, S., Okada, Y., and Amatsu, M. (1992). *Neurosci. Lett.*, **139**, 126–129.

# Index


(+)-(S)-202-791 104
(+)-PN200-110 104
α-adrenergic receptor 282
α-scorpion toxins 24, 25, 38, 39
α-tocopherol (vitamin E) 48, 50
$\alpha_2$-adrenergic 115
β-scorpion toxins 24, 39
μ-conotoxin 24, 25
ω-Aga-IIIA 130
ω-Agatoxin-IVB 130
ω-Agatoxin-IVA 104, 130, 137, 138, 460
ω-CgTx GVIA 133, 134, 459
ω-conotoxin (CTx)-GVIA-sensitive channels(N-types) 104
ω-conotoxin-GVIA 103, 104, 130
ω-conotoxin-MVIIA 130
ω-conotoxin-MVIIC 107, 130
1-aminocyclopropanecarboxylic acid; ACPC 439
2-amino-7-phosphonoheptanoic acid; AP-7 438
4-aminopyridine (4-AP) 173, 174, 333, 439
5-hydroxypropafenone 90
5, 7-dichlorokynurenic acid 443
6-cyano-7-nitroquinoxaline-2, 3-dione (CNQX) 460
7-chloro-kynurenic acid 402
9-aminoacridine 35
9-aminoacridine 172, 173

$A_1$-adenosine receptor 458
aconitine 24, 43
A-current($I_A$) 6
adenine nucleotide translocator 334
adenosine 457
ajmaline 75
allethrin 46
aluminium 153
amiloride 130, 131, 273
aminoglycoside 273
amiodarone 83, 86, 329
amitriptyline 449
AMPA receptors 407
amphipathic 275
amyloid β protein (AβP[1-40]) V149
amyloid [AβP(1-40)] channels 159
amyloid channels 153, 155
  ion selectivity of amyloid channels 152
antiarrhythmic drugs 74
  fast kinetic (class IB) blockers 85
  open channel blockers 86
antidepressants 438
antimycin A 331
$Ap_4A$ 464
$Ap_5A$ 464
apamin 175
aprindine 88, 90
arachidonic acid 275

$Ba^{2+}$ currents 7, 107, 170, 175
batrachotoxin (BTX) 24, 40
Bay K 8644 104, 107, 130, 133, 450
BDNF 9, 10, 13
benzocaine 90
benzothiazepines 133
brevetoxins 24, 25, 44
BTX 86, 87
bupivacaine 90

$Ca^{+}$ dependent inactivation 100
$Ca^{2+}$ buffering systems 299
$Ca^{2+}$-sensitive $Ca^{2+}$ release 303, 313
  channel open probability 313
  open state probability of $Ca^{2+}$-sensitive 307
$Ca^{2+}$ channel gatings 97, 98
$Ca^{2+}$ channels 7, 97
  pertussis toxin-sensitive 110
caffeine 285, 286, 302
calcium channels
  cardiac L-type channels 101
  DHP antagonists 104, 115
  DHP-agonists 104
  DHPs 103, 107
  dihydropyridine receptor 284
  dihydropyridines 104, 132
  gating modes 100

high-threshold channels 103
inactivation kinetics 99
L-type $Ca^{2+}$ channels 9, 108, 132
low threshold channels 101
LVA channel 101
modulation of high-threshold $Ca^{2+}$ channels 110
modulation by neurotransmitters 117
N-type 104, 106, 108, 111
N-type calcium channels 133
N-type channels 459
neuronal N-type channel 101
non-L-type channels 106
other types of calcium channels 140
P-type 104, 115, 136
Q-type 104, 106, 108, 141, 142
T-type calcium channels 102, 115, 131
time-dependent inactivation 99
voltage dependence of dihydropyridine block 132
voltage-dependent calcium channels 129
voltage-dependent facilitation of $Ca^{2+}$ currents 113
voltage-induced 'facilitation' 113
calcium homeostasis 453
calcium signal 299
calmodulin 287
calreticulin 300
calsequestrin 300, 302
CaM-dependent protein kinase 287
cation channels 343
CCCP 331
cessium 172, 175
ceteidil 175
cGMP-gated channel 383
charybdotoxin 175, 355
chick cochlear neurones 7
chloramine-T 45
chlorpromazine 35, 448
chlorprothixene 448
cholera toxin-sensitive G-proteins 110
chromaffin cells 343
chromaffin granule channels 371
channel gating 356
flicker 350
high open-probability (hop) 353, 361
lop mode 356
permeability ratio $P_K/P_{Na}$ 350
cignatoxin 25
ciguatoxin 24, 45
citalopram 445
clotrimazole 175
cocaine 34
cochlear ganglion 4
coelenterate peptide α toxins 86
conductance levels 424, 426
conductivity 390
cono-peptides 104
conotoxins 32, 33
Crabtree effect 324
cyclic ADP-ribose 313
cyclopenthyltheophylline (CTP) 459
cyclopiasonic acid 302
cyclosporin A 331, 335
cyphenothrin 46
cytochrome P-450 91

D-(–)-2-amino-5-phosphonovalerate 405
D-tubocurarine 175
dantrolene 286, 304, 312
DDT 46
decamethonium 172
deltamethrin 46
dendrotoxin 172
desensitization 389, 400
desipramine 90, 449
development 3
diacylglycerol 281, 305
diadenosine polyphosphates 464
digitoxin 312
diltiazem 133, 393
diprafenone 86
disopyramide 75, 83, 86, 87, 89
dizocilpine binding 443, 450
dizocilpine; MK-801 438
DL-2-amino-5-phosphonovaleric acid (APV) 460
domoate 409
dose response relation 388
Down's syndrome 149
doxorubicine 312
DPI 201-106 79, 86, 87

EGTA 355
electron transport 321
elevation of intracellular $Ca^{2+}$ 273
eliprodil 441, 442
encainide 83
endoplasmic reticulum 281
enkephalin 110
ethacizin 86, 90
ethosuximide 130, 131
excitotoxic cell death 401

fatty acids 275
fenvalerate 46
fibroblast growth factors 12

flecainamide 83, 89, 90
flufenamic acids 407
flunarizine 130
fluspirilene 130, 143
forskolin 172, 173
FPL 64176 130
FTX 130
furosemide 407

G protein 366, 458
GABA 115
$GABA_β$ 115
Gaboon viper venom 172
gadolinium 271, 272
GDP-β-S 115, 116, 344
glibenclamide 333
glutamate 399
glycine receptor 401, 423, 450
glycylxylidide 90
Golgi apparatus 281
goniopora toxin 25
grayanotoxin (GTX) 24, 40
GTP-γ-S 115, 116, 344

haloperidol 175
heparin 313
heterotrimetric GTP-binding proteins 344
hexamethonium 172
human depression 453
HVA 103

iberiotoxin 175
(IC) drugs 85
ifenprodil 403
imipramine 75, 90, 445
imperator toxin 172, 173
inositol triphosphate 281
inositol 1, 4, 5-triphosphate 300
  $InsP_3$-sensitive channels 306
inositol 1, 4, 5-triphosphate receptor 281
inositol-phosphoglycan (IPG) 10
input conductance 168, 169
insecticide 45
insulin-like growth factors (IGFs) 12
insulin-secreting RINm5F cells 115
intermediate (IA) and slow kinetic antiarrythmic
ischaemia–reperfusion 326
isoproterenol 358

kainic acid 150
kainate receptors 407
kaliotoxin 175
kynurenate 403

lidocaine 35, 36, 83, 87, 88, 89, 90
light-sensitive channel 384
  channel gating 388
  flicker 388
  molecular properties 393
  open probability 386
  pharmacology 391
liposomes 322
local anaesthesics 25, 34, 89
long-term potentiation 315, 457
LY 97241 172

mag-fura-2 technique 302
malignant hyperthermia 285
margatoxin 172, 173
MCD (mast cell degranulating peptide) 172
mechano-gated 271
membrane tension 274
mepacrine 175
methoxuflurane 130
mexiletine 83, 88, 89, 90
miniature EPSC 458
mitochondrial channels 281, 327
  inner mitochondrial membrane 326
  multiple conductance channel 329
  permeability transition 326
  pH-dependent 323
  uncoupling protein 325
  voltage-dependent anion-selective channel 322
mitoplasts 327
MK-801 406
monoamine oxidase inhibitors 446
MTX 33
muscarinic 112

*N*-acetylprocainamine 91
*N*-bromoacetamide 25, 37
*N*-bromosuccinide 37
nerve growth factor (N) 3
neuronal differentiation 3
neurotrophins 9, 11
neurotrophins 3, 4, 5 , 9
NGF 9, 13
nickel 103
nifedipine 104, 132
niflumic 407
nimodipine 132, 134, 450

nitrendipine 104, 132, 156, 175
NMDA antagonists 438, 442
NMDA receptor channels 401, 423, 450
  expression cloning 399
  expression in *Xenopus* oocyte 430
  extrasynaptic receptors 422
  low-conductance channels 427
  outward rectification 391
  recombinant NMDA receptor 425
  regulatory sites 401
  stoichiometry 426
non-selective organic blocker 142
noradrenaline 10, 112, 356
noxiustoxin 175
nuclear envelope 281

oxotremorine 110, 114

palmitoyl carnitine 313
pancreatic β cells 167
pancuronium 35
pargyline 446
PbTX-1 44
PbTX-8 44
PC12 13
penticainide 83, 87
phalloidin 173
phencyclidine 172, 173, 401, 405
phenothrin 46
phenylalkylamines 133
phenytoin 130
phosphodiesterase 384
phospholipase 275, 282
photoreceptors 383
pimozide 130, 393
planar bilayers 152, 321, 344
plasticity 401
polyamines 172, 174, 401, 403
polyanions 323
porin 322
porin/VDAC system 151
potassium channels 4
  after-hyperpolarization 314
  ATP-sensitive $K^+$ channels 313, 332
  $Ca^{2+}$-activated $K^+$ channels 175, 176, 314
  delayed rectifier current ($K_v$) 5, 173
  fast transient outward current ($I_{to,\ f}$) 6
  inward (anomalous) rectifier K current ($K_{IR}$) 170, 173
  $K^+$ channel 167, 344
  $K_{ATP}$ 167, 172
  pore segment (P) 170
  shaker-like channels 170
  transient outward current 173
prajmaline 86
prajmalium 87
presynaptic facilitation 122
procainamide 75, 83, 91
procaine 34, 286, 309
pronase 25, 37, 86
propafenone 83, 86, 87, 89, 90
propoxyphene 90
propranolol 329, 358
protein kinase C 282, 287, 305
protein kinase C (PKC) 115
proton pump 370
pyrazinecarboxyamides 273
pyrethroids 45, 52

quinidine 75, 83, 86, 89, 90, 172
quinine 175
quisqualate 409
QX-314 35

R-type 142, 145
rapidly inactivating outward current 5
ropitoin 90
ruthenium red 284, 286, 309
ryanodine 175, 283, 285, 286, 304, 308
  binding 86
  -binding site 85
ryanodine receptor 281, 284
  molecular cloning 284

sarcoplasmic-endoplasmic reticulum $Ca^{2+}$ ATPase-pumps 300
saxitoxin(STX) 23, 24
scorpion α toxins 86
scorpion toxins 38
sea anemone toxins 4, 25, 37, 38
secretory ranules 343
sensory neurones 3
smooth muscle 81
sodium channel 4, 13, 23, 28, 175
  cardiac $Na^+$ channels 77, 81, 82
  epithelial $Na^+$ channel 74
  flicker 87
  frequency-dependent blockade of $Na^+$ channels 82
  kinetics 77
  $Na^+$ channel α-subunits 75
  pharmacological properties 28
  resistant to TTX 75
  sodium channel modulators 37
  sodium channel density 26, 27

stoichiometry for TTX binding 26
subclassification of $Na^+$ channel blockers 85
tonic $I_{Na}$ block 84
use dependent blockade 35, 84
voltage-dependent blockade of $Na^+$ channels 35, 82
voltage-gated sodium channels 74
somatostatin 121
soyasaponins 175
spermidine 286, 403
spermine 403
$Sr^{2+}$ 172
staurosporine 305
stereocilia 274
stretch-induded channels 273
striatoxin 25, 44
structure of AβP in solution 151
strychnine 35
STX conotoxin 24
sulphonylurea derivatives 167
sulphonylurea receptor 172
synaptic function 315
synaptic plasticity 315
synaptic receptors 422
synaptic vesicles 281

T-tubules 283
tacrine 157
tetraethylammonium (TEA) 172, 173, 174, 175
tetra-aminoacridine 172
tetramethrin 46
tetrodotoxin(TTX) 23, 25, 74
thapsigargin 300, 302
thermogenin 325
thioridazine 130
thromethamine 155
tocainide 83, 88
toxin γ 39
trifluoroperazine 175
trimeric G proteins 367
*Trk* proto-oncogene family 11
tromethamine 162
trypsin 86
TTX binding 28
TTX conotoxin 24
TTX-R 75
TTX-resistant sodium channels 51, 52
TTX-sensitive sodim channels 51, 52
tyrosine kinase 286

valinomycin 325
vascular smooth muscle 167
verapamil 133
veratridine 24, 41
vitamin E 48

*Xenopus* oocytes 274

zinc 172, 173, 406